Cage Metal Complexes

Yan Voloshin • Irina Belaya •
Roland Krämer

Cage Metal Complexes

Clathrochelates Revisited

Springer

Yan Voloshin
Kurnakov Institute of General and Inorganic Chemistry
Russian Academy of Sciences
Moscow, Russia

Gubkin Russian State University of Oil and Gas
Moscow, Russia

Irina Belaya
Nesmeyanov Institute of Organoelement Compounds
Russian Academy of Sciences
Moscow, Russia

Roland Krämer
Anorganisch Chemisches Institut
Ruprecht-Karls-Universität Heidelberg
Heidelberg, Germany

ISBN 978-3-319-56419-7 ISBN 978-3-319-56420-3 (eBook)
DOI 10.1007/978-3-319-56420-3

Library of Congress Control Number: 2017940403

Printed on acid-free paper

This Springer imprint is published by Springer Nature
The registered company is Springer International Publishing AG
The registered company address is: Gewerbestrasse 11, 6330 Cham, Switzerland

Preface

Compounds with encapsulated species were found to have unique chemical, physical, and physicochemical properties. Those that contain organic ions, nonmetal inorganic cations, and neutral molecules as guests have been proposed for the efficient recognition, separation, carrying, storage, and detection of various encapsulated caged species, thus entering the field of molecular and supramolecular devices and modern material. Those include efficient catalysts, carriers of therapeutically active compounds, and many others; main results in these fields have been recently summarized in our book "The encapsulation phenomenon—synthesis, reactivity and applications of caged ions and molecules," 2016 [1]. On the other hand, the encapsulation of metal cations has been skipped in the above book; the situation in this field, as it appeared before 2002, has been highlighted in the monograph "Clathrochelates: synthesis, structure and properties" by Yan Z. Voloshin, Nina A. Kostromina, and Roland Krämer [2]. The presence of encapsulated metal ion(s) in these molecules (so-called clathrochelates) suggests many other fields of potent applications, associated with their metal-centered radioemitter, redox, and magnetic activities, thus allowing to design radiodiagnostic and radiotherapeutic cage compounds, mediators of electron transfer, biomimetics of metal enzymes, paramagnetic probes, and magnetic materials.

Depending on the nature of chemical bonds between their ligand synthons (precursors), metal clathrochelates can be divided into two main classes: covalent and coordination capsules. Chapter 1 describes the scope and goal of this book, classification of such clathrochelates and their caging ligands, as well as main approaches for their synthesis and functionalization.

Chapter 2 highlights main synthetic approaches for preparation of cage metal complexes and their reactivity. Such complexes include the clathrochelate derivatives of tetra- and pentadentate encapsulating ligands (Sect. 2.1). The pioneering works of A. Bencini, A. Bianchi, and M. Micheloni in this field should be emphasized. Chemistry of hexadentate polyamine caging ligands, their analogs, and cage complexes (Sect. 2.2) has been mainly developed by A. M. Sargeson, J. M. Harrowfield, and coworkers. Various types of polyazomethine pseudo- and clathrochelates, the derivatives of pyrazoloximate, oximehydrazonate, and α-dioximate ligand synthons, as well as hybrid systems based on them have been mostly prepared in our research group. Pioneering efforts in the synthesis and physical chemistry of Schiff-base imine capsules and their polyamine derivatives (Sect. 2.3) have been made by J. Nelson and coworkers. Most polynuclear clathrochelate complexes, the derivatives of dinucleating encapsulating iminophenol ligands (including those with terminal donor groups), have been synthesized and characterized by P. Chaudhuri, K. Severin, and coworkers. K. Severin is also a pioneer in the use of cage metal complexes with terminal donor groups as ligand synthons for preparation of clathrochelate-based coordination cages, polymers, metallomacrocycles, and MOFs (Sect. 2.4).

Main data on spatial and electronic structure of the above types of cage metal complexes and their derivatives are highlighted in Chap. 3. Chapter 4 describes magnetic properties of the paramagnetic clathrochelates and the related spectral data as well.

Chapter 5 summarizes practical applications of cage metal complexes, including their interaction with biosystems and the design of prodrugs, clathrochelates, radioemitters, and target compounds for radiodiagnostics and radiotherapy and paramagnetic probes for NMR and MRI studies (Sect. 5.1). The potential of these macropolycyclic complexes as catalysts and precatalysts of various chemical reactions (including HER) is highlighted in Sect. 5.2; their other practical applications in molecular (photo)electronics and modern materials science are summarized in Sect. 5.3.

Section 5.4 describes the straightforward design of the encapsulating macropolycyclic ligands for stabilization of unusual (including catalytically active) oxidation states of an encapsulated metal ion.

The authors would like to thank Dr. Ekaterina Lebed, Dr. Yulia Nelubina, and Dr. Anna Vologzhanina (INEOS RAS) for their valuable contribution in writing the book. Artworks for this book were prepared with the support of Russian Science Foundation (grant 16-13-10475).

Moscow, Russia — Yan Z. Voloshin
Moscow, Russia — Irina G. Belaya
Heidelberg, Germany — Roland Krämer
December, 2016

References

1. Voloshin Y, Belaya I, Krämer R (2016) The encapsulation phenomenon: synthesis, reactivity and applications of caged Ions and molecules. Springer International Publishing, Switzerland
2. Voloshin YZ, Kostromina NA, Krämer R (2002) clathrochelates: synthesis, structure and properties. Elsevier, Amsterdam

Contents

List of Abbreviations

ADMA	4-[3-(9-anthryl)propyl]-*N*,*N*-dimethylaniline
Ac	Acetate
acac	Acetylacetone
ADH	Alcohol dehydrogenase
ADP	Anisotropic displacement parameters
AIBN	Azobisisobutyronitrile
AscH	Ascorbic acid
bpy	2,2′-bipyridine
bpz	2,2′-bipyrazole
Bn	Benzyl radical
BSA	Bovine serum albumin
Bz	Benzene sulfonate radical
Bu	*n*-butyl radical
CD	Circular dichroism
cod	Cyclooctadiene
Cp	Cyclopentadiene
Cp*	Pentamethyl cyclopentadiene
CT	Charge transfer
CTB	Charge transfer band
CV	Cyclic voltammetry
cytochrome c(III)	Oxidized form of cytochrome *c*
DED	Deformation electron density
DBA	Dibenzylideneacetone
DBU	1,8-diazabicyclo[5.4.0] undec-7-ene
DBN	1,5-diazabicyclo(4.3.0)non-5-ene
DCC	*N*,*N*′-dicyclohexylcarbodiimide
DDQ	2,3-dichloro-5,6-dicyano-1,4-benzoquinone
DDT	Dodecanethiol
DFT	Density functional theory
DLS	Dynamic light scattering
DMA	Dimethylacetamide
DMAP	4-dimethylaminopyridine
DMF	Dimethylformamide
DMSO	Dimethylsulfoxide
dppp	1,3-bis(diphenylphosphino)propane

ECD	Electronic circular dichroism
EDD	Electron density distribution
EIE	Equilibrium isotope effect
EIS	Electrochemical impedance spectroscopy
en	Ethylenediamine
ESP	Electron spin polarization
ET	Electron transfer
Fc	Ferrocenyl radical
Fe-SOD	Iron superoxide dismutase
FT-IR	Fourier transform infrared spectroscopy
FRET	Fluorescence resonance energy transfer
GFC	Gel filtration chromatography
Hal	Halogen
Hipip	The oxidized form of the high-potential iron-sulfur protein
HMDSO	Hexamethyldisiloxane
HDXO	Diacetylmonooxime hydrazone
HOMO	Highest occupied molecular orbital
HVOR	Hydroxycarboxylate viologen oxireductase
ICP	Inductively coupled plasma
ICP-MS	Inductively coupled plasma mass spectrometry
IEC	Ion-exchange chromatography
Im	Imidazole
IS	Isomeric shift
ITO	Tin-doped indium oxide
L	Ligand
LIESST	Light-induced excited spin state trapping
LIR	Lanthanide-induced relaxation
LIS	Lanthanide-induced shift
M	Metal ion
MD	Molecular dynamics
Me	Methyl radical
MM	Molecular modeling
Ms	Mesityl radical
NBS	*N*-bromosuccinimide
NCT	Neutron capture therapy
NMP	*N*-methylpyrrolidone
Nu^-	Nucleophilic agent
OTf	Trifluoromethanesulfonate
PBN	Phenyl-tert-butyl nitrone
PEG	Polyethylene glycol
PET	Photoinduced electron transfer
PGSE	Pulse gradient spin-echo
tht	Tetrahydrothiophene
PNA	Peptidonucleic acids
PTC	Phase-transfer conditions
Py	Pyridine

PzOxH	2-acetylpyrazoloxime
QS	Quadrupole splitting
RAFT	Reversible addition-fragmentation chain transfer
RDA	Radial distributions of atoms
RTP	Rotating tube processor
r_i	Physical (Shannon) ionic radius
SAM	Self-assembled monolayer
SCO	Spin-crossover
SDP	Spinning disk processing
SECM	Scanning electrochemical microscopy
SPECT	Single-photon emission computed tomography
STM	Scanning tunneling microscopy
TAP	Trigonal antiprism
TBP	Trigonal bipyramid
TDDFT	Time-dependent DFT
tempo	2,2,6,6-tetramethylpiperidine-1-oxyl
TEBAC	Triethylbenzylammonium chloride
THF	Tetrahydrofuran
TOF	Triethyl orthoformate
Tol	Tolyl radical
TP	Trigonal prism
tren	tris(2-aminoethylamine)
TR-FRET	Time-resolved fluorescence resonance energy transfer
Ts	Tosyl radical
VCD	Vibrational circular dichroism

1 General Considerations

1.1 Scope and Goal

The following main criteria for the formation of a cage complex with encapsulated species [organic or inorganic anion(s), cation(s), and neutral molecule(s)] have been provided [1]. They include "(1) the presence of three-dimentional cavity in either covalent, or coordination, or supramolecular caging ligand (capsule) and (2) encapsulation of the above guest anion(s), cation(s) or neutral molecule(s), or their combination within this cavity *via* formation of intracavitary donor–acceptor (coordination) or supramolecular (hydrogen, electrostatic dipole–dipole, van der Walls etc.) bonds" [1]. On the other hand, the term "clathrochelate" (derived from Latin "clāthrātus" to furnish with lattice, encaged), introduced by D. H. Busch [2], suggests the formation of a complex with an encapsulated metal ion in which (1) a three-dimensional cavity in its macropolycyclic ligand is present, and (2) the metal ion coordinates heteroatoms of this ligand within its cavity, isolating such encapsulated ion from the external factors.

This book also describes the clathrochelate-containing and clathrochelate-based hybrid systems, polynuclear compounds, coordination cages, polymers, and MOFs as well as the modern materials and the nanodevices based on them.

1.2 Encapsulating Ligands and Their Cage Metal Complexes: Main Classes and Subclasses; General Approaches for the Synthesis and Functionalization

Fundamental concepts of clathrochelate chemistry [i.e., cage complexes with encapsulated metal ion(s)] as well as general classification of their encapsulating (caging) ligands (based, first of all, on the nature of donor groups) have been provided in our book [3] in 2002, which highlighted the chemistry of "classical" clathrochelate compounds, such as polyamine hexadentate macrobicyclic ligands and the corresponding cage metal complexes (sarcophaginates and sepulchrates), and their thio-containing analogs as well as that of polyazomethine metal clathrochelates, such as macropolycyclic α-dioximates, oximehydrazonates, hydrazonates, and azineoximates and their heterocyclic analogs (i.e., cage metal bipyridinates and phenanthrolinates, and the related compounds). On the other hand, the following types of encapsulating ligands (capsules) and their cage metal complexes were not covered in [3]:

1. Tetradentate covalent capsules
2. Pentadentate covalent capsules

Y. Voloshin et al., *Cage Metal Complexes*, DOI 10.1007/978-3-319-56420-3_1

3. Schiff-base imine capsules and their polyamine derivatives
4. Coordination capsules and some other types of mono- and polynucleating caging ligands

Since 2002, following new types of the clathrochelate complexes, their analogs, hybrid systems, and derivatives have been prepared and characterized:

5. Tris-pyrazoloximate pseudoclathrochelates
6. Hybrid dioximatopyrazoloximates (clathrochelatoscorpionates) and polytopic phthalocyanino- and porphirinoclathrochelates
7. Hybrid polyhedral compounds, such as *closo*-borato- and carboranoclathrochelates
8. Clathrochelate-based polynuclear complexes, metallomacrocycles, coordination cages, polymers, and MOFs, using cage metal complexes with terminal donor group(s) as ligand synthons

In addition to classical synthetic approaches of organic, elementoorganic, and macrocyclic chemistry (including nucleophilic substitution with highly active nucleophiles, transmetallation, and template self-assembly on an appropriate metal ion as a matrix), the elaborated new methods for their preparation have been developed until 2002:

1. Free-radical substitution of the clathrochelate precursors
2. Their metal-promoted nucleophilic substitution with low-active nucleophiles
3. Other types of their metal-promoted (catalyzed) substitution reactions
4. Metal-promoted hydrodehalogenation, halogen exchange, and reductive homocoupling reactions of clathrochelate complexes
5. Electrophilic addition to suitable ribbed-functionalized clathrochelate precursors
6. Ligand-centered redox reactions of the redox-active cage complexes
7. Their metal-centered redox reactions

The above synthetic approaches allowed obtaining a plethora of cage metal complexes, their derivatives, and modern materials based on them with unusual chemical and physical properties for a wide range of various practical applications (see Chap. 5).

References

1. Voloshin Y, Belaya I, Krämer R (2016) The encapsulation phenomenon: synthesis, reactivity and applications of caged Ions and molecules. Springer International Publishing, Switzerland
2. Busch DH (1964) Rec Chem Progr 25:107
3. Voloshin YZ, Kostromina NA, Krämer R (2002) Clathrochelates: synthesis, structure and properties. Elsevier, Amsterdam

2 Synthesis and Reactivity of Cage Metal Complexes

Synthesis of cage metal complexes of different symmetry and functionality is one of the most important tasks in the chemistry of these compounds with unique or uncommon chemical, physicochemical, and physical properties. Until publication of our first book on clathrochelate chemistry in 2002 [1], main synthetic approaches for preparation of various types of clathrochelates have included:

1. Template self-assembly on a suitable metal ion as a matrix (for all the classes of such cage complexes)
2. Coordination of an initially prepared encapsulating ligand toward an appropriate metal ion (for the conformationally labile polyamine and polyimine ligands)
3. Nucleophilic substitution with highly active nucleophiles (for the reactive halogenoclathrochelates)
4. Transmetallation (a capping group exchange) reaction [for the cage complexes with labile cross-linking group(s)]
5. Chemical transformation of apical substituent(s) at a cage framework [for the clathrochelates with reactive terminal group(s)]
6. Condensation of an appropriate macrocyclic precursor with a suitable ligand synthon
7. Macrobicyclization (cross-linking) of a reactive semiclathrochelate precursor (first of all, for the hydrazonate, oximehydrazonate, and azineoximate cage complexes)
8. Imine condensation (for the Schiff-base clathrochelates)

During last decade, the above new approaches (see Sect. 1.2) have been used for their preparation.

In most cases, the clathrochelate *d*-metal tris-dioximates with equivalent ribbed and apical fragments (Scheme 2.1a) have been synthesized either using direct template self-assembly on a metal ion as a matrix of the symmetric α-dioximes or by functionalization of their hexahalogenoclathrochelate precursors with highly active nucleophilic agents. The apical functionalization of cage complexes of this type has been performed using a suitable Lewis-acidic capping agent containing either such functionalizing substituent(s) or the reactive group(s) allowing their further functionalization by well-known organic reactions, such as imine condensation, metal-catalyzed cross-coupling, nucleophilic substitution, etc.

In the case of non-symmetric α-dioximes, a mixture of *fac*- and *mer*-isomers of the corresponding metal clathrochelates (Scheme 2.1c, d) in their statistical ratio 1:3 (Scheme 2.2) has been formed in the course of their template self-assembly.

The clathrochelates of this type with non-equivalent capping fragments (Scheme 2.1b) have been prepared either by a stepwise "assembling" using an elementoxide surface as a matrix

Y. Voloshin et al., *Cage Metal Complexes*, DOI 10.1007/978-3-319-56420-3_2

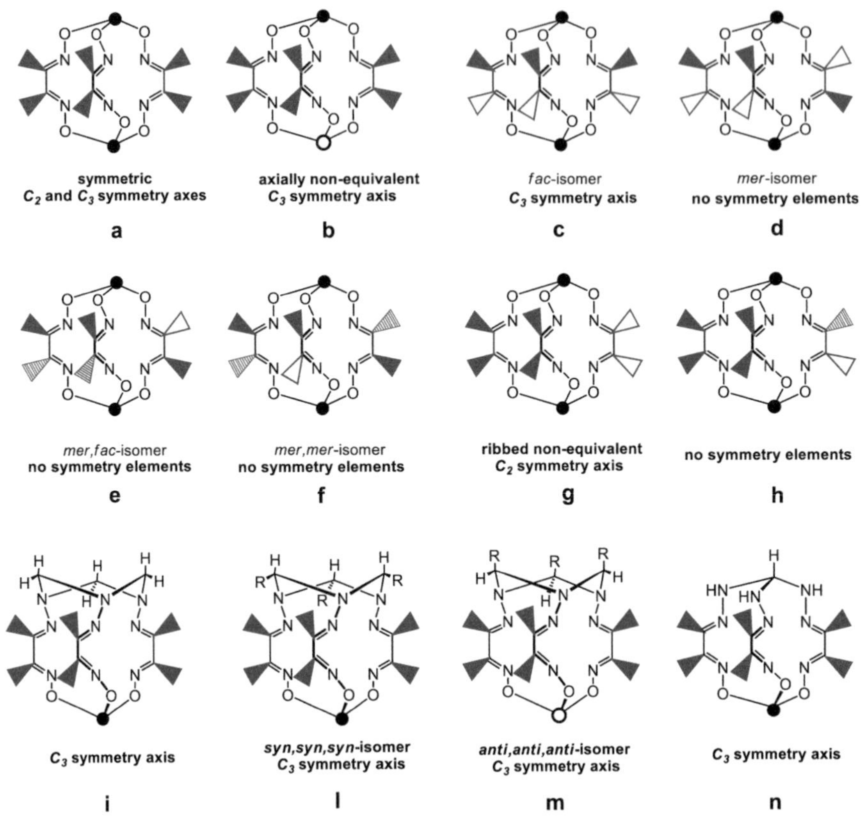

Scheme 2.1 Main types of clathrochelate frameworks [2]

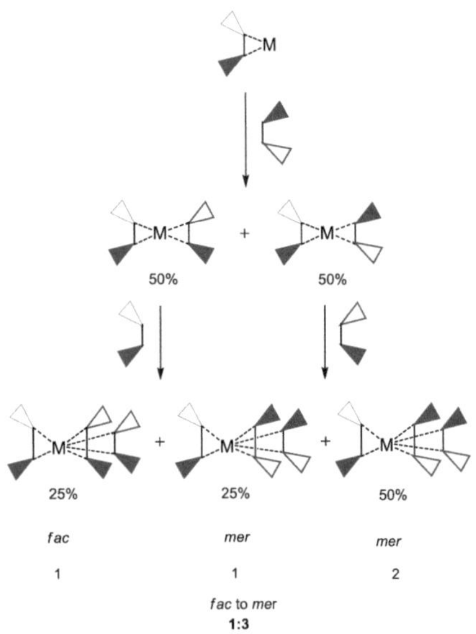

Scheme 2.2 Formation of *mer*- and *fac*-isomers of the metal clathrochelates [2]

or by capping group exchange (transmetallation) reactions. Lability of the apical cross-linking group(s) of the trialkylantimony-capped iron(II) semiclathrochelates and clathrochelates, easily undergoing a transmetallation reaction with various types of Lewis-acidic compounds, the clathrochelate derivatives of which could not be synthesized in reasonable yields using standard synthetic approaches (i.e., the template self-assembly and condensation reactions), has been used for preparation of the mono- and bis-clathrochelates with nonequivalent capping groups as well as of hybrid complexes, including the phthalocyanino- and porphirinoclathrochelates. High reactivity of the above antimony-capped clathrochelate precursors can be explained by their specific structures that are intermediate between those of the macrobicyclic complexes and those of the corresponding molecular adducts (Scheme 2.3).

In most cases, clathrochelate complexes with nonequivalent ribbed chelate fragment (Scheme 2.1g, h) have been synthesized by condensation (cycloaddition) of the corresponding α-dioxime or pyrazoloxime to a difluoroboron-cross-linked metal bis-dioximate as a reactive macrocyclic precursor. In the case of mono- or dihalogenodioximes, the obtained fluoroboron-capped mono- or dihalogenoclathrochelates undergo nucleophilic, free-radical, or metal-catalyzed (promoted) substitution, thus giving mono- or disubstituted monoribbed-functionalized cage complexes. In several cases, the ribbed functionalization of such fluoroboron-capped cage complexes has been performed with a suitable α-dioxime with functionalizing substituent(s) as a ligand synthon. On the other hand, mono- and diribbed-functionalized clathrochelates with other types of apical substitute(s) [first of all, those with functionalizing group(s)] could not be prepared using the above synthetic pathway.

The feasible pathways for the synthesis of such mono- and diribbed-functionalized iron(II) clathrochelates with apical functionalizing substituents, shown in Scheme 2.4, have been compared [3]. The dihalogeno-α-dioximate ligand synthons are substantially more weaker donors than their aliphatic and aromatic analogs, and the use of mixture of the corresponding α-dioximes in the direct template condensation (Pathway I) is described to give mainly the C_3-symmetric clathrochelate products with equivalent ribbed fragments but not the cage complexes with nonequivalent chelate moieties. Condensation of the macrocyclic bis-α-dioximates with these weakly donor dihalogenodioximes (Pathway II) typically proceeds under vigorous reaction conditions, also mainly giving the boron-capped C_3-symmetric aliphatic or aromatic tris-dioximates as the clathrochelate products of a symmetrization reaction. The halogenation of the apically functionalized glyoxime macrobicyclic complexes (Pathway III) is complicated by side reactions of their apical and ribbed substituents as

Scheme 2.3 Equilibrium between the macrobicyclic and "onic" forms of the antimony-capped clathrochelates

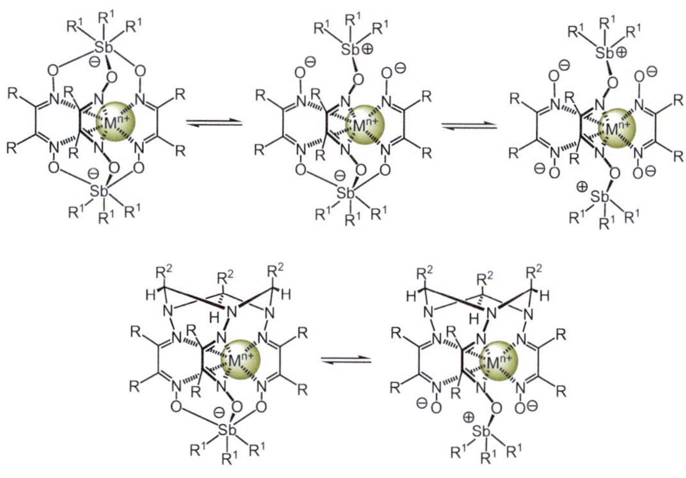

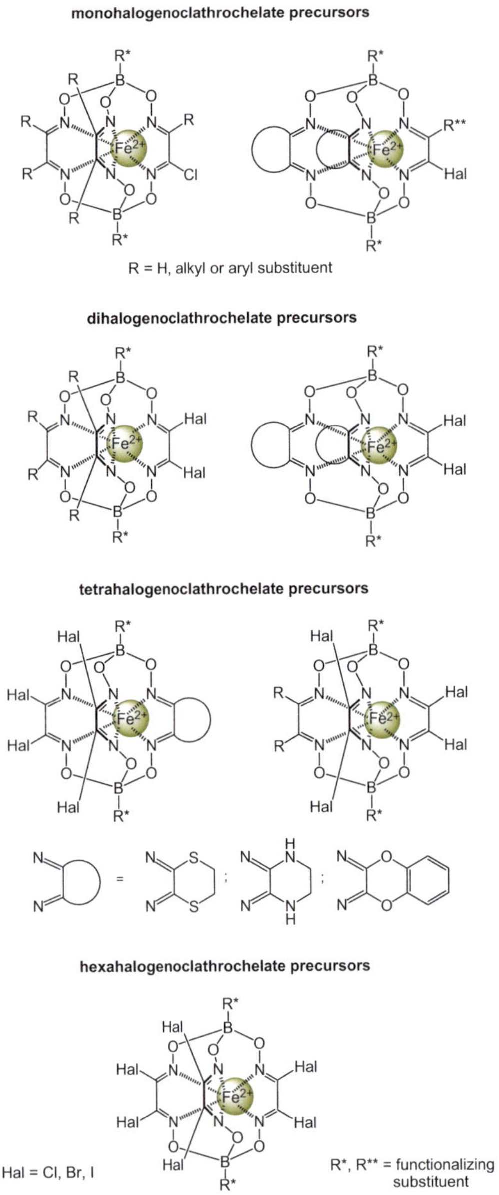

Scheme 2.4 Main types of halogenoclathrochelate precursors

well as by complete destruction of their cage frameworks under these vigorous reaction conditions. So, Pathway IV is described to be most convenient for this purpose: the hexahalogenoclathrochelate precursors can be obtained in high yields using known synthetic procedure, and their nucleophilic substitution with several bis-nucleophiles forming the stable cyclic six-membered ribbed fragments can be performed stepwise, thus giving the corresponding di- and tetrahalogenoclathrochelates shown in Schemes 2.5 and 2.6. Nucleophile substitution of the *vic*-dichloroclathrochelates with secondary aliphatic amines as *N*-nucleophiles is known to give mono-amine cage complexes as the macrobicyclic products of a substitution of only one of the two halogen atoms in the *vic*-position of the same ribbed fragment. This partial functionalization allows obtaining the corresponding monohalogenoclathrochelates (Scheme 2.5), which can undergo further nucleophilic substitution with more active nucleophiles, such as primary aliphatic amines and thiolate anions. Analogous synthetic pathway can be used for the synthesis of the tetrahalogenoclathrochelates by Scheme 2.6. So, stepwise nucleophilic substitution using different bis-nucleophilic agents has been proposed [3] for the synthesis of the di- and tetrachloroclathrochelate complexes with non-equivalent ribbed substituents, such as N_2- and S_2-containing six-membered cycles. On the other hand, the inherent electron-donating amine groups at a quasiaromatic polyazomethine cage framework are known to deactivate the remaining halogen substituents, and they can not undergo a further nucleophilic substitution even at high temperatures and with large excess of a nucleophilic agent. In general, the anionic forms of pyrocatechol, ethylenediamine, and 1,2-ethanedithiol as O_2-, N_2-, and S_2-bis-nucleophiles, generated in situ in the presence of triethylamine, are described [3] to be appropriate nucleophiles suitable for the synthesis of such clathrochelate precursors due to the formation of stable cyclic six-membered ribbed substituents at a macrobicyclic framework.

Mononuclear tris-oximehydrazonate cage complexes have been prepared using a two-step synthetic procedure that includes an isolation of the corresponding monocapped semiclathrochelate precursor with reactive hydrazonate NH_2 groups and its further macrobicyclization with active carbonyl compounds, such as formaldehyde or TOF. So, the use of either an appropriate carbonyl ligand synthon for such macrobicyclization or a suitable

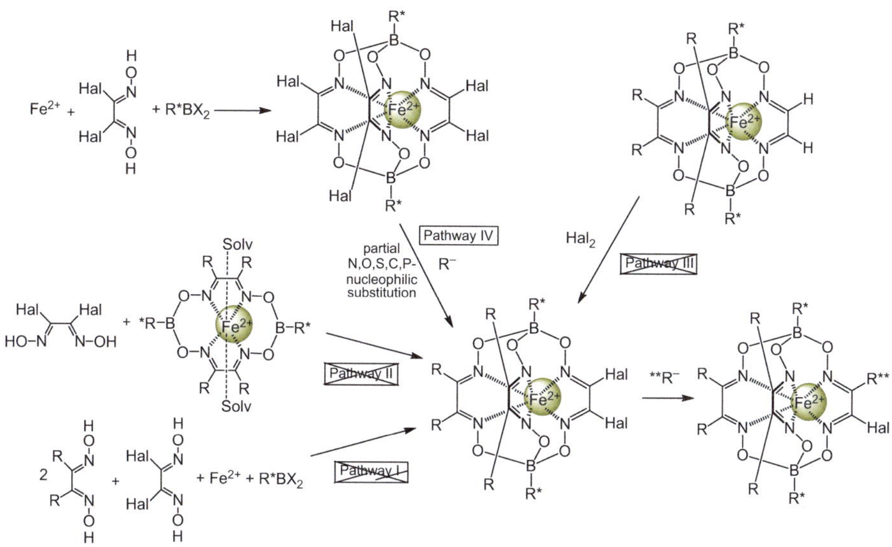

R = aliphatic or aromatic substituent; (R,R) = alicyclic fragment or its N_2-, O_2- or S_2- containing analog
R*, R** = functionalizing substituents

Scheme 2.5 Main synthetic pathways to the iron(II) dihalogenoclathrochelates [3]

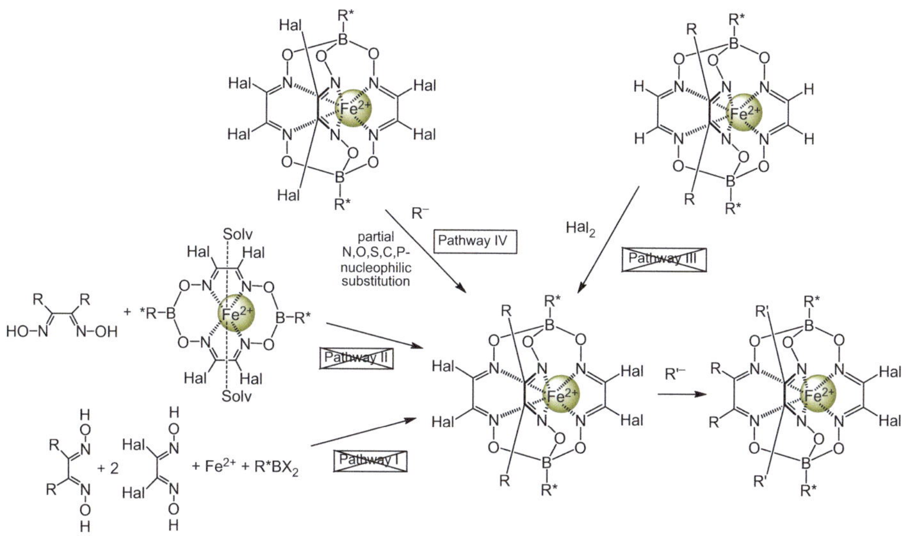

R,R' = aliphatic or aromatic substituent; (R,R), (R',R') = alicyclic fragment or its N_2-, O_2- or S_2-containing analog
R* = functionalizing substituent

Scheme 2.6 Main synthetic pathways to the iron(II) tetrahalogenoclathrochelates [3]

Lewis-acidic capping agent with functionalizing substituent(s) allows their apical functionalization. The latter approach has been also used for apical functionalization of the pseudoclathrochelate metal tris-pyrazoloximates, which have been prepared by template self-assembly of a pyrazoloxime ligand synthon and an appropriate boronic acid on the corresponding metal ion as a matrix. On the other hand, condensation of this pyrazoloximate ligand synthon with a macrocyclic bis-dioximate precursor allowed obtaining the first representative of hybrid dioximatopyrazoloximate cage complexes, the clathrochelatoscorpionates. Lability of the single triethylantimony-capping group of the corresponding macrobicyclic iron(II) tris-oximehydrazonates has been used for preparation of hybrid oximehydrazonate phthalocyaninoclathrochelates.

Apical functionalization of metal-free polyamine macrobicyclic ligands with one or two terminal NH_2 group(s) at their capping carbon atom(s) is reported [4] to be a real challenge, as they have either seven or eight reactive centers (one or two apical primary and six ribbed secondary amino groups) resulting in a poorly separable mixture of the products. However, in their cage complexes, the donor secondary amino groups are deactivated by their coordination to an encapsulated metal ion, and therefore, their apical functionalization can be easily performed using classical organic reactions. Lability of the polysaturated macrobicyclic ligands of these cage complex allowed for their demetallation, thus giving metal-free caging chelators with functionalizing apical substituent(s). These ligands can encapsulate a series of suitable metal ions (including the radioactive ones) to form their polyamine clathrochelates. Another pathway to such apically functionalized cage complexes (in particular, those with long-chain aliphatic substituent(s)) is based on the use of an appropriate ligand synthon in their template self-assembly. The same synthetic approaches have been developed for apical functionalization of their azathio-containing $N_{6-X}S_X$-macrobicyclic analogs. Ribbed functionalization of the above polyamine clathrochelates has been performed in only a few cases and used the reactivity of their secondary donor amino groups.

In most cases, the Schiff-base covalent capsules have been synthesized by imine condensation of the corresponding aldehyde and amine ligand synthons (in several cases, on an appropriate template metal ion as a matrix); their further reduction gave the corresponding polyamine caging ligands.

Other types of polyamine, polyoxaamine, and pyrocatechol-based encapsulating ligands as well as their sulfur-containing analogs have been prepared using classical methods of organic and macrocyclic chemistry (including, in particular, the condensations in high dilution conditions). It should be noted that this synthetic approach cannot be used in the case of the above pseudo- and clathrochelate dioximates, oximehydrazonates, and pyrazoloximates, as their caging ligands do not exist in their metal-free states due to thermodynamical reasons.

2.1 Tetra- and Pentadentate Encapsulating Ligands and Their Complexes

The *tren*-capped calix[4]arene-based caging ligands **1** and **2** are reported [5] to encapsulate the first-row transition metal cations, such as cobalt, nickel, and copper(II) ions by Scheme 2.7, thus giving their 1:1 cage complexes; those with an encapsulated copper(II) ion were found to be the most stable. The relatively small Zn^{2+} ion is described to form both the host–guest 1:1 and 1:2 complexes; a binding affinity of these ligands to the second zinc(II) ion is affected by the cavity size of an encapsulating ligand.

The calix[6]arene-based caging ligand **3** with a tripodal amine-capping fragment was found [6] to form its heteroguest 1:1:1 cage complex with the encapsulated zinc(II) ion and co-encapsulated acetonitrile molecule by Scheme 2.8. The same encapsulating ligand also gave a zinc(II) clathrochelate with co-encapsulated ethanol molecule [7]. Coordination of **3** toward copper(II) ion has been used [8] for preparation in a high yield of the corresponding 1:1 cage complex also shown in Scheme 2.8.

The copper(II) cage complex of the tris-pyridine encapsulating ligand **4** (Scheme 2.9) is reported [9] to undergo the solvent exchange reactions of its coordinated water molecule with organic guests (first of all, with π-acceptor molecules of acetonitrile and DMF).

The N_5-donor macrobicyclic ligand **5** with one ribbed 2,2′-bipyridine fragment has been prepared [10] using condensation of the

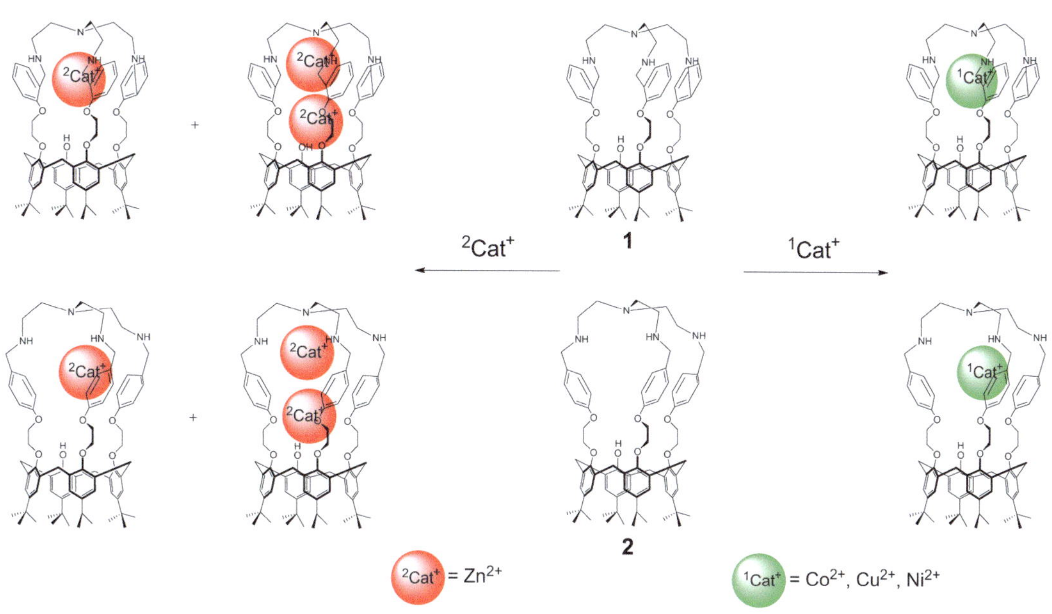

Scheme 2.7 Preparation of calix[6]arene-based cage complexes

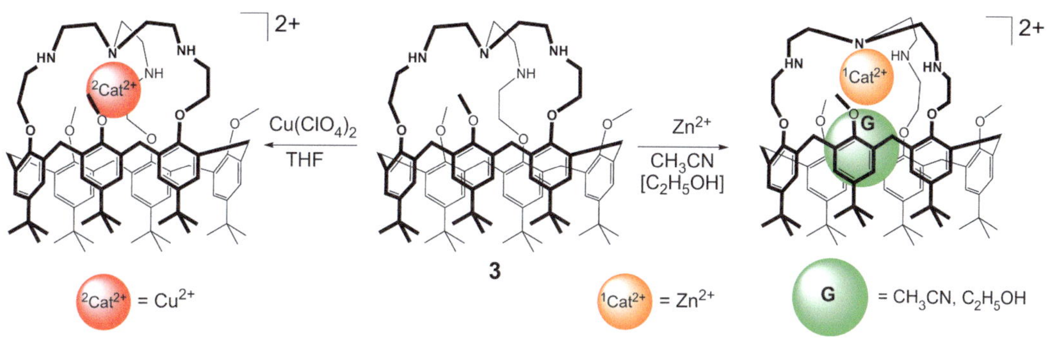

Scheme 2.8 Synthesis of calix[6]arene-based cage complexes

corresponding protected terminal diol with a suitable diamine macrocyclic ligand synthon *1* by Scheme 2.10; this ligand was found to encapsulate copper(II) ion giving the corresponding 1:1 cage complex.

Condensation of the tetraazamacrocyclic ligand synthon *2* with 2,6-bis-(bromomethyl)-pyridine in the presence of Na_2CO_3 by Scheme 2.11 has been used [11] for preparation of the pyridine-strapped amine macrobicyclic ligand **6** that efficiently binds copper(II) ion, thus forming its 1:1 cage complex.

A series of the bis-N_5-macrobicyclic ligands **7–11** with cross-linking pyridine groups is reported [12] to form 1:1 copper(II) and cobalt(III) cage complexes by Scheme 2.12.

1:1 europium, terbium, and gadolinium(III) cage complexes of the macrobicyclic ligand **12** have been synthesized [13] by Scheme 2.13 using an ionic exchange reaction of the encapsulated lithium(I) ion in its initially prepared cage complex as a clathrochelate precursor.

A series of the macropolycyclic cage compounds **13–29** have been prepared [14] by simple and straightforward direct reactions between the corresponding diamines and the suitable bis-(bromomethyl) ligand synthons using

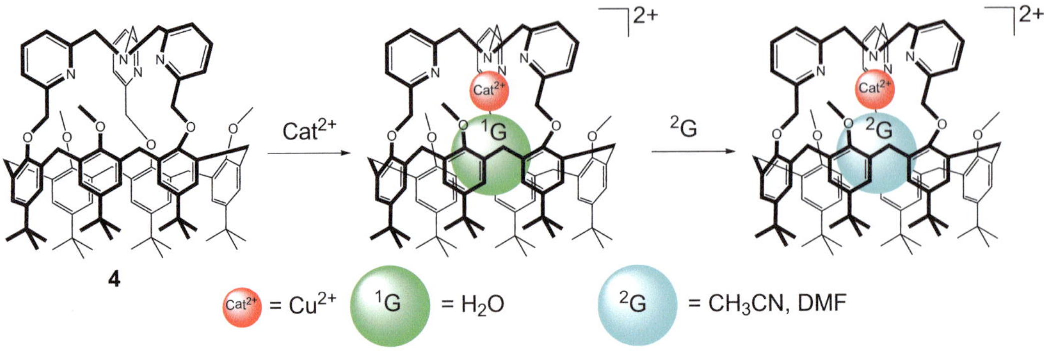

Scheme 2.9 Preparation of the tris-pyridine clathrochelates

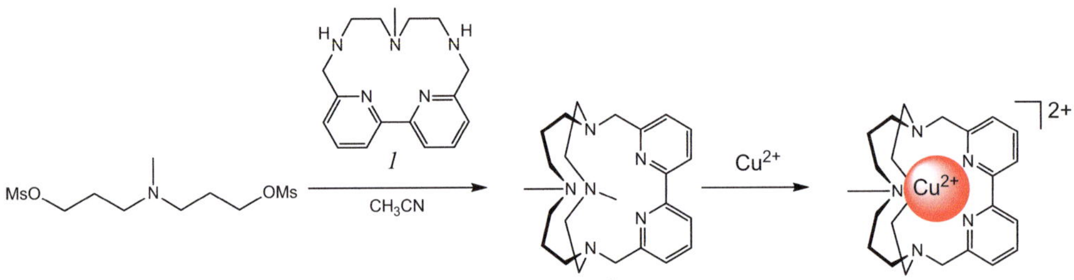

Scheme 2.10 Synthesis of the 2,2′-bypyridine-containing ligand **5** and its copper(II) complex

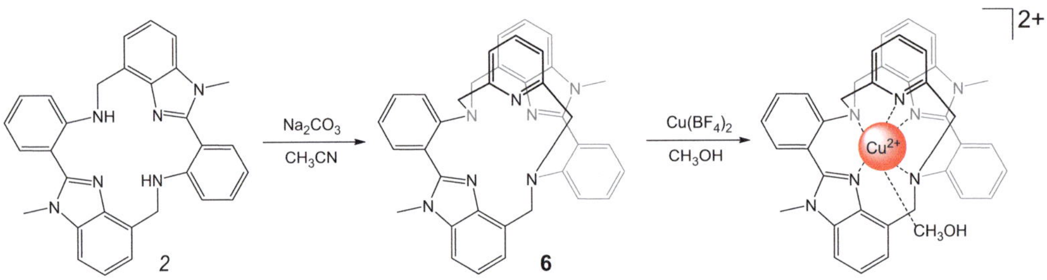

Scheme 2.11 Preparation of the pyridine-strapped macrobicyclic ligand **6** and its copper(II) complex

two synthetic approaches shown in Schemes 2.14, 2.15, and 2.16; their hexanitro-containing analog **31** has been synthesized using direct nitration of **30** by Scheme 2.17. Template effect of a metal ion has not been found in the above coupling reactions: the yields of their metal-free forms and those of the corresponding 1:1 potassium(I) clathrochelates are reported [14] to be almost the same.

Mono- and homodinuclear 1:1 and 1:2 copper and silver(I) cage complexes of the encapsulating ligands **32** and **33** have been synthesized [15] using exchange reactions of an encapsulated metal ion of their potassium(I)-encapsulating clathrochelate precursors by Schemes 2.18 and 2.19, respectively. Demetallation of these metal (I) complexes allowed to obtain the corresponding metal-free macropolycyclic compounds.

Classical approaches of macrocyclic chemistry shown in Scheme 2.20 have been used [16] for preparation of the macrobicyclic compounds **34** and **35**; template effect of alkali metal cations has not been observed. Smaller of them, the

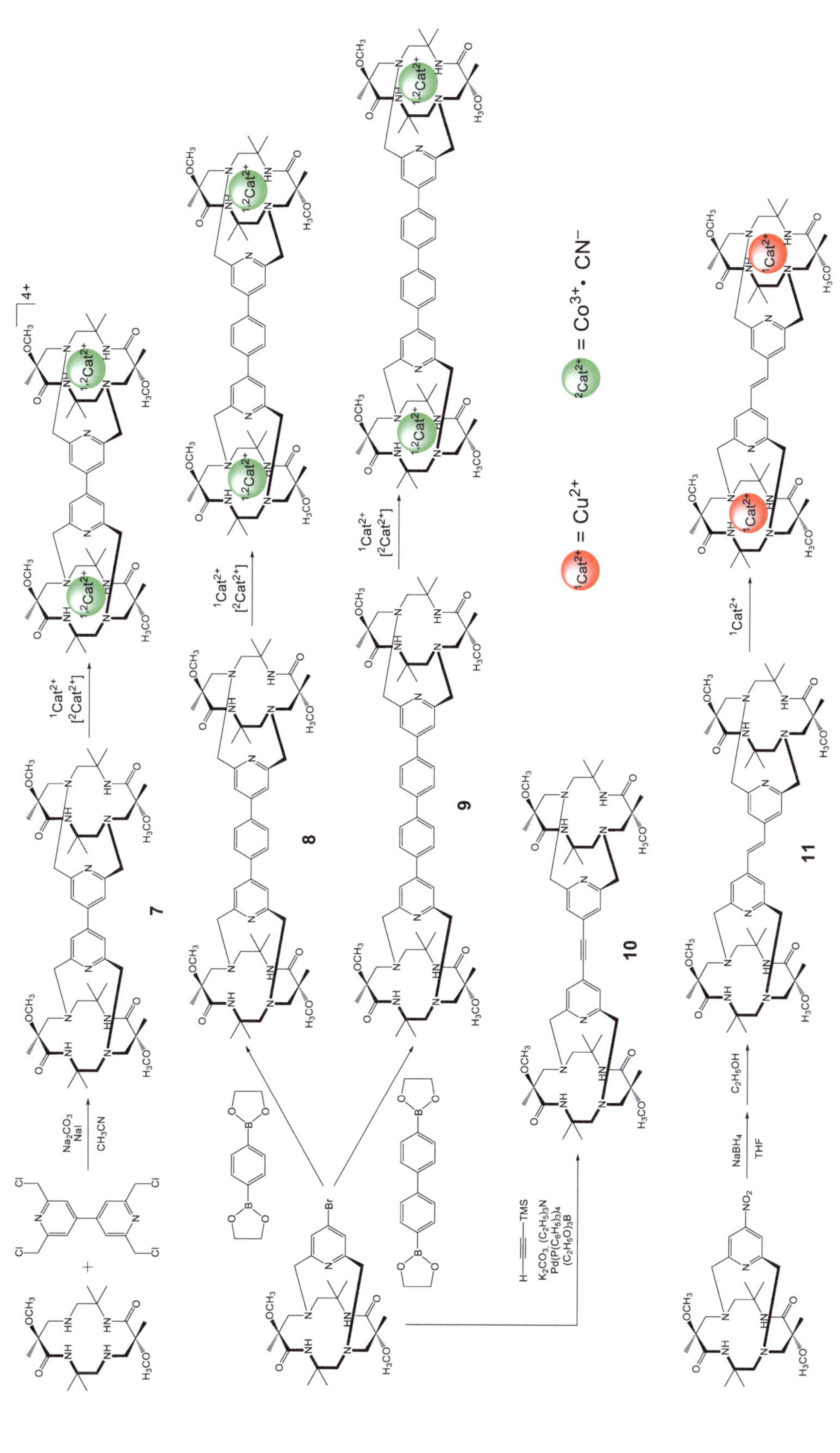

Scheme 2.12 Synthesis of the bis-macrobicyclic ligands **7–11** and their cage complexes

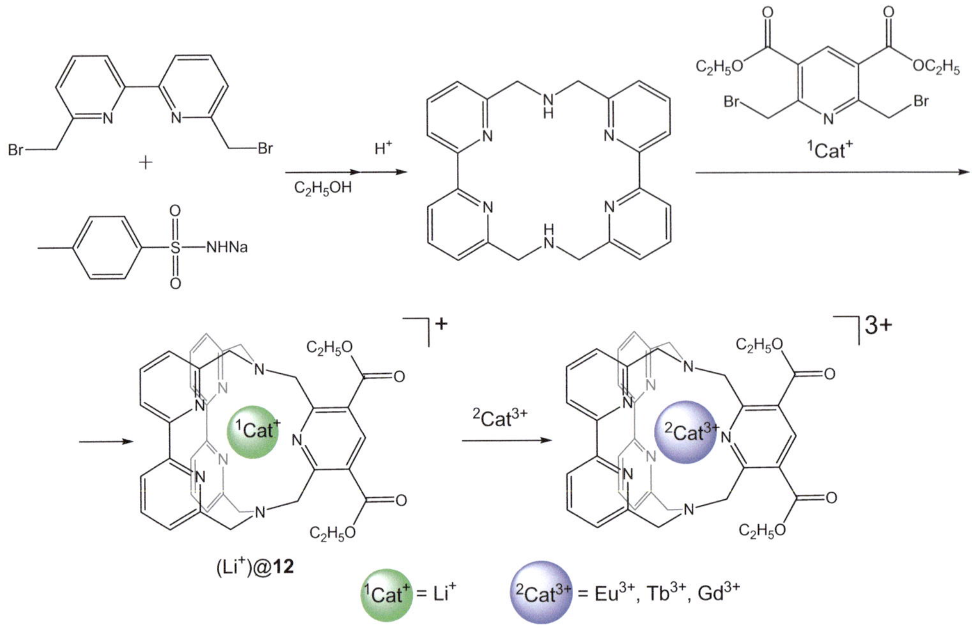

Scheme 2.13 Preparation of the bis-pyridine macrobicyclic ligand **12** and its cage complexes

macrobicycle **34**, selectively encapsulates the lithium(I) ion, whereas its homolog **35** with a bigger cavity is able also to bind Na^+ and K^+ cations, thus giving their 1:1 cage complexes [16].

The N_5-pentadentate macrobicyclic ligand **36** (Scheme 2.21) was found [17] to encapsulate the lithium(I) and copper(II) cations, thus forming their 1:1 cage complexes. Its analog **37** (Scheme 2.22) is described [18] to bind a wide range of metal(I) and metal(II) cations, but only 1:1 sodium(I) and zinc(II) cage complexes of **37** have been isolated and X-ray structurally characterized.

The N_4O-macrobicyclic ligand **38** (Scheme 2.23) is reported [19] to encapsulate one lithium(I) or magnesium(II), or zinc(II) ion, forming their 1:1 cage complexes.

The covalent capsules **32**, **33**, and **39** with donor pyridyl groups and their 1:1 potassium (I) cage complexes have been synthesized [20] by Scheme 2.24.

1:1 rubidium(I) cage complex of the fluorinated caging ligand **27** (Scheme 2.25) has been prepared [21] and characterized by single-crystal X-ray diffraction.

The N_4O-macrobicyclic compound **40** has been prepared [22] by Scheme 2.26 starting from the tetraazamacrocyclic precursor *3*. Its pentaaza macrobicyclic analog **41** (Scheme 2.27) is reported [23] to selectively encapsulate the lithium(I) or copper(II) cation, thus giving the corresponding 1:1 cage complexes.

The dibenzofurane- and diphenyl ether-containing N_4O-caging ligands **42** and **43** are described [24] to selectively bind one copper or zinc(II) ion, forming their 1:1 clathrochelates shown in Scheme 2.28.

2.2 Mononucleating Hexadentate Ligands and Their Cage Complexes

2.2.1 Polyamine Caging Ligands and Their *O*-, *S*-containing Analogs

2.2.1.1 Template and Non-template Synthesis

A series of the hexaoxaamine macrobicyclic compounds **44–49** with pendant functionalizing

Scheme 2.14 Synthesis of the macropolycyclic compounds **13–20**

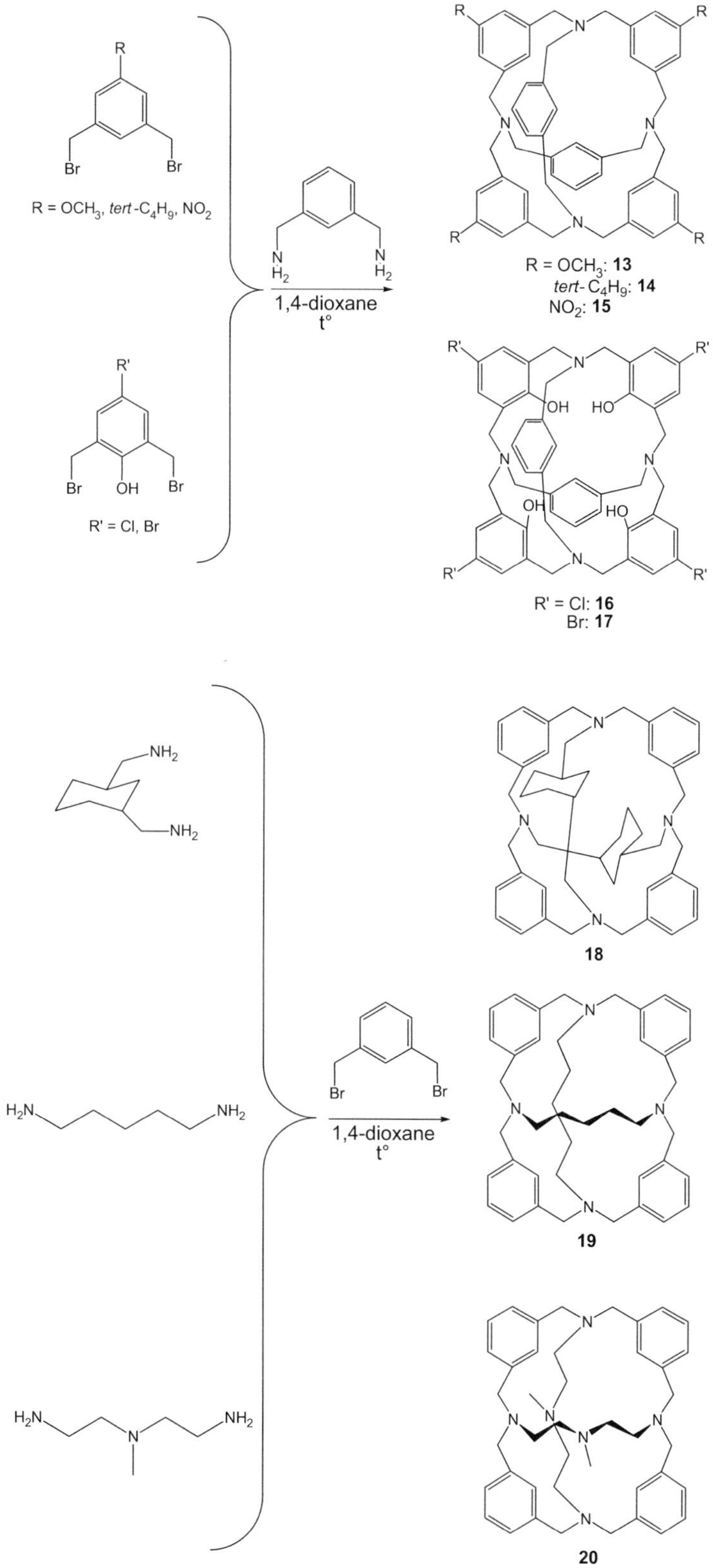

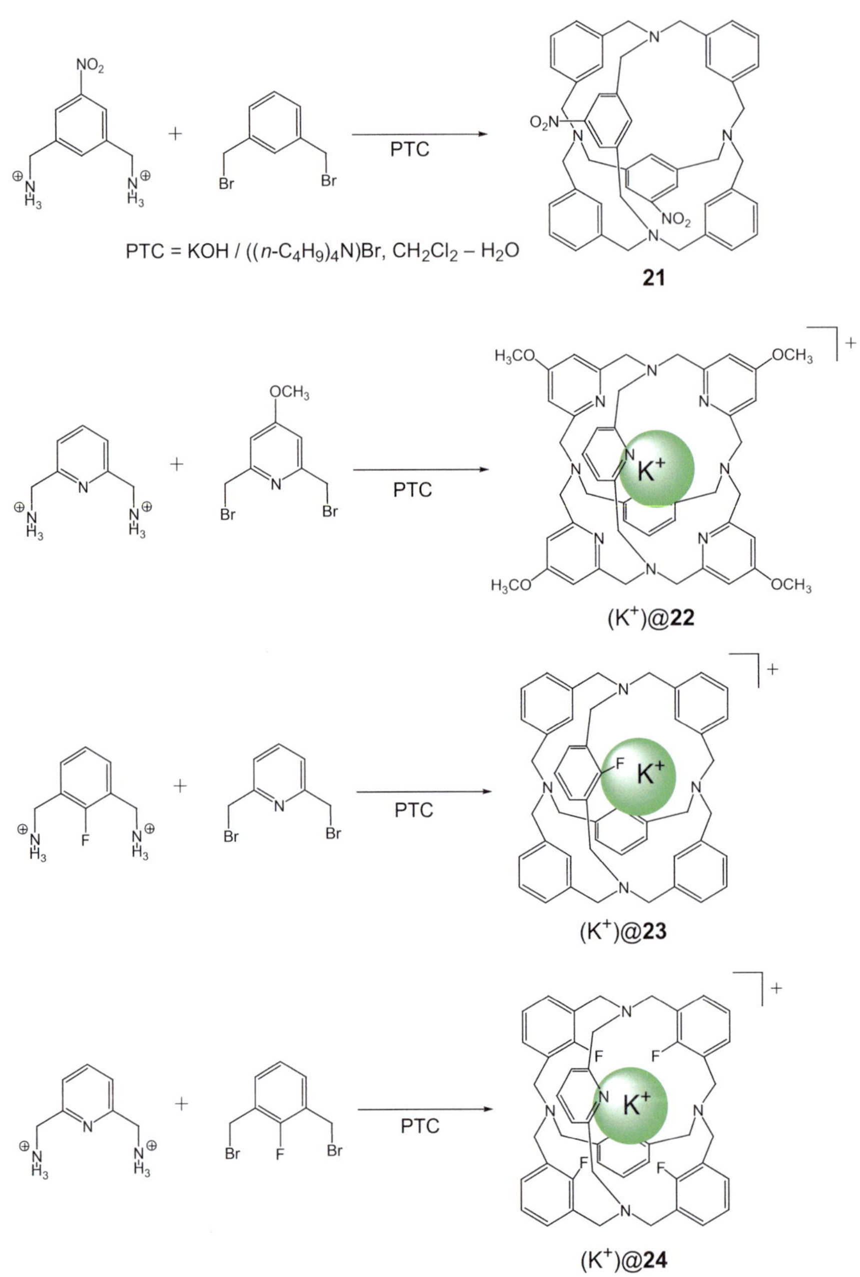

Scheme 2.15 Preparation of the macropolycyclic compound **21** and potassium(I) complexes of the ligands **22–24**

substituents has been prepared [25] by Scheme 2.29. The same macrobicyclic ligands and their analogs shown in Scheme 2.30 are reported [26] to form their 1:1 cage complexes with encapsulated *3d*-transition and main groups metal ions (including alkali and rare-alkali cations, Scheme 2.31).

The nonsymmetric oxaaza macrobicyclic caging ligand **58** is reported [27] to easily encapsulate one silver(I), zinc, cadmium, or lead(II) ion, forming their 1:1 cage complexes by Scheme 2.32.

Extraction capacity of the macrobicyclic ligand **59** to bind alkali metal halides has been studied [28] using ^{1}H NMR and ESI-MS

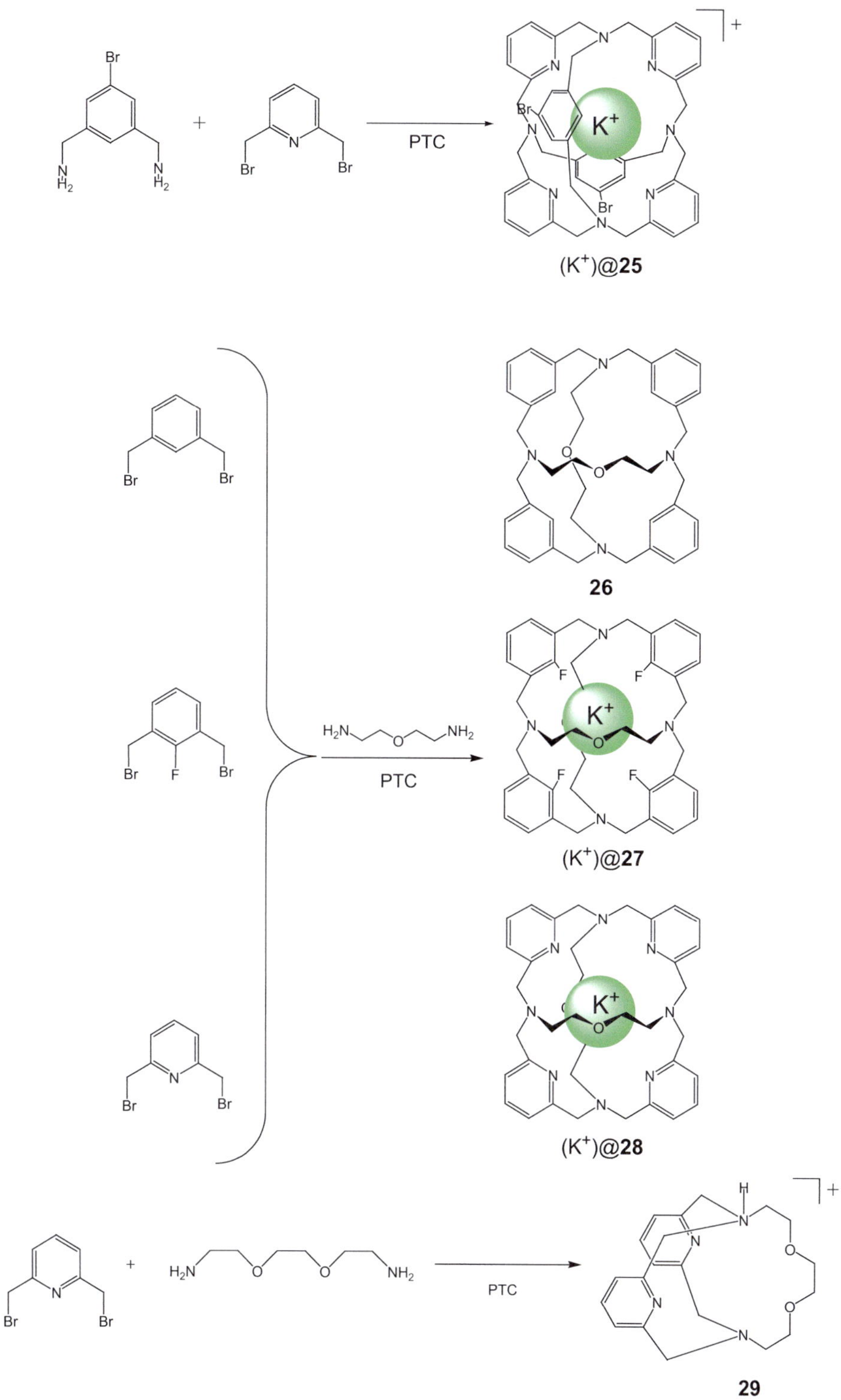

Scheme 2.16 Synthesis of potassium(I) complexes of the ligands **25**, **27** and **28** and its analogs **26** and **29**

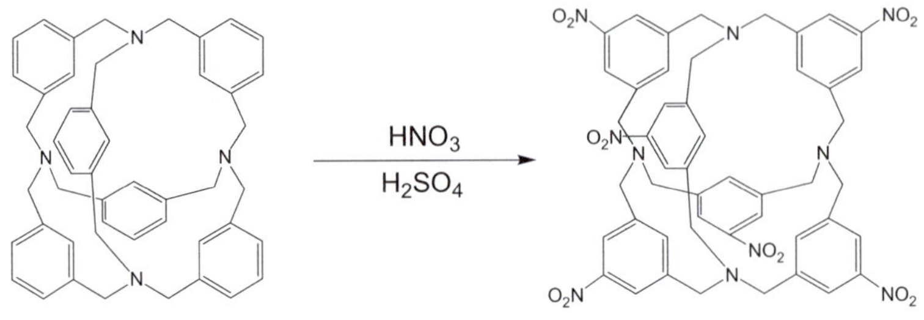

30 **31**

Scheme 2.17 Preparation of nitro-containing ligand **31**

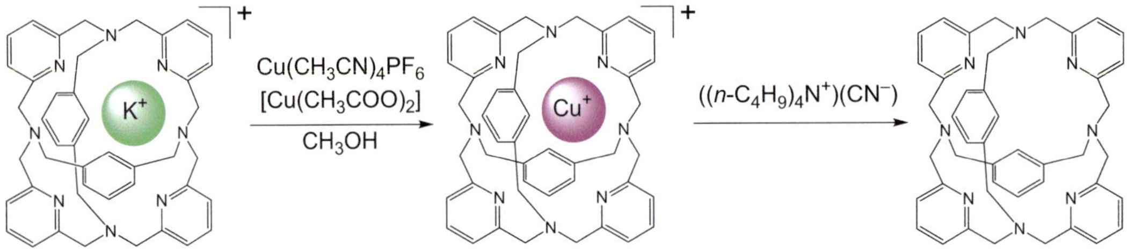

32

Scheme 2.18 Synthesis of metal-free macropolycyclic compound **32**

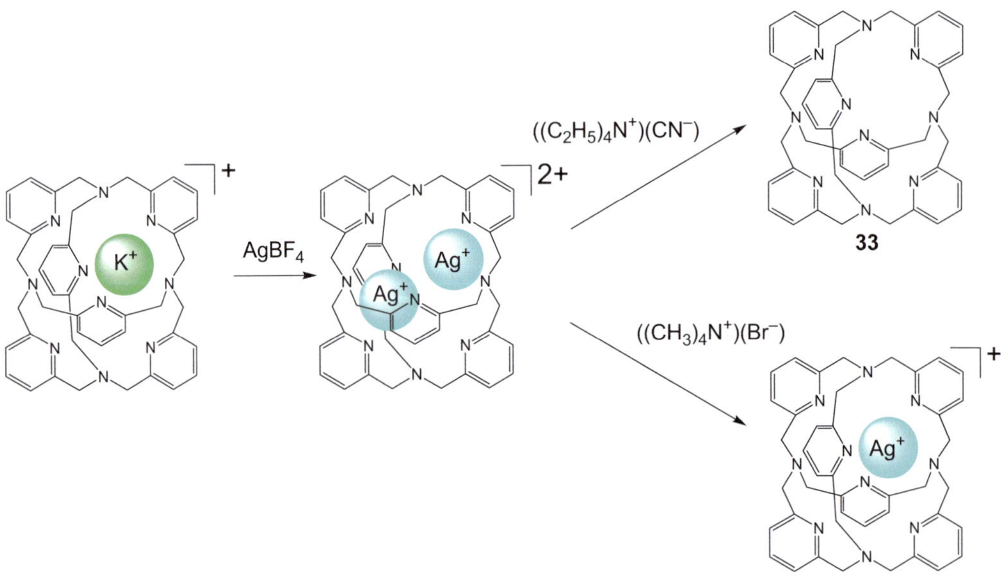

Scheme 2.19 Preparation of the ligand **33** and its silver(I) cage complexes

methods. This ligand was found to extract potassium halides from their aqueous solutions to chloroform phase, thus giving 1:1 potassium(I) cage complex (K$^+$)@**59** shown in Scheme 2.33.

The N_3O_3-macrobicyclic ligand **60** is reported [29] to form 1:1 cage complexes with $3d$-transition metal ions: in particular, it easily encapsulates copper(II) ion, forming by Scheme

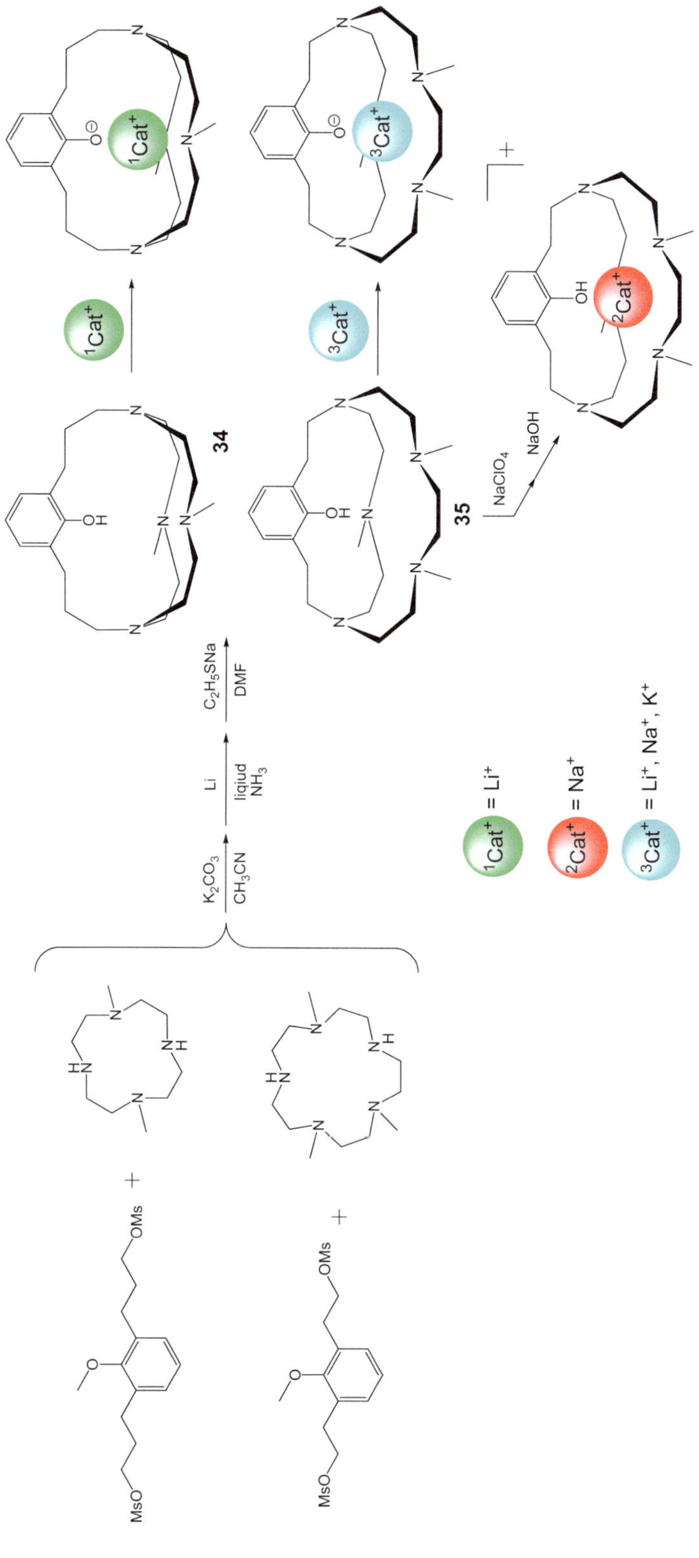

Scheme 2.20 Synthesis of the macrobicyclic ligands **34** and **35** and their cage complexes

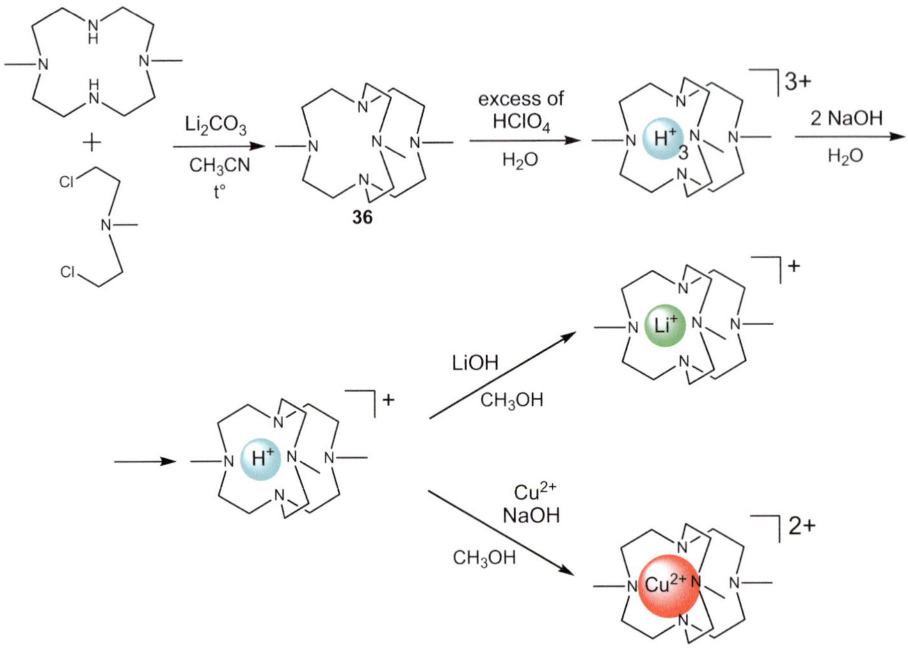

Scheme 2.21 Preparation of the macrobicycle **36** and its complexes

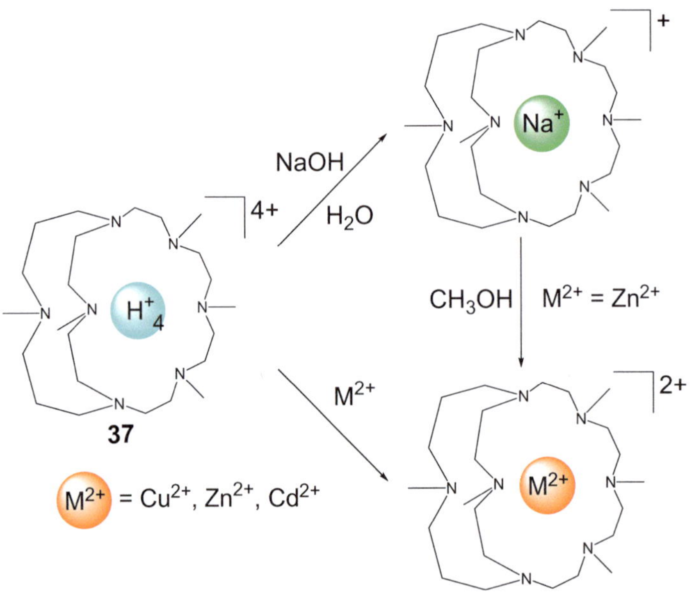

Scheme 2.22 Synthesis of clathrochelate complexes of the macrobicyclic ligand **37**

2.34 its clathrochelate with pseudotetrahedral CuN_4-coordination polyhedron.

The laterally nonsymmetric N_5O_3-macrobicyclic ligand **60**, obtained [29] by condensation of tripodal amine *tren* with tris{[2-(3-(oxomethyl)phenyl)oxy]ethyl}amine (Scheme 2.35), is reported [30] to readily form 1:1 cage complexes with encapsulated transition and main group II metal ions. These caged metal ions were found to occupy a *tren*-end of the cavity of **60**: their MN_4-coordination polyhedra are formed by three secondary

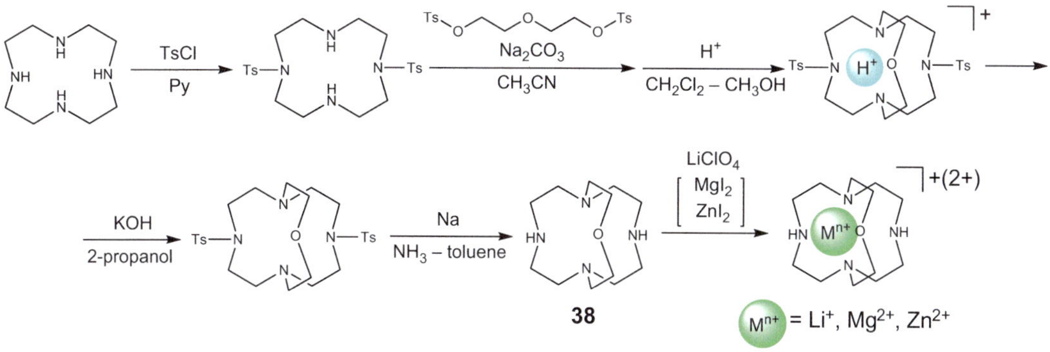

Scheme 2.23 Preparation of the N_4O-macrobicyclic ligand **38** and its cage complexes

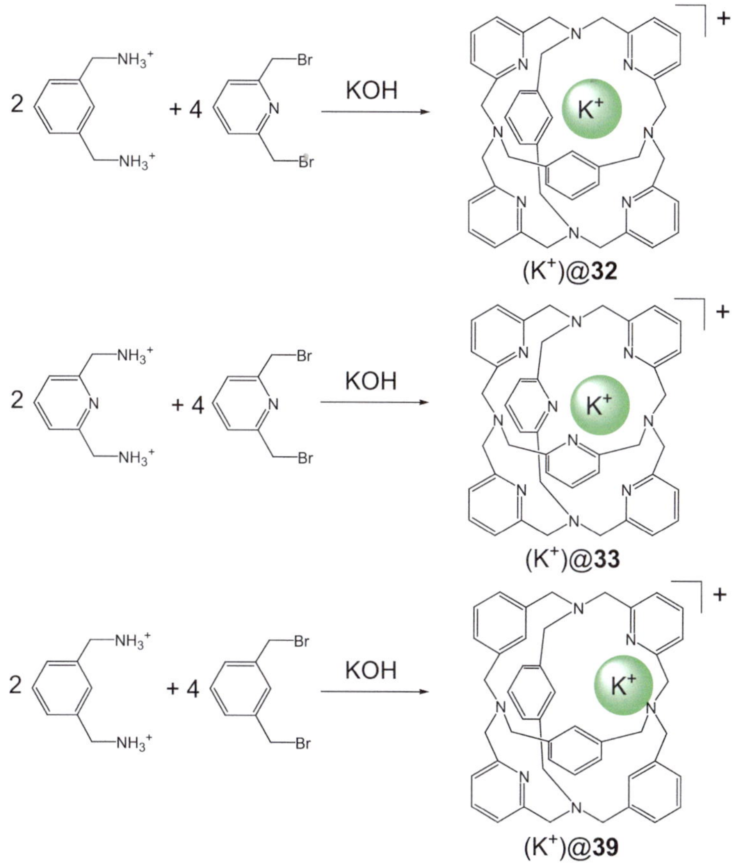

Scheme 2.24 Synthesis of potassium(I) complexes of the ligands **32**, **33** and **39**

amino and one bridgehead nitrogen atoms. The functionalization of **60** with π-accepting 2,4-dinitrobenzene substituent by Scheme 2.35 gave the macrobicyclic ligand **50**. Such functionalization caused the conjugation of a lone pair of the corresponding amine nitrogen atom of **50** with this π-acceptor group, thus reducing its availability. When nickel, copper, zinc, or cadmium(II) perchlorates or tetrafluoroborates have been used in their coordination-driven self-assembly with **50**, the corresponding metal ion was found [30] to be encapsulated within its

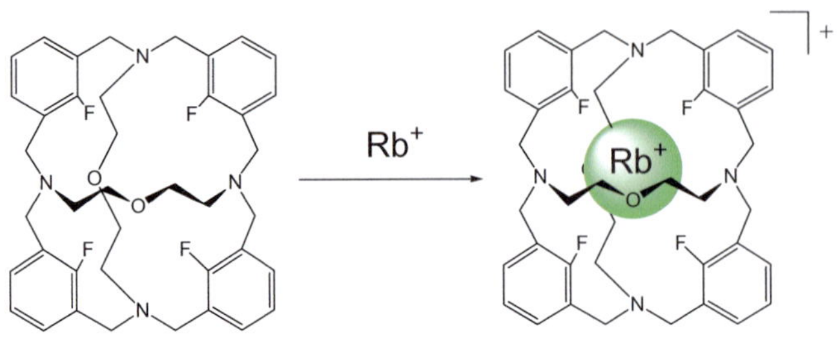

Scheme 2.25 Preparation of the clathrochelate (Rb$^+$)@**27**

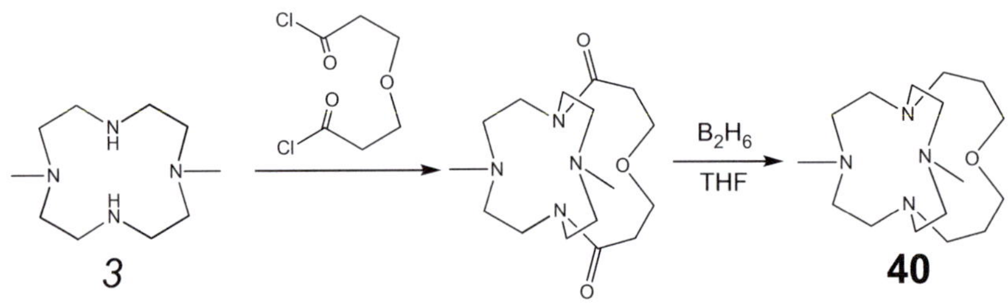

Scheme 2.26 Synthesis of the N_4O-macrobicyclic compound **40**

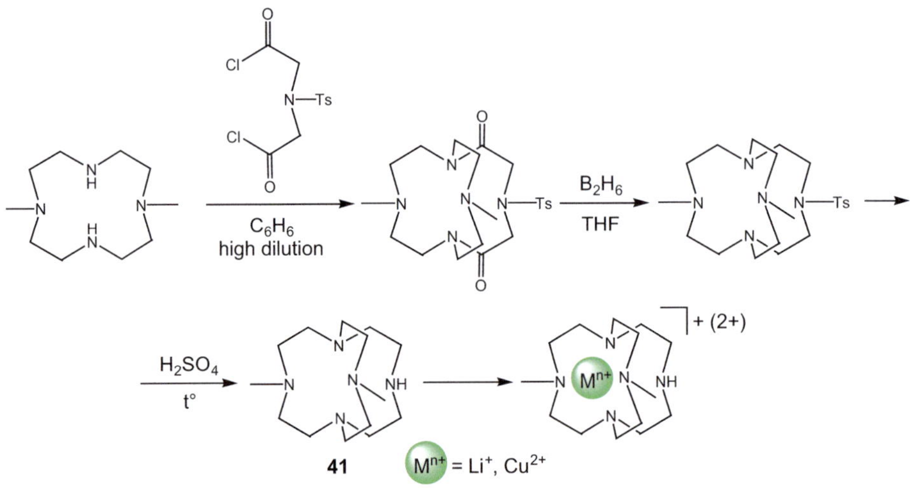

Scheme 2.27 Preparation of the N_5-macrobicyclic ligand **41** and its cage complexes

cavity, while in the presence of more strong donors (Cl$^-$, N_3^-, and SCN$^-$ anions), this metal ion came out of this cavity, binding to **50** outside of its cavity at a site away from the bulky dinitrobenzene substituent.

The same laterally nonsymmetric N_5O_3-macrobicyclic platform but with crown ether NO_4-substituent at 9- and 10-positions of the anthracenyl pendant group has been used [31] for the design and preparation by Scheme 2.36 of the polytopic caging ligand **61**, showing a fluorescence enhancement in the case of simultaneous binding of alkali and transition metal ions.

Redox-driven translocation of the encapsulated copper ion within the cavities of N_5S_3-caging ligands **62** and **63** with two distinct metal binding sites (Scheme 2.37) is reported [32]. Upon addition of one equivalent of NaBH

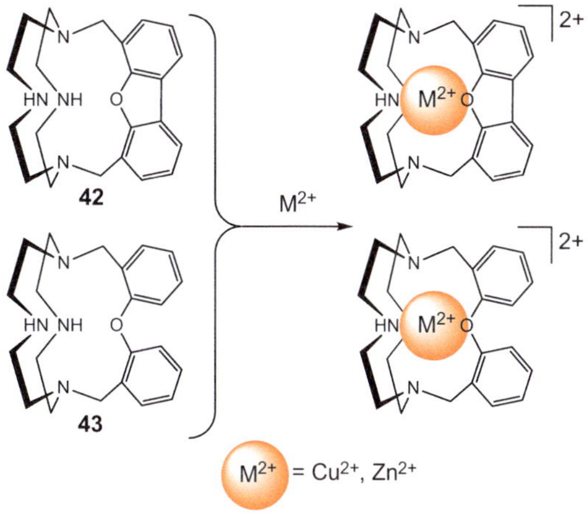

Scheme 2.28 Encapsulation of copper and zinc(II) ions by the ligands **42** and **43**

$(CH_3COO)_3$ to THF solution of 1:1 copper(II) complex of **62**, its encapsulated metal ion underwent a reduction to the copper(I) cation. This resulted in a quenching of the fluorescence of **62**: such caged ion moved away from N_4-tripodal fragment of **62**, which is preferable for the copper(II) ion, to its NS_3-cap, which is preferable for binding of the copper(I) cation as a soft Pearson acid, thus allowing PET. Upon exposure to air, the above encapsulated copper(I) ion was found [32] to undergo reoxidation to the caged copper(II) cation, moving back to the above N_4-fragment of **62**, thus blocking FRET and, therefore, causing the recovery of its fluorescence.

The biphenyl-based N_6S_2-macrobicyclic ligands **64** and **65** have been synthesized [33] using classical organic reactions and cyclization methods of macrocyclic chemistry shown in Scheme 2.38: the conversion of dialdehydes to the corresponding tetraaldehydes has been performed by treatment with ethanedithiol, while their macrobicyclic derivatives have been prepared by [2+1] condensation with two equivalents of diethylenetriamine in medium-dilution reaction conditions followed by reduction with $NaBH_4$. The followed reductive methylation of these macrobicycles with formaldehyde and formic acid under Eschweiler–Clarke conditions is reported [33] to give the above permethylated macrobicycles.

Stepwise synthesis of the pyridinophane N_4O_2-caging ligand **66** and its 1:1 lithium and sodium(I) cage complexes by Scheme 2.39 has been performed [34]. This caging ligand is reported to encapsulate the above alkali metal cations without selectivity but not the potassium(I) ion.

The bis-N_5O_3-macrobicyclic compounds **67**, **68**, and **69** with *para*-xyloyl, benzene-1,4-dicarbolyol, and 9,10-dimethylene anthracene spacers have been synthesized [35] by Schemes 2.40, 2.41, and 2.42. This allowed to evaluate the distance effect on their single- and two-step FRET processes; these macrobicycles have been probed for FRET in the presence of a metal ion as an input. The compounds **67** and **68** were found to exhibit poor performance in single-step FRET, while in the case of **69**, forming 1:1 and 1:2 cage complexes with encapsulated copper and mercury(II) ions, respectively, as an input, a large two-step FRET process is reported [35] to be operational.

Facile and efficient synthetic procedure for preparation of the nonsymmetric N_7-macrobicyclic ligand **70** in a moderate yield by Scheme 2.43 using a one-pot palladium(II)-catalyzed amidocarbonylation reaction has been evaluated [36]; this ligand was found to encapsulate copper(II) ion, thus forming its 1:1 cage complex.

The laterally nonsymmetric macrobicyclic precursors **60** and **71** have been derivatized [37] with electron-withdrawing functionalized fluorophoric 7-nitrobenz-2-oxa-1,3-diazole group(s) to give the corresponding mono-, bis-, and tris-substituted caging ligands **58** and **72–76** (Scheme 2.44), which form the fluorescent 1:1 mercury(II) cage complexes.

The N_5O_3-macrobicyclic ligand **77** (Scheme 2.45) was found [38, 39] to encapsulate one cobalt or copper(II) ion, thus giving their 1:1 cage complexes.

The crown ether-based lateral macrobicyclic ligands **78** and **79** are described [40] to bind lead(II) cation, forming its 1:1 cage complexes by Scheme 2.46.

The N_2O_6-macrobicyclic compound **80** (Scheme 2.47) has been proposed [41] as a potent

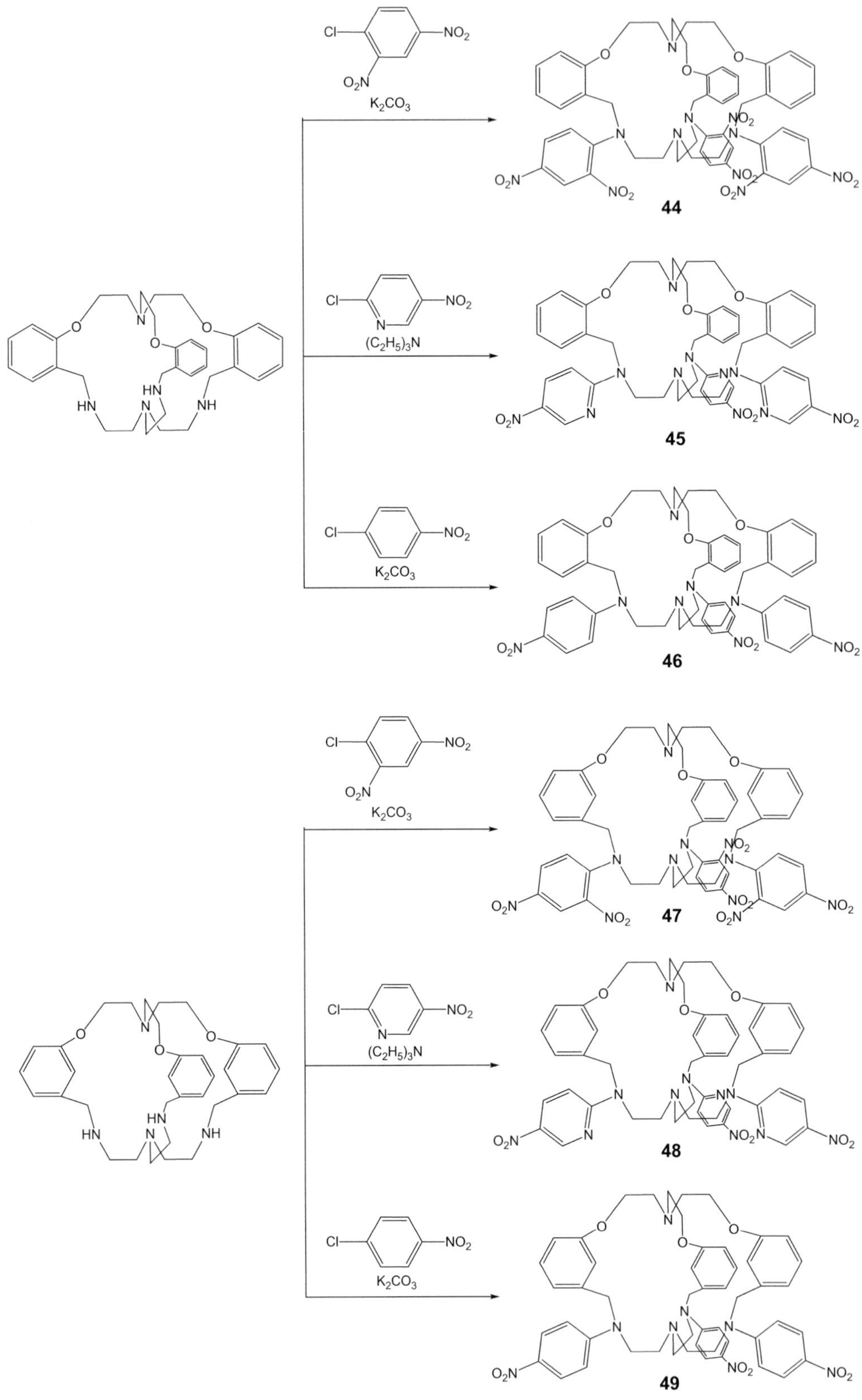

Scheme 2.29 Preparation of the nitro-containing macrobicycles **44**–**49**

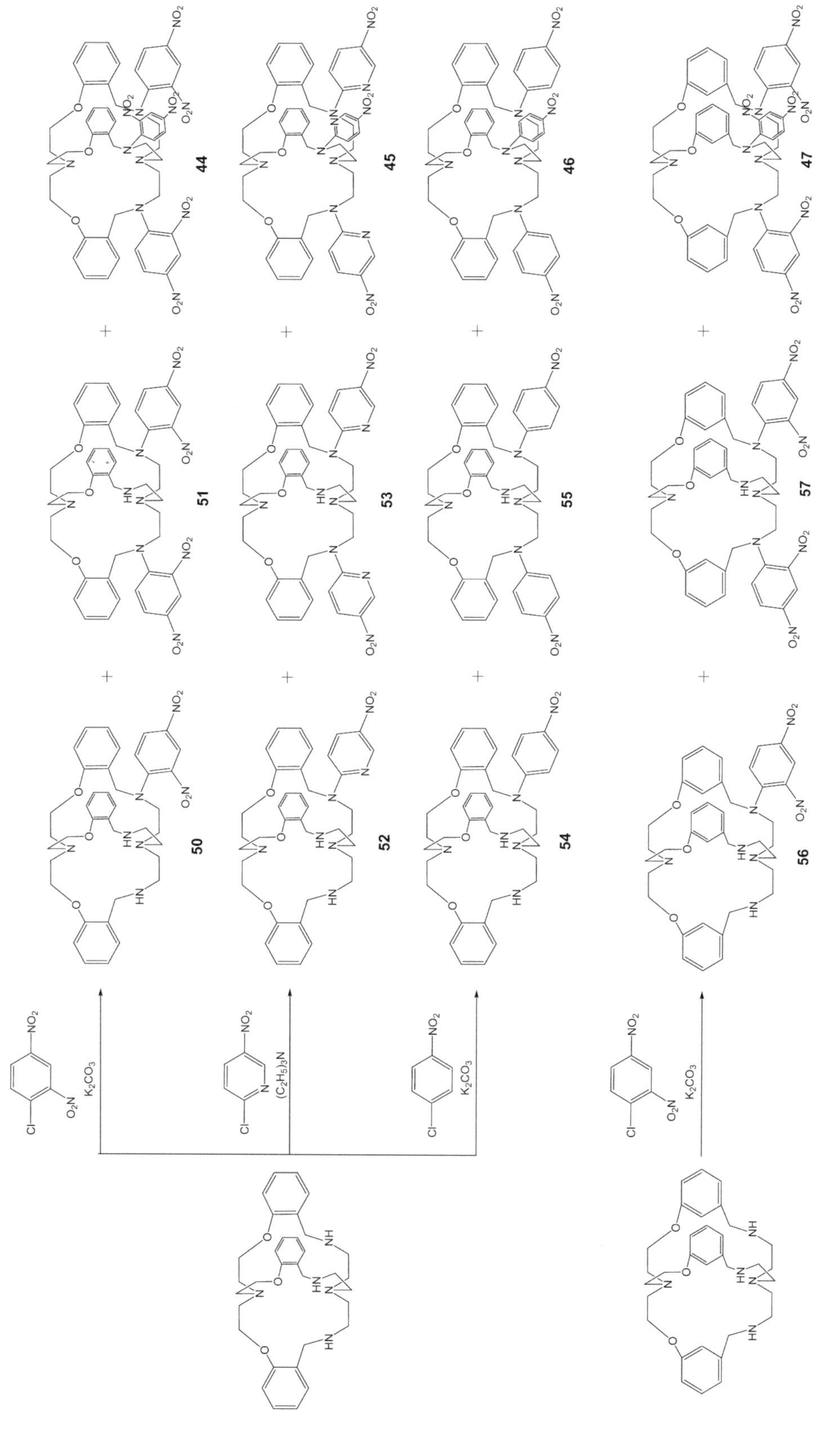

Scheme 2.30 Synthesis of a series of the nitro-containing macrobicyclic compounds

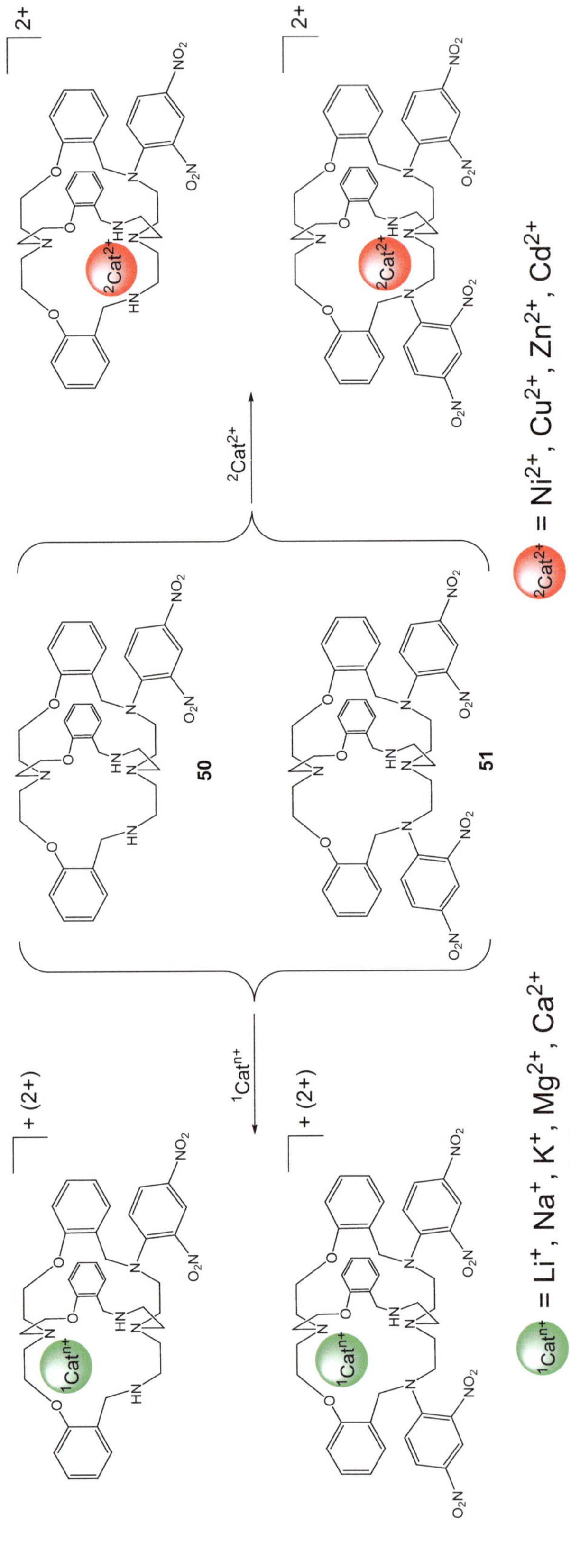

Scheme 2.31 Encapsulation of metal(I) and metal(II) ions by nitro-containing ligands **50** and **51**

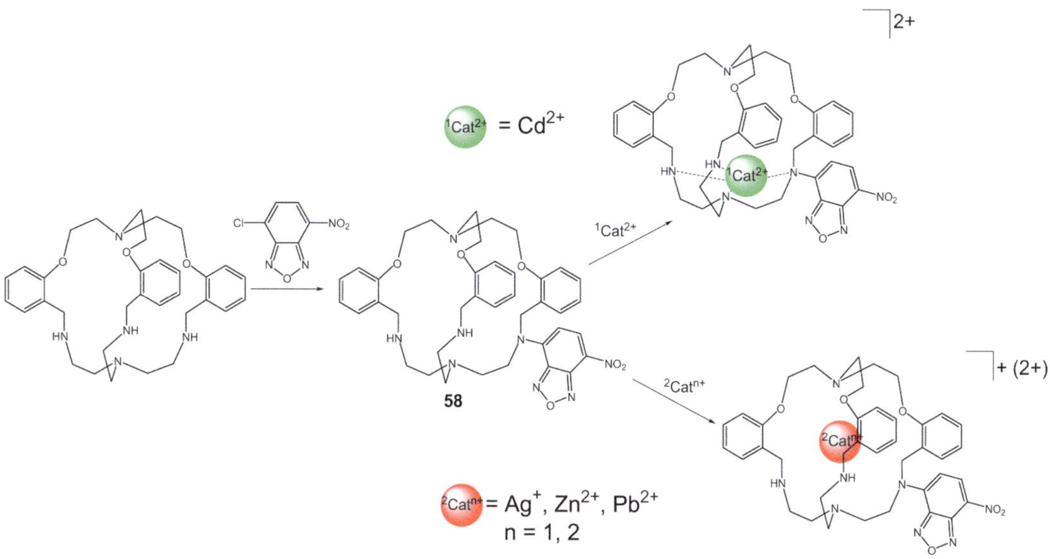

Scheme 2.32 Preparation of the nonsymmetric ligand **58** and its cage complexes

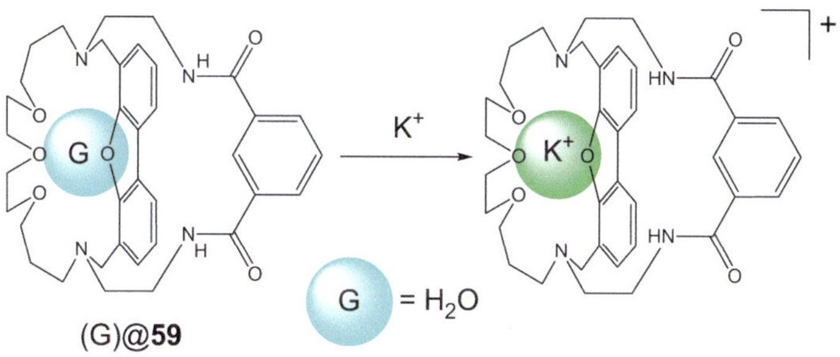

Scheme 2.33 Synthesis of the clathrochelate (K^+)@**59**

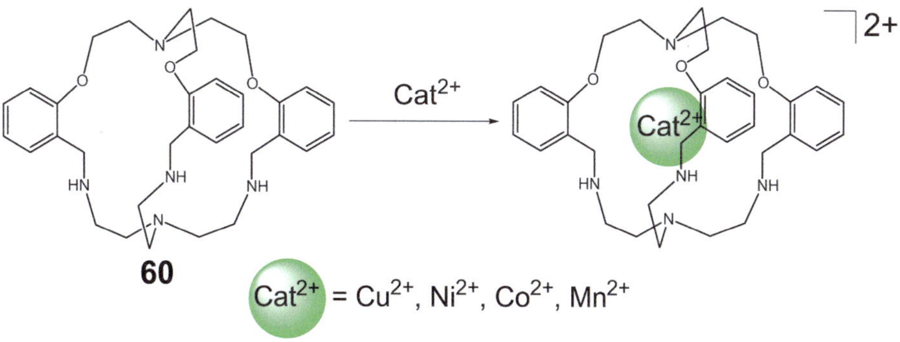

Scheme 2.34 Preparation of cage complexes of the encapsulating ligand **60**

caging ligand for encapsulation of various metal cations.

A series of the N_2O_6-macrobicyclic compounds **80–85** (Scheme 2.48) has been prepared [42] using a common stepwise synthetic approach by dialkylation of an appropriate macrocyclic precursor; the procedure for **80** is shown in Scheme 2.49. The compounds **81** and **83** (but

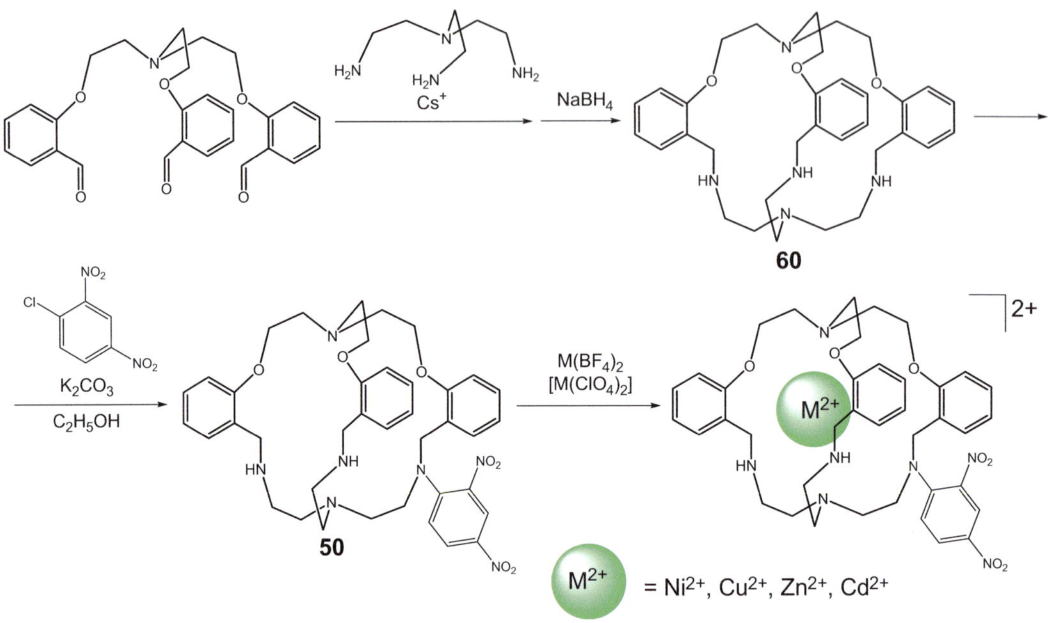

Scheme 2.35 Synthesis of the macrobicycle **60** and its chemical transformations

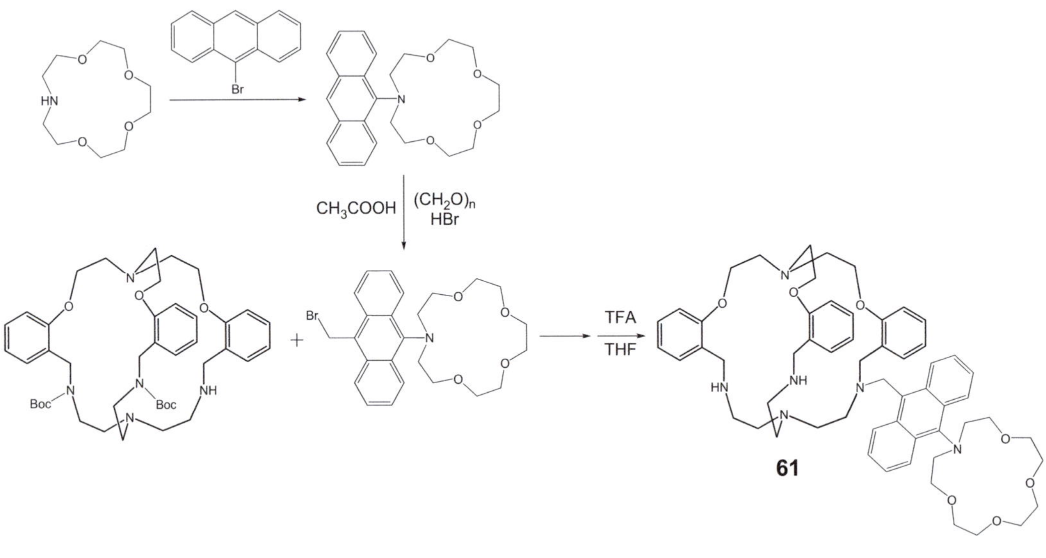

Scheme 2.36 Preparation of the crown ether-containing ligand **61**

not **80**) are described to efficiently bind alkali metal cations, forming their 1:1 cage complexes by Scheme 2.50.

1:1 cobalt(II) cage complex of the octaazamacrobicyclic ligand **86** has been synthesized [43] by Scheme 2.51 using direct template condensation of tripodal amine *tren* and paraformaldehyde on a cobalt(II) ion as a matrix.

The polyamine macrobicyclic compounds **87**, **88**, and **89** have been prepared [44] by Scheme 2.52 using ammonolysis of their tetraazamacrocyclic precursor *5* with the corresponding terminal aliphatic diamines.

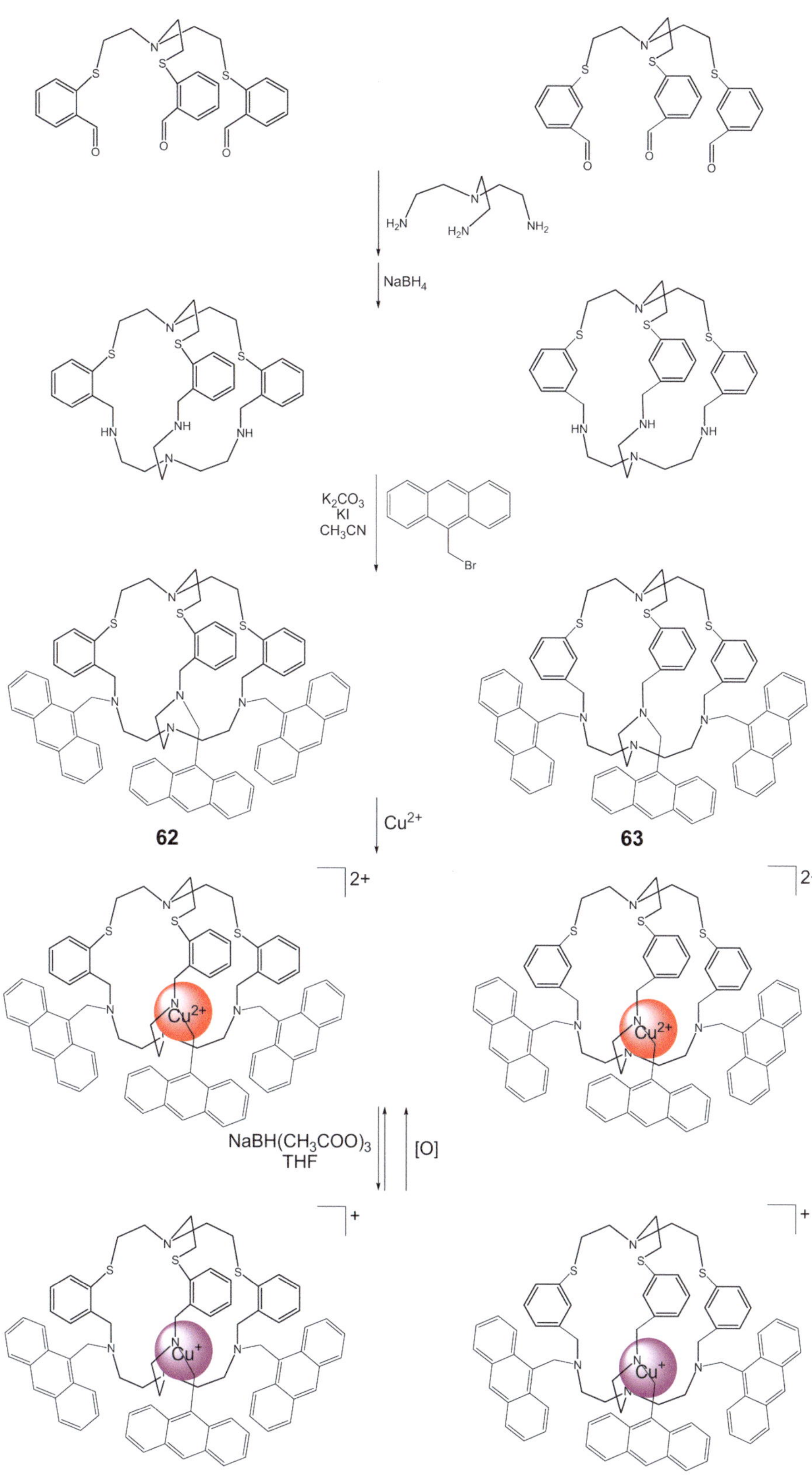

Scheme 2.37 Synthesis of the anthracenyl-appended macrobicyclic ligands **62** and **63**

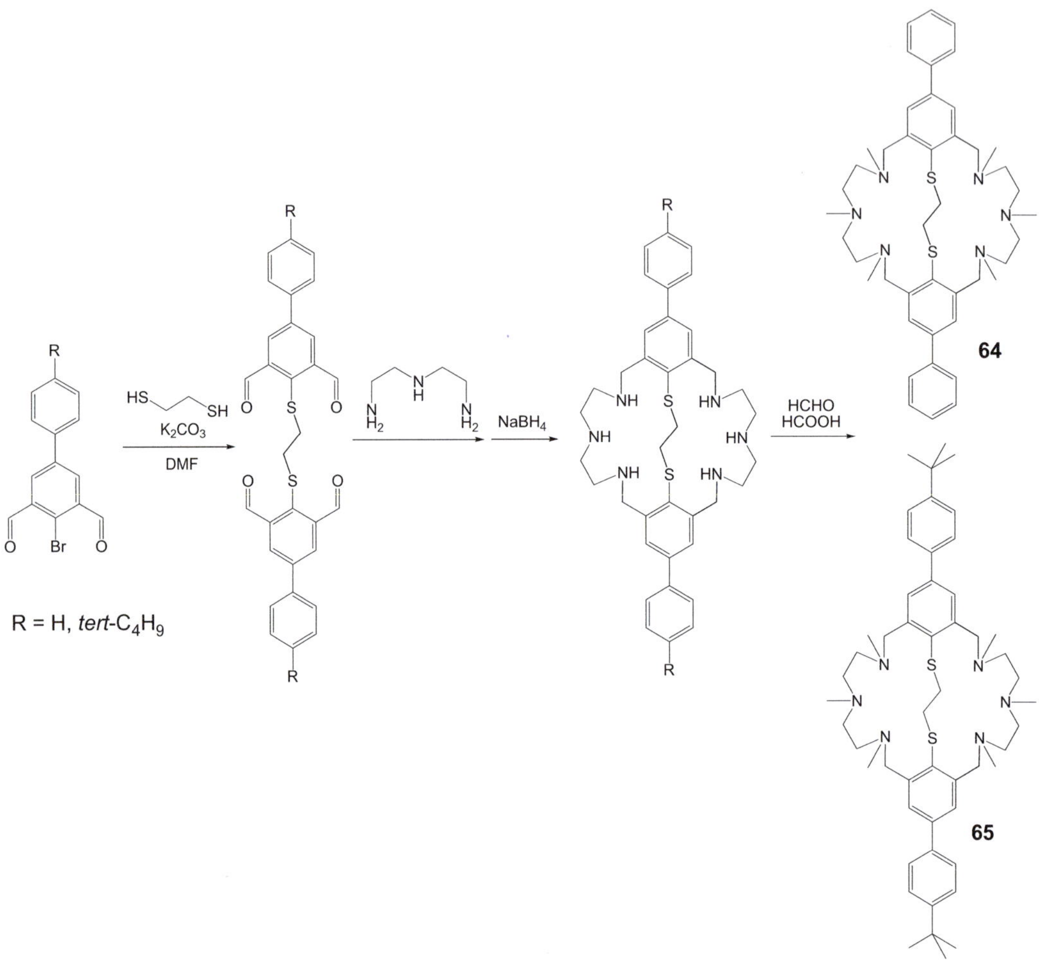

Scheme 2.38 Preparation of the macrobicycles **64** and **65**

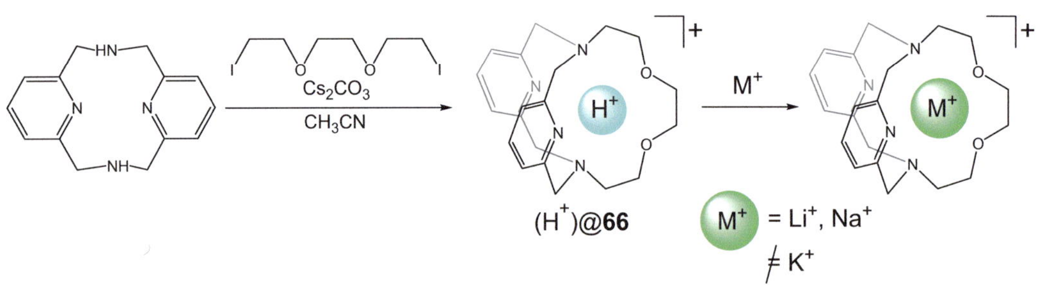

Scheme 2.39 Synthesis of cage complexes of the ligand **66**

The synthesis of 1:1 manganese, iron, cobalt, nickel, and zinc(II) cage complexes of the octaamine caging ligand **90** has been performed [45] by Scheme 2.53.

Template condensation of tripodal amine *tren* and formaldehyde in the presence of potassium(I) and cadmium(II) iodides by Scheme 2.54 has been used [46] for preparation of host–guest 1:1

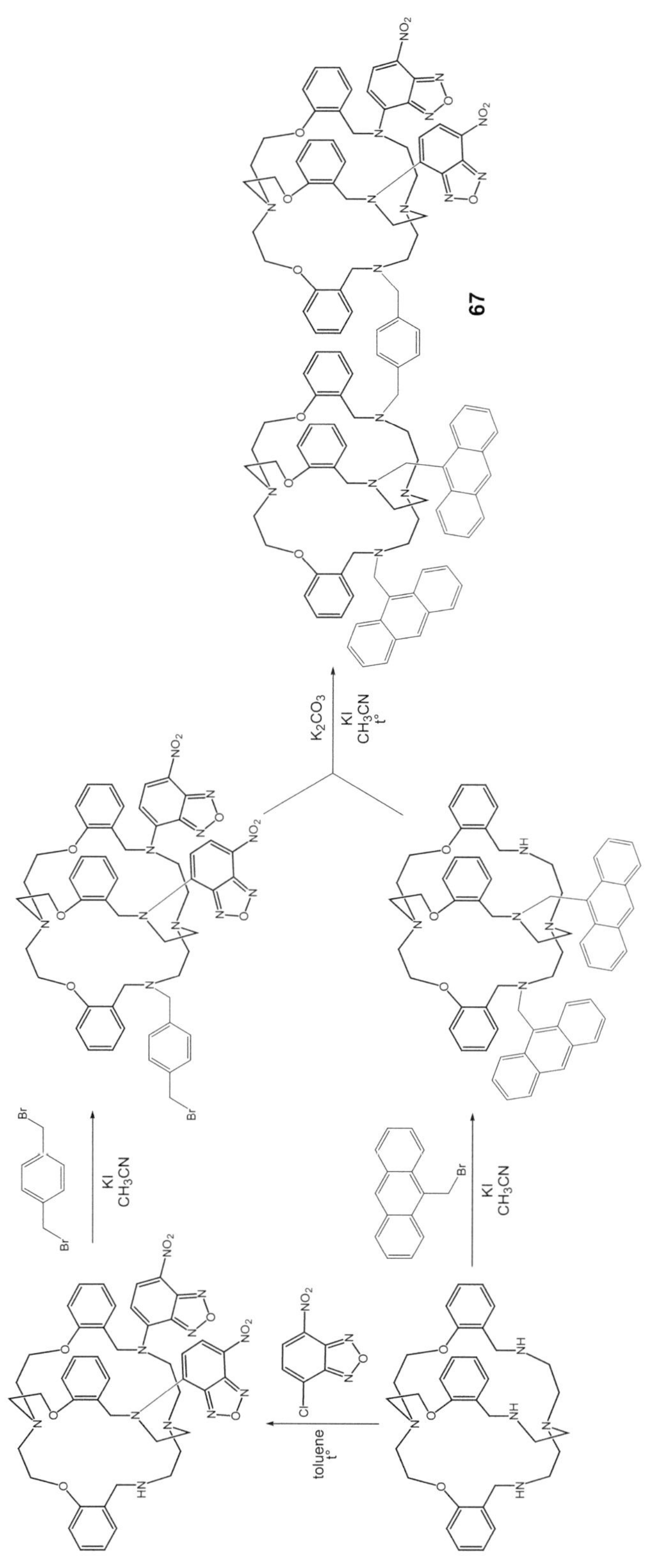

Scheme 2.40 Preparation of the bis-cage compound **67**

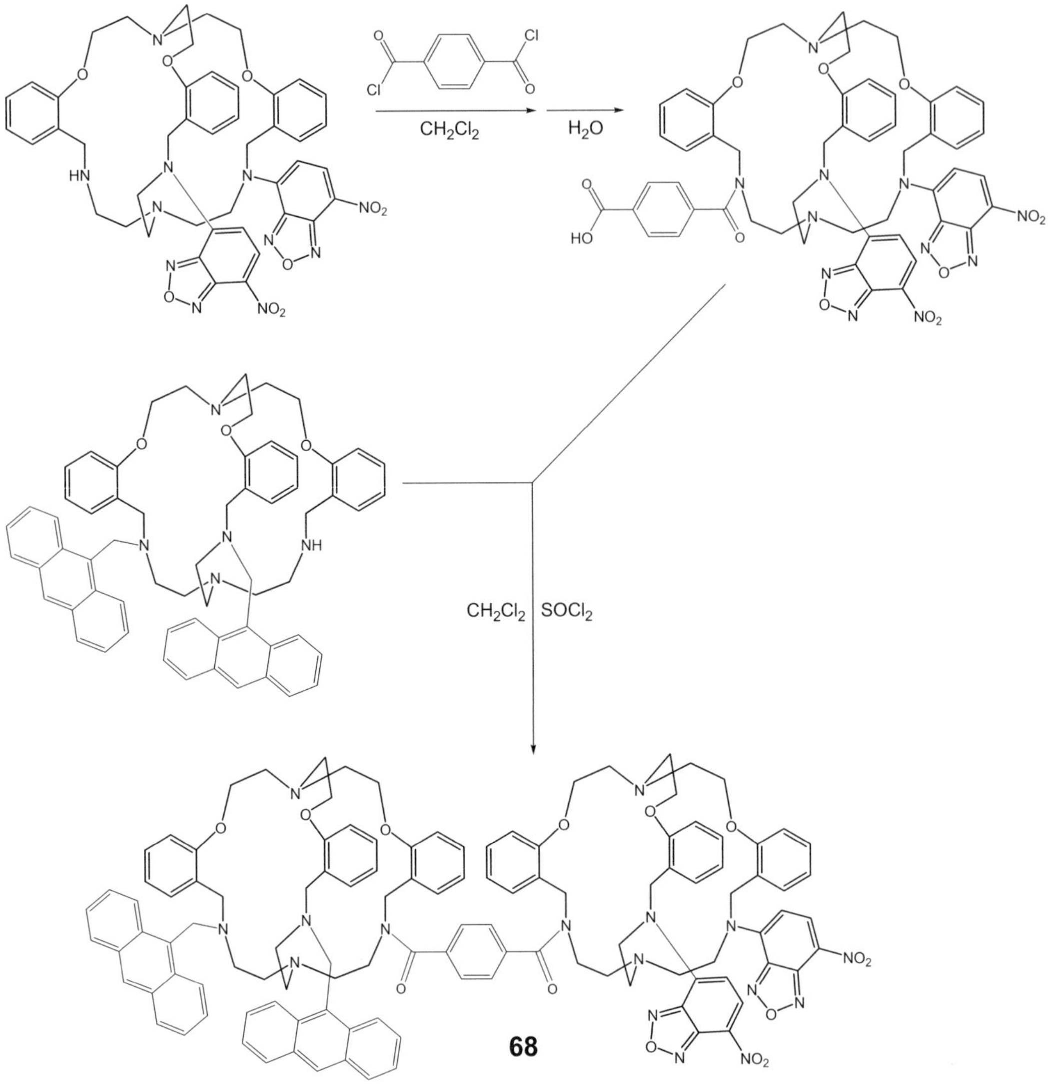

Scheme 2.41 Synthesis of the bis-cage compound **68**

complex of the 16-amine caging ligand **91** with encapsulated guest trication $[K_3(H_2O)]^{3+}$.

One-pot condensation of propanal, formaldehyde, and cobalt(II) tris-1,2-cyclohexanediaminate followed by a reduction with $NaBH_4$ has been used [47] for the synthesis of the macropentacyclic cobalt(III) polyamine clathrochelate (Co^{3+})@**92** by Scheme 2.55.

A series of the amphiphilic polyamine cobalt(III) clathrochelates with apical long-chain paraffin substituents, such as octyl, dodecyl, and hexadecyl groups, has been prepared [48] by Scheme 2.56 using co-condensation of formaldehyde and the corresponding aliphatic aldehydes with a tripodal hexaamine cobalt(III) semiclathrochelate; the use of its apically active Λ-$(-)_D$-stereoisomer allowed us to obtain 1:1 cobalt(III) cage complex of the corresponding dodecyl-functionalized chiral clathrochelate ligand Λ-$(-)_D$-**94**. The monoimine intermediate macrobicyclic products of these reactions have been reduced [48] to the target hexaamine clathrochelates (Co^{3+})@**93**, (Co^{3+})@**94**, and (Co^{3+})@**95** with $NaBH_4$.

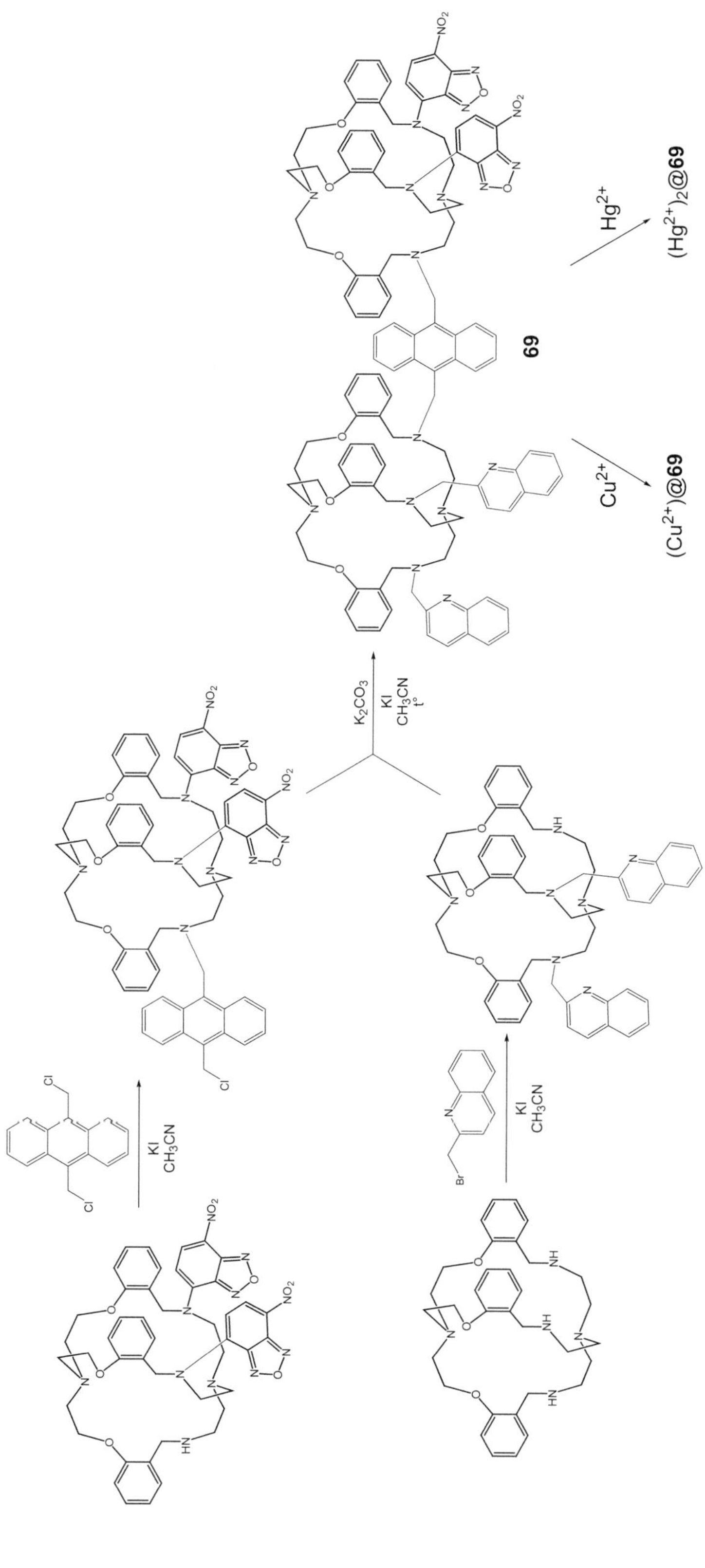

Scheme 2.42 Preparation of the bis-cage ligand **69** and its complexes

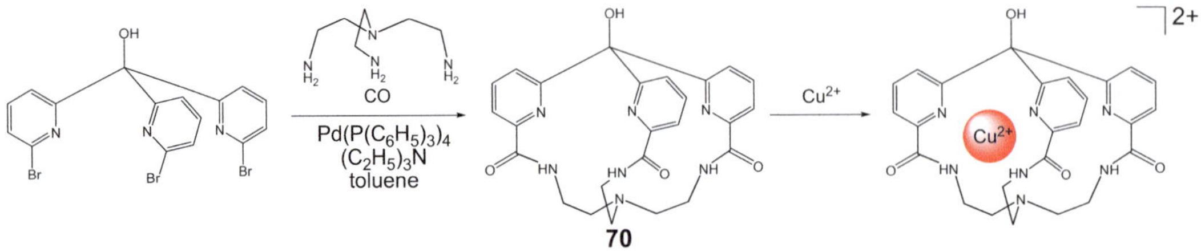

Scheme 2.43 Preparation of the ligand **70** and its copper(II) complex

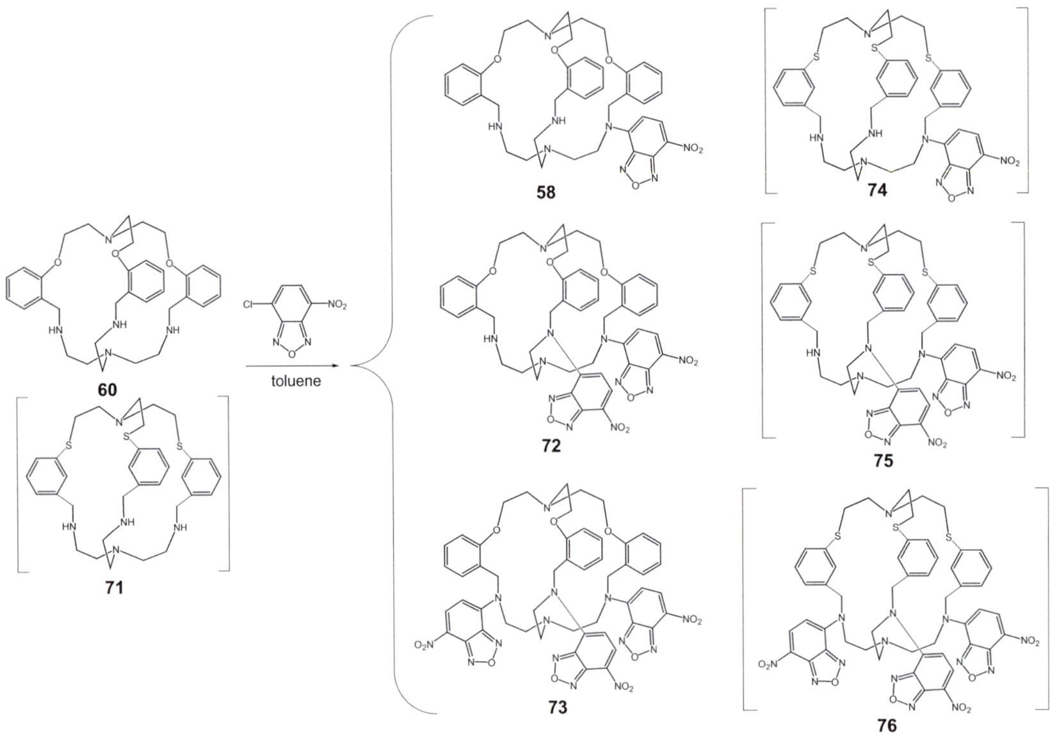

Scheme 2.44 Derivatization of the macrobicyclic precursors **60** and **71**

The racemic C_3-symmetric platinum(IV) clathrochelate (Pt^{4+})@**96** has been stereospecifically and regiospecifically prepared [49] in a moderate yield by Scheme 2.57 using template condensation of its bis-triamine precursor with formaldehyde and then propanal under basic reaction conditions. The reduction of this triimine cage complex with $NaBH_4$ gave its hexaamine macrobicyclic derivative (Pt^{4+})@**97**.

This expanded hexaamine ligand **97** has been isolated [50] in the form of its hexahydrochloride salt by demetallation of the parent cobalt(III) cage complex in strongly acidic aqueous solution using zinc dust as a reductant. It is reported [50] to form 1:1 zinc, cadmium, and mercury(II) cage complexes by Scheme 2.58.

Chiral resolution of the cobalt(III) dinitro- and diammoniumclathrochelate (Co^{3+})@**98** and (Co^{3+})@(**99** · $2H^+$) (Scheme 2.59) has been performed [51] by selective crystallization of the macrobicyclic polycations with optically active bis-μ-(*R*),(*R*)-tartratodiantimonate(III) ($[Sb_2(R,R\text{-tart})_2]^{2-}$) dianion as well as using column chromatography with $Na_2[Sb_2(R,R\text{-tart})_2]$ as an eluent.

Cobalt(III) cage complex of the expanded imine N_4S_2-macrobicyclic ligand **100** has been prepared [52] using template cross-linking of its tetraamine precursor by Scheme 2.60.

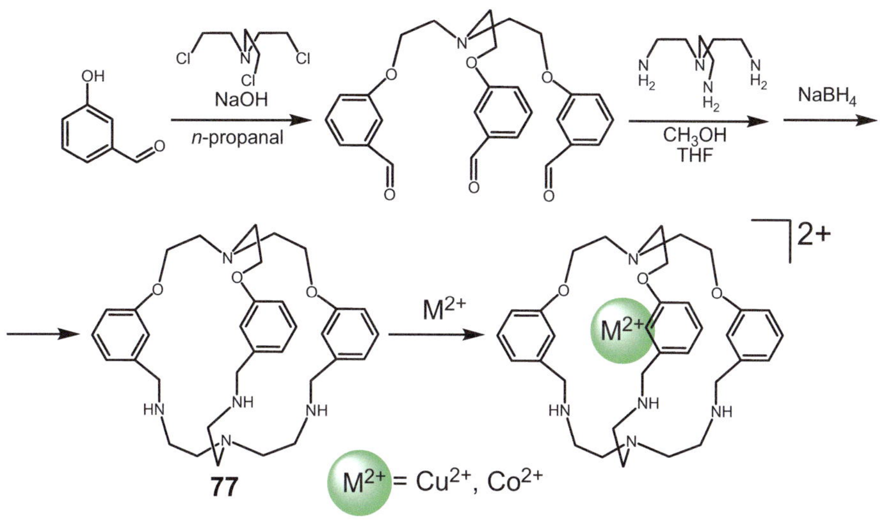

Scheme 2.45 Synthesis of the ligand **77** and its cage complexes

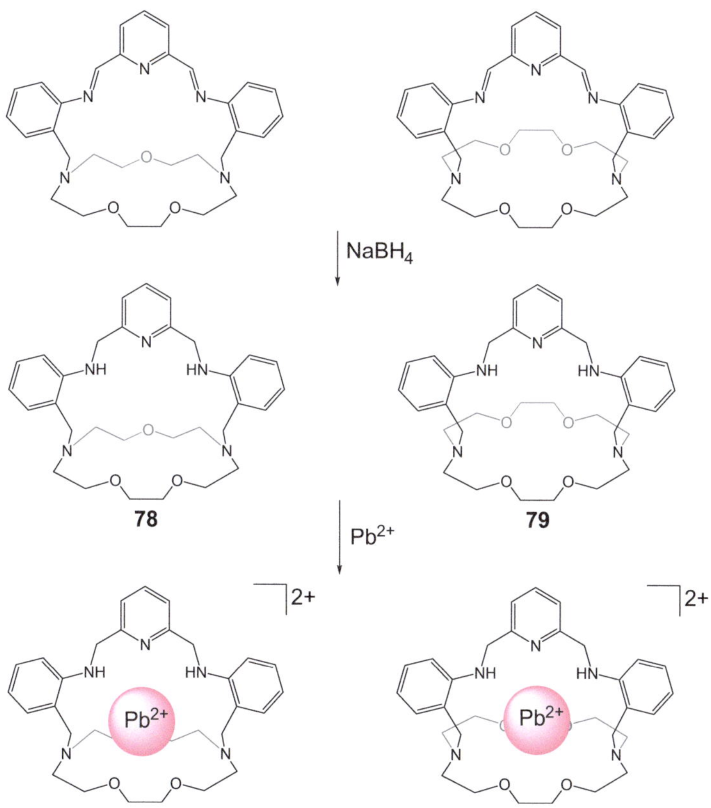

Scheme 2.46 Preparation of the macrobicycles **78** and **79** and their complexes

Coordination-driven self-assembly of the monoaminomethyl N_3S-macrobicyclic ligand **101** with iron(II) ions has been used [53] for the synthesis of the corresponding 1:1 cage complex, which is stable toward an oxidation both in solution and in the solid state (Scheme 2.61). On the other hand, iron(III) ions are reported to not form the similar cage complex with this encapsulating

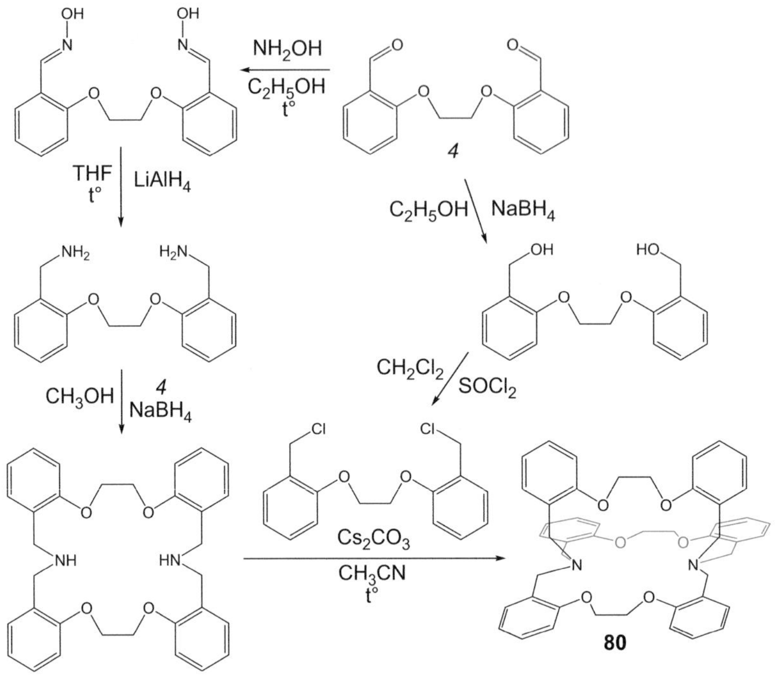

Scheme 2.47 Synthesis of the macrobicyclic compound **80**

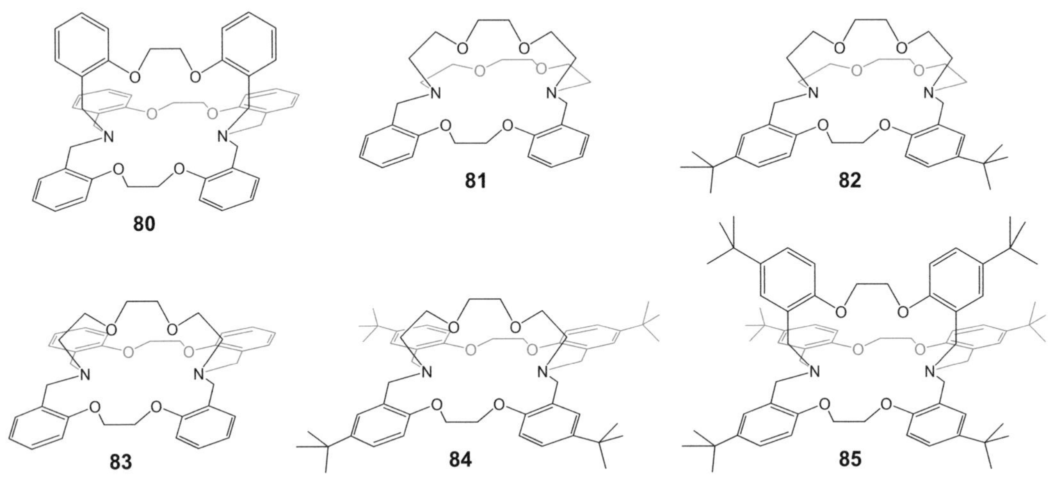

Scheme 2.48 Chemical drawings of the macrobicycles **80–85**

ligand. The products of its coordination toward copper(II) ion were found to be affected by the nature of the initial copper(II) salt: this reaction gave the corresponding 1:1 cage complex with encapsulated copper(II) ion (Scheme 2.62) in the case of copper(II) perchlorate, while $Cu(BF_4)_2$ forms the macrocyclic copper(II) complex of **101** that plays the role of a tetradentate ligand [54]. In the case of chloride or bromide counterions in DMSO media, an equilibrium between these macrobicyclic (clathrochelate) and monomacrocyclic forms has been detected [55], while in aqueous solution only the copper(II) cage complex (Cu^{2+})@**101** has been formed. These results are described to be undoubtedly important for a practical use of 64-copper(II)

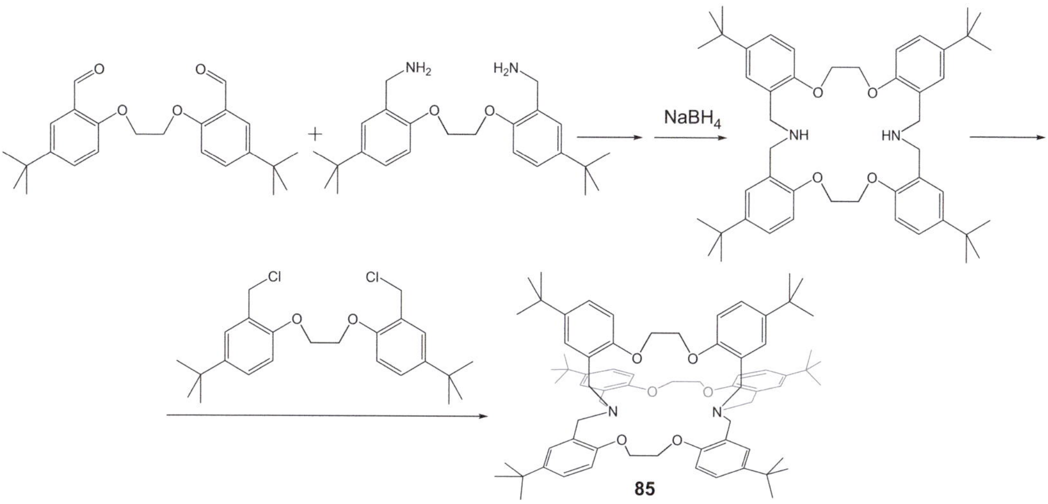

Scheme 2.49 Preparation of the *tert*-butyl-containing macrobicycle **85**

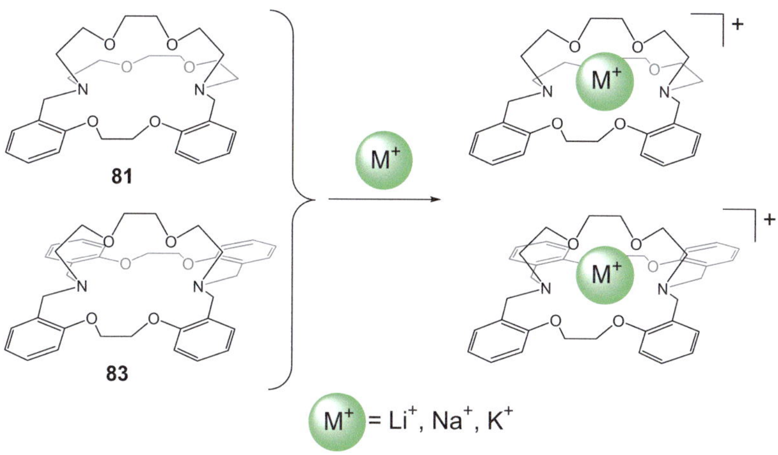

Scheme 2.50 Synthesis of cage complexes of the ligands **81** and **83**

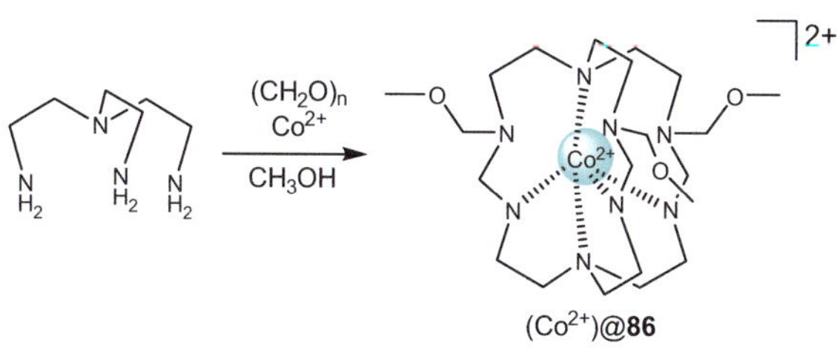

Scheme 2.51 Template synthesis of the clathrochelate (Co^{2+})@**86**

cage complexes of **101** in PET diagnostics (see Sect. 5.1.4.1).

Coordination of the macrobicyclic ligand **97** with an extended inner cavity (see above), obtained [56] by demetallation of its cobalt(III) cage complex by Scheme 2.63, toward copper(II) ion has been used [57] for preparation of their cage complex (Cu^{2+})@**97**. Its octamethyl-containing macrobicyclic analog **102** has been prepared [58] by Scheme 2.64 using template

Scheme 2.52 Preparation of the macrobicycles **87**–**89**

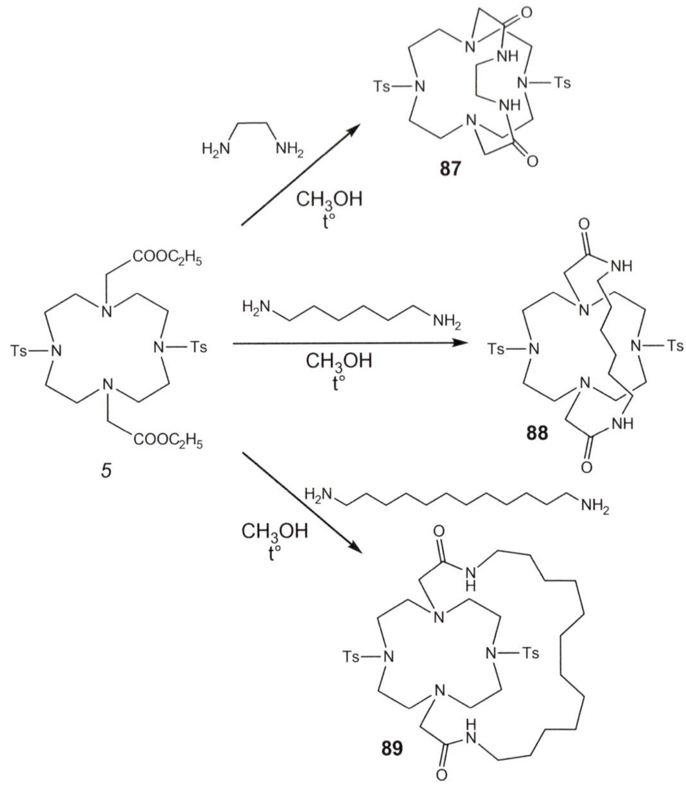

Scheme 2.53 Synthesis of the clathrochelate complexes of **90**

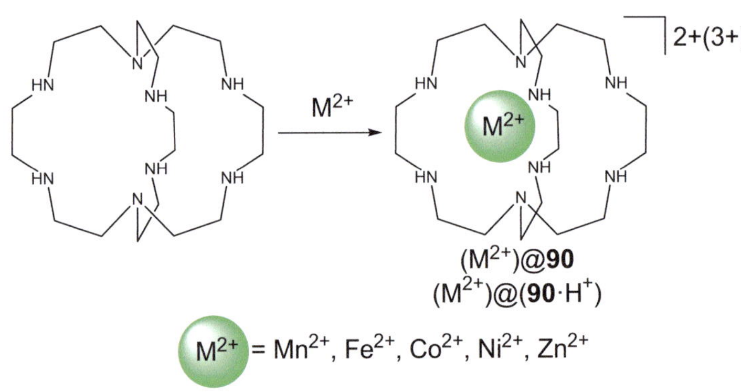

condensation of the corresponding non-macrocyclic bis-triamine precursor with formaldehyde and 2-methylpropanal followed by reduction with $NaBH_4$ and demetallation with zinc dust. As follows from the UV-vis, EPR, and single crystal X-ray diffraction data of [58], coordination of this ligand toward copper(II) ion both in solution and in the solid state gives the penta- and hexacoordinate macrobicyclic complexes of this ion also shown in Scheme 2.64.

The racemic and resolved forms of the hydroxyl-terminated apically functionalized cobalt(III) clathrochelates (Co^{3+})@**103** and (Co^{3+})@**104** have been synthesized [59] by Scheme 2.65.

McMurry coupling of tris(5-formyl-2-thienyl)-methane ligand precursors by Scheme 2.66 has been used [60] for preparation of the S_6-macrobicyclic compounds **105** and **106** with tris-(2-thienyl)methane-capping fragments. All

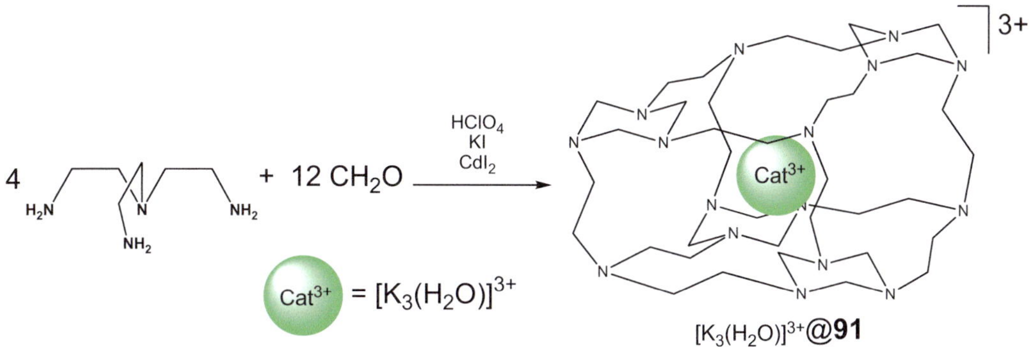

Scheme 2.54 Template synthesis of the clathrochelate $[K_3(H_2O)]^{3+}$@**91**

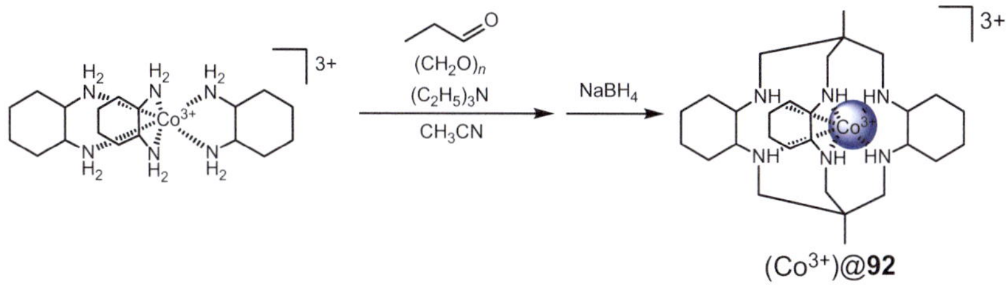

Scheme 2.55 Preparation of the cobalt(III) clathrochelate (Co^{3+})@**92**

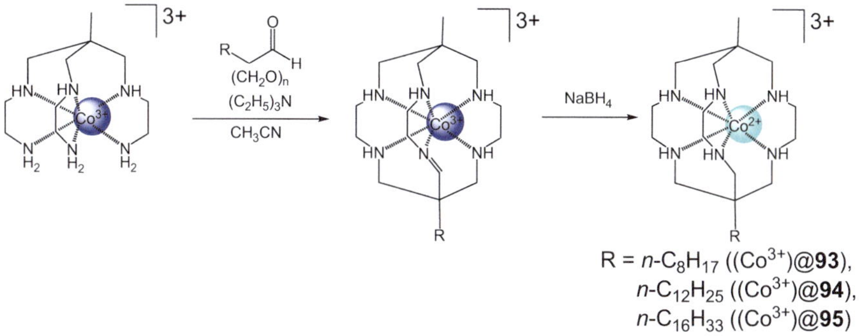

Scheme 2.56 Synthesis of a series of polyamine cobalt(III) clathrochelates

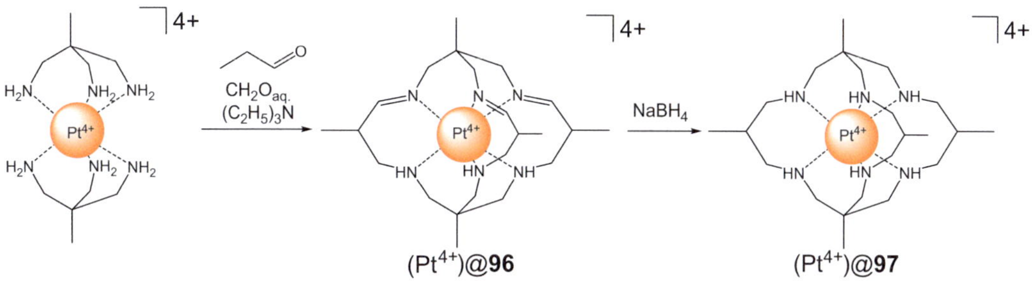

Scheme 2.57 Preparation of the platinum(IV) cage complex of **97**

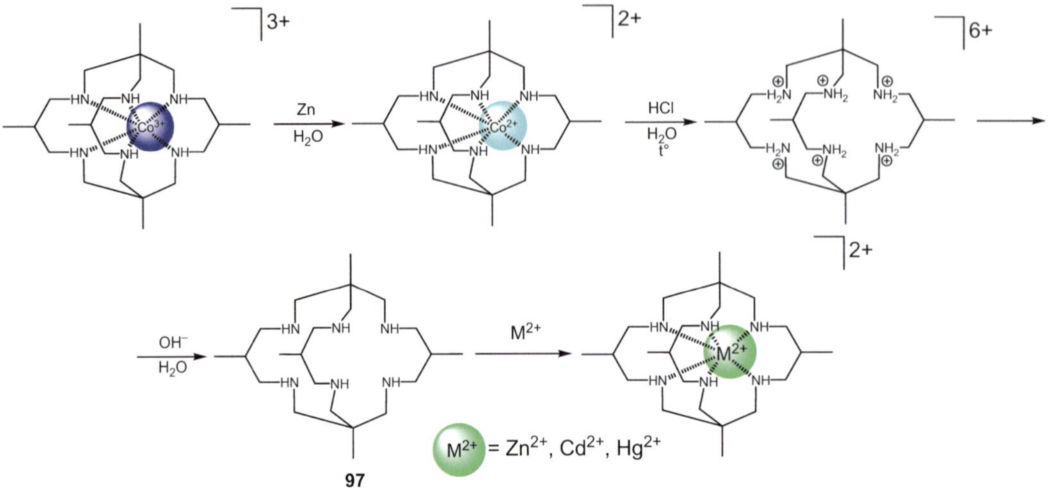

Scheme 2.58 Synthesis of the macrobicycle **97** and its cage complexes

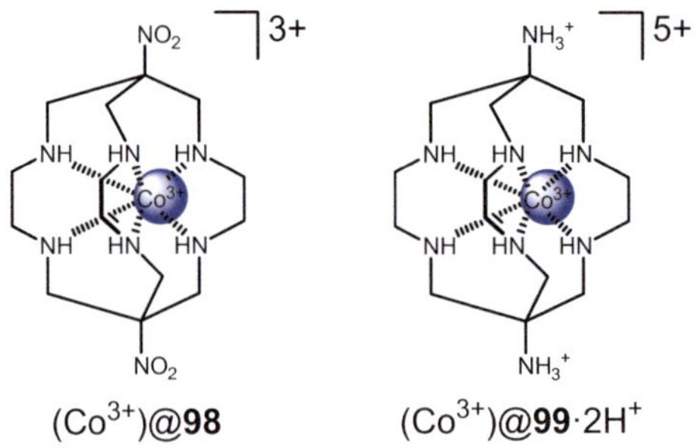

Scheme 2.59 Chemical drawings of the cobalt(III) complexes (Co^{3+})@**98** and (Co^{3+})@(**99**·$2H^+$)

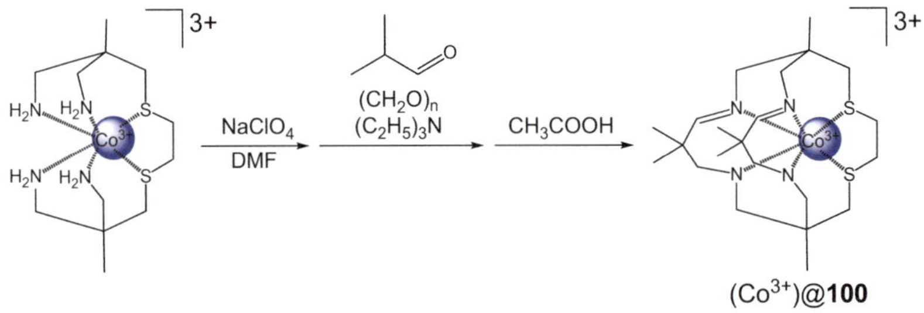

Scheme 2.60 Preparation of the clathrochelate (Co^{3+})@**100**

the sulfur atoms of their thienyl groups are directed inward, thus allowing to obtain 1:1 copper(II) cage complex of the encapsulating ligand **105**. The same synthetic approach has been used [61] for the synthesis of four neutral, cationic, and anionic cage compounds **107–110** by Scheme 2.67. The dianionic macrobicycle **110** underwent its further transformations with methyl iodide and chlorotrimethylsilane, thus giving the cage compounds **105** and **111** also shown in this Scheme.

A series of the S_6-macrobicyclic compounds **112–119** has been prepared [62] using stepwise synthetic procedures shown in Schemes 2.68, 2.69, and 2.70.

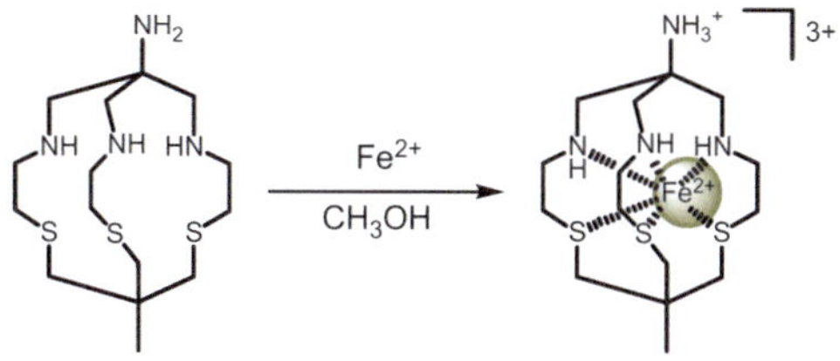

101

Scheme 2.61 Encapsulation of iron(II) by the ligand **101**

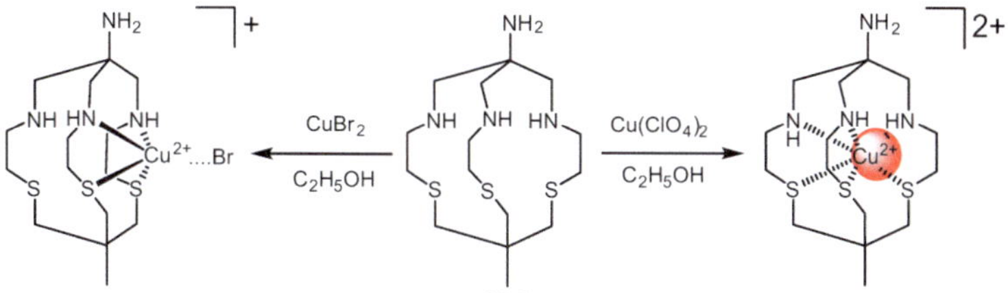

101

Scheme 2.62 Preparation of copper(II) complexes of **101**

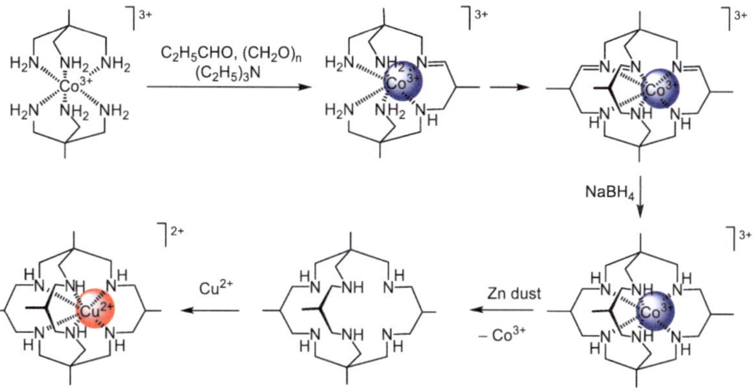

97

Scheme 2.63 Synthesis of the macrobicycle **97** and its copper(II) complex

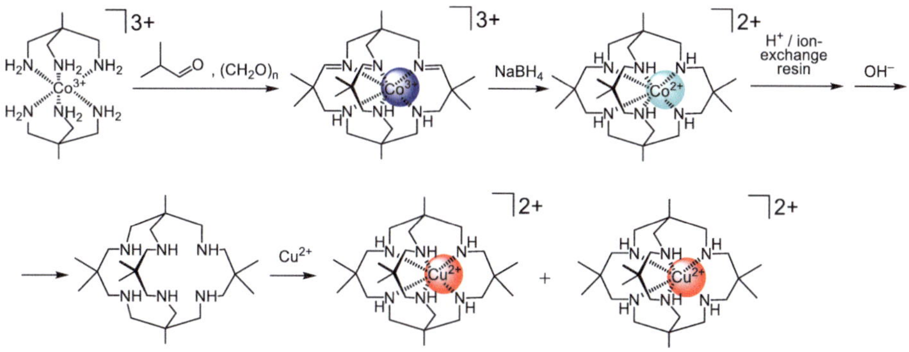

102

Scheme 2.64 Preparation of the macrobicycle **102** and its copper(II) complexes

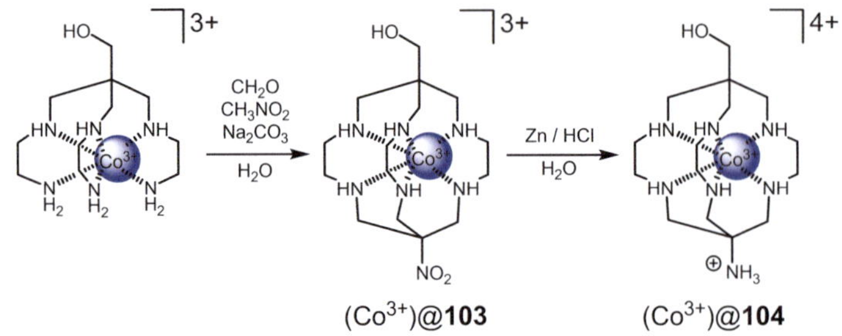

Scheme 2.65 Synthesis of cobalt(III) cage complexes of **103** and **104**

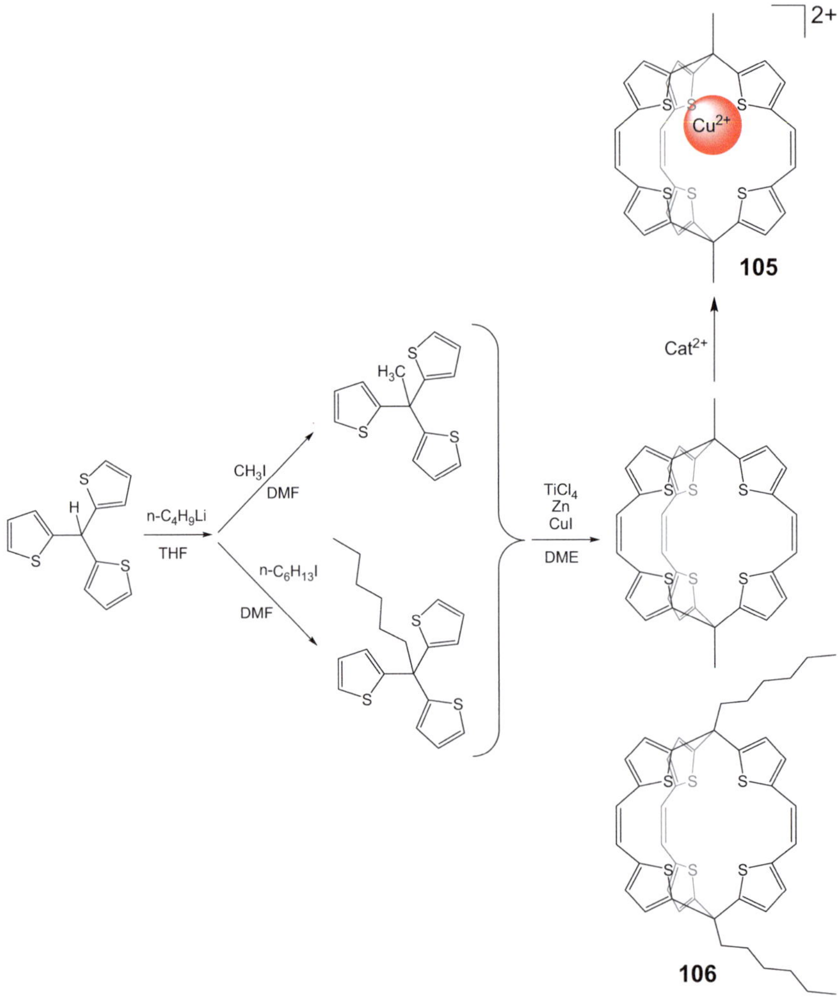

Scheme 2.66 Preparation of the macrobicycles **105** and **106**

2.2.1.2 Apical Functionalization of Polyamine Ligands and Cage Complexes

A metal-free apically monofunctionalized polyamine precursor **120** of the peptide-terminated macrobicyclic chelator **121** has been prepared [4] by Scheme 2.71 using the copper(II) diammoniumclathrochelate (Cu^{2+})@(**99** · $2H^+$) as a reactive cage intermediate with secondary amino groups, deactivated by their coordination

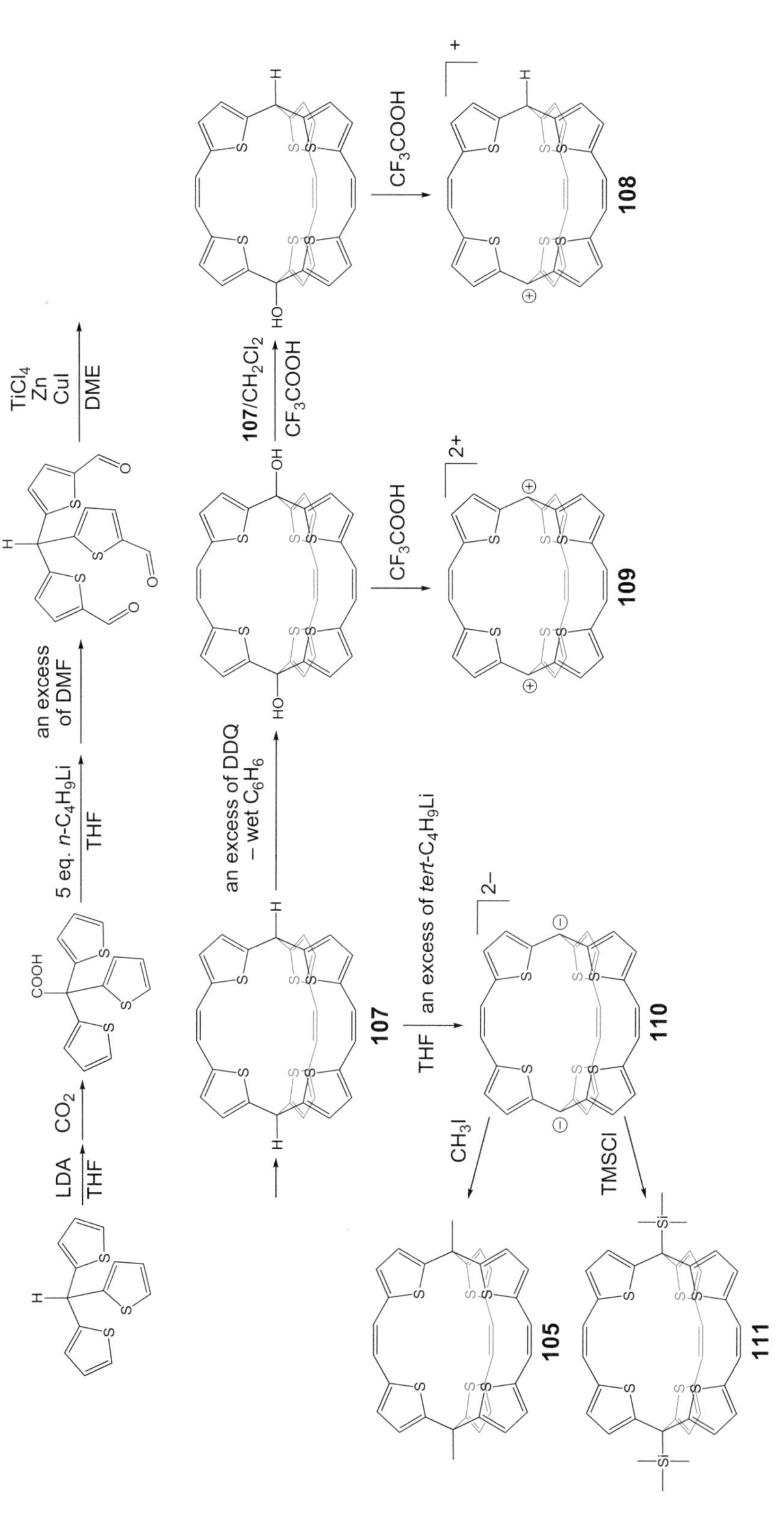

Scheme 2.67 Synthesis of a series of the S_6-macrobicyclic compounds

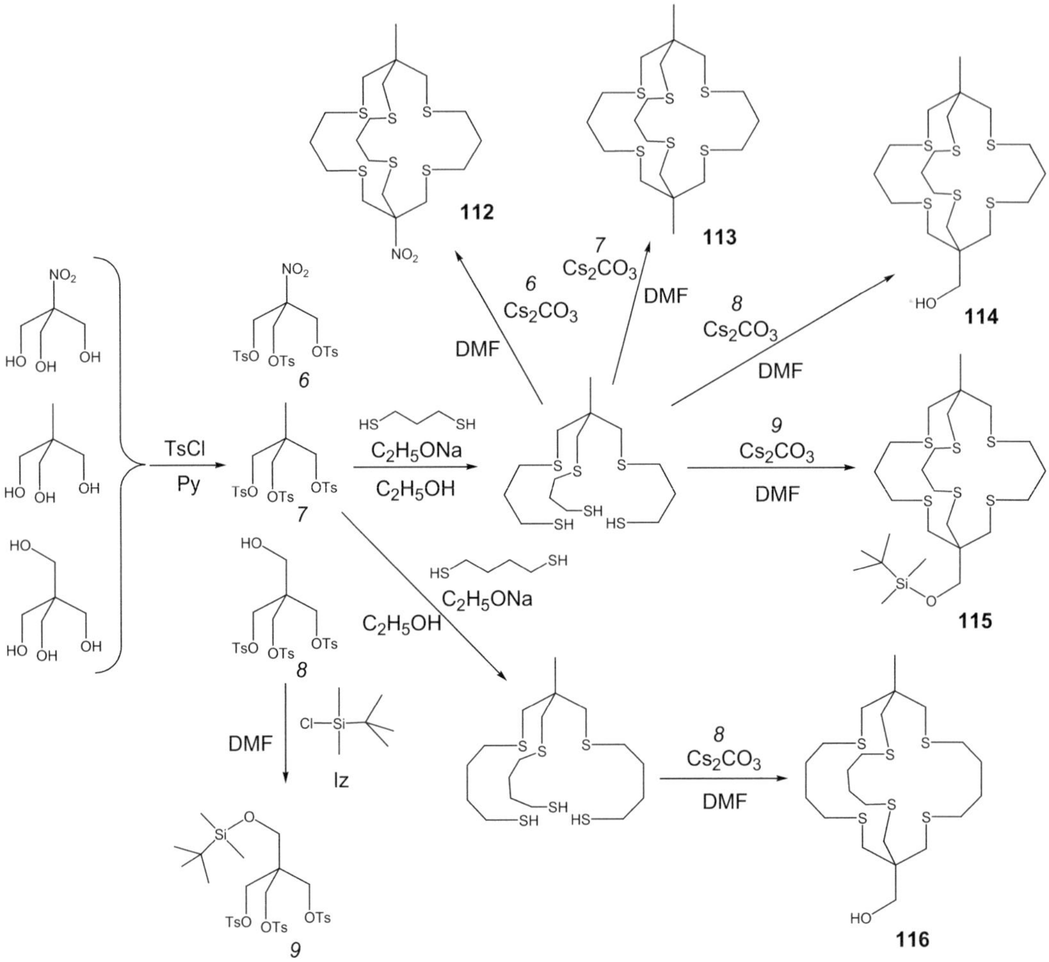

Scheme 2.68 Preparation of the S_6-macrobicycles **112–116**

to an encapsulated metal ion. The carboxyl-terminated macrobicycle **120** has been then modified [4] by its condensation with cyclic peptide *12*, containing Ala-Gly-Asp fragment and pendant lysine residue for covalent binding with this clathrochelate platform.

Other synthetic pathways for preparation of this monoaminomacrobicyclic precursor without isolation of the above copper(II) clathrochelate intermediate are described [63]. **120** has been prepared by Scheme 2.72 using either a demetallation of 1:1 cobalt(III) cage complex of the macrobicyclic ligand **122** with NaOH followed by a basic hydrolysis with K_2CO_3 or, vice versa, by the basic hydrolysis of (Co^{3+})@**122** with K_2CO_3 followed by a demetallation with NaOH.

The analogous apically difunctionalized macrobicyclic compound **123** with two terminal carboxylic groups has been prepared [64] by Scheme 2.73 using direct alkylation of **99** with 4-bromomethylbenzoic acid. Its functionalization with the same peptide *12* [64] gave the clathrochelate chelator **124** with improved biodistribution properties (see Sect. 5.1.4).

Condensation of (Cu^{2+})@**99** with glutaric and malic anhydrides by Scheme 2.74 has been used [65] for preparation of the corresponding mono- and dicarboxyl-terminated copper(II) clathrochelates with various linker fragments.

Condensation of the mono- and dicarboxymethylated cobalt(III)-encapsulating macrobicyclic precursors (Co^{3+})@**129** and (Co^{3+})@**130** with

Scheme 2.69 Synthesis of the S_6-macrobicycles **117** and **118**

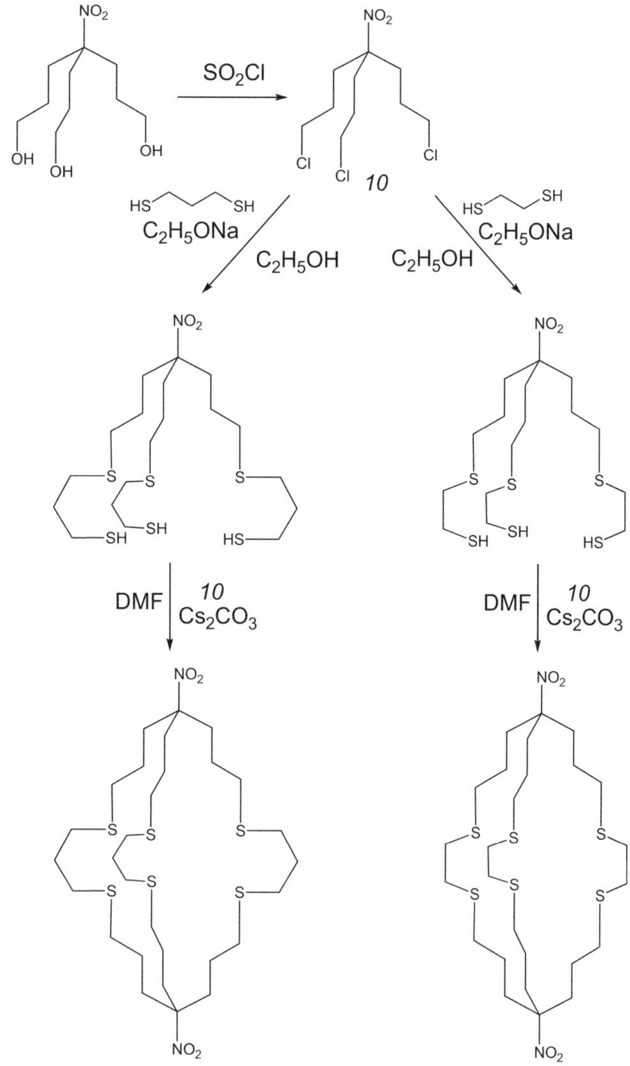

linear polyamines (such as ethylenediamine, spermine, and *N,N′*-bis(3-aminopropyl)-1,2-ethanediamine), phenylalanine and tripodal amine *tren*, has been used [66] for the synthesis of a series of apically functionalized cobalt(III) clathrochelates with amine and peptide substituents by Schemes 2.75, 2.76, and 2.77.

Amphiphilic cobalt, zinc, nickel, copper, and magnesium(II) cage complexes of the apically functionalized mono- and difunctionalized macrobicyclic ligands **144–150** have been synthesized [67] using reductive alkylation of their metal-free mono- and diaminomacrobicyclic precursors **143** and **99** with the corresponding long-chain aliphatic aldehydes by Scheme 2.78. The use of the magnesium(II)-encapsulating clathrochelate precursors (Mg^{2+})@**149** and (Mg^{2+})@**150** allowed to easily isolate the corresponding metal-free apically functionalized covalent capsules, which are suitable for encapsulation of a series of metal(II) ions [67].

Copper(II) and cobalt(III) macrobicyclic complexes with terminal long-chain paraffin substituents have been synthesized [68] by Scheme 2.79. A series of the polyamine cobalt(III) clathrochelates with terminal mercaptoethylamide

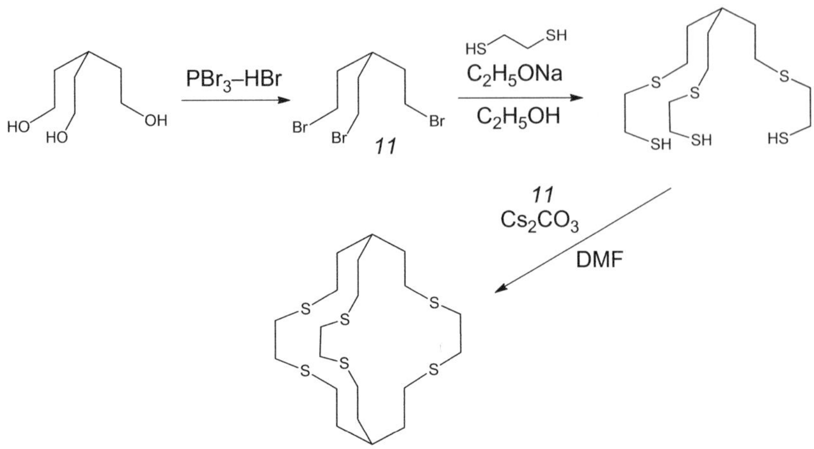

Scheme 2.70 Preparation of the S_6-macrobicycle **119**

groups has been prepared [69] using multistep synthetic procedure shown in Scheme 2.80.

Apical functionalization of the cobalt(III) mono- and diaminoclathrochelates (Co^{3+})@**99** and (Co^{3+})@**143** has been performed [70] by Schemes 2.81 and 2.82, thus giving the mono- and bis-clathrochelates with methylamide (from the reactions with acetic anhydride), Schiff-base arylimine (from imine condensations), benzylamine, phthalimide, and tosylate terminal substituents.

The apically mono- and difunctionalized cobalt(III) mono- and diaminoclathrochelates (Co^{3+})@**177**, (Co^{3+})@**178**, and (Co^{3+})@**179** with terminal 2- and 3-thienylmethyl groups have been prepared [71] by Scheme 2.83.

Reductive alkylation of the diaminomacrobicyclic compound **99** and its cobalt(III) 1:1 cage complex by Schemes 2.84 and 2.85 has been used [72] for preparation of cobalt(III) and copper(II) cage complexes with lipophilic apical substituents; the coordination of the secondary amino groups of these macrobicyclic ligands toward an encapsulated metal ion protected them from side alkylation reactions (see, e.g., macrobicycle **180**).

Efficient carboxymethylation of the aminoclathrochelate precursor (Co^{3+})@**143** has been performed [73] by Scheme 2.86 using a large excess of chloroacetate, thus giving 1:1 cobalt(III) cage complexes of the macrobicyclic ligands **184** and **185**. The reactions of (Co^{3+})@**143** with $CuCO_3 \cdot Cu(OH)_2$ and $NiCO_3 \cdot 2Ni(OH)_2 \cdot 4H_2O$ in aqueous solutions afforded their trinuclear 2:1 complexes also shown in Scheme 2.86.

2.2.2 Macrobicyclic Metal Tris-dioximates and Other Types of Polyazomethine Clathrochelates

2.2.2.1 Apical Functionalization of Quasiaromatic Macrobicyclic Compounds

Template Synthesis

Apically functionalized iron(II) tris-nioximates (Fe^{2+})@**186**, (Fe^{2+})@**187**, (Fe^{2+})@**188**, (Fe^{2+})@**189**, (Fe^{2+})@**190**, and (Fe^{2+})@**191** have been prepared [74] in high yields by direct template reaction on an iron(II) ion as a matrix (Scheme 2.87) using Lewis acidity of the corresponding monofunctionalized boronic acids with their boron atom in trigonal coordination environment; the tetrahedral anion $[B(C{\equiv}C(C_6H_5))_4]^-$ has been also used as a cross-linking agent. The choice of nioxime as a ligand synthon for preparation of these apically functionalized clathrochelates is explained [74]

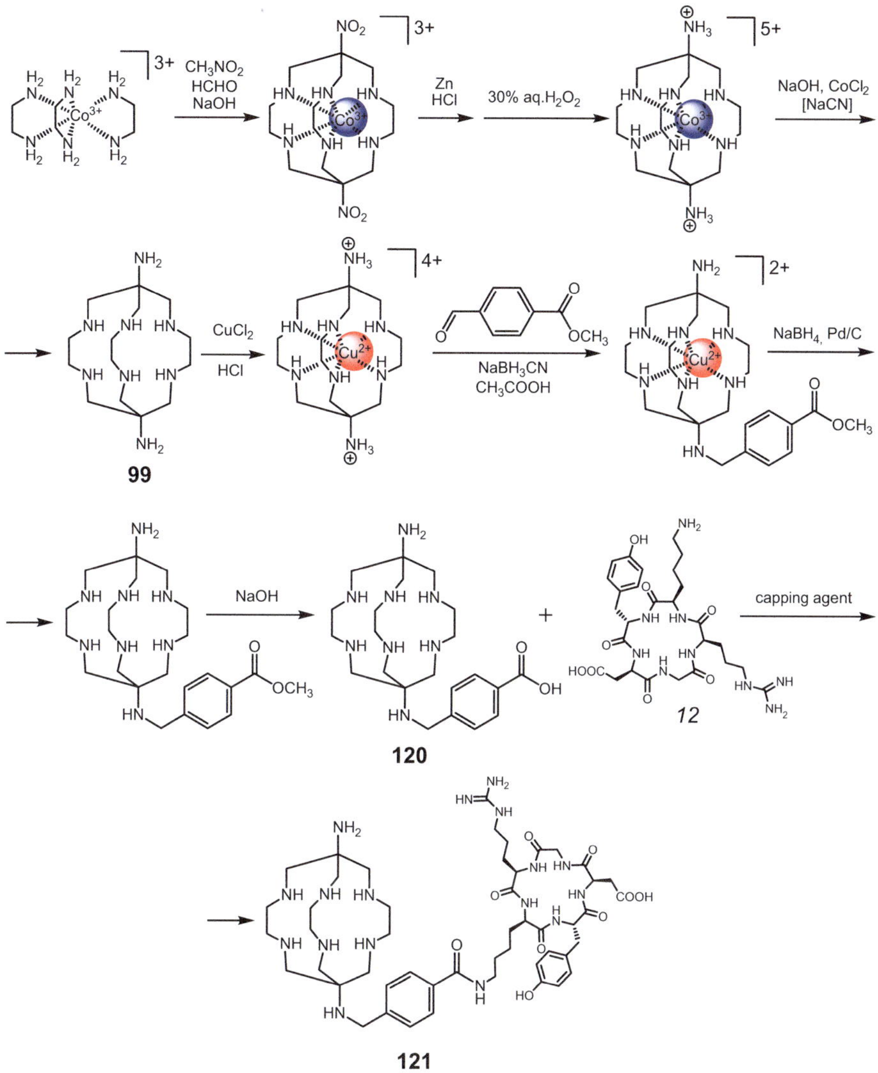

Scheme 2.71 Synthesis of apically functionalized macrobicycles **120** and **121**

by higher stability of the macrobicyclic complexes formed by six- and octa-membered alicyclic dioximes as compared to their acyclic analogs; in the latter cases, the coordination toward a metal ion is accompanied by a *trans*-to-*cis* conformation transition. Their hexachloroclathrochelate analog (Fe^{2+})@**192**, the derivative of a weakly donor dichloroglyoxime, has been synthesized [74] under vigorous reaction conditions by Scheme 2.88.

The diolefin-terminated aliphatic and aromatic iron(II) clathrochelates (Fe^{2+})@**193**, (Fe^{2+})@**194**, (Fe^{2+})@**195**, (Fe^{2+})@**196**, and (Fe^{2+})@**197** have been synthesized [75] by Scheme 2.89 using template cross-linking of the corresponding α-dioxime with the *iso*-propyl ester of a given diolefin boronic acid on an iron(II) ion as a matrix. Typically, such condensations with protolytically stable alkyl- and arylboronic acids have been earlier performed in polar protic solvents (first of all, in alcohol media) without the addition of bases for neutralization of H^+ ions releasing in the course of such self-assembly [75], but these reaction conditions are reported to be unacceptable for the above diolefin boron-containing esters

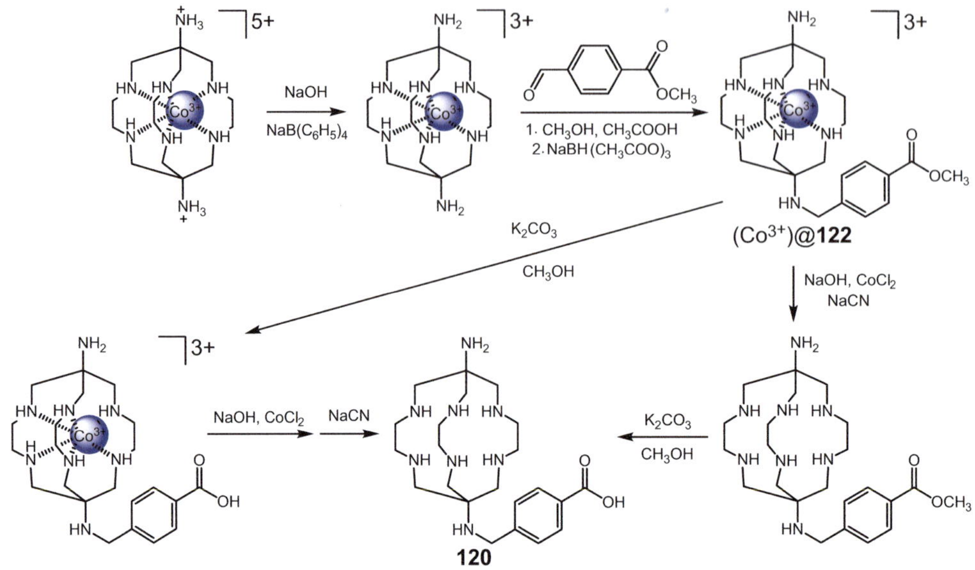

Scheme 2.72 Synthetic pathways to the macrobicycle **120**

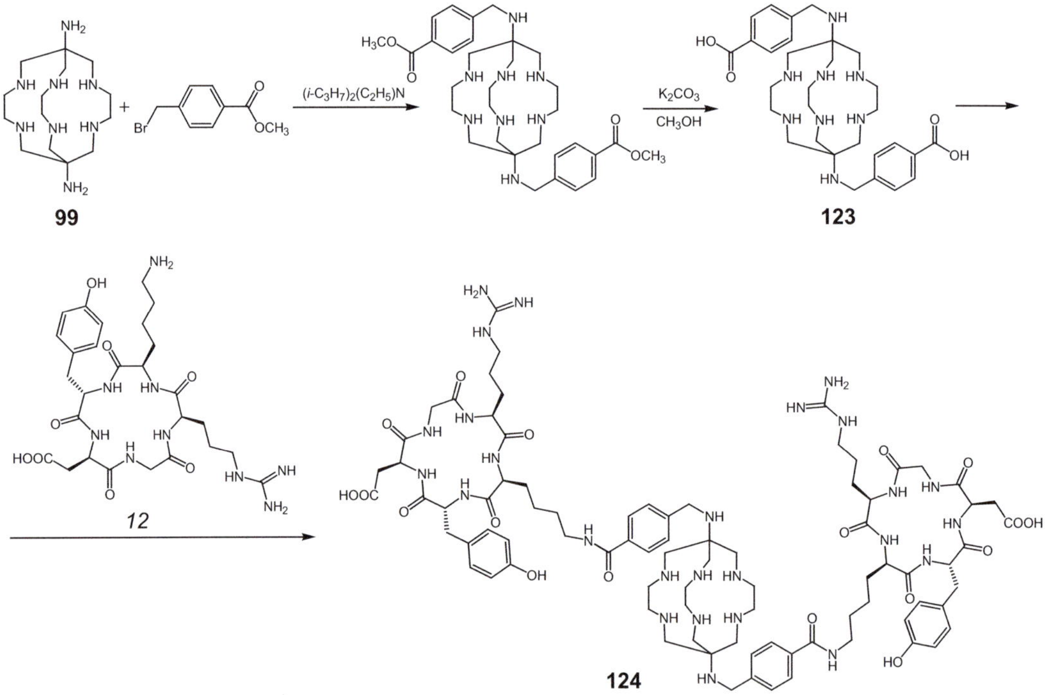

Scheme 2.73 Preparation of the macrobicycles **123** and **124**

because of the protolytic cleavage of their B–C bonds. As a result, polar aprotic acetonitrile has been used [75] as a solvent for macrobicyclization (capping) of the initially prepared non-macrocyclic iron(II) tris-nioximate, and the equimolar amount of triethylamine has been added to neutralize the releasing H^+ ions.

A series of the ferrocenylboron-capped iron(II) tris-dioximates (Fe^{2+})@**198**, (Fe^{2+})@**199**, and (Fe^{2+})@**200** has been prepared [76] by template

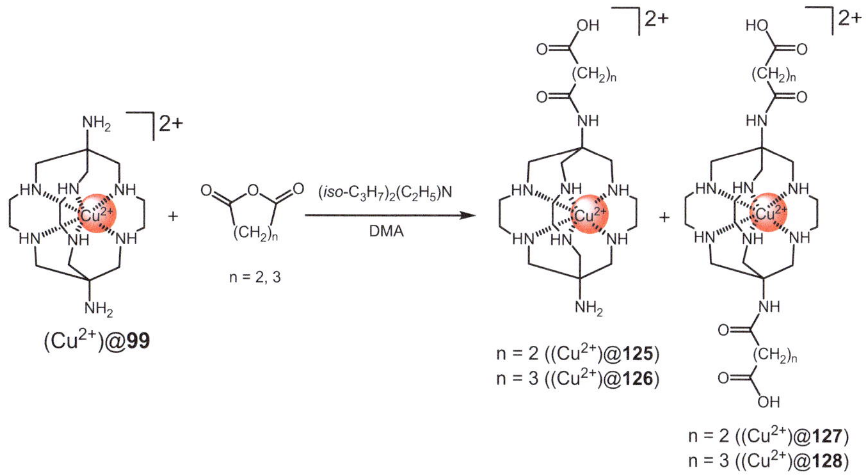

Scheme 2.74 Synthesis of copper(II) complexes of the ligands **125–128**

cross-linking of three molecules of the corresponding aliphatic α-dioxime with ferrocenylboronic acid on an iron(II) ion (Scheme 2.90).

The boron-capped aliphatic alicyclic and acyclic iron(II) tris-*α*-dioximates (Fe^{2+})@**201**, (Fe^{2+})@**202**, (Fe^{2+})@**203**, and (Fe^{2+})@**204** with apical adamantyl substituents have been self-assembled [77] by Scheme 2.91 under mild reaction conditions on an iron(II) ion as a matrix using dimethyl ester of adamantylboronic acid as a capping agent. This ester easily underwent an oxidation by air oxygen in the course of such self-assembly. As a result, the corresponding monohydroxymonoadamantyl-terminated iron(II) cage complexes have been also isolated [77]. Their diadamantylboron-capped hexachloroclathrochelate analog (Fe^{2+})@**205** has been synthesized in a high yield by Scheme 2.92 under vigorous reaction conditions.

Template condensation of nioxime with dithiopheneboronic acid by Scheme 2.93 on the corresponding metal(II) ion as a matrix has been performed [78], thus giving the macrobicyclic iron, cobalt, and ruthenium(II) tris-nioximates with redox-active apical substituents. The same synthetic approach has been used [79] for preparation of their dithiophene-terminated iron and cobalt(II) tris-α-benzildioximate analogs by Scheme 2.94.

The 2-thiopheneboron-capped cobalt(II) dimethylglyoximate and α-benzildioximate clathrochelates (Co^{2+})@**208** and (Co^{2+})@**209** and their hexachloromacrobicyclic analog (Co^{2+})@**210** have been synthesized [80] by Scheme 2.95 using direct template condensation of the corresponding α-dioximate ligand synthon with 2-thiopheneboronic acid on a cobalt(II) ion as a matrix. The macrobicyclic tris-dimethylglyoximate (Co^{2+})@**208** has been easily formed under mild reaction conditions in the presence of triethylamine as organic base, whereas its α-benzildioximate and dichloroglyoximate analogs could be prepared under vigorous conditions only [80].

Their tris-nioximate iron- and cobalt(II)-encapsulating analogs (Fe^{2+})@**211** and (Co^{2+})@**211**, resulting from direct template condensation of three molecules of nioxime with 2-thiopheneboronic acid on the corresponding metal ion as a matrix by Scheme 2.96, are described [81].

The synthesis of the boron-capped iron and cobalt(II) tris-nioximates (M^{2+})@**212**, (M^{2+})@**213**, (M^{2+})@**214**, (M^{2+})@**215**, and (M^{2+})@**216** with terminal reactive hydroxyl, carboxyl, and amino groups has been performed [82] in moderate yields by Scheme 2.97 using template condensation of nioxime with monofunctionalized OH-, NH_2-, and COOH-substituted phenylboronic acids on the

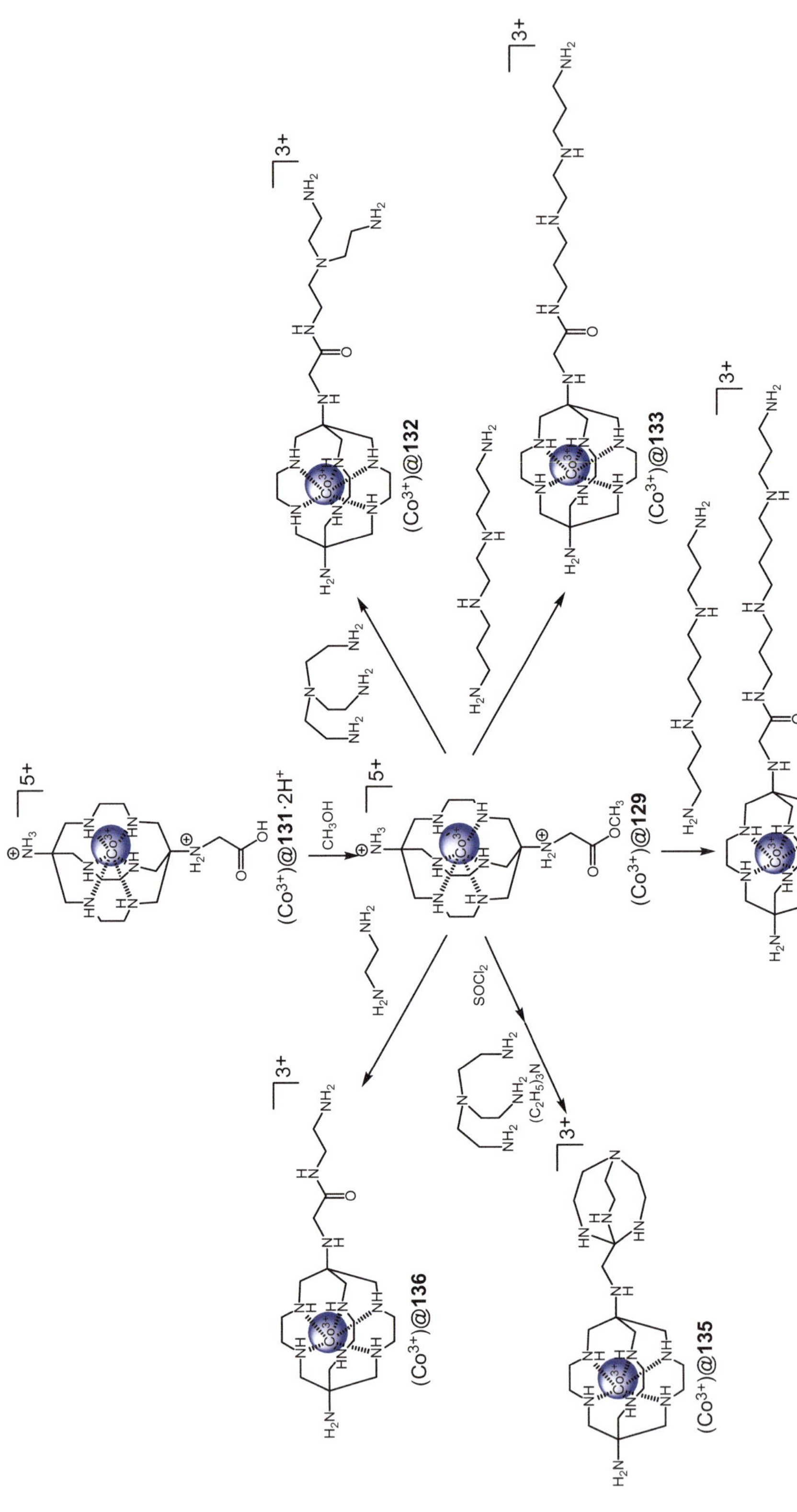

Scheme 2.75 Synthesis of a series of the apically monofunctionalized cobalt(III) clathrochelates

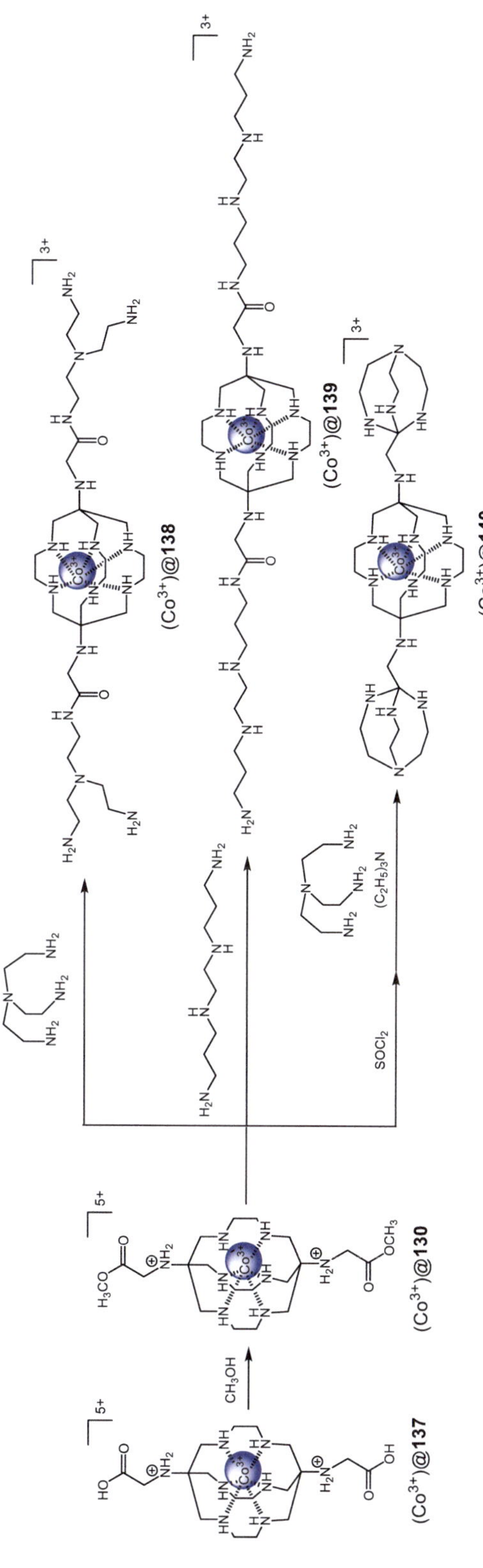

Scheme 2.76 Synthesis of a series of the apically difunctionalized cobalt(III) clathrochelates

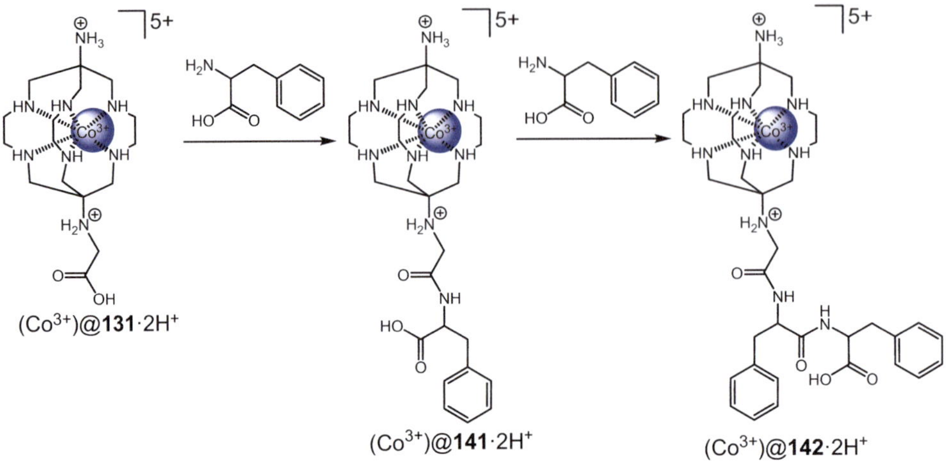

Scheme 2.77 Synthesis of the cobalt(III) clathrochelates (Co^{3+})@(**141**·$2H^+$) and (Co^{3+})@(**142**·$2H^+$)

corresponding metal ion as a matrix. In the case of 3-aminophenylboron-capped cage complex, triethylamine has been added to prevent a protonation of the terminal amino groups by H^+ ions releasing in the course of self-assembly reaction [82]. The preparation and single crystal X-ray structure of the same *para*-carboxyphenylboron-capped tris-nioximate iron(II) clathrochelate (Fe^{2+})@**212** are described [83].

Direct template cross-linking of three molecules of the corresponding aliphatic α-dioxime with 4- and 3-formylphenylboronic acids on an iron(II) ion as a matrix under mild reaction conditions in methanol or acetonitrile media by Schemes 2.98 and 2.99 has been used [84] for preparation of 4- and 3-formylphenylboron-capped iron(II) tris-nioximates and tris-dimethylglyoximates (Fe^{2+})@**217**, (Fe^{2+})@**218**, (Fe^{2+})@**219**, (Fe^{2+})@**220**, (Fe^{2+})@**221**, (Fe^{2+})@**222**, (Fe^{2+})@**223**, (Fe^{2+})@**224**, (Fe^{2+})@**225**, (Fe^{2+})@**226**, (Fe^{2+})@**227**, and (Fe^{2+})@**228** in high yields. With methanol as a solvent, such one-pot reactions gave the mixtures of these cage complexes and their acetals; the latter have been converted to their formyl-terminated macrobicyclic analogs using H^+-catalyzed hydrolysis. One-step self-assembly reactions in acetonitrile medium afforded the formyl-terminated iron(II) clathrochelates in the yields from a low to a moderate, and the overall yields of the above two-step synthetic procedures are reported [84] to be higher than those of this one-step condensation in acetonitrile as a solvent. On the other hand, the iron(II) hexachloroclathrochelate (Fe^{2+})@**229** with terminal formyl groups has been prepared under vigorous reaction conditions (boiling nitromethane as a solvent with distillation of nitromethane–water azeotrope) using template capping of three molecules of a weakly donor dichloroglyoxime with 3-formylphenylboronic acid by Scheme 2.100 on an iron(II) ion as a matrix.

The *n*-butyl- and *n*-hexadecylboron-capped iron and cobalt(II) tris-dioximates (M^{2+})@**230**, (M^{2+})@**231**, (M^{2+})@**232**, (M^{2+})@**233**, (M^{2+})@**234**, (M^{2+})@**235**, (M^{2+})@**236**, (M^{2+})@**237**, (M^{2+})@**238**, and (M^{2+})@**239** shown in Schemes 2.101 and 2.102 have been synthesized [85] by direct template condensation of three molecules of the corresponding aliphatic or aromatic α-dioxime with an appropriate alkylboronic Lewis acid on either an Fe^{2+} ion or a Co^{2+} cation as a matrix under mild or vigorous reaction conditions, depending on the nature of an α-dioximate ligand synthon.

The detailed synthetic protocols for preparation of the hexadecylboron-capped iron and cobalt(II) hexachloroclathrochelates in moderate yields by Scheme 2.103 using direct template condensation of three dichloroglyoxime

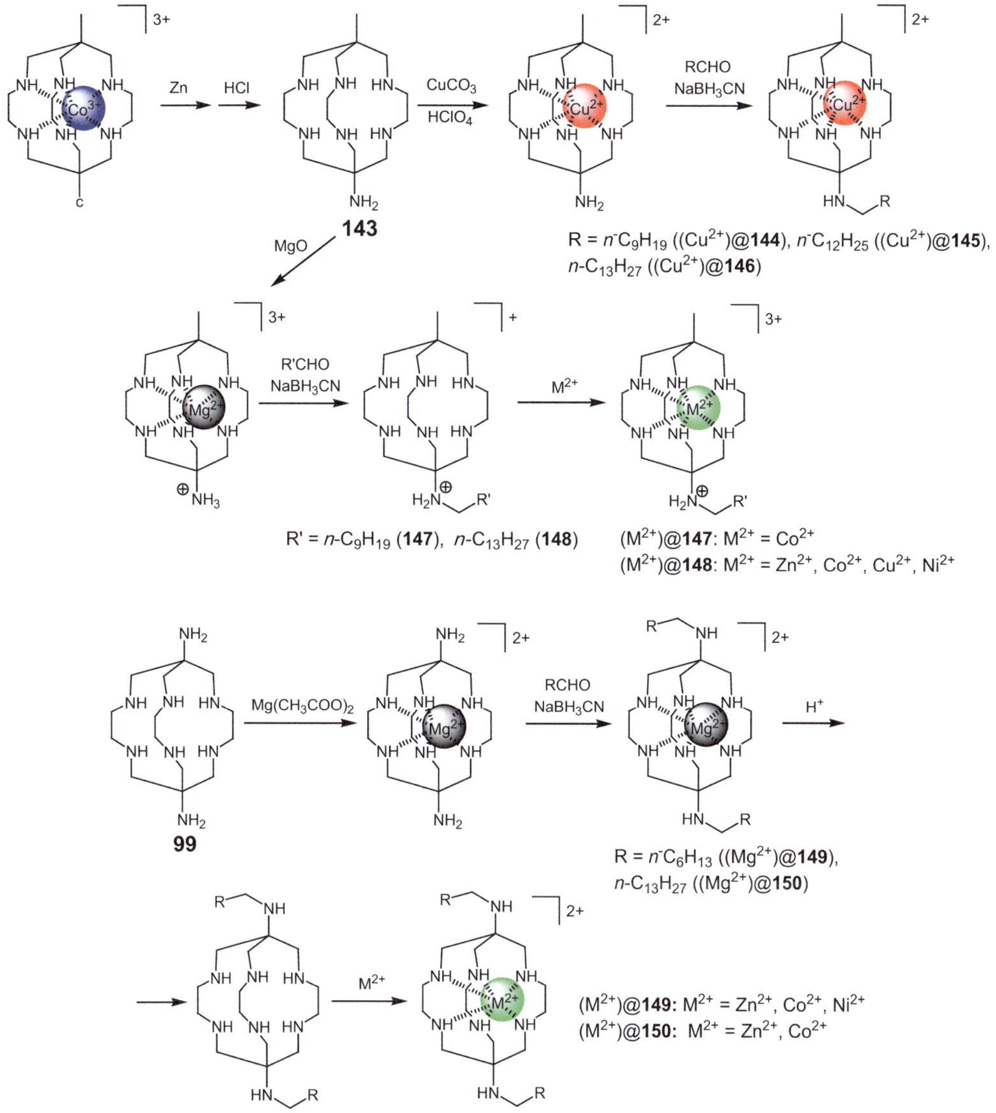

Scheme 2.78 Preparation of a series of polyamine metal(II) cage complexes

molecules with *n*-hexadecylboronic acid on the corresponding metal ion as a matrix are reported [86].

The 4-pyridylboron-capped clathrochelate iron(II) tris-dioximates have been prepared [87] in moderate yields by Scheme 2.104 using template cross-linking of aliphatic and aromatic α-dioximes [(Fe^{2+})@**241**, (Fe^{2+})@**242**, and (Fe^{2+})@**243**] and dichloroglyoxime [(Fe^{2+})@**244**] with 4-pyridylboronic acid on an iron(II) ion as a matrix. The aliphatic macrobicyclic iron(II) tris-dimethylglyoximate and tris-nioximate have been synthesized in methanol as a solvent, whereas in the case of aromatic α-benzildioxime, the reaction in methanol medium gave the Chugaev-type iron(II)bis-dioximate with 4-pyridylboronic acid as an axial N-donor ligand. So, the 4-pyridylboron-capped iron(II) tris-α-benzildioximate (Fe^{2+})@**243** has been synthesized under vigorous reaction conditions using a boiling nitromethane as a solvent. The use of trifluoroacetic acid as a solvent in

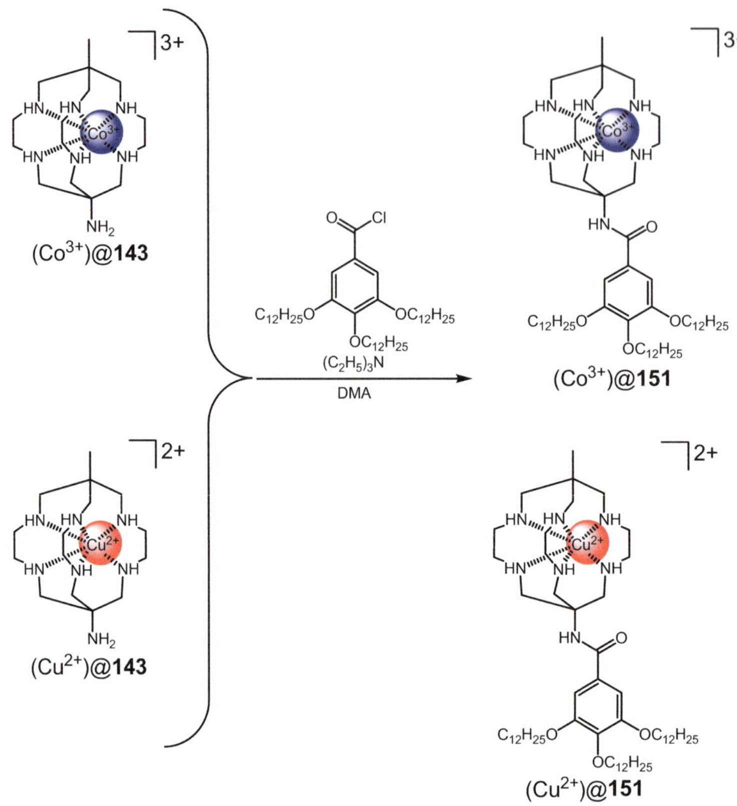

Scheme 2.79 Synthesis of cage complexes of the ligand **151**

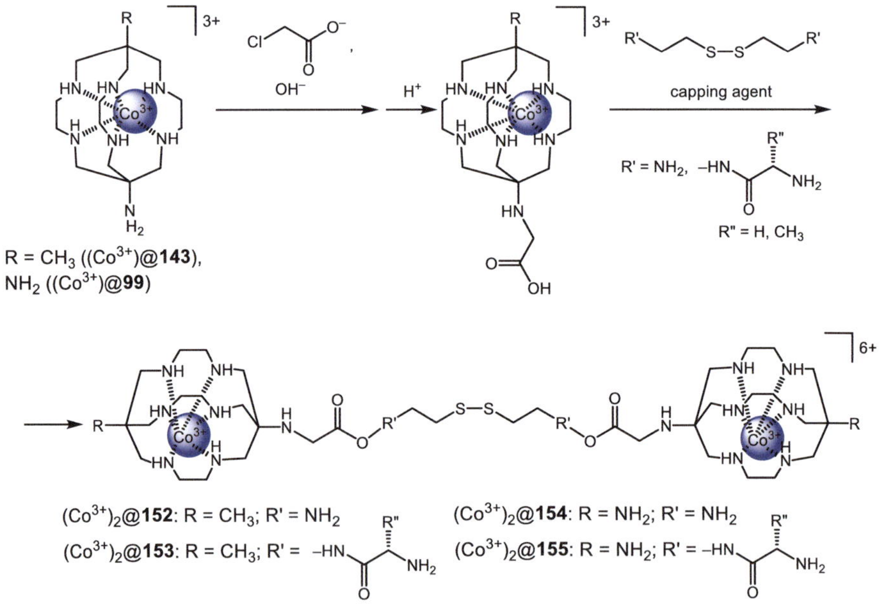

Scheme 2.80 Preparation of a series of cobalt(III) bis-clathrochelates

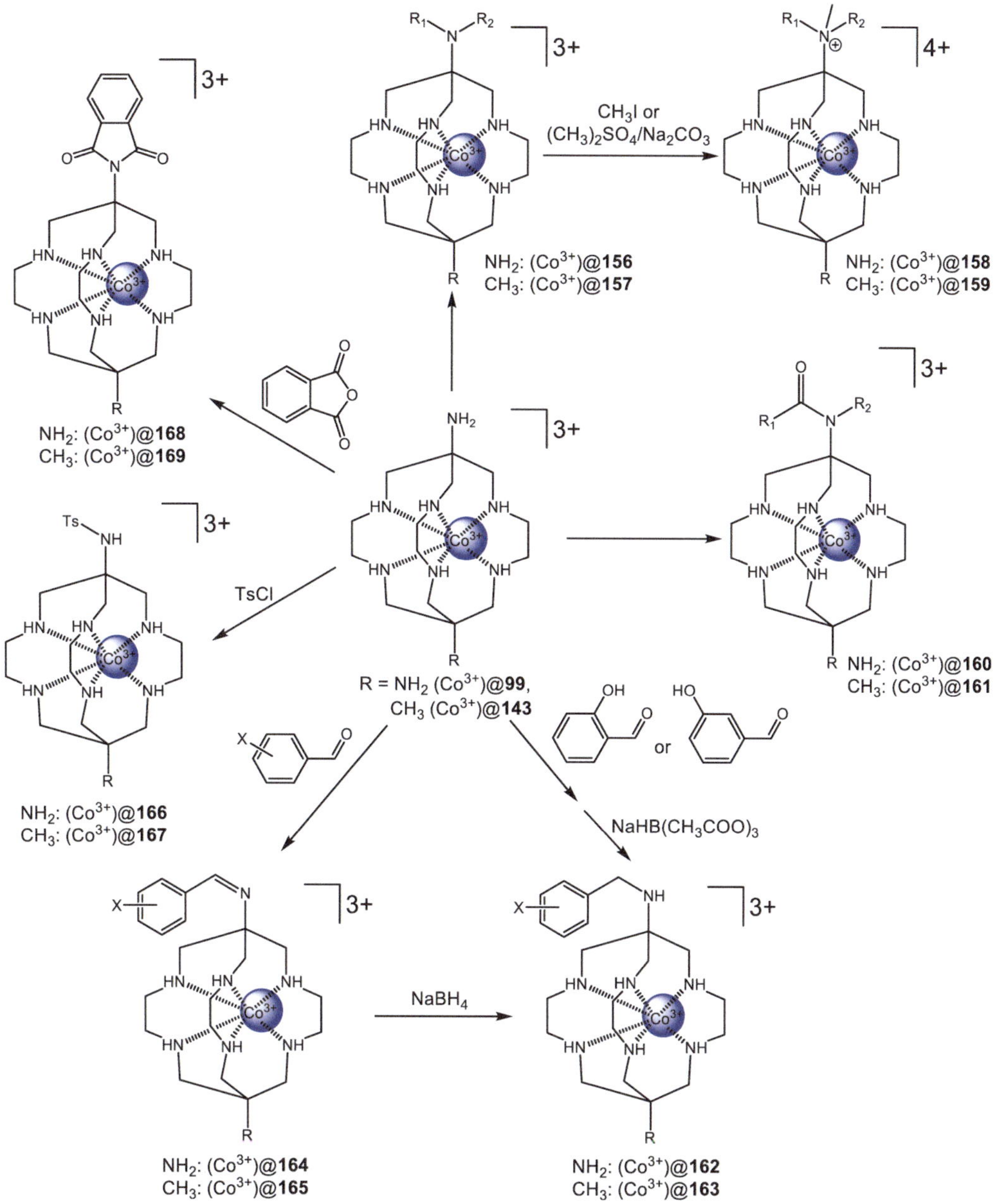

Scheme 2.81 Synthesis of a series of polyamine cobalt(III) cage complexes

the case of a weakly donor dichloroglyoxime allowed obtaining [87] the bis-trifluoroacetate salt of the corresponding diprotonated clathrochelate dication, which has been converted into the macrobicyclic intracomplex (Fe^{2+})@**244** with triethylamine.

Template condensation of a non-macrocyclic iron(II) tris-nioximate precursor with *ortho*-carboranyl-based capping agents in the presence of triethylamine, preventing the protolytic cleavage of their reactive B–C bonds, has been used [88] for the synthesis of the C-carboranylboron-capped iron(II) clathrochelates (Fe^{2+})@**245** and (Fe^{2+})@**246** by Scheme 2.105.

The iron(II) tris-dimethylglyoximate and tris-nioximate clathrochelates with apical

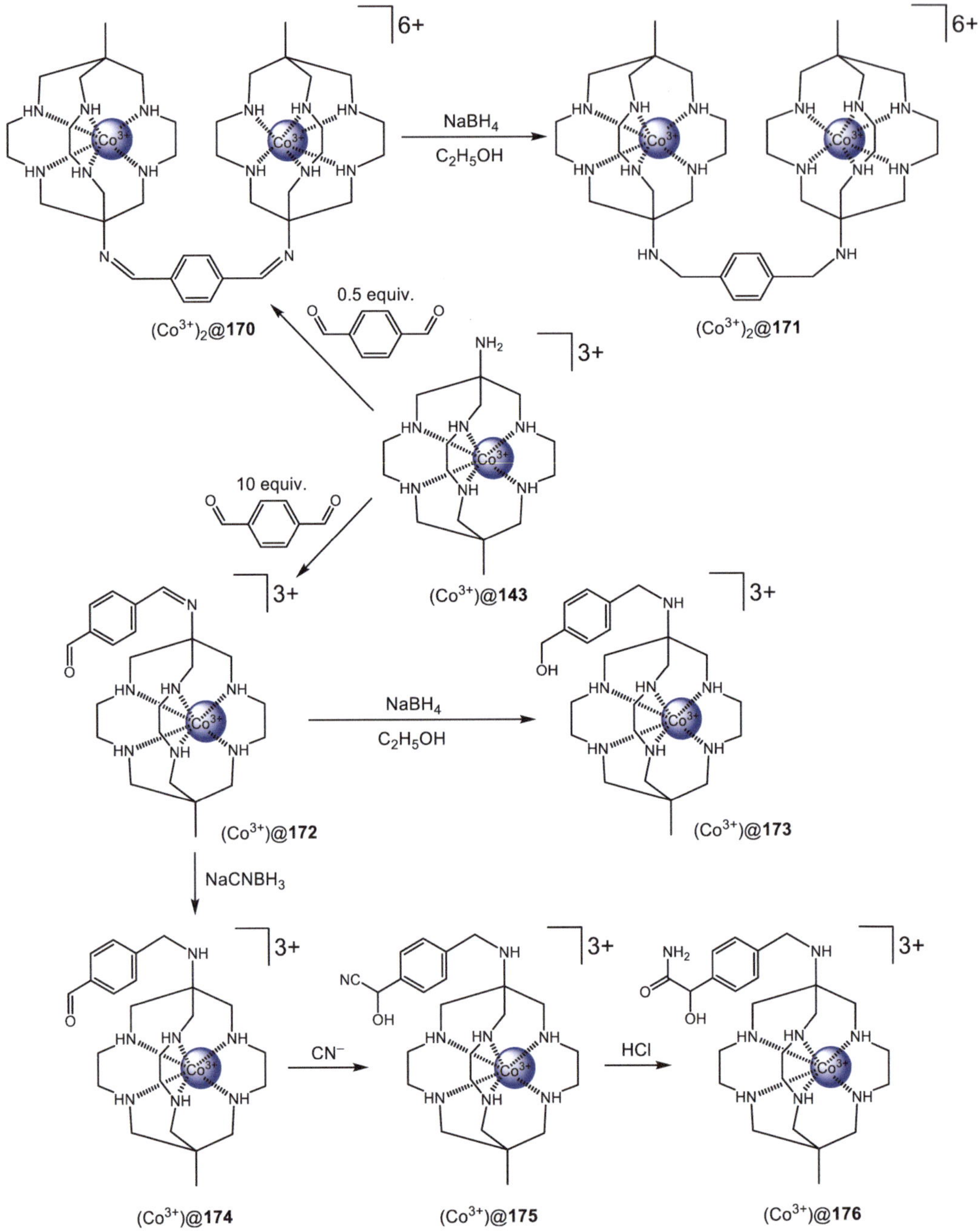

Scheme 2.82 Preparation of the cobalt(III) mono- and bis-clathrochelates

formylpyridine groups have been synthesized [89] by Scheme 2.106 using an appropriate acetal-protected boronic acid as a capping agent. The deprotection of them has been achieved with 4-toluenesulfonic acid using microwave heating to give the target formyl-terminated cage complexes (Fe^{2+})@**247** and (Fe^{2+})@**248**. Their 4-aminophenyl-terminated macrobicyclic analogs have been prepared by Scheme 2.107 using Lewis acidity of an appropriate benzyl-protected aromatic boronic acid. The macrobicyclic intermediates of these

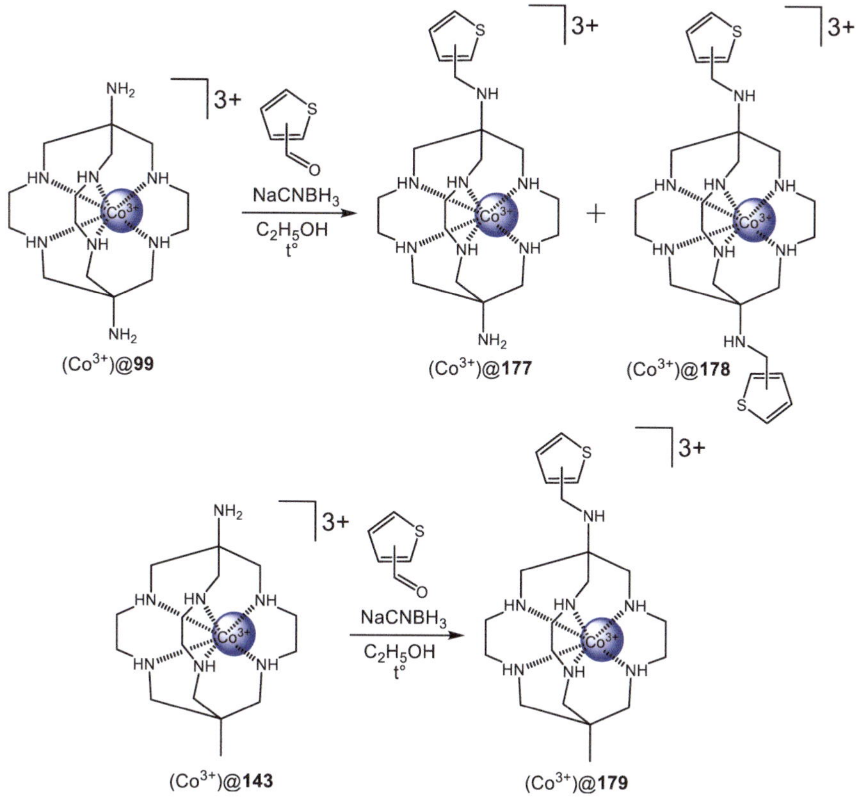

Scheme 2.83 Synthesis of the thienyl-terminated polyamine cobalt(III) clathrochelates

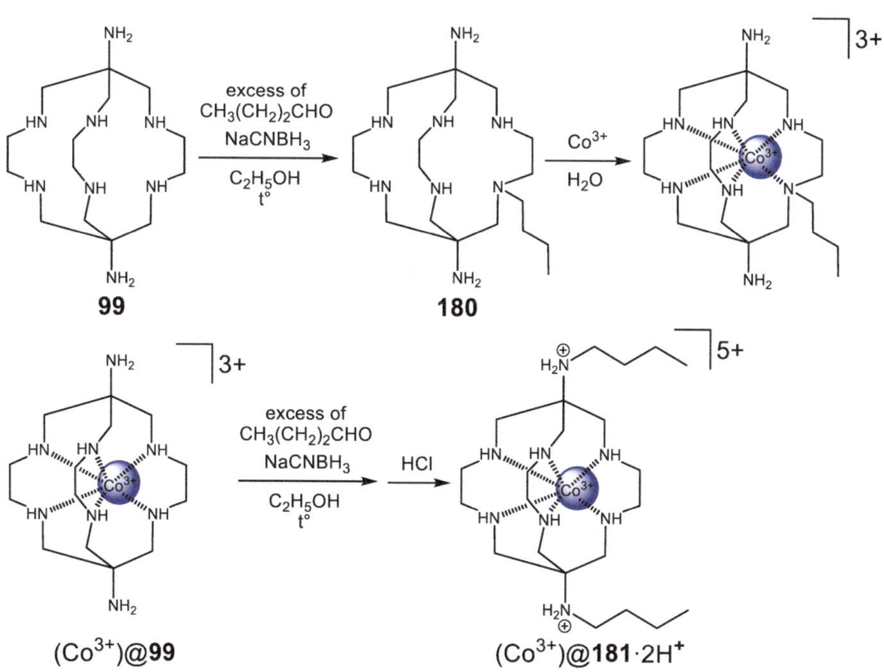

Scheme 2.84 Preparation of cobalt(III) complexes of the ligands **180** and **181**·2H⁺

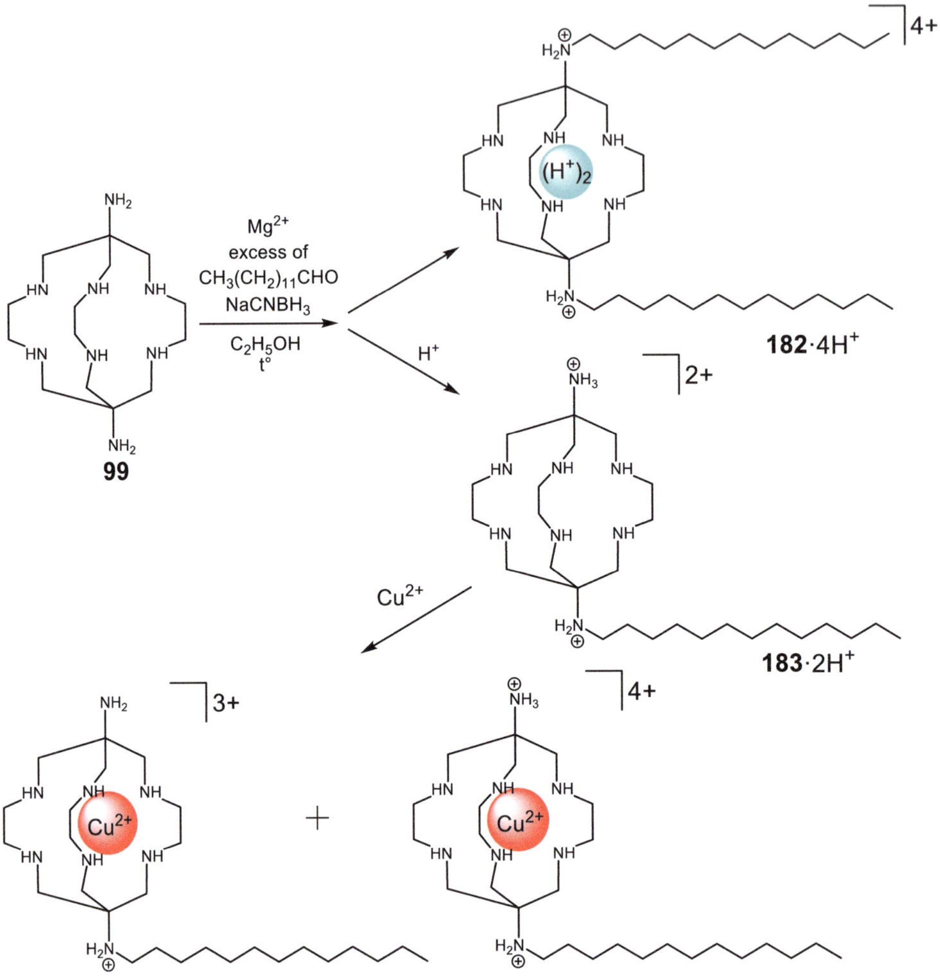

Scheme 2.85 Synthesis of the macrobicycles **182**·2H$^+$ and **183**·2H$^+$

template reactions have been deprotected through hydrogenation, thus giving the clathrochelates (Fe^{2+})@**249** and (Fe^{2+})@**250** with terminal amine groups. The 3-pyridyl-terminated glyoximate, diethylglyoximate, and nioximate iron(II) cage complexes (Fe^{2+})@**251**, (Fe^{2+})@**252**, and (Fe^{2+})@**253** have been synthesized by template self-assembly of 3-pyridylboronic acid and the corresponding α-dioxime on an iron(II) ion as a matrix. These complexes with donor 3-pyridyl terminal groups (Scheme 2.108) are reported [89] to form various types of the clathrochelate-based coordination cages (see Sect. 2.4).

Their di- and tetracarboxyl-terminated iron(II)-tris-dimethylglyoximate and tris-nioximate analogs (Scheme 2.109) have been used [90] as the macrobicyclic ligand synthons for preparation of the corresponding clathrochelate-based 1D, 2D, and 3D coordination polymers (see Sect. 2.4).

Template cross-linking of the corresponding α-dioxime with 3-pyridylboronic acid on an iron(II) ion afforded the iron(II) tris-dimethylglyoximate and tris-nioximate clathrochelates (Fe^{2+})@**251** and (Fe^{2+})@**258** (Scheme 2.110) with terminal 3-pyridyl groups [91].

A series of the 4-pyridylboron-capped iron(II) nioximate mono- and bis-clathrochelates (Fe^{2+})@**242**, (Fe^{2+})$_2$@**259**, and (Fe^{2+})$_2$@**260** has been self-assembled [92] using one-pot template synthetic procedures shown in Scheme 2.111 on

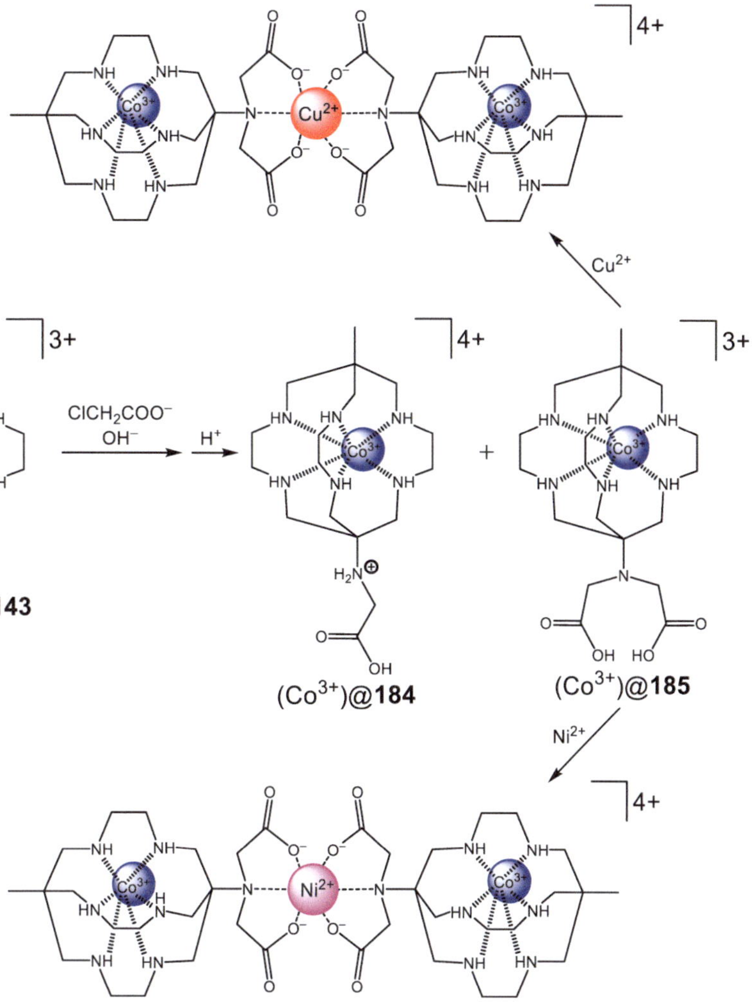

Scheme 2.86 Copper and nickel(II) complexes of the clathrochelate ligand (Co^{3+})@**185**

an iron(II) ion as a matrix. The boron-capped clathrochelate iron(II) complexes with two non-equivalent alkyne- or iodine- and 4-pyridyl-terminated apical substituents have been prepared by one-step synthetic procedure shown in Scheme 2.112 using the mixtures of two different arylboronic (4-pyridylboronic and 4-ethynylphenylboronic or 4-iodophenylboronic) acids for template cross-linking of the corresponding α-dioxime; the macrobicyclic products of these reactions [(Fe^{2+})@**261**, (Fe^{2+})@**262**, and (Fe^{2+})@**263**] with equivalent and nonequivalent capping groups have been chromatographically isolated [92] as individual compounds.

The triethylantimony-capped iron(II) tris-dimethylglyoximate (Fe^{2+})@**264** has been prepared [93] by Scheme 2.113 using direct template cross-linking of dimethylglyoxime with triethylantimony(V) dibromide as a capping Lewis-acidic agent on an iron(II) ion as a matrix; the attempted using of the same synthetic pathway was unsuccessful in the case of its cobalt(III)-encapsulating analog (Co^{3+})@**264**. So, the latter cage complex has been synthesized [93] in a moderate yield by macrobicyclization of the initially obtained potassium salt of non- macrocyclic cobalt(III) tris-dimethylglyoximate by Scheme 2.114.

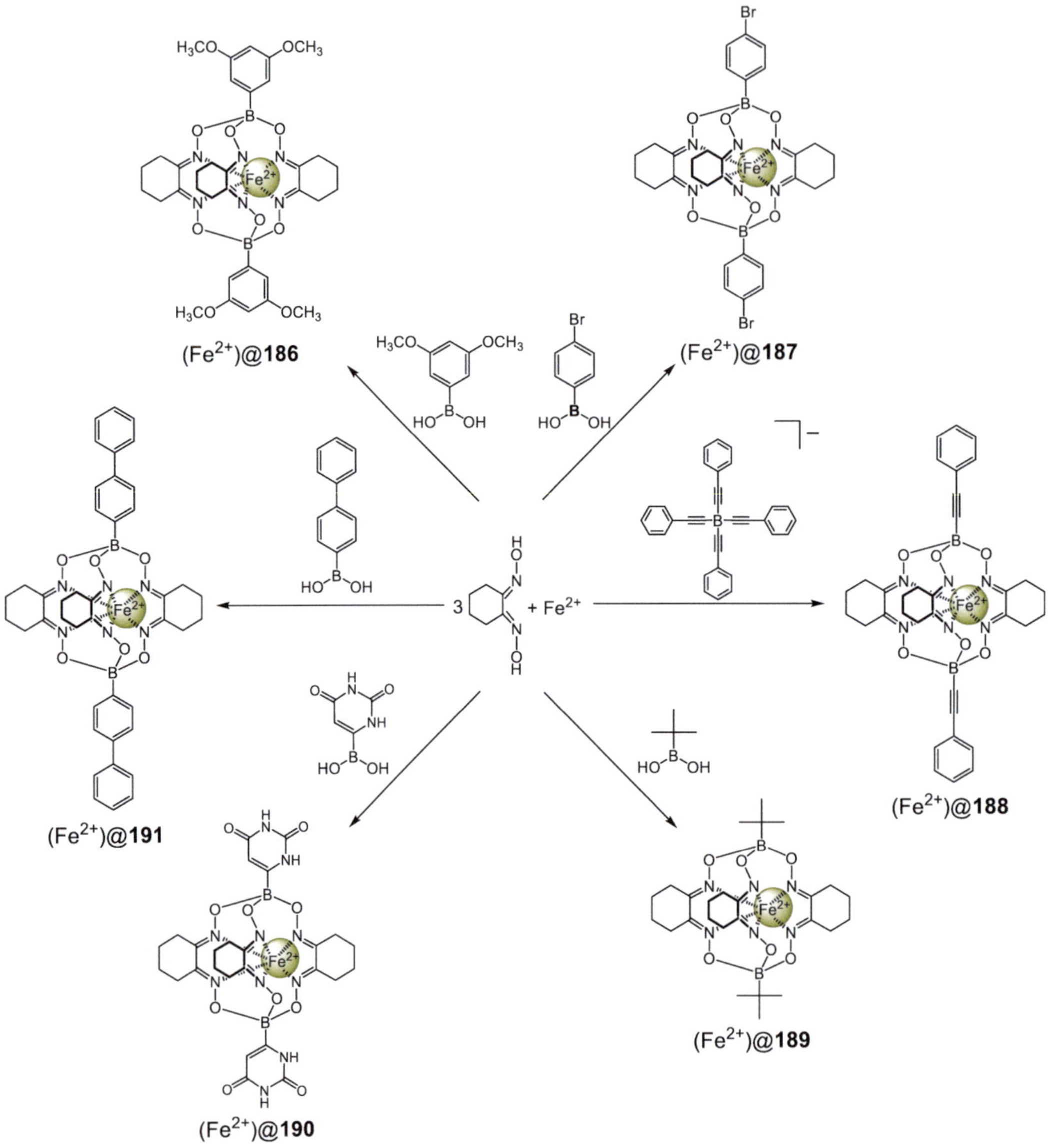

Scheme 2.87 Template self-assembly of apically functionalized iron(II) clathrochelates

Reactivity of Apical Groups

Capping of the semiclathrochelate (Fe^{2+})(**265**) (see Sect. 2.2.4) with mono- and bifunctional Lewis-acidic cross-linking agents, other than $(C_6H_5)_3Sb(OH)_2$ and its derivatives, has been used [94] for the synthesis of the mono- and -bis-clathrochelates with non-equivalent capping fragments shown in Scheme 2.115. Clathrochelates with labile triethylantimony capping group(s) [94] undergo a transmetallation in the presence of silica gel as a heterogeneous catalyst; this process did not occur under mild reaction conditions in the absence of SiO_2 and yielded a mixture of the products under more vigorous conditions. The first stage of this reaction gives the surface-immobilized mixed SbSi-capped macrobicyclic intermediates, further undergoing a desorption with other Lewis-acidic capping agents (Scheme 2.116) to form the mono- or bis-clathrochelate products (Fe^{2+})@**270**,

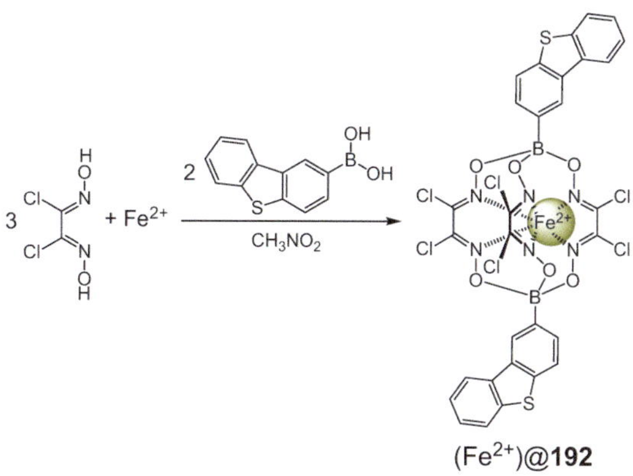

Scheme 2.88 Template self-assembly of the clathrochelate (Fe^{2+})@**192**

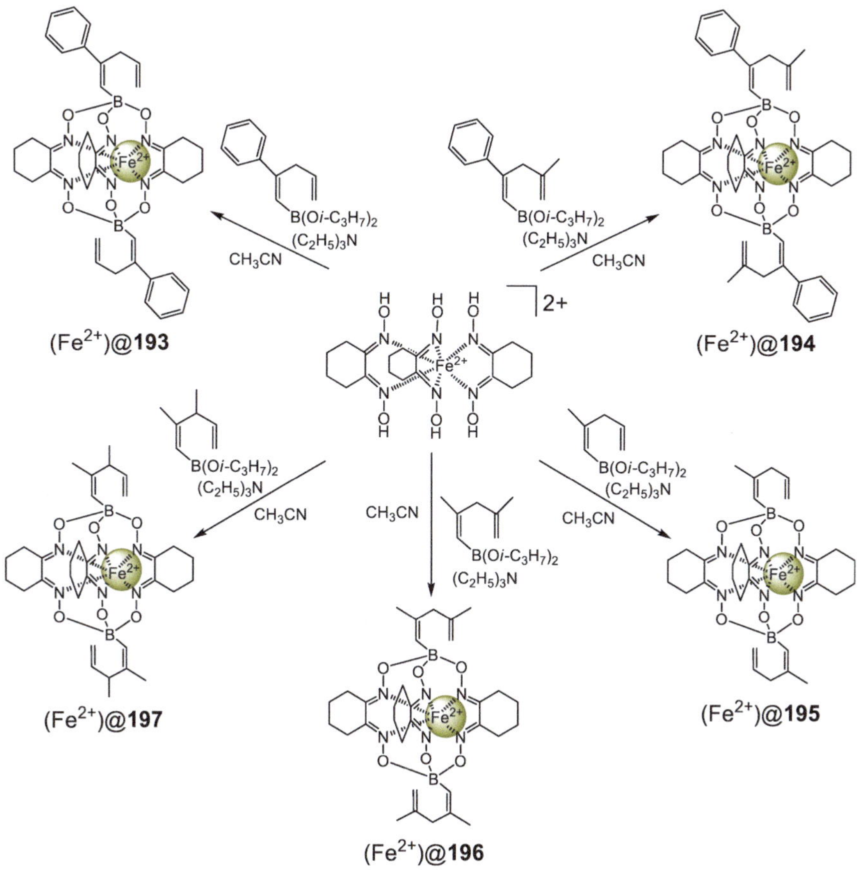

Scheme 2.89 Synthesis of diolefin-terminated iron(II) clathrochelates

(Fe^{2+})@**271**, (Fe^{2+})@**272**, $(Fe^{2+})_2$@**273**, and $(Fe^{2+})_2$@**274** with mono- and bifunctional Lewis acids, respectively. Such synthetic approach can be used for the straightforward synthesis of the mono- and bis-clathrochelates with two non-equivalent capping groups [94].

Transmetallation of the macrobicyclic boron, antimony-cross-linked tris-nioximate

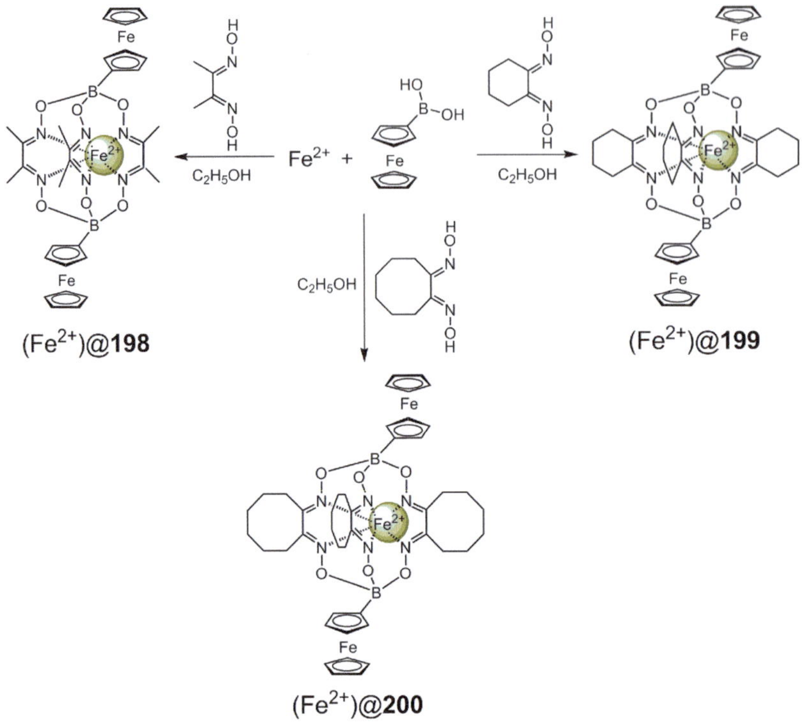

Scheme 2.90 Template synthesis of the ferrocenylboron-capped iron(II) complexes

(Fe^{2+})@**271** by Scheme 2.117 with tris(trifluoromethyl)germanium(IV) iodide as an efficient capping agent, three electron-withdrawing substituents of which enhance the Lewis acidity of the capping germanium atom and prevent the possible polymerization giving the coordination polymers, has been used [95] for preparation of a mixed boron, germanium-capped clathrochelate anion (Fe^{2+})@**275**. In contrast to their iron(II)-containing analogs, the bis(triethylantimony)-capped cobalt(III) tris-dimethylglyoximate (Co^{3+})@**264** (see section “Template Synthesis”) did not undergo either transmetallation with various boronic acids or stepwise reaction with $IGe(CF_3)_3$, giving on its first stage a mixed germanium, antimony-capped cobalt(III) cage intracomplex. Only the bis-germanium-capped clathrochelate monoanion(Co^{3+})@**276** has been formed by Scheme 2.118 with an excess of this capping agent [95].

The apically linked iron(II) α-dioximate bis-clathrochelates $(Fe^{2+})_2$@**277** and $(Fe^{2+})_2$@**278** have been synthesized [96] by Scheme 2.119 using transmetallation of a mixed boron, antimony-capped iron(II) tris-nioximate (Fe^{2+})@**271** as their macrobicyclic precursor with the corresponding bifunctional Lewis-acidic capping agent (i.e., 1,4-phenylenediboronic acid and the butyl ester of 1,3-propanediboronic acid). Transmetallation of the same clathrochelate precursor with phenylboronic acid in the presence of silica gel as a heterogeneous catalyst gave the diboron-capped tris-nioximate iron(II) monoclathrochelate (Fe^{2+})@**279** with non-equivalent capping fragments [96].

The formyl-terminated iron(II) tris-dimethylglyoximate and tris-nioximate clathrochelates (see section “Template Synthesis”) are reported [84] to undergo H^+-catalyzed condensation with isoniazid as hydrazone component, thus giving the corresponding hydrazonate cage complexes (Fe^{2+})@**280** and (Fe^{2+})@**281** shown in Scheme 2.120 [84].

Sonogashira coupling by Scheme 2.121 has been used [92] for the synthesis of the 4-pyridyl-terminated iron(II) bis- and tris-clathrochelates $(Fe^{2+})_2$@**282**, $(Fe^{2+})_2$@**283**, and $(Fe^{2+})_3$@**284** with a rigid, rod-shaped geometry.

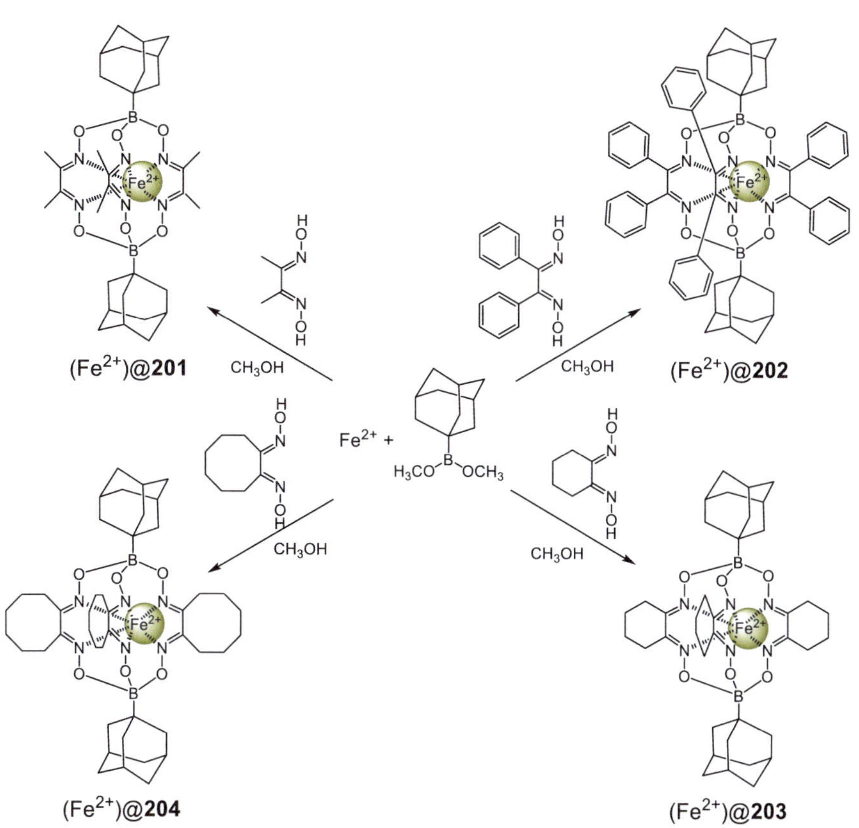

Scheme 2.91 Template synthesis of the adamantylboron-capped iron(II) complexes

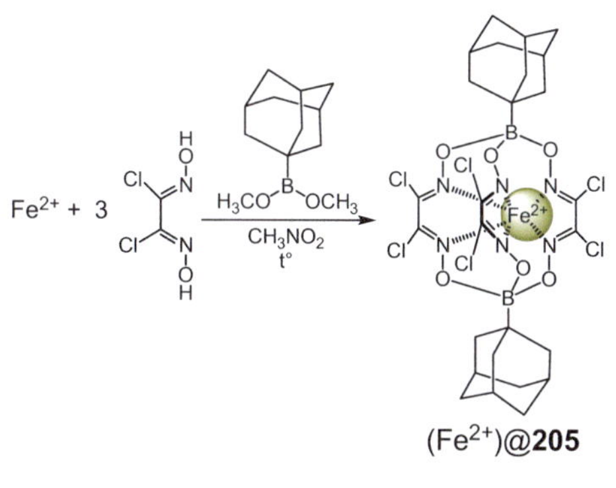

Scheme 2.92 Preparation of the hexachloroclathrochelate (Fe^{2+})@**205**

2.2.2.2 Ribbed Functionalization of Quasiaromatic Macrobicyclic Compounds

Template Synthesis and Condensation Reactions

Template condensation of dichloroglyoxime with methaneboronic acid on the corresponding metal ion as a matrix by Scheme 2.122 under vigorous reaction conditions has been used [97] for preparation of the methylboron-capped iron and cobalt(II) hexachloroclathrochelates (M^{2+})@**285**. Their analogs (M^{2+})@**286** (Scheme 2.123) have been prepared [98] using the same synthetic approach.

The iron and cobalt(II) hexabromoclathrochelates (M^{2+})@**287** have been self-assembled [99] by

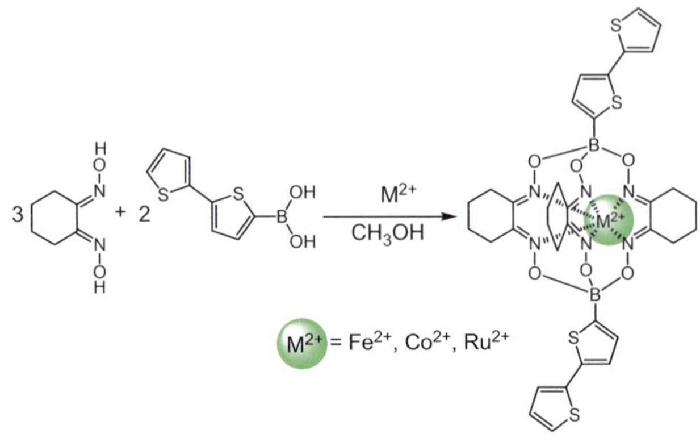

(M^{2+})@**206**

Scheme 2.93 Self-assembly of the dithienyl-terminated complexes (M^{2+})@**206**

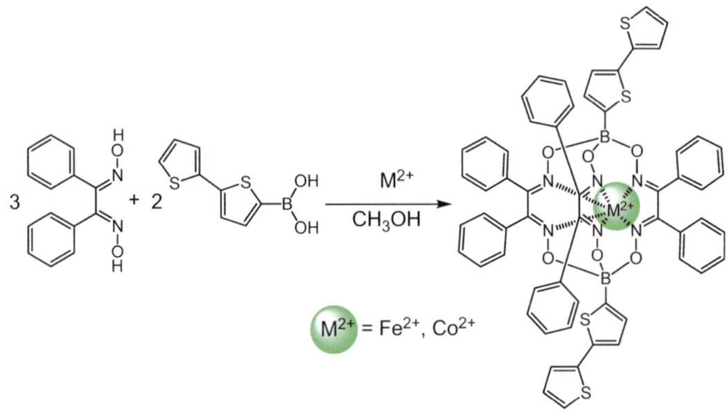

(M^{2+})@**207**

Scheme 2.94 Template synthesis of the dithienyl-terminated iron and cobalt(II) tris-α-benzildioximates

Scheme 2.124 under vigorous reaction conditions using template cross-linking of three molecules of dibromoglyoxime with *n*-butylboronic acid on the corresponding metal (II) ion as a matrix. Their *n*-butylboron-capped iron(II) hexaiodoclathrochelate analog (Fe^{2+})@**288** has been prepared [100] by Scheme 2.125 using direct condensation of diiodoglyoxime with *n*-butylboronic acid on an iron(II) ion under vigorous reaction conditions.

Template synthesis of the iron and cobalt(II) (per)fluoroclathrochelates with superhydrophobic exterior is described [101]. For the first time, the cage complex of this series, (Fe^{2+})@**289**, has been prepared [100] by two-step procedure, including perfluoroarylation of its hexaiodoclathrochelate precursor with Cu(C_6F_5) (see section "Metal-Promoted (Catalyzed) Reactions"). This synthetic approach needs an isolation of the metal hexaiodoclathrochelate precursor for each type of the apical capping groups and could not be performed for less thermodynamically stable and more redox and chemically reactive cobalt(II) cage complexes [101]. So, the *n*-butyl-, phenyl-, and pentafluorophenylboron-capped iron and cobalt(II) tris-perfluoro-α-benzildioximates have been synthesized by Scheme 2.126 using template condensation of three molecules of this weakly donor α-dioxime with an appropriate boronic acid on the corresponding metal(II) ion as a matrix. Such self-assemblies have been performed under vigorous reaction conditions, thus giving the target cage complexes in moderate yields [101].

The pentafluorophenylboron-capped iron and cobalt(II) hexachloroclathrochelates have been synthesized [102] using one-pot template condensation of dichloroglyoxime with

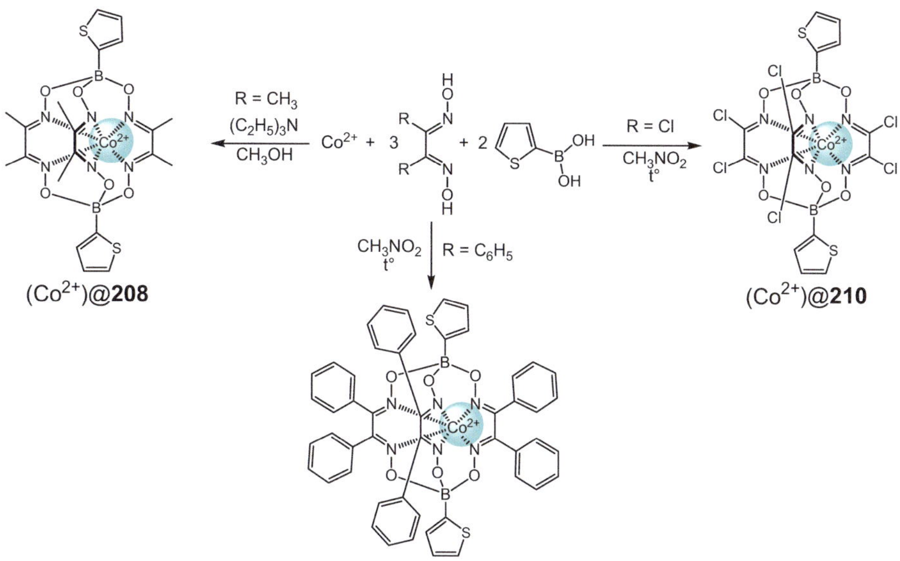

Scheme 2.95 Preparation of the macrobicyclic cobalt(II) tris-α-benzildioximate (Co^{2+})@**209**

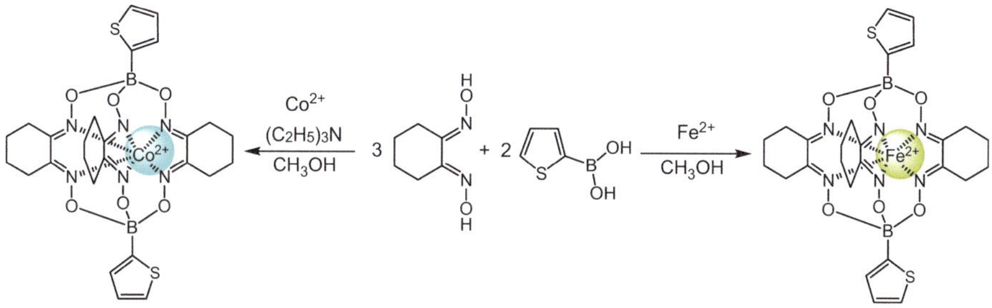

Scheme 2.96 Synthesis of the thienyl-terminated iron and cobal(II) clathrochelates

pentafluorophenylboronic acid on an iron or cobalt(II) ion as a matrix by Scheme 2.127 under vigorous reaction conditions (i.e., trifluoroacetic acid as a solvent). Such unusual solvent was used for increasing of Lewis acidity of $C_6F_5B(OH)_2$ and for avoiding its side deborylation reaction. The formed pentafluorophenylboron-capped hexachloroclathrochelates are perfectly stable in such strongly acidic media [102].

Template condensation of dichloroglyoxime with *n*-butylboronic acid on a cobalt(II) ion as a matrix by Scheme 2.128 under vigorous reaction conditions has been used [103] for preparation of the *n*-butylboron-capped cobalt(II) hexachloroclathrochelate (Co^{2+})@**293** in a high yield. The detailed synthetic procedures for preparation of the *n*-butyl-, phenyl-, and fluoroboron-capped cobalt(II) hexachloro-clathrochelates (Co^{2+})@**243**, (Co^{2+})@**294**, and (Co^{2+})@**295** using template self-assemblies on a cobalt(II) ion by Scheme 2.129 are described [104]. Co^{2+} ion is reported to be less efficient template than Fe^{2+} cation, and the yields of these cobalt(II) cage complexes were substantially lower than those for their iron(II)-encapsulating analogs. This fact is explained [104] by their different Shannon radii and electronic configurations: the hexacoordinate cobalt(II) ion with electronic configuration d^7 has $r_i = 0.79$ and 0.885 Å for its low- and high-spin states, respectively; so it is

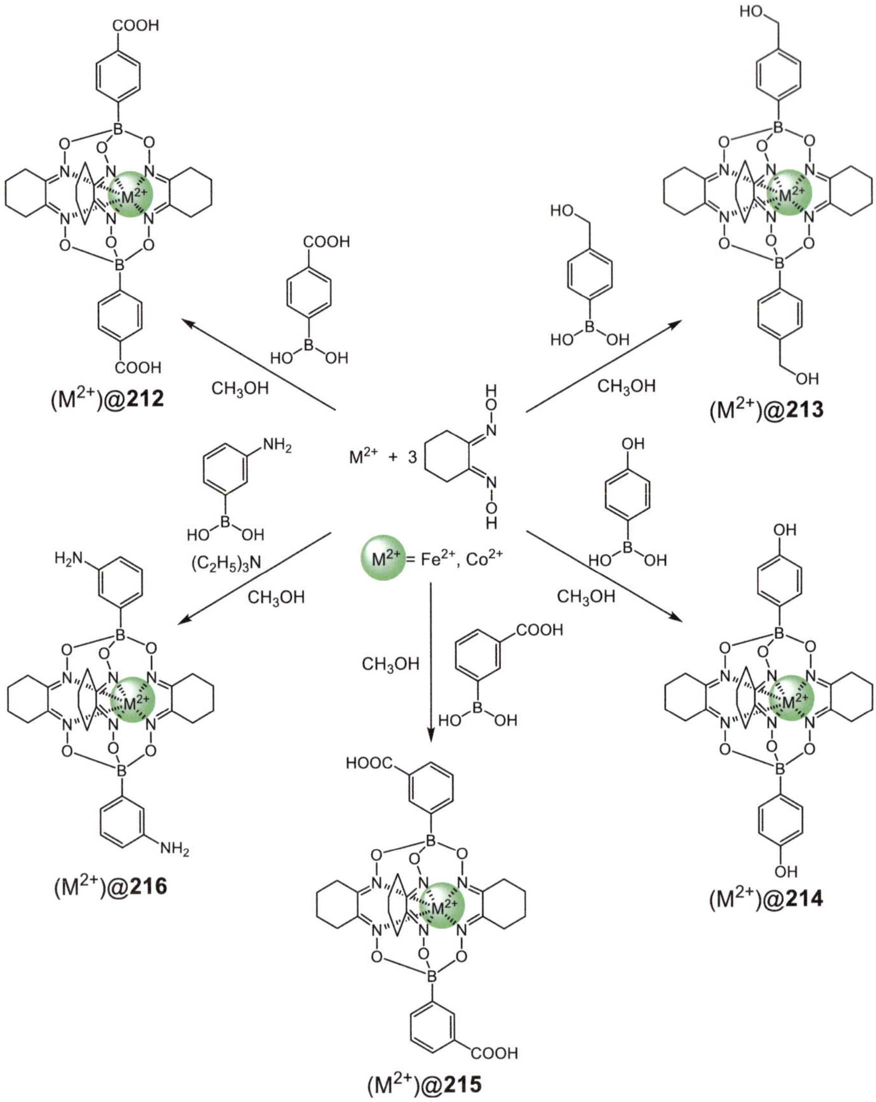

Scheme 2.97 Preparation of the iron and cobalt(II) tris-nioximates with terminal reactive groups

substantially larger than the low-spin iron(II) ion (r_i = 0.75 Å) with electronic configuration d^6. Moreover, the cobalt(II) ion is known to easily form the stable square-planar macrocyclic bis-dioximates, and these side reactions are described [104] to decrease the yields of the above cobalt(II) hexachloroclathrochelates.

The tris-dichloroglyoximate clathrochelate dianion (Fe^{2+})@**296** has been prepared [105] by Scheme 2.130 using $((CH_3)_4N)_2SnCl_6$ as a capping agent.

The improved synthetic procedure for preparation of a mixed α-benzildioximate–dimethylglyoximate clathrochelate (Fe^{2+})@**297** by condensation of dimethylglyoxime with an iron(II) macrocyclic complex *13* by Scheme 2.130 under vigorous reaction conditions has been elaborated [106] (Scheme 2.131).

The monochloroclathrochelate (Fe^{2+})@**298** has been synthesized [107] using condensation of the same macrocyclic precursor with methylchloroglyoxime by Scheme 2.132.

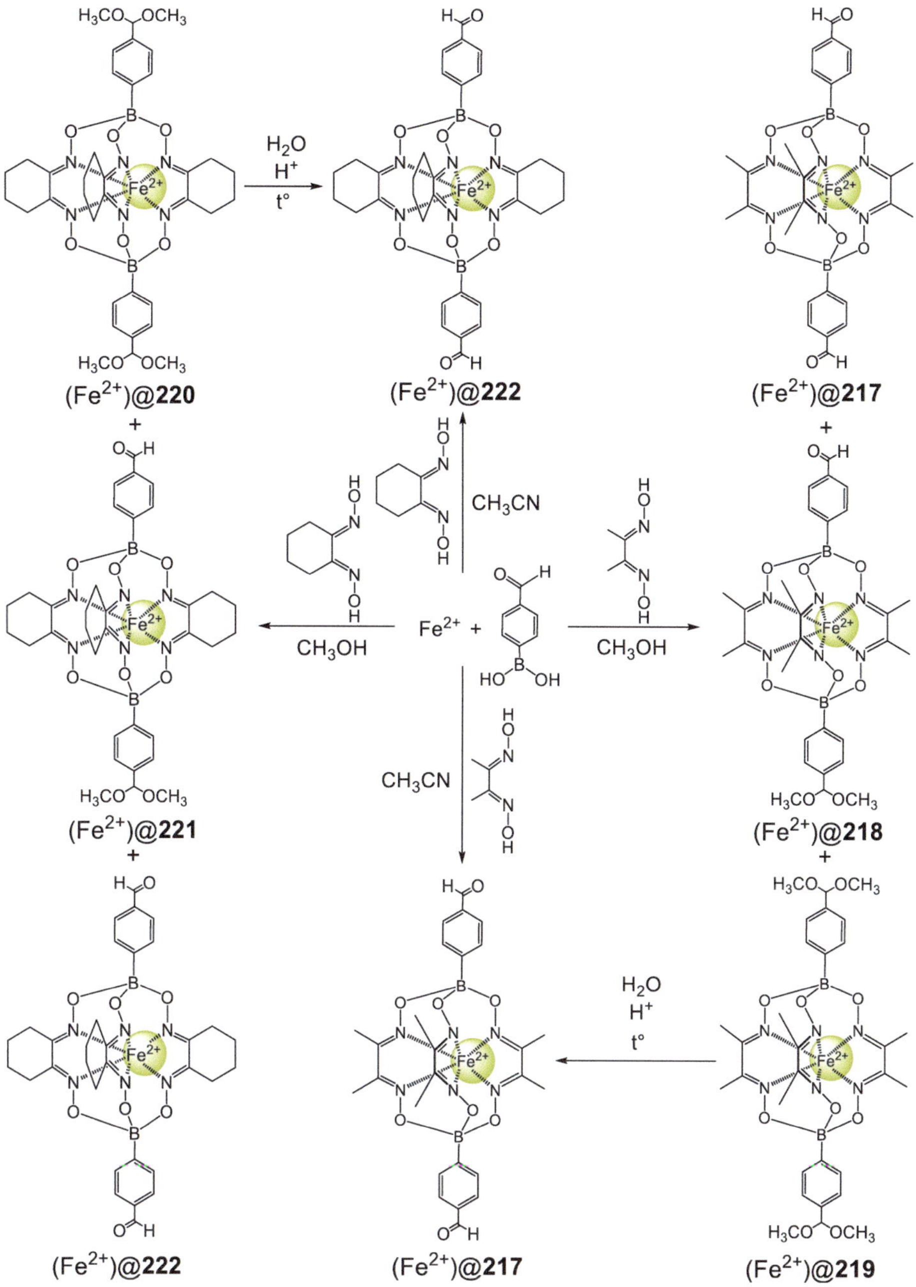

Scheme 2.98 Synthesis of the pare-formyl(acetal)-terminated iron(II) dimethylglyoximates

The *mer*- and *fac*-isomers of the iron(II) trichloroclathrochelates (Fe^{2+})@**299** and (Fe^{2+})@**300** have been synthesized [107] by Schemes 2.133 and 2.134 using direct template cross-linking of three methylchloroglyoxime molecules with phenylboronic acid and $BF_3 \cdot O(C_2H_5)_2$ as the capping Lewis-acidic agents, respectively, on an iron(II) ion as a matrix.

Condensation of mono- and dichloroglyoximes with a cobalt(II) bis- α- benzildioximate *14* by Scheme 2.135 under vigorous reaction conditions

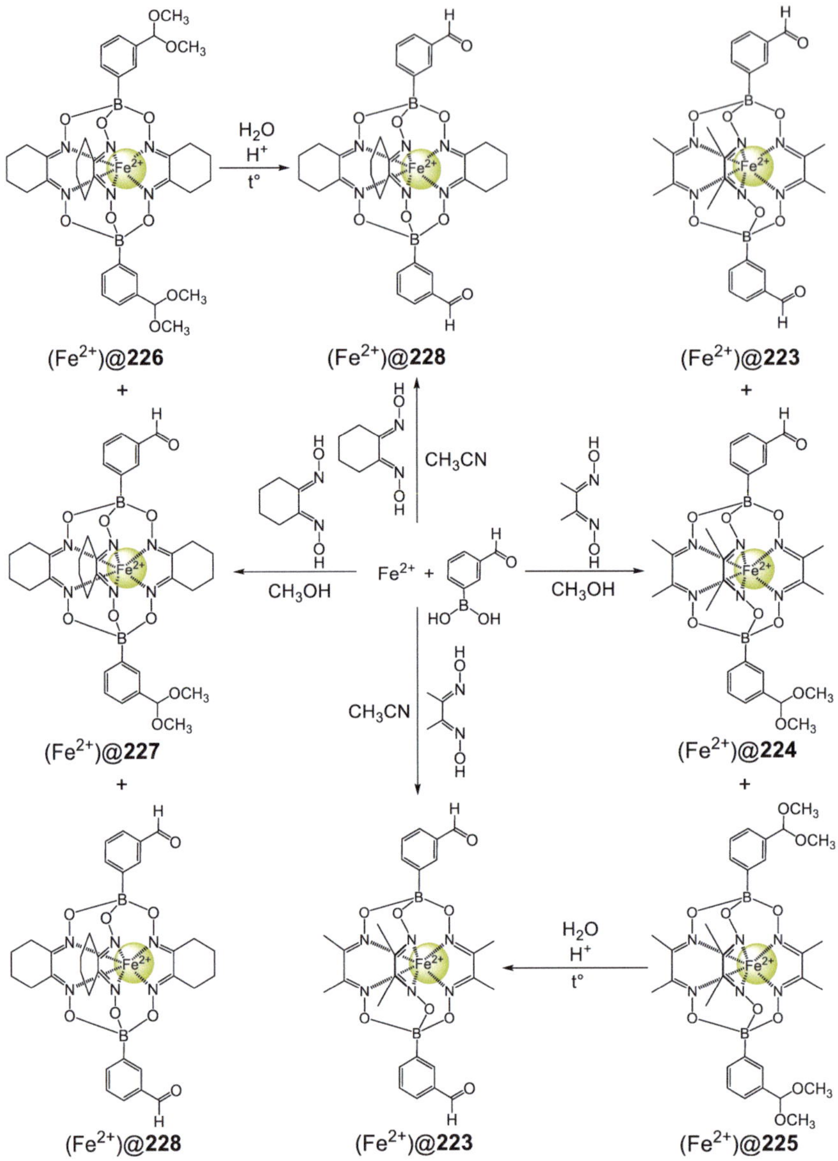

Scheme 2.99 Synthesis of the meta-formyl(acetal)-terminated iron(II) dimethylglyoximates

has been used [108] for preparation of the cobalt(II) mono- and dichloroclathrochelates (Co^{2+})@**301** and (Co^{2+})@**302** in moderate yields.

The phenyl-, fluoro-, and *n*-butylboron-capped trichloroglyoximate cage complexes (Fe^{2+})@**303**, (Fe^{2+})@**304**, and (Fe^{2+})@**305** have been prepared [2] by template cross-linking on an iron(II) ion (Scheme 2.136) using Lewis acidity of the corresponding trigonal boron compounds as cross-linking agents. The *fac*-to-*mer* isomer ratio for these tris-chloroglyoximate clathrochelates was found to be close to the statistically predicted one. Condensation of a macrocyclic iron(II) bis-α-benzildioximate *13* with chloroglyoxime has been performed [2] by Scheme 2.137 to give the monochloroclathrochelate (Fe^{2+})@**301**. That of a monophenol-terminated α-dioxime *15* with the same macrocyclic precursor by Scheme 2.138 has been used [109] for preparation of the iron(II)

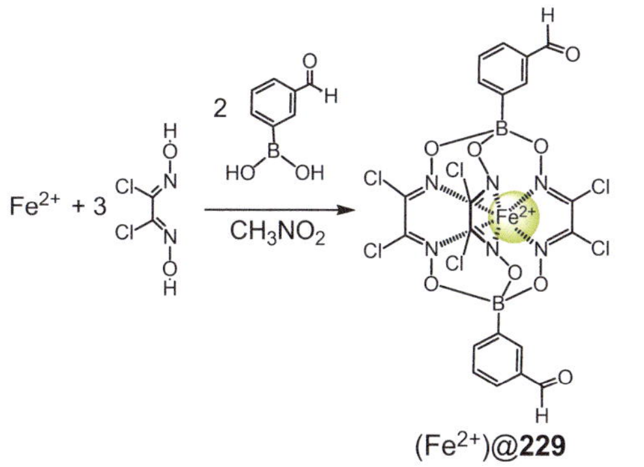

Scheme 2.100 Template synthesis of the hexachloroclathrochelate (Fe^{2+})@**229**

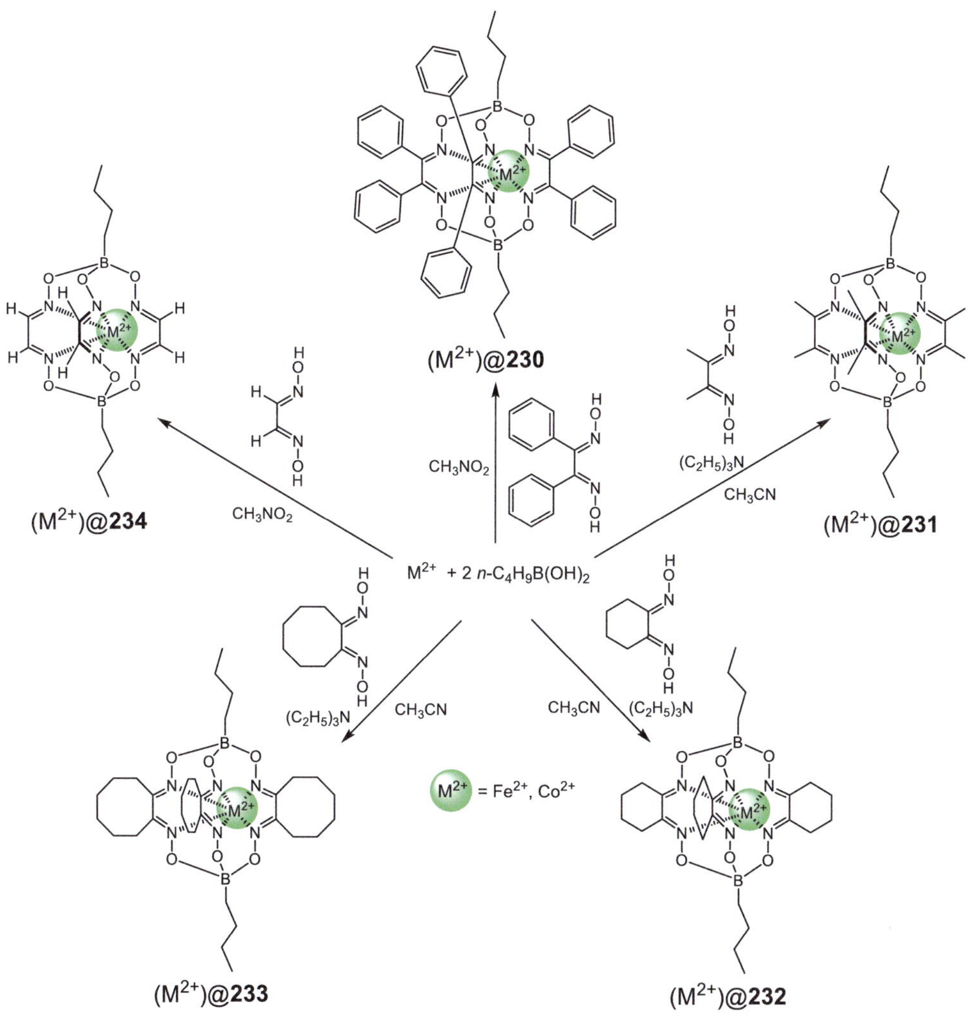

Scheme 2.101 Self-assembly of *n*-butylboron-capped iron and cobalt(II) clathrochelates

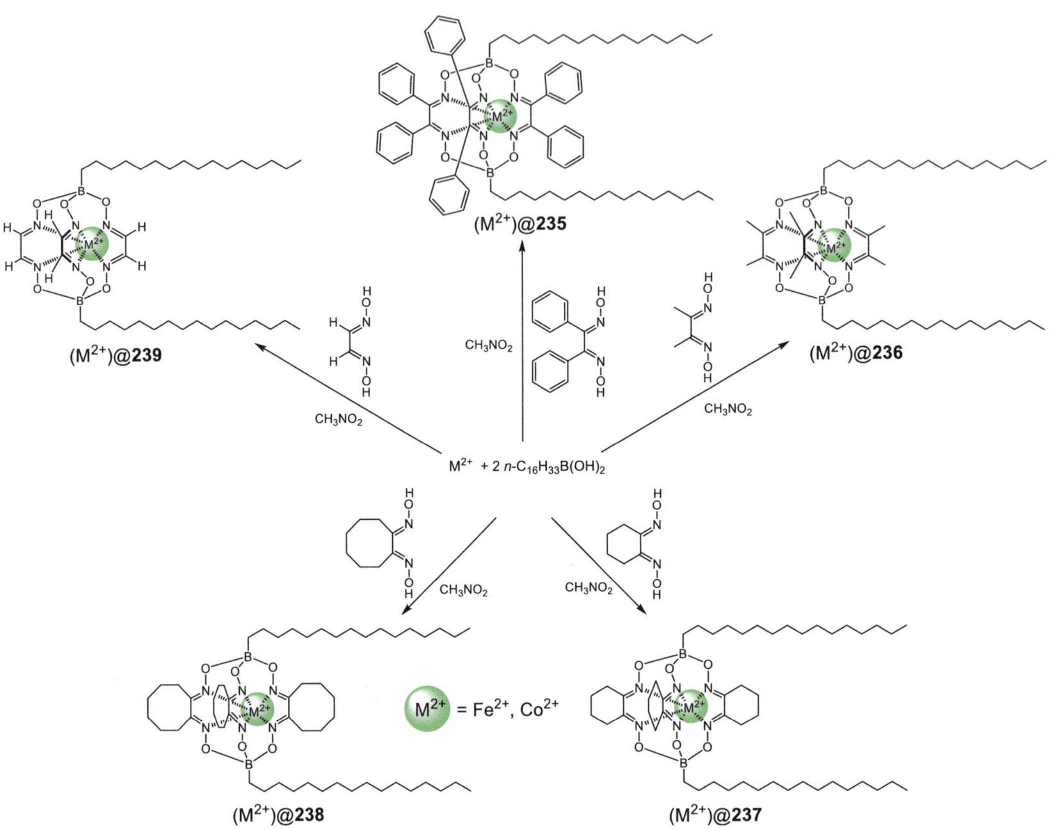

Scheme 2.102 Self-assembly of *n*-hexadecylboron-capped iron and cobalt(II) clathrochelates

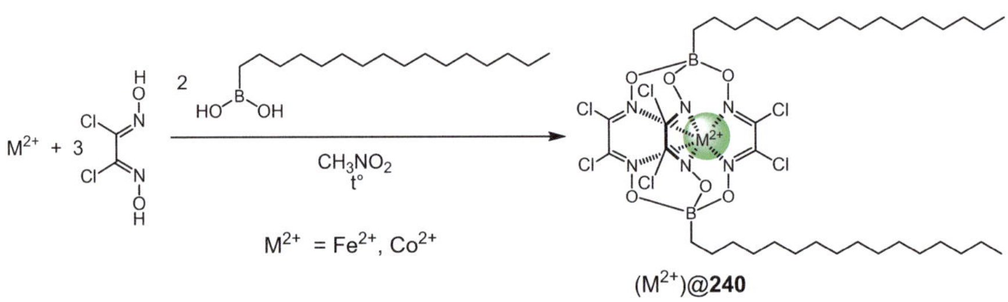

Scheme 2.103 Preparation of the *n*-hexadecylboron-capped iron and cobalt(II) hexachloroclathrochelates

clathrochelate (Fe^{2+})@**306** with one inherent ribbed substituent of this type.

Synthetic procedures for preparation of the cobalt(II) clathrochelates (Co^{2+})@**296** and (Co^{2+})@**307** with non-equivalent chelate ribbed fragments (Scheme 2.139) are reported [110] to have several peculiarities as compared with those for their iron(II)-encapsulating analogs. The attempts of their synthesis starting from the macrocyclic cobalt(II) bis-dioximates are complicated by easy oxidation and reversible binding of O_2 by these planar complexes. The cobalt(III) clathrochelates with non-equivalent ribbed fragments could not be obtained starting from the kinetically inert cobalt(III) bis-dioximates, whereas in the presence of a reductant only the macrobicyclic cobalt(II)-containing products with equivalent ribbed fragments have been isolated [110]. On the other hand, in the course of one-pot synthetic

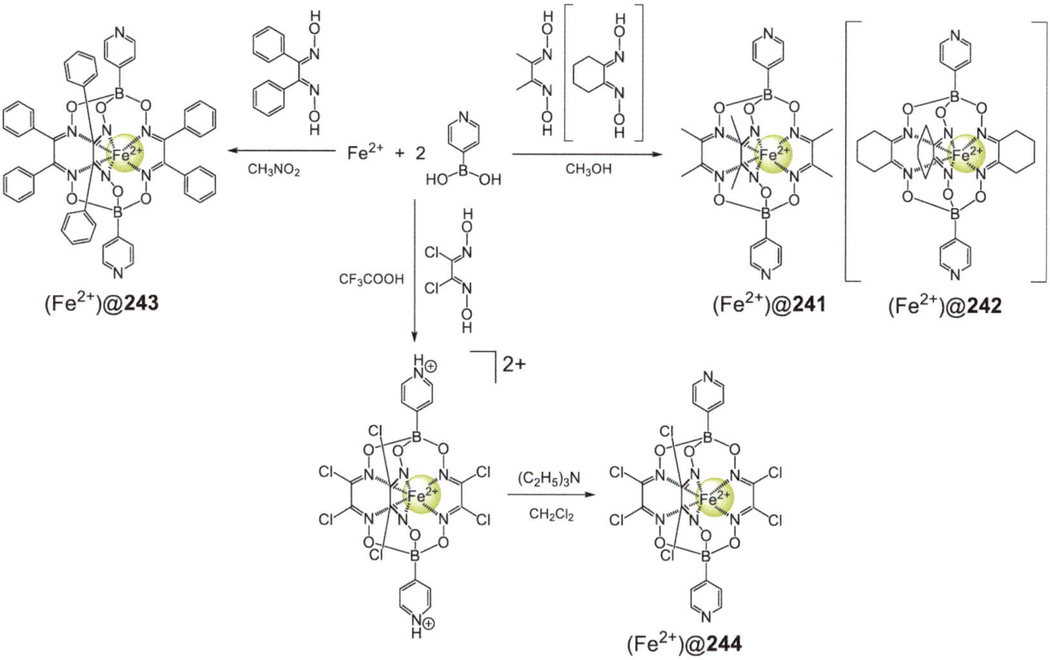

Scheme 2.104 Template synthesis of the pyridyl-terminated iron(II) clathrochelates

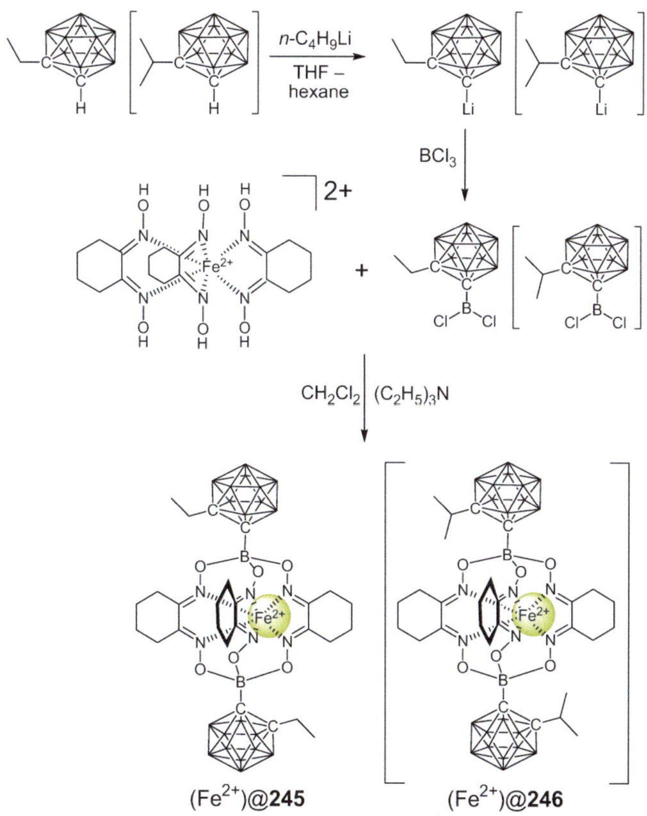

Scheme 2.105 Preparation of the carboranylboron-capped iron(II) complexes

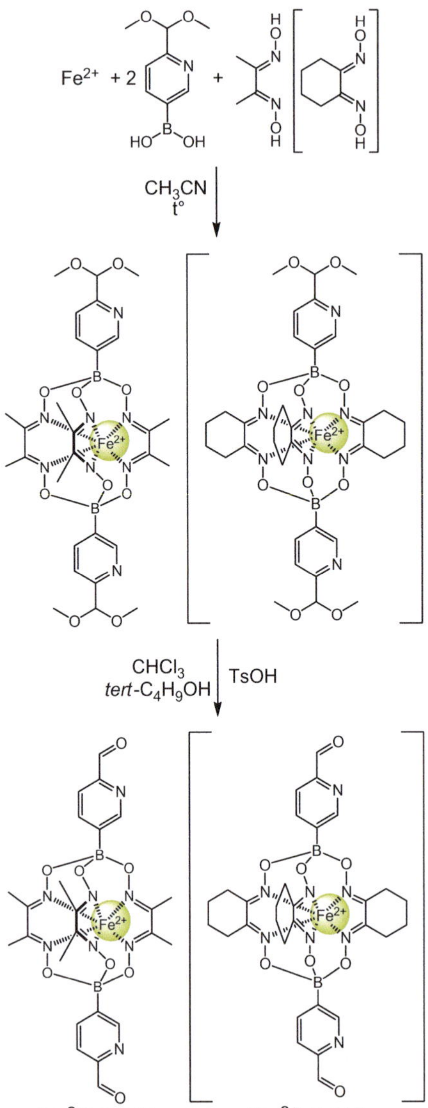

Scheme 2.106 Synthesis of the iron(II) clathrochelates with terminal reactive formyl group

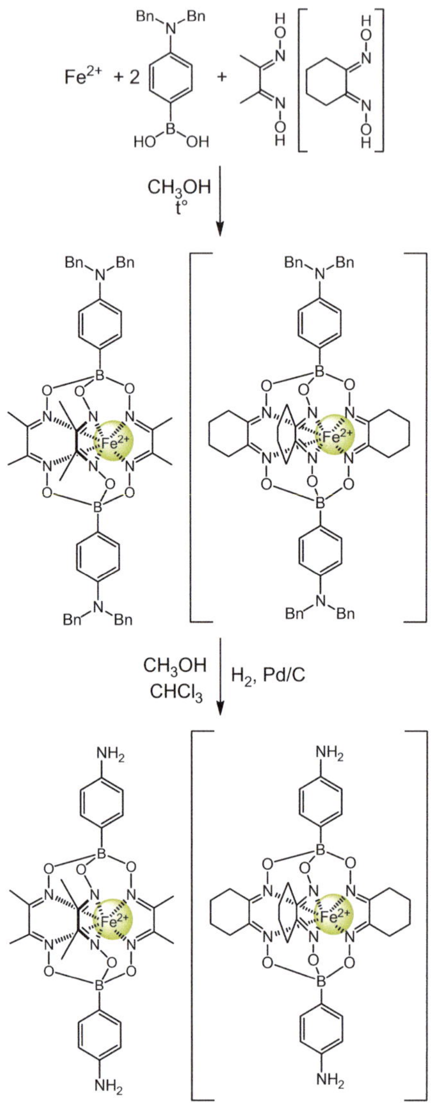

Scheme 2.107 Synthesis of the iron(II) clathrochelates with terminal reactive amino group

procedure shown in Scheme 2.139, an initial kinetically inert non-macrocyclic cobalt(III) bis-α-benzildioximate *16* was found to undergo its simultaneous macrocyclization with $BF_3 \cdot O(C_2H_5)_2$ and a reduction with metallic silver, thus giving the intermediate difluoroboron–cross-linked macrocyclic cobalt(II) complex. Its condensation with dimethylglyoxime gave the target cobalt(II) clathrochelate (Co^{2+})@**296** with non-equivalent α-dioximate fragments. Br^- and H^+ ions releasing on the initial stage of this synthetic procedure have been eliminated from the reaction mixtures with Ag_2CO_3. In the case of its analog (Co^{2+})@**307**, the poorly soluble "pre-clathrochelate" intermediate was found [110] to be formed on the first stage of this tandem reaction; its reprecipitation from DMSO solution with water caused the further macrocyclization yielding the target cage complex.

The same synthetic strategy, based on condensation of α-furyldioxime with a BF_2-cross-linked

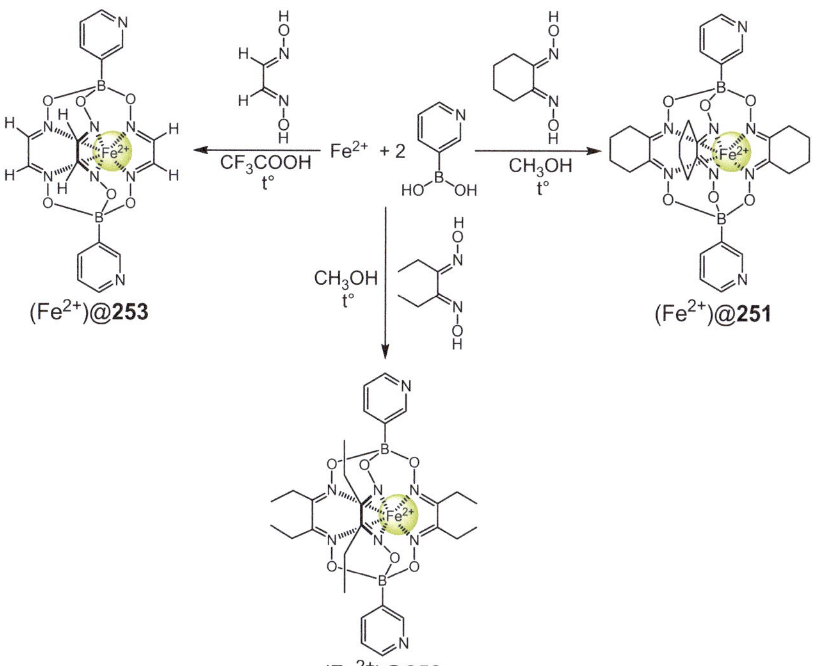

Scheme 2.108 Template self-assembly of the 3-pyridyl-terminated iron(II) complexes

macrocyclic iron(II) bis-α-benzildioximate *13*, has been used [111] for preparation of the corresponding 2-furyl-containing clathrochelate (Fe^{2+})@**308** with non-equivalent aromatic chelate fragments by Scheme 2.140. Monoribbed-functionalized iron(II) diiodo- and dibromoclathrochelates (Fe^{2+})@**309** and (Fe^{2+})@**310** have been synthesized [112, 113] using condensation of the corresponding dihalogenodioxime with the same macrocyclic precursor *13* by Scheme 2.141.

Nucleophilic Substitution

The tris-dichloroglyoximate tin-capped clathrochelate dianion (Fe^{2+})@**297** (see section "Template Synthesis and Condensation Reactions") is reported [106] to be a substantially less reactive macrobicyclic precursor than its boron-capped analogs. It did not undergo nucleophilic substitution neither with sterically unhindered primary aliphatic amines, such as *n*-butylamine and cyclohexylamine, nor with alkylthiolate anions in DMF media. With a phenolate anion as *O*-nucleophile, the substitution gave only the mixture of partially substituted (mainly the tri- and tetra-functionalized) macrobicyclic products even after prolonged heating of the reaction mixture. On the other hand, with thiophenolate anion as a *S*-nucleophile, the hexaphenylsulfide iron(II) clathrochelate dianion (Fe^{2+})@**311** shown in Scheme 2.142 has been isolated [106].

Nucleophilic substitution of the phenylboron-capped iron(II) hexachloroclathrochelate with the corresponding thiolate anions in various media has been used [114] for preparation of the hexasulfide tris-dioximate iron(II) complexes with pendant *n*-butyl, *n*-octylsulfide, silatrane, and captopril functionalizing groups in high yields; minimal amounts of by-products of the side decomposition reactions have been detected in aprotic solvents with a small donor number, such as chloroform and dichloromethane (Scheme 2.143).

Reaction of the dichloroclathrochelate precursor (Fe^{2+})@**302** with quinoxaline-2,3-dithiol in the presence of triethylamine as organic base by Scheme 2.144 has been used [115] for preparation of the monoribbed-functionalized clathrochelate (Fe^{2+})@**316** with annelated disulfide quinoxaline

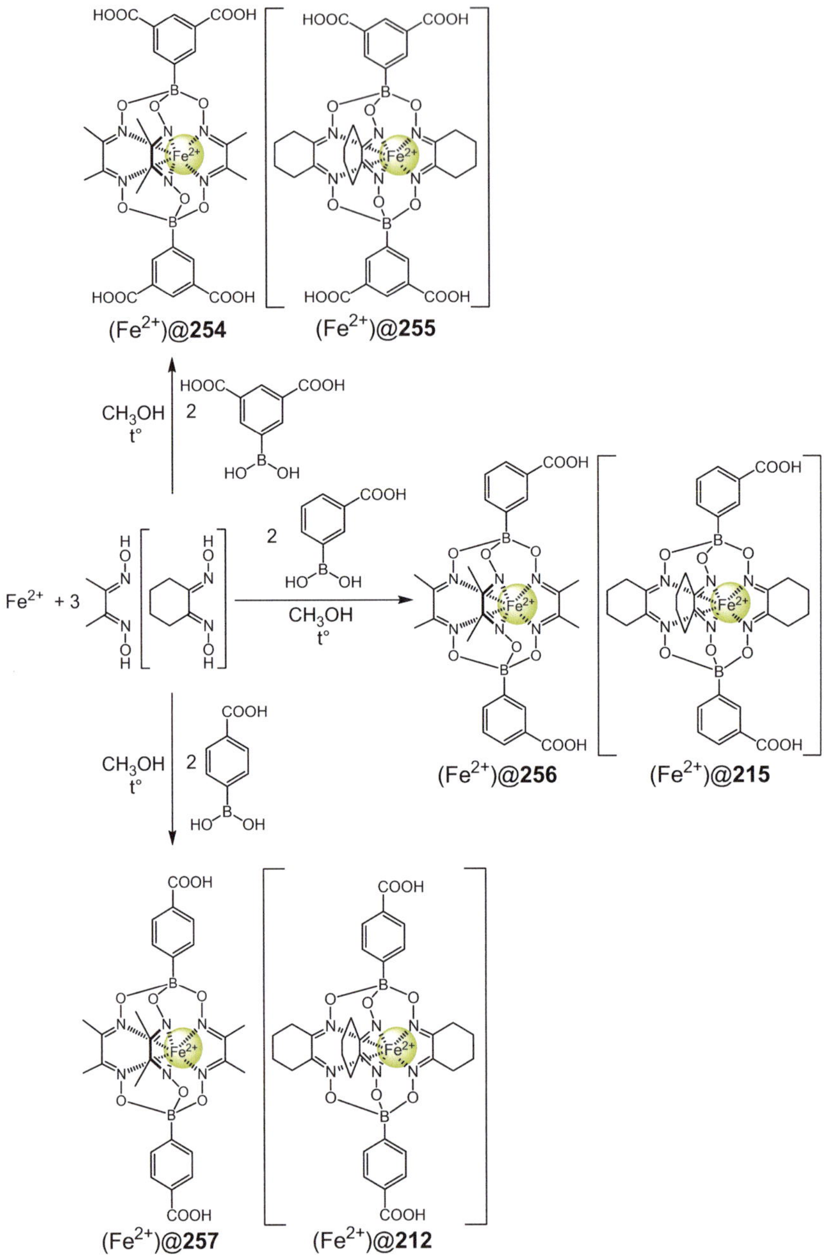

Scheme 2.109 Synthesis of the carbonyl-terminated iron(II) clathrochelates

substituent [115]. Its diallylsulfide analog (Fe^{2+})@**317** has been synthesized [116] by Scheme 2.145 using nucleophilic substitution of the same dichloroclathrochelate precursor.

Mono- and biradical mono- and dinitroxide macrobicyclic iron(II) complexes (Fe^{2+})@**318** and (Fe^{2+})@**319** have been prepared [117] by Scheme 2.146 using nucleophilic substitution of (Fe^{2+})@**302** with an appropriate aminonitroxide as a *N*-nucleophile.

The attempted nucleophilic substitution of the same diclathrochelate precursor with monofunctionalized *closo*-borates with inherent HO-, HS-, and NH_2-groups in the presence of strong bases

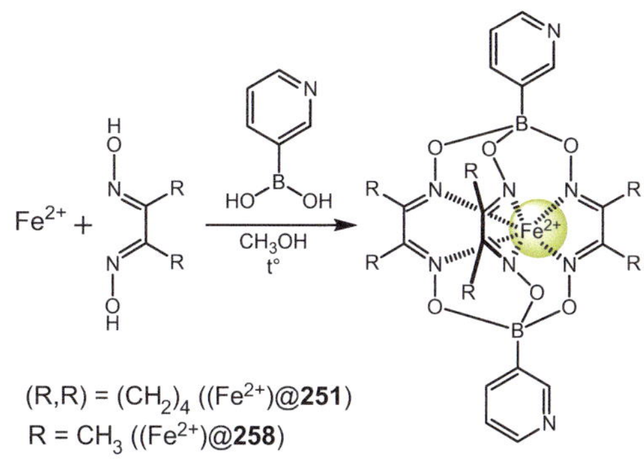

Scheme 2.110 Template synthesis of the 3-pyridylboron-capped iron(II) complexes

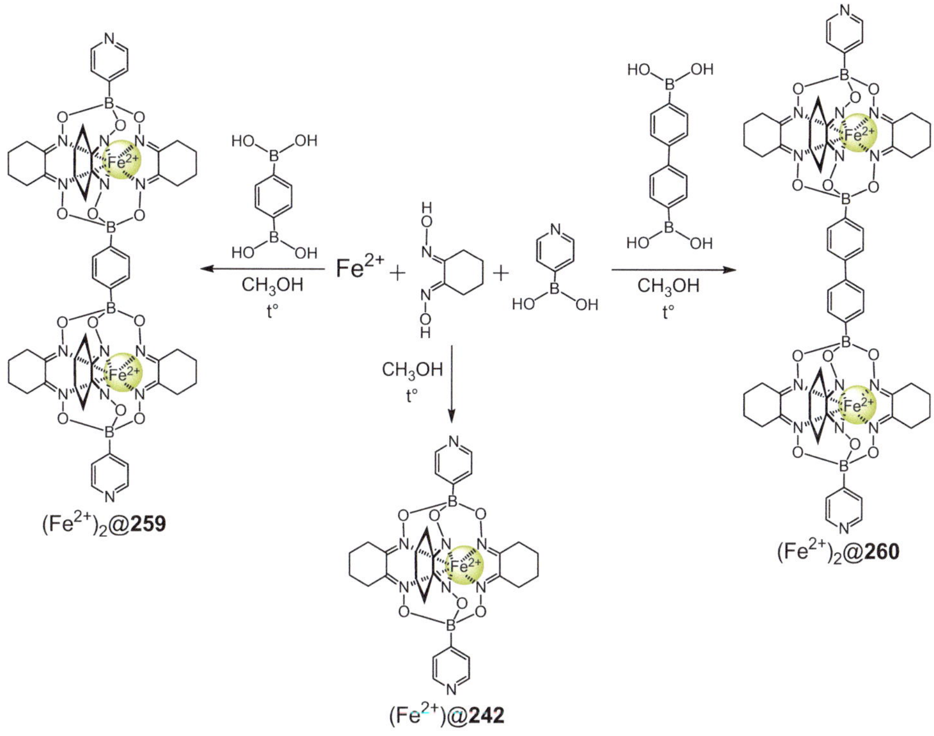

Scheme 2.111 Preparation of the 4-pyridyl-terminated iron(II) mono- and bis-clathrochelates

(in particular, potassium *tert*-amylate) has been successfully performed [118] only in the case of amino-*closo*-borate anion $B_{12}H_{11}NH^{3-}$ (Scheme 2.147), thus giving the mono-*closo*-borate-functionalized clathrochelate dianion (Fe^{2+})@**320**, the first representative of the hybrid *closo*-boratoclathrochelates.

Functionalization of an iron(II)-encapsulating clathrochelate framework with biorelevant pharmacophoric dopamine-ribbed substituents has been performed [119]. Most common method of the synthesis of such ribbed-functionalized cage complexes is based on nucleophilic substitution of the reactive chlorine atom(s) of an appropriate halogenoclathrochelate precursor with a suitable nucleophilic agent (see above). However, the reaction of (Fe^{2+})@**302** with dopamine as polynucleophilic agent having two

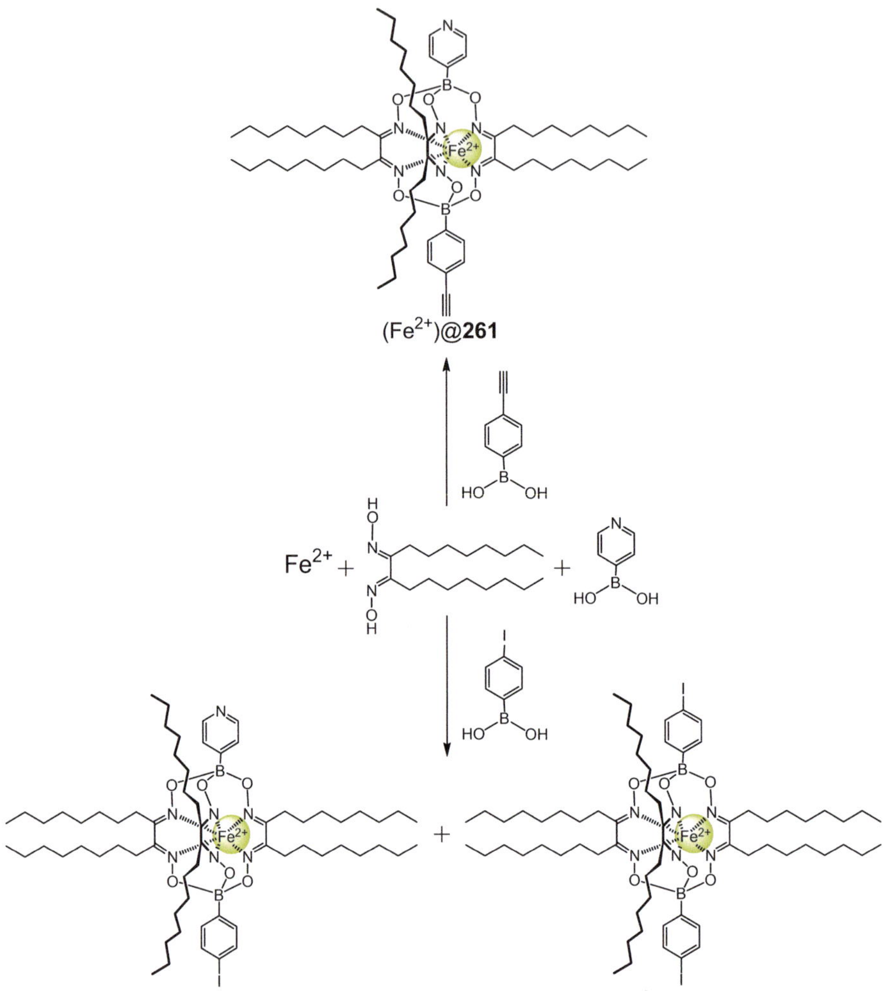

Scheme 2.112 Synthesis of the iron(II) clathrochelates with ribbed octyl substituents

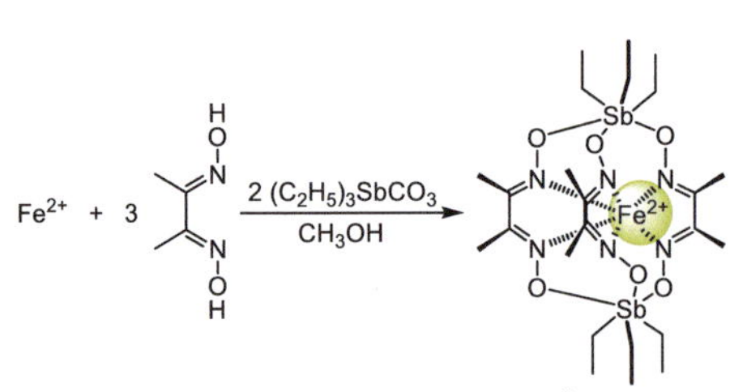

Scheme 2.113 Preparation of the triethylantimony-capped iron(II) tris-dimethylglyoximate

non-equivalent reactive centers (i.e., the terminal primarily aliphatic amino group and the hydroxyl substituents of its pyrocatechol fragment) gave the mixture of the macrobicyclic products. Trying to perform a selective nucleophilic substitution only with amino group of this dopamine nucleophile, its pyrocatechol fragment has been protected [119] by coordination toward

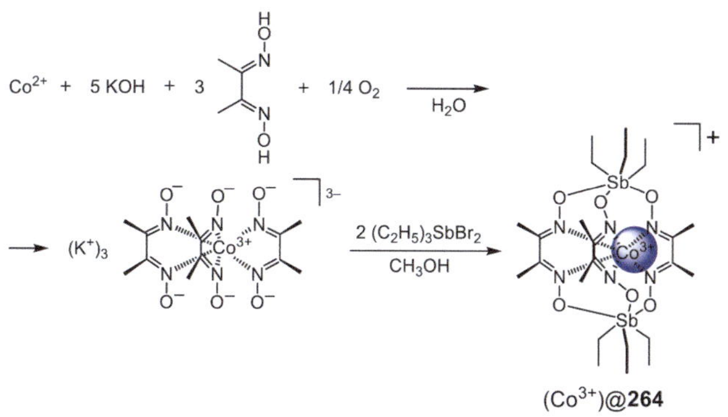

Scheme 2.114 Synthesis of the triethylantimony-capped cobalt(III) tris-dimethylglyoximate

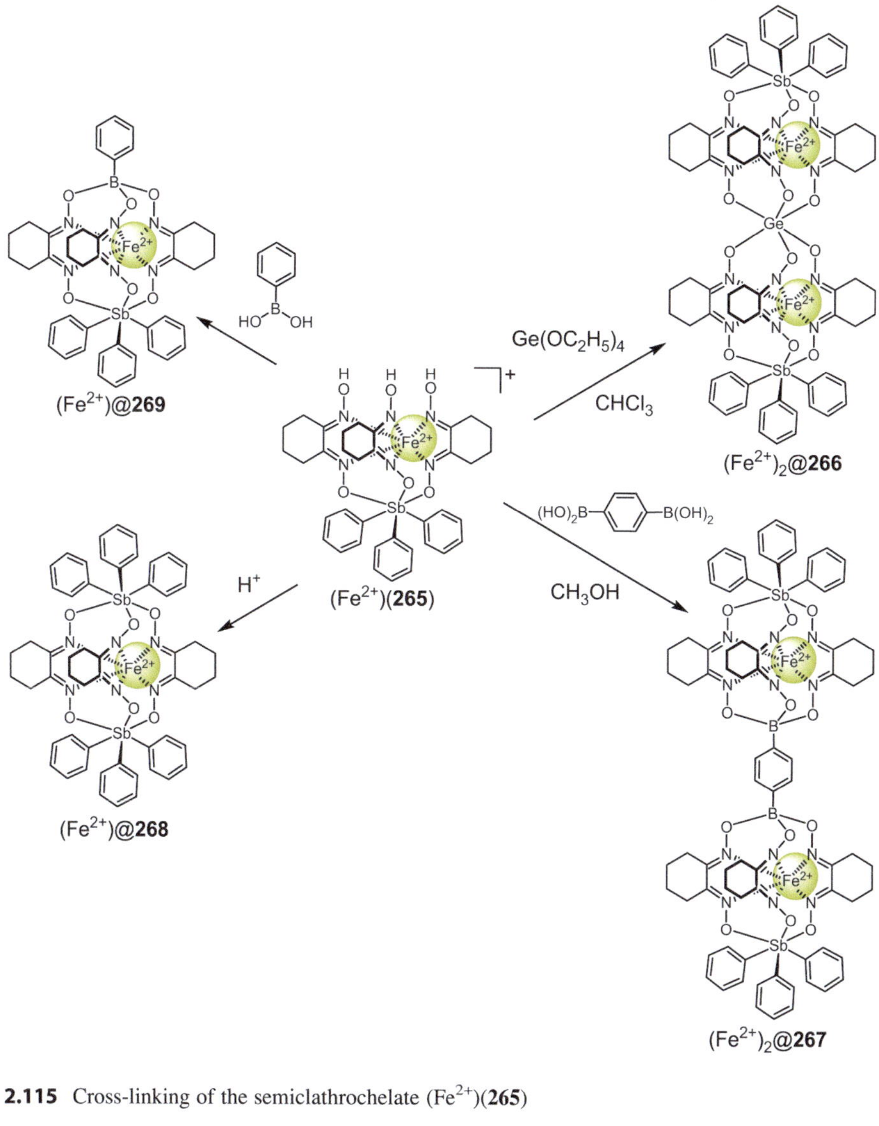

Scheme 2.115 Cross-linking of the semiclathrochelate (Fe^{2+})(**265**)

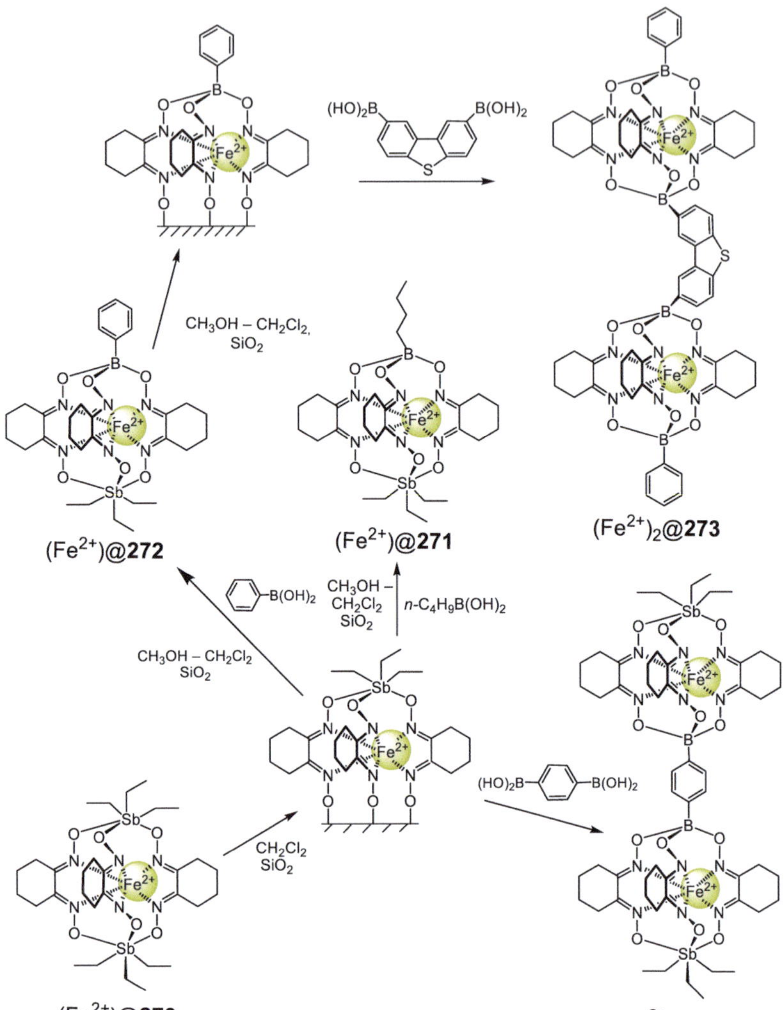

Scheme 2.116 Synthesis of the iron(II) mono- and bis-clathrochelates with non-equivalent capping groups

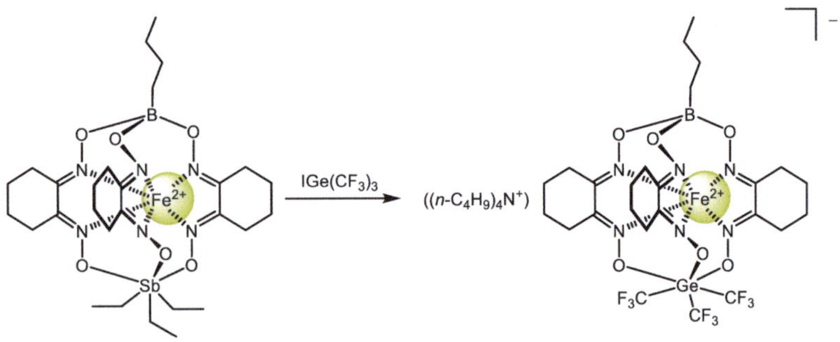

Scheme 2.117 Transmetallation of the triethylantimony-capped iron(II) clathrochelate with $IGe(CF_3)_3$

phenylboronic acid. So, the monoribbed-functionalized clathrochelate (Fe^{2+})@**321** with two dopamine substituents in a *vic*-position of the same chelate α-dioximate fragment has been synthesized [119] by Scheme 2.148 using nucleophilic substitution of (Fe^{2+})@**302** with

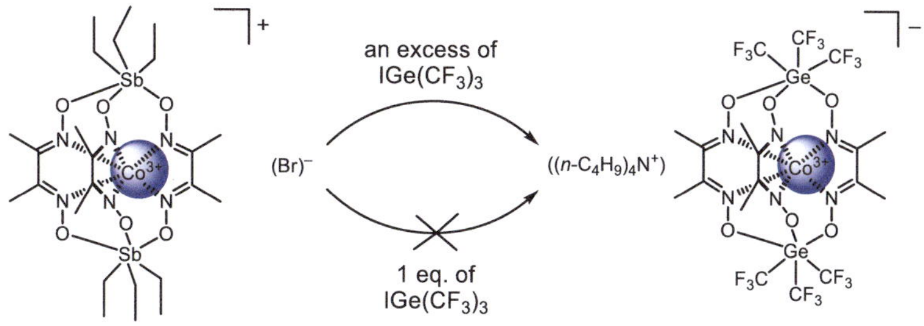

(Co3+)@**264** (Co3+)@**276**

Scheme 2.118 Transmetallation of the bis-triethylantimony-capped cobalt(III) complex with $IGe(CF_3)_3$

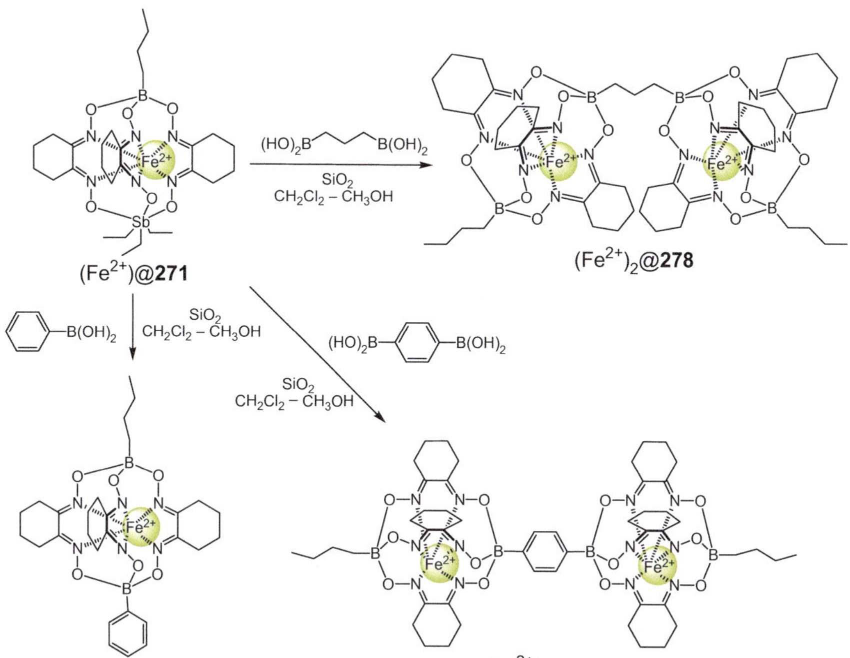

(Fe2+)@**279** (Fe2+)2@**277**

Scheme 2.119 Transmetallation of the antimony-capped iron(II) complex with mono- and diboronic acids

phenylboron-protected dopamine in the presence of triethylamine.

Reaction of this macrobicyclic precursor with *N-tert*-butylhydroxylamine in crude DMF by Scheme 2.149 unexpectedly gave the cage complex (Fe^{2+})@**322** with one dimethylamine-ribbed substituent [120]. This result was explained by contamination of the crude DMF with an admixture of dimethylamine. The same reaction in the presence of ammonia in DMF, purified from dimethylamine, afforded the (amino)dimethylaminoclathrochelate (Fe^{2+})@**323**, the derivative of dimethylamine, which resulted from a transamination of DMF, occurring in the course of this reaction. Partial nucleophilic substitution of (Fe^{2+})@**302** with ammonia in toluene has been used [120] for preparation of the monochloroclathrochelate (Fe^{2+})@**324** with one inherent ribbed amine substituent in a *vic*-position to its chlorine atom.

The bis-benzildioximate iron(II) clathrochelate (Fe^{2+})@**325** with two inherent amino groups in one of the three ribbed chelate fragments has been prepared [121] in a high yield by Scheme 2.150 using nucleophilic substitution of (Fe^{2+})@**302** with liquid ammonia. These amino groups have an amide character, thus undergoing a deprotonation only in the presence

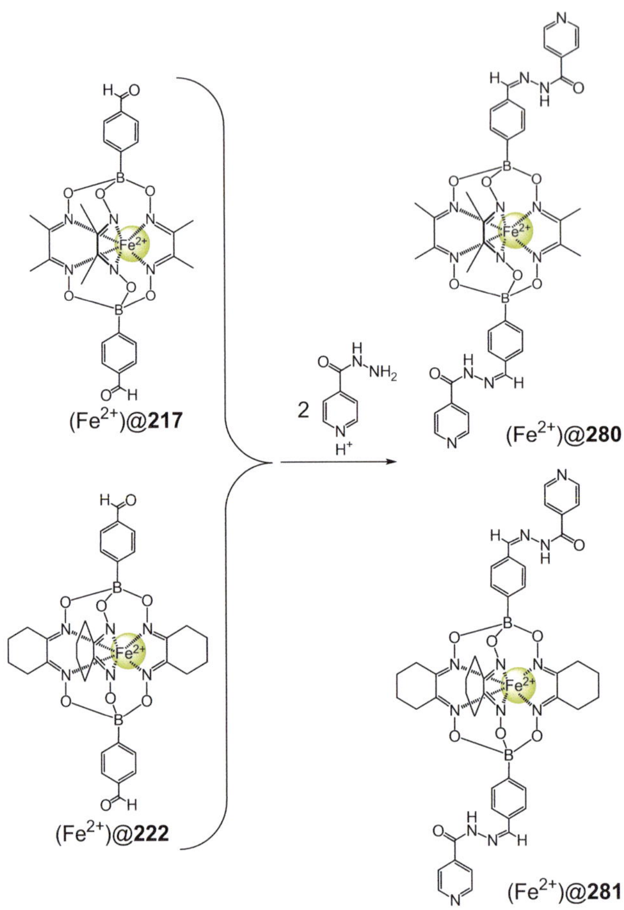

Scheme 2.120 Preparation of the hydrazonate iron(II) clathrochelates

of strong bases, such as *tert*-C_4H_9OK or KOH; the solutions of (Fe^{2+})@**325** in aprotic polar solvents (1,4-dioxane or DMSO) rapidly change their color from red to deep violet because of such deprotonation process [121].

The monoribbed-functionalized mono- and dialkoxy-substituted iron(II) clathrochelates (Fe^{2+})@**326** and (Fe^{2+})@**327** have been prepared [122] by Scheme 2.151. The authors failed to obtain the clathrochelates with inherent alkoxyl group(s) using alkaline metal alkoxides in alcohol media as *O*-nucleophiles because of the complete destruction of a macrobicyclic framework of (Fe^{2+})@**302** by highly basic and low nucleophilic (in protonogenic media) alkoxide anions. So, nonpolar and aprotic solvents have been used to increase a nucleophilicity of these anions and to decrease their basicity. In particular, TOF, known to be stable in alkaline media and the pyrolysis of which is a simple method to generate the highly reactive diethoxycarbene, has been proposed [122] for this purpose. Indeed, the reaction of (Fe^{2+})@**302** with TOF in the presence of metallic sodium gave the iron(II) cage complexes (Fe^{2+})@**326** and (Fe^{2+})@**327** with inherent ethoxyl substituent(s) in their ribbed fragments (mainly, the difunctionalized cage complex). TOF is reported [122] to in situ generate the *O*-nucleophilic ethoxide anions with metallic sodium as it shown in Scheme 2.151.

Nucleophilic substitution of the same iron(II) dichloroclathrochelate (Fe^{2+})@**302** giving the dialkylsulfide cage complexes (Fe^{2+})@**328**, (Fe^{2+})@**329**, (Fe^{2+})@**330**, (Fe^{2+})@**331**, (Fe^{2+})@**332**, and (Fe^{2+})@**333** by Scheme 2.152 has been performed [123] with RSH/TEA system

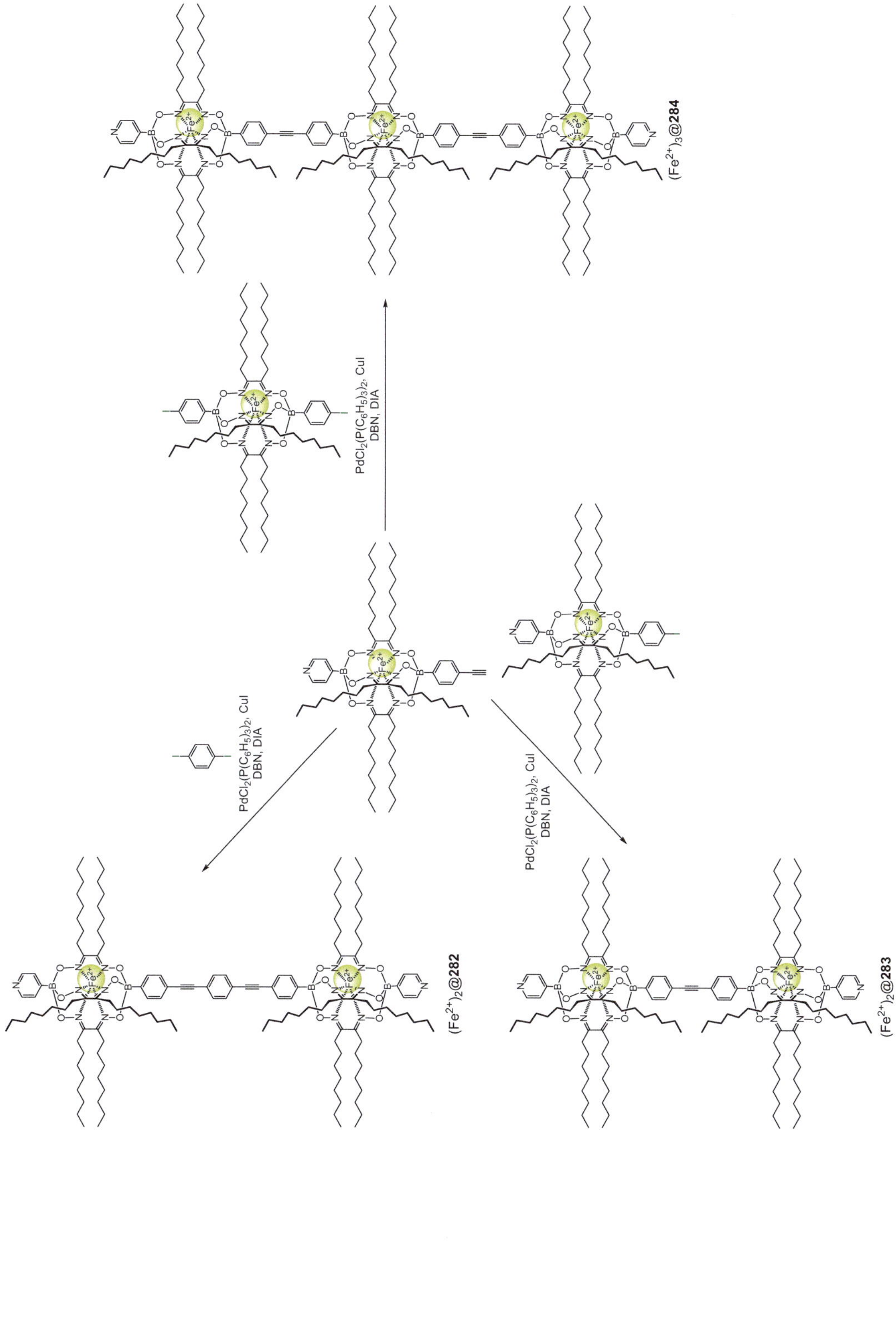

Scheme 2.121 Synthesis of the 4-pyridyl-terminated iron(II) polyclathrochelates with ribbed octyl substituents

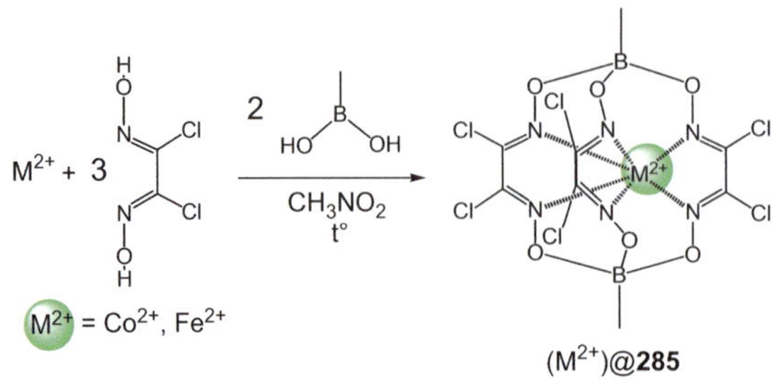

Scheme 2.122 Template synthesis of the methylboron-capped iron and cobalt(II) hexachloroclathrochelates

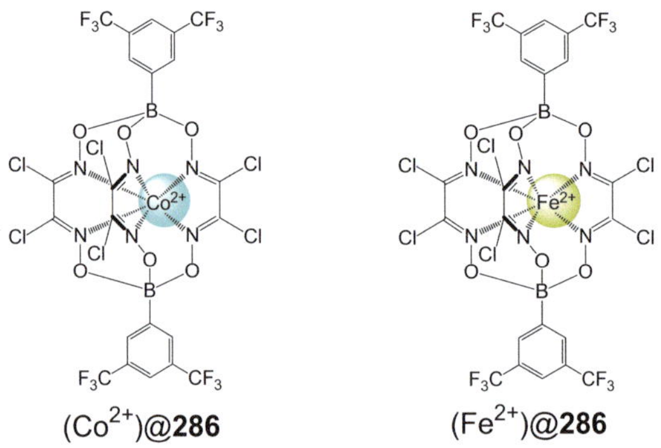

Scheme 2.123 Chemical drawings of iron and cobalt(II) complexes of the hexachloroclathrochelate **286**

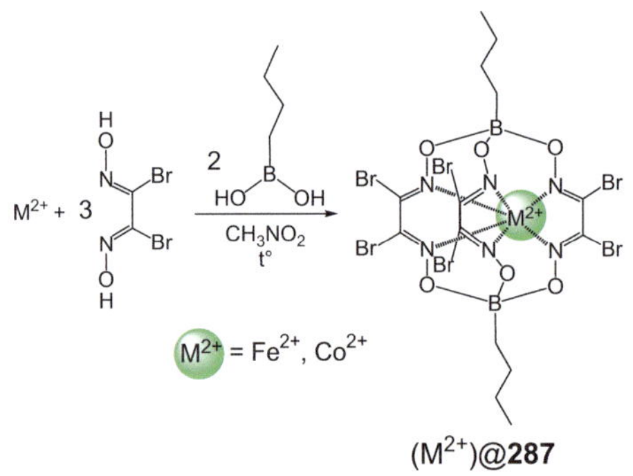

Scheme 2.124 Preparation of the iron and cobalt(II) hexabromoclathrochelates

in dichloromethane as a solvent instead of most common at the moment RSH/alkaline metal alcoholate system in donor media (DMF, THF, or 1,4-dioxane). In the case of the monothiolate anions as *S*-nucleophiles, the target products have been prepared in high yields under mild reaction conditions. The use of this RSH/TEA/CH_2Cl_2 system in the case of S_2-dinucleophilic dithiolates allowed to decrease the formation of by-products that resulted from the side intramolecular self-dealkylation reactions of the type "dragon biting its own tail" (Scheme 2.153) because thiolate anion–thiol equilibrium is shifted in the direction of a protonated thiol form. In particular, the authors of [123] failed to obtain the iron(II) clathrochelates with annelated thiocrown ribbed substituent by condensation of (Fe^{2+})@**302** with 3,5-dithiaoctane-1,8-dithiol and 3,6-dithianonane-1,9-dithiol in their clathrochelate-to-dithiol 1:1 molar ratio

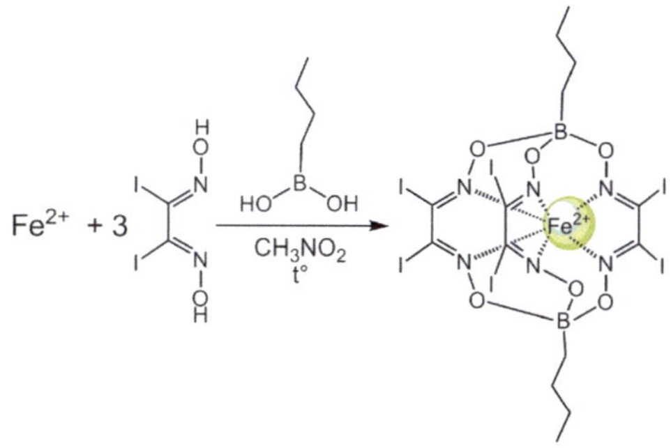

Scheme 2.125 Synthesis of the *n*-butylboron-capped iron(II) hexaiodoclathrochelate

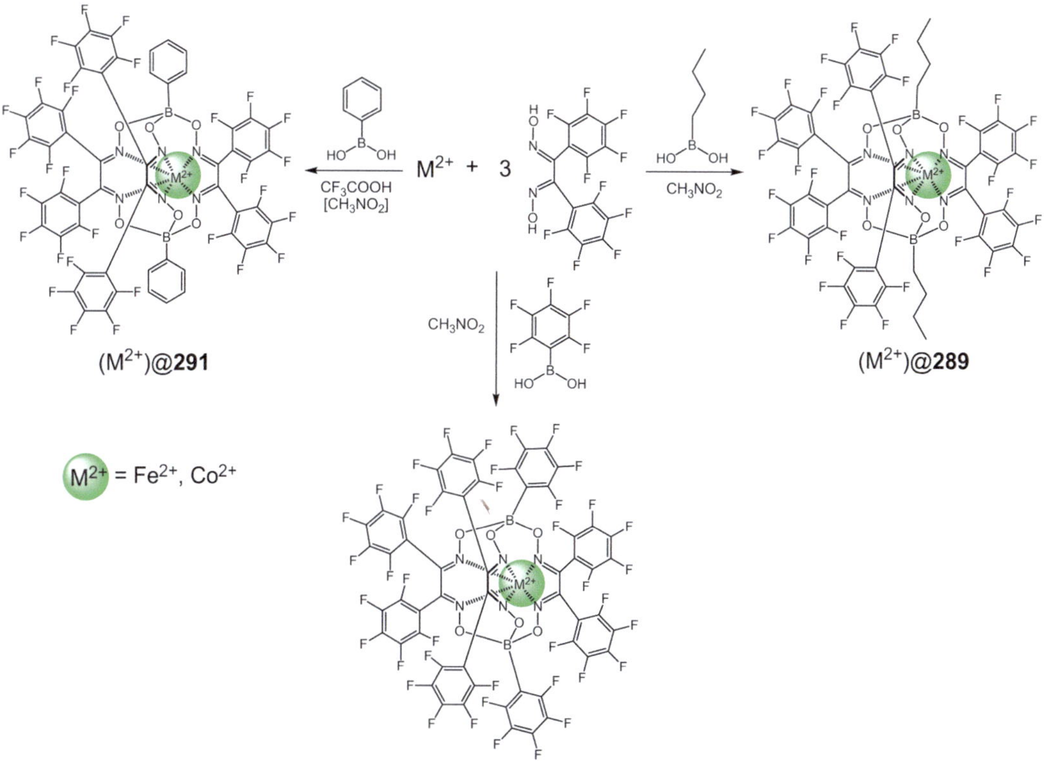

Scheme 2.126 Preparation of the boron-capped iron and cobalt(II) tris-perfluoro-α-benzildioximates

due to occurrence of the above side intramolecular reactions. Template condensation of the thiodithiol $(HSCH_2CH_2)_2S$ with this clathrochelate precursor in the presence of Cd^{2+} ions and triethylamine afforded a poorly separable mixture of the S_3-thiocrown- and S_6-thiocrown-containing iron(II) mono- and bis-clathrochelates (Fe^{2+})@**329** and $(Fe^{2+})_2$@**330**. The attempted preparation of the above S_6-thiocrown ether clathrochelate using a stepwise synthetic procedure shown in Scheme 2.152 gave an unexpected result: if the clathrochelate (Fe^{2+})@**328** with terminal thiol groups easily underwent its alkylation with ethyl iodide/TEA/

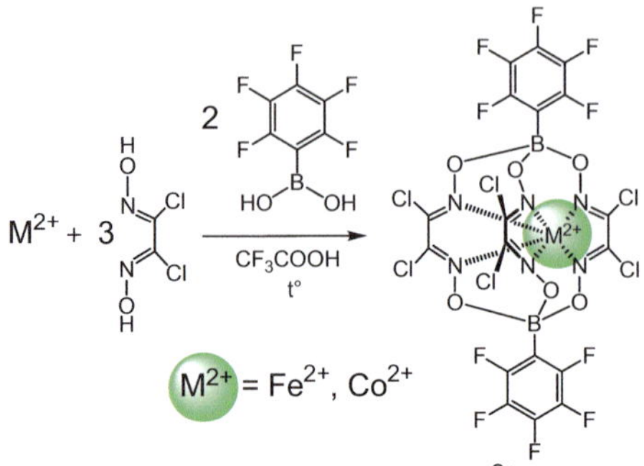

(M^{2+})@**292**

Scheme 2.127 Synthesis of the pentafluorophenylboron-capped iron and cobalt(II) hexachloroclathrochelates

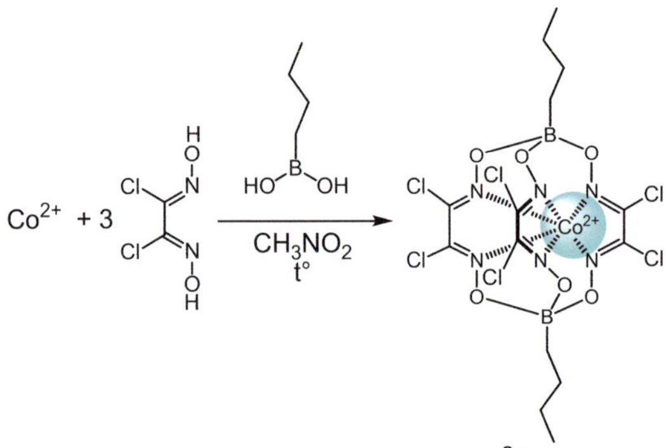

(Co^{2+})@**293**

Scheme 2.128 Template self-assembly of the hexachloroclathrochelate (Co^{2+})@**293**

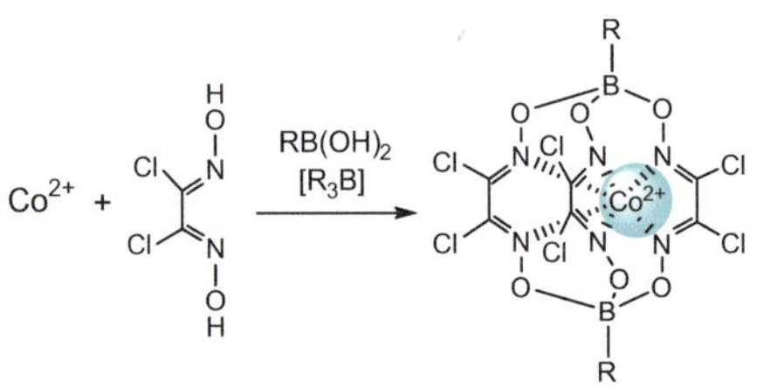

R = n-C_4H_9 (Co^{2+})@**293**,
C_6H_5 (Co^{2+})@**294**,
F (Co^{2+})@**295**

Scheme 2.129 Preparation of a series of the boron-capped cobalt(II) hexacloroclathrochelates

DMF system to form the macrobicyclic product (Fe^{2+})@**334**, template condensation of (Fe^{2+})@**328** with 1,2-diiodoethane in the presence of Cs_2CO_3 in DMF medium mainly resulted in the formation of the S_3-thiocrown ether macrobicyclic complex (Fe^{2+})@**329**. The authors of [123] suggested that the deprotonated terminal group of (Fe^{2+})@**328** attacks the alkylsulfide substituent in a *vic*-position of the same chelate ribbed fragment, thus causing its dealkylation (Scheme 2.153) and the formation of nine-membered S_3-alicyclic substituent in this fragment by Scheme 2.154. The above side process substantially decreases the yield of the target

Scheme 2.130 Synthesis of the tin-capped iron(II) hexachloroclathrochelate

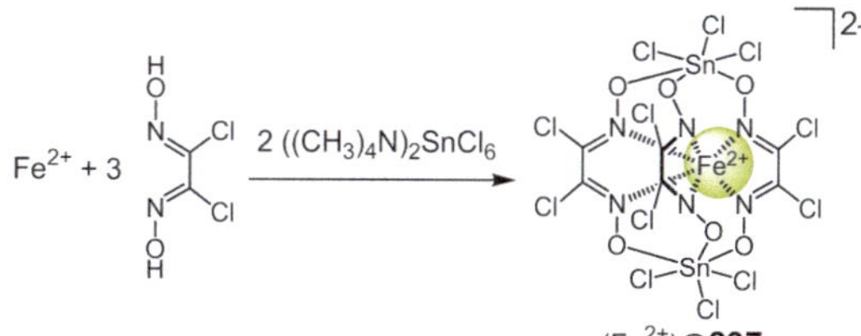

Scheme 2.131 Preparation of the iron (II) clathrochelate with non-equivalent ribbed fragments

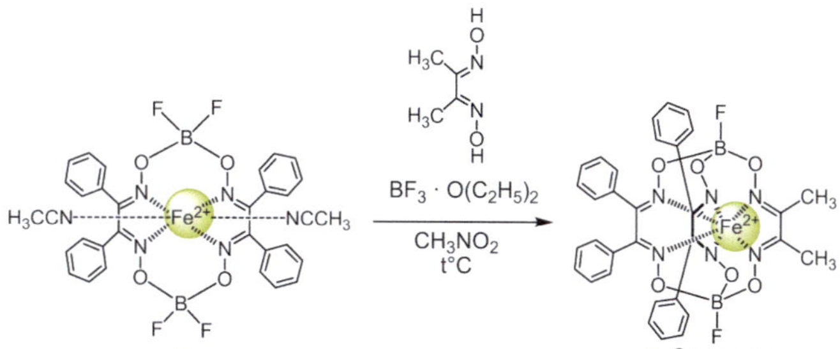

Scheme 2.132 Synthesis of the iron(II) monochloroclathrochelate (Fe^{2+})@**298**

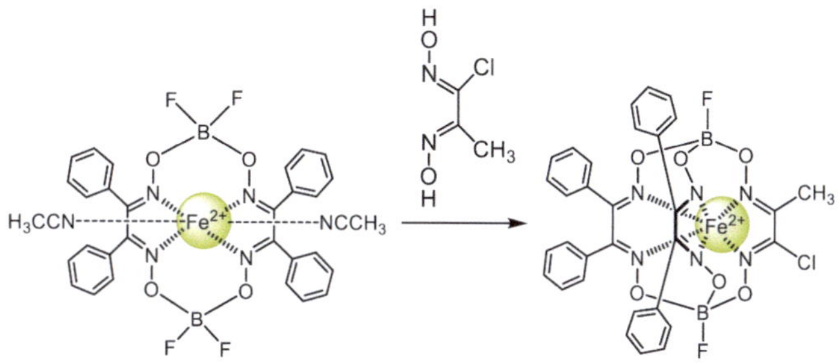

Scheme 2.133 Preparation of the *mer*- and *fac*-isomers of (Fe^{2+})@**299**

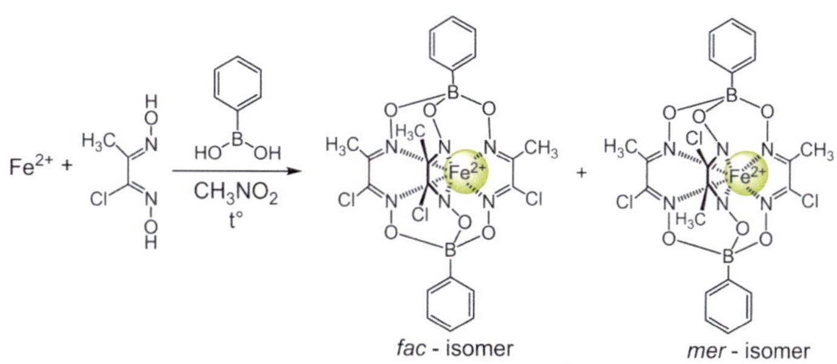

S_6-crown ether cage complex, which only a trace amount has been detected [123].

The iron(II) cage complex (Fe^{2+})@**335** with terminal thiol group has been synthesized [124] by Scheme 2.155 using nucleophilic substitution of its monochloroclathrochelate precursor (Fe^{2+})@**301** with bis(2-mercaptoethyl) sulfide in the presence of triethylamine. Trying to prevent the formation of a bis-clathrochelate by-product, this dithiol has been used [124] in its fivefold excess, while triethylamine has been added in practically stoichiometric amount.

Nucleophilic substitution of (Fe^{2+})@**302** with potassium salt of dimercaptomaleodinitrile,

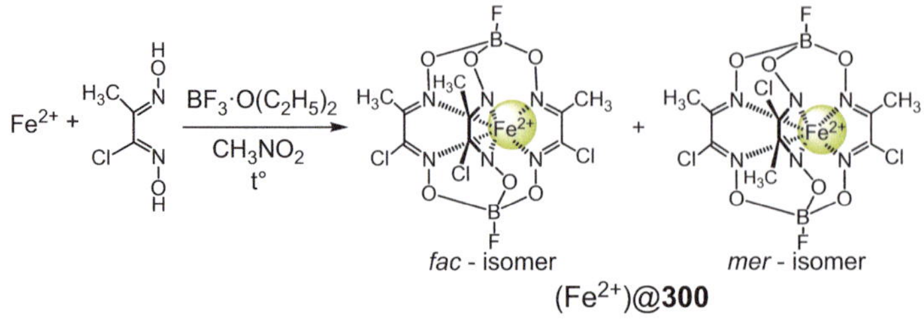

Scheme 2.134 Preparation of the *mer*- and *fac*-isomers of (Fe^{2+})@**300**

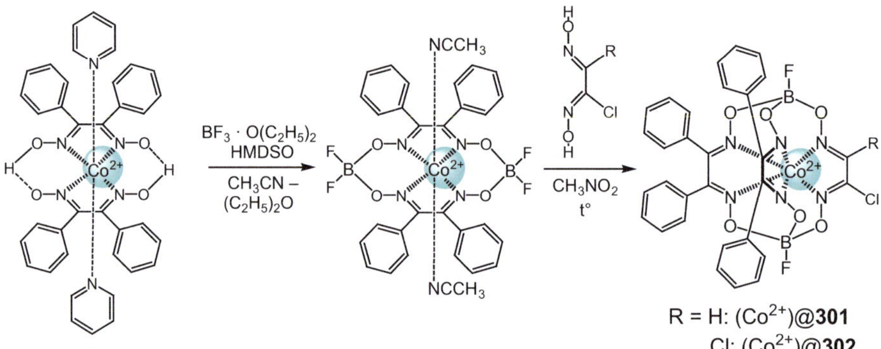

Scheme 2.135 Synthesis of the cobalt(II) mono- and dichloroclathrochelates

giving the cage complex (Fe^{2+})@**336** with annelated ribbed substituent by Scheme 2.156, is reported [125] to proceed in a low yield due to the formation of a mixture of by-products (including polymeric ones) of the side reactions of (Fe^{2+})@**302** with a *trans*-form of this S_2-dinucleophile. Moreover, the highly reactive and thiophilic clathrochelate product (Fe^{2+})@**336** was found [125] also to undergo its side chemical transformations under these reaction conditions.

Monoribbed-functionalized iron(II) clathrochelate (Fe^{2+})@**337** with two ribbed inherent imidazolyl groups has been synthesized [126] by Scheme 2.157 using nucleophilic substitution of (Fe^{2+})@**302** with a Grignard reagent, the organomagnesium derivative of imidazole, which was initially prepared from imidazole with *n*-butyl magnesium bromide. Coordination of pyridine toward Mg^{2+} ion is reported [126] to cause the destruction of $ImMg^{+}\ldots Br^{-}$ ionic pairs, while tetra-*n*-butyl phosphonium bromide played a role of phase-transfer compound that forms the corresponding soluble ionic associate.

Controlled hydrolysis of the same dichloroclathrochelate precursor has been used [127] for preparation of the iron(II) clathrochelates with inherent ribbed hydroxyl substituent(s). Tetra-*n*-butylammonium salt of the macrobicyclic anion (Fe^{2+})@**338** has been synthesized by hydrolysis of this reactive precursor with 10% aqueous tetra-*n*-butylammonium hydroxide by Scheme 2.158; its dihydroxy-containing analog (Fe^{2+})@**339** has been also formed in the course of such nucleophilic substitution with OH^{-} anion as an *O*-nucleophile, but it could not be isolated as an individual compound. The ionic associate (Fe^{2+})@**338** gave the corresponding macrobicyclic acid in strongly acidic media, and the derivatives of this clathrochelate anion have been commonly formed as by-products of the side hydrolysis reactions in the course of nucleophilic substitution of (Fe^{2+})@**302** with various nucleophiles under basic conditions [127].

The bis-propargylamine macrobicyclic iron(II) complex (Fe^{2+})@**340** has been prepared [128] in a high yield by Scheme 2.159 using the

Scheme 2.136 Template synthesis of the iron(II) trichloroclathrochelates

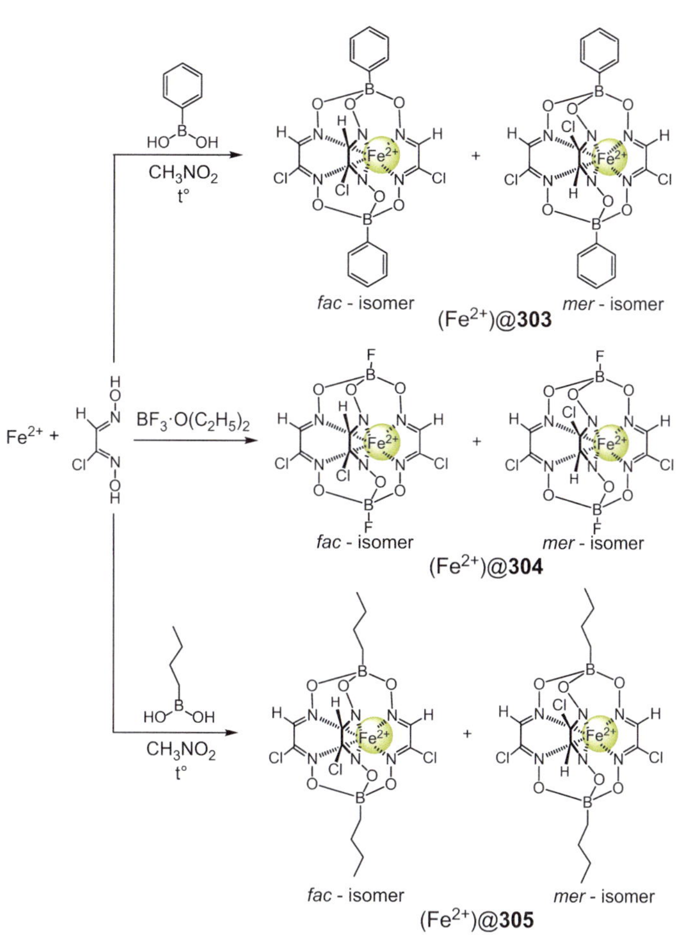

reaction of (Fe^{2+})@**302** with an excess of propargylamine. This diyne-terminated cage complex underwent a copper(I)-catalyzed "click" reaction with benzyl azide in dichloromethane–water biphase medium. Unexpectedly, main product of this 1,3-dipolar cycloaddition reaction was a copper(II) complex of the clathrochelate ligand (Fe^{2+})@**340** with donor groups in its ribbed substituents: the central Cu^{2+} ion coordinates their two deprotonated amino groups and two azaheterocyclic fragments as well [128].

Nucleophilic substitution of the same iron(II) dichloroclathrochelate with *ortho*-phenylenediamine by Scheme 2.160 in the presence of triethylamine is reported [129] to easily proceed even under mild reaction conditions, thus giving the macrobicyclic complex (Fe^{2+})@**341** in a high yield.

The monochloromorpholine cage complex (Fe^{2+})@**342** has been prepared [130] by Scheme 2.161 using the reaction of (Fe^{2+})@**302** with morpholine as a secondary amine *N*-nucleophile: only one of the two chlorine atoms of this dichloroclathrochelate precursor underwent nucleophilic substitution. The clathrochelate (Fe^{2+})@**342** has been used as a monochloroclathrochelate precursor for its further functionalization by Scheme 2.161 with more active nucleophiles, such as primary 2-ethanolamine as a *N*-nucleophile and deprotonated forms of 3- and 4-mercaptobenzoic acids as a *S*-nucleophiles, thus giving the heterodifunctionalized cage complexes (Fe^{2+})@**343**, (Fe^{2+})@**344**, and (Fe^{2+})@**345** in high yields under mild reaction conditions [130].

The analogous synthetic approach has been implemented [131] for preparation of the

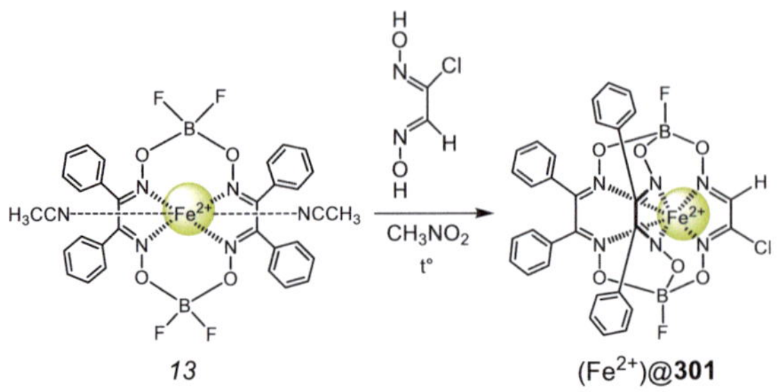

Scheme 2.137 Preparation of the iron(II) monochloroclathrochelate (Fe^{2+})@**301**

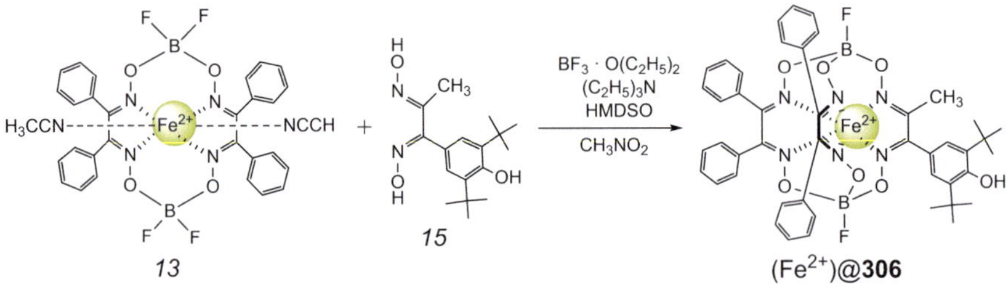

Scheme 2.138 Synthesis of the phenol-terminated cage complex (Fe^{2+})@**306**

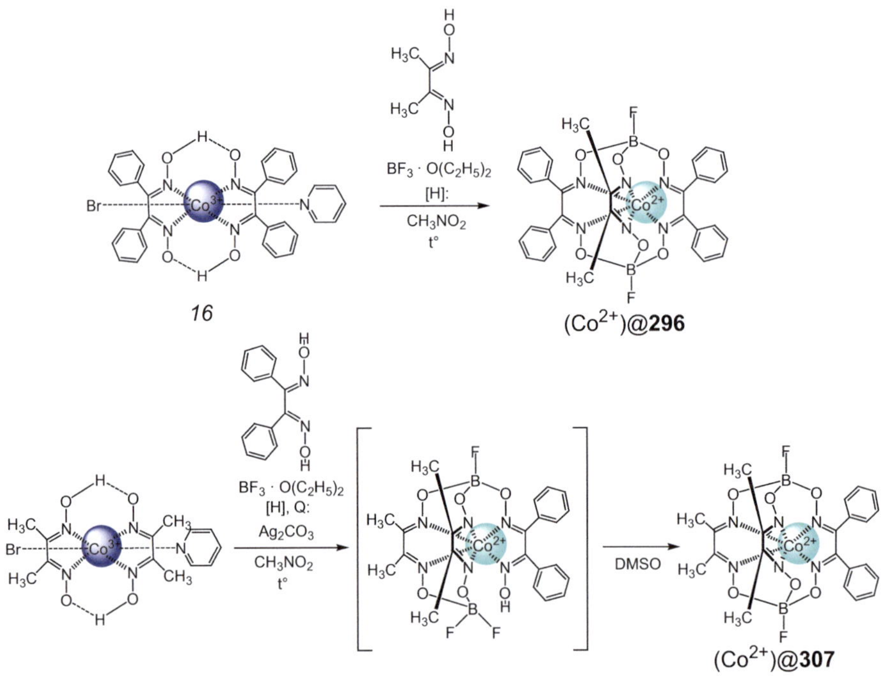

Scheme 2.139 Preparation of the cobalt(II) clathrochelates with non-equivalent ribbed fragments

monoribbed-functionalized *tert*-butylsulfide iron(II) clathrochelate (Fe^{2+})@**346** by Scheme 2.162 using an excess of sodium *tert*-butylthiolate as a *S*-nucleophile.

4,5-Dimercapto-[1, 3]dithiol-2-thione *17* has been tested [132] as a potent S_2-binucleophilic agent for ribbed functionalization of cage metal complexes. This compound has been usually

Scheme 2.140 Synthesis of a mixed α-furyl-bis-α-benzildioximate iron(II) clathrochelate

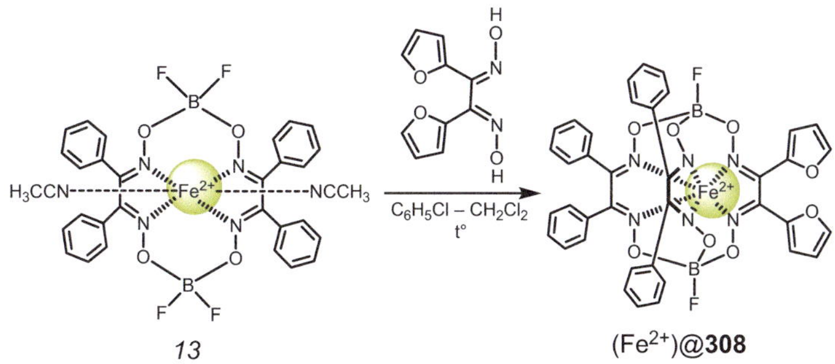

Scheme 2.141 Preparation of the iron (II) dibromo- and diiodoclathrochelates

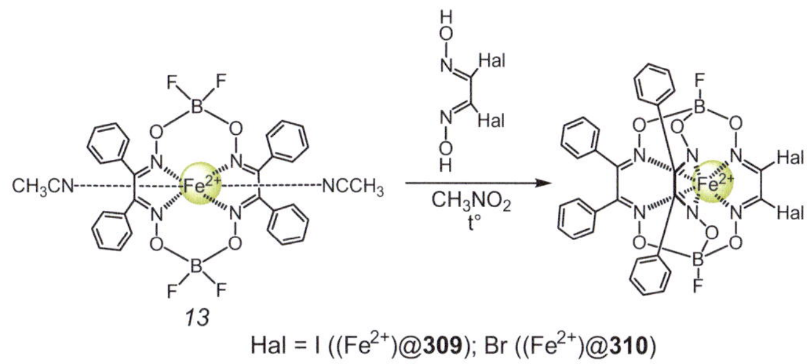

Scheme 2.142 Nucleophilic substitution of tin-capped iron(II) hexachloroclathrochelate

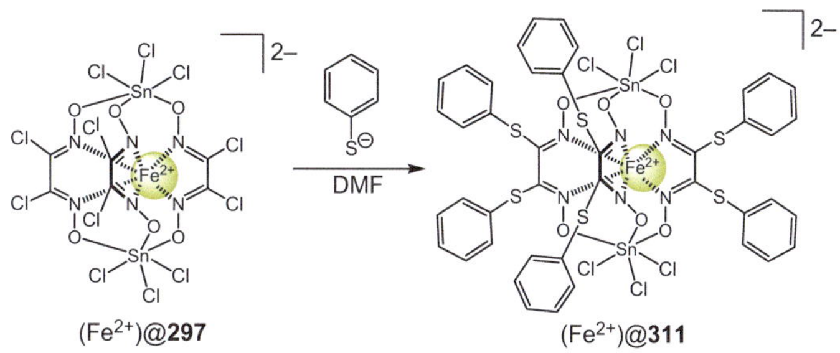

isolated as a complex of zinc(II) ion with its anionic form (because of a low chemical stability and a high reactivity of the protonated form of *17*). Aiming to in situ generate this anionic form, the competing coordination of a low nucleophilic 2,2′-bipyridine, forming the stable chelate complex with zinc(II) ion, has been used [132]. As a result, nucleophilic substitution of (Fe^{2+})@**302** with generated in situ 4,5-dimercapto-[1, 3] dithiol-2-thione dianion by Scheme 2.163 gave the cage complex (Fe^{2+})@**347** with annelated disulfide heterocyclic fragment in a relatively low yield due to an occurrence of the side dealkylation reactions [132].

The clathrochelate complexes with aliphatic amine ribbed substituent(s) can be easily obtained by nucleophilic substitution of the corresponding halogenoclathrochelate precursors with relatively highly basic primary or secondary amines as *N*-nucleophiles (see above). However, in the case of aromatic amines, the basicity of their amino groups often is insufficient to perform such a reaction. On the other hand, the presence of electron-withdrawing groups in the molecules of the aromatic amines substantially increases the acidity of their amino group, thus allowing the use of strong bases for generation of the corresponding anions with higher nucleophilic

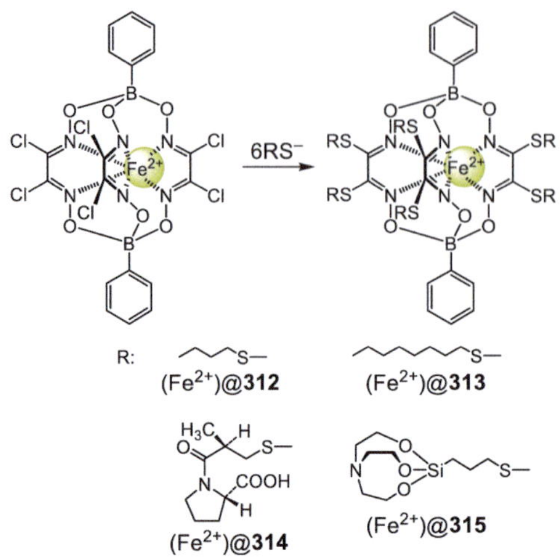

Scheme 2.143 Synthesis of the ribbed-functionalized iron(II) clathrochelates (Fe^{2+})@**312** – (Fe^{2+})@**315**

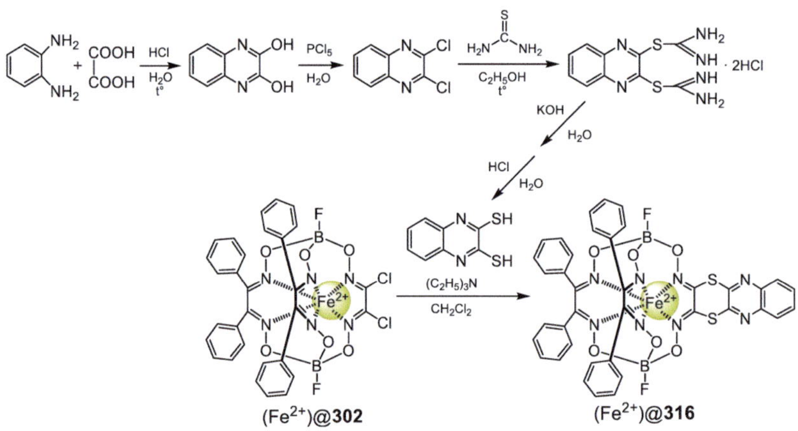

Scheme 2.144 Preparation of the quinoxaline-functionalized clathrochelate (Fe^{2+})@**316**

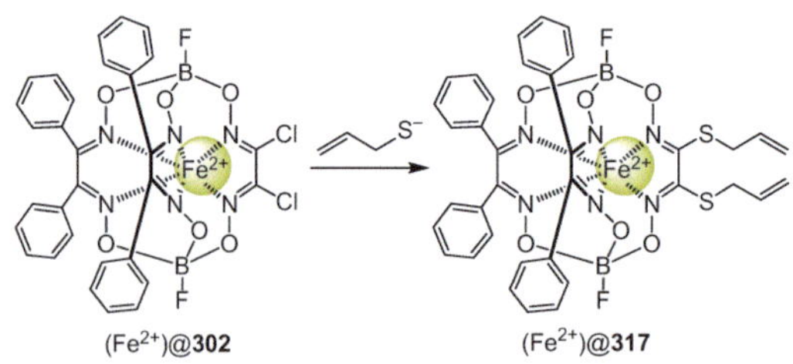

Scheme 2.145 Synthesis of the diallylsulfide cage complex (Fe^{2+})@**317**

activity [133, 134]. Indeed, nucleophilic substitution of the iron(II) dichloroclathrochelate (Fe^{2+})@**302** with 2-nitroaniline has been performed [133] by Scheme 2.164 in the presence of DBU as a strong organic base. Unexpected macrobicyclic product of this reaction resulted from nucleophilic addition of DBU to (Fe^{2+})@**302**; the plausible pathway of this process is also shown in Scheme 2.164.

Two alternative approaches to the synthesis of the iron(II) clathrochelate (Fe^{2+})@**348** with annelated phenylimidazole fragment (i.e.,

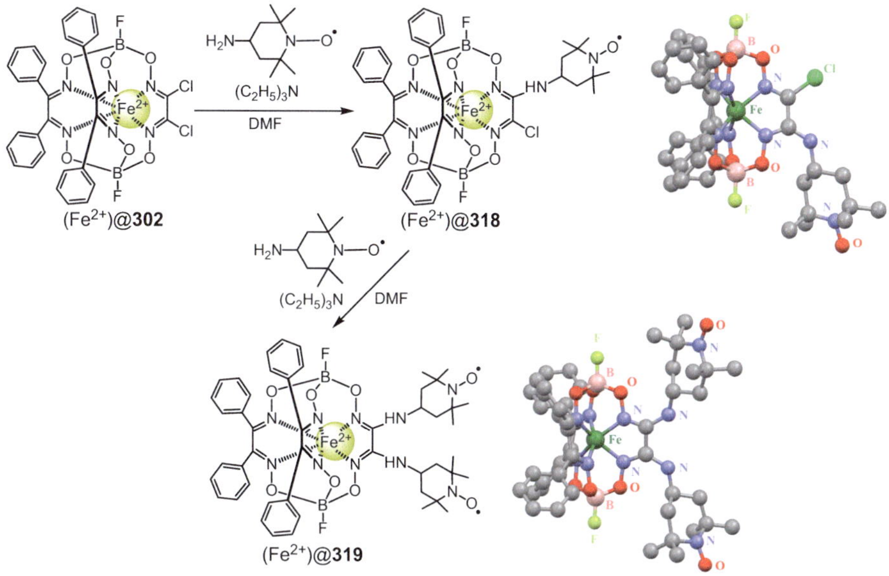

Scheme 2.146 Synthesis of the mono- and dinitroxide iron(II) clathrochelates

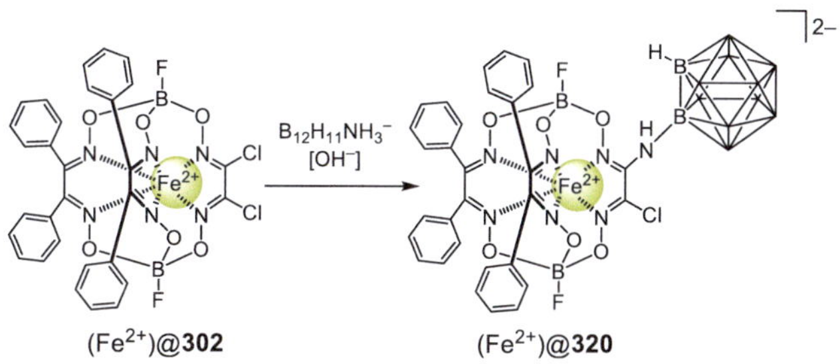

Scheme 2.147 Preparation of the mono-*closo*-borate clathrochelate (Fe^{2+})@**320**

nucleophilic substitution and electrophilic addition) have been compared [134]. Most obvious synthetic approach to the target complex (Fe^{2+})@**348** is based on nucleophilic substitution of (Fe^{2+})@**302** with benzamidine hydrochloride by Scheme 2.165 in the presence of the same strong organic base.

The synthesis of the first phosphorylated iron (II) monochloroclathrochelate (Fe^{2+})@**349** and the products of its nucleophilic substitution are described [135]. The reactions of (Fe^{2+})@**302** with organophosphorus(III) compounds (di- and trialkyl(aryl)phosphines and their halogen-containing derivatives) did not give the target phosphine cage complexes shown in Scheme 2.166, but mainly resulted in an occurrence of the side reduction—hydrodehalogenation processes leading to the mono- and dimethine clathrochelate by-products. The plausible scheme of this reaction includes the formation of clathrochelate anion radicals by a reduction of (Fe^{2+})@**302** with the corresponding phosphine as a reductant and an abstraction of hydrogen atom from a solvent or reactants by these very reactive species [135]. Most feasible pathway to the target mono- and difunctionalized thio- and phosphorylated iron(II) clathrochelates is based on a stepwise nucleophilic substitution of their dihalogenoclathrochelate precursors with the corresponding phosphorus-containing nucleophiles. Such a phosphorylation did not proceed under the typical conditions of Arbuzov and Michaelis–Becker reactions (inert solvent and the temperature range 0–150°C) due to the complete destruction of a cage framework under these reaction conditions. So, nucleophilic

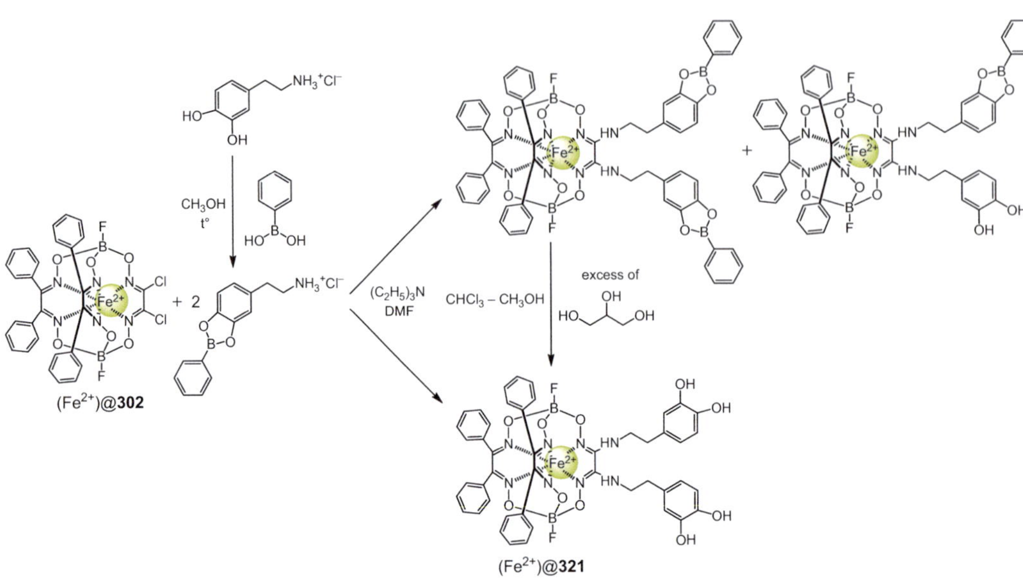

Scheme 2.148 Preparation of the dopamine-containing complex (Fe^{2+})@**321** and its derivatives

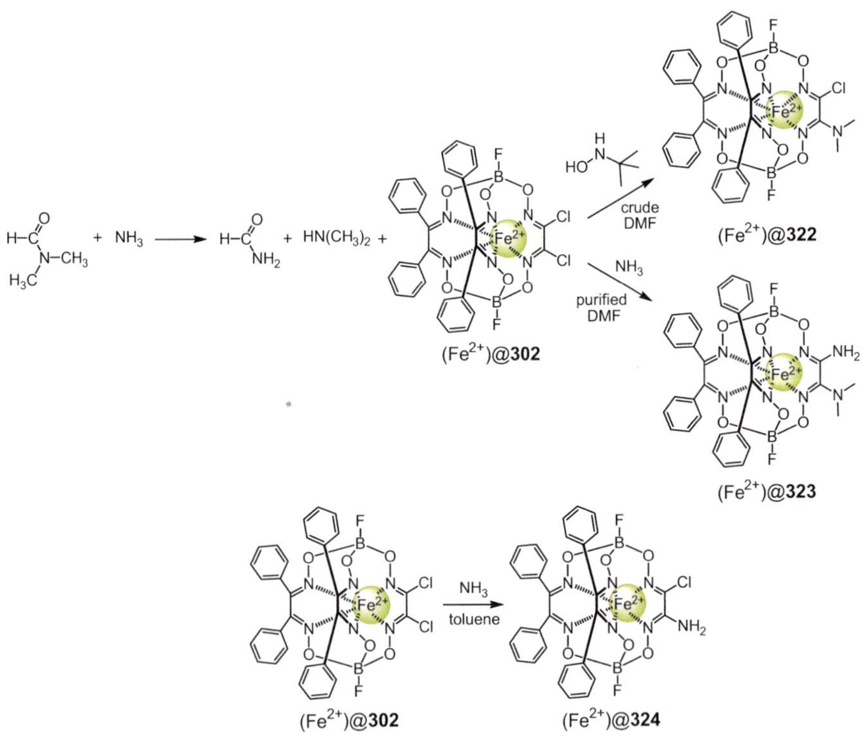

Scheme 2.149 Preparation of the mono- and diamine iron(II) clathrochelates (Fe^{2+})@**322** – (Fe^{2+})@**324**

Scheme 2.150 Synthesis and deprotonation of the diamine iron(II) clathrochelate

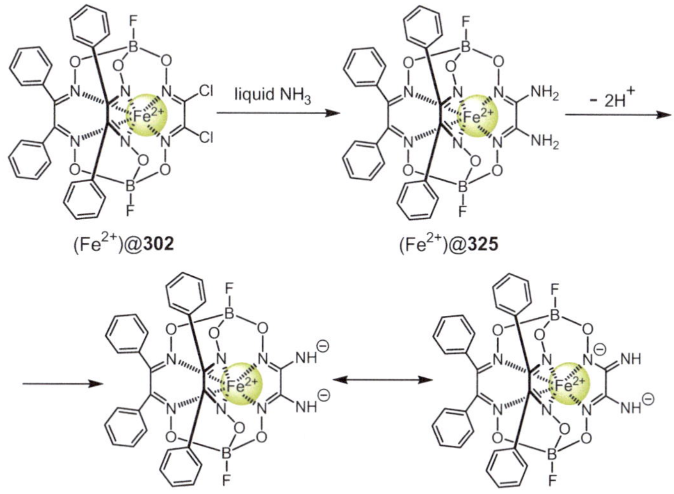

Scheme 2.151 Preparation of the diethoxyl-containing iron(II) cage complex

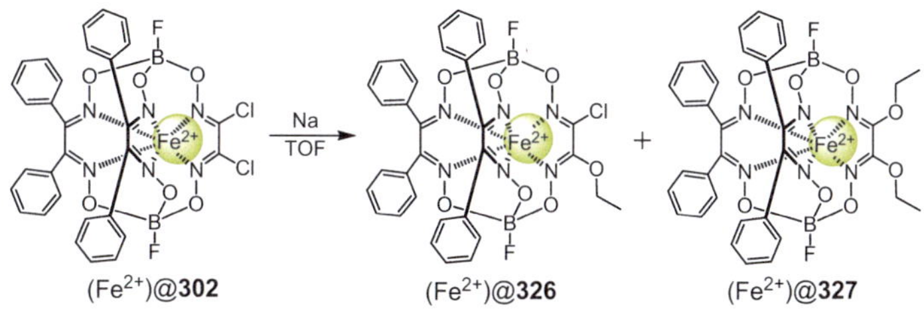

$$HC(OC_2H_5)_3 \longrightarrow (C_2H_5O)_2C: + C_2H_5OH$$
$$(C_2H_5O)_2C: \longrightarrow (C_2H_5O)_2C = C(OC_2H_5)_2$$
$$C_2H_5OH + Na \longrightarrow NaOC_2H_5 + 1/2\ H_2$$

substitution of (Fe^{2+})@**302** with diphenylphosphine sulfide has been performed [135] by Scheme 2.167 under PTC, giving the monochloromonothiophosphoryl cage complex (Fe^{2+})@**349** in a high yield. This thiophosphorylated iron(II) monochloroclathrochelate was found to undergo no further thiophosphorylation, leading to the dithiophosphorylated cage complex, even with an excess of the above thiophosphorylating agent and under the classical phosphorylation reaction conditions (Scheme 2.167) as well. This fact is explained [135] by sterical clashes between two bulky diphenylthiophosphine ribbed substituents in a *vic*-position of the same chelate fragment of a clathrochelate framework. On the other hand, this monochloroclathrochelate (Fe^{2+})@**349** easily underwent nucleophilic substitution with sterically unhindered primary and secondary amines under mild reaction conditions by Scheme 2.168, thus giving *P*,*N*-substituted monoribbed-functionalized iron(II) cage complexes in high yields. For the synthesis of its bis-clathrochelate derivative (Fe^{2+})$_2$@**355**, two synthetic strategies shown in Scheme 2.169 have been used [135]: the first one is based on the reaction of an excess of (Fe^{2+})@**302** with cadaverine, while the other used a nucleophilicity of the clathrochelate (Fe^{2+})@**354** with a terminal amino group. Both these synthetic approaches allowed to obtain [135] the target iron(II) bis-clathrochelate in high yields under mild reaction conditions.

The analogous phosphorylation reactions of (Fe^{2+})@**302** by Scheme 2.170 with diphenylphosphine oxide and diethyl thiophosphite have been performed [136] under PTC. A choice of PTC system for efficient (thio)phosphorylation of this macrobicyclic precursor was governed

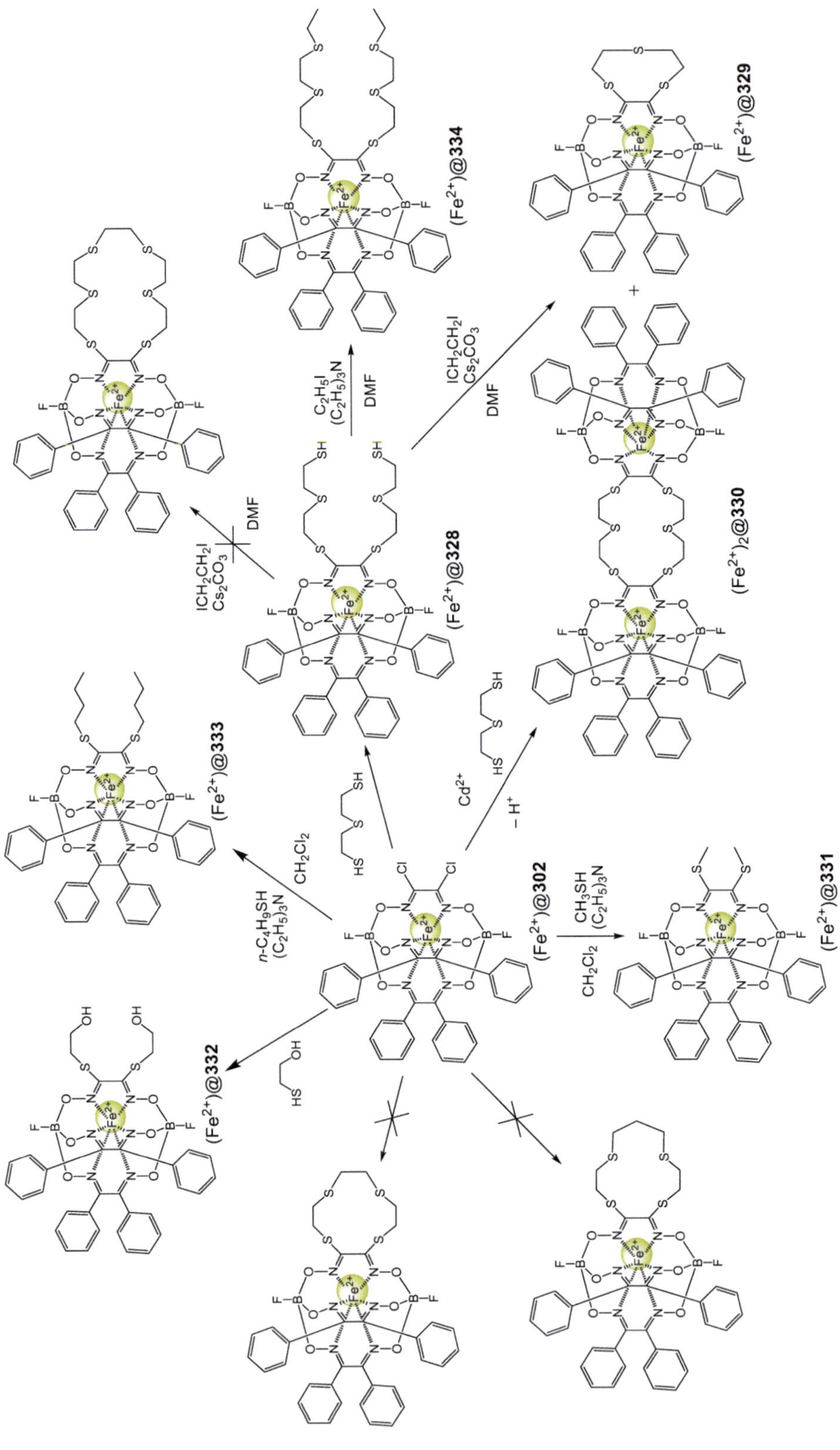

Scheme 2.152 Synthesis of a series of the alkylsulfide iron(II) mono- and bis-clathrochelates

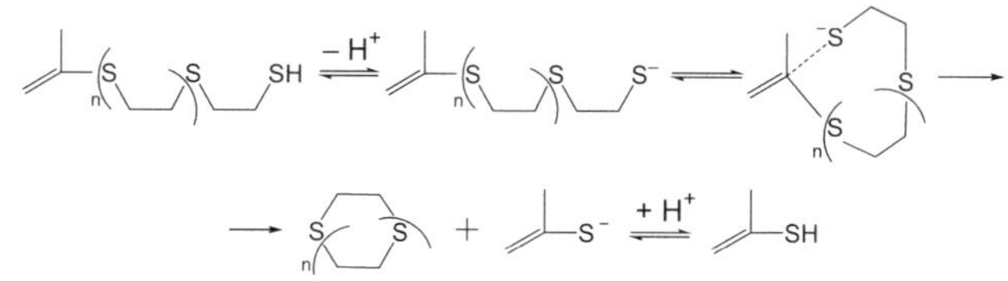

Scheme 2.153 Chemical transformations of the HS-terminated ribbed substituent

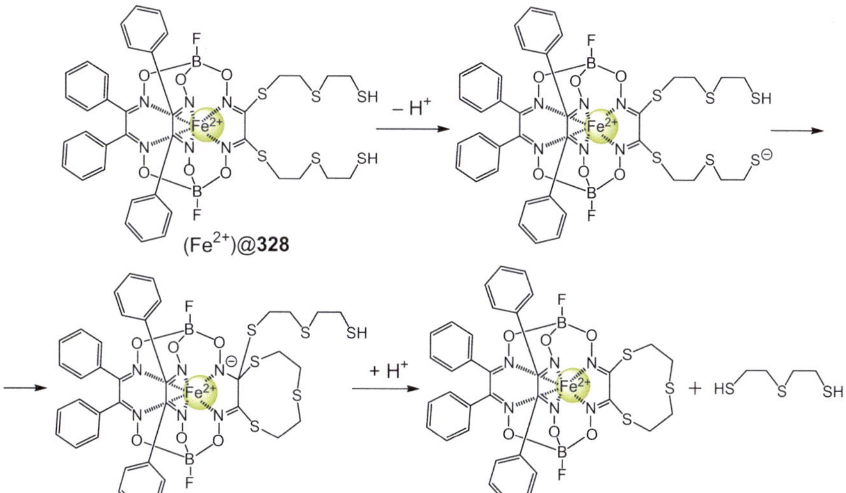

(Fe^{2+})@**328**

(Fe^{2+})@**329**

Scheme 2.154 Intramolecular dealkylation reaction of the complex (Fe^{2+})@**328**

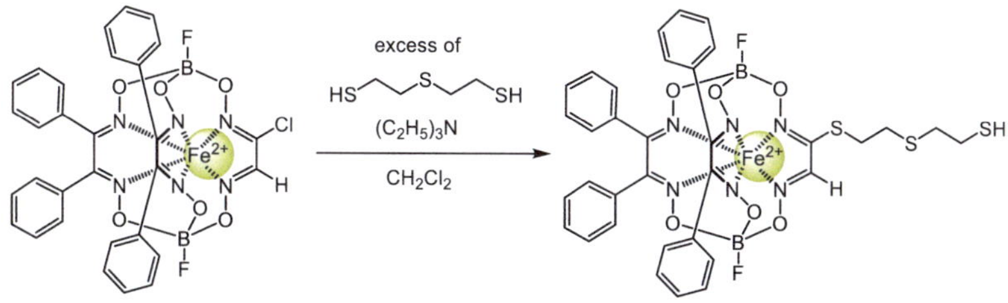

(Fe^{2+})@**301** (Fe^{2+})@**335**

Scheme 2.155 Preparation of the thiol-terminated clathrochelate (Fe^{2+})@**335**

mainly by the nature of an initial (thio) phosphorylating agent. In the case of diphenylphosphine sulfide and diethyl thiophosphite as *P*-nucleophiles, the best yields have been observed in dichloromethane–50% NaOH aqueous solution–5 mol.% TEBAC system; nucleophilic substitution with diphenylphosphine oxide under the same conditions gave a poorly separable mixture of the clathrochelate complexes and by-products. The selectivity of phosphorylation reaction in this case has been achieved in K_2CO_3–acetonitrile–5 mol.% TEBAC system, thus giving the target monophosphorylated clathrochelate (Fe^{2+})@**356** in a high yield. The use of different dichloroclathrochelate-to-diethyl thiophosphite molar ratios allowed obtaining both the mono- and diphosphorylated macrobicyclic complexes (in contrast to the reactions of (Fe^{2+})@**302** with diphenylphosphine oxide or with its sulfur-containing analog) [136].

Nucleophilic substitution of the macrobicyclic iron(II) trichlorotrimethylglyoximates (Fe^{2+})@**299**

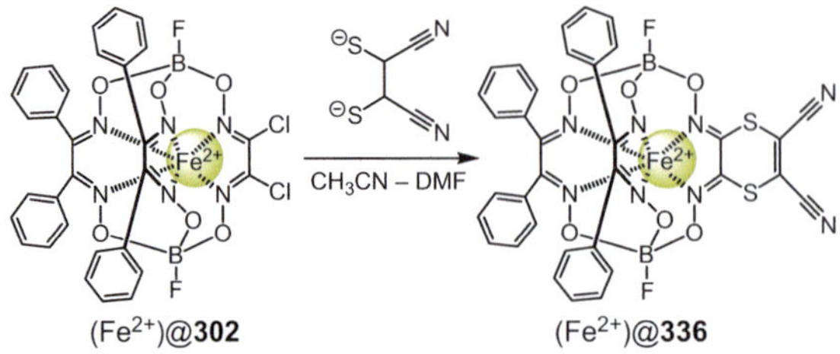

Scheme 2.156 Synthesis of the dinitrile cage complex (Fe^{2+})@**336**

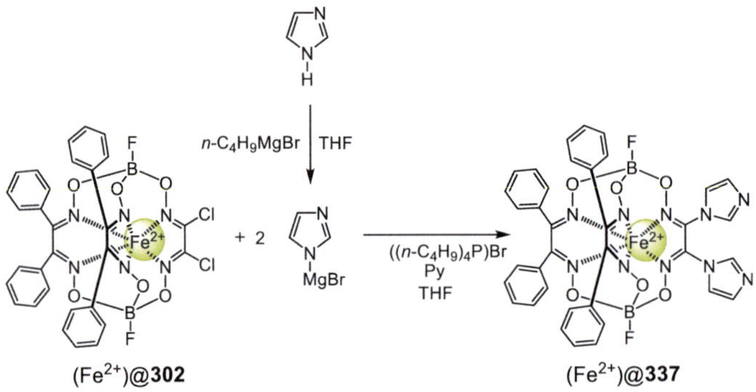

Scheme 2.157 Preparation of the diimidazolyl clathrochelate (Fe^{2+})@**337**

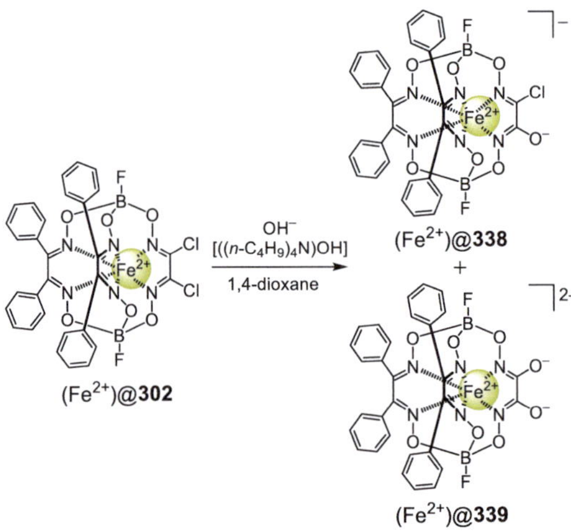

Scheme 2.158 Stepwise hydrolysis of the iron(II) dichloroclathrochelate (Fe^{2+})@**302**

and (Fe^{2+})@**300** (see section "Template Synthesis and Condensation Reactions") with primary unhindered aliphatic amines, such as *n*-butylamine and cyclohexylamine, by Schemes 2.171 and 2.172 is reported [107] to give the products of their partial substitution, the difunctionalized clathrochelates (Fe^{2+})@**359** and (Fe^{2+})@**361**; these reactions affected only two of their three α-dioximate chelate fragments. An unexpected result has been observed [107] in the case of nucleophilic substitution of the fluoroboron-capped clathrochelate precursor (Fe^{2+})@**300** with *n*-butylamine. Unlike the above phenylboron-cross-linked trichloroclathrochelate (Fe^{2+})@**299**, main product of its reaction with this N-nucleophile in DMF medium by Scheme 2.172 has been identified [107] as the constitutional isomers of the

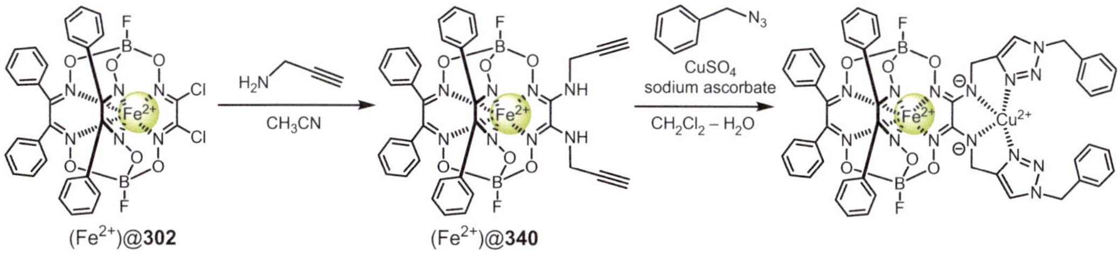

Scheme 2.159 Preparation of copper(II) complex of the clathrochelate ligand (Fe^{2+})@**340**

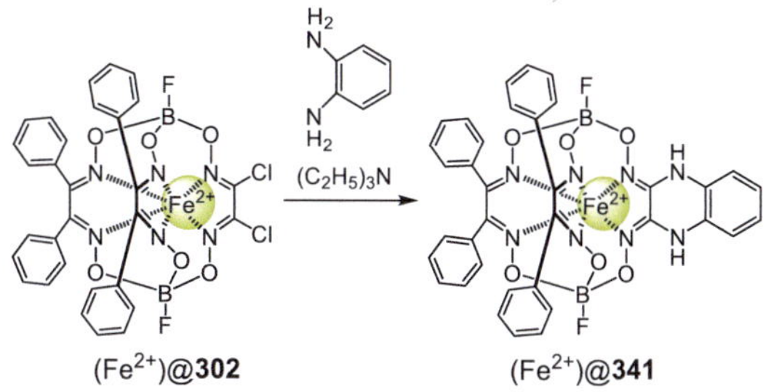

Scheme 2.160 Synthesis of the phenylenediamine cage complex (Fe^{2+})@**341**

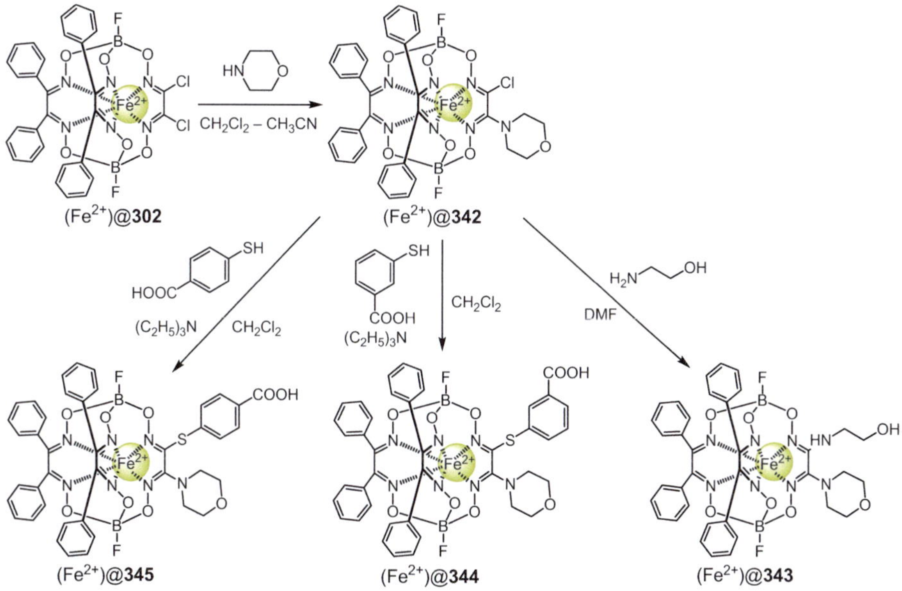

Scheme 2.161 Preparation and reactivity of the monochloromorpholine clathrochelate (Fe^{2+})@**342**

triamine clathrochelate (Fe^{2+})@**362** with their functionalizing substituents in its three α-dioximate chelate fragments. Typically, the nature of an apical substituent at a cage framework does not affect the products of nucleophilic substitution of the iron(II) halogenoclathrochelates with amines. In this case, however, passing from the phenyl group to the strong electron-withdrawing fluorine substituent is enough for activation of the electrophilic chloroglyoximate fragment for its nucleophilic substitution with *n*-butylamine as

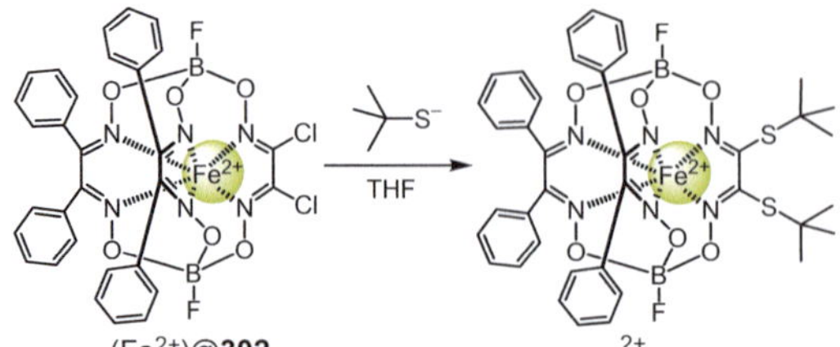

Scheme 2.162 Synthesis of the di-*tert*-butylsulfide iron(II) clathrochelate (Fe^{2+})@**346**

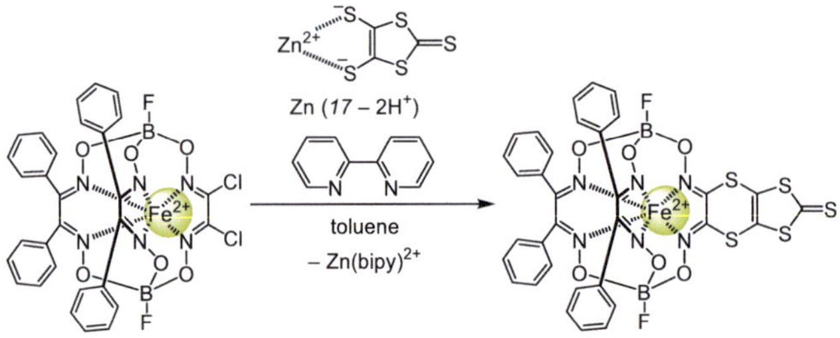

Scheme 2.163 Preparation of the iron(II) clathrochelate (Fe^{2+})@**347** with annelated disulfide ribbed fragment

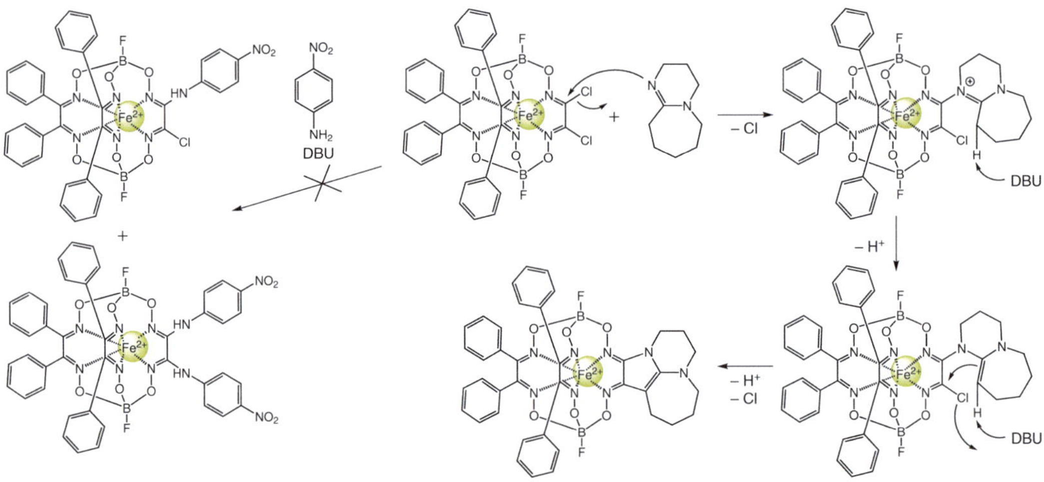

Scheme 2.164 Nucleophilic addition of DBU to an iron(II) dichloroclathrochelate

a *N*-nucleophile. An attempted chromatographical isolation of the *fac*- and *mer*-isomers of (Fe^{2+})@**362** is described [107] to be unsuccessful, but allowed to obtain the diamine macrobicyclic product (Fe^{2+})@**364**, resulting from the functionalization of its OH-containing dichloroclathrochelate precursor (Fe^{2+})@**363** as an admixture to the trichloromacrobicyclic complex (Fe^{2+})@**300**.

Nucleophilic substitution of the monochloroclathrochelate (Fe^{2+})@**298** by Scheme 2.173 has been used [107] for the synthesis of its monofunctionalized macrobicyclic derivatives (Fe^{2+})@**366**, (Fe^{2+})@**367**, (Fe^{2+})@**368**, and (Fe^{2+})@**369**.

Reactivity of the tris-chloroglyoximate clathrochelates (Fe^{2+})@**303** and (Fe^{2+})@**304** is reported [2] to be substantially higher than that of the above methylchloroglyoximate analogs. In particular, their triamine macrobicyclic derivatives have been isolated as major products of nucleophilic substitution of these precursors

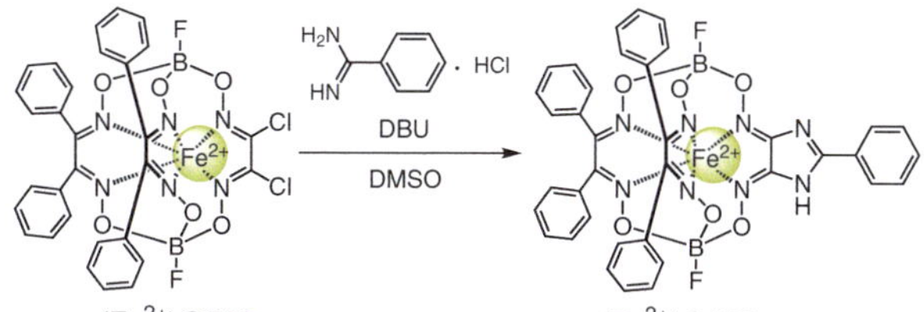

Scheme 2.165 Synthesis of an iron(II) clathrochelate with annelated phenylimidazole fragment

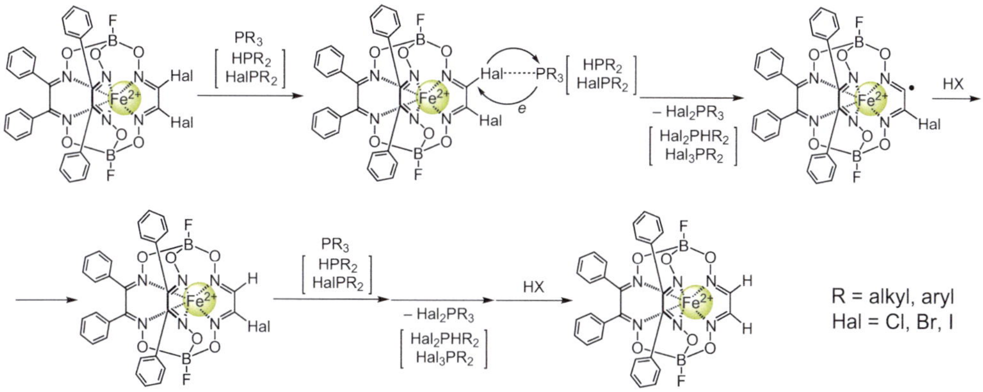

Scheme 2.166 Reactions of an iron(II) dihalogenoclathrochelates with organophosphorus(III) compounds

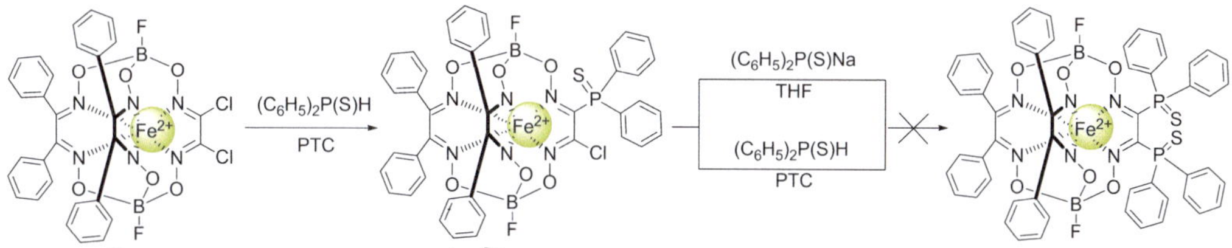

Scheme 2.167 Thiophosphorylation of an iron(II) dichloroclathrochelate precursor

with *n*-butylamine (Schemes 2.174 and 2.175), whereas the reactions of the phenylboron-capped clathrochelate (Fe^{2+})@**299** with *n*-butylamine gave only its diamine macrobicyclic derivative (see above). On the other hand, the mixtures of isomers of the corresponding diamine clathrochelate have been formed in the case of nucleophilic substitution of (Fe^{2+})@**303** and (Fe^{2+})@**305** with less reactive cyclohexylamine (Scheme 2.174); those with in situ generated with alkyl- and arylthiolate anions have proceeded more easily, thus giving the trisulfide iron(II) clathrochelate derivatives (Fe^{2+})@**372** and (Fe^{2+})@**374** by Schemes 2.174 and 2.176.

The monochloroclathrochelate precursor (Fe^{2+})@**301** also underwent nucleophilic substitution of its single reactive chlorine atom with amines and thiolate anions by Scheme 2.177, thus giving in particular the clathrochelate complexes (Fe^{2+})@**375**, (Fe^{2+})@**376**, (Fe^{2+})@**377**, and (Fe^{2+})@**378** with reactive terminal HO-, H_2N-, and HS-groups [2]. An attempt of [2] to obtain an iron(II) bis-clathrochelate by nucleophilic substitution of (Fe^{2+})@**301** with an anionic derivative of the monothiol-terminated complex (Fe^{2+})@**378** gave the iron(II) clathrochelate (Fe^{2+})@**379** with inherent

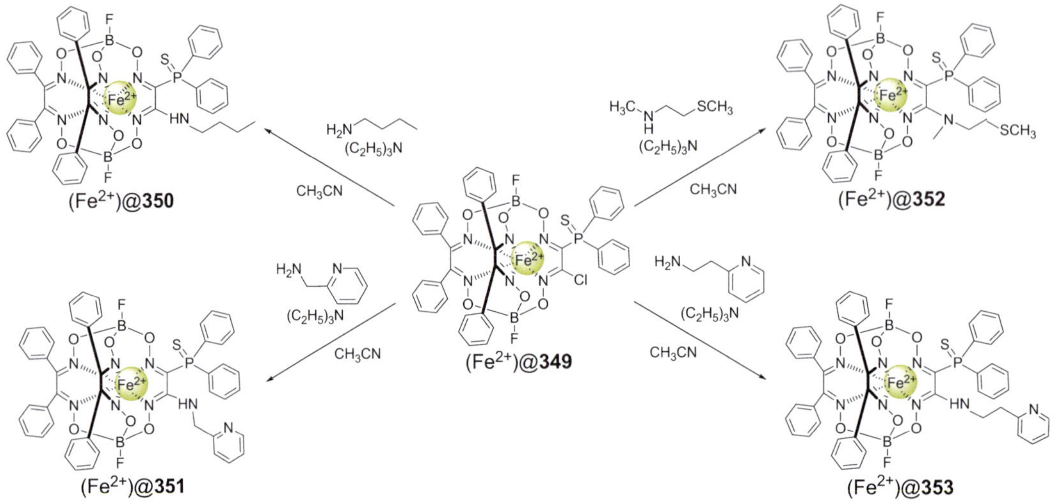

Scheme 2.168 Nucleophilic substitution of the phosphorus-containing monochloroclathrochelate (Fe^{2+})@**349**

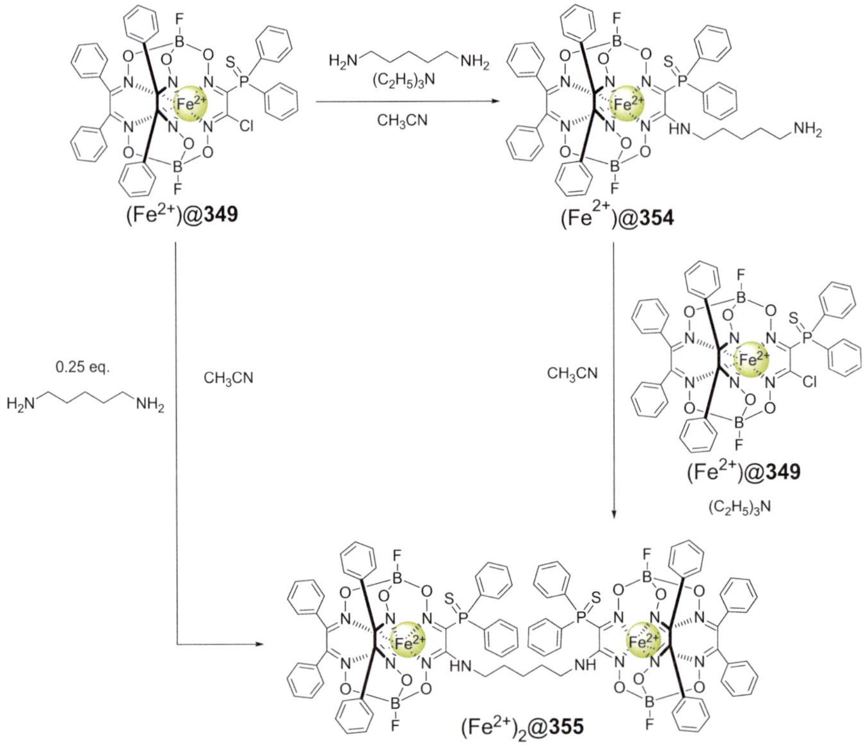

Scheme 2.169 Preparation of the amine-terminated clathrochelate (Fe^{2+})@**354** and its bis-macrobicyclic derivative $(Fe^{2+})_2$@**355**

HS-group in one of its three chelate α-dioximate fragments as a major product of this reaction by Scheme 2.177. The detailed study of this reaction [2] evidenced that the clathrochelate (Fe^{2+})@**378** underwent an intramolecular elimination of ethylene sulfide in basic media to yield the above macrobicyclic product with inherent HS-group; the plausible

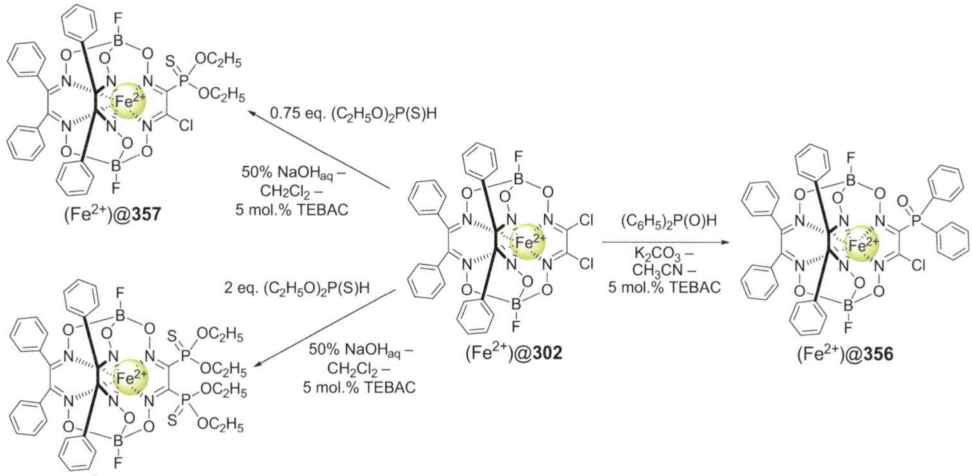

Scheme 2.170 (Thio)phosphorylation of the dichloroclathrochelate (Fe^{2+})@**302**

pathway of this process is shown in Scheme 2.178. The above terminal HS-group easily underwent its alkylation with methyl iodide in the presence of K_2CO_3 in DMF media, thus giving the cage complex (Fe^{2+})@**380** (Scheme 2.177) [2].

Complete nucleophilic substitution of six reactive chlorine atoms of the adamantylboron-capped hexachloroclathrochelate precursor (Fe^{2+})@**205** with in situ generated adamantylthiolate anions by Scheme 2.179 under vigorous reaction conditions (using potassium *tert*-amylate as strong organic base in homogeneous DMF medium) gave the octaadamantyl cage complex (Fe^{2+})@**381** in a moderate yield [77]. In contrast, nucleophilic substitution of the phenylboron-capped iron(II) hexachloroclathrochelate (Fe^{2+})@**294** by Scheme 2.180 with the same nucleophilic anion proceeds under mild reaction conditions, allowing to obtain the target hexasulfide macrobicyclic product (Fe^{2+})@**382** in a much higher yield [77].

The iron(II) cage complexes with sulfide ribbed substituents containing terminal iono- and protonogenic groups have been prepared [137] by Schemes 2.181 and 2.182 using nucleophilic substitution of their mono-, di-, and hexachloroclathrochelate precursors with thiolate ions, the derivatives of *ortho*-, *meta*-, and *para*-mercaptobenzoic acids and sodium 3-mercapto-1-propanesulfonate, as *S*-nucleophiles. The reactions of the hexachloroclathrochelate precursor (Fe^{2+})@**293** with anionic derivatives of these mercaptobenzoic acids, generated in situ in the presence of triethylamine, afforded their hexasulfide macrobicyclic complexes shown in Scheme 2.181. In the case of *para*-mercaprobenzoic acid, the corresponding clathrochelate hexaanion has been isolated [137] in the form of its triethylammonium salt, the reprecipitation of which from DMF solution with $NaClO_4$ gave the water-soluble sodium salt of (Fe^{2+})@**385**.

With *meta*-mercaptobenzoic acid, a protonated form of the hexasulfide iron(II) clathrochelate (Fe^{2+})@**384** has been formed on the first stage of this reaction. In the case of a clathrochelate derivative of *ortho*-mercaptobenzoic acid, its fully protonated form has been also isolated [137]. Nucleophilic substitution of the mono- and dichloroclathrochelate precursors (Fe^{2+})@**301** and (Fe^{2+})@**302** with the above aromatic thiolate anions afforded the corresponding mono- and disulfide monoribbed-functionalized cage complexes shown in Scheme 2.181. Functionalization of the above hexa- and dichloroclathrochelate precursors (Fe^{2+})@**293**, (Fe^{2+})@**294**, and (Fe^{2+})@**302** with a thiolate derivative of sodium 3-mercapto-1-propanesulfonate under the same reaction conditions has been used [137] for preparation of sodium salts of the hexa- and disulfide

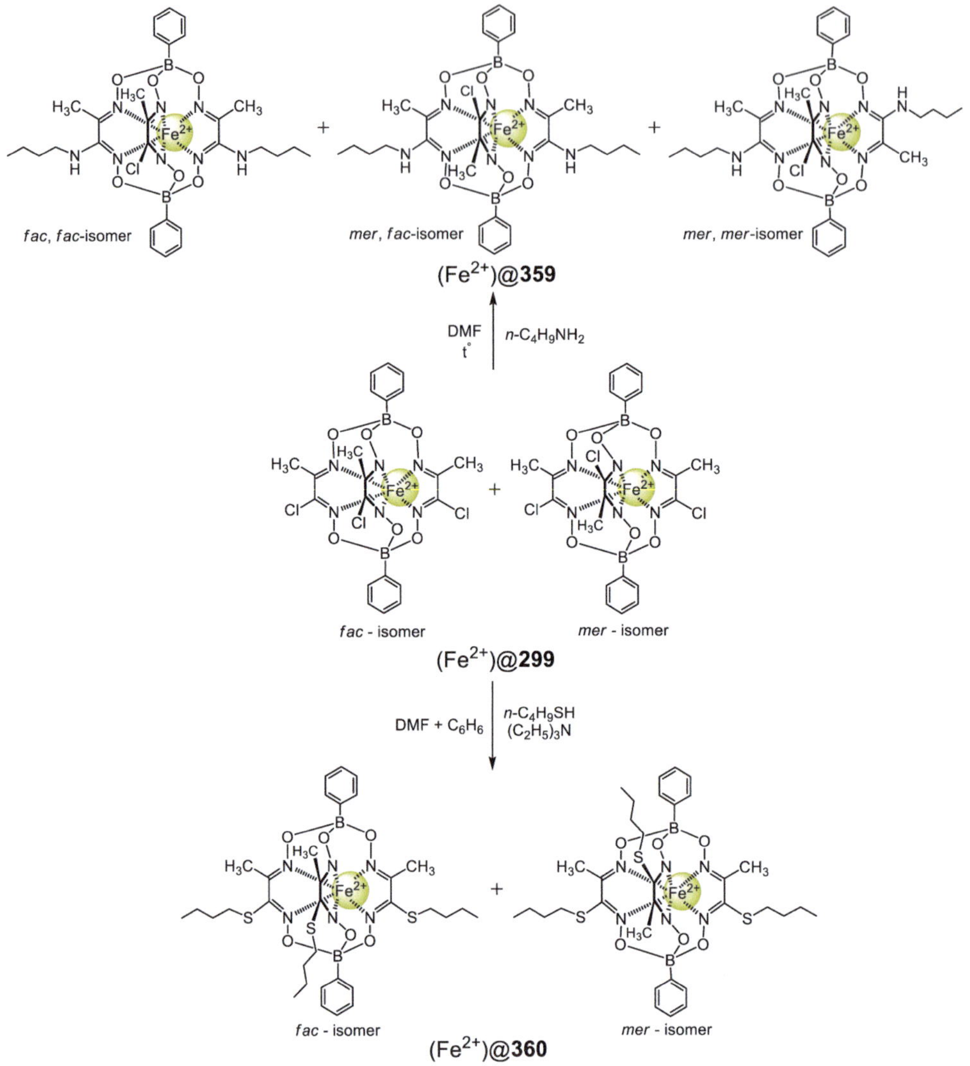

Scheme 2.171 Nucleophilic substitution of the trichloroclathrochelate precursor (Fe^{2+})@**299**

macrobicyclic hexa- and dianions by Scheme 2.182. The water solubility of the above ribbed-functionalized iron(II) clathrochelates was found [137] to increase with the number of iono- and protonogenic functionalizing substituents at their macrobicyclic frameworks.

The similar reaction of the 3-formylphenylboron-capped iron(II) hexachloroclathrochelate precursor (Fe^{2+})@**229** (see section "Template Synthesis and Condensation Reactions") by Scheme 2.183 with thiolate anions generated in situ from *para*-mercaptobenzoic acid in the presence of triethylamine as organic base has been used [84] for preparation of the hexasulfide iron(II) clathrochelate (Fe^{2+})@**395** with six ribbed *para*-mercaptobenzoic substituents and terminal reactive groups in its apical substituents.

Mono-, di-, and hexaphenol iron(II) clathrochelates (Fe^{2+})@**396**, (Fe^{2+})@**397**, (Fe^{2+})@**398**, (Fe^{2+})@**399**, and (Fe^{2+})@**400** have been synthesized [109] by Scheme 2.184 using nucleophilic substitution of the corresponding iron(II) mono-, di-, and hexachloroclathrochelate precursors with 2,6-di-*tert*-butyl-4-mercaptophenolate anions, generated in situ in the presence of triethylamine in acetonitrile as a solvent: the

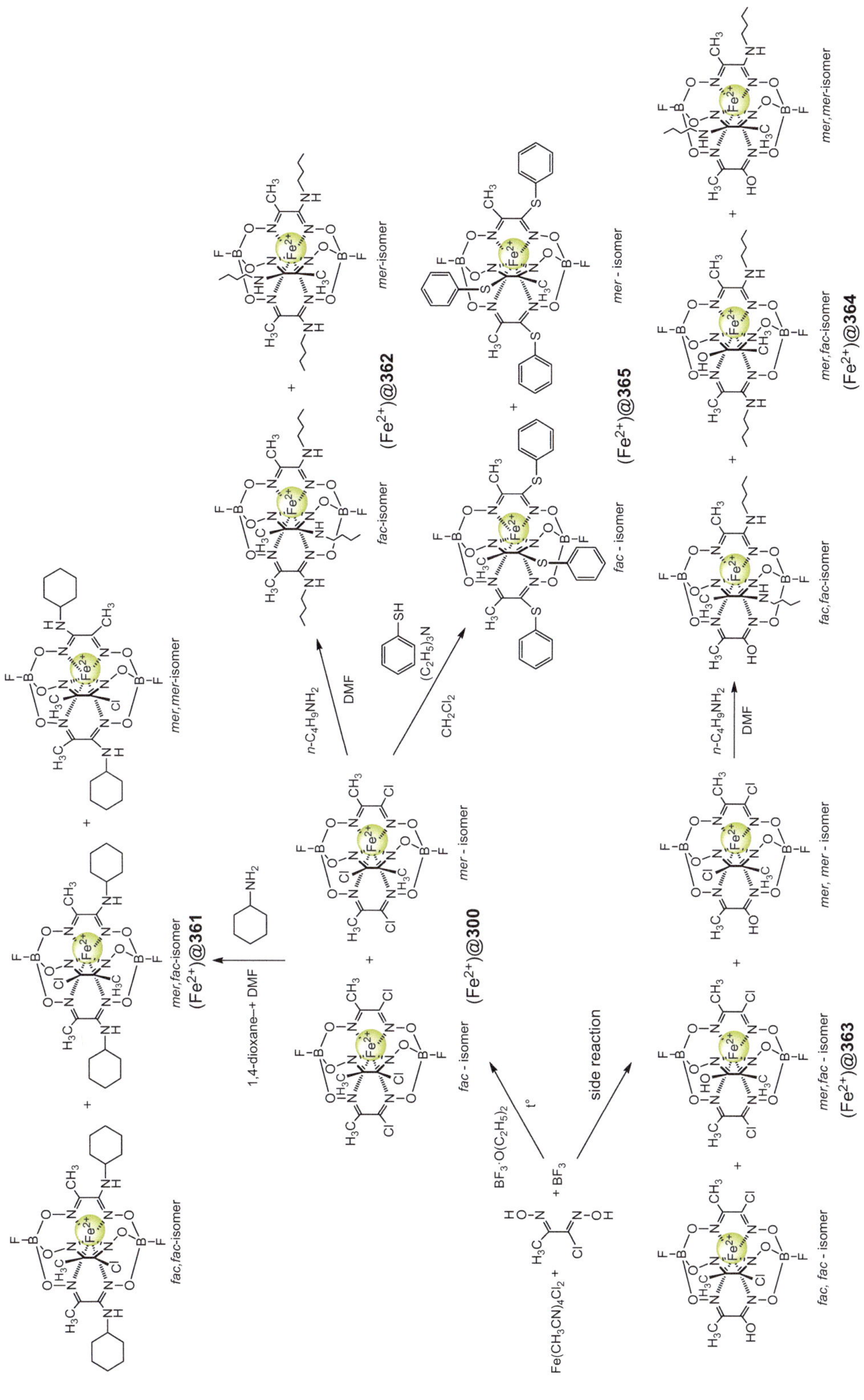

Scheme 2.172 Preparation and nucleophilic substitution of the chloroclathrochelates (Fe^{2+})@**300** and (Fe^{2+})@**363**

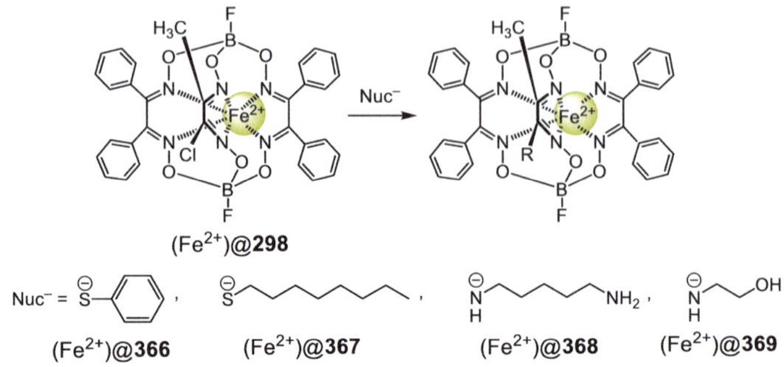

Scheme 2.173 Nucleophilic substitution of the monochloroclathrochelate (Fe^{2+})@**298**

same reactions in dichloromethane media gave only the partially functionalized macrobicyclic products. When DMF has been used as a solvent, a complete functionalization of the clathrochelate molecules has been observed, but together with simultaneous increase in the formation of by-products of the complete destruction of their cage frameworks and those of an oxidation of the initial thiol as well [109]. The synthesis of the hexaphenolsulfide cobalt(II) clathrochelate (Co^{2+})@**397** by Scheme 2.185 is reported to be complicated by the side redox and chemical reactions, such as the formation of the corresponding cobalt(I) cage complex and the complete destruction of a macrobicyclic framework, thus decreasing the yield of the target clathrochelate. The above reduction did not occur in benzene medium in the case of a slow addition of triethylamine to the reaction mixture, but this reaction is described [109] to give mainly the tetrasubstituted clathrochelate products; their complete functionalization has been performed by Scheme 2.185 using 1,4-dioxane as a solvent.

Mono-, di-, and hexaphenylsulfide iron(II) clathrochelate analogs (Fe^{2+})@**401** and (Fe^{2+})@**402** of the above phenol-terminated cage complexes have been also prepared [109] using nucleophilic substitution by Scheme 2.186 of the same iron(II) halogenoclathrochelates with thiophenolate anion as an active *S*-nucleophilic agent, thus giving the target macrobicyclic products in high yields.

Nucleophilic substitution of the boron- and tin-capped iron(II) hexa- and dichloroclathrochelate precursors with S_2-dinucleophiles, generated in situ from the corresponding α-dithiols in the presence of triethylamine as organic base, has been used [138] for preparation of the mono- and triribbed-functionalized iron(II) clathrochelates (Fe^{2+})@**404**, (Fe^{2+})@**405**, (Fe^{2+})@**406**, (Fe^{2+})@**407**, (Fe^{2+})@**408**, and (Fe^{2+})@**409** with six-membered S_2-cyclic substituent(s) by Scheme 2.187 in high yields under mild reaction conditions. The attempted synthesis of their tetra- and hexaamine analogs using nucleophilic substitution of the iron(II) hexachloroclathrochelates (Fe^{2+})@**293** and (Fe^{2+})@**294** with ethylenediamine under the same reaction conditions was unsuccessful, while the reaction of their dichloroclathrochelate analog (Fe^{2+})@**302** with an excess of this N_2-dinucleophile gave the target diamine iron(II) clathrochelate (Fe^{2+})@**408** in a low yield [138].

Analogous reactions of the same S_2- and N_2-dinucleophiles with the boron-capped cobalt and ruthenium(II) hexachloroclathrochelate precursors by Scheme 2.188 have been performed [139]. In the case of the bifunctional dithiolate nucleophiles, the target triribbed-functionalized hexasulfide cage complexes have been isolated in high yields using the pseudo-high dilution conditions: this allowed to avoid the side reactions giving the bis-clathrochelate by-products. The reaction of the cobalt(II) hexachloroclathrochelate (Co^{2+})@**294** with ethylenediamine in DMF medium by Scheme 2.188 afforded the target cobalt(III) tris-ethylenediamine macrobicyclic monocation (Co^{3+})@**410** in a low yield only due to a side reaction of the complete destruction of its macrobicyclic framework with this highly basic nucleophile [139].

The clathrochelates (Fe^{2+})@**411**, (Fe^{2+})@**412**, (Fe^{2+})@**413**, (Fe^{2+})@**414**, (Fe^{2+})@**415**, (Fe^{2+})@**416**, and (Fe^{2+})@**417** have been

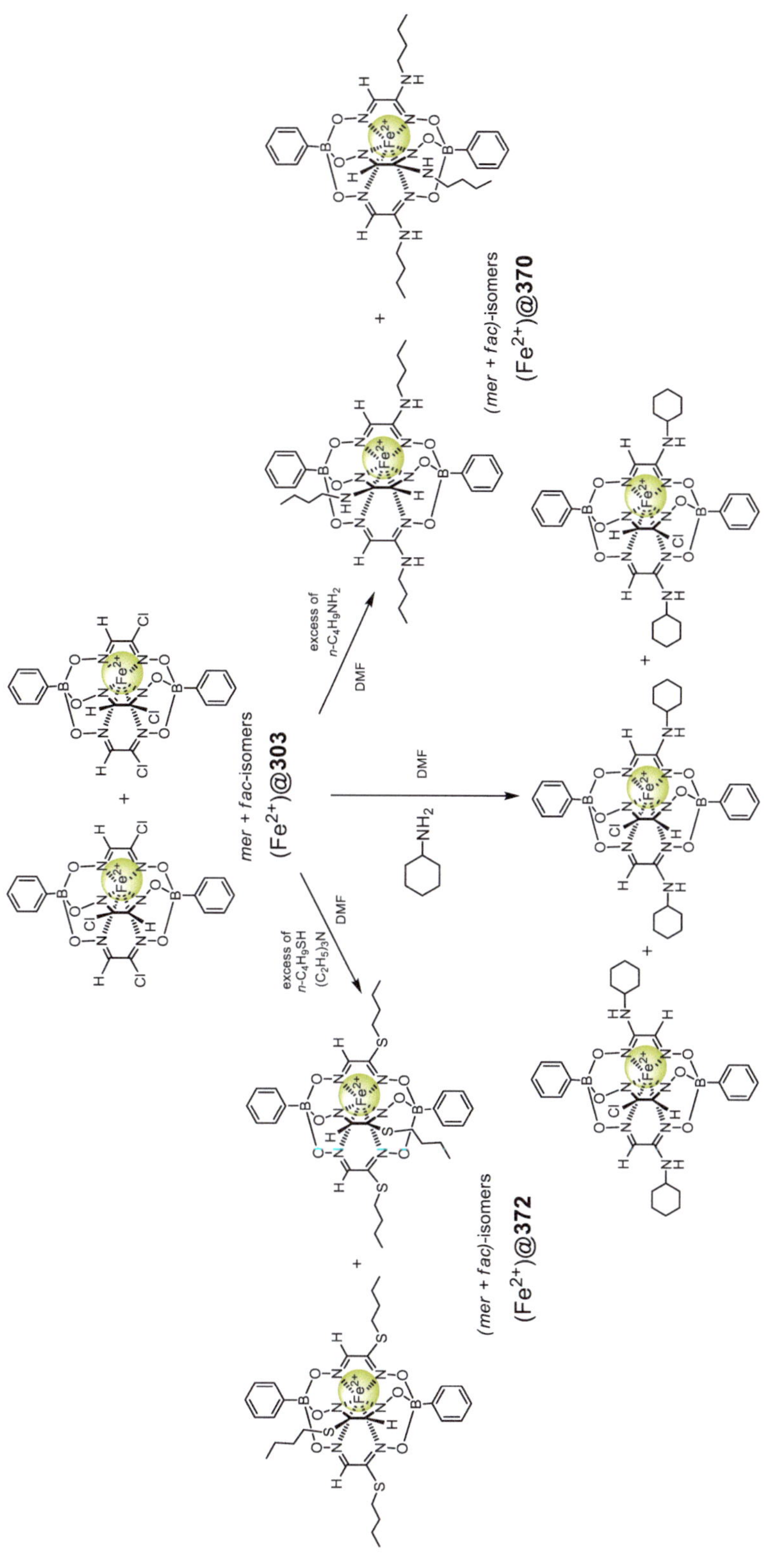

Scheme 2.174 Nucleophilic substitution of the trichloroclathrochelate precursor (Fe^{2+})@**303**

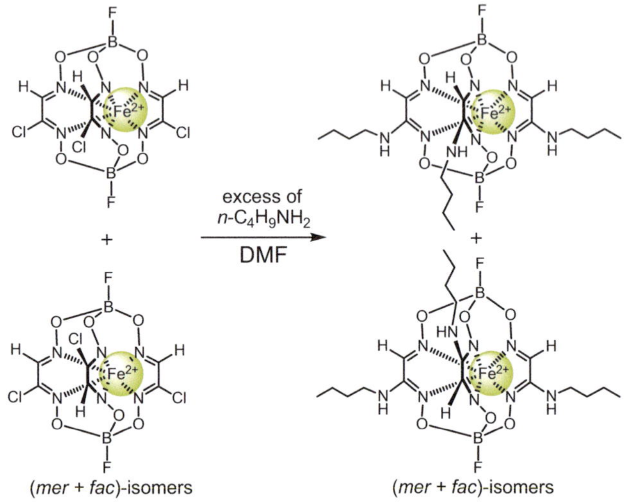

(Fe^{2+})@**304** (Fe^{2+})@**373**

Scheme 2.175 Synthesis of the constitutional isomers of the triamine complex (Fe^{2+})@**373**

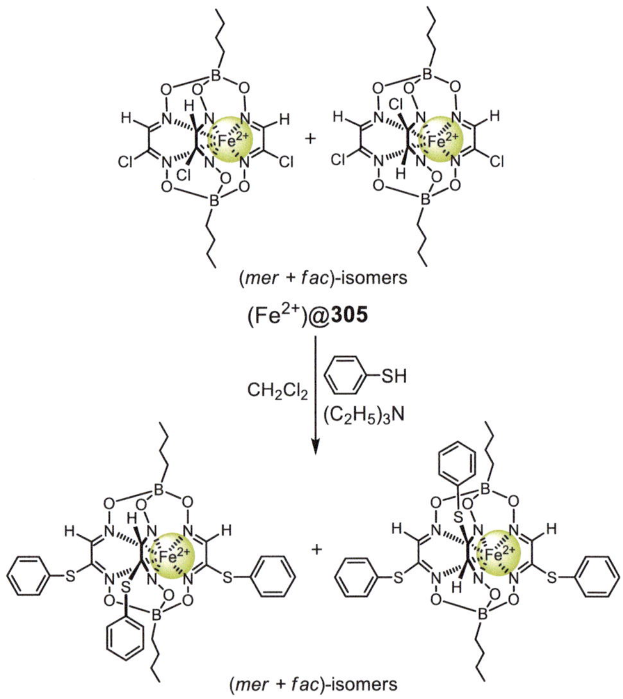

(Fe^{2+})@**305**

(Fe^{2+})@**374**

Scheme 2.176 Preparation of the constitutional isomers of the trisulfide clathrochelate (Fe^{2+})@**374**

prepared [3] in high yields by Schemes 2.189 and 2.190 using nucleophilic substitution of the n-butylboron-capped iron(II) hexachloroclathrochelate (Fe^{2+})@**293** with anionic derivatives of pyrocatechol, ethylenediamine, and 1,2-ethanedithiol, generated in situ in the presence of triethylamine. This intramolecular cyclization, giving O_2-, N_2-, and S_2-six-membered ribbed substituent(s) at a cage framework, has been performed under pseudo-high dilution conditions, thus limiting a formation of various macrobicyclic by-products. In the case of anionic

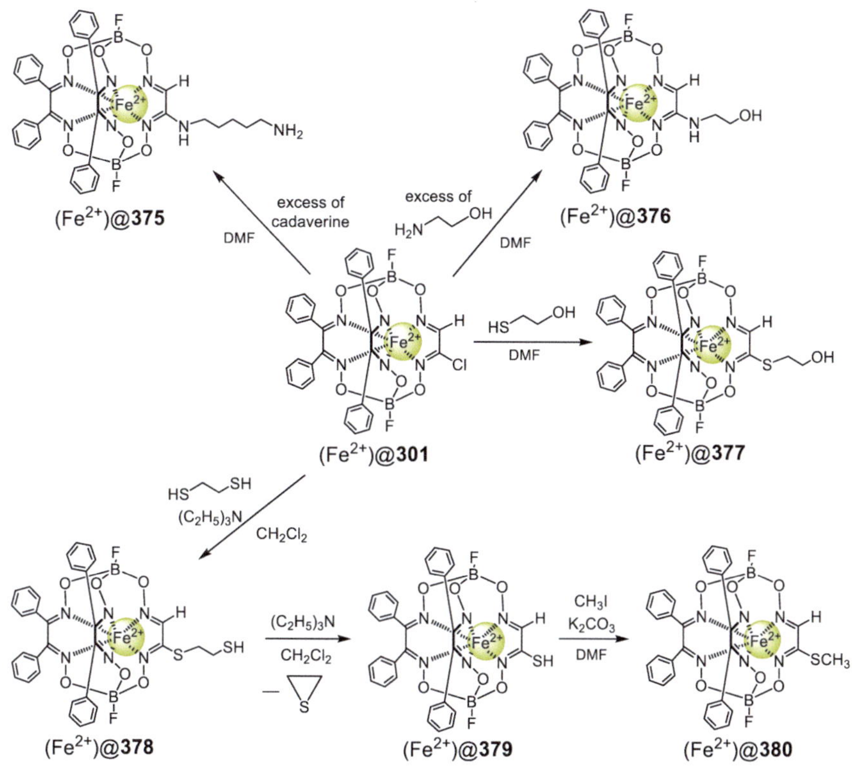

Scheme 2.177 Nucleophilic substitution of the monochloroclathrochelate (Fe^{2+})@**301**

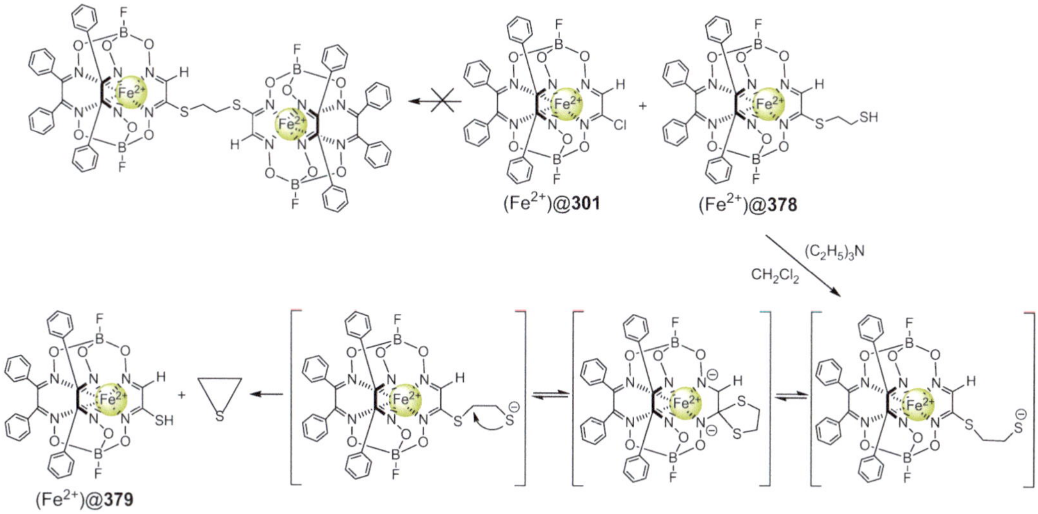

Scheme 2.178 Chemical transformation of the thiol-terminated complex (Fe^{2+})@**378** with elimination of ethylene sulfide

derivatives of ethylenediamine as a N_2-dinucleophile, only the mono-N_2-alicyclic iron(II) tetrachloroclathrochelate (Fe^{2+})@**411** has been formed in a moderate yield, while its bis-N_2-alicyclic dichloroclathrochelate analog (Fe^{2+})@**418** did not form even with an excess

Scheme 2.179 Synthesis of the octaadamantyl-containing clathrochelate (Fe^{2+})@**381**

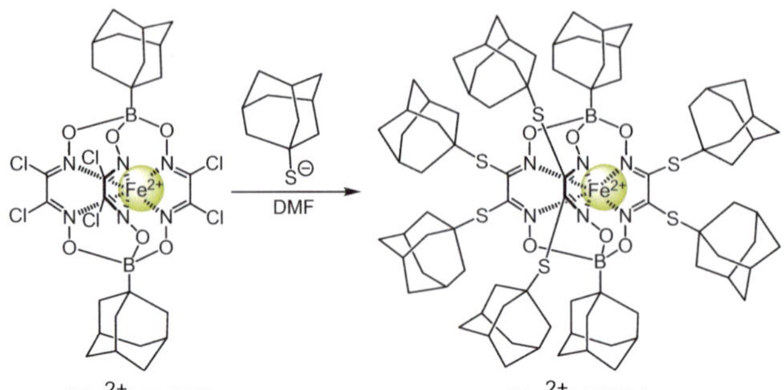

Scheme 2.180 Preparation of the hexaadamantyl-containing cage complex (Fe^{2+})@**382**

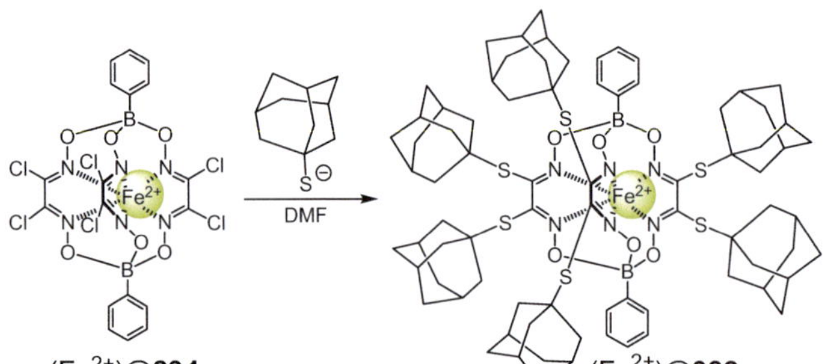

of ethylenediamine [3]. The tetrachloroclathrochelate (Fe^{2+})@**411** underwent further nucleophilic substitution of one of its two dichloroglyoximate fragments by Scheme 2.189, thus giving the N_2, S_2-alicyclic dichloroclathrochelate (Fe^{2+})@**412** with three non-equivalent ribbed chelate fragments. Stepwise nucleophilic substitution with anionic derivatives of 1,2-ethanedithiol has been used [3] for the synthesis of mono- and diribbed-substituted S_2-alicyclic tetra- and dichloroclathrochelates (Fe^{2+})@**413** and (Fe^{2+})@**414** also shown in Scheme 2.189. Nucleophilic substitution by Scheme 2.190 with anionic derivatives of pyrocatechol as an O_2-dinucleophile allowed obtaining iron(II) di- and tetrachloroclathrochelate precursors (Fe^{2+})@**415** and (Fe^{2+})@**416** with six-membered O_2-cyclic ribbed substituent(s). Such mono- and bispyrocatecholate macrobicyclic complexes have been prepared in relatively high yields by stepwise nucleophilic substitution of (Fe^{2+})@**293** with pyrocatecholate dianions using clathrochelate-to-nucleophile molar ratios 1:1 and 1:2, respectively. At the same time, the tris-pyrocatechol cage complex (Fe^{2+})@**417** with equivalent ribbed fragments has not been found among the clathrochelate products of such nucleophilic substitutions under the standard conditions; it has been isolated [3] in a low yield with a large excess of pyrocatechol at high temperature using a boiling nitromethane as a solvent. To avoid the formation of the bis-pyrocatechol iron(II) macrobicyclic complex, the synthesis of the monoribbed-functionalized tetrachloroclathrochelate (Fe^{2+})@**416** has been performed at low temperature (−50 °C) in dichloromethane medium to increase its yield. On the other hand, the above diribbed-functionalized complex (Fe^{2+})@**415** has been isolated [3] in a relatively high yield using carbon tetrachloride as a solvent and prolonged heating of the reaction mixture.

Nucleophilic substitution of the reactive chlorine atoms of the cobalt(II) hexachloroclathrochelates (Co^{2+})@**293**, (Co^{2+})@**294**, and (Co^{2+})@**295** by Scheme 2.191 [104] considerably differs from the similar reactions of their iron(II)- and ruthenium(II)-encapsulating analogs. In the latter cases, their hexaaryl- and hexaalkylsulfide clathrochelate derivatives have been synthesized

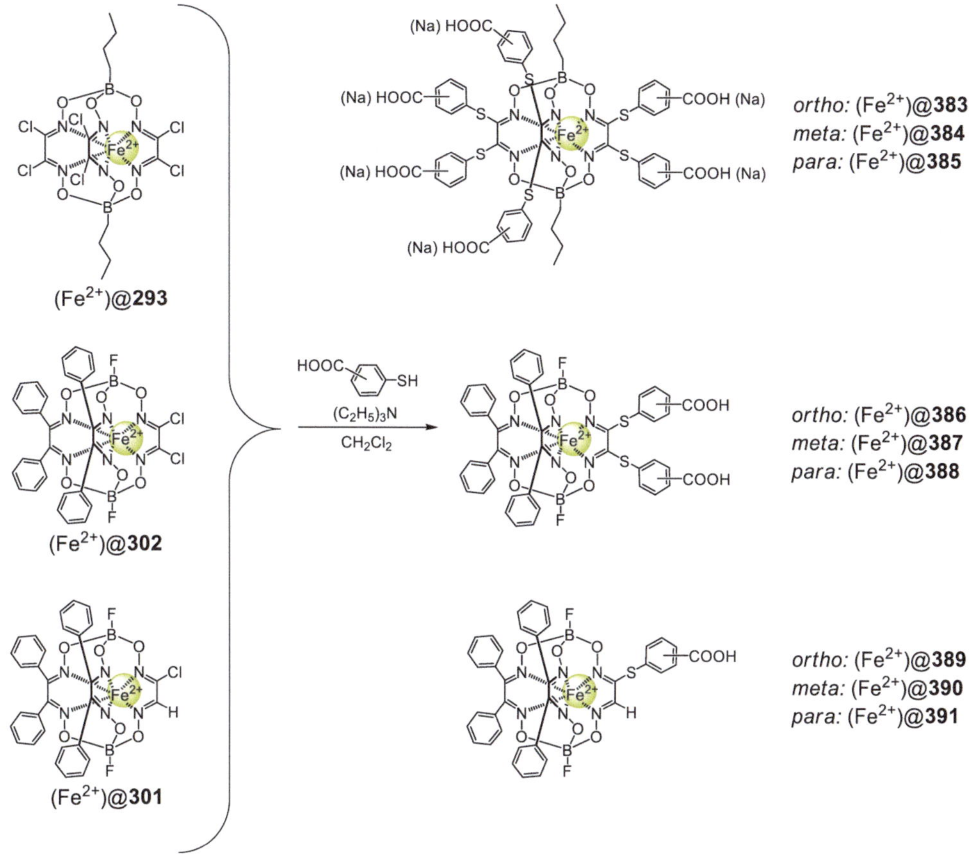

Scheme 2.181 Synthesis of a series of the mono-, di- and hexacarboxyl-containing iron(II) clathrochelates

both with alkaline metal thiolates in the solvents with a high donor number and with a RSH/triethylamine system in aprotonic solvents with a low donor number (see above). In the case of the cobalt(II) hexachloroclathrochelates, the use of alkaline metal thiolates as nucleophilic agents caused the complete destruction of their cage frameworks [104]. So, the system RSH/triethylamine in dichloromethane medium with in situ generated thiolate anions has been used to obtain the hexasulfide cobalt(II) clathrochelates under mild reaction conditions. An unusual trinuclear $Co^{III}Co^{II}Co^{III}$ bis-clathrochelate $(Co^{3+})_2$@**425** has been isolated [104] as a by-product of nucleophilic substitution of the phenylboron-capped cobalt(II) hexachloroclathrochelate (Co^{2+})@**294** with *n*-butylthiolate anion by Scheme 2.192 in 1,4-dioxane as a solvent. The plausible pathway of its formation is as follows: this precursor has been partially reduced with a thiolate ion to produce the corresponding cobalt(I) clathrochelate, a cage framework of which is relatively unstable because of a larger Shannon radius of Co^{+} ion as compared with that of Co^{2+} ion. The cavity size of this framework is optimal for low-spin Fe^{2+} and Co^{3+} ions, and this size fits, to a lesser extent, with the sizes of low- and high-spin Co^{2+} ions. It is evident that high-spin Co^{+} ion with electronic configuration d^8 poorly fits this cavity. As a result, one of the two capping fragments of its clathrochelate detaches with a formation of the corresponding semiclathrochelate intermediate, thus revealing the rigidity of a macrobicyclic framework. Then three possibilities exist: (1) this hexachlorosemiclathrochelate undergoes nucleophilic substitution

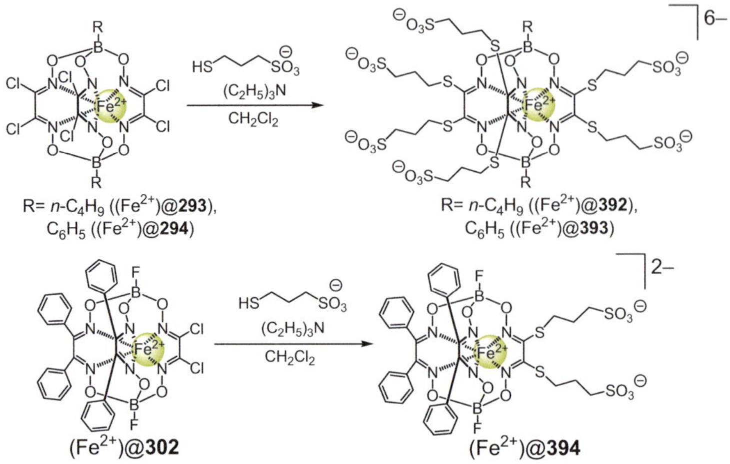

Scheme 2.182 Preparation of the SO_3^--terminated iron(II) clathrochelates

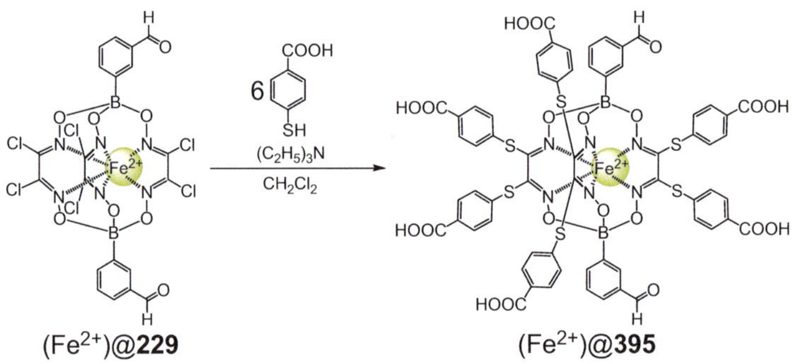

Scheme 2.183 Synthesis of the hexacarboxyl-containing iron(II) clathrochelate with terminal formyl groups

with thiolate ions, followed by oxidation of an encapsulated cobalt ion, and by cross-linking of the formed hexasulfide cobalt(III) semiclathrochelate with a Lewis-acidic cobalt(II) ion, which is present in the reaction mixture because of the partial complete destruction of a clathrochelate framework; (2) the second pathway includes a cross-linking two molecules of the above hexachlorosemiclathrochelate intermediate with a cobalt(II) ion to produce the dodecachloro-bis-clathrochelate, which then undergoes nucleophilic substitution and a oxidation of an encapsulated cobalt ion; or (3) nucleophilic substitution of an admixture of the hypothetical homotrinuclear $Co^{II}Co^{II}Co^{II}$ dodecachloro-bis-clathrochelate. The first of these pathways to the above heterotrinuclear $Co^{III}Co^{II}Co^{III}$ bis-clathrochelate $(Co^{3+})_2$@**425** (Scheme 2.192) is described [104] to be most probable.

Nucleophilic substitution of the cobalt(II) hexachloroclathrochelates with primary aliphatic amines, such as *n*-butylamine, *tert*-butylamine, and cyclohexylamine, gave the hexaamine cobalt(III) clathrochelates shown in Scheme 2.191 in a wide range of the reaction conditions: their formation has been observed in different solvents with an excess and with low molar amounts of the corresponding amine as well as using $RNH_2/RNH_3^+Cl^-$ systems to decrease its reactivity [104]. In the case of *tert*-butylamine in dichloromethane medium, a mixture of the constitutional isomers of the corresponding tetra-*tert*-butylamine cobalt(II) clathrochelate has been formed. Its reaction with excess of *n*-butylamine in DMF gave the mixture of the constitutional isomers of the hexaamine cobalt(III) clathrochelate (Co^{3+})@**420** with four *tert*-butylamine and two *n*-butylamine ribbed substituents. In the case of

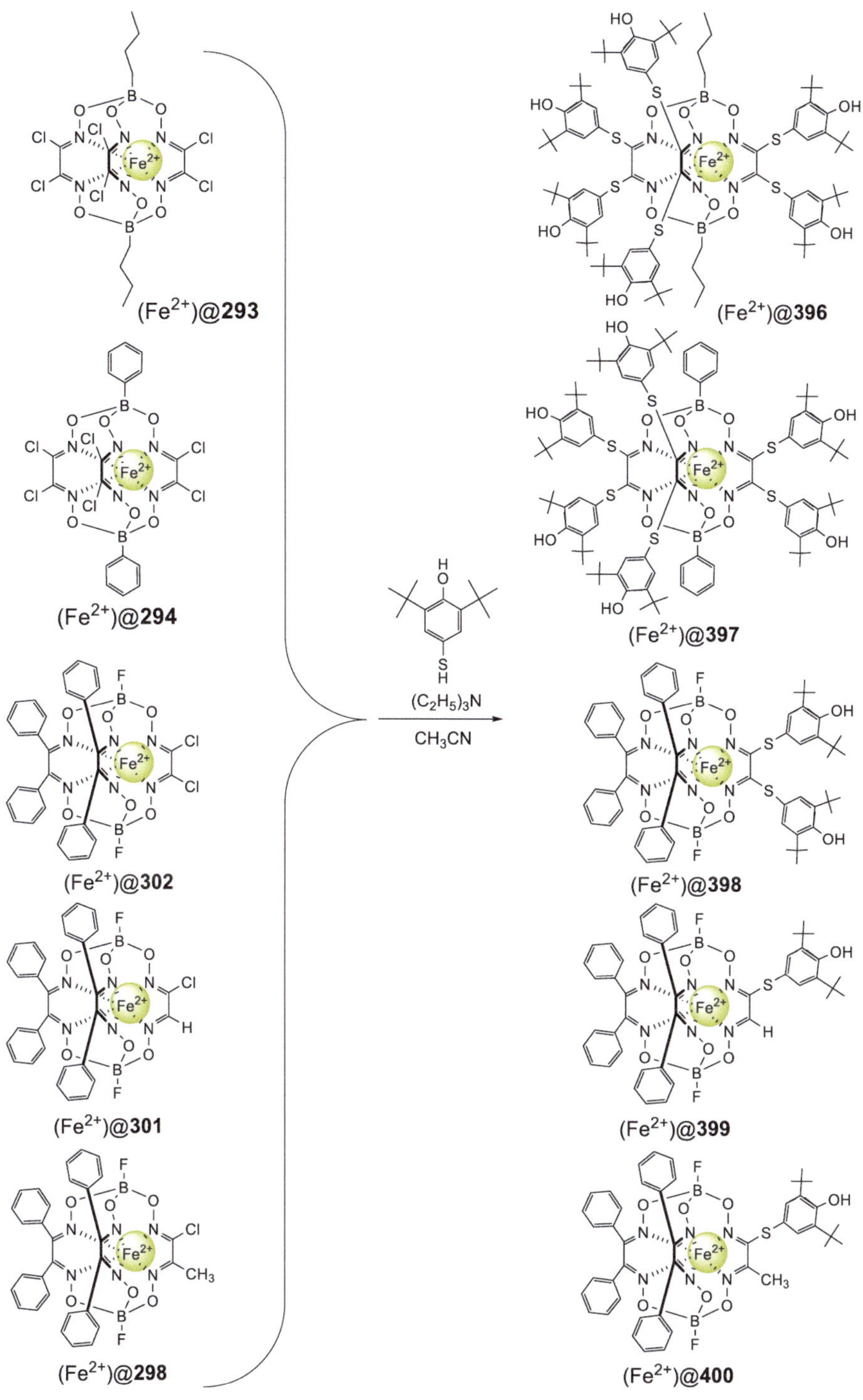

Scheme 2.184 Preparation of a series of the phenol-terminated iron(II) clathrochelates

diethylamine as a *N*-nucleophile in DMF medium, the formation of the trifunctionalized cobalt(III) clathrochelates has been observed [104]; their reactions with primary aliphatic amines in DMF media resulted in the formation of the constitutional isomers of the

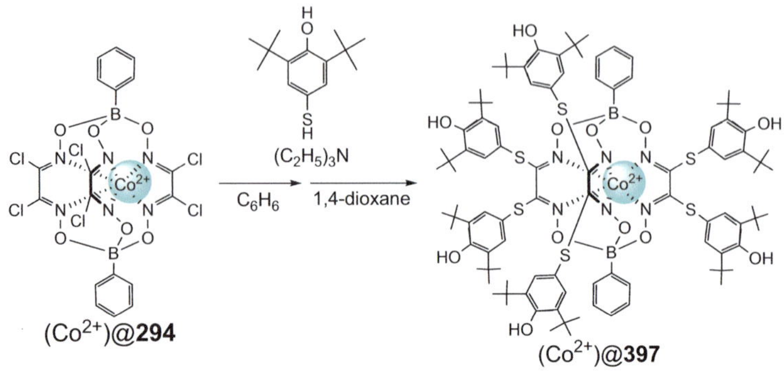

Scheme 2.185 Synthesis of a hexaphenolsulfide cobalt(II) cage complex

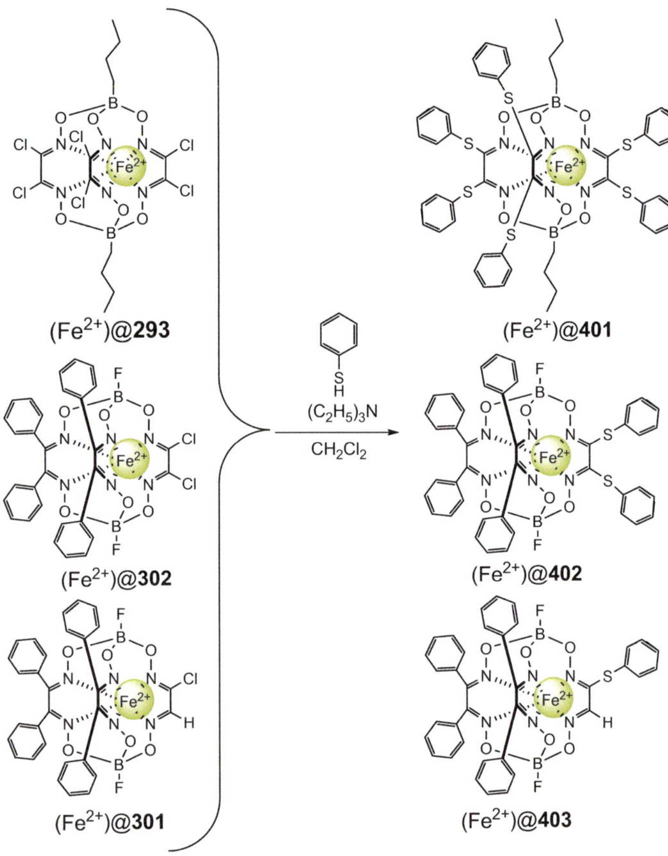

Scheme 2.186 Preparation of the mono-, di- and hexasulfide iron (II) clathrochelates

corresponding hexaamine cobalt(III) clathrochelate (Co^{3+})@**419** also shown in Scheme 2.191 [104].

Unusual $Co^{III}Co^{II}Co^{III}$ bis-clathrochelate by-product $(Co^{3+})_2$@**426** of a crystallization of the 2-thiopheneboron-capped cobalt(II) tris-dimethylglyoximate (Co^{2+})@**208** (see section "Template Synthesis") from chloroform–heptane mixture that was deposited on air atmosphere under the sunlight has been isolated and X-ray structurally characterized [140]: in this complex, its two tris-dioximate cobalt(III)-encapsulating macrobicyclic entities are cross-linked by the cobalt(II) ion as a bifunctional Lewis acid (Scheme 2.193). The attempts to obtain this trinuclear bis-cage complex using template

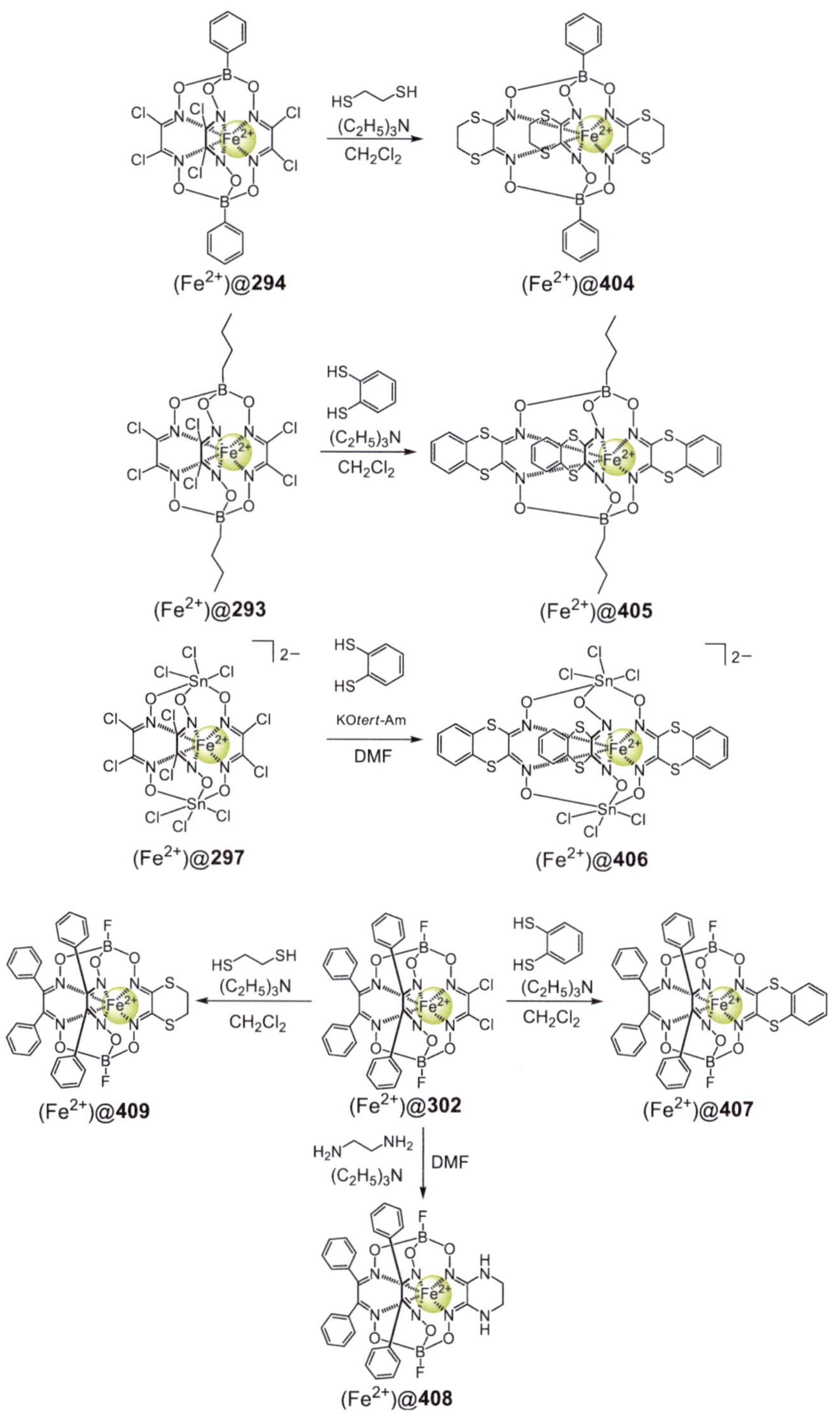

Scheme 2.187 Synthesis of the iron(II) clathrochelates with S_2- or N_2-containing six-membered ribbed fragment(s)

condensation with an excess of $CoCl_2$ under aerobic conditions in the presence of triethylamine as organic base gave only the cobalt(II) monoclathrochelate (Co^{2+})@**208**. The above trinuclear complex was not found also among the products of the regular template reaction,

Scheme 2.188
Preparation of the cobalt and ruthenium clathrochelates with S_2- or N_2-containing six-membered ribbed fragments

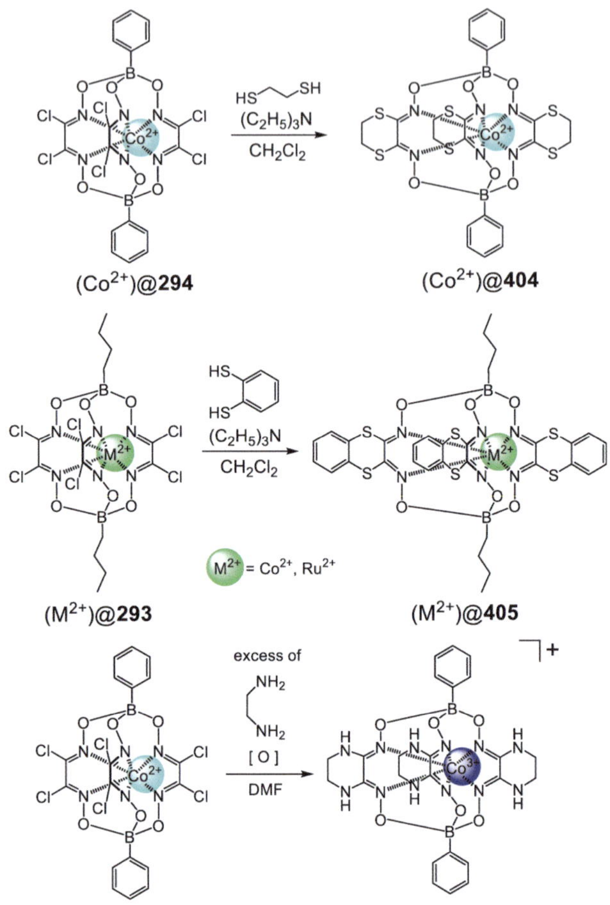

used for preparation of (Co^{2+})@**208**. So, such bis-clathrochelate has been initially suggested [140] to result from the side chemical transformation of the monoclathrochelate (Co^{2+})@**208** with $COCl_2$ as the product of photo-induced oxidation of chloroform during a slow evaporation of the chloroform solution of (Co^{2+})@**208** in the presence of air oxygen. However, the model experiment with halogen lamp irradiation using either a crude chloroform or a purified solvent showed no formation of $(Co^{3+})_2$@**426** and resulted only in a partial oxidation of its monoclathrochelate analog (Co^{2+})@**208** followed by complete destruction of its macrobicyclic framework. The photochemical experiments with high-pressure Hg-lamp filtered and non-filtered irradiation of the chloroform solution of this cobalt(II) cage complex either under inert atmosphere or with air bubbling have been performed [140]. The treatment of its solution with non-filtered Hg-lamp irradiation caused a fast decomposition of (Co^{2+})@**208**, and the registered MALDI-TOF mass spectra contained the characteristic peaks of the trinuclear bis-clathrochelate $(Co^{3+})_2$@**426**. These peaks disappeared after few minutes of such irradiation due to the complete decomposition of the above cage and bis-cage complexes under these irradiation conditions. In the case of the glassy-filtered Hg-lamp irradiation of an initial chloroform solution, the complete decomposition of (Co^{2+})@**208** proceeded more slowly and only its gradual decomposition has been observed under inert atmosphere; no formation of its bis-clathrochelate derivative $(Co^{3+})_2$@**426** has been detected [140]. In the case of air bubbling

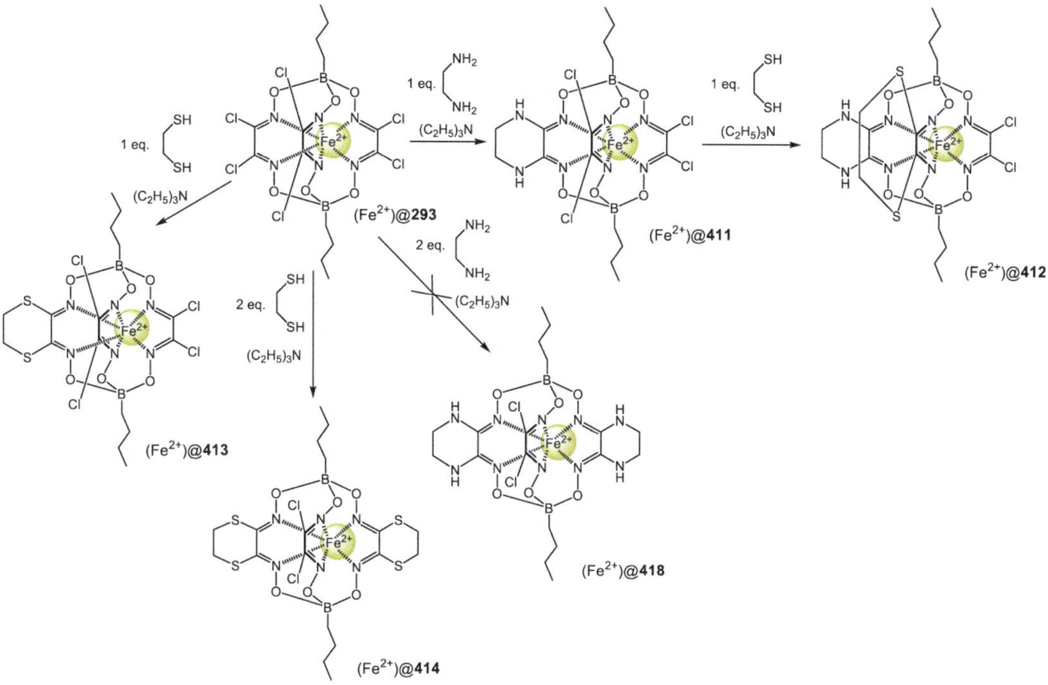

Scheme 2.189 Synthesis of the iron(II) clathrochelates with S_2,N_2-alicyclic ribbed gragment(s)

through this chloroform solution, its decomposition proceeded much slowly and the trinuclear bis-clathrochelate $(Co^{3+})_2$@**426** has been detected among the reaction products using mass spectrometry. Therefore, most probable pathway to this bis-cage complex includes the photoinitiated radical oxidation of chloroform with air oxygen leading to the reactive species ($COCl_2$, HCl, and the radicals) that cause the opening, rearrangement, and recapping of a macrobicyclic framework of (Co^{2+})@**208** by Scheme 2.193, thus giving the trinuclear $Co^{III}Co^{II}Co^{III}$ bis-cage complex $(Co^{3+})_2$@**426** [140].

The hexaperfluoroarylsulfide ribbed-functionalized iron and cobalt(II) tris-dioximates (M^{2+})@**427**, (M^{2+})@**428**, and (M^{2+})@**429** have been prepared [141] by Scheme 2.194 under mild reaction conditions by nucleophilic substitution of six reactive chlorine atoms of the *n*-butylboron-capped iron and cobalt(II) hexachloroclathrochelates (Fe^{2+})@**293** and (Co^{2+})@**293** with perfluoroarylthiolate anions generated in situ in the presence of triethylamine as organic base. Such nucleophilic substitutions in the case of the cobalt(II) complex (Co^{2+})@**293** have been complicated by its side reduction reactions with thiolate anions resulting in the formation of the navy-blue cobalt(I)-encapsulating macrobicyclic monoanions. The single crystals of the dodecasulfide $Co^{III}Co^{II}Co^{III}$ bis-clathrochelate $(Co^{3+})_2$@**430** (Scheme 2.195), which is a by-product of nucleophilic substitution of (Co^{2+})@**293** with pentafluorothiophenolate anion, have been also isolated [102]. The plausible pathway to this trinuclear cage complex is similar to that for its *n*-butylsulfide analog $(Co^{3+})_2$@**425** (see Scheme 2.196). The suggestion that the hexachloroclathrochelate precursor (Co^{2+})@**293** may contain an admixture of the corresponding homotrinuclear $Co^{II}Co^{II}Co^{II}$ dodecachloro-bis-clathrochelate and its nucleophilic substitution in the presence of air oxygen gives the dodecapentafluorophenylsulfide bis-clathrochelate $(Co^{3+})_2$@**430** has been excluded [141] basing on the results of additional experiments.

The pentafluorophenylboron-capped clathrochelate precursors (Fe^{2+})@**292** and (Co^{2+})@**292** (see section "Template Synthesis

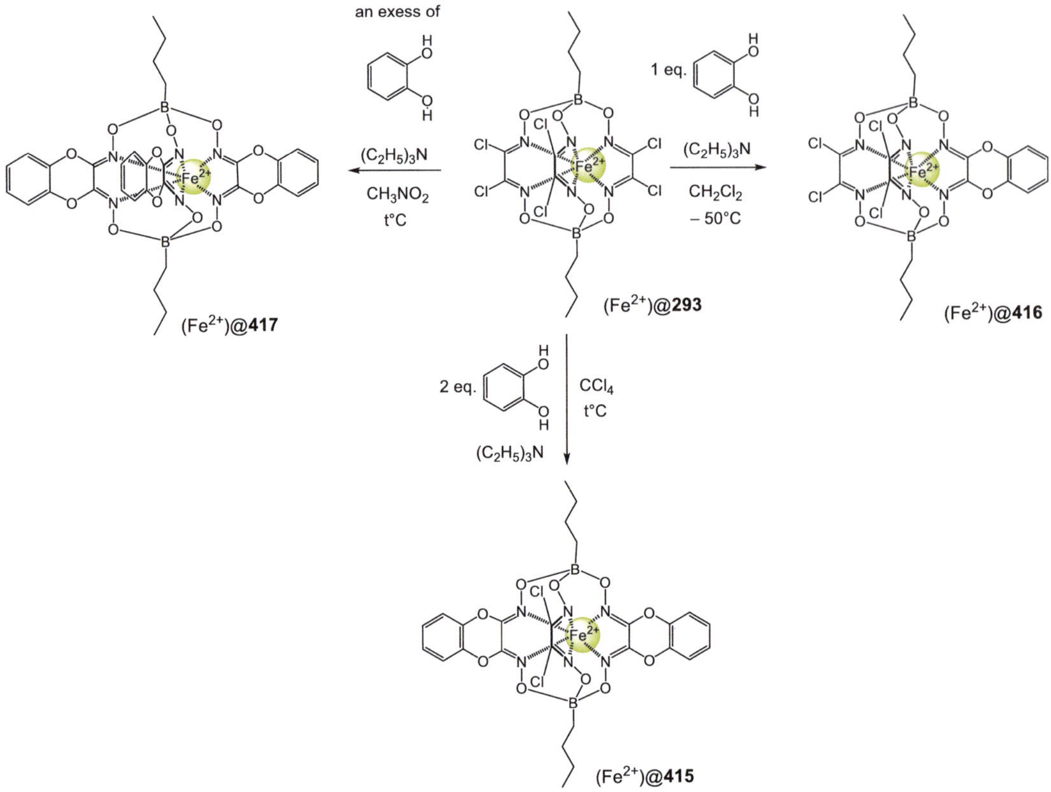

Scheme 2.190 Preparation of the pyrocatecholate iron(II) clathrochelate

and Condensation Reactions") easily underwent nucleophilic substitution by Scheme 2.197 with (per)fluoroarylthiolate anions generated in situ from the corresponding thiols with triethylamine [102]. This gave the metal(II) macrobicyclic complexes, the second superhydrophobic shell of which is formed by fluorine atoms of both the apical and the ribbed (per)fluoroaryl substituents. In the case of functionalization of (Co^{2+})@**292** with pentafluorophenylthiolate anion, along with the target hexasulfide monomacrobicyclic product (Co^{2+})@**433**, a mixed-valence trinuclear $Co^{III}Co^{II}Co^{III}$ bis-clathrochelate by-product $(Co^{3+})_2$@**434** (Scheme 2.197) has been isolated and X-ray structurally characterized [102]. Two plausible pathways for its formation are shown in Scheme 2.196. They include either (1) the formation of a $Co^{II}Co^{II}Co^{II}$ dodecachloro-bis-clathrochelate precursor as a by-product of template condensation of dichloroglyoxime with pentafluorophenylboronic acid on a cobalt(II) ion: its nucleophilic substitution could give this dodeca (pentafluorophenylsulfide) bis-clathrochelate, or (2) a reduction of the encapsulated cobalt(II) ion of (Co^{2+})@**292** with pentafluorophenylthiolate anion followed by detachment of one of the two capping pentafluorophenylboron fragments; the formed cobalt(I) semiclathrochelate undergoes its cross-linking with a Lewis-acidic Co^{2+} cation, followed by nucleophilic substitution of a bis-clathrochelate intermediate and its oxidation giving the bis-clathrochelate $(Co^{3+})_2$@**434** [102].

Mono- and dichloroclathrochelate precursors (Co^{2+})@**301** and (Co^{2+})@**302** are reported [108] to easily undergo nucleophilic substitution of their reactive chlorine atom(s) with *N*,*S*-nucleophiles (i.e., the deprotonated aliphatic mono- and diamines and mono- and dithiolate anions) by Scheme 2.198. In the case of secondary aliphatic amines as the sources of *N*-nucleophiles, only one of the two chlorine atoms of (Co^{2+})@**302** underwent nucleophilic substitution. That with ethylenediamine and

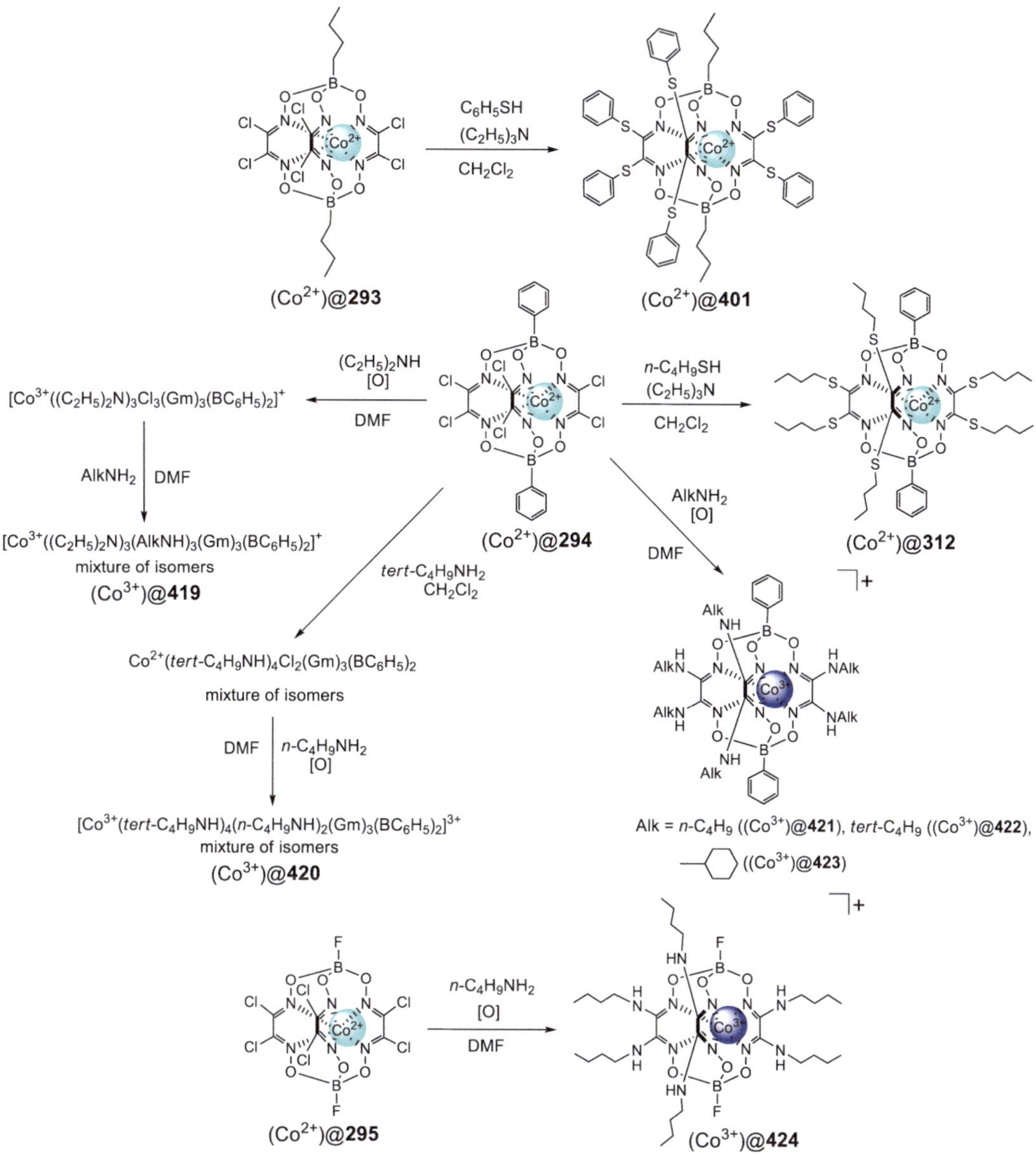

Scheme 2.191 Nucleophilic substitution of the boron-capped cobalt(II) hexachloroclathrochelates

1,2-ethanedithiol as dinucleophilic agents having short spacer fragments gave only their macrobicyclic derivatives with α,α′-N_2 and α,α′-S_2 alicyclic six-numbered ribbed fragments (Scheme 2.198) have been isolated [108], whereas no formation of the corresponding ribbed-functionalized cage complexes with terminal amine or mercapto group(s) has been observed. On the other hand, the cobalt(II) clathrochelates with terminal mercapto group have been synthesized [108] by Scheme 2.198 using the cobalt(II) monochloroclathrochelate (Co^{2+})@**301** as a macrobicyclic precursor and the corresponding aliphatic dithiol as a nucleophilic agent.

Nucleophilic substitution of the reactive chlorine atoms of the hexachloroclathrochelate (Fe^{2+})@**293** by Scheme 2.199 with spacer-containing thiol-terminated *closo*-decaborate dianion $[B_{10}H_9OCH_2CH_2OCH_2CH_2SH]^{2-}$ in the

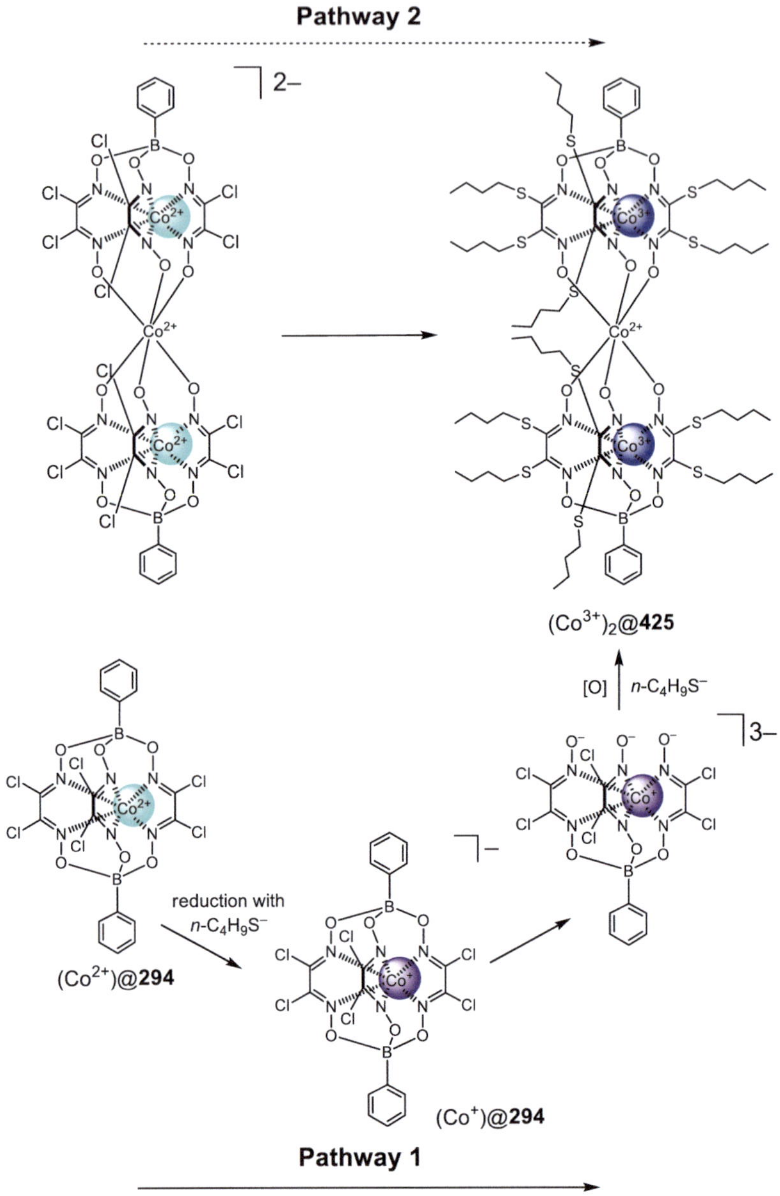

Scheme 2.192 Formation of the bis-clathrochelate $(Co^{3+})_2$@**425**

presence of triethylamine as organic base for in situ deprotonation of its thiol group gave a hybrid hexafunctionalized cage complex, the hexa-*closo*-decaborate iron(II) clathrochelate dodecaanion (Fe^{2+})@**438** [142].

Aromatic nucleophilic substitution of the dibromoclathrochelate precursor (Fe^{2+})@**439** (see section "Electrophilic Addition and Other Reactions of Inherent Ribbed Substituents and Clathrochelate Frameworks") with the derivative of piperazine as a *N*-nucleophile in the presence of triethylamine has been used [111] for preparation of the monofunctionalized iron(II) monobromoclathrochelate (Fe^{2+})@**440** in a moderate yield by Scheme 2.200; the corresponding difunctionalized cage complex has not been detected even with a threefold excess of this nucleophilic agent at high temperature.

The C–C conjugated iron(II) dihalogeno-bis-clathrochelates have been used [143] as the reactive bis-macrobicyclic precursors that easily underwent their nucleophilic substitution with

Scheme 2.193 Formation of the bis-clathrochelate $(Co^{3+})_2$@**426**

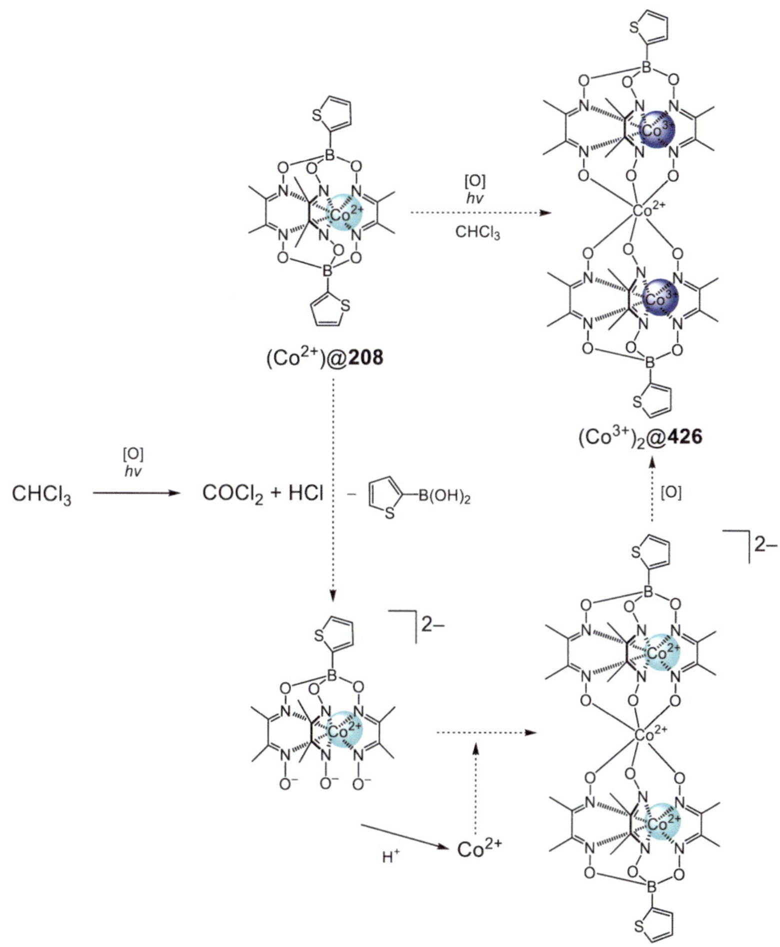

Scheme 2.194 Synthesis of the iron and cobalt(II) hexa(polyfluorophenylsulfide) clathrochelates

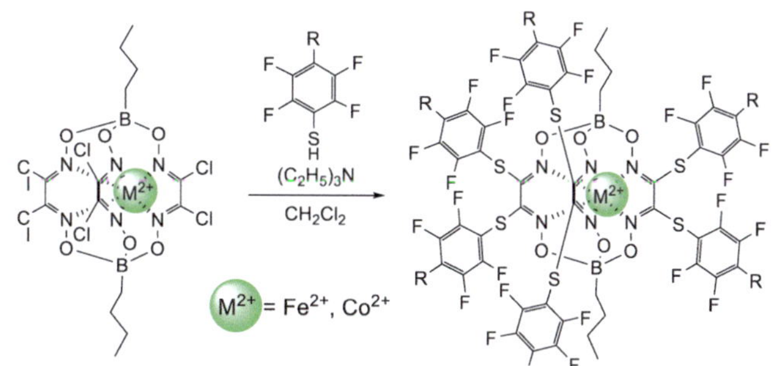

(M^{2+})@**293**

R = F ((M^{2+})@**427**), Cl ((M^{2+})@**428**, CF_3 ((M^{2+})@**429**)

various *N*,*S*-nucleophiles, thus giving the ribbed-functionalized bis-cage complexes; the performed reactions of the dibromo-bis-clathrochelate $(Fe^{2+})_2$@**441** are shown in Scheme 2.201.

The hybrid calixarenoclathrochelate conjugates with alkylsulfide linkers have been prepared [144, 145] by Scheme 2.202 using nucleophilic substitution of (Fe^{2+})@**302** with thiolate anions, the derivatives of calix[4]arene

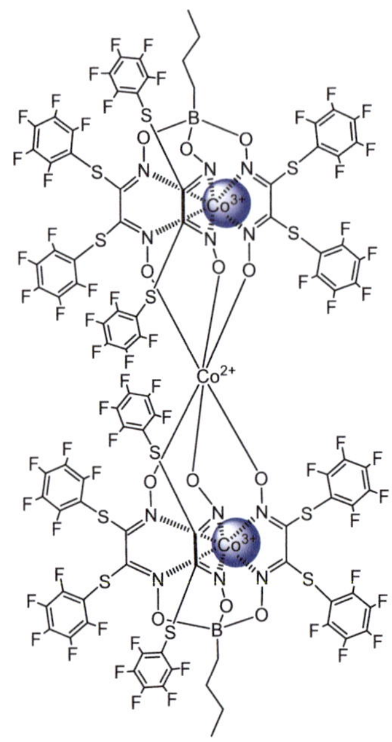

$(Co^{3+})_2$@**430**

Scheme 2.195 Chemical drawing of the bis-clathrochelate $(Co^{3+})_2$@**430**

ligand synthons with pendant thiol-terminated functionalizing substituents; the yields of the above hybrid complexes are affected by a length of their methylene spacers [145].

Free-Radical Substitution

Substitution of chlorine atom(s) of the iron(II) dichloroclathrochelate (Fe^{2+})@**302** by 1,4-dioxan-2-yl group(s) has been unexpectedly observed [146] in the course of an attempted alkylation of this macrobicyclic precursor with acetylacetone by Scheme 2.203 in the presence of metallic sodium in boiling 1,4-dioxane. Its major macrobicyclic product, the difunctionalized clathrochelate (Fe^{2+})@**452**, is a by-product of several syntheses that used a reflux of (Fe^{2+})@**302** in 1,4-dioxane medium. The monofunctionalized analog (Fe^{2+})@**453** has been isolated [146] from the products of an attempted hydroxyamination of (Fe^{2+})@**302** with $NH_2OH \cdot HCl$ – metallic sodium system in 1,4-dioxane. Metallic sodium or other components were found [146] to not play any significant role in this reaction, having a radical nature. It is triggered by radical derivatives of 1,4-dioxane hydroperoxide: the absence of nucleophilic agents in the reaction mixture excludes the possible functionalization of this precursor by nucleophilic substitution. The clathrochelates (Fe^{2+})@**452** and (Fe^{2+})@**453** have been formed during a reflux of (Fe^{2+})@**302** in crude 1,4-dioxane or in purified 1,4-dioxane that has been exposed on air, but they were not found in a freshly purified 1,4-dioxane under inert atmosphere, thus suggesting that the formation of the intermediate radical species resulted from the decomposition of the peroxide derivatives of 1,4-dioxane. The plausible pathway of this radical reaction [146] is shown in Scheme 2.204. On its first stage, the 1,4-dioxane hydroperoxide (or *tert*-C_4H_9OOH that can be also used as a radical initiator) gave the reactive radical species upon their thermal decomposition, abstracting the hydrogen atom from the solvent and producing the nucleophilic 1,4-dioxan-2-yl radicals. They attack an azomethine C=N fragment of a clathrochelate framework, thus giving the macrobicyclic intermediate (Fe^{2+})@**454** (Scheme 2.204, stage III; only its dichloroglyoximate chelate fragment is

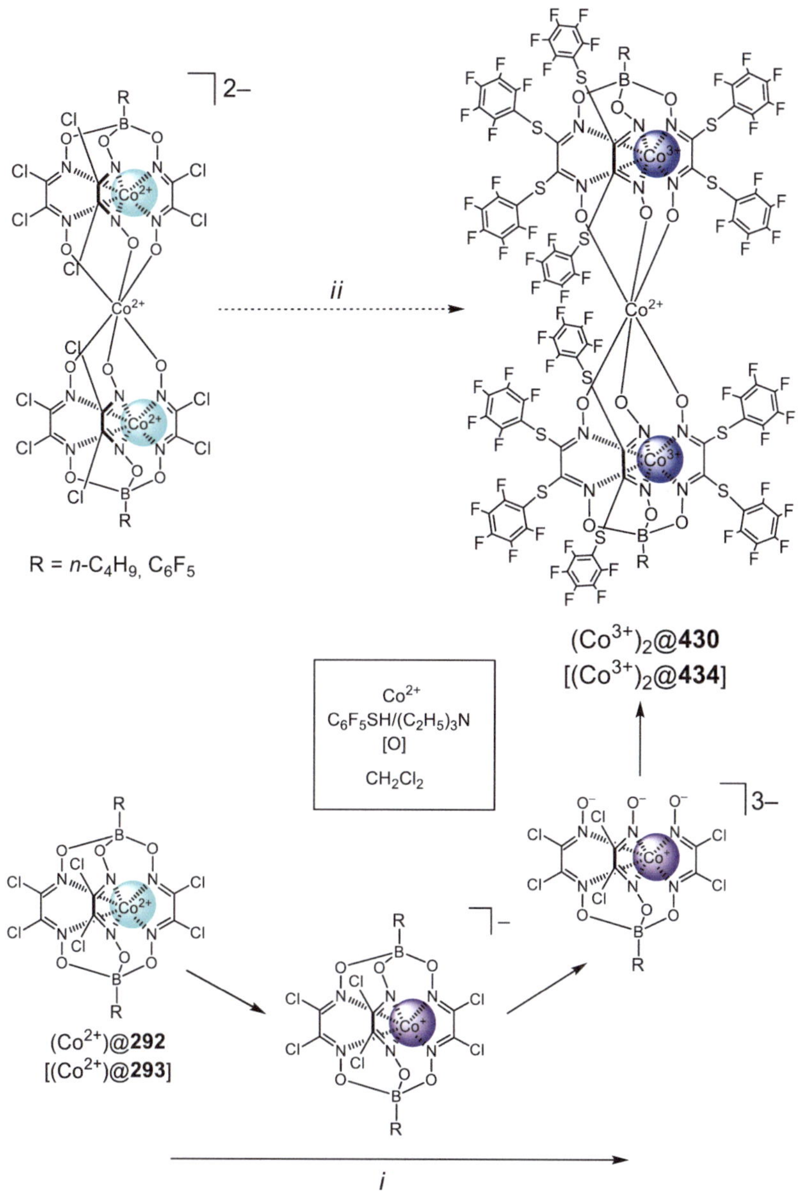

Scheme 2.196 Formation of the bis-clathrochelates $(Co^{3+})_2$@**430** and $(Co^{3+})_2$@**434**

shown). An addition of the radical species to the corresponding carbon atom is followed by β-elimination, resulting in the chlorine radical and the monoribbed-functionalized clathrochelate product (Fe^{2+})@**453**. So, this reaction [146] proceeds as a free-radical addition–elimination process that formally resulted in the above substitution. Moreover, a prolonged reflux of (Fe^{2+})@**302** in crude 1,4-dioxane in the presence of metallic sodium gave not only the above cage complexes (Fe^{2+})@**452** and (Fe^{2+})@**453**, the products of its substitution with 1,4-dioxan-2-yl radicals, but also the clathrochelate by-product (Fe^{2+})@**455** (Scheme 2.205), the ribbed substituent of which resulted from 1,4-dioxane ring opening through

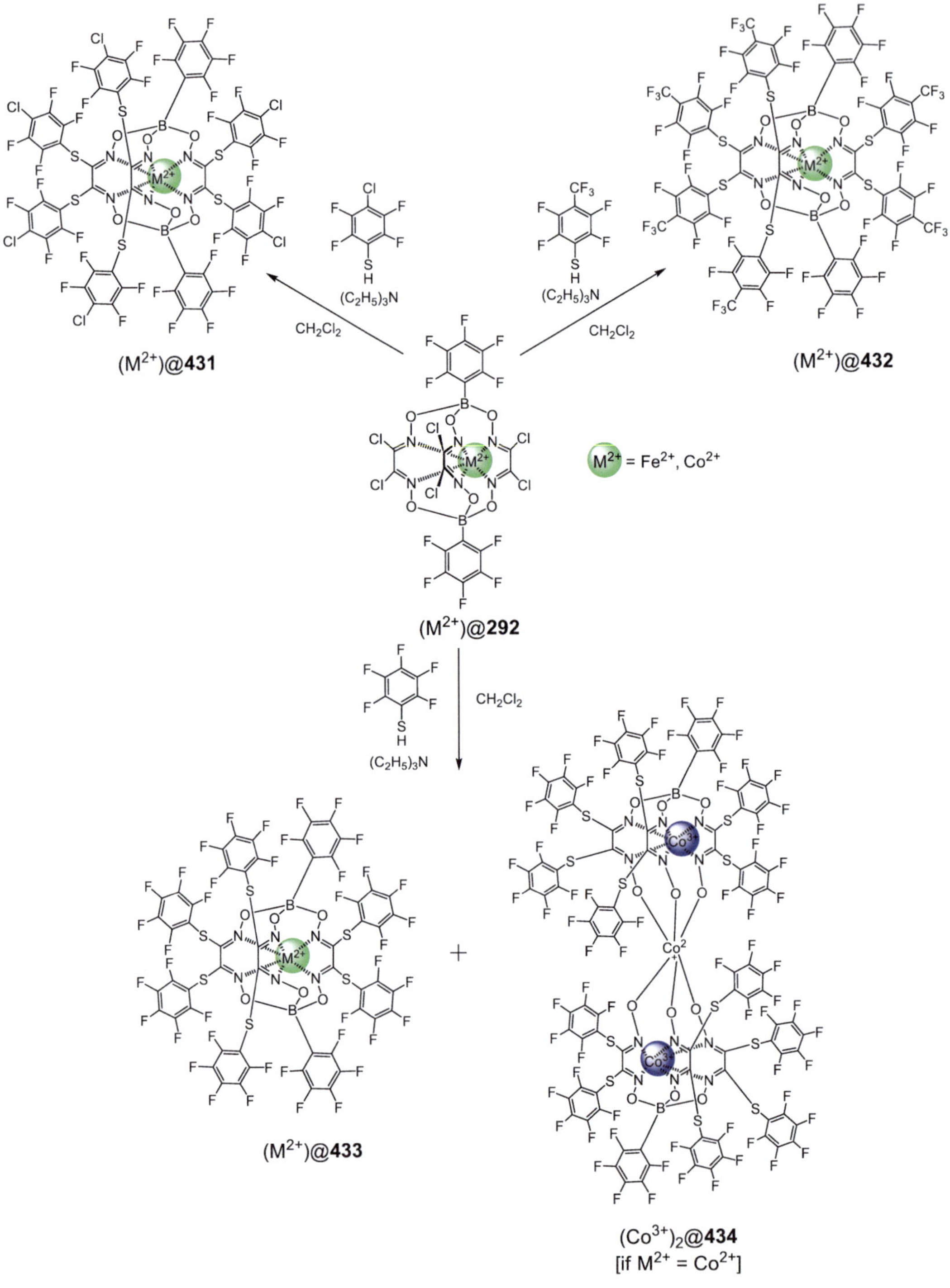

Scheme 2.197 Nucleophilic substitution of the pentafluorophenylboron-capped iron and cobalt(II) hexachloroclathrochelates

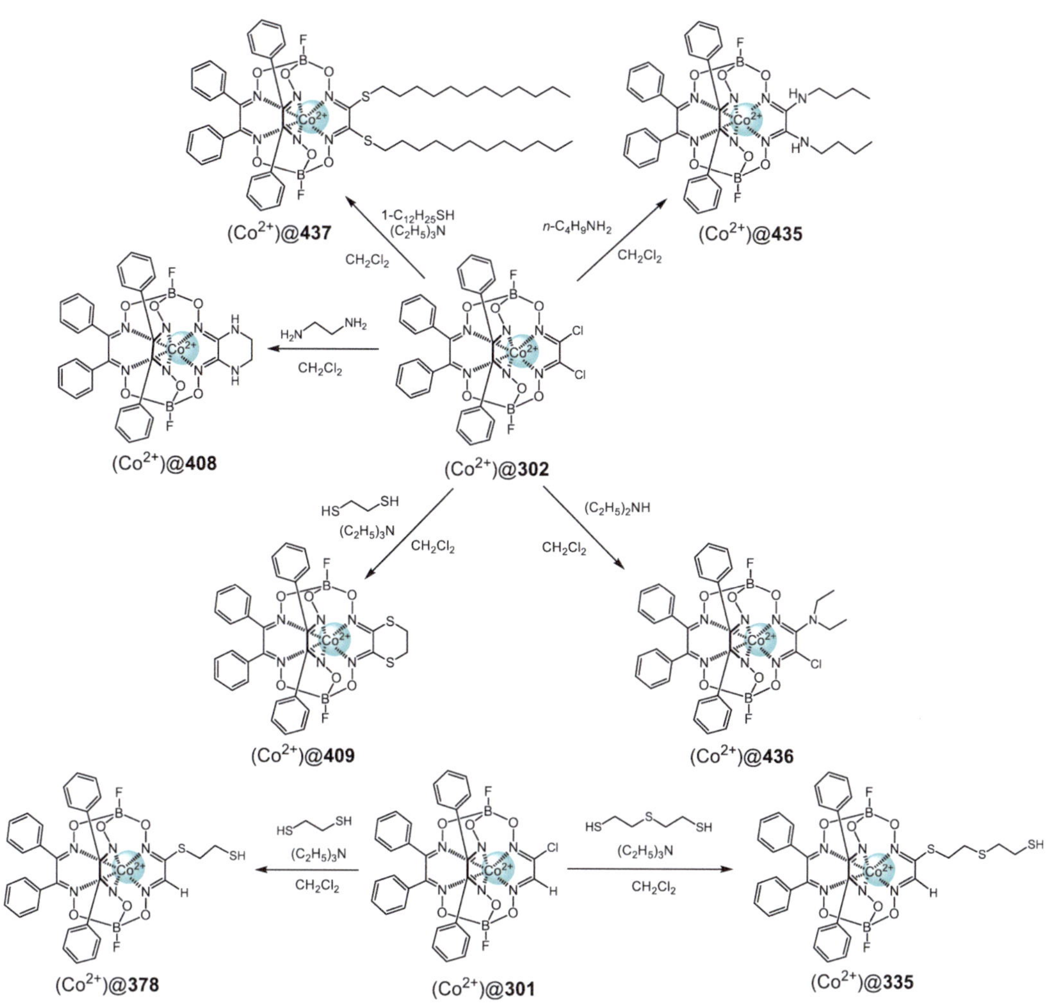

Scheme 2.198 Nucleophilic substitution of the cobalt(II) mono- and dichloroclathrochelates

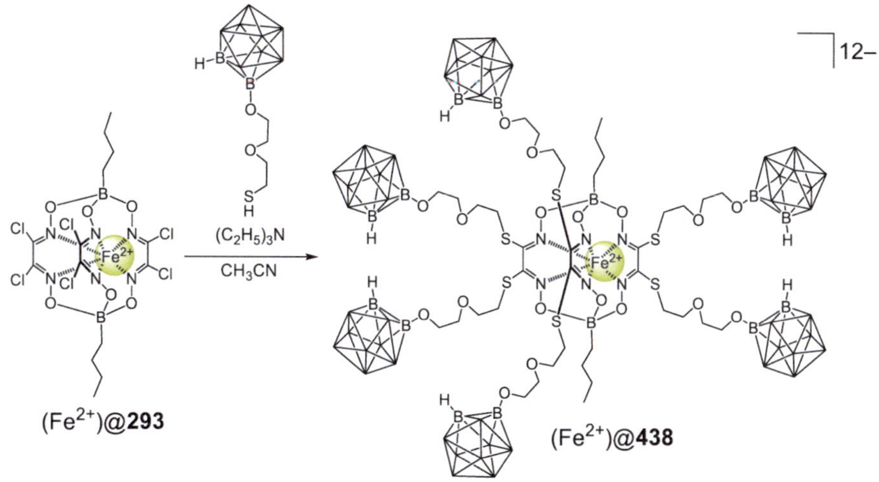

Scheme 2.199 Synthesis of the hexa-*closo*-borate iron(II) cage complex

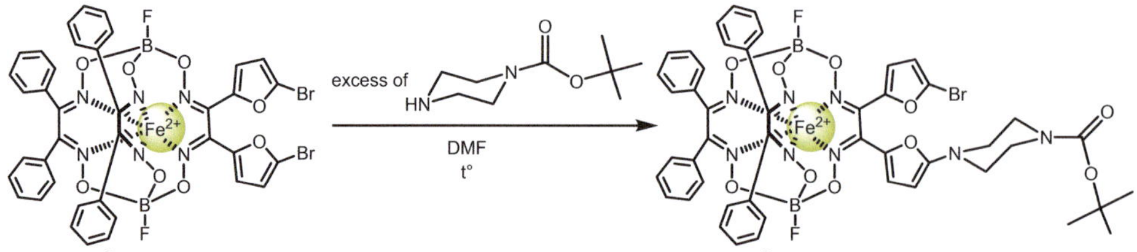

Scheme 2.200 Nucleophilic substitution of the dibromoclathrochelate precursor (Fe^{2+})@**439**

an oxidative cleavage of the carbon–carbon bond in the functionalized ribbed fragment of a macrobicyclic encapsulating ligand [147].

Radical substitution of one chlorine atom of (Fe^{2+})@**302** with cyclohexyl radical using *tert*-butyl hydroperoxide as a free-radical initiator has been performed [148] by Scheme 2.206, thus giving the cage complex (Fe^{2+})@**456** in a moderate yield.

The monotetrahydrofuryl-functionalized clathrochelate (Fe^{2+})@**457** has been prepared [149] in a high yield by Scheme 2.207 using a reflux of (Fe^{2+})@**302** in THF in the presence of *tert*-butylhydroperoxide as a radical initiator; benzoyl peroxide and 1,1'-azobis (cyclohexanecarbonitrile) can also be used as the radical initiators. This reaction proceeded even in the presence of air oxygen, thus suggesting that a quasiaromatic cage complex (Fe^{2+})@**302** successfully competes as a radical scavenger with O_2 for the radical derivatives of THF. The experiments in inert atmosphere with substoichiometric amounts of the above radical initiators showed that this reaction has a chain character, its average propagation length is approximately 2.5, and it proceeds through 2-C atom of tetrahydrofuryl substituent with a high regioselectivity due to the stabilization of a tetrahydrofur-2-yl radical by its adjacent oxygen atom [149]. Radical substitution of the same iron(II) dichloroclathrochelate with 1,3-dioxolane by Scheme 2.208 gave the corresponding monofunctionalized cage complex (Fe^{2+})@**458** in a high yield [150]. This reaction with 1,3-dioxanyl radical also proceeded through its 2-C atom with a high regioselectivity due to the stabilization of 1,3-dioxan-2-yl radical species by two adjacent oxygen atoms [150].

Free-radical substitution of (Fe^{2+})@**302** with generated in situ radical derivatives of ethanol and *iso*-propanol by Scheme 2.209 using *tert*-C_4H_9OOH as a thermal radical initiator has been used [151] for preparation of the monofunctionalized iron(II) clathrochelates (Fe^{2+})@**459** and (Fe^{2+})@**460** with a single inherent alcohol C-substituent. These alcohols interacted with reactive radical species through the hydrogen abstraction at their carbon atoms adjacent to the hydroxyl group; in this case, the formed radical species are stabilized by α-oxygen atom. These radicals $(CH_3)_2(OH)C^{\bullet}$ and $(CH_3)(H)(OH)C^{\bullet}$ are highly nucleophilic and underwent an addition–elimination reaction with a quasiaromatic macrobicyclic framework of (Fe^{2+})@**302** at azomethine group in its dichloroglyoximate ribbed fragment, thus giving the above monofunctionalized clathrochelate products in the yields from a moderate to a high. Higher yield of an *iso*-propanol clathrochelate derivative, the complex (Fe^{2+})@**460**, is explained [151] by higher nucleophilicity and stability of the corresponding radical species; the radical trapping with air oxygen did not severely compete with their addition to a polyazomethine cage framework.

The synthesis of the monoribbed-difunctionalized macrobicyclic derivative of ethanol, the cage complex (Fe^{2+})@**461**, has been performed [152] by Scheme 2.210.

Reaction of the monodimethylglyoximate bis-α-benzildioximate iron(II) clathrochelate (Fe^{2+})@**296** with $(C_2H_5)_3B–O_2$ system as a "soft" radical source by Scheme 2.211 [105] gave mainly the macrobicyclic product (Fe^{2+})@**462** of a reductive addition of two carbon-centered radicals to azomethine group of the

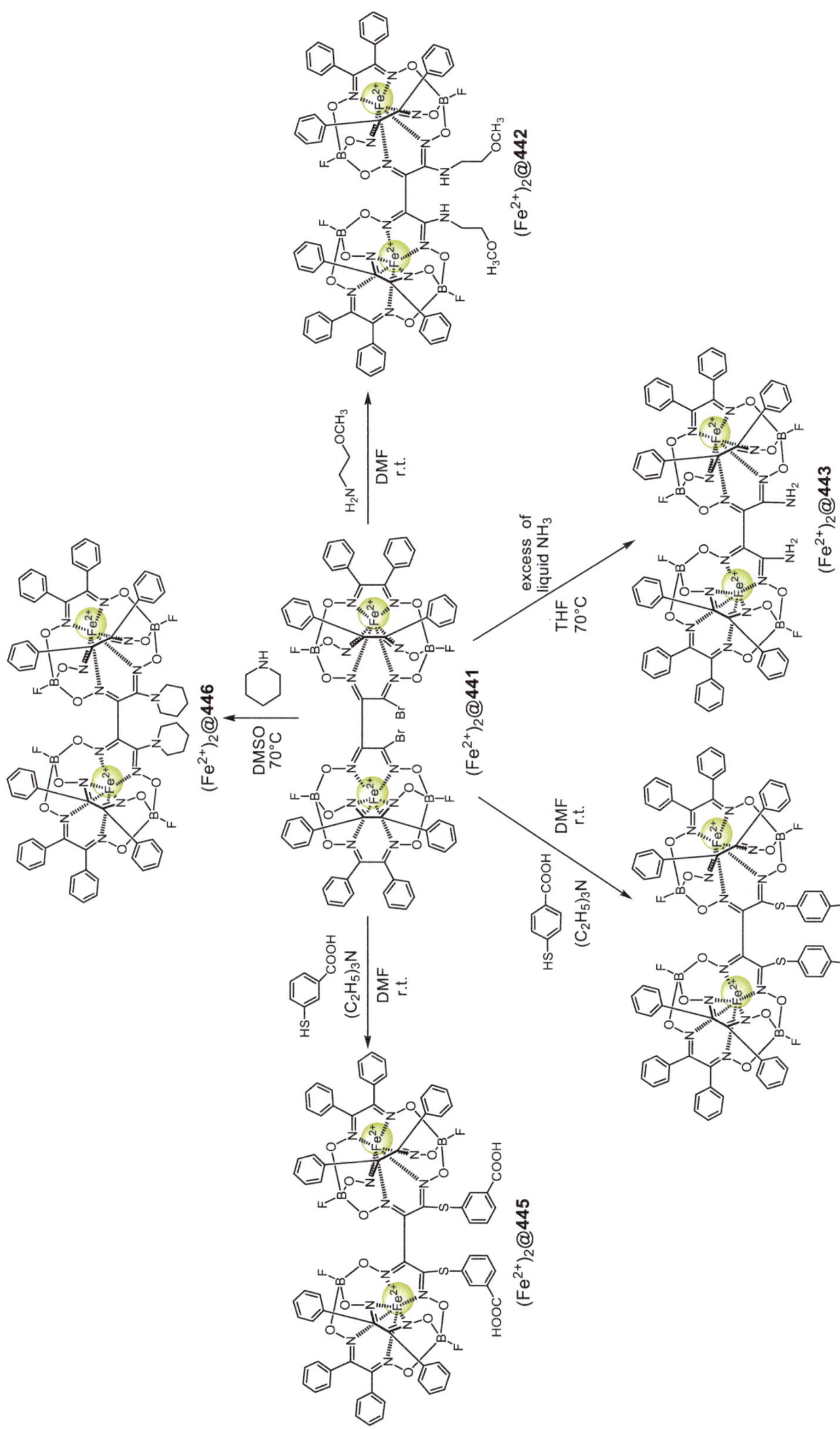

Scheme 2.201 Nucleophilic substitution of the dibromo-bis-clathrochelate precursor $(Fe^{2+})_2$@**441**

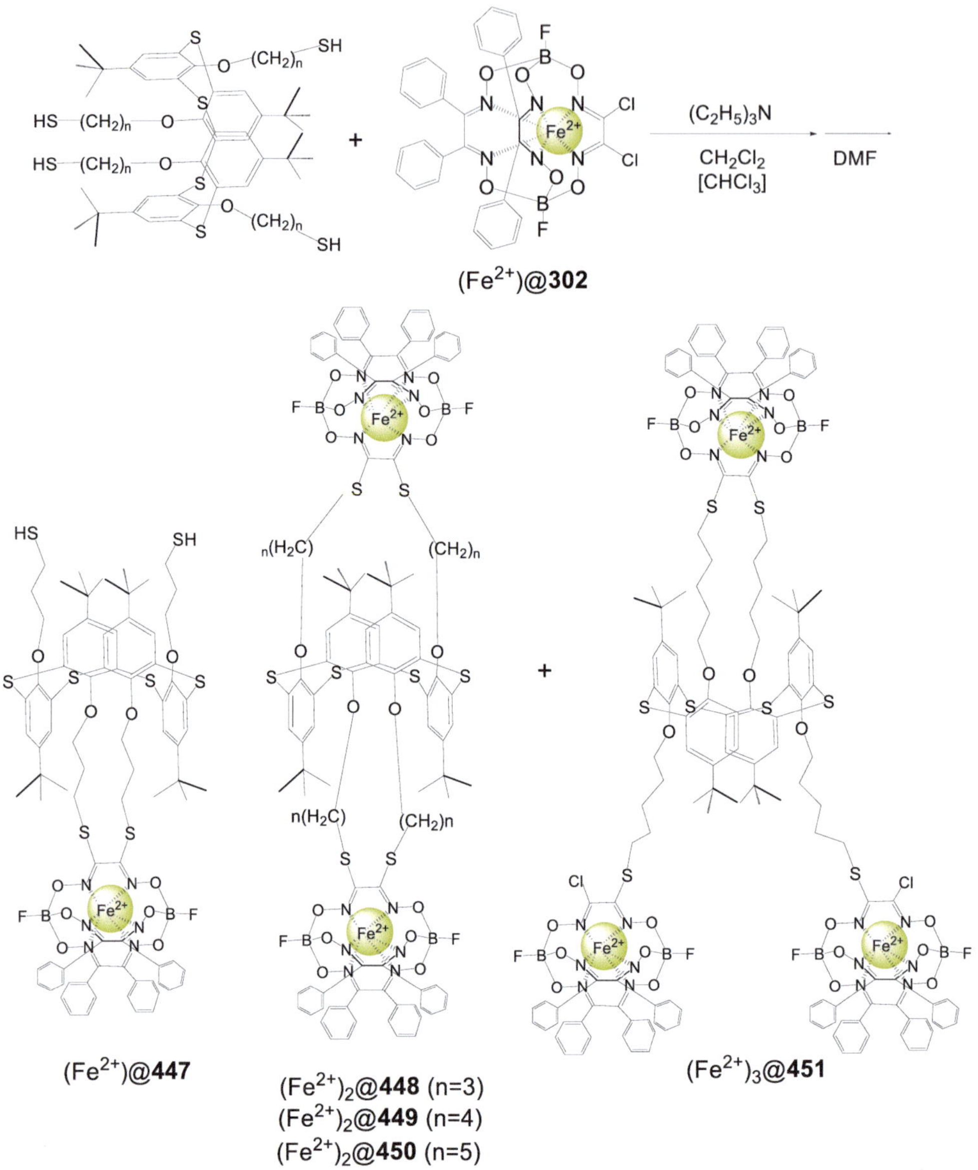

Scheme 2.202 Synthesis of the hybrid iron(II) calixarenoclathrochelates

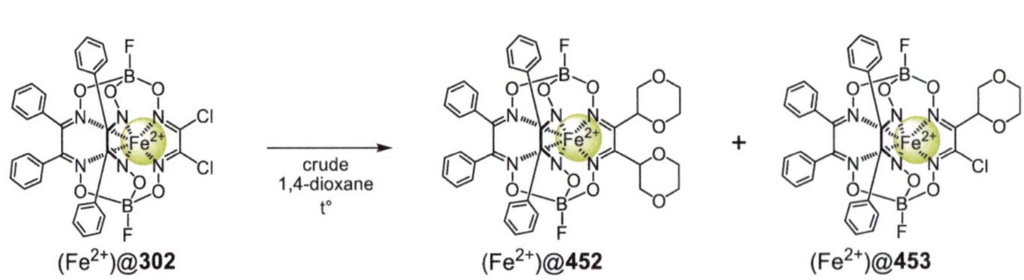

Scheme 2.203 Free-radical substitution of an iron(II) dichloroclathrochelate with 1,4-dioxan-2-yl radical(s)

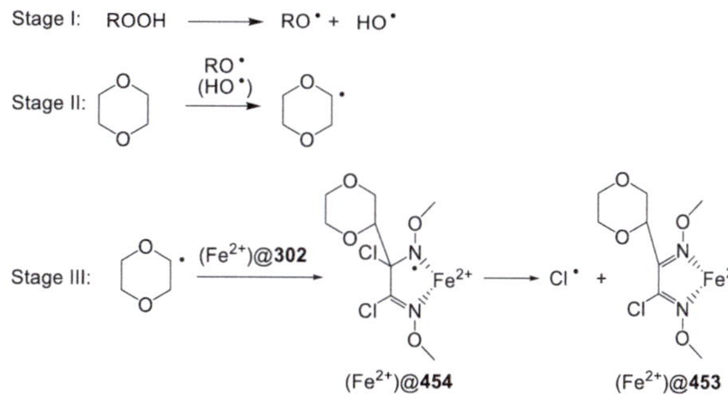

Scheme 2.204 Plausible pathway of a free-radical substitution of (Fe^{2+})@**302**

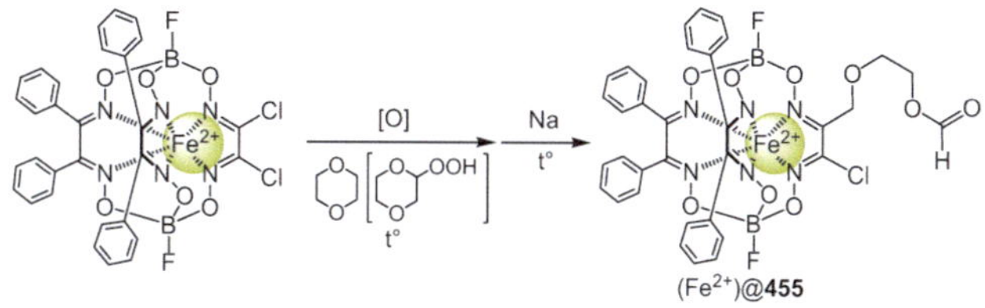

Scheme 2.205 Formation of the clathrochelate product (Fe^{2+})@**455** of a ring opening reaction

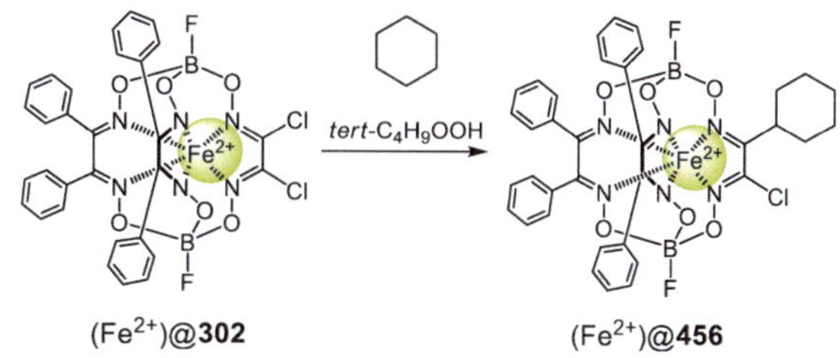

Scheme 2.206 Free-radical substitution of an iron(II) dichloroclathrochelate with cyclohexyl radical

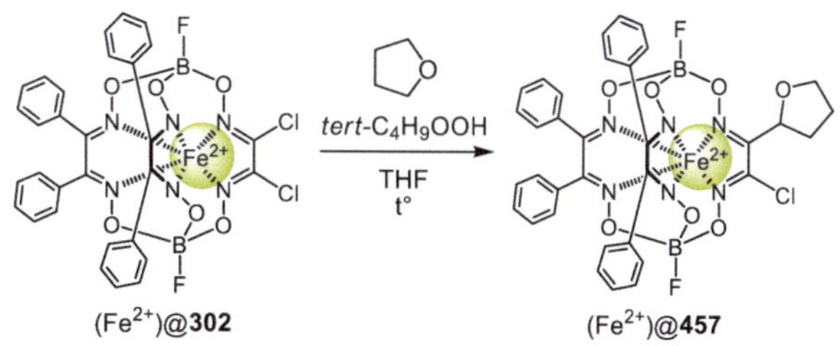

Scheme 2.207 Free-radical substitution of an iron(II) dichloroclathrochelate with tetrahydrofuryl radical

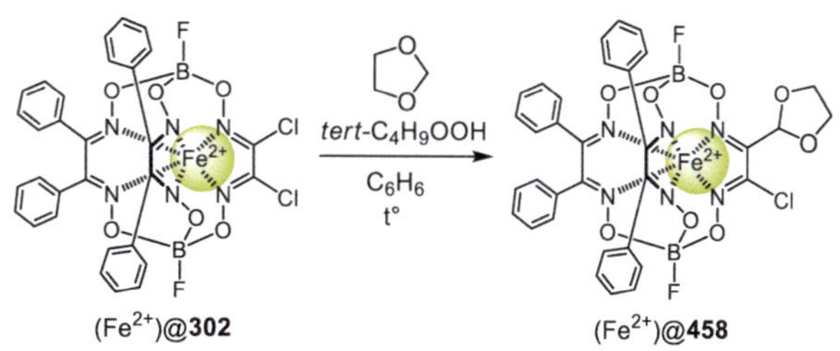

Scheme 2.208 Free-radical substitution of an iron(II) dichloroclathrochelate with 1,3-dioxanyl radical

dimethylglyoximate chelate fragment of the above clathrochelate precursor. The resulting macrobicyclic ligand **462** contains both the oxime and hydroxylamine donor groups, thus suggesting an unusual reduction of one of the donor oxime groups of (Fe^{2+})@**296** to the

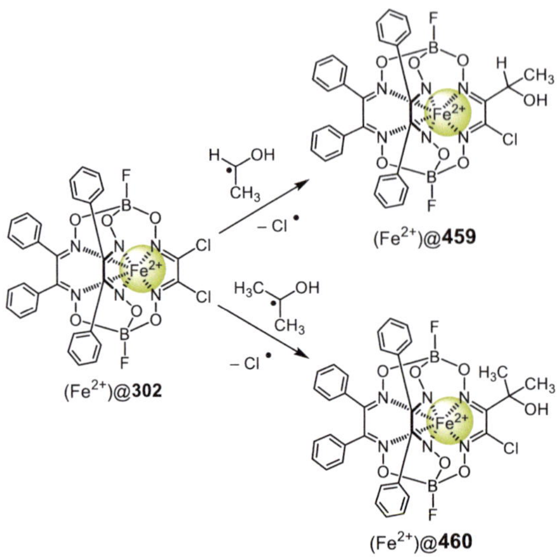

Scheme 2.209 Free-radical substitution of an iron(II) dichloroclathrochelate with radical derivatives of ethanol and *iso*-propanol

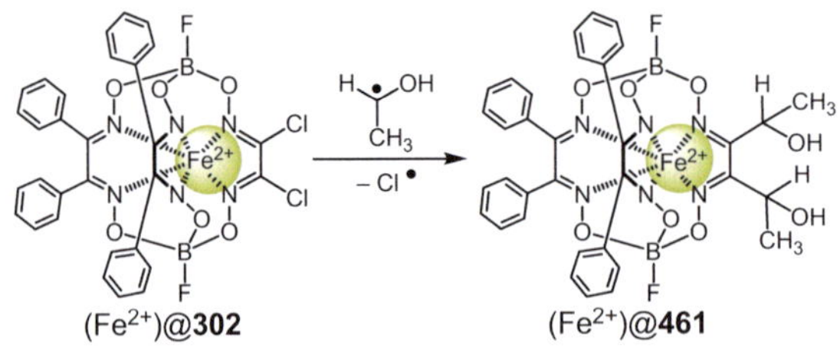

Scheme 2.210 Synthesis of the difunctionalized iron(II) cage complex (Fe^{2+})@**461**, the derivative of ethanol

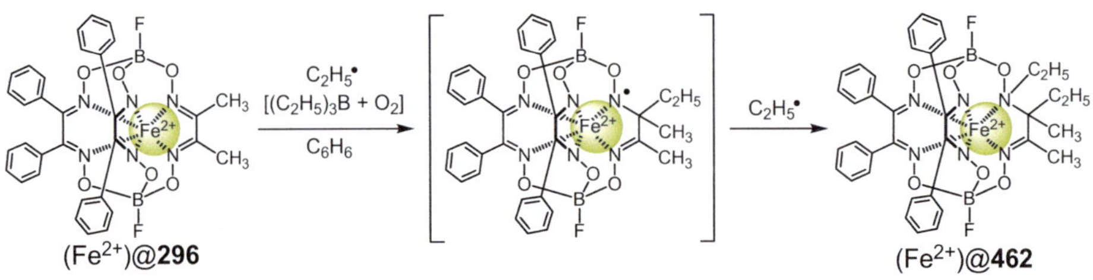

$$(C_2H_5)_3B + O_2 \longrightarrow (C_2H_5)_2BOO^\bullet + C_2H_5^\bullet$$

Scheme 2.211 Reaction of the monodimethylglyoximate clathrochelate (Fe^{2+})@**296** with radical derivatives of triethylboron

hydroxylamine one. The plausible pathway to (Fe^{2+})@**462** includes (1) the reaction of triethylborane with air oxygen yielding the ethyl radicals (see Scheme 2.211), (2) their trapping by azomethine group of (Fe^{2+})@**296** occurring exclusively at its dimethylglyoximate chelate fragment, and (3) as methyl radical is a poor leaving group, the radical adduct formed is reported [105] to be long-lived enough to trap the second ethyl radical, thus giving the macrobicyclic product (Fe^{2+})@**462**.

Macrobicyclic products of the reactions of (Fe^{2+})@**302** with a simplest carbon-centered methyl radical, generated in situ in DMSO–H_2O–H_2O_2–Fe^{2+} system, have been studied [153]. These reactions (Scheme 2.212) gave

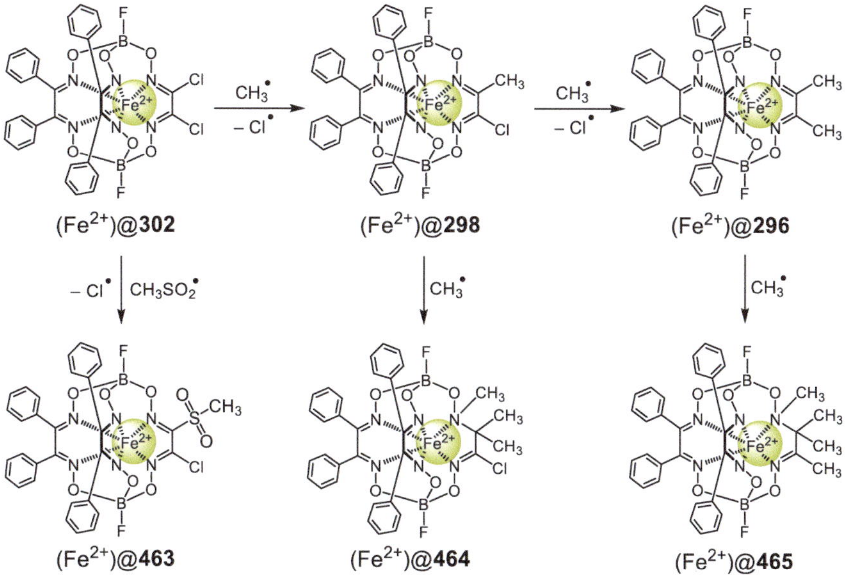

Scheme 2.212 Reactions of the mono- and dichloroclathrochelate precursors with methyl radical(s) and their dimethylsulfonyl analog

mainly the clathrochelate products resulting from methylation of the dichloroglyoximate chelate fragment of its caging ligand. Both their nature and yields are reported [153] to be affected by concentration of H_2O_2. At its low concentrations, a degree of the conversion of such macrobicyclic precursor was found to be a rather small, giving two major macrobicyclic products, the methylchloroglyoximate complex (Fe^{2+})@**298** and the clathrochelate (Fe^{2+})@**463** with methylsulfonyl ribbed substituent. At high H_2O_2 concentrations, the yields of the macrobicyclic products of this free-radical methylation reaction, dominating over nucleophilic substitution, were found to be higher; the initial macrobicyclic products (Fe^{2+})@**298** and (Fe^{2+})@**296**, resulting from the subsequent attacks by methyl radicals on the chloroglyoximate group(s) of (Fe^{2+})@**302**, underwent further methylation of their already substituted azomethine fragment, thus giving the oximatohydroxylamine cage complexes (Fe^{2+})@**464** and (Fe^{2+})@**465** [153].

Metal-Promoted (Catalyzed) Reactions

Nucleophilic substitution of the halogenoclathrochelate precursors with sterically unhindered aliphatic amines and thiolate anions typically proceeds easily and in high yields (see section "Nucleophilic Substitution"). On the other hand, such substitution with soft nucleophilic agents requires an activation of the leaving halogen atoms and an increase of the nucleophilicity of these agents without the use of the strong bases, causing the complete destruction of a clathrochelate framework [154]. So, the convenient pathways of the ribbed functionalization of the tris-dioximate clathrochelates, based on nucleophilic substitution of these macrobicyclic precursors with soft *N*- and *O*-nucleophiles in the presence of cadmium diamides as the promoters, have been developed [154]. Two cadmium (II) diamides, obtained from one-pot reactions of diisopropylamine or hexamethyldisilazane with *n*-butyl lithium followed by the reactions of their lithium derivatives with $CdCl_2$ by Scheme 2.213, have been used [154] for activation of nucleophilic substitution of the reactive chlorine atoms of the di- and hexachloroclathrochelates, such as (Fe^{2+})@**302** and (Fe^{2+})@**293**. These cadmium(II) derivatives activated their C–Hal bond both by coordination of their halogen atoms toward cadmium(II) ion (see Scheme 2.214) and through the formation of anionic

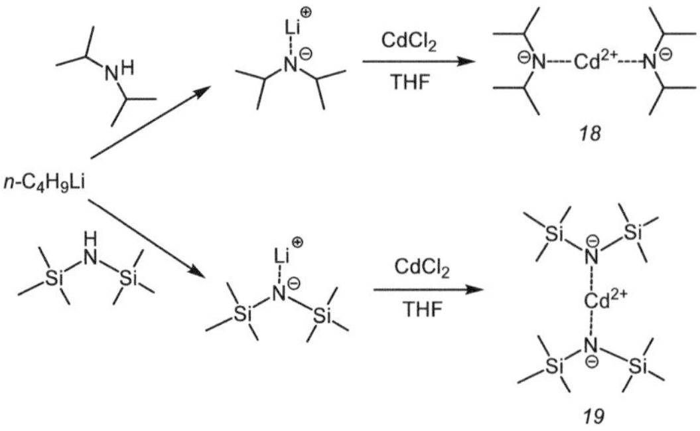

Scheme 2.213 Preparation of cadmium(II) amides

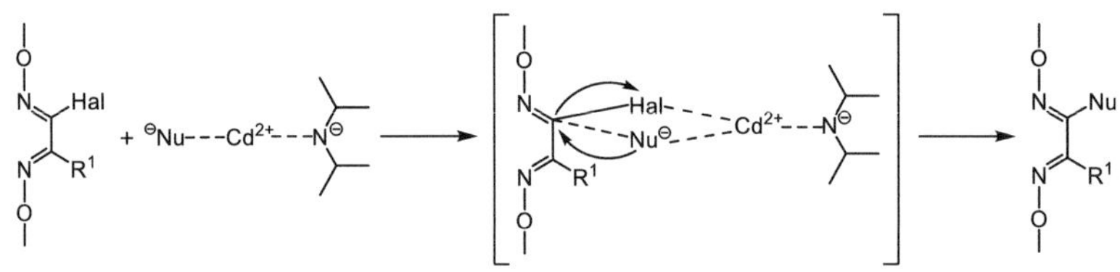

Scheme 2.214 Activation of the clathrochelate's C–Hal bond by cadmium(II) amide

forms of the corresponding nucleophilic agent with cadmium amides, thus increasing an electron density on its reactive site. The in situ generated cadmium amides and alcoholates are low-basic reagents allowing to obtain the clathrochelate derivatives of soft nucleophilic agents, such as alcohols (including the fluorinated ones) and aromatic amines (including those with electron-withdrawing substituents), in moderate yields using one-pot synthetic procedures under mild reaction conditions. In particular, this synthetic approach has been used [154] for preparation of the triribbed-functionalized clathrochelates with six fluorinated alkoxyl groups and the monoribbed-functionalized cage complexes with *n*-butoxyl, crown ether, aromatic amine, carboxymethyl- and nitro-containing chelate substituents shown in Schemes 2.215, 2.216, and 2.217.

Cadmium-promoted *C*-nucleophilic substitution has been used [155] for preparation of the macrobicyclic complex (Fe^{2+})@**474** with pendant ribbed dicyanopyrazine substituent by Scheme 2.218. Carbanion condensation of the dichloroclathrochelate (Fe^{2+})@**302** with 2,3-dicyano-5,6-dimethylpyrazine in the presence of the cadmium(II) amide *19* as a promoter gave the monofunctionalized clathrochelate (Fe^{2+})@**474** as its major product. Only one pair of the carbanionic and electrophilic reactive centers has been involved in this reaction, whereas the macrobicyclic product of double condensation of a *vic*-dihalogenoglyoximate fragment with *ortho*-dimethylsubstituted pyrazine molecule has not been detected. This result is explained [155] by formation on its first stage of an intermediate dicyanopyrazine carbanion, which is stabilized by coordination of its heterocyclic nitrogen atom toward cadmium(II) ion. The clathrochelate (Fe^{2+})@**474** is a strong CH-acid: the addition of aliphatic amines to its acetonitrile solution caused the change of color from yellowish-brown to violet due to the formation of a macrobicyclic anion with highly conjugated polyazomethine quasiaromatic system [155].

The same synthetic approach using cadmium(II) amides as promoters of nucleophilic substitution has been successfully used [108] for functionalization of the cobalt(II) dichloroclathrochelate (Co^{2+})@**302**: its nucleophilic substitution with *n*-butanol in the presence of *19* by Scheme 2.219 gave the monoribbed-functionalized cobalt(II) macrobicyclic complex (Co^{2+})@**471** with two inherent di-*n*-butoxyl substituents.

Copper(I)-promoted halogen exchange reaction of the ribbed chlorine atoms of the iron(II)

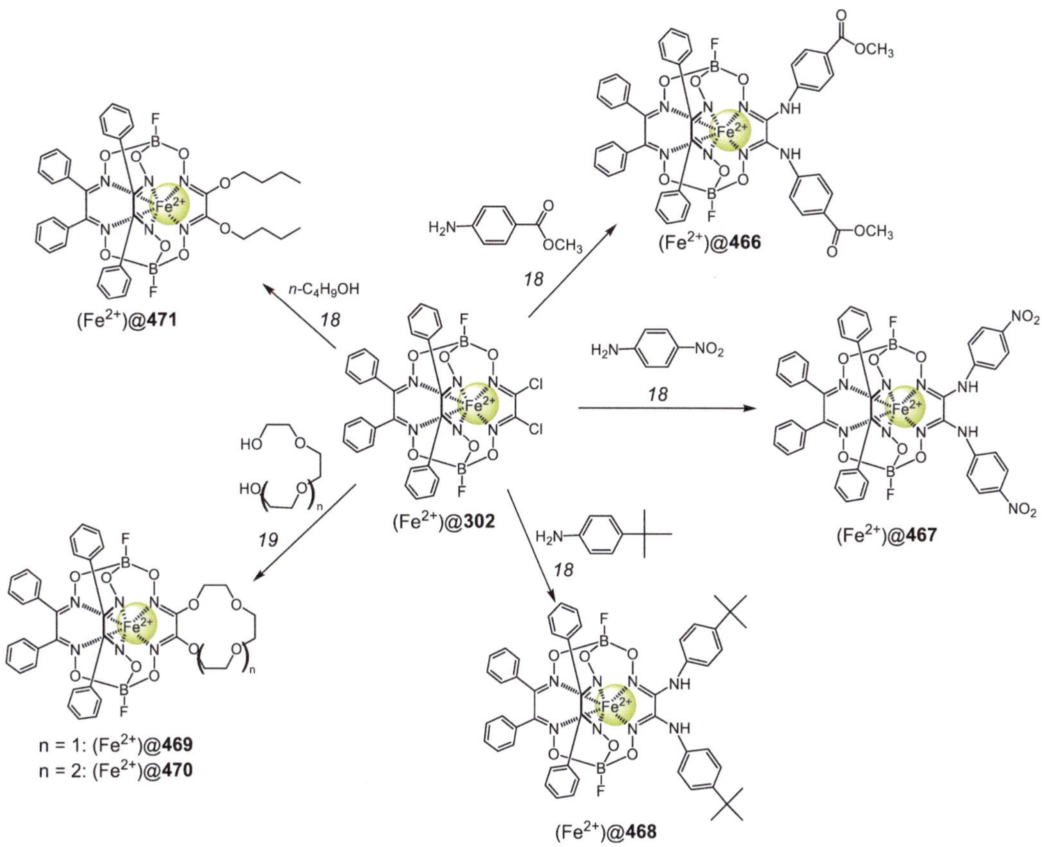

Scheme 2.215 Cadmium-promoted nucleophilic substitution of the dichloroclathrochelate (Fe^{2+})@**302**

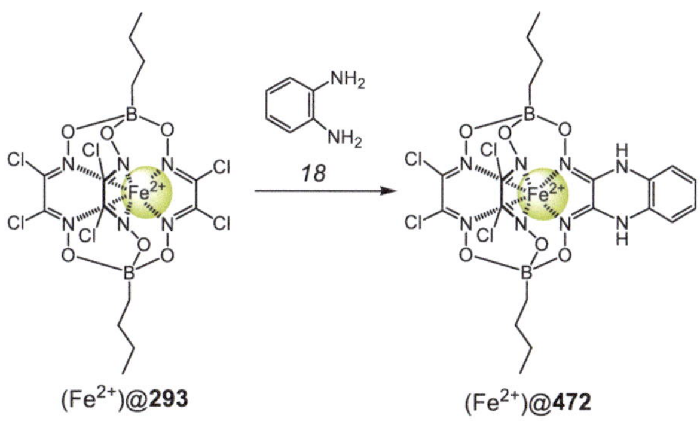

Scheme 2.216 Cadmium-promoted nucleophilic substitution of the hexachloroclathrochelate (Fe^{2+})@**293**

dichloroclathrochelate (Fe^{2+})@**302** has been studied [156]. Its reaction with twofold excess of CuI by Scheme 2.220 gave the mixture of the diiodo- and iodochloroclathrochelates, which have been chromatographically isolated as individual compounds. This reaction occurred with CuI playing the roles of both the catalyst and the reagent, and only in the case of the *vic*-dihalogen-substituted macrobicyclic complexes. Using the complex $Na[CuI_2(NMP)_n]$, generated in situ, with its iodide anions coordinated toward copper(I) ion is reported [156] to be important for easy proceeding of the above halogen exchange reaction and for the formation of the target

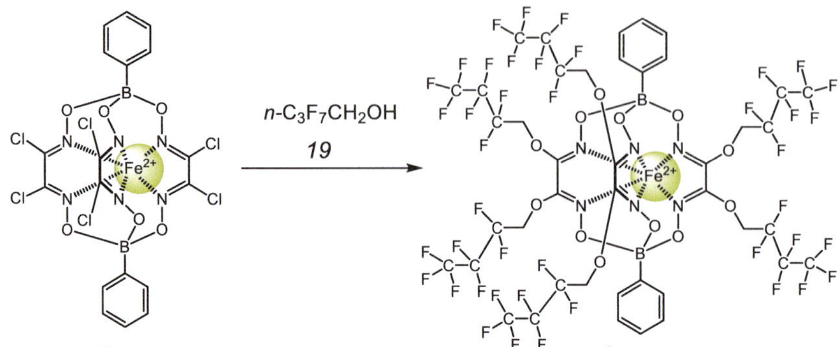

Scheme 2.217 Cadmium-promoted nucleophilic substitution of the hexachloroclathrochelate (Fe^{2+})@**294**

clathrochelate product as well. The plausible pathway of this process is shown in Scheme 2.221; it includes the formation of the intermediate anion-radical macrobicyclic species, which have been effectively generated by the above CuI–*N*-methylpyrrolidone complex [156].

2,5-dithiahexane-assisted perfluoroalkylation of the iron(II) diiodoclathrochelate (Fe^{2+})@**309** (see section "Template Synthesis and Condensation Reactions") by Scheme 2.222 has been performed [157]. Trifluoromethylation of this quasiaromatic macrobicyclic precursor with trifluoromethyl copper(I) using 2,5-dithiahexane as a co-ligand gave the mixture of the mono- and bis-trifluoromethylated clathrochelate products (Fe^{2+})@**476** and (Fe^{2+})@**477** in a moderate yield.

Perfluoroarylation of the iron(II) di- and hexaiodoclathrochelates (Fe^{2+})@**309** and (Fe^{2+})@**288** has been performed [100]. Their reactions with pentafluorophenyl copper(I) by Schemes 2.223 under mild reaction conditions afforded the mono-, di-, and hexaperfluoroarylated cage complexes (Fe^{2+})@**478**, (Fe^{2+})@**479**, and (Fe^{2+})@**289** with inherent pentafluorophenyl substituents [100].

Reaction of the same iron(II) diiodoclathrochelate (Fe^{2+})@**309** with a copper(I) cyanide–triphenylphosphine complex by Scheme 2.224 in acetonitrile medium under mild reaction conditions has been used [112] for preparation of the iron(II) clathrochelates (Fe^{2+})@**480** and (Fe^{2+})@**481** with one and two inherent cyano group(s), respectively. The dinitrile cage complex (Fe^{2+})@**481** was only a minor product of this reaction, whereas its major monomethine mononitrile macrobicyclic product resulted from the side dehalogenation–reduction process. The plausible pathway of this tandem process that caused the substitution of one of the two iodine atoms by cyano group and the second one by hydrogen atom is shown in Scheme 2.225. At its first stage, the iron(II) diiodoclathrochelate undergoes an anion-radical hydrodehalogenation with acetonitrile molecule as a source of the hydrogen atom. The monomethine monoiodine iron(II) clathrochelate (Fe^{2+})@**482**, the product of this hydrogen abstraction, is less reactive in a reductive rupture of its C–I bond because of a lower reduction potential of this clathrochelate intermediate, and therefore, it undergoes only the substitution of its iodine atom with a cyano group, but not the dehalogenation–reduction tandem reaction [112].

C-carboranylation of the same quasiaromatic iron(II) cage complex (Fe^{2+})@**309** has been performed [158] using a palladium-catalyzed cross-coupling reaction with lithium (1-*ortho*-carboranyl)trimethoxyborate as a carboranylating agent, generated in situ from equimolar amounts of lithium 1-*ortho*-carboranyl and trimethoxyborate. This one-pot reaction in the presence of catalytic amounts of the corresponding palladium(II)-based catalyst by Scheme 2.226 gave the ribbed-functionalized mono-*ortho*-carboranylclathrochelate (Fe^{2+})@**483**; no formation of its bis(carboranylated) analog has been observed

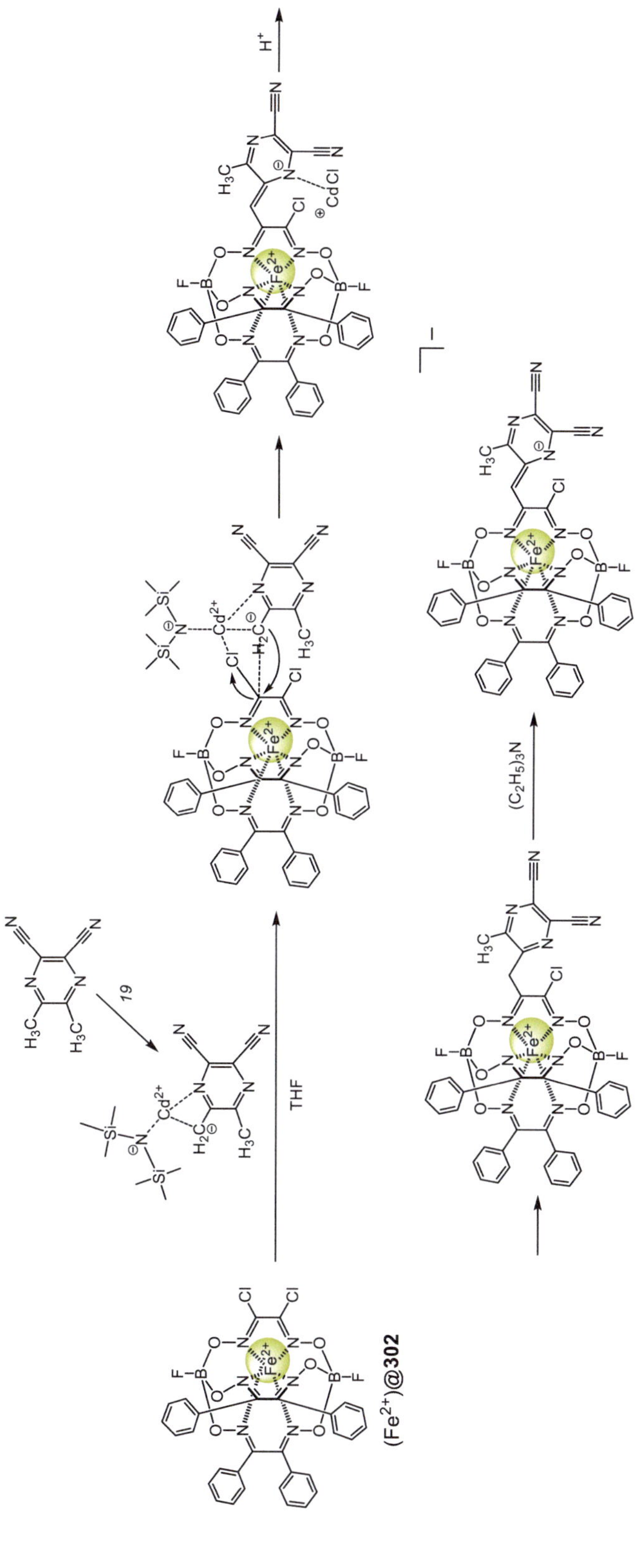

Scheme 2.218 Cadmium-promoted *C*-nucleophilic substitution of (Fe^{2+})@**302**

Scheme 2.219 Cadmium-promoted *O*-nucleophilic substitution of the cobalt(II) dichloroclathrochelate

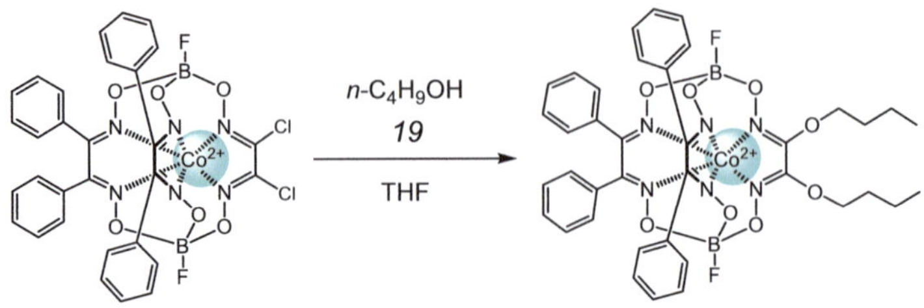

(Co^{2+})@**302** (Co^{2+})@**471**

Scheme 2.220 Halogen exchange reaction of an iron(II) dichloroclathrochelate

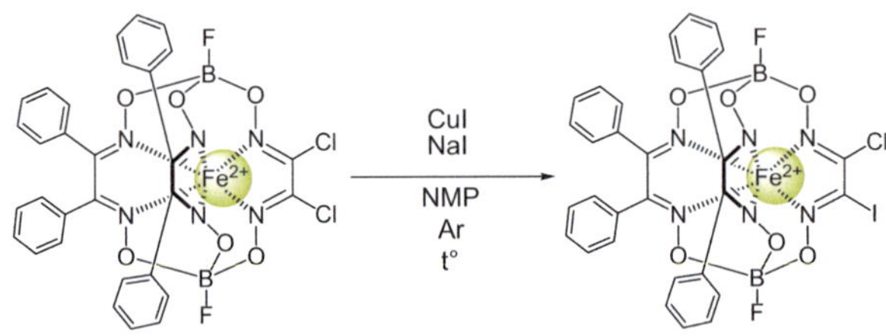

(Fe^{2+})@**302** (Fe^{2+})@**475**

[158] even in the presence of an excess of the above carboranylating agent due to sterical clashes between the bulky carboranyl substituents in a *vic*-position of the same chelate-ribbed fragment of a macrobicyclic framework. The above formation of the C(carborane)–C(clathrochelate) bond has been accompanied by hydrodeiodination in its vicinal position, thus giving the hybrid monomethine monocarboranyl macrobicyclic complex (Fe^{2+})@**483**. The plausible pathway of this tandem process, resulting in substitution of one of the two iodine atoms by carboranyl group and the second one by hydrogen atom (Scheme 2.227), includes either (1) the anion-radical hydrodeiodination of (Fe^{2+})@**309** with a solvent molecule as a source of this hydrogen atom followed by carboranylation of the clathrochelate intermediate or (2) the substitution of one of the two iodine atoms of this diiodoclathrochelate by carboranyl group followed by hydrodehalogenation of the monoiodo monocarboranyl macrobicyclic intermediate [158].

Chemical transformations of the iron(II) dihalogenomonoclathrochelates in the presence of an acetonitrile copper(I) solvato-complex are reported [143] to be affected both by the nature of the halogen atoms in their ribbed chelate fragments and by the reaction conditions (i.e., the solvent and the temperature used). The dibromo- and diiodoclathrochelates (Fe^{2+})@**310** and (Fe^{2+})@**309** (see section "Template Synthesis and Condensation Reactions") in HMPA solution at 70–80 °C underwent a copper(I)-promoted reductive homocoupling by Scheme 2.228 giving the C–C conjugated dibromo- and diiodo-bis-clathrochelates $(Fe^{2+})_2$@**441** and $(Fe^{2+})_2$@**484** in high yields. Their dichloromonoclathrochelate analog (Fe^{2+})@**302** did not undergo such homocoupling under the same reaction conditions, and the corresponding dichloro-bis-clathrochelate $(Fe^{2+})_2$@**486** has been prepared in a high yield using copper(I)-promoted dimerization of its heterodihalogenoclathrochelate precursor (Fe^{2+})@**475** (see above) by Scheme 2.229. The use of NMP as a solvent at high temperatures gave the mixture of the bis-clathrochelate products resulted from tandem homocoupling–hydrodehalogenation reactions of their dihalogenomonoclathrochelate precursors; these reactions are also shown in Scheme 2.228. An initial acetonitrile copper(I) solvato-complex in its NMP solution at high temperatures underwent a re-solvatation and a disproportionation, leading to the copper(II) ions and metallic nano-copper, which promote the above hydrodehalogenation process even at room temperature. Two plausible pathways of this tandem process can be illustrated by Scheme 2.230 [143]. Pathway I includes the

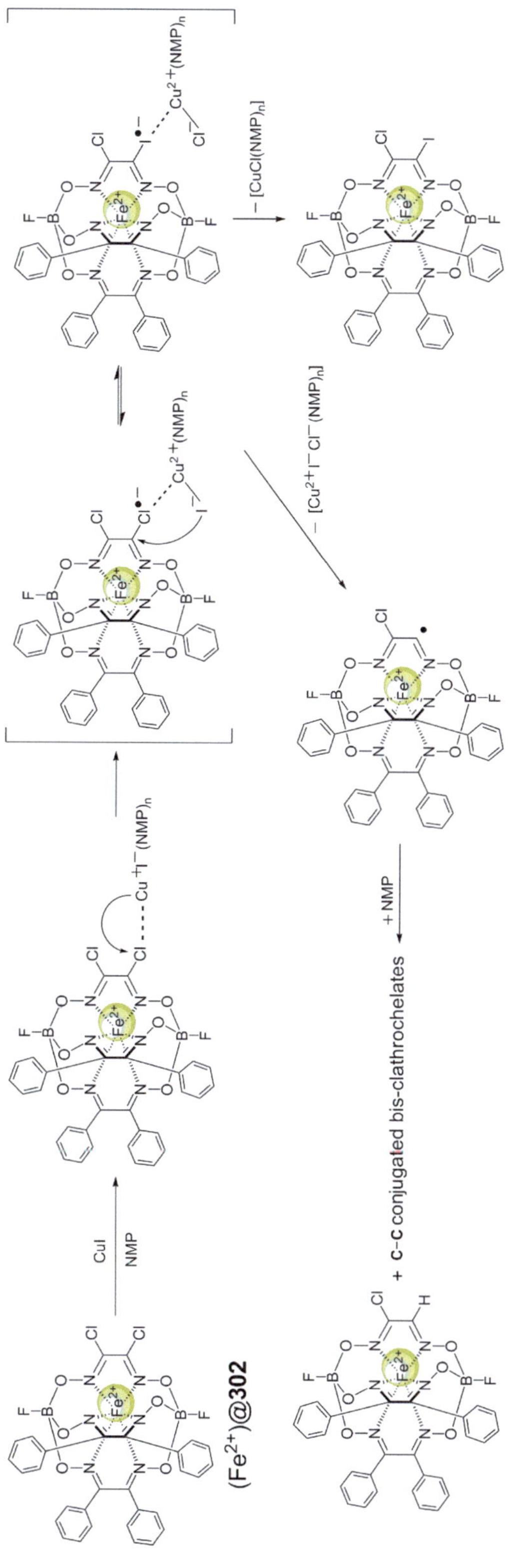

Scheme 2.221 Plausible pathway of copper(I)-promoted halogen exchange reaction of (Fe^{2+})@**302**

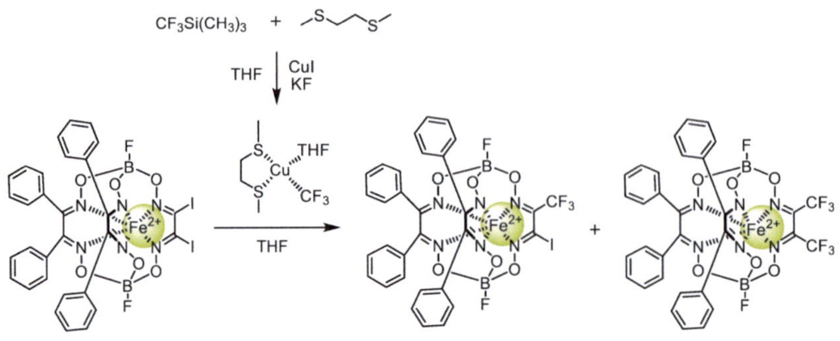

(Fe^{2+})@**309** (Fe^{2+})@**476** (Fe^{2+})@**477**

Scheme 2.222 2,3-thiahexane-assisted perfluoromethylation of an iron(II) diiodoclathrochelate

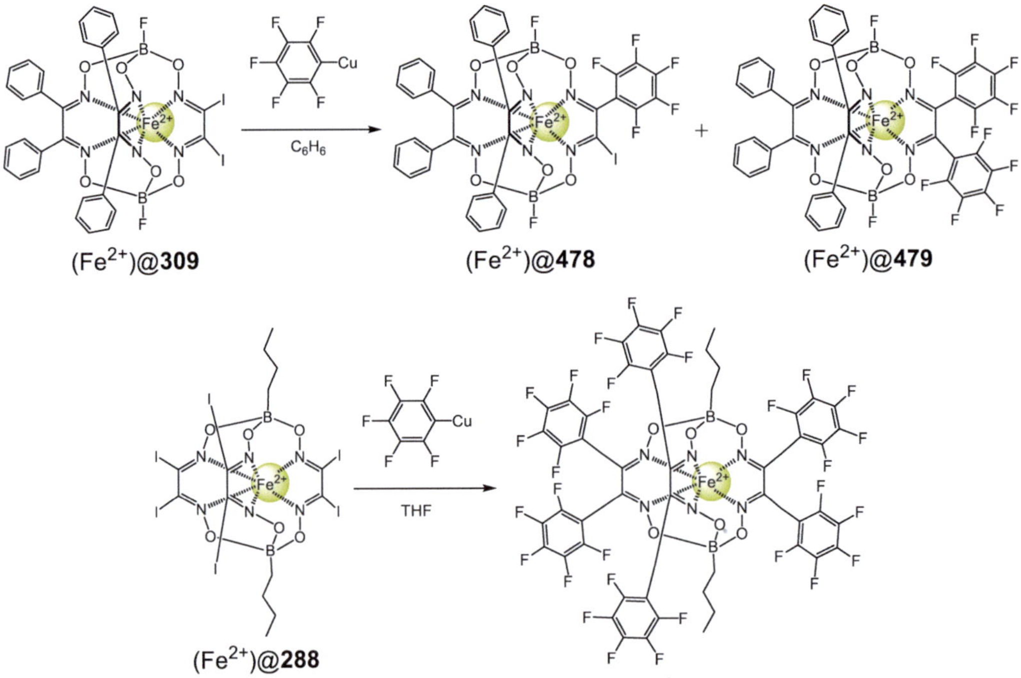

(Fe^{2+})@**309** (Fe^{2+})@**478** (Fe^{2+})@**479**

(Fe^{2+})@**288** (Fe^{2+})@**289**

Scheme 2.223 Perfluoroarylation of the iron(II) di- and hexaiodoclathrochelates

formation of a very reactive monobromoclathrochelate anion-radical intermediate that abstracts the hydrogen atom from the solvent or reactants, thus giving the corresponding iron(II) monomethine monobromoclathrochelate. The following reduction–hydrodehalogenation of this intermediate with nano-copper affords the monomethine anion-radical macrobicyclic species which easily undergo a dimerization to form a C–C conjugated bis-clathrochelate. Such a highly reactive radical species may hypothetically also abstract the hydrogen atom from the solvent or reactants, thus giving the corresponding dimethine monoclathrochelate, but such cage complexes have not been detected in appreciable amounts among the clathrochelate products of this reaction [143]. Therefore, pathway II, including in situ hydrodebromination of an already formed

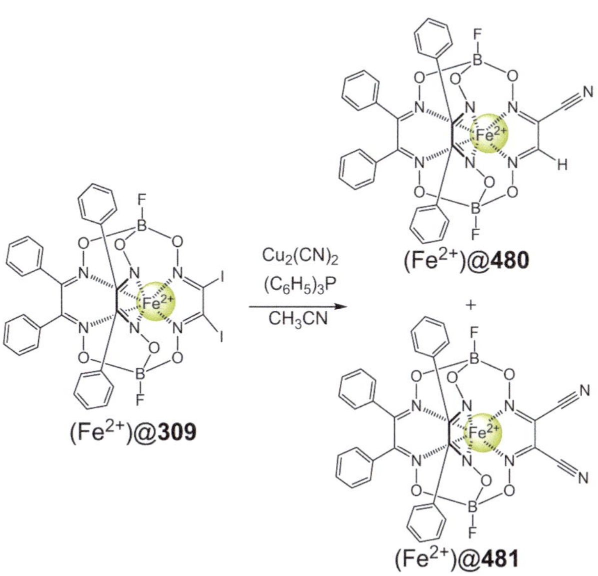

Scheme 2.224 Preparation of the iron(II) clathrochelates with one and two inherent cyano group(s)

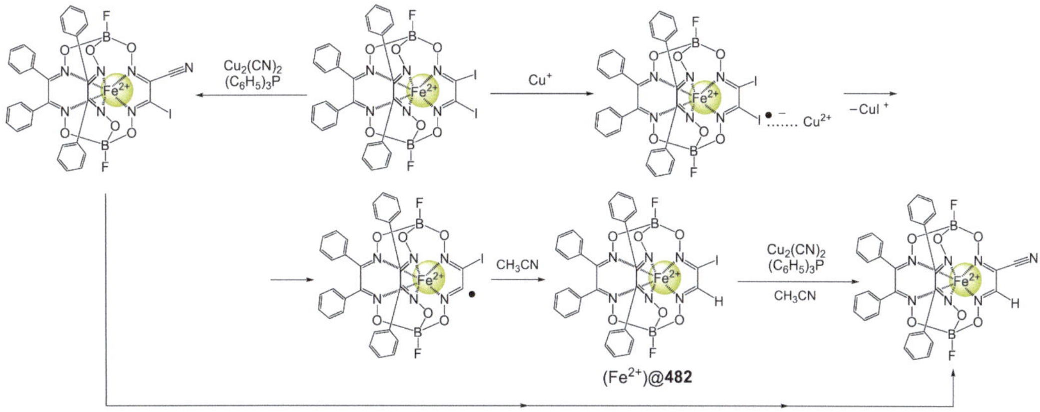

Scheme 2.225 Plausible pathway of the side dehalogenation–reduction process

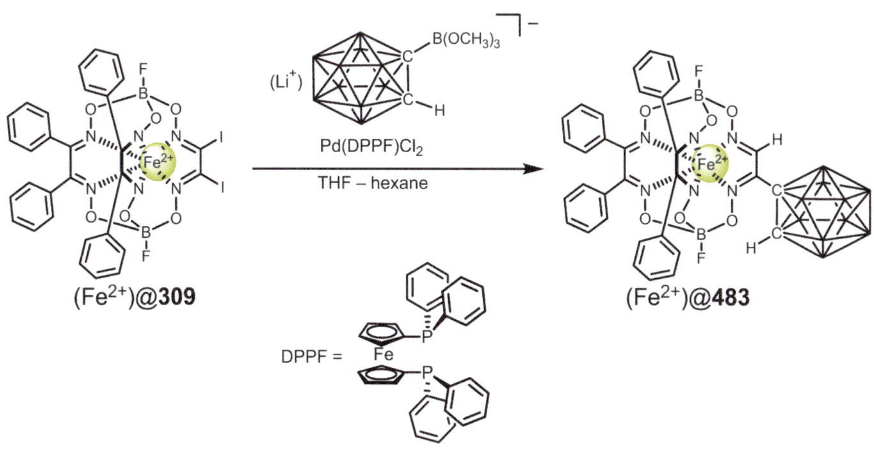

Scheme 2.226 Palladium-catalyzed *C*-carboranylation reaction of an iron(II) diiodoclathrochelate

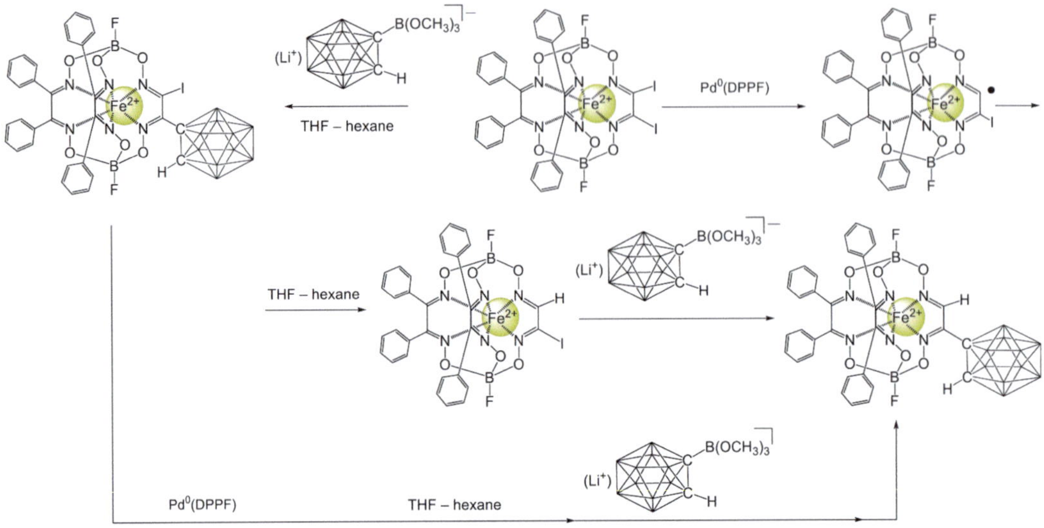

Scheme 2.227 Plausible pathway of the tandem *C*-carboranylation–hydrodeiodination reaction of an iron(II) diiodoclathrochelate

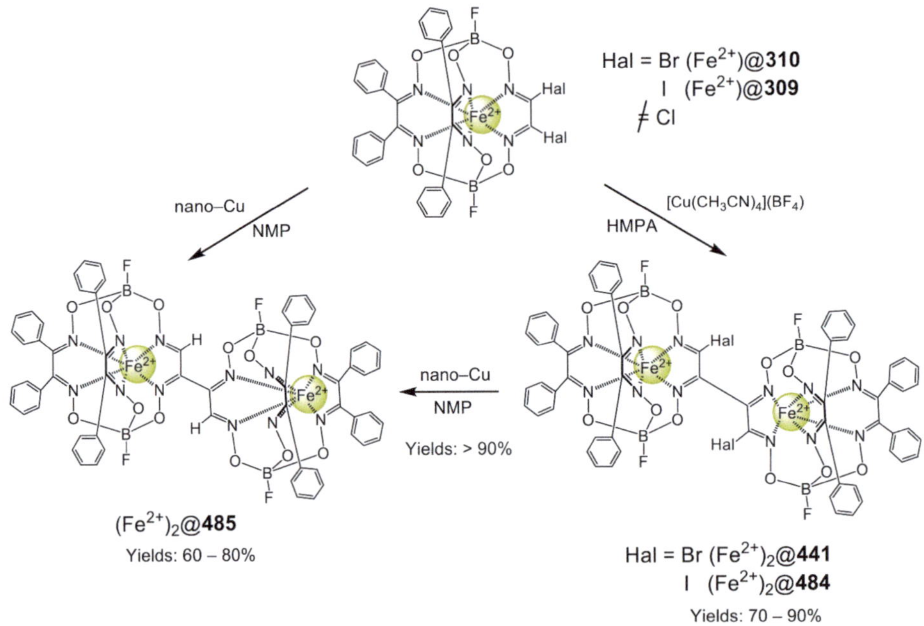

Scheme 2.228 Copper(0)- and copper(I)-promoted reactions of the iron(II) dihalogenoclathrochelates

dibromo-bis-clathrochelate $(Fe^{2+})_2$@**441**, is reported [143] to be the most probable. This suggestion has been confirmed by almost quantitative conversion of $(Fe^{2+})_2$@**441** into $(Fe^{2+})_2$@**485** in the presence of nano-copper and copper(II) ions in NMP medium at high temperatures (Scheme 2.231).

Suzuki–Miyaura and Sonogashira reactions of the reactive 5-bromo-2-furyl substituents of the monoribbed-functionalized clathrochelate (Fe^{2+})@**439** (see section "Electrophilic Addition and Other Reactions of Inherent Ribbed Substituents and Clathrochelate Frameworks") with 4-carboxyphenylboronic acid and

Scheme 2.229 Synthesis of the iron(II) dichloro-bis-clathrochelate $(Fe^{2+})_2$@**486**

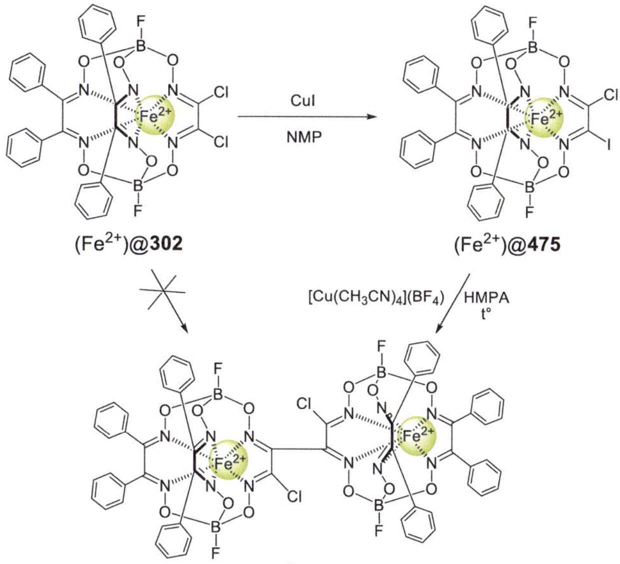

trimethylsilylacetylene, respectively, by Scheme 2.232 in the presence of a palladium(II)-containing catalyst *20*, generated in situ, have been used [111] for preparation of the cage complexes (Fe^{2+})@**487** and (Fe^{2+})@**488** with inherent aromatic ribbed substituents.

Electrophilic Addition and Other Reactions of Inherent Ribbed Substituents and Clathrochelate Frameworks

Unexpected result has been obtained [125] when the reaction of the iron(II) dichloroclathrochelate (Fe^{2+})@**302** with potassium salt of dimercaptomaleodinitrile has been performed in nitromethane as a solvent: major product of this reaction was found to be the cage complex (Fe^{2+})@**489** with annulated thiazine thiophene heterocycle in its ribbed chelate fragment (Scheme 2.233). The formation of this macrobicyclic product under PTC conditions is explained [125] by hindering transfer of the dimercaptomaleodinitrile dianion–tetra-*n*-butylammonium cation ionic pair, while monocharged and lipophilic 2-cyano-3-mercapto-4-amino-5-nitrothiophene anion, resulting from the condensation of dimercaptomaleodinitrile with nitromethane, has been efficiently transferred to the organic phase in the form of its ionic associate with $(n\text{-}C_4H_9)_4N^+$ cation. Following nucleophilic substitution of (Fe^{2+})@**302** with the above nucleophilic anion gave the clathrochelate (Fe^{2+})@**489**; the pathway of a formation of such substituted nitrothiophene is shown in Scheme 2.233. The second plausible pathway to this cage complex with annelated heterocyclic ribbed fragment includes nucleophilic substitution with one of the two thiolate groups of mercaptomaleodinitrile at its first stage [125].

Attempted [128] demetallation of a copper(II) complex of the macrobicyclic ligand (Fe^{2+})@(**490**–$2H^+$) with potassium cyanide by Scheme 2.234 unexpectedly gave the target bis-triazolyl clathrochelate (Fe^{2+})@**490** as only its minor product, while major macrobicyclic product of this reaction, the complex (Fe^{2+})@**491**, has one inherent amine group and one benzyl-1,2,3-triazolylmethylamine ribbed substituent at its clathrochelate framework. So, the latter cage complex has resulted from the side dealkylation reaction of (Fe^{2+})@**490**; the plausible pathway of this side process is also shown in Scheme 2.234. The intermediate macrobicyclic product, formed by coordination of two cyanide ions toward the parent copper(II) complex of (Fe^{2+})@**490**, undergoes an intermolecular ET from its quasiaromatic caging ligand to the

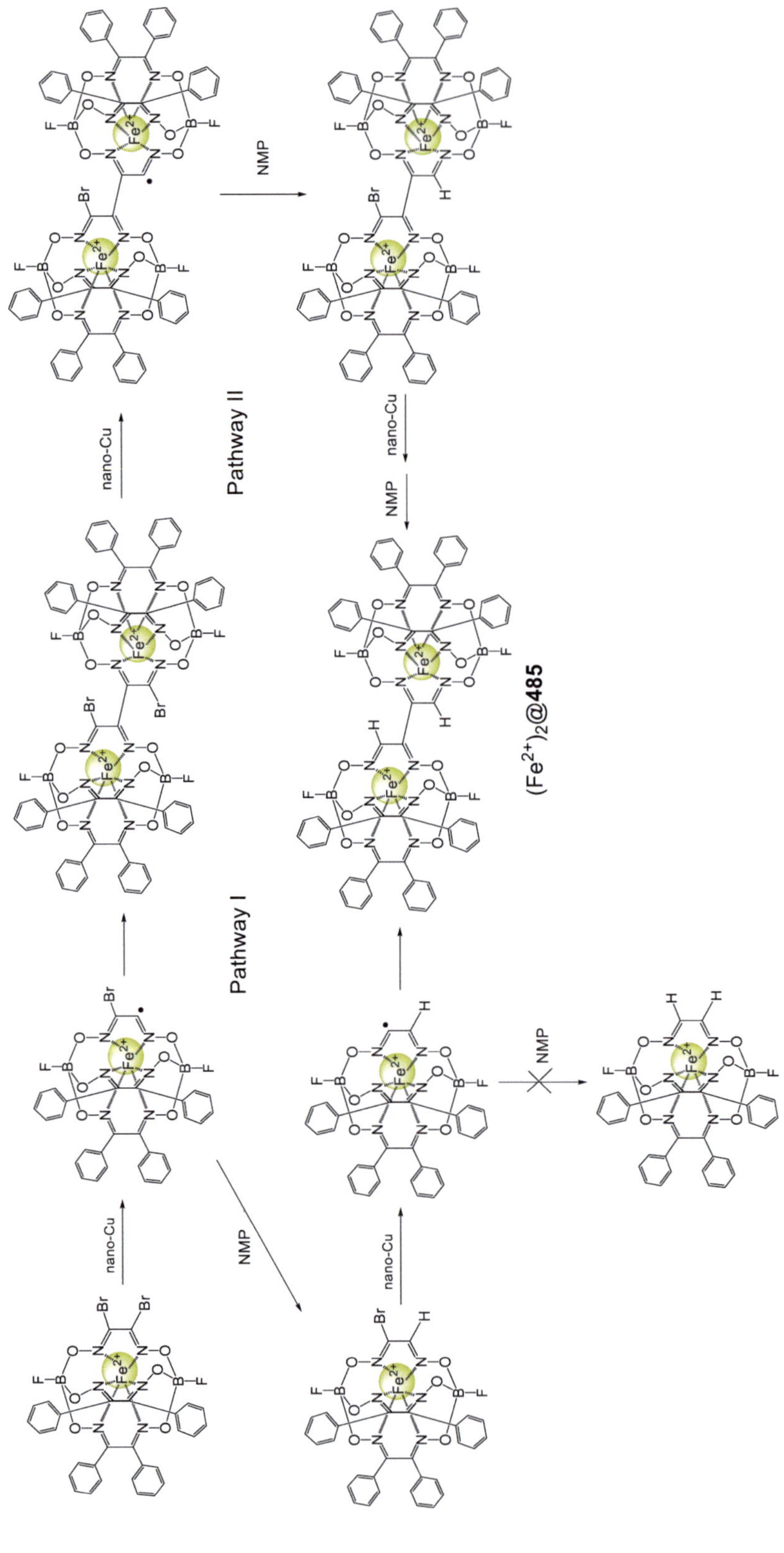

Scheme 2.230 Plausible pathway of the tandem homocoupling–hydrodehalogenation reaction

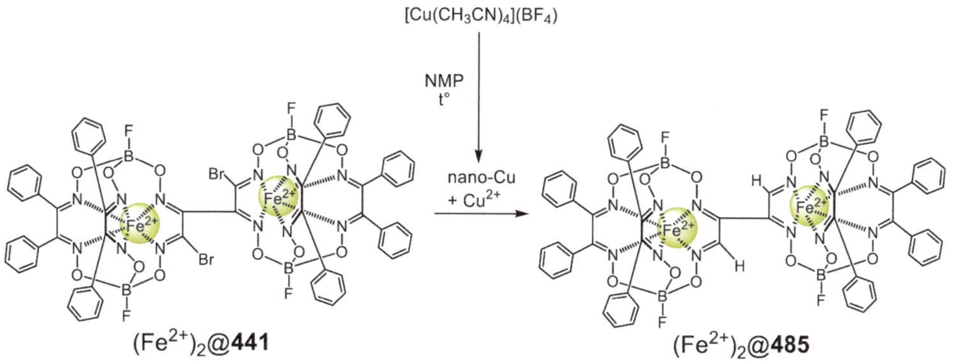

Scheme 2.231 Copper(0)-promoted hydrodehalogenation of an iron(II) dibromo-bis-clathrochelate

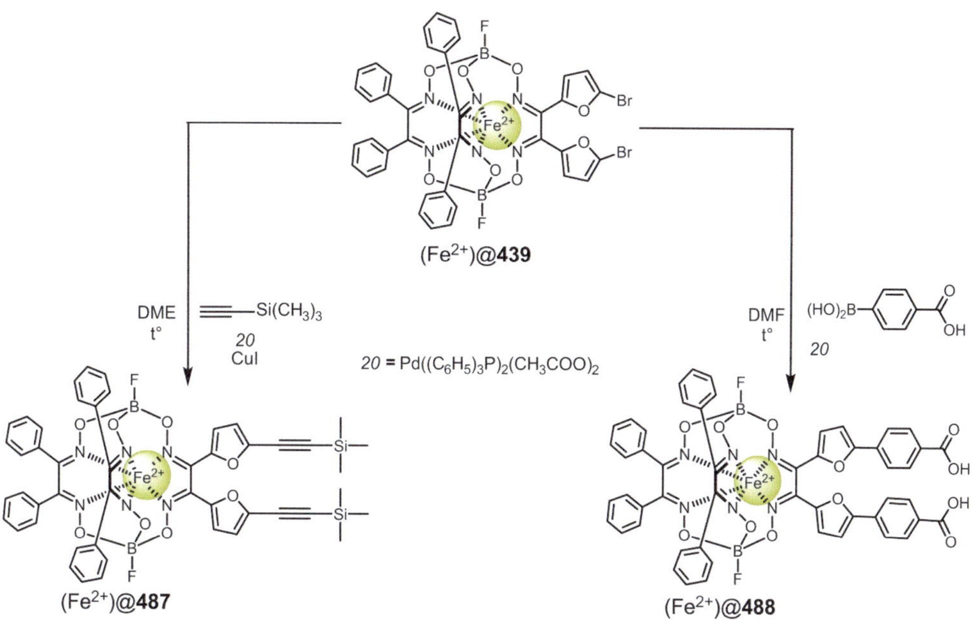

Scheme 2.232 Palladium-catalyzed reactions of the dibromoclathrochelate (Fe^{2+})@**439**

central metal ion, thus giving the anion-radical clathrochelate species.

The following intermolecular oxidation of these anion-radicals with the second molecule of the above cyanide complex led to the corresponding copper(I)-containing derivative of the same clathrochelate ligand. The copper(I) complex of (Fe^{2+})@**490** with one dehydrogenated benzyl-1,2,3-triazolyl-methineimine ribbed substituent easily undergoes a basic hydrolysis, thus giving the monoamine monotriazol clathrochelate (Fe^{2+})@**491** and the corresponding aldehyde as well [128].

The clathrochelate (Fe^{2+})@**341** (see section "Nucleophilic Substitution") easily underwent an oxidation with air oxygen in solution by Scheme 2.235, leading to its oxidized derivative (Fe^{2+})@**492** [129]. The latter cage complex has been also prepared using one-pot synthetic procedure avoiding an isolation of the macrobicyclic intermediate (Fe^{2+})@**341**.

The phosphorus(V)-containing clathrochelate (Fe^{2+})@**356** has been reduced [136] with an

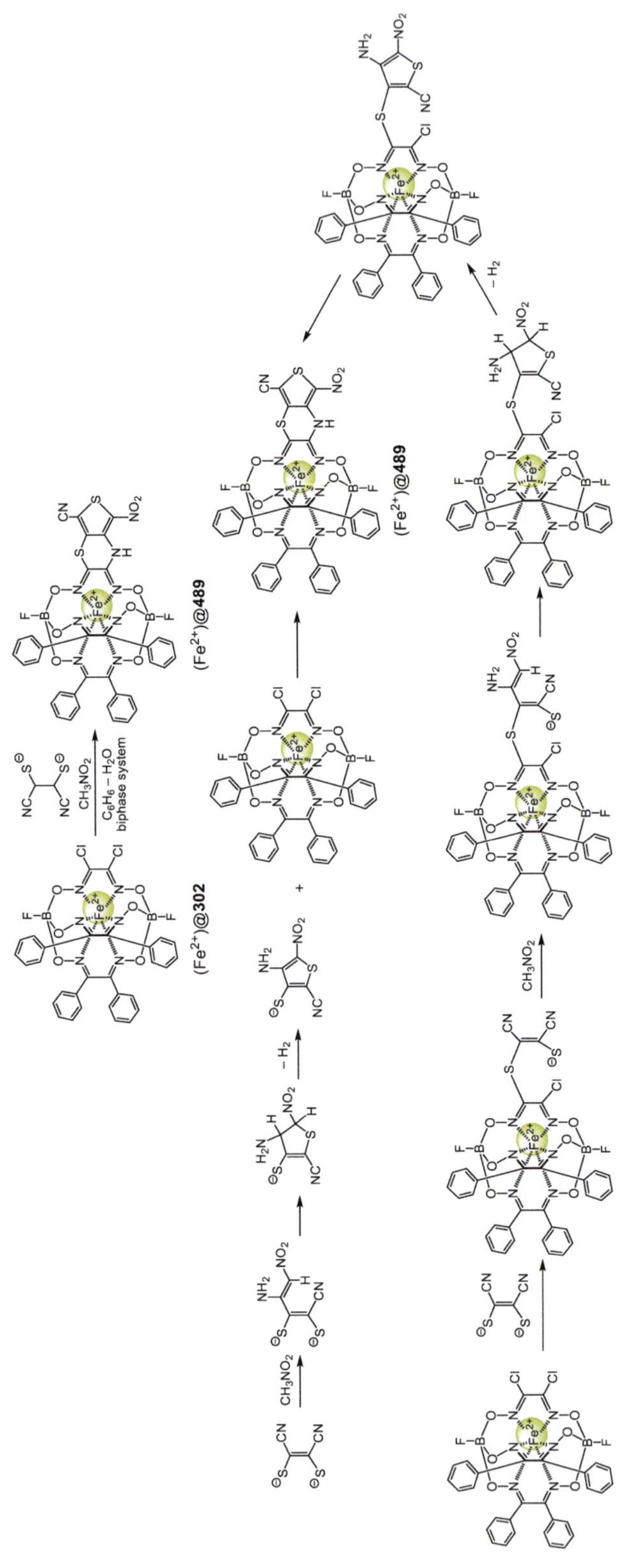

Scheme 2.233 Plausible pathways of the formation of the iron(II) clathrochelate (Fe^{2+})@489

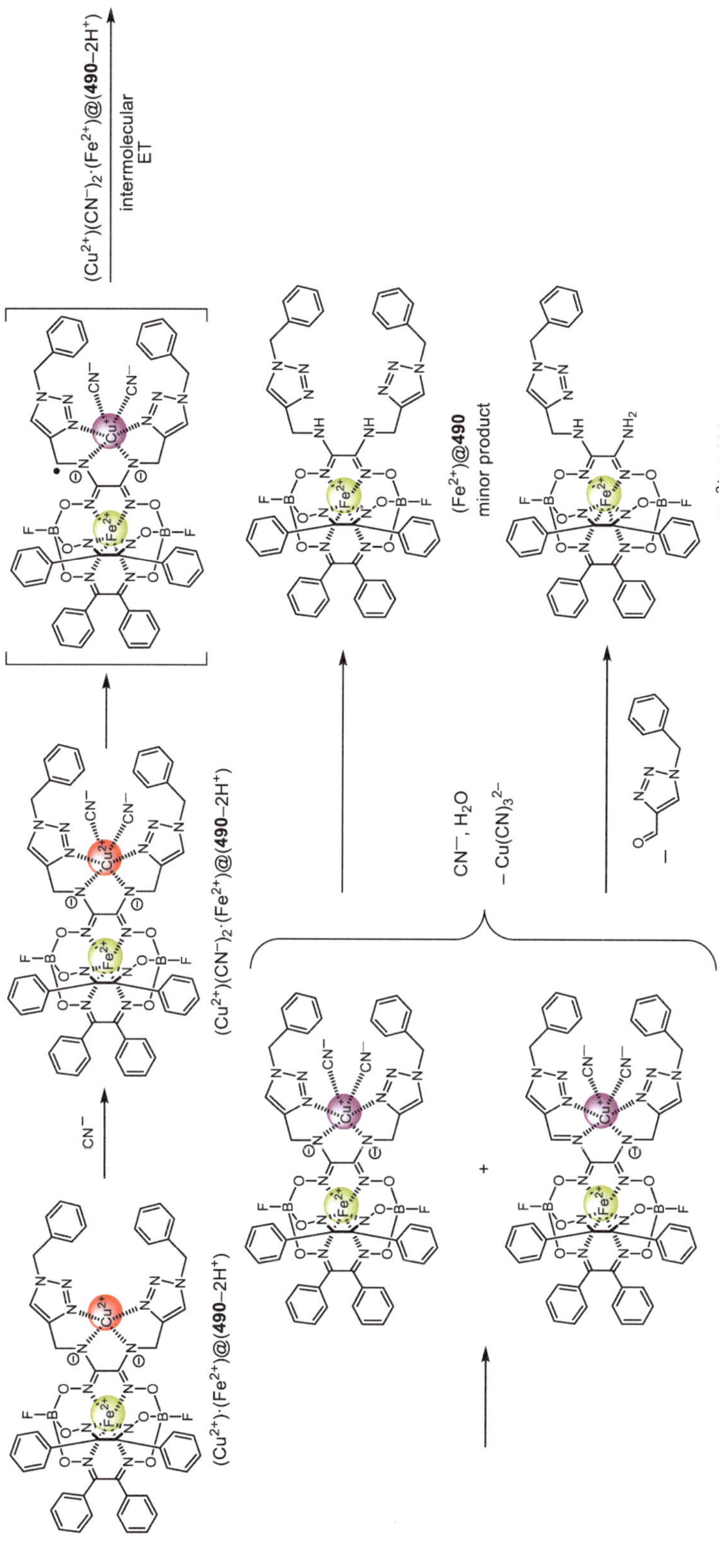

Scheme 2.234 Formation of the clathrochelates (Fe^{2+})@**490** and (Fe^{2+})@**491**

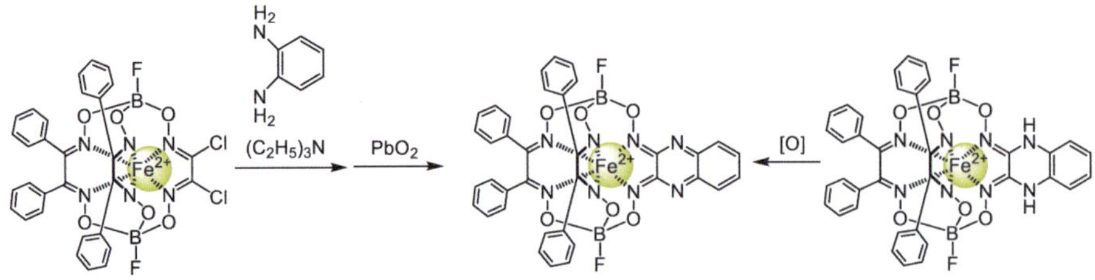

Scheme 2.235 Synthetic pathways to the complex (Fe^{2+})@**492**

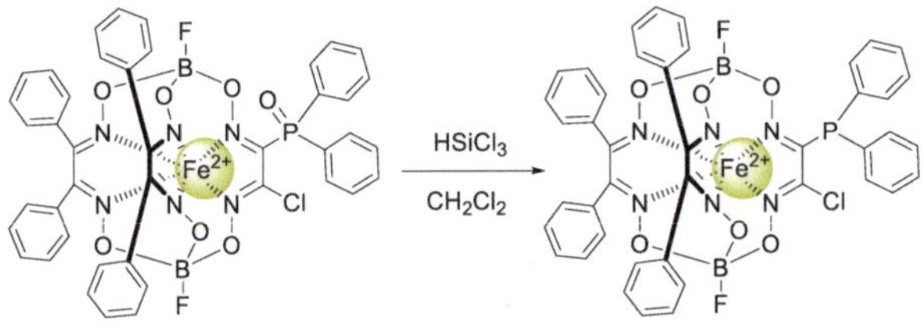

Scheme 2.236 Reduction of the phosphorus(V)-containing clathrochelate (Fe^{2+})@**356** to its phosphorus(III)-containing derivative

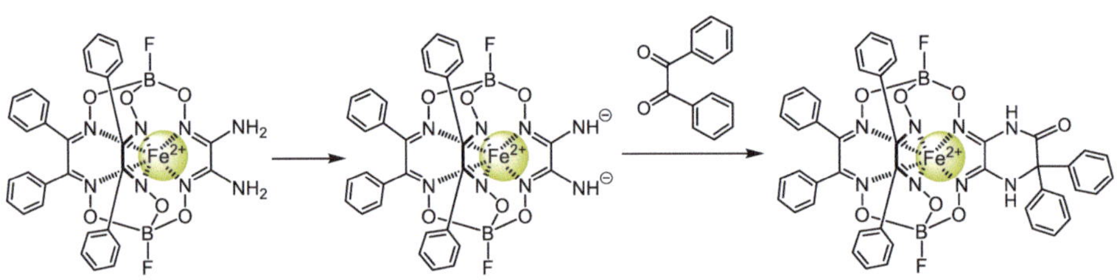

Scheme 2.237 Electrophilic addition of benzil to the diaminoclathrochelate (Fe^{2+})@**325**

excess of silicoform by Scheme 2.236, thus giving the iron(II) cage complex (Fe^{2+})@**493** with inherent diphenylphosphorus(III) group in a quantitative yield. The macrobicyclic product of this reaction easily underwent a reoxidation to the parent cage complex (Fe^{2+})@**356**.

Coordination of the anionic forms of the diaminoclathrochelate (Fe^{2+})@**325** toward the electrophilic Cu^{2+} ion allowed to study [121] the reactivity of these anionic macrobicyclic species against the organic electrophiles, such as α-diketones. The choice of benzil as an electrophilic agent is explained by its stability against a side self-condensation process under basic conditions. The reaction of (Fe^{2+})@**325** with benzil in the presence of potassium *tert*-butoxide by Scheme 2.237 gave the clathrochelate (Fe^{2+})@**494** in a moderate yield. This macrobicyclic product results not from a plain addition of (Fe^{2+})@(**325**–$2H^+$) to benzil but from an 1,2-shift of its phenyl group [121]. Such a migration of phenyl group in the macrobicyclic intermediate *21* caused a formation of the intermediate *22*, following cyclization of which afforded the morpholinone *23* (Scheme 2.238). The plausible pathway of this reaction includes an elimination of hydroxide ion from the molecule *21*, thus giving the intermediate *24* (Scheme 2.237); its following transformations (i.e, a ring closure with a migration of the phenyl substituent to the azomethine carbon atom) gave the clathrochelate (Fe^{2+})@**494** [121]. So, the above clathrochelate dianion (Fe^{2+})@(**325**–$2H^+$)

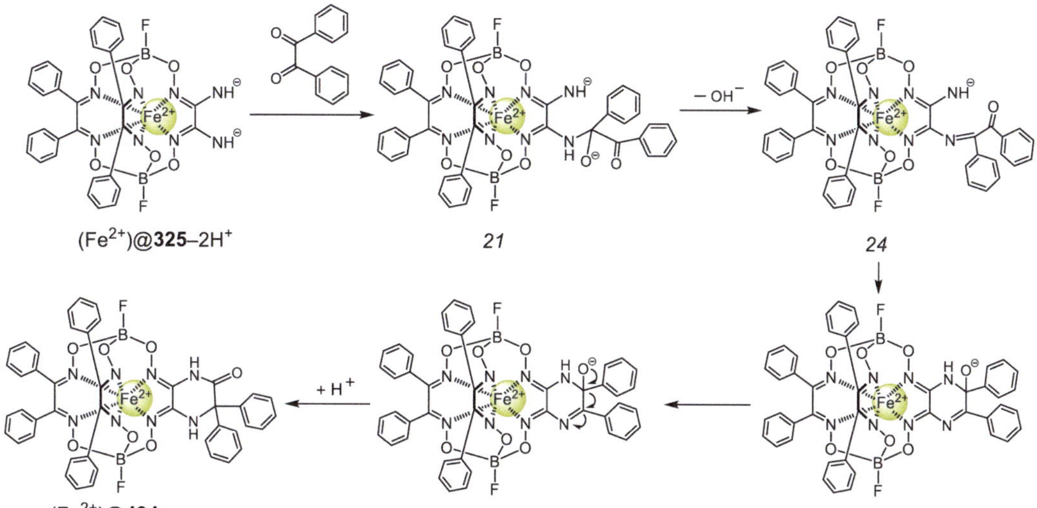

Scheme 2.238 Plausible pathway of the reaction of (Fe^{2+})@**325**–2H$^+$ with benzil

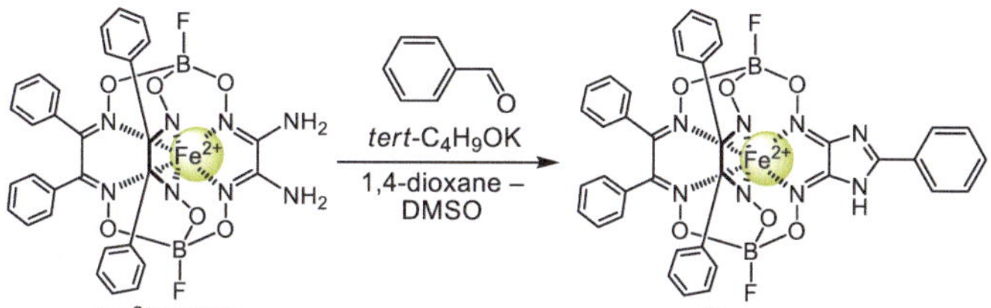

Scheme 2.239 Electrophilic addition of benzaldehyde to the diaminoclathrochelate (Fe^{2+})@**325**

has been reacted with benzil as an organic electrophile, thus giving the macrobicyclic product with annulated heterocyclic ribbed piperazinone fragment, resulting from a benzilic-type rearrangement with 1,2-shift of its phenyl substituent.

Electrophilic addition by Scheme 2.239 has been used [134] as an alternative to nucleophilic substitution (see section "Nucleophilic Substitution") synthetic approach to the cage complex (Fe^{2+})@**348**. The electrophilic carbonyl group of benzaldehyde attacks the corresponding macrobicyclic dianion, formed *via* deprotonation of the amine ribbed substituents at a quasiaromatic polyazomethine cage framework of (Fe^{2+})@**325** with potassium *tert*-butoxide DMSO solution. The subsequent elimination of water molecule and an oxidation with air oxygen gave the annelated phenylimidazole substituent at this framework [134].

Selective bromination of the furyl-ribbed substituents of the complex (Fe^{2+})@**308** (see section "Template Synthesis and Condensation Reactions") with *N*-bromosuccinimide by Scheme 2.240 has been performed [111], thus giving the dibromoclathrochelate (Fe^{2+})@**439** in a high yield.

2.2.3 Tris-oximehydrazonate Cage Complexes and Their Semiclathrochelate Precursors

The ferrocenylboron-capped semiclathrochelate precursor *25* of the tris-oximehydrazonate cage complex (Fe^{2+})@**495** has been prepared [159] using two synthetic approaches. First of them is based on an initial isolation of the mixture of *fac*- and *mer*-isomers of the non-macrocyclic iron(II)

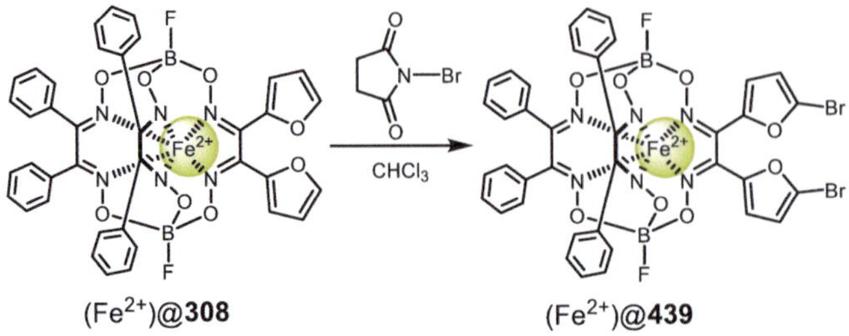

Scheme 2.240 Preparation of the dibromoclathrochelate precursor (Fe^{2+})@**439**

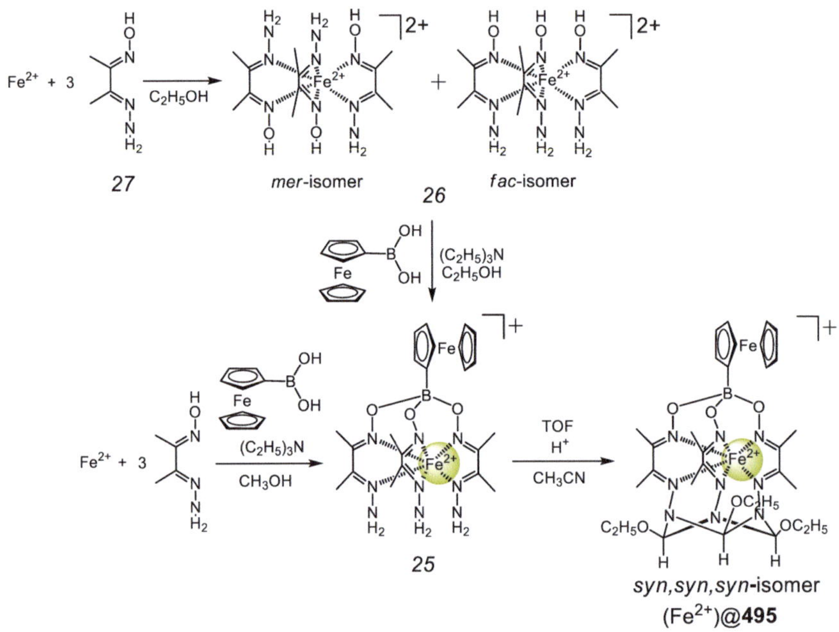

Scheme 2.241 Synthesis of the ferrocenylboron-capped oximehydrazonate clathrochelate (Fe^{2+})@**495**

tris-oximehydrazonate *26*, which then underwent a cross-linking with ferrocenylboronic acid by Scheme 2.241. The overall yield of this semiclathrochelate was a relatively low. So, another approach, based on template condensation of three molecules of its oximehydrazonate ligand synthone *27* and ferrocenylboronic acid on an iron(II) ion as a matrix, giving *25* in a moderate yield, has been elaborated [159]. Its macrobicyclization with TOF proceeded with unexpected difficulties. Under the standard reaction conditions (i.e., at room temperature under air atmosphere in the presence of catalytic amounts of trifluoroacetic anhydride), this reaction mainly gave the partially cross-linked complexes, while a reflux of the reaction mixture caused the formation of the macrocyclic and open-chain iron(II) oximehydrazonates due to an oxidation of the ferrocenylboron-capping fragment of *25* and its clathrochelate derivative (Fe^{2+})@**495** by air oxygen. So, this H^+-catalyzed macrobicyclization with TOF has been successfully performed [159] under the anaerobic conditions at room temperature.

The authors of [160] suggested to use the methyl ester of *ortho*-carboranyldiboronic acid as a bifunctional Lewis acid for preparation of the corresponding binuclear bis-semiclathrochelate iron(II) hexaoximehydrazonate with *ortho*-position of its semiclathrochelate fragments at a carborane framework. However, template condensation of

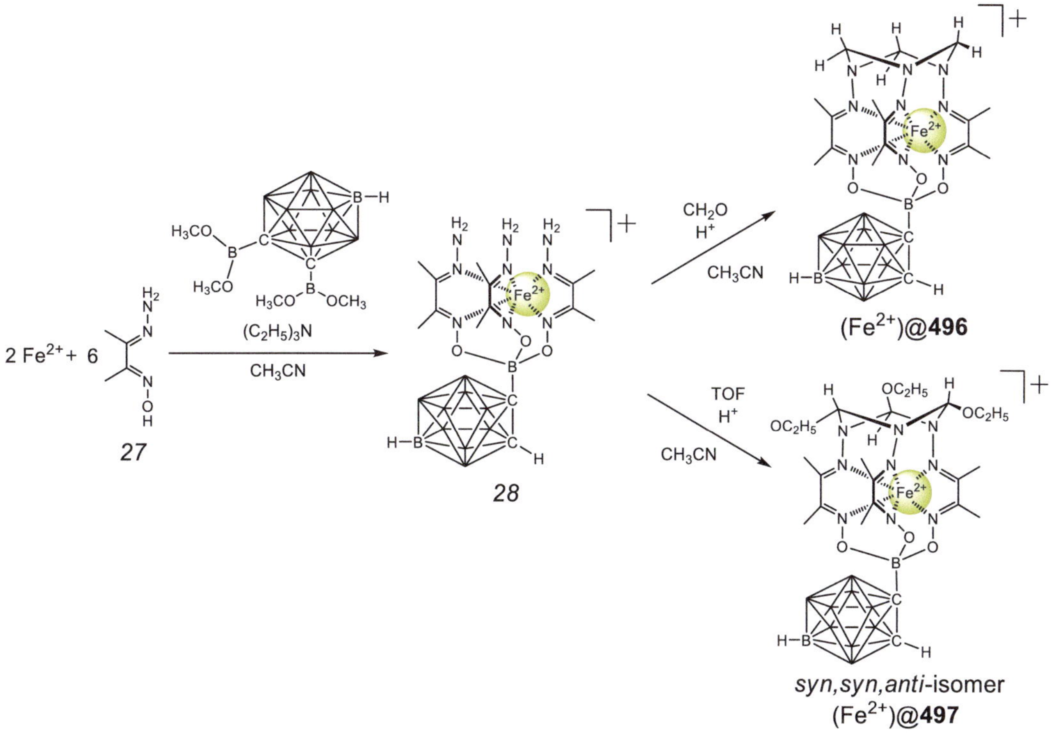

Scheme 2.242 Preparation of the carboranylboron-capped iron(II) oximehydrazonates

the oximehydrazonate ligand synthone *27* with this cross-linking agent on an iron(II) ion as a matrix in the presence of triethylamine has unexpectedly resulted in the formation of the iron(II) monosemiclathrochelate *28* shown in Scheme 2.242. This product resulted from a deborylation (an elimination of one of the two $B(OCH_3)_2$ groups) of the above capping agent in the course of self-assembly process. This fact is explained [160] by sterical clashes in *ortho*-position of its carborane framework. H^+-catalyzed condensation of *28* with formaldehyde or TOF as the active carbonyl components has been used [160] for the synthesis of the tris-oximehydrazonate clathrochelate cations (Fe^{2+})@**496** and (Fe^{2+})@**497** (its *syn,syn,anti*-isomer) with apical cross-linking 1,3,5-triazacyclohexane fragment (Scheme 2.242). Macrobicyclization of their *n*-butylboron-capped semiclathrochelate analog *29* with TOF by Scheme 2.243 has been also performed [160], thus giving a *syn,syn,syn*-isomer of the macrobicyclic cation (Fe^{2+})@**498**.

The semiclathrochelate iron(II) complexes *30–34* have been prepared [161] by Scheme 2.244 using template condensation of *27* with the *iso*-propyl ester of the corresponding 1,4-pentadienylboronic acid on an iron(II) ion as a matrix in presence of triethylamine for neutralization of H^+ ions releasing in the course of this reaction. These ions caused a protolytic cleavage of the reactive $B–C_{vinyl}$ bonds, thus decreasing the yields of the target iron(II) semiclathrochelates. H^+-catalyzed macrobicyclization of these semiclathrochelate precursors with an excess of TOF has been used [161] for preparation of the tris-oximehydrazonate clathrochelates (Fe^{2+})@**499**, (Fe^{2+})@**500**, (Fe^{2+})@**501**, (Fe^{2+})@**502**, and (Fe^{2+})@**503**.

The synthesis of the oximehydrazonate iron(II) clathrochelates (Fe^{2+})@**504** and (Fe^{2+})@**505** with a single triethylantimony-capping group has been accomplished [162] by Scheme 2.245. The greater lability of this apical group, explained by low Lewis acidity of the trialkylantimony(V) compounds, has

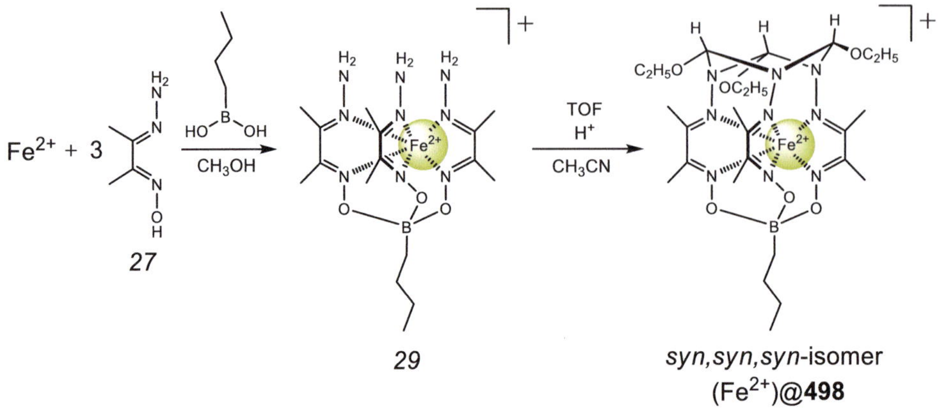

Scheme 2.243 Synthesis of the *n*-butylboron-capped iron(II) oximehydrazonate (Fe^{2+})@**498**

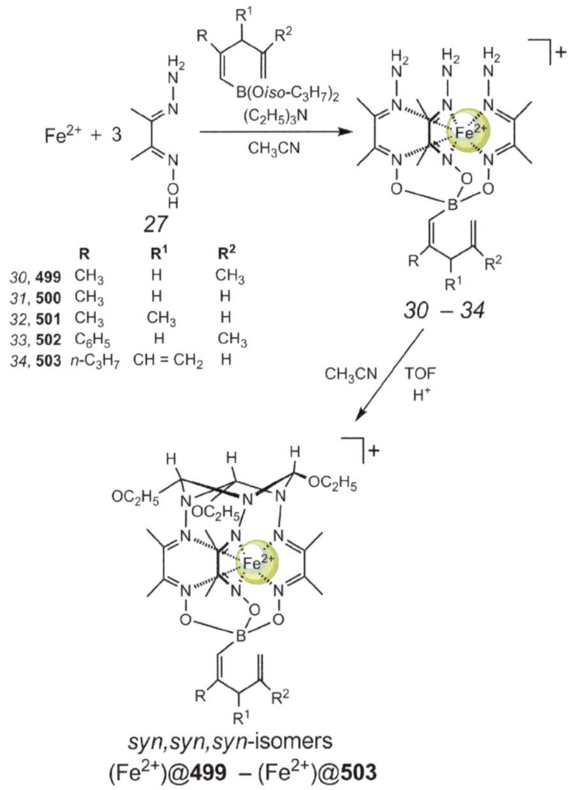

Scheme 2.244 Preparation of the diolefinboron-capped oximehydrazonates (Fe^{2+})@**499** – (Fe^{2+})@**503**

complicated an isolation of the target semiclathrochelate intermediate *35*. Trying to obtain it in a higher yield, the azeotropic distillation of water from the reaction mixture has been used [162]. The synthesis of its clathrochelate derivatives (Fe^{2+})@**504** and (Fe^{2+})@**505** has been also complicated by the side reactions (in particular, by reduction of their capping antimony(V) group): a relatively low yield of the macrobicyclization of *35* with TOF is also explained [162] by high lability of its triethylantimony-capping fragment.

The homodinuclear oximehydrazonate iron(II) bis-clathrochelate $(Fe^{2+})_2$@**506** has

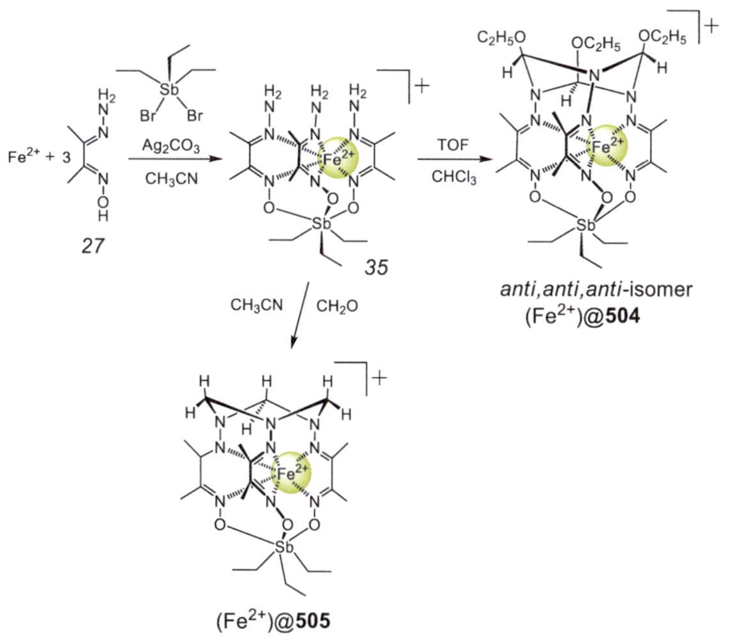

Scheme 2.245 Synthesis of the triethylantimony-capped iron(II) oximehydrazonates

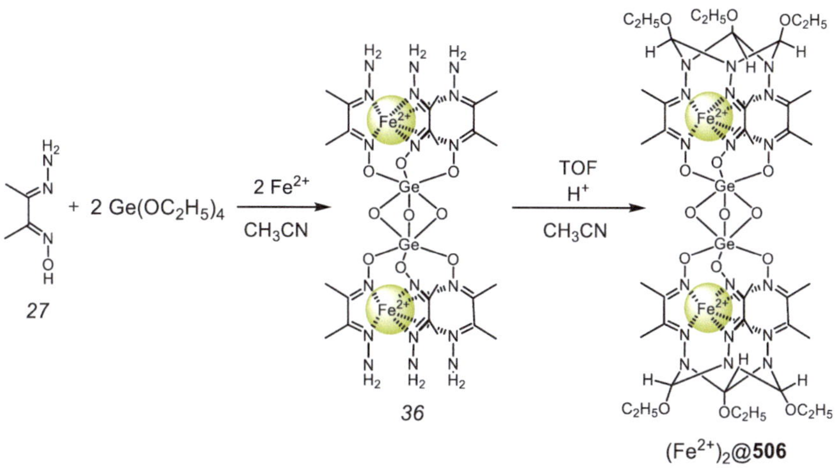

Scheme 2.246 Preparation of the oximehydrazonate bis-clathrochelate $(Fe^{2+})_2$@**506**

been prepared [94] by two-step synthetic procedure shown in Scheme 2.246 starting from *27*, the oxime and hydrazonate groups of which have substantially different reactivities, and using germanium(IV) tetraethoxide as a bifunctional Lewis acid. The bis-semiclathrochelate intermediate *36* is practically insoluble in organic solvents, but it underwent H^+-catalyzed macrobicyclization through its hydrazonate groups with TOF, thus giving the bis-cage complex $(Fe^{2+})_2$@**506**.

The analogous apically linked diboron-capped bis-clathrochelates $(Fe^{2+})_2$@**507** and $(Fe^{2+})_2$@**508** has been prepared [96] by two-step synthetic procedure shown in Scheme 2.247. It includes (1) an isolation of the bis-semiclathrochelate iron(II) oximehydrazonate *37* by template condensation of six molecules of *27* with the butyl ester of 1,3-propanediboronic acid on an iron(II) ion as a matrix and (2) its following H^+-catalyzed

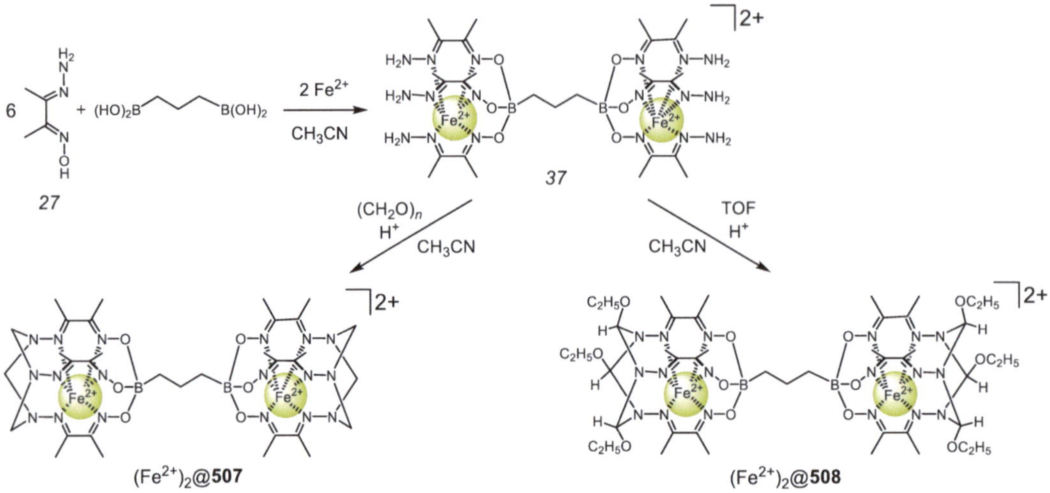

Scheme 2.247 Synthesis of the oximehydrazonate iron(II) bis-clathrochelates

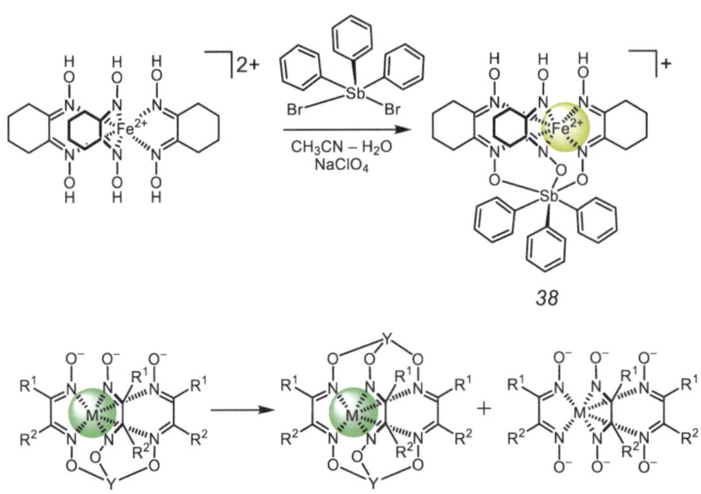

Scheme 2.248 Formation of the monoantimony-capped iron(II) semiclathrochelate *38* and a disproportion reaction of a semiclathrochelate complex

macrobicyclization with formaldehyde or TOF, respectively, thus giving the bis-cage iron(II) complexes with 1,3,5-triazacyclohexane-capping fragments [96].

2.2.4 Tris-pyrazoloximate and Tris-dioximate Pseudoclathrochelates

An unusual semiclathrochelate product *38* has been isolated [94] in the course of self-assembly reaction by Scheme 2.248. This semiclathrochelate tris-dioximates readily underwent a disproportionation, thus yielding its more kinetically and thermodynamically stable clathrochelate derivative.

The zinc, cobalt, iron, and manganese(II) tris-pyrazoloximate pseudoclathrochelates (M^{2+})@**509** have been prepared [163] by Scheme 2.249 in high yields under mild reaction conditions using one-pot template condensation of a mixed pyrazoloxime ligand synthon *39* with phenylboronic acid on the corresponding metal ion as a matrix. Hydrogen atoms of three terminal pyrazolyl NH groups form a pseudomacrobicyclic framework of these complexes *via* their hydrogen bonding with

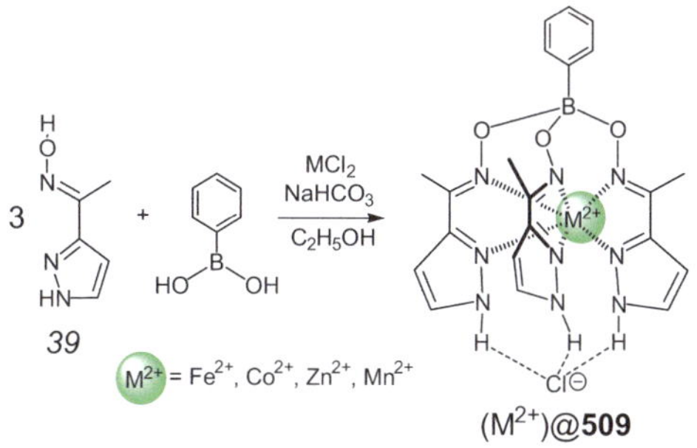

Scheme 2.249 Template self-assembly of the phenylboron-capped metal(II) pseudoclathrochelates

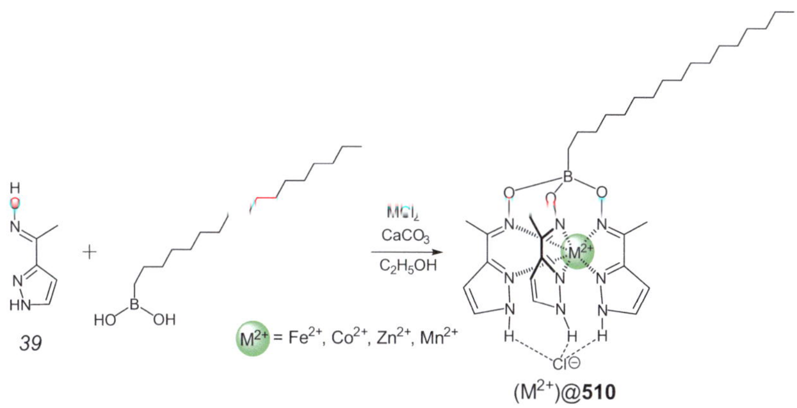

Scheme 2.250 Synthesis of the hexadecylboron-capped metal(II) pseudoclathrochelates

chloride counterion, thus giving the pseudocapping μ_3–Cl^- ... $(HN)_3$ apical group and forming three additional six-membered pseudochelate cycles in its tripodal tris-pyrazolyl fragment [163]. Efficient self-assembly of the monohexadecylboron-capped zinc, cobalt, iron, and manganese(II) tris-pyrazoloximate pseudoclathrochelates (M^{2+})@**510** has been performed [164] by Scheme 2.250 under mild reaction conditions using one-pot template condensation of the same ligand synthon with *n*-hexadecylboronic acid as a capping Lewis-acidic agent.

2.2.5 Hybrid Systems and Other Types of Macropolycyclic Metal Complexes

The luminescent bipyridine and pyrazole europium, terbium and gadolinium(III) clathrochelates have been prepared [165] by Scheme 2.251 using the ionic exchange reactions of 1:1 cage complexes of the macrobicyclic ligands **511** and **512** with an encapsulated alkali metal ion.

One-pot condensation of the corresponding disaccharide ligand synthons *40* and *41* with the N_2O_4-macrocycle *42* by Scheme 2.252 on a sodium(I) cation as a matrix under the conditions of tandem Staudinger–aza-Wittig reaction has been used [166] for the synthesis of 1:1 sodium(I) cage complexes of the monocellobiosyl macrobicyclic ligands **513** and **514**. The plausible pathway of this reaction includes the formation of 1,6-diisocyanatocellobiose intermediate from 1,6-diazidocellobiose with simultaneous double nucleophilic addition of isocyanate to the sodium(I) complex of *42* that is reported [166] to be preorganized for its further macrobicyclization stage.

The cobalt(II)-capped polyether caging ligands **515** and **516** have been prepared [167] using a coordination-driven self-assembly of

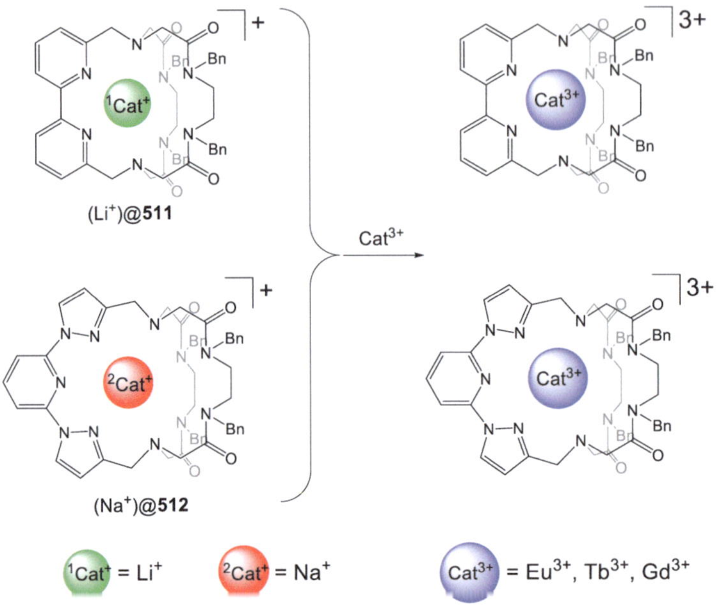

Scheme 2.251 Preparation of the bipyridine and pyrazole metal(III) clathrochelates

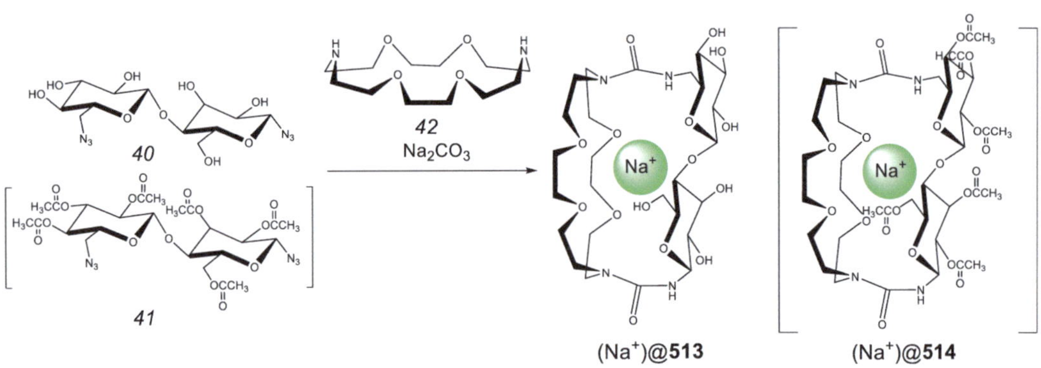

Scheme 2.252 Synthesis of potassium(I) complexes of the monocellobiosyl ligands **513** and **514**

alkali metal polyether bis-xanthates with cobalt(II) ions by Scheme 2.253. According to the ESI-MS data of [167], the caging ligand **516** with a bigger cavity efficiently encapsulates alkali metal cations, thus forming their 1:1 cage complexes, while the cavity of its analog **515** is too small for accommodation of these ions.

The fluorinated caging ligand **517** has been synthesized [168] by Scheme 2.254; it has been tested as an extractant for alkali metal and silver(I) cations. This ligand highly selectively forms 1:1 potassium and silver(I) cage complexes. In the clathrochelate cation (K^+)@**517**, its six fluorine atoms are coordinated toward an encapsulated metal cation, thus giving a distorted octahedral KF_6-coordination polyhedron [168, 169].

The formation of the mono- and homodinuclear cage complexes of the polypyridine encapsulating ligands **33** and **518** (Scheme 2.255) with caged alkali and *3d*, *4f*-transition metal cations (in the former case, with a selectivity toward of Rb^+ ion) is reported [170].

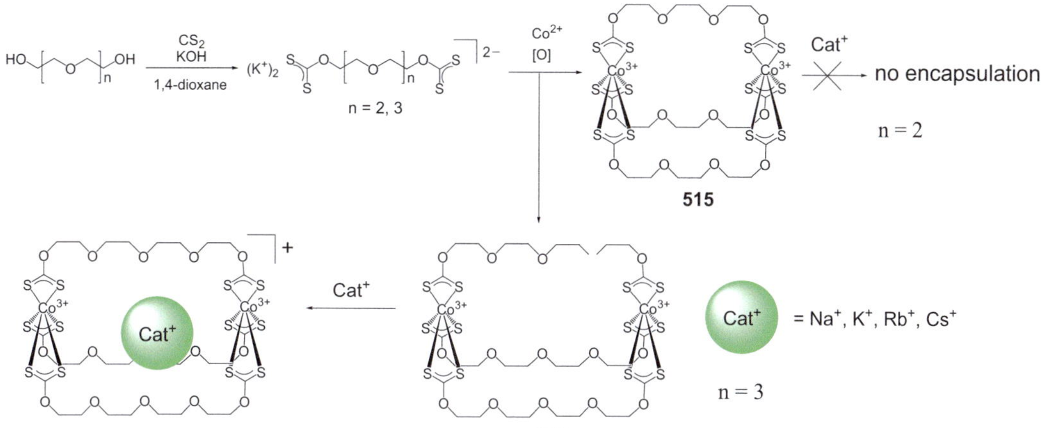

Scheme 2.253 Preparation of the cobalt(II)-capped polyether macrobicyclic ligands and their cage complexes

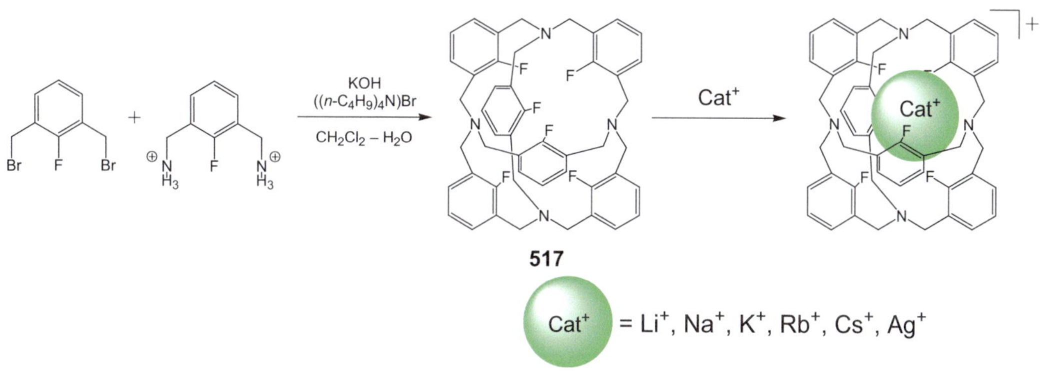

Scheme 2.254 Synthesis of the fluorinated ligand **517** and its cage complexes

1:1 europium and terbium(III) cage complexes of the tris-bipyridine macrobicyclic ligand **519** have been synthesized [171] by Scheme 2.256 using the exchange reactions of a caged sodium(I) cation in its clathrochelate (Na^{+})@**519**. The same synthetic approach (i.e., the ionic exchange of an encapsulated metal ion) has been used [172] for preparation of their gadolinium(III)-encapsulating analogs (Gd^{3+})@**519** and (Gd^{3+})@**520** by Scheme 2.257.

The presence of photoactive pendant polyaromatic substituents in the molecule of the N_4O_4-macrobicyclic ligand **521** has been used [173] for fluorescent detection of the formation of its 1:1 calcium(II) cage complex by Scheme 2.258.

The reactions of metallic indium and thallium with 1,2-dicyanobenzene by Scheme 2.259 under low nitrogen pressure at 210 °C have been used [174, 175] for the synthesis of 1:1 indium and thallium(III) cage complexes of the N_6-hexadentate macrobicyclic ligand **522**.

Coordination-driven self-assembly of the cesium salt of tin(IV) tris-4-(diphenylphosphino) benzene-1,2-diolate *43* with $AgNO_3$ or AgOTf in their molar ratio of 2:3 by Scheme 2.260 has been used [176] for preparation of 1:1 cesium(I) cage complex of the encapsulating ligand **523**. Cesium(I) cation is reported [176] to play a crucial role in this reaction: the attempt using a series of other metal cations in such self-assembly instead of Cs^{+} ion gave no cage complexes.

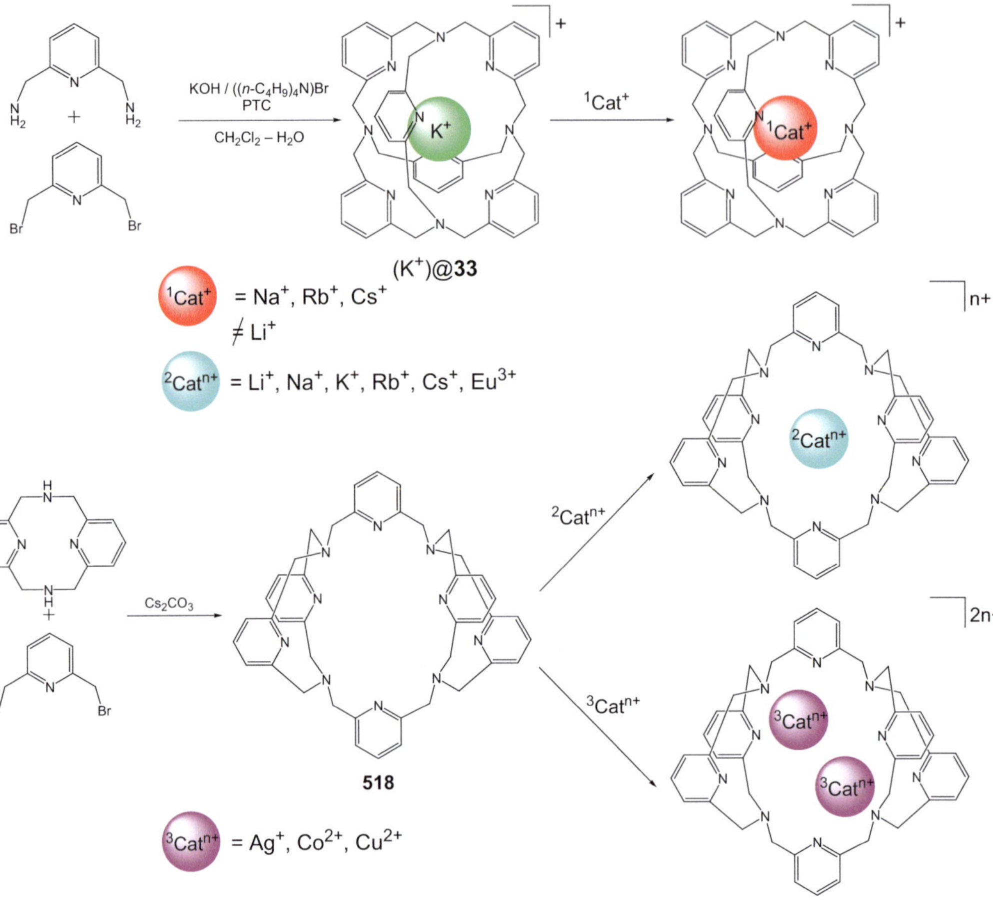

Scheme 2.255 Preparation of the polypyridine ligands **33** and **518** and their cage complexes

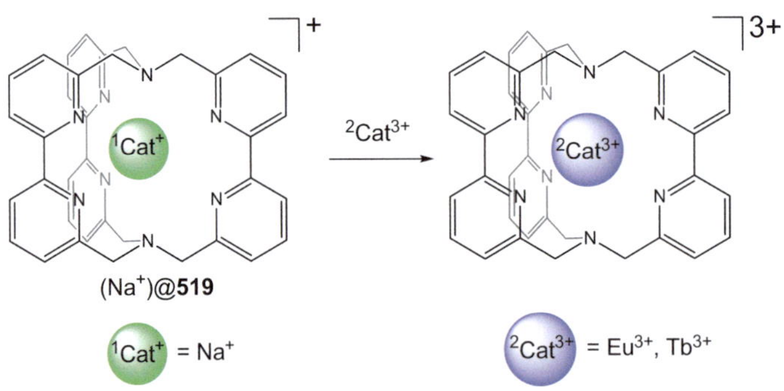

Scheme 2.256 Synthesis of the complexes of the tris-bipyridine caging ligand **519**

Lewis acidity of zirconium and hafnium(IV) phthalocyanines has been used [177] for the synthesis of hybrid phthalocyaninoclathrochelates with phthalocyanine-containing capping group by Scheme 2.261. The boron, antimony-capped clathrochelate precursors (Fe^{2+})@**271** and (Fe^{2+})@**524** with labile triethylantimony cross-linking group underwent a transmetallation (a capping group exchange reaction) at room temperature with these Lewis acids under mild

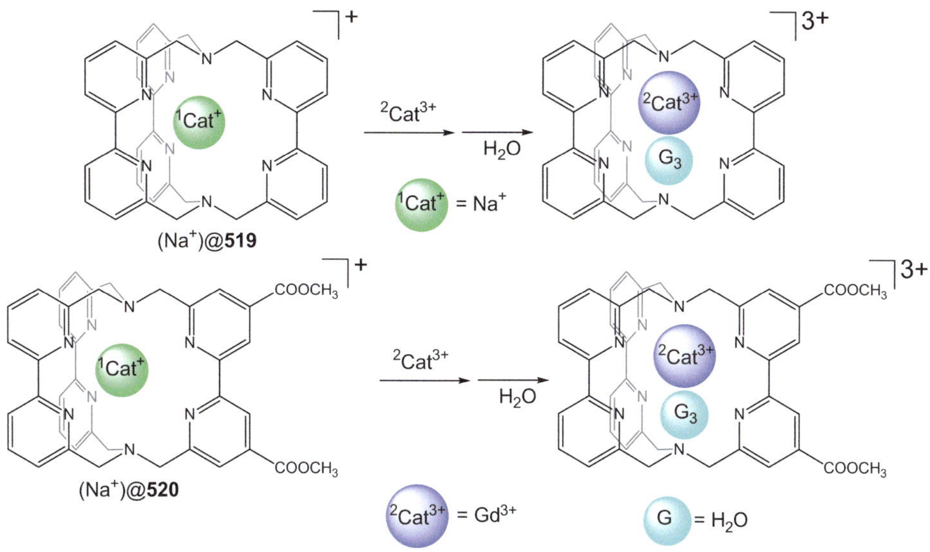

Scheme 2.257 Preparation of gadolinium(III) complexes of the macrobicycles **519** and **520**

reaction conditions, thus giving the hybrid compounds in practically quantitative yields.

Analogous oximehydrazonate phthalocyaninoclathrochelates, the lutetium(III) phthalocyanine-capped iron(II) cage complexes (Fe^{2+})@**528** and (Fe^{2+})@**529**, have been synthesized [162] by Scheme 2.262 using a transmetallation of the corresponding triethylantimony-capped tris-oximehydrazonate precursors (see Sect. 2.2.4), allowing to obtain the target hybrid complexes in moderate yields.

Ditopic hybrid monoporphyrinoclathrochelates (Fe^{2+})@**530** and (Fe^{2+})@**531** have been prepared [178] in moderate yields using a transmetallation of a mixed boron, antimony-capped iron(II) tris-nioximate clathrochelate (Fe^{2+})@**271** as a macrobicyclic precursor with -Lewis-acidic dichlorozirconium or dichlorohafnium(IV) 5,10,15,20-tetraphenylporphyrins (*44* and *45*, respectively) by Scheme 2.263 under mild reaction conditions. Their heterotrinuclear analogs (Fe^{2+})@**532** and (Fe^{2+})@**533** with two apical porphyrin fragments have been also synthesized using the same synthetic approach starting from the bis-triethylantimony-cross-linked iron(II) clathrochelate precursor (Fe^{2+})@**270** with two reactive apical groups (Scheme 2.263).

Condensation of the pyrazoloxime ligand synthon *39* with a BF_2-cross-linked macrocyclic iron(II) bis-α-benzildioximate *13* by Scheme 2.264 [179] gave the mixture of the hybrid clathrochelatoscorpionate (Fe^{2+})@**534** and its pyrazoloxime-armed macrocyclic intermediate *46*: the latter complex, forming in the range 70–80°C, underwent its further macrobicyclization at approximately 100°C.

The hybrid porphyrin–clathrochelate system (Co^{2+})@**535** has been prepared [180] in a high yield by Scheme 2.265 using a coordination-driven self-assembly of the 4-pyridylboron-capped cobalt(II) clathrochelate (Co^{2+})@**241** with zinc(II) porphyrin in the presence of triethylamine.

2.3 Polynucleating Macrobicyclic Ligands and Their Metal Complexes

2.3.1 Schiff-Base Imine Capsules and Their Polyamine Derivatives

The large octaamine macrobicycle **536** with six pendant $Si(OC_2H_5)_3$ groups, the derivative of

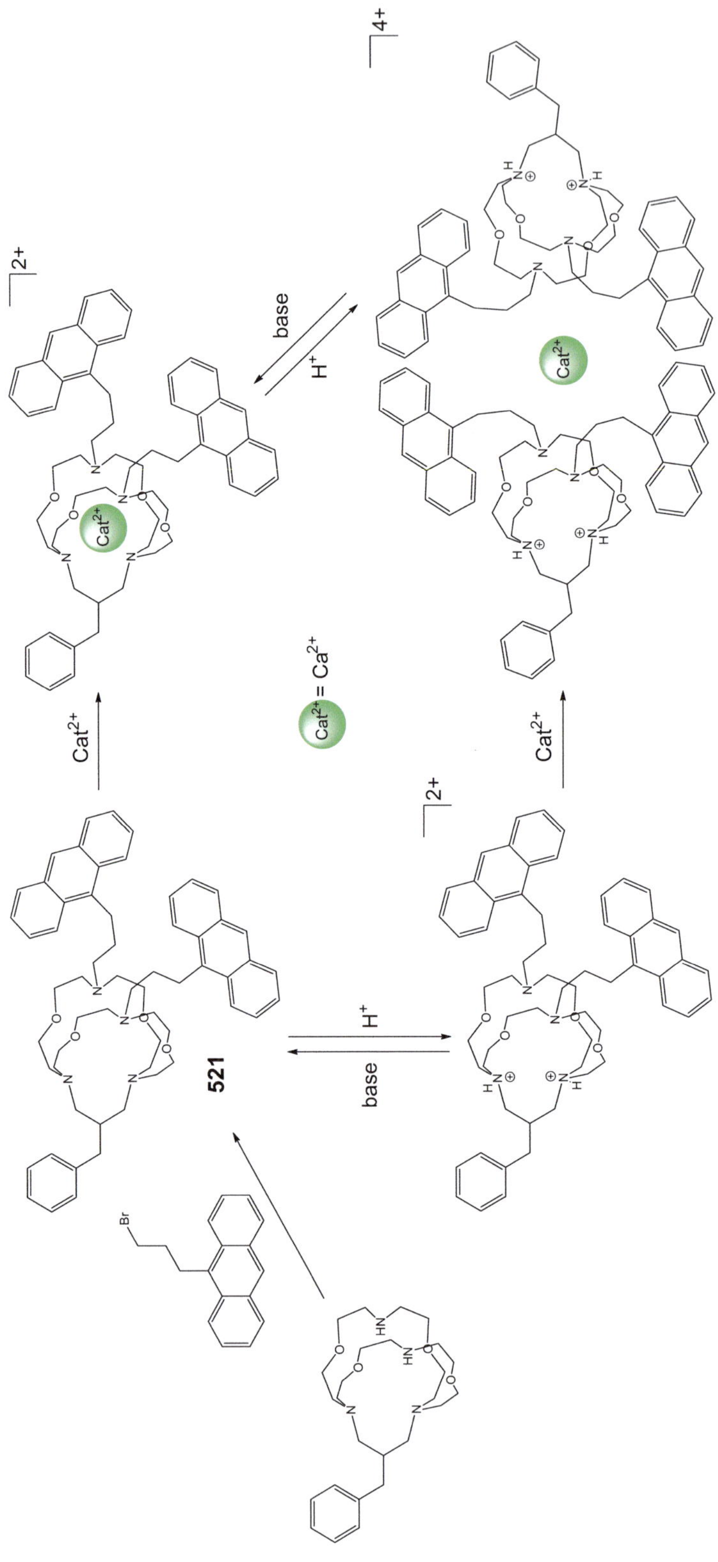

Scheme 2.258 Preparation of the anthracenyl-appended ligand **521** and its calcium(II) complex

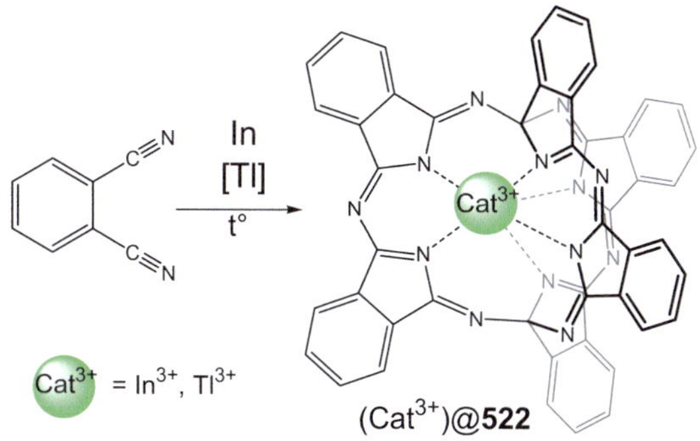

Scheme 2.259 Template self-assembly of the indium and thalium(III) clathrochelates (Cat^{3+})@**522**

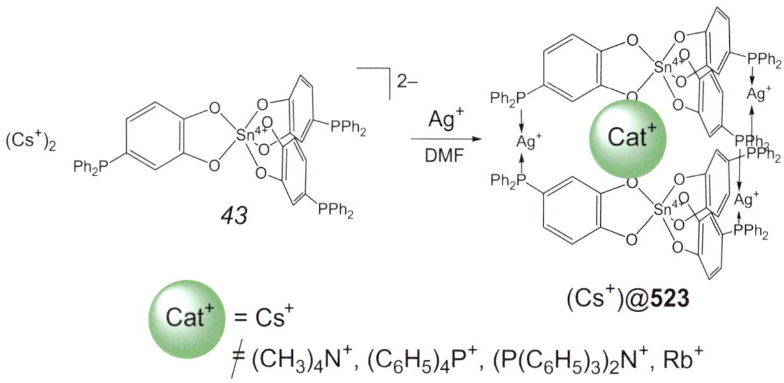

Scheme 2.260 Self-assembly of cesium(I) complex of the coordination capsule **523**

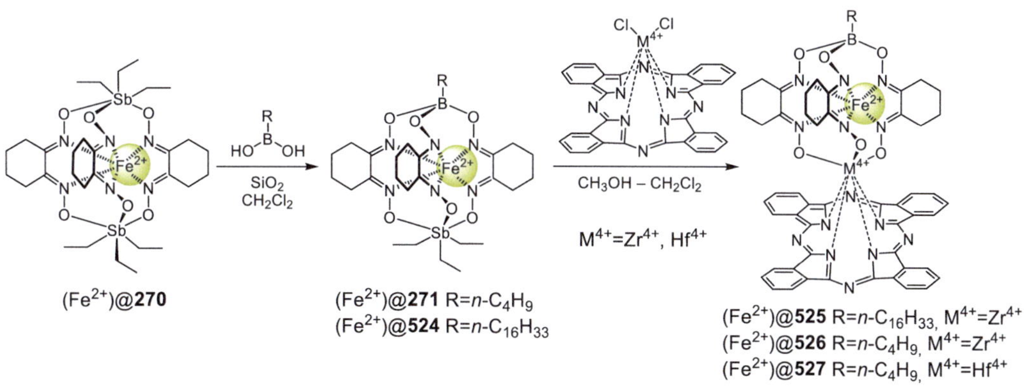

Scheme 2.261 Preparation of the hybrid α-dioximate iron(II) phthalocyaninoclathrochelates

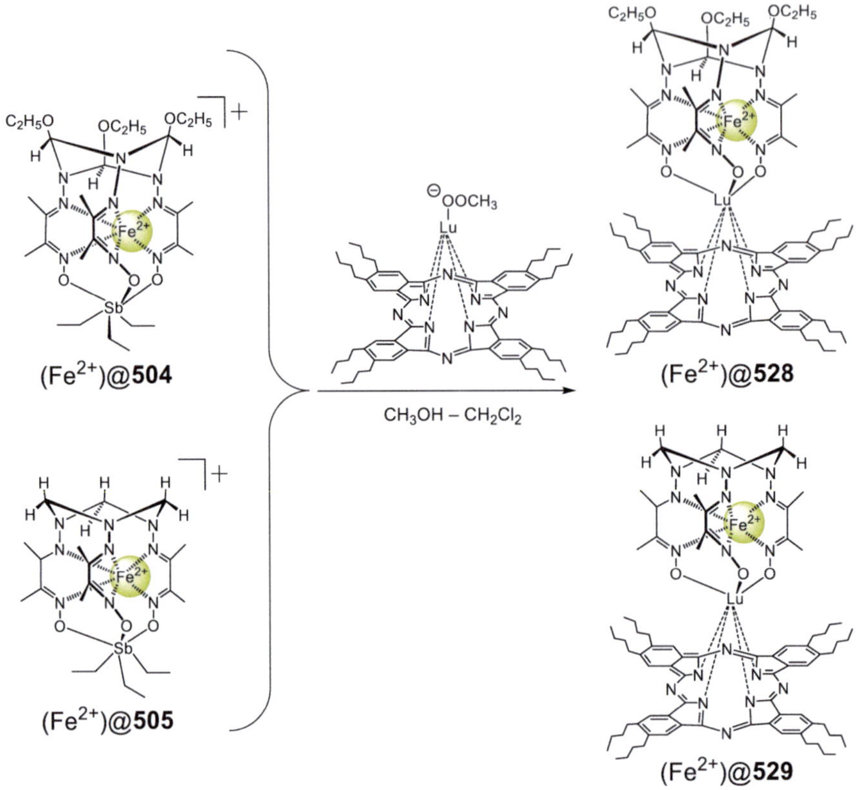

Scheme 2.262 Synthesis of the hybrid oximehydrazonate iron(II) phthalocyaninoclathrochelates

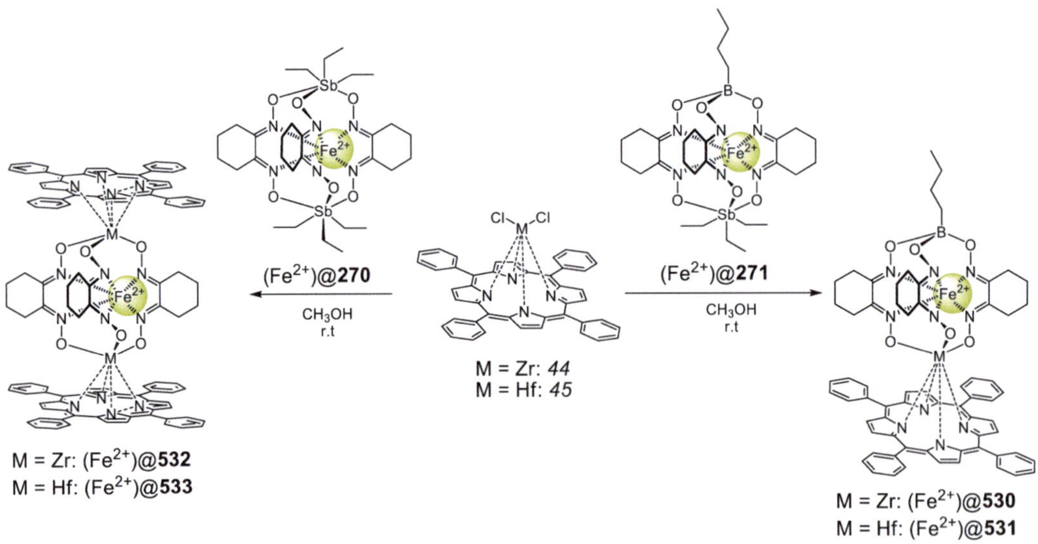

Scheme 2.263 Preparation of the hybrid iron(II) mono- and bis-porphyrinoclathrochelates

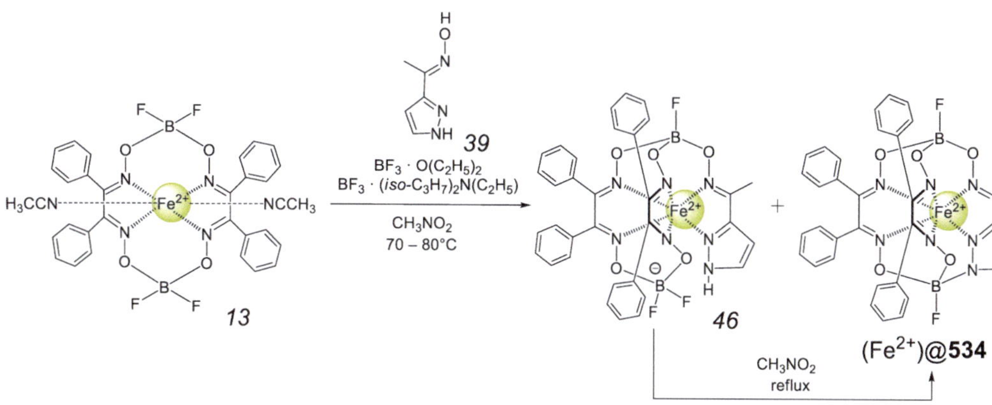

Scheme 2.264 Synthesis of the hybrid iron(II) clathrochelatoscorpionate and its macrocyclic analog

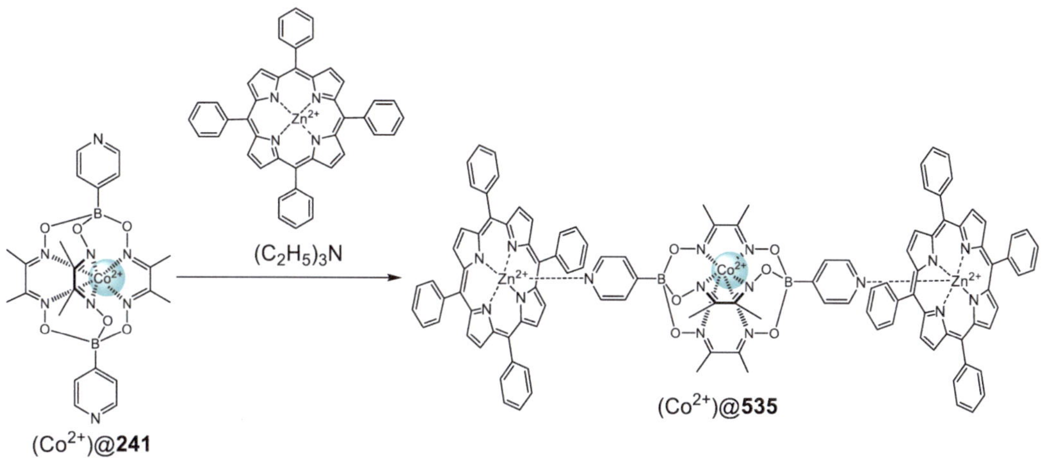

Scheme 2.265 Self-assembly of zinc(II) porphyrin with clathrochelate ligand (Co^{2+})@**241**

tripodal amine *tren*, has been prepared [181] by Scheme 2.266. It encapsulates the copper(II) ions both in solution and their trapped into sol–gel matrixes, thus giving the corresponding 1:2 cage complex.

1:1:1 heterodinuclear cage complex of the extended Schiff-base macrobicyclic ligand **537** with co-encapsulated silver and copper(I) cations has been prepared [182] by Scheme 2.267 using the exchange reaction of an encapsulated metal ion of the homodinuclear silver(I) clathrochelate $(Ag^+)_2$@**537**.

The hexaimine macrobicyclic ligand **538**, prepared by imine condensation of its dialdehyde synthon *47* with tripodal amine *tren* by Scheme 2.268, co-encapsulates two copper(I) cations, thus forming the corresponding homodinuclear 1:2 cage complex [183].

1:2 homodinuclear copper(I) cage complex of the Schiff-base tris-pyrrole macrobicyclic ligand **539** has been synthesized [184] by Scheme 2.269 using template imine condensation of its dialdehyde synthon *48* with tripodal amine *tren* on a copper(I) ion as a matrix. This clathrochelate underwent the chemical redox transformations also shown in Scheme 2.269: its one-electron oxidation yielded a localized mixed-valence $Cu^{I}Cu^{II}$ cage complex of a monodeprotonated form of **539**, while two-electron oxidation gave the homodinuclear 1:2 copper(II) clathrochelate, the derivative of a triply deprotonated form of **539** [184].

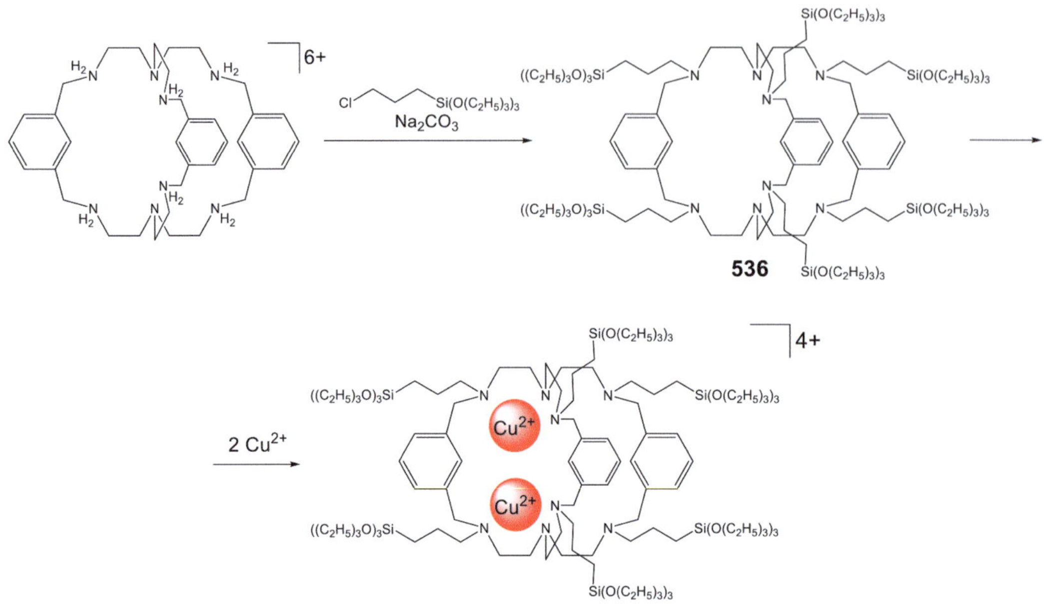

Scheme 2.266 Preparation of the octaamine macrobicycle **536** and its copper(II) cage complex

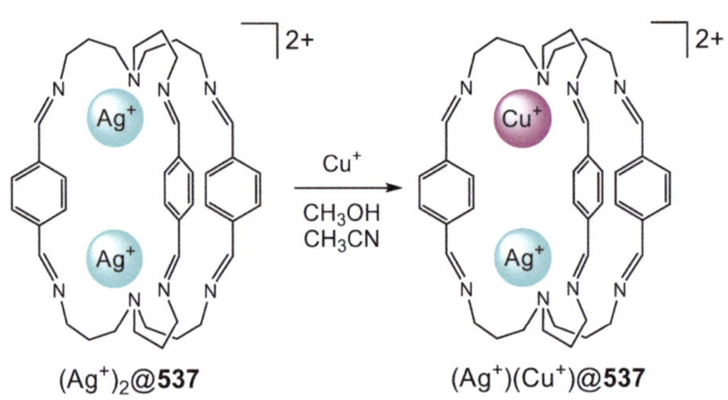

Scheme 2.267 Synthesis of the heterodinuclear clathrochelate $(Ag^+)(Cu^+)$@**537**

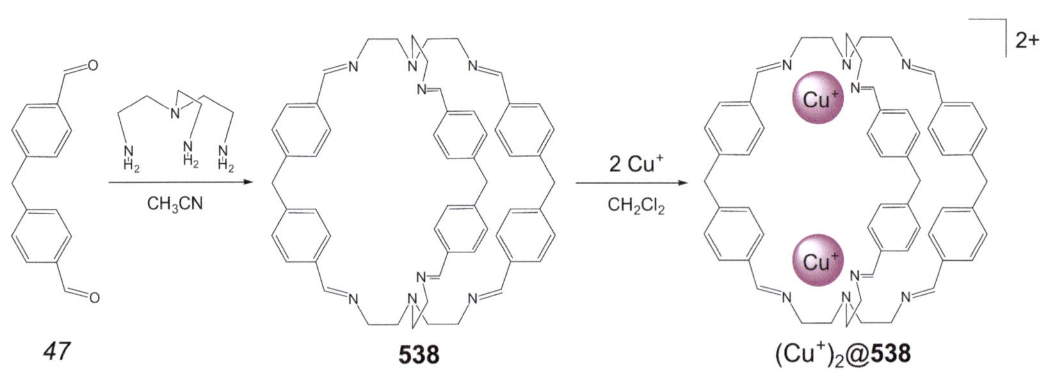

Scheme 2.268 Preparation of the hexaimine macrobicycle **538** and its dicopper(I) cage complex

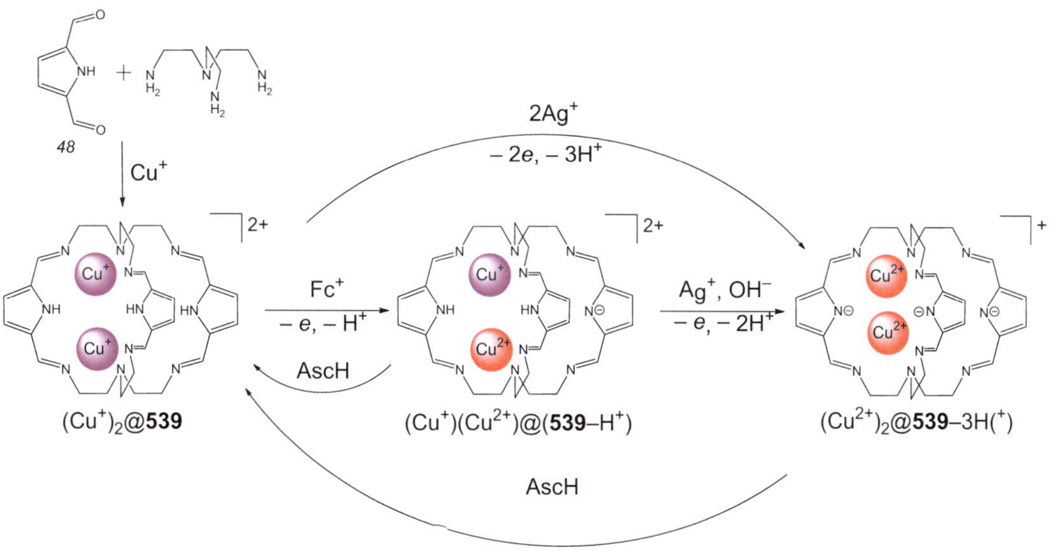

Scheme 2.269 Synthesis of copper cage complexes of **539** and its deprotonated form

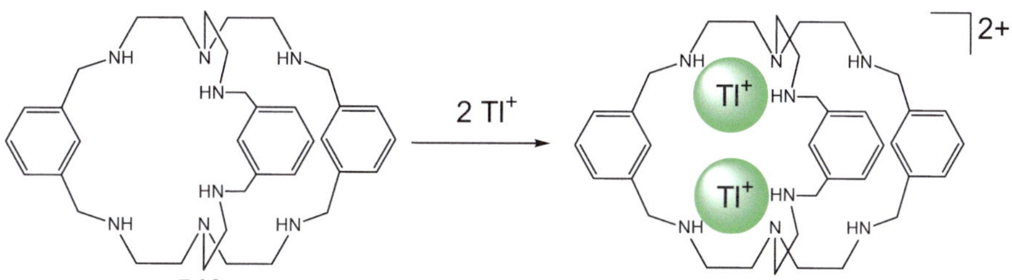

Scheme 2.270 Preparation of homodinuclear thallium(I) clathrochelate $(Tl^{+})_2$@**540**

Coordination-driven self-assembly of the Schiff-base *meta*-xylyl caging ligand **540** with thallium(I) ions has been used [185] for preparation of 1:2 homodinuclear cage complex $(Tl^{+})_2$@**540** by Scheme 2.270.

1:1:1 heteronuclear cage complexes of the europium(III) ion with co-encapsulated cadmium, nickel, and zinc(II) cations have been prepared [186] using two-step synthetic procedure (Scheme 2.271) *via* the formation of an intermediate mixed europium(III), calcium(II)-encapsulating clathrochelate $(Eu^{3+})(Ca^{2+})$@**541**.

The synthesis of the homo- and heterodinuclear 1:2 and 1:1:1 cage complexes of the macrobicyclic ligand **545** with encapsulated copper(I,II) and cobalt(II) ions has been performed [187] by Scheme 2.272.

Preparation of the homo- and heterobinuclear 1:2 and 1:1:1 cage complexes of the Schiff-base encapsulating ligands **543** and **544** has been accomplished by Scheme 2.273 [188].

The octaamine macrobicyclic ligand **540** is described [189] to form its homodinuclear cage complex with co-encapsulated copper(II) ions (Scheme 2.274).

A mixed-valence dinuclear silver(I,III) cage complex of the extended Schiff-base encapsulating ligand **537** has been synthesized [190] using template [2+3] imine condensation by Scheme 2.275.

A series of 1:2 homo- and 1:1:1 heterodinuclear copper(I) and copper(II) cage complexes of the encapsulating tris-pyridazine ligand **545** and its macrobicyclic amine

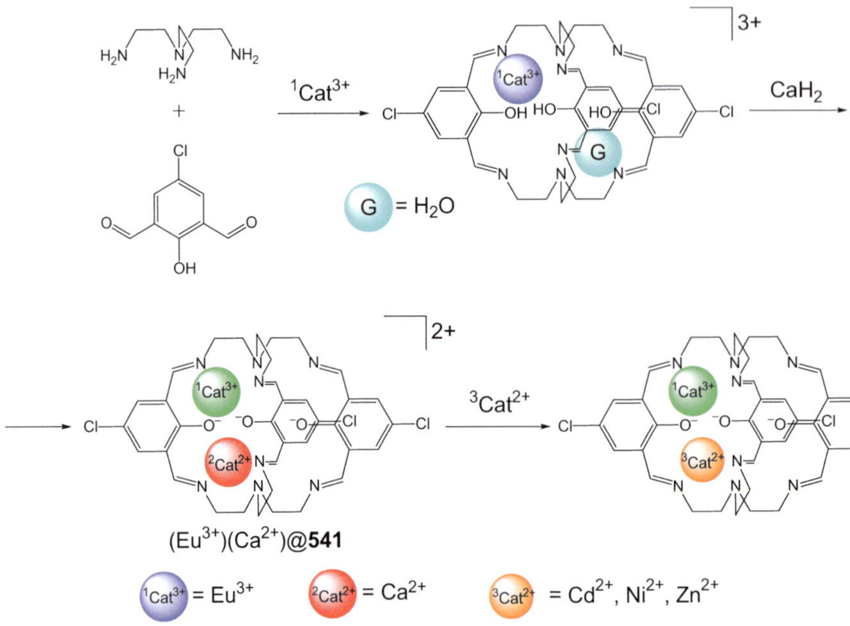

Scheme 2.271 Synthesis of the heterodinuclear cage complexes of the ligand **541**

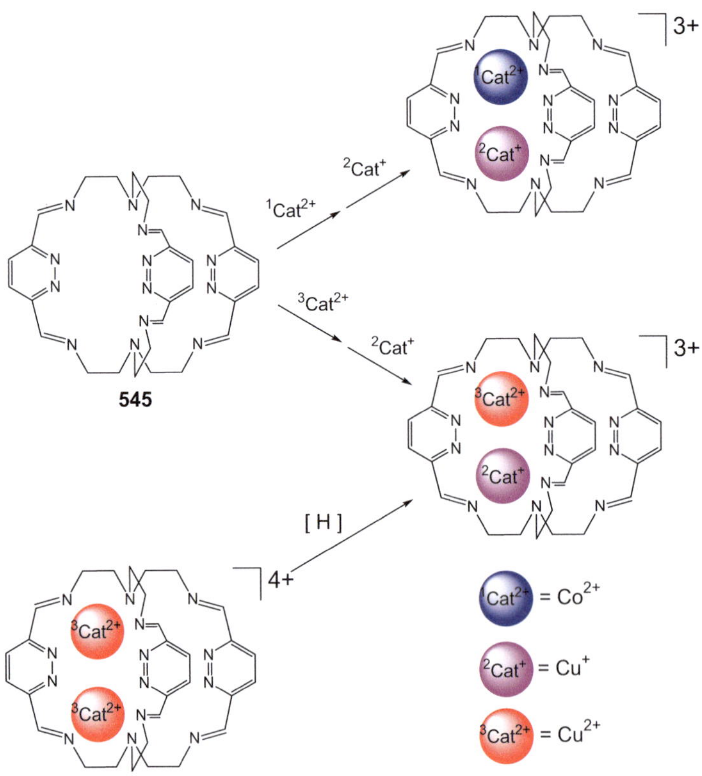

Scheme 2.272 Preparation of homo- and heterodinuclear complexes of the macrobicycle **545**

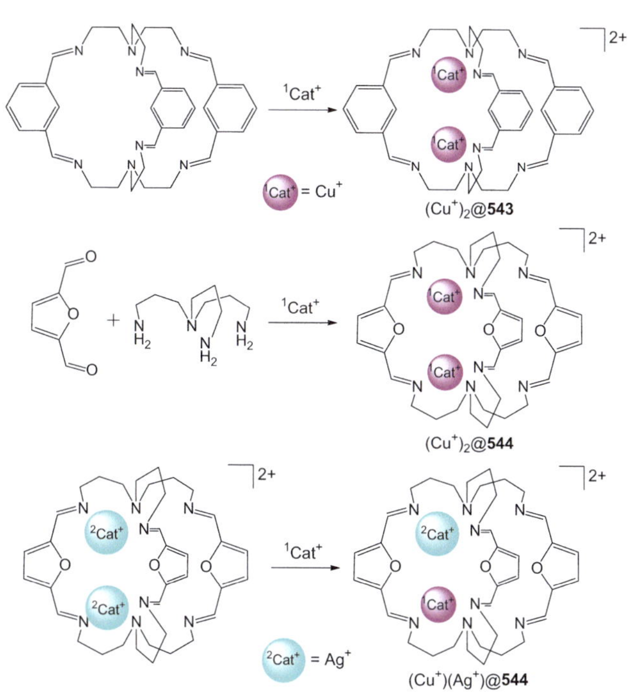

Scheme 2.273 Synthesis of homo- and heterodinuclear complexes of the ligand **544**

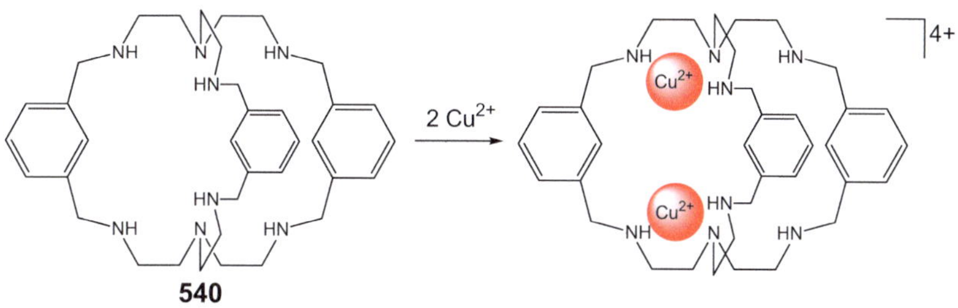

Scheme 2.274 Preparation of copper(II) complex of the caging ligand **540**

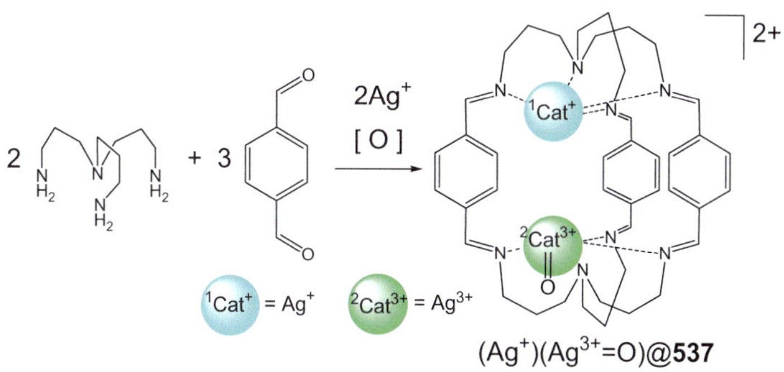

Scheme 2.275 Template synthesis of silver(I),silver(III) cage complex of **537**

derivative **546** has been prepared [191] by Scheme 2.276. The hexaimine ligand **545** encapsulates two copper(I) ions, thus forming its homodinuclear 1:2 copper(I) cage complex, or one copper(II) ion to give its 1:1 clathrochelate. An addition of copper(I) ions to this mononuclear clathrochelate resulted in the formation of 1:1:1 heterodinuclear mixed-valence copper(I)- and copper(II)-encapsulating cage complex $(Cu^{+})(Cu^{2+})$@**545**. The polyamine macrobicyclic ligand **546** co-encapsulates two copper(II) ions, thus giving the corresponding homodinuclear 1:2 clathrochelate $(Cu^{2+})_2$@**546** also shown in Scheme 2.276 [191].

Template imine condensation of the dialdehyde ligand synthon 49 with tripodal amine tren by Scheme 2.277 on a silver(I) ion as a matrix has been used [192] for the synthesis of the homodinuclear 1:2 cage complex of the Schiff-base hexaimine macrobicyclic ligand **542** with two co-encapsulated silver(I) ions.

Template-free condensation of tripodal amine *tren* with terephthalaldehyde *49* has been used [193] for preparation of the hexaimine binucleating ligand **542**; its reduction with $NaBH_4$ gave the polyamine macrobicycle **547**. The ligand **542** co-encapsulates two silver or two copper(I) ions, thus giving the corresponding homodinuclear 1:2 cage complexes shown in Scheme 2.278.

Coordination-driven self-assembly of the macrobicyclic ligand **548** with copper(I) ions has been used [194] for the synthesis of 1:2 homodinuclear cage complex $(Cu^{+})_2$@**548** by Scheme 2.279.

The tris-pyrazolate polyamine macrobicyclic ligand **549** has been synthesized [195, 196] by Scheme 2.280 using [3+2] imine condensation of 3,5-pyrazoledicarbaldehyde with tripodal amine *tren* followed by a reduction with $NaBH_4$. A monodeprotonated form of this clathrochelate ligand co-encapsulates two copper(II) ions, thus giving 1:2 homodinuclear cage complex $(Cu^{2+})_2$@(**549**–H^{+}) [196].

A series of 1:2 homodinuclear silver and copper(I) cage complexes of the thiophene-containing Schiff-base imine macrobicyclic ligand **550**, its extended analog **551**, and its reduced amine derivative **552** has been prepared [197] using the common synthetic approaches shown in Scheme 2.281.

The homodinuclear silver(I) cage complex of the N_8O_6-macrobicyclic ligand **553** has been synthesized [198] by Scheme 2.282.

The authors of [199] reported their unsuccessful attempts to obtain the macrobicyclic compound **554** using a metal-free [2+3] imine condensation of tripodal amine tren with 2,5-diformyl-4-methylphenol, whereas in the presence of a suitable template, main group 1 or 2 cation as a matrix, this compound has been isolated in a high yield. Because its cage complexes with encapsulated group 2 metal ions are poorly soluble, 1:1 sodium(I) cage complexes of **554** with various counteranions have been used [199] as clathrochelate precursors for the corresponding exchange reactions of an encapsulated metal ion. Such reactions with metal groups 12 and 14 cations in neutral solutions by Scheme 2.283 gave only the mononuclear 1:1 cage complexes of this potentially binucleating caging ligand. Similarly, the kinetically stable 1:1 lanthanide(III) cage complexes of **554** and its analog **555** shown in Scheme 2.284 have been prepared [200]. The exchange reactions of an encapsulated metal ion of 1:1 sodium(I) cage complexes of these encapsulating ligands with iron(II) and iron(III) cations under aerobic conditions afforded only their iron(III) clathrochelates [201]. In contrast, the analogous reactions with manganese(II) and manganese(III) ions gave only their 1:1 manganese(II) cage complexes.

A series of 1:1 lanthanide(III) cage complexes of the encapsulating hexaimine ligand **554** has been prepared [202] by Scheme 2.285 using one-step [2+3] imine condensation of tripodal amine *tren* with 2,6-diformyl-4-methylphenol *50* in the presence of the corresponding lanthanide(III) cation.

The mono- and homodinuclear lanthanide(III) cage complexes of the macrobicyclic ligand **554** and its triple deprotonated form, respectively, have been prepared [203] by Scheme 2.286 using template imine condensation of triethylammonium salt of *50* with tripodal amine *tren* on a lanthanide(III) ion as a matrix.

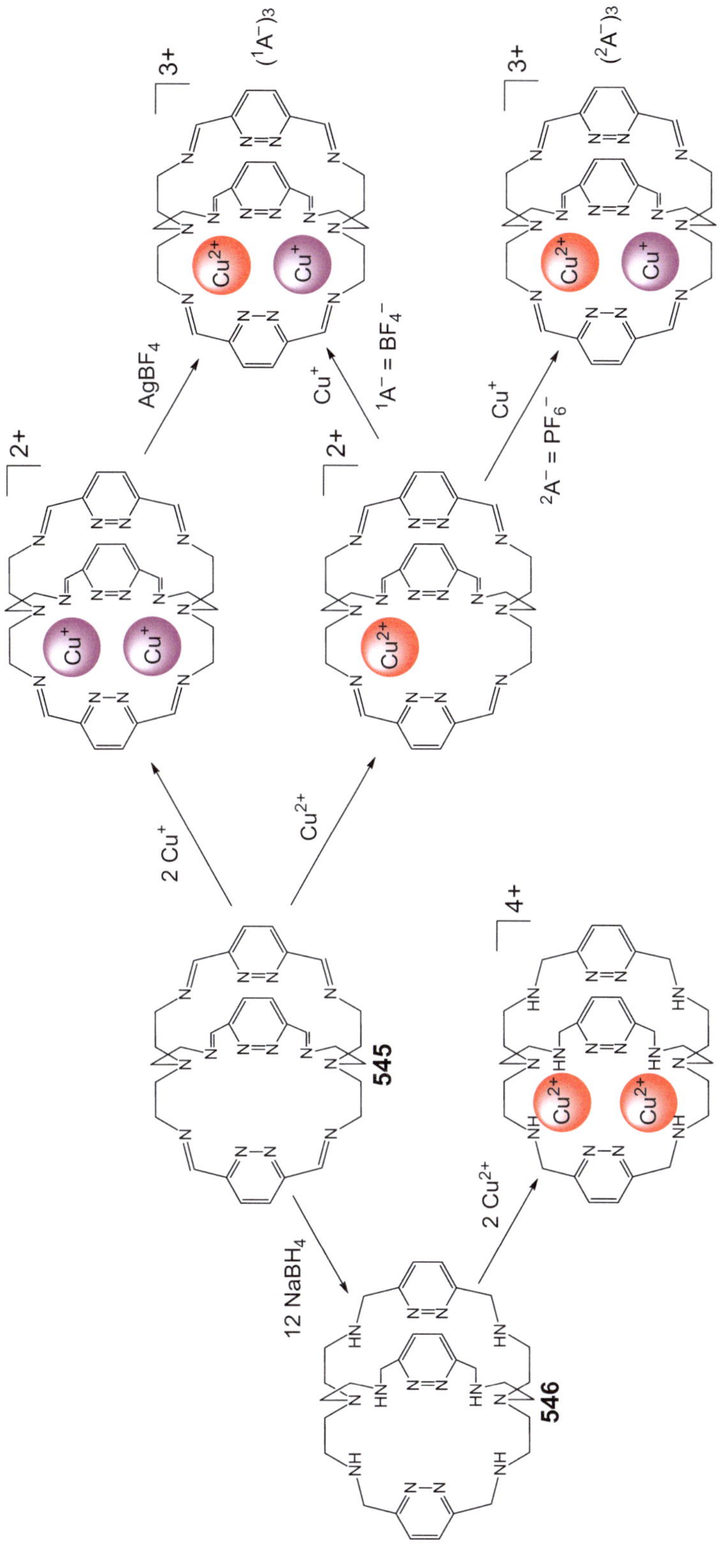

Scheme 2.276 Synthesis of mono- and dinuclear cage complexes of the hexaimine ligand **545** and its amine deivative **546**

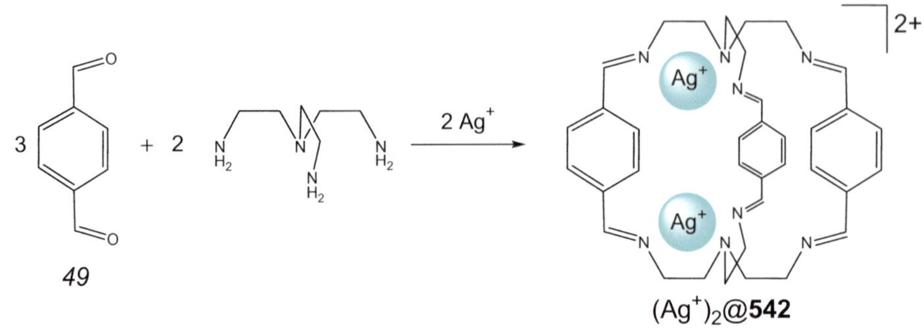

Scheme 2.277 Template synthesis of the homodinuclear silver(I) cage complex of the ligand **542**

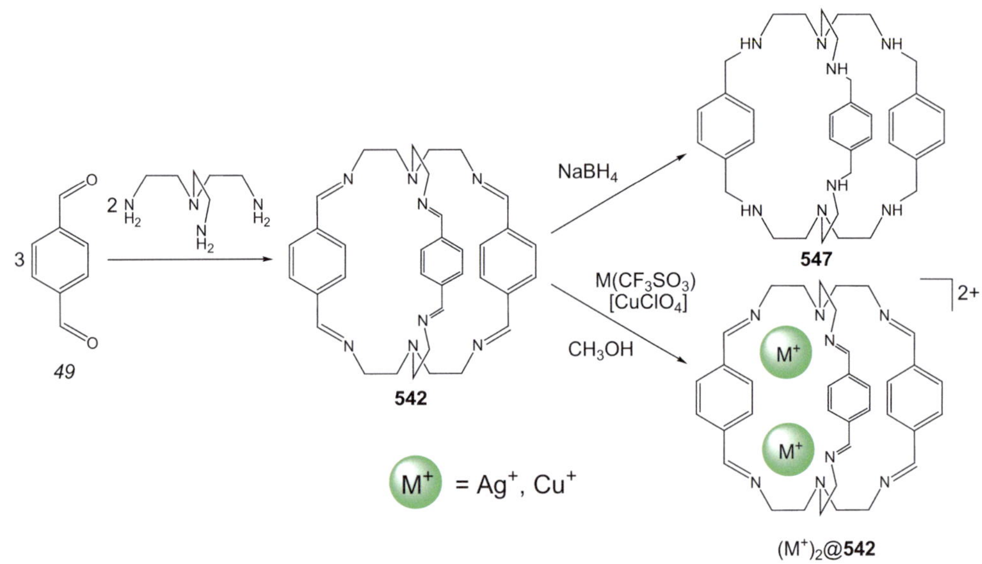

Scheme 2.278 Preparation of the imine macrobicyclic ligand **542**, its amine derivative **547** and homodinuclear cage complexes of **542**

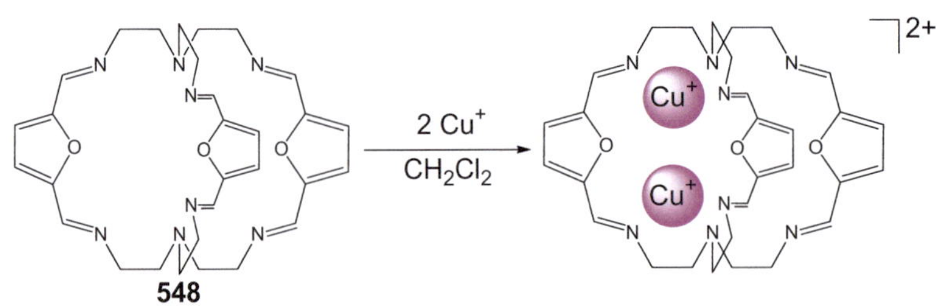

Scheme 2.279 Encapsulation of copper(I) ions by the binucleating ligand **548**

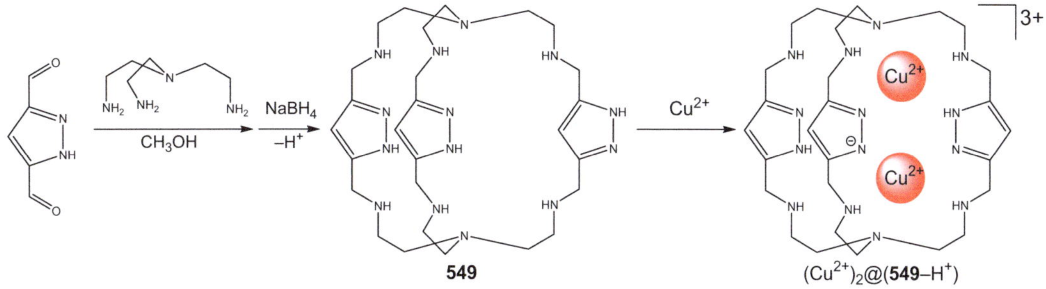

Scheme 2.280 Synthesis of homodinuclear copper(II) clathrochelate $(Cu^{2+})_2$@(**549**–H^+)

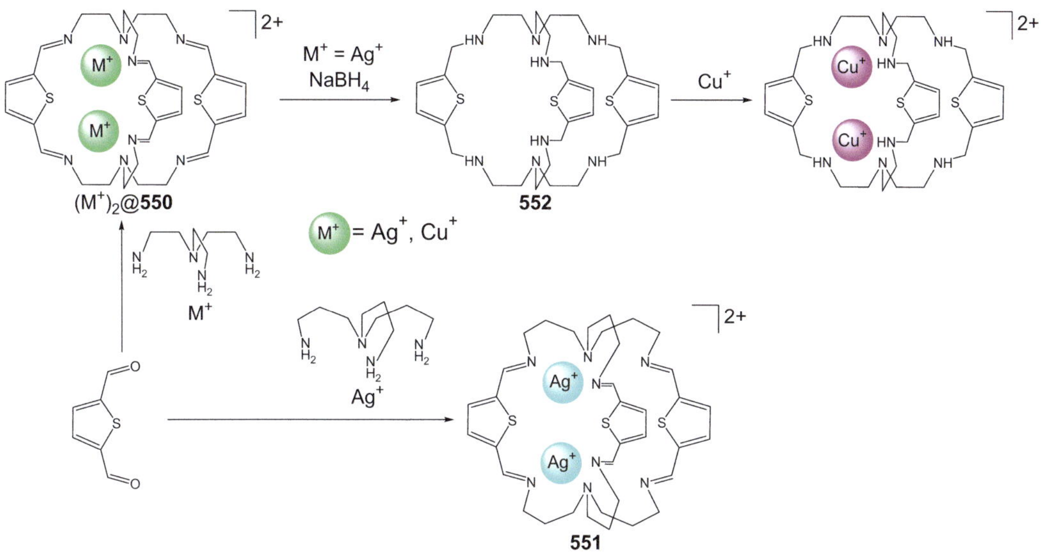

Scheme 2.281 Preparation of homodinuclear cage complexes of the ligands **550–552**

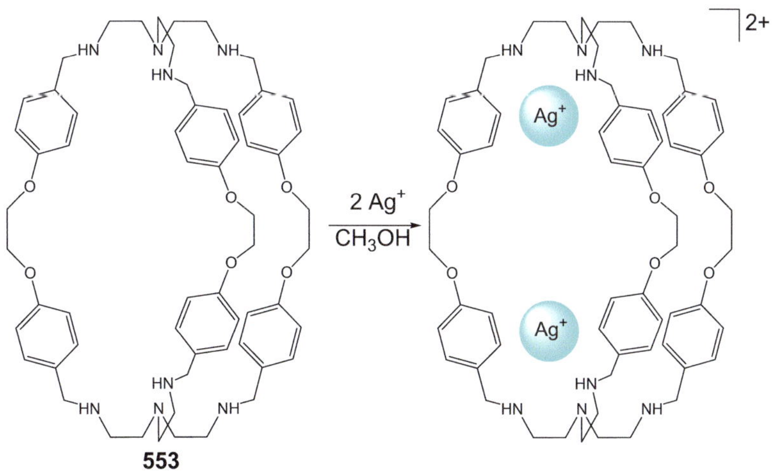

Scheme 2.282 Synthesis of homodinuclaer silver(I) complex of the ligand **553**

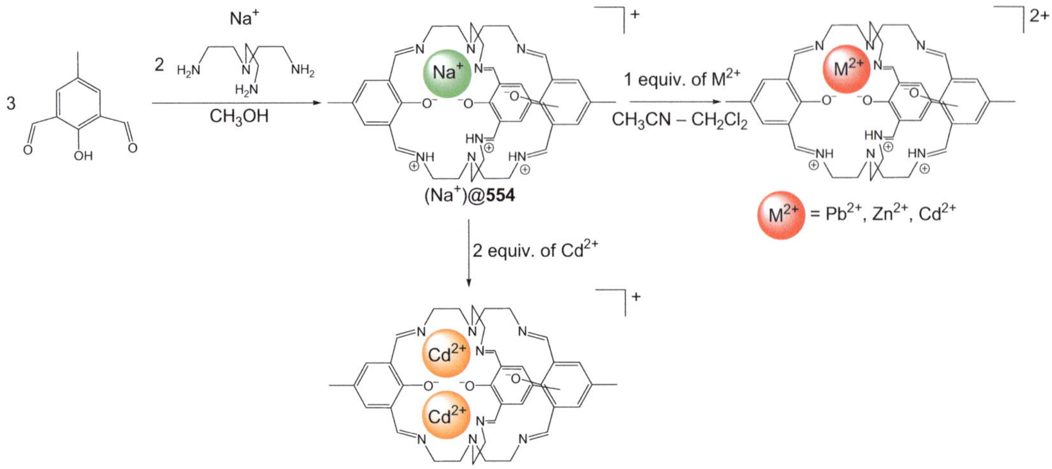

Scheme 2.283 Synthesis of mono- and dinuclear cage complexes of the ligand **554**

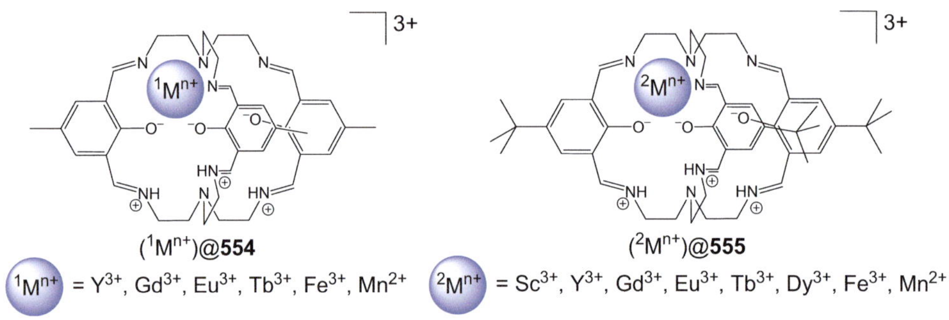

Scheme 2.284 Chemical drawings of metal(II) and metal(III) cage complexes of the ligands **554** and **555**

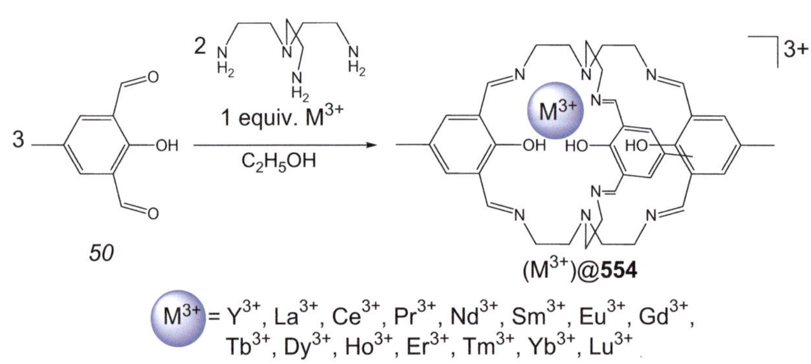

Scheme 2.285 Preparation of mononuclear lanthanide(III) cage complexes of **544**

[2+3] imine condensation of tripodal amine *tren* with *50* on the corresponding lanthanide(III) ion by Scheme 2.287 has been used [204] for preparation of a series of the homodinuclear 1:2 cage complexes of the encapsulating ligand **554**.

The above mononuclear lanthanide(III) cage complexes of the triple deprotonated form of the Schiff-base macrobicyclic ligand **554** are reported [205] to co-encapsulate zinc(II) ion in the presence of calcium hydride by Scheme 2.288, thus giving the corresponding

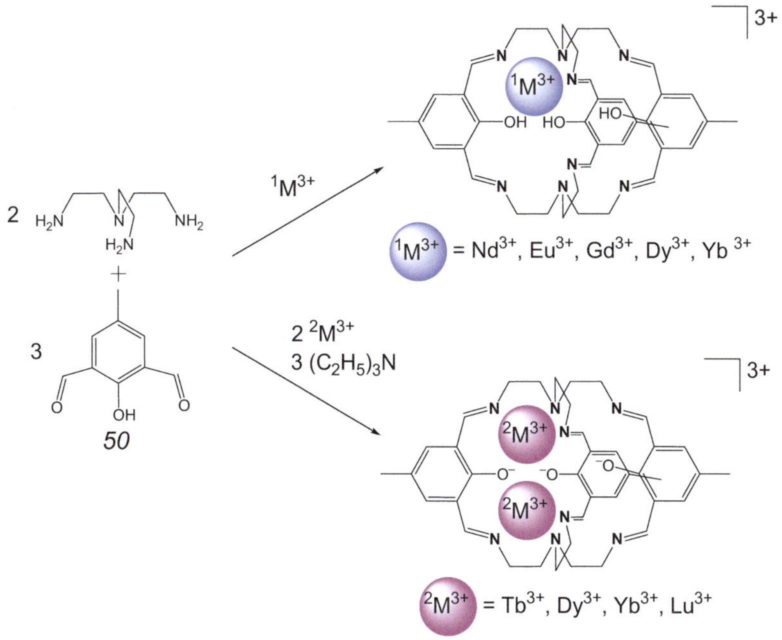

Scheme 2.286 Synthesis of mono- and dinuclear lanthanide(III) complexes of **554**

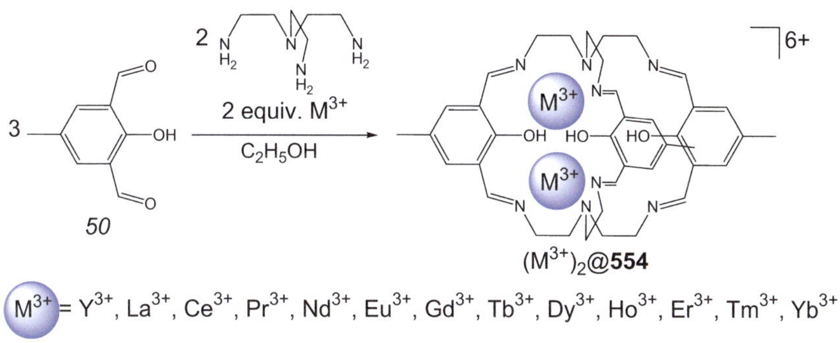

Scheme 2.287 Preparation of the homodinuclear lanthanide clathrochelates $(M^{3+})_2$@**554**

heterodinuclear 1:1:1 host–guest compounds in moderate yields; the synthetic pathway to the homodinuclear 1:2 zinc(II) clathrochelate $(Zn^{2+})_2$@**554** is also shown in this Scheme.

1:1 cage complexes of the macrobicyclic imine ligand **556** with an encapsulated high-spin iron(II) ion ($s = 2$) and that of **555** with a caged iron(III) cation in an intermediate spin state ($s = 3/2$) have been synthesized [206] by Scheme 2.289 *via* coordination of the initially prepared macrobicycle **556** toward iron(II) cation and by direct template imine condensation on an iron(III) ion as a matrix, respectively. The exchange of an encapsulated sodium(I) ion in its 1:1 cage complex of the ligand **554** has been used [206] for preparation of the iron(III) clathrochelate (Fe^{3+})@**554**.

A series of 1:1 sodium(I) cage complexes, the clathrochelate derivatives of four Schiff-base encapsulating ligands **539**, **548**, **554**, and **556**, has been synthesized [207] by Scheme 2.290.

The octaamine caging ligands **540**, **547**, **557**, and **558** shown in Scheme 2.291 are reported [208] to form 1:1 and 1:2 cage complexes with

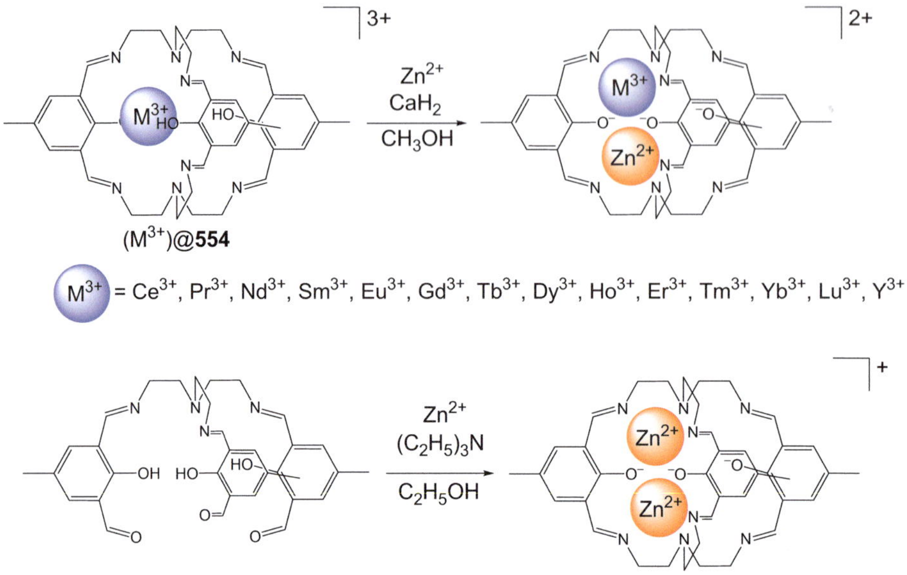

Scheme 2.288 Synthesis of the zinc(II)-encapsulating homo- and heterodinuclear cage complexes of **554**

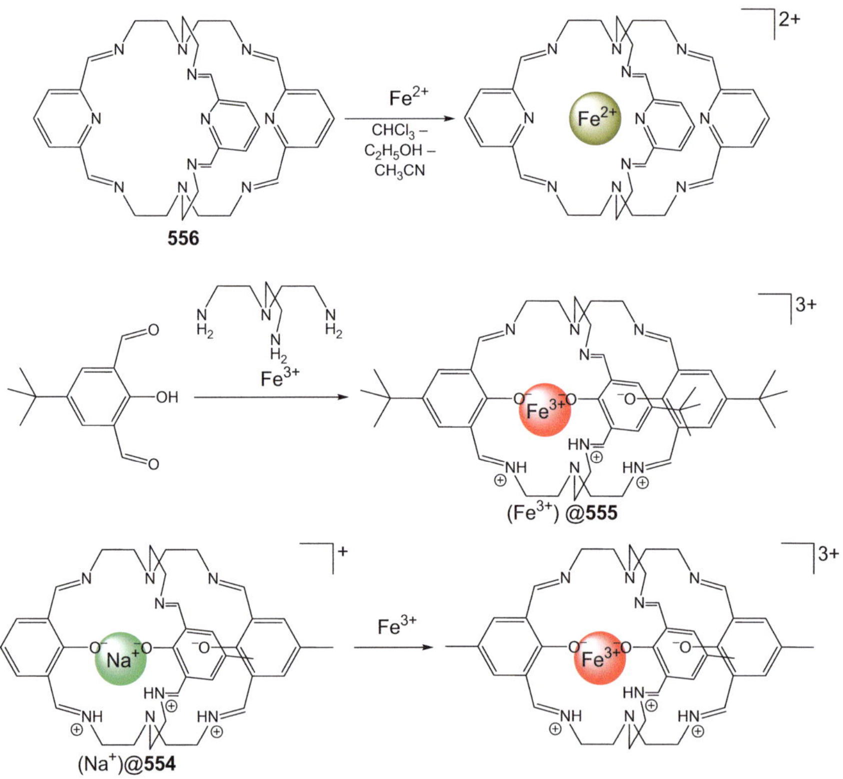

Scheme 2.289 Preparation of the iron(III) cage complexes of **554** and **555**

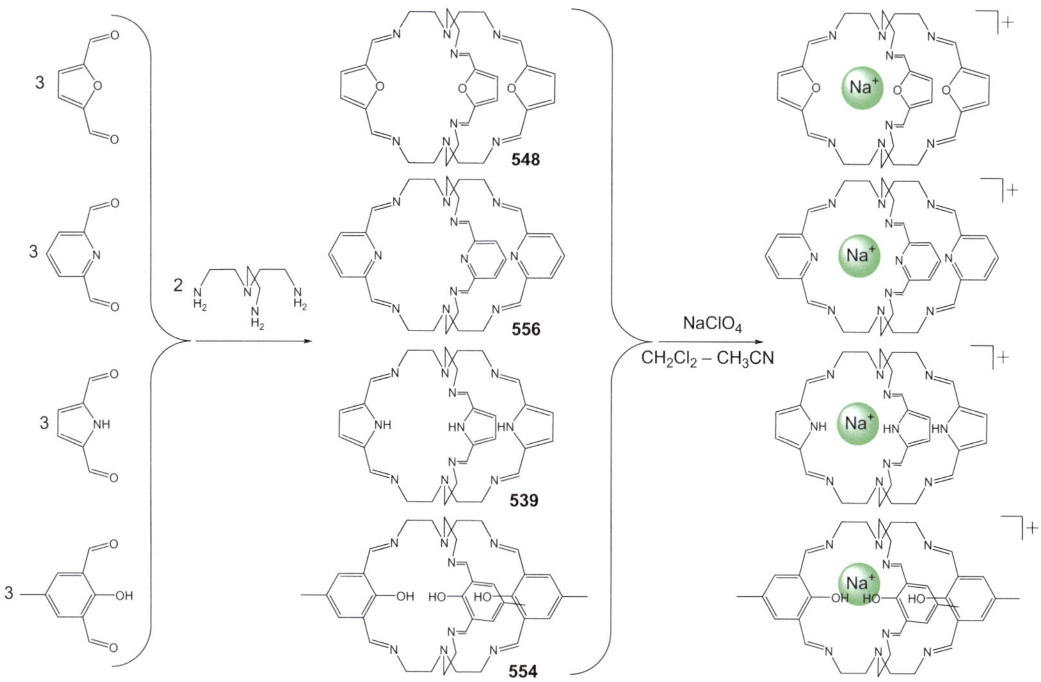

Scheme 2.290 Synthesis of a series of the hexaimine macrobicycles and their sodium(I) cage complexes

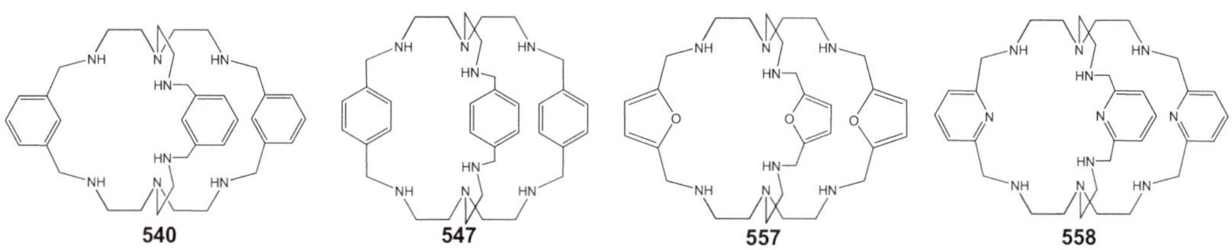

Scheme 2.291 Chemical drawimgs of a series of the macrobicyclic polyamine compounds

encapsulated cobalt, nickel, and zinc(II) cation(s). Thermodynamic stabilities of the mononuclear cage complexes of these macrobicyclic ligands and that of the corresponding homodinuclear metal(II) clathrochelates, the derivatives of their analogs of **557** and **558**, change in a row: $Ni^{2+} < Co^{2+} << Cu^{2+} >> Zn^{2+}$ [208].

The *tren*-capped binucleating macrobicyclic ligand **559** is reported [209, 210] to encapsulate zinc, copper, nickel, and cobalt(II) ions in a stepwise manner, thus giving their 1:1 and 1:2 cage complexes by Scheme 2.292; its protonated form formed the analogous mono- and dinuclear clathrochelates shown in Scheme 2.293. Their stability constants decrease in a row: $Cu^{2+} > Zn^{2+}$, $Ni^{2+} > Co^{2+}$, and those for binding of the second guest metal ion are lower by four or five orders of magnitude.

The octaamine macrobicyclic ligand **547** has been effectively encapsulated silver(I), copper, cobalt, or nickel(II) ions, giving their 1:2 cage complexes by Scheme 2.294 [211]. In the case of iron and manganese(II) triflates, a tetraprotonated form of **547**, possessing the intramolecular hydrogen bonds, has been formed, thus preventing it from an encapsulation of the solvent guest molecules and the metal ions as well [211].

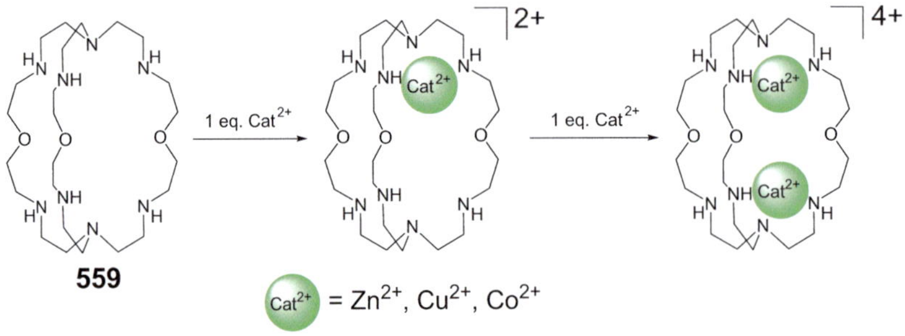

Scheme 2.292 Stepwise encapsulation of metal(II) ions by the ligand **559**

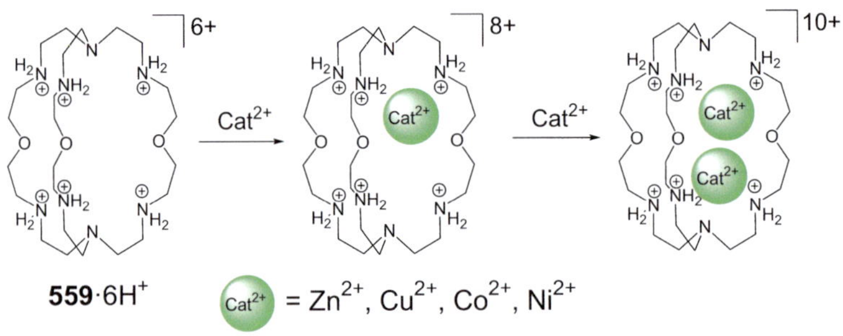

Scheme 2.293 Stepwise encapsulation of metal(II) ions by the ligand **559**·6H$^+$

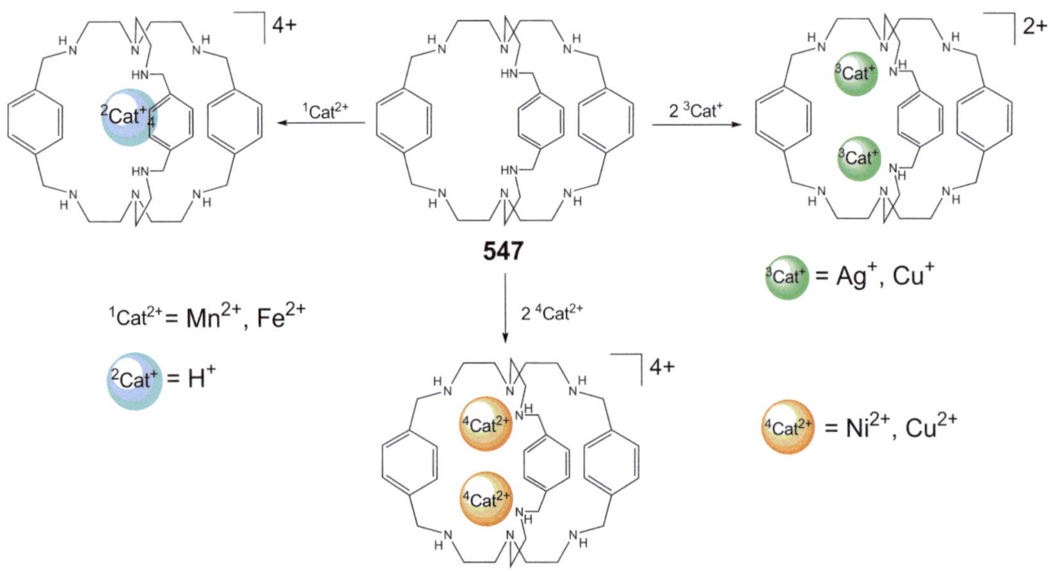

Scheme 2.294 Preparation of mono- and dinuclear cage complexes of the ligand **547**

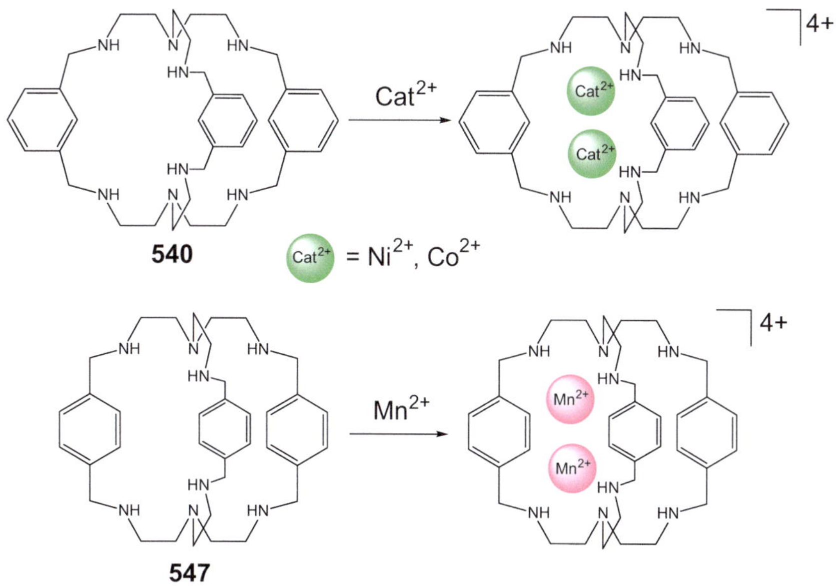

Scheme 2.295 Encapsulation of metal(II) ions by the ligands **540** and **547**

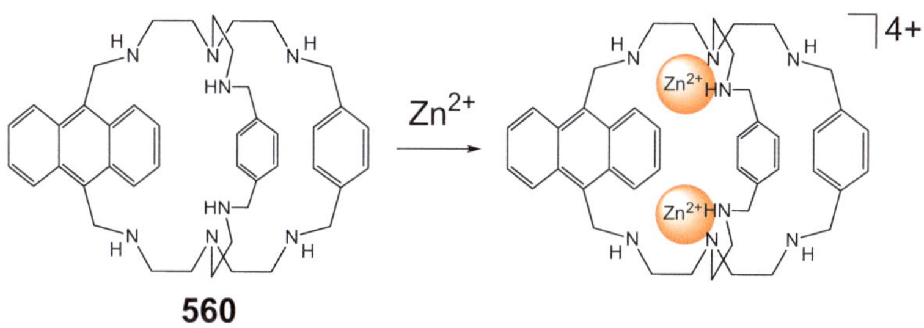

Scheme 2.296 Preparation of the homodinuclear zinc(II) cage complex of the ligand **560**

The homodinuclear 1:2 nickel and cobalt(II) cage complexes of the polyamine macrobicyclic ligand **540** have been prepared [212] by Scheme 2.295. A manganese(II) ion is too large to form its clathrochelate complex with this *meta*-phenylene ligand, whereas the *para*-phenylene isomer **547** did this.

The fluorescent macrobicyclic ligand **560** with an extended cavity and 9,10-anthracene-ribbed fragment is described [213] to accommodate two zinc(II) ions, thus forming the corresponding 1:2 cage complex by Scheme 2.296.

Potentiometric titration has been used [214] to study a coordination-driven self-assembly of the furan-based macrobicyclic ligand **557** with copper(II) ions. Their homodinuclear 1:2 cage complex shown in Scheme 2.297 has been formed in the range of pH 4–7.

The thiophene-based macrobicyclic ligand **552** also formed its homodinuclear copper(II) cage complex by Scheme 2.298 [215]. In contrast, the *N*-methylated macrobicyclic ligand **561** (Scheme 2.299) encapsulates only one copper(II) cation in the presence of oxygen-centered cationic co-guest H_3O^+ species, forming the strong hydrogen bonds with three *N*-methylated amino groups of this caging ligand [216]. Such hydrogen bonding is reported [217] to strongly compete with an encapsulation of the second Cu^{2+}ion by **561**, thus suppressing the formation of its homodinuclear 1:2 copper(II) cage complex.

Scheme 2.297 Synthesis of the homodinuclear copper(II) cage complex of the ligand **557**

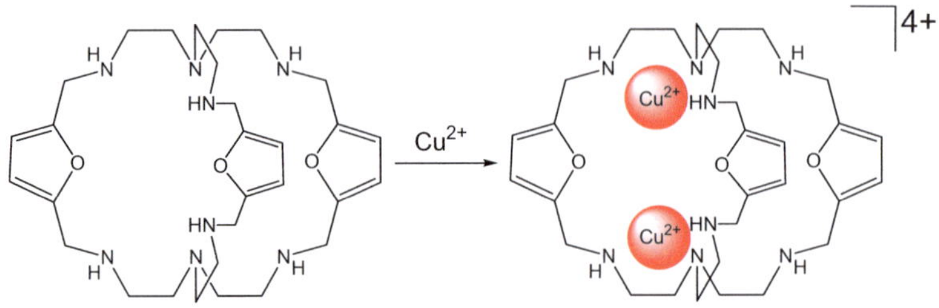

Scheme 2.298 Preparation of the homodinuclear copper(II) cage complex of the ligand **552**

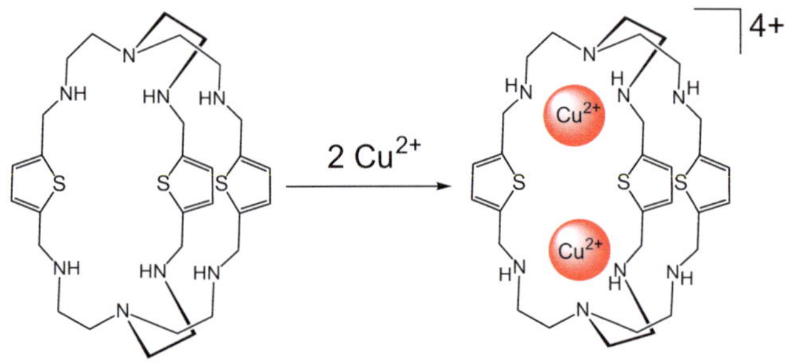

Scheme 2.299 Synthesis of the copper(II) cage complex of the ligand **561**

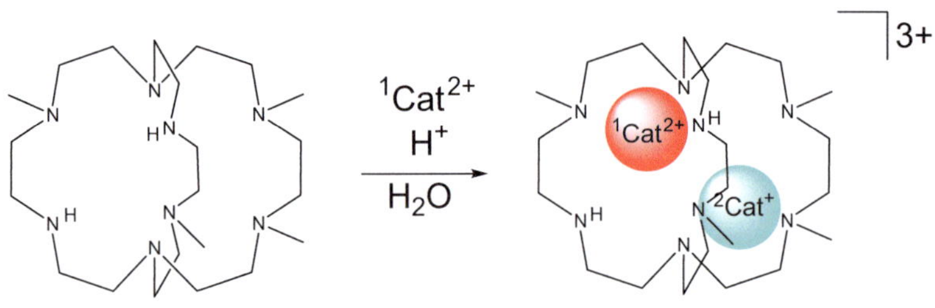

The homodinuclear 1:2 copper(II) cage complexes of the octaamine macrobicyclic ligands **547**, **540**, and **557** have been synthesized [218] by Scheme 2.300.

The macrobicyclic *tren*-based ligand **562**, which has been self-assembled [219] by Scheme 2.301, forms its homodinuclear cage complex $(Cu^+)_2$@**562** with two co-encapsulated copper(I) ions.

The binucleating hexacarboxamide caging ligand **563** (Scheme 2.302) with C_3-symmetric macrobicyclic framework easily accommodates cobalt, iron, nickel, and zinc(II) ions, thus forming their homodinuclear 1:2 cage complexes [220, 221].

2.3.2 Other Types of Polynucleating Caging Ligands

The labile octaaminophenolic macrobicyclic ligand **564** is described [222] to encapsulate zinc and copper(II) cations, thus giving the corresponding homodinuclear 1:2 cage complexes by Scheme 2.303.

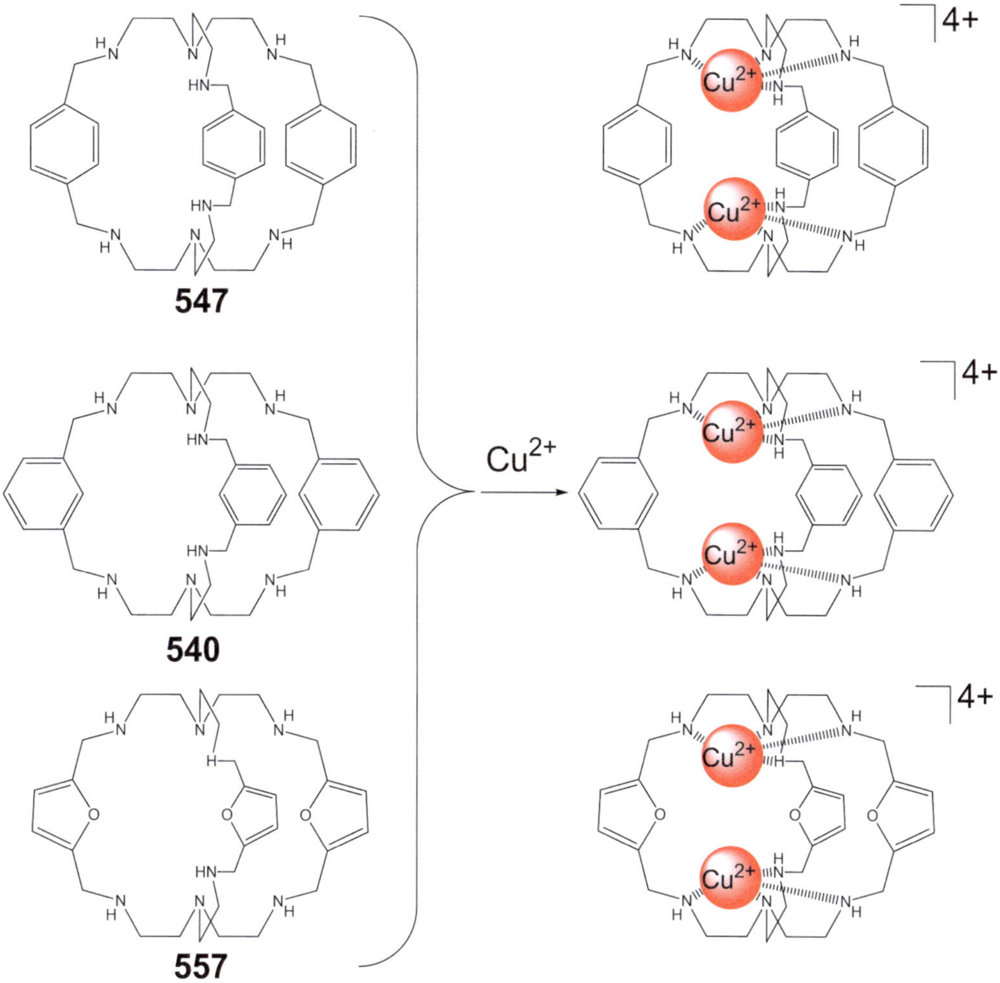

Scheme 2.300 Preparation of the homodinuclear copper(II) cage complexes of the ligands **540**, **547** and **557**

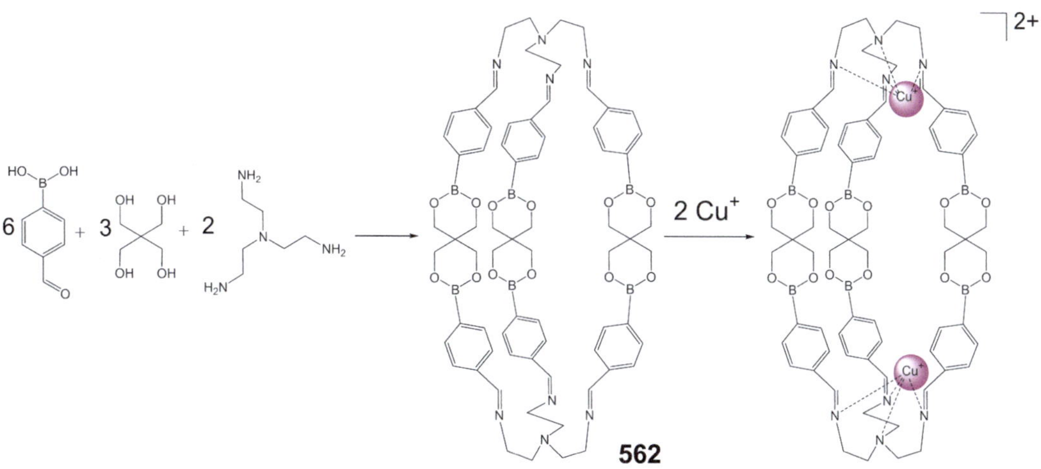

Scheme 2.301 Self-assembly of the macrobicyclic ligand **562** and its homodinuclear copper(I) cage complex

Scheme 2.302 Synthesis of the homodinuclear cage complexes of the ligand **563**

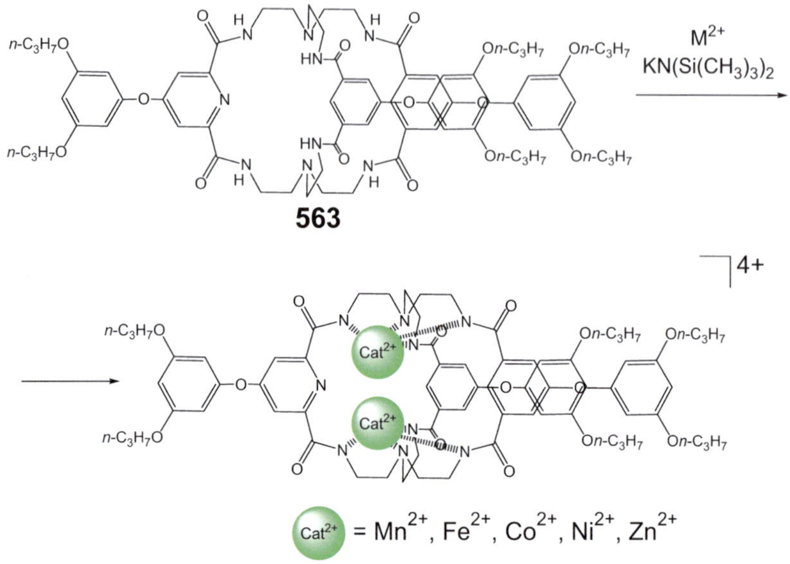

Scheme 2.303 Preparation of the homodinuclear copper(II) complex of the ligand **564**

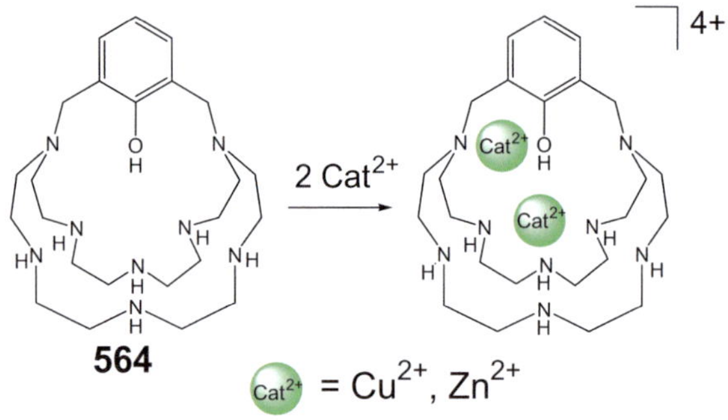

Scheme 2.304 Synthesis of the homotrinuclear copper(I) and copper(II) cage complexes of the ligand **565**

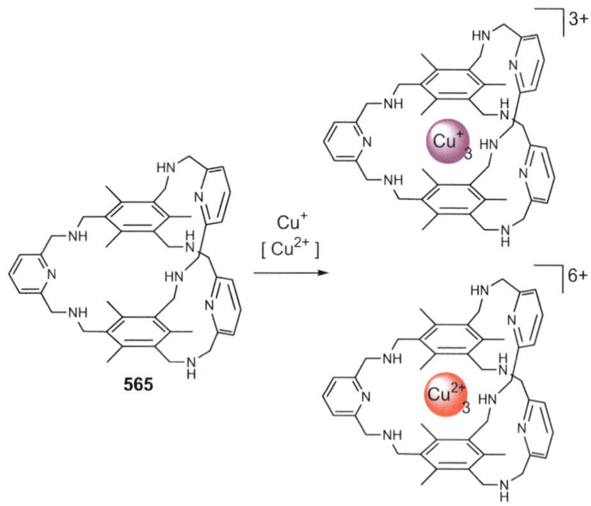

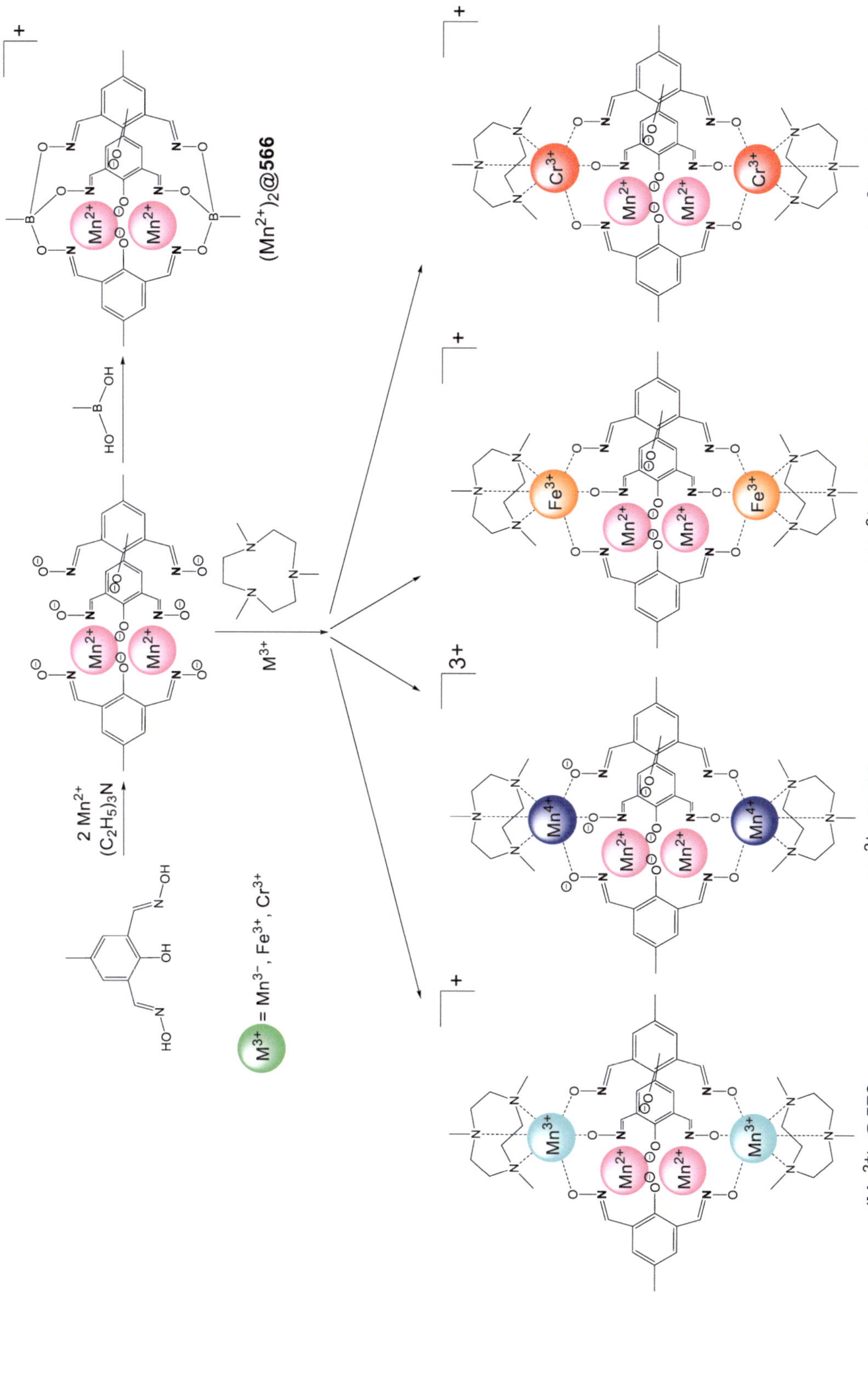

Scheme 2.305 Preparation of the homodinuclear manganese(II) cage complexes of the ligands **566–570**

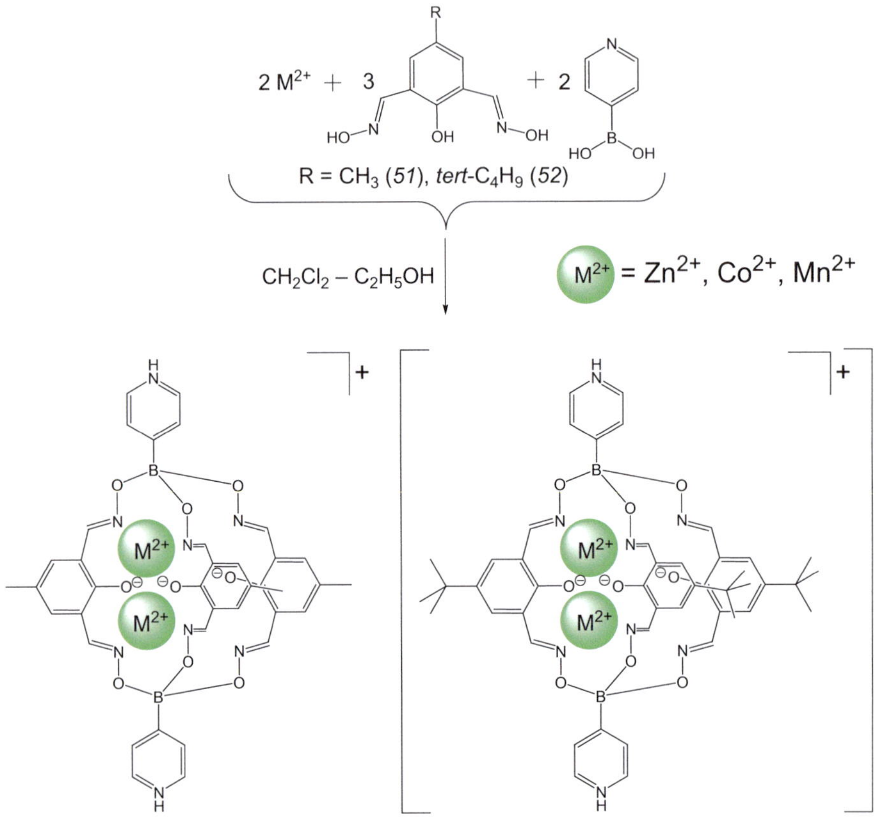

Scheme 2.306 Synthesis of the homodinuclear metal(II) cage complex of the ligands **571** and **572**

Preparation of Cu^{I}_{3}- and Cu^{II}_{3}-homotrinuclear cage complexes of the encapsulating ligand **565** by Scheme 2.304 is reported [223].

A series of 1:2 homodinuclear manganese(II) cage complexes of the boron-, chromium, iron, and manganese(III)- and manganese(IV)-capped coordination capsules **566–570**, the derivatives of 2,6-diformyl-4-methylphenol oxime as a ligand synthon, has been synthesized [224] using two-step synthetic procedure shown in Scheme 2.305.

1:2 homodinuclear 4-pyridylboron-capped zinc, cobalt, and manganese(II) cage complexes of the macrobicyclic ligands **571** and **572**, the derivatives of 2,6-diformyl-4-methylphenol oxime *51* and 2,6-diformyl-4-*tert*-butylphenol oxime *52*, respectively, have been self-assembled [225] by Scheme 2.306. The formed cationic clathrochelates underwent a deprotonation of their encapsulating ligands with organic bases, such as triethylamine or tetraethylammonium hydroxide: the corresponding macrobicyclic anions have been prepared in high yields by Scheme 2.307 [225].

Capping of the ligand synthon *51* with 4-formylphenylboronic acid on the corresponding metal(II) ion as a matrix has been used [83] for preparation of 1:2 homodinuclear manganese and cobalt(II) cage complexes of the formyl-terminated encapsulating ligand **573** by Scheme 2.308.

The anionic macrobicyclic homodinuclear cobalt and zinc(II) complexes of the binucleating ligand **574** and **575** with two terminal carboxylic groups have been synthesized [90] by template condensation of 4-carboxyphenylboronic acid

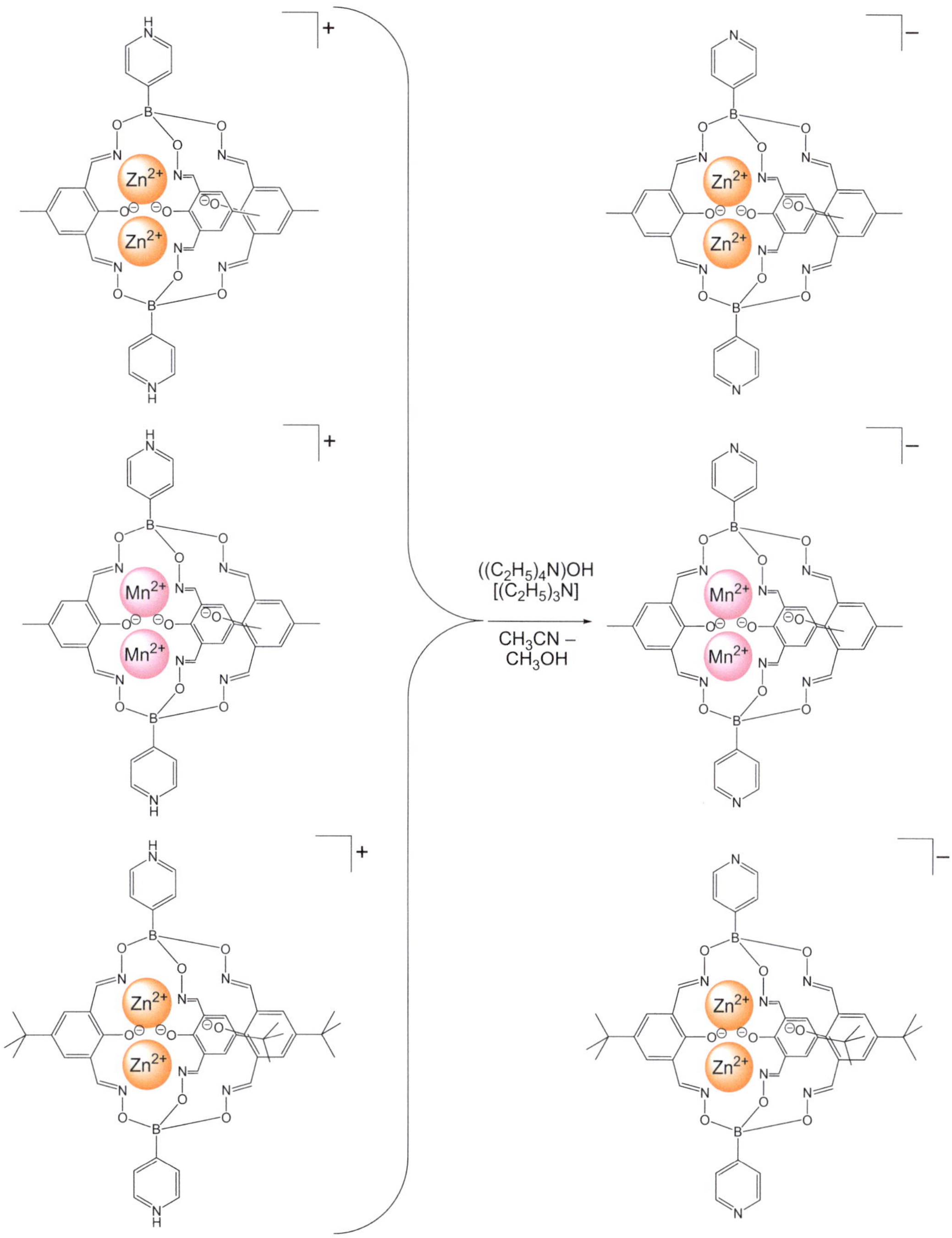

Scheme 2.307 Deprotonation of a series of the homodinuclear metal(II) clathrochelates

with *51* and *52* as the ligand synthons, respectively, on the corresponding metal(II) ion as a matrix by Scheme 2.309; tetraethylammonium hydroxide has been used as organic base for deprotonation of these synthons. The similar synthetic strategy (Scheme 2.310) has been used [90] for preparation of their clathrochelate analog $(Zn^{2+})_2$@**576** with four terminal carboxylic

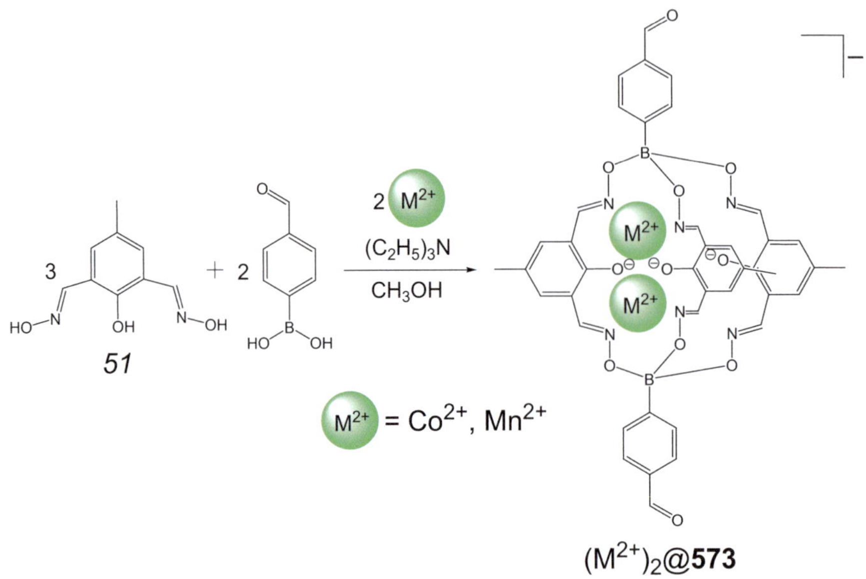

Scheme 2.308 Preparation of the homodinuclear cage complexes of the ligand **573**

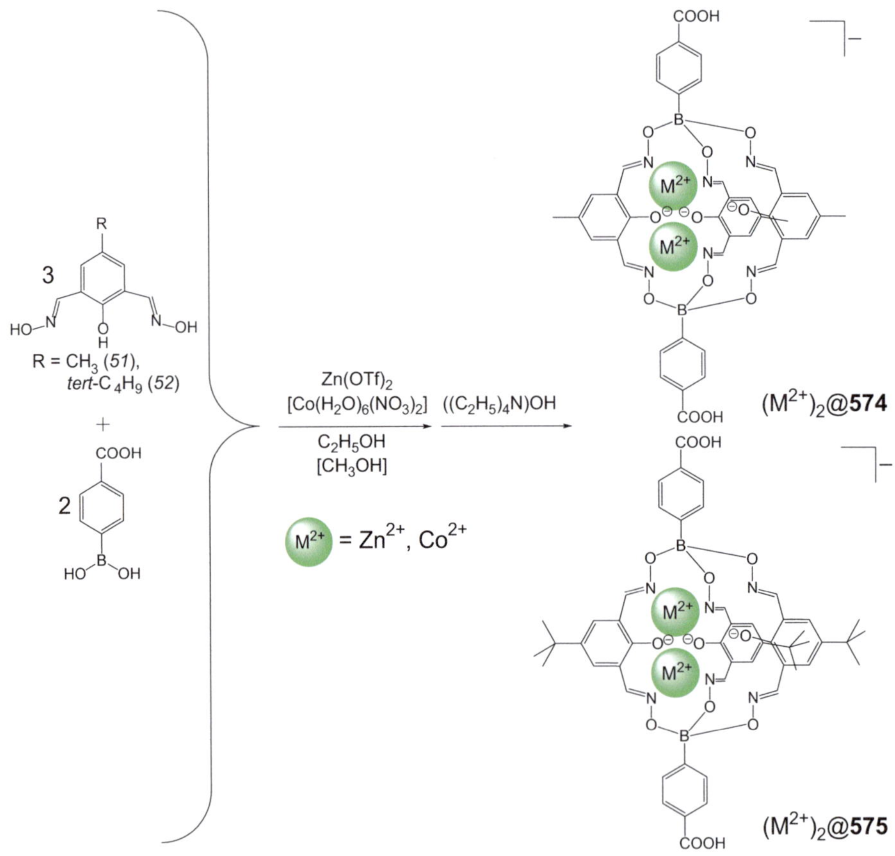

Scheme 2.309 Synthesis of the homodinuclear cage complexes of the ligands **574** and **575**

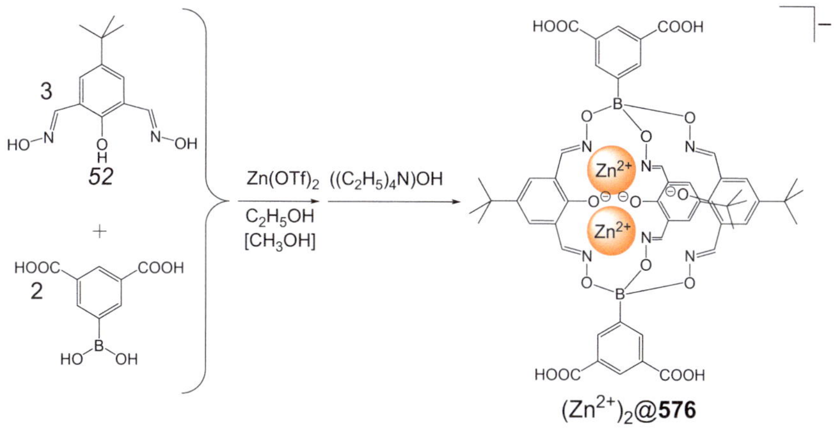

Scheme 2.310 Self-assembly of the homodinuclear zinc(II) cage complex $(Zn^{2+})_2$@**576**

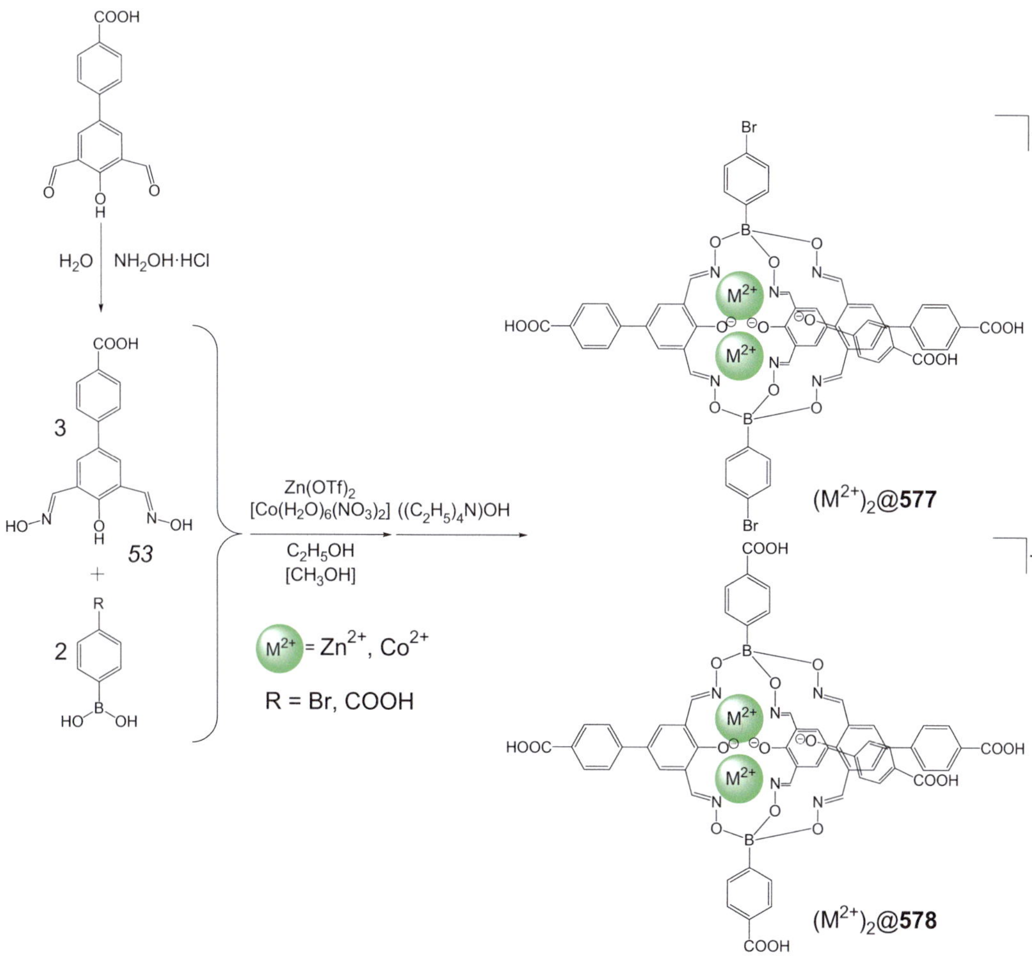

Scheme 2.311 Synthesis of the homodinuclear zinc and cobalt(II) cage complexes of the ligands **577** and **578**

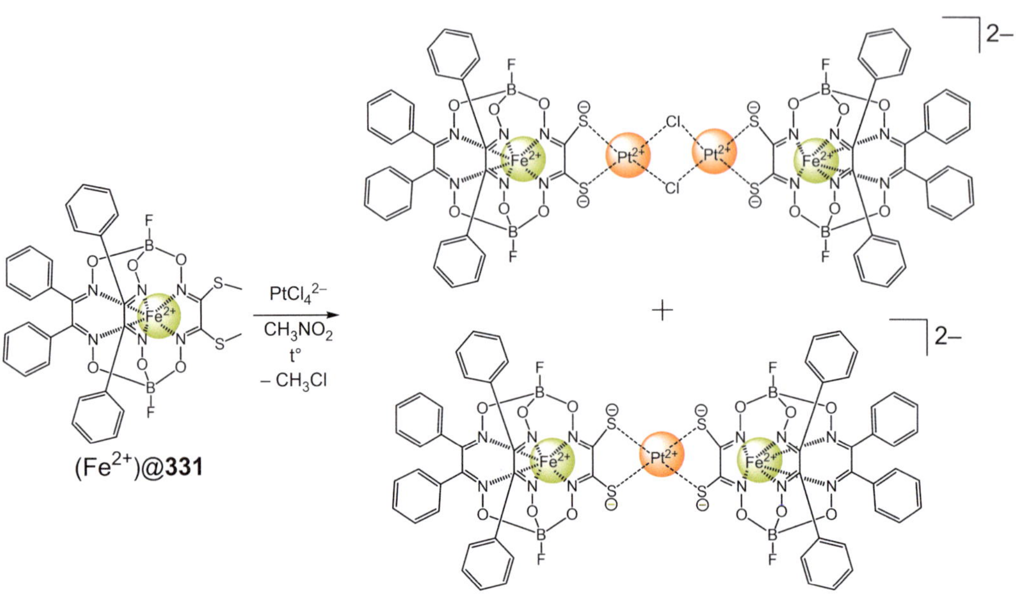

(Fe^{2+})@**331**

Scheme 2.312 Preparation of platinum(II) complexes of a clathrochelate dianion

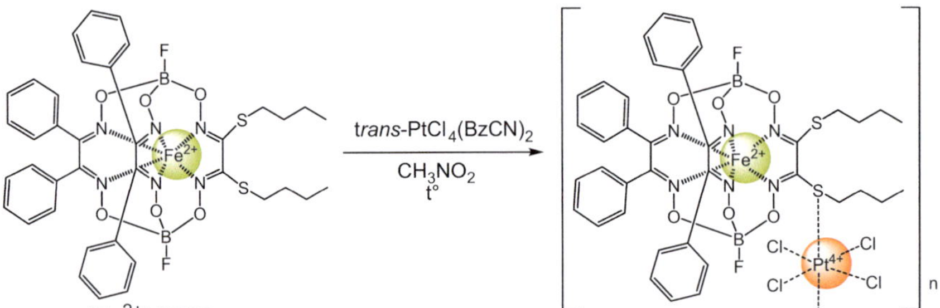

(Fe^{2+})@**333**

Scheme 2.313 Synthesis of platinum(IV) complex of the clathrochelate ligand (Fe^{2+})@**333**

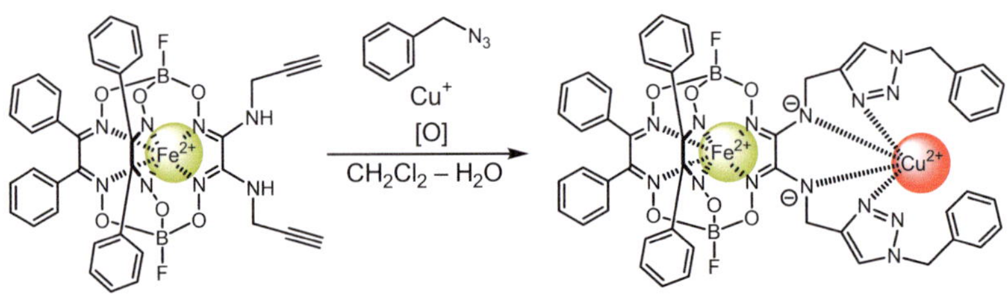

(Fe^{2+})@**340** $(Cu^{2+})\cdot(Fe^{2+})$@(**490**–$2H^+$)

Scheme 2.314 Self-assembly of copper(II) complex of the clathrochelate ligand (Fe^{2+})@(**490**–$2H^+$)

groups attached to its apical arylboronate capping moieties. The use of the carboxyl-terminated ligand synthon *53* allowed obtaining the cage complexes of the encapsulating ligand **577**, containing three carboxylic groups in its equatorial positions and two bromide atoms in its axial positions. Template self-assembly of the same synthon with 4-carboxyphenylboronic acid on

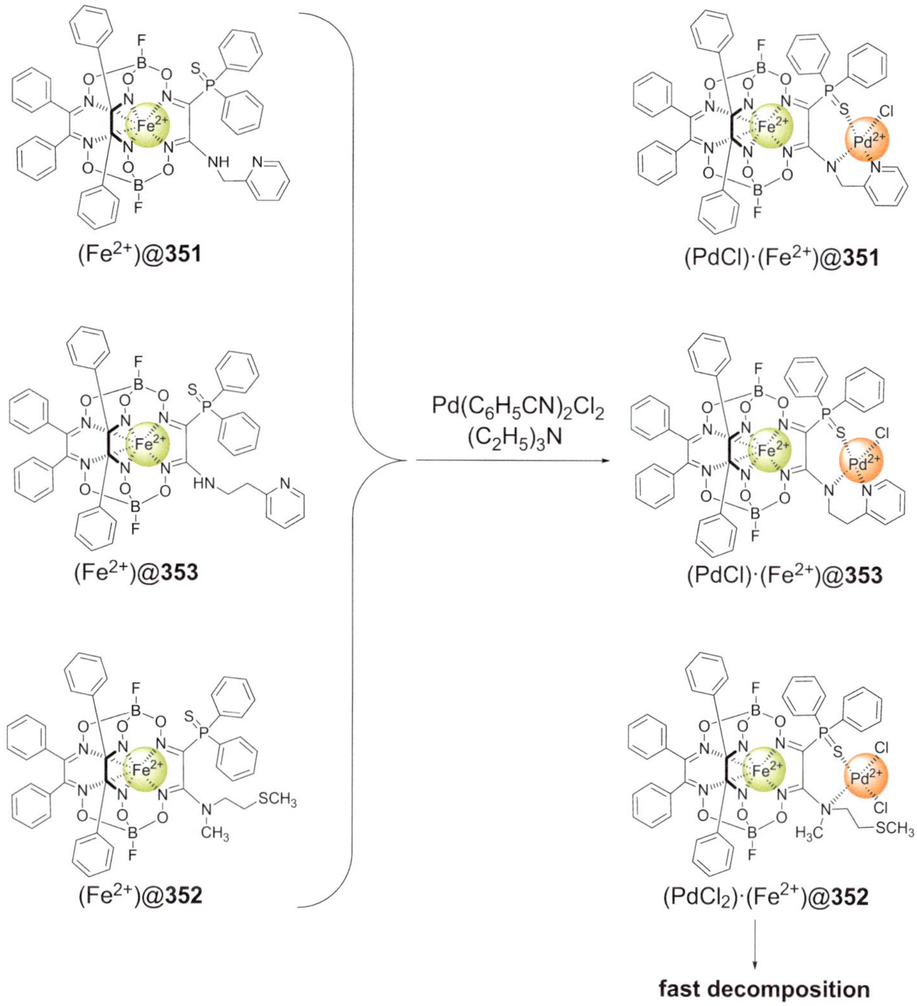

Scheme 2.315 Synthesis of palladium(II) complexes of the phosphorus(V)-containing chelate ligands

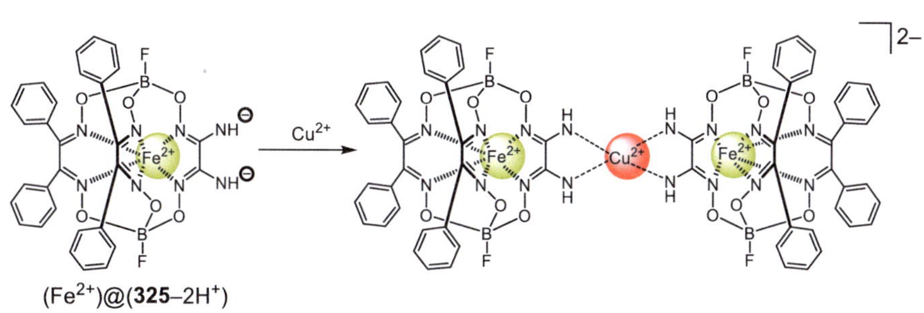

Scheme 2.316 Self-assembly of copper(II) complex of the clathrochelate ligand (Fe^{2+})@(**325**–$2H^+$)

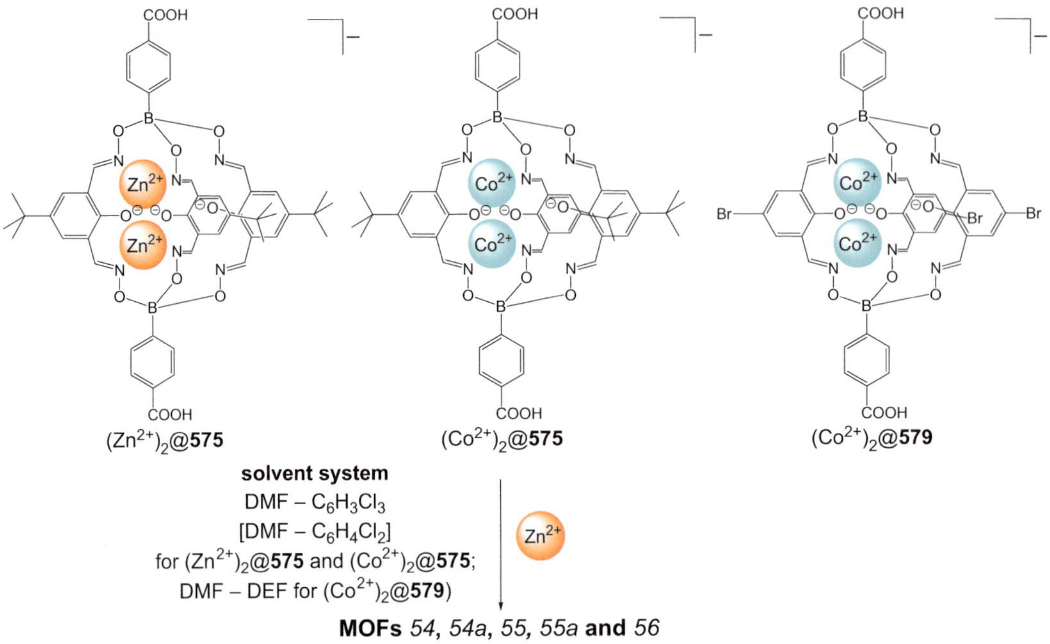

Scheme 2.317 Self-assembly of a series of the clathrochelate-based MOFs

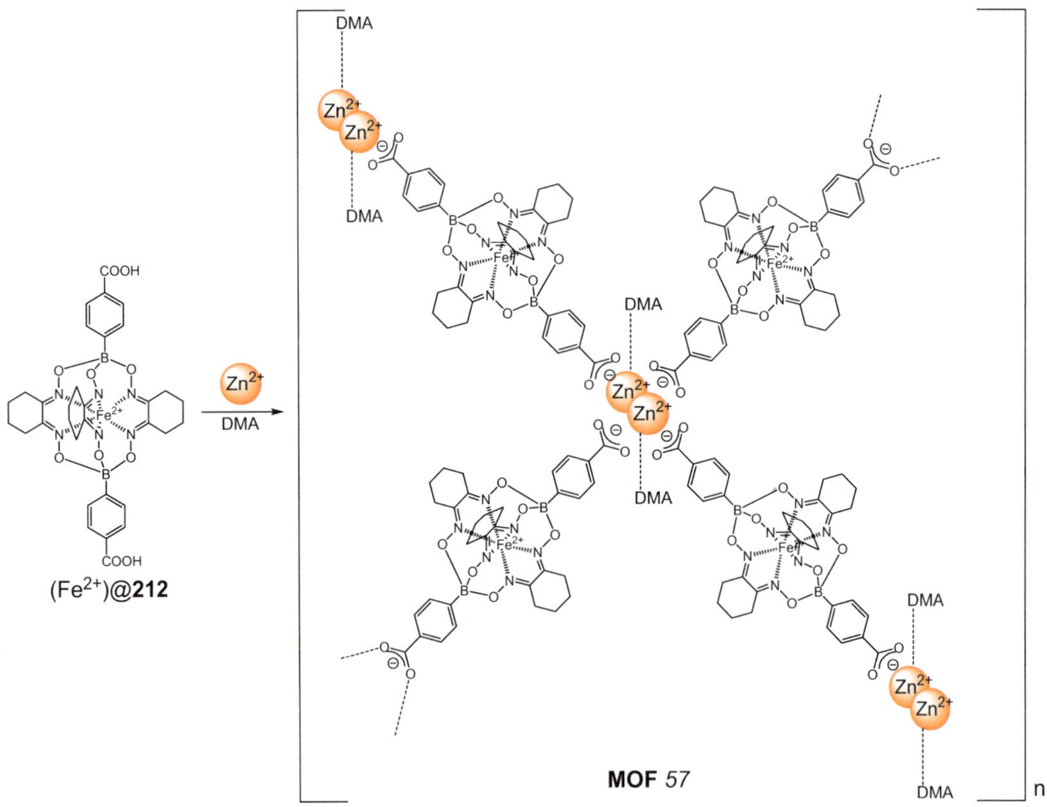

Scheme 2.318 Preparation of the iron(II) clathrochelate-based **MOF** *57*

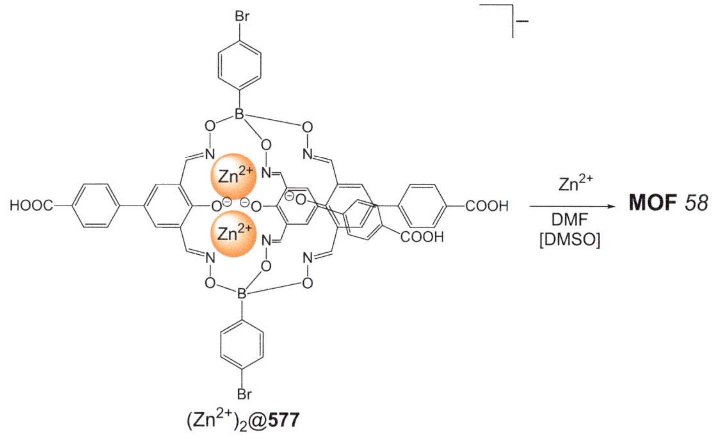

Scheme 2.319 Self-assembly of the zinc(II) clathrochelate-based **MOF** *58*

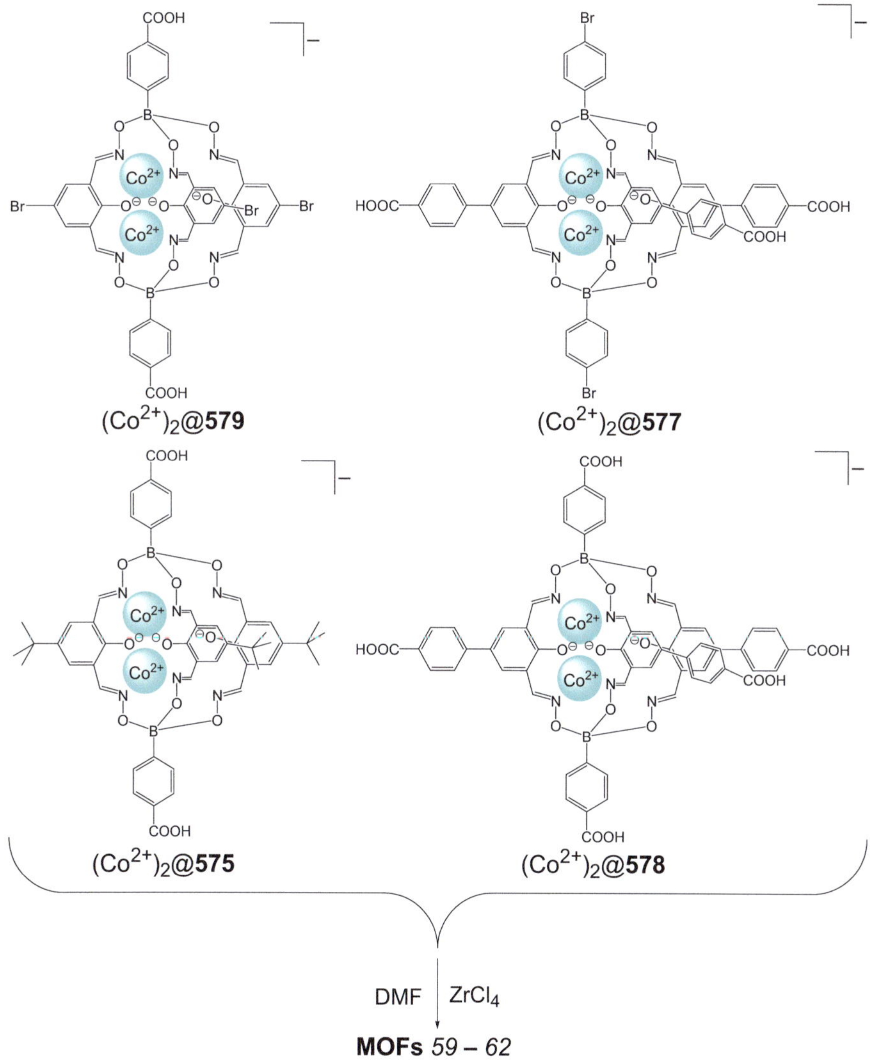

Scheme 2.320 Synthesis of the cobalt(II) clathrochelate-based **MOFs** *59–62*

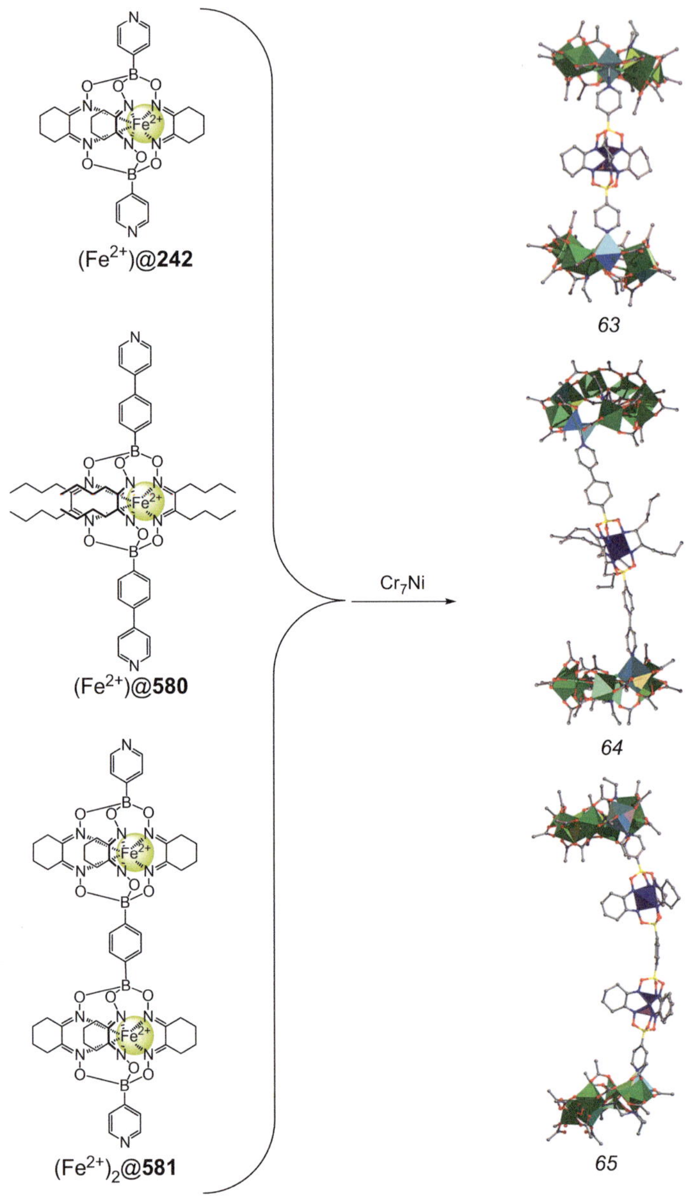

Scheme 2.321 Formation of polynuclear Cr_7Ni complexes of the 4-pyridyl-terminated iron(II) clathrochelate ligands

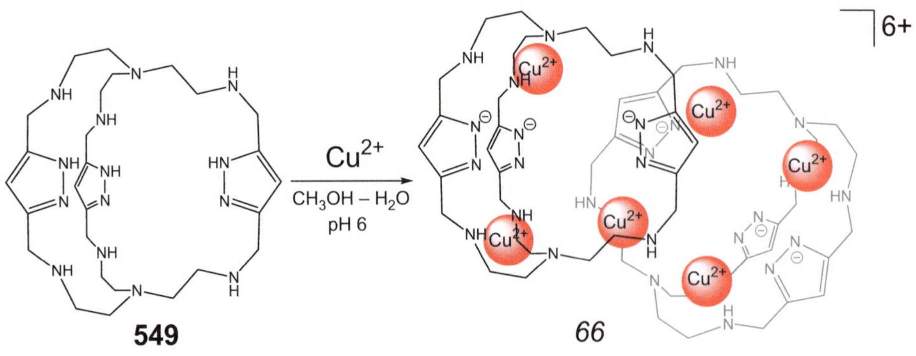

Scheme 2.322 Formation of the hexanuclear copper(II) cage complex *66*

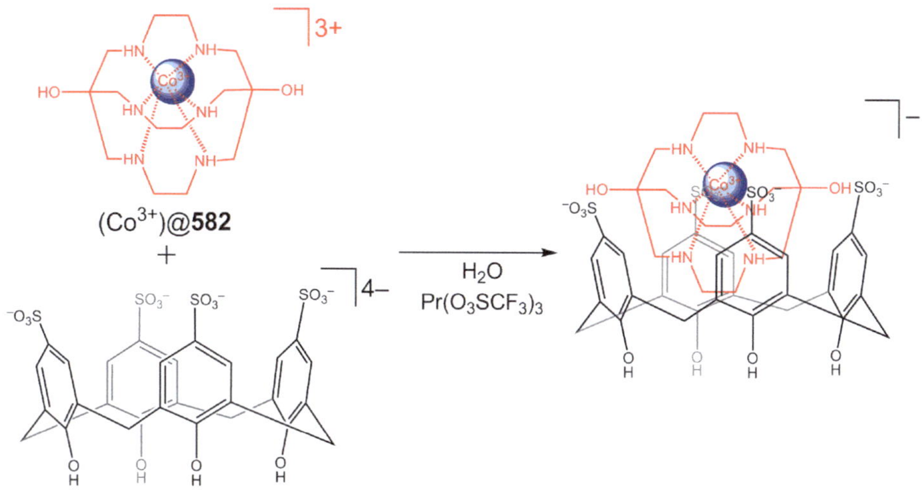

Scheme 2.323 Supramolecular complex of the clathrochelate (Co^{3+})@**582** with *para*-sulfonatocalix[4]arene

the corresponding metal(II) ion as a matrix gave the cage complexes of the pentatopic clathrochelate ligand **578** also shown in Scheme 2.311.

2.4 Clathrochelate-Based Polynuclear Metal Complexes, Coordination Cages, and MOFs

Reaction of the dimethylsulfide clathrochelate (Fe^{2+})@**331** (see section "Nucleophilic Substitution") with $PtCl_4^{2-}$ dianion by Scheme 2.312 afforded the dimeric tetranuclear complexes with and without two bridging chlorine ions, the derivatives of the corresponding dianionic macrobicyclic dithiolate ligand resulting from elimination of methyl chloride [123]. Coordination of the di-*n*-butylsulfide clathrochelate (Fe^{2+})@**333** toward the platinum(IV) *trans*-complex $PtCl_4(C_6H_5CH_2CN)_2$ has been used [123] for preparation of the corresponding 1D coordination polymer shown in Scheme 2.313 with this disulfide iron(II) macrobicycle playing the role of a monodentate *S*-donor ligand.

1,3-Dipolar cycloaddition of benzyl azide to the propargylamine clathrochelate (Fe^{2+})@**340** (see section "Nucleophilic Substitution") in the presence of copper(I) ions by Scheme 2.314 gave the copper(II) complex of the clathrochelate ligand (Fe^{2+})@**490**. This complex is formed *via* coordination of two deprotonated amino groups and two donor azaheterocyclic fragments of its ribbed substituents toward the central metal (II) ion [128].

The monoribbed-functionalized *P,N*-substituted cage complexes (Fe^{2+})@**351** and (Fe^{2+})@**353** (see section "Nucleophilic Substitution"), containing two strongly donor centers per a clathrochelate molecule (i.e., the sulfur atom of the diphenylthiophosphine group and the nitrogen atom of its pyridyl moiety), have been used [135] as *N,S*-donor macrobicyclic ligands toward palladium(II) ion: they form the corresponding Pd,Fe-dinuclear 1:1 complexes (Scheme 2.315) in high yields. The relative stability of these complexes is explained [135] by high donor ability of the above ribbed substituents as well as by disposition of their donor centers allowing them to give the thermodynamically favorable five- and six-membered chelate cycles. Their macrobicyclic analog (Fe^{2+})@**352** also gives its 1:1 complex with palladium(II) ion in solution, but this complex underwent a fast decomposition: the donor amino group of **352** has an amide character due to the conjugation with a quasiaromatic clathrochelate framework and, therefore, is

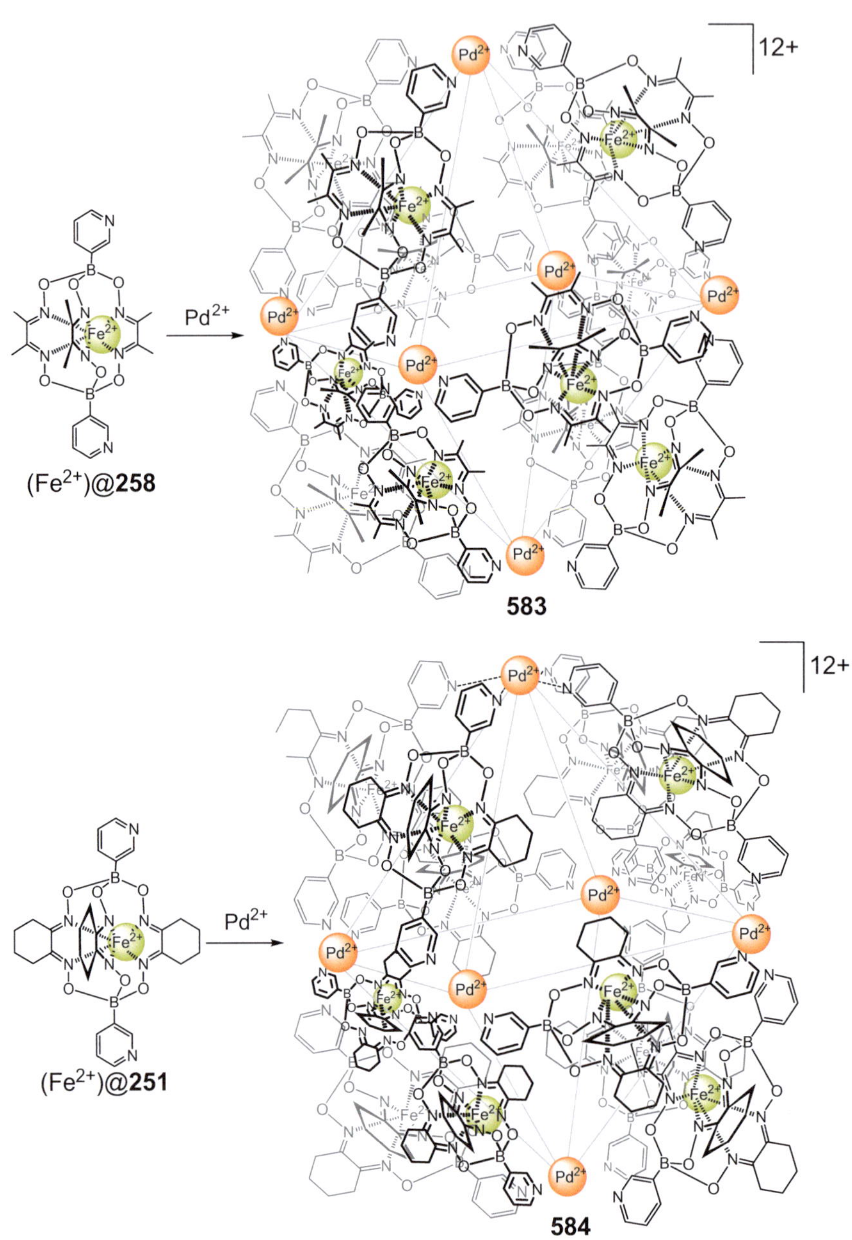

Scheme 2.324 Self-assembly of the coordination cages **583** and **584**

poor donor, while two sulfur atoms of the ribbed substituents in a *vic*-position of the same ribbed fragment of **352** are far from each other and cannot form a stable chelate cycle [135].

1:2 heterodinuclear copper(II) complex of a dianionic form of the clathrochelate ligand (Fe^{2+})@**325**, generated in situ in the presence of large excess of potassium *tert*-butoxide, has been prepared [121] by Scheme 2.316.

Solvothermal reactions of the ditopic clathrochelate ligands $(Zn^{2+})_2$@**575**, $(Co^{2+})_2$@**575**, and $(Co^{2+})_2$@**579** with zinc(II) nitrate by Scheme 2.317 at 120 °C have been used [90] for preparation of the corresponding MOFs *54*, *55*, and *56*. The

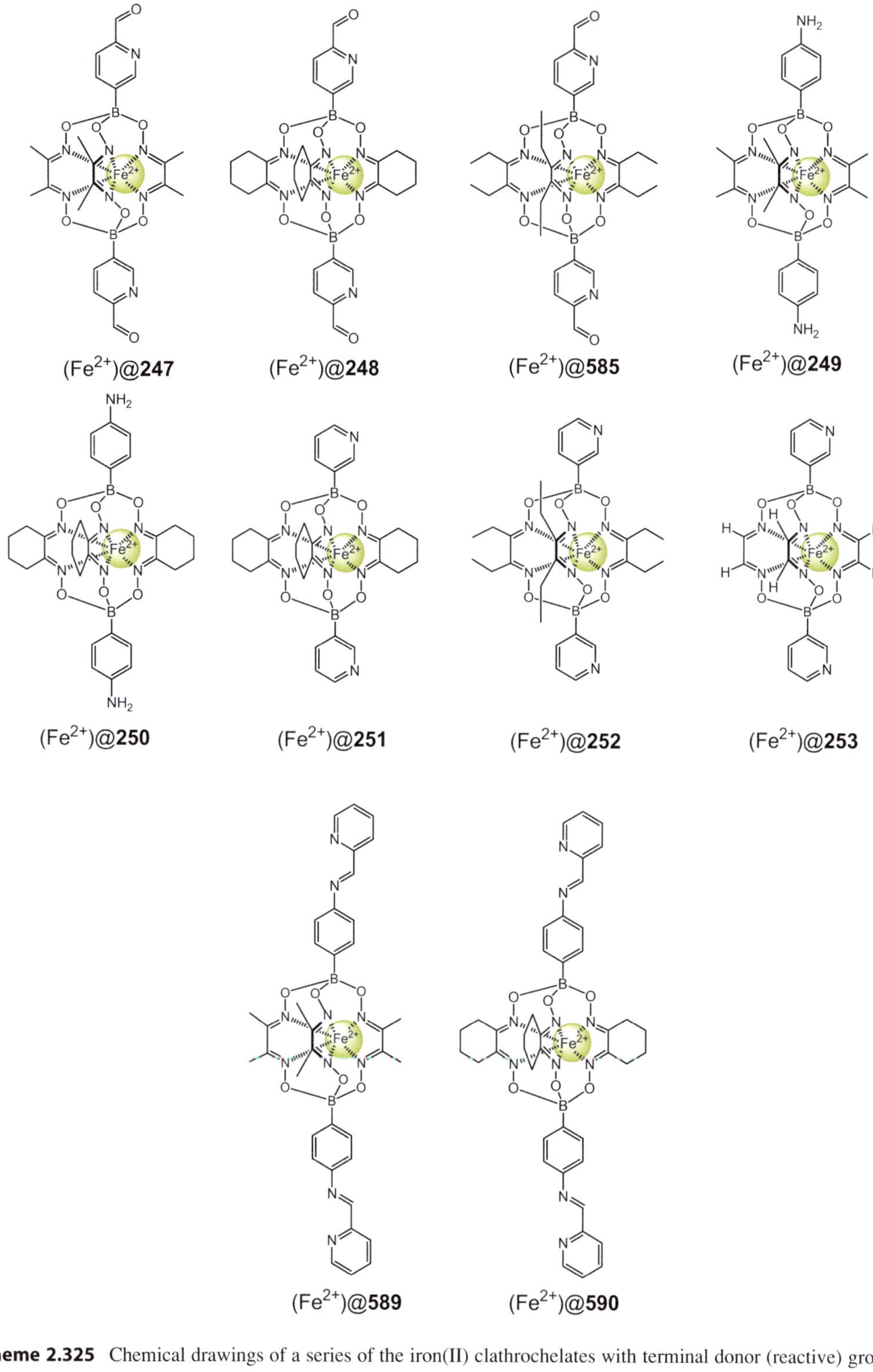

Scheme 2.325 Chemical drawings of a series of the iron(II) clathrochelates with terminal donor (reactive) groups

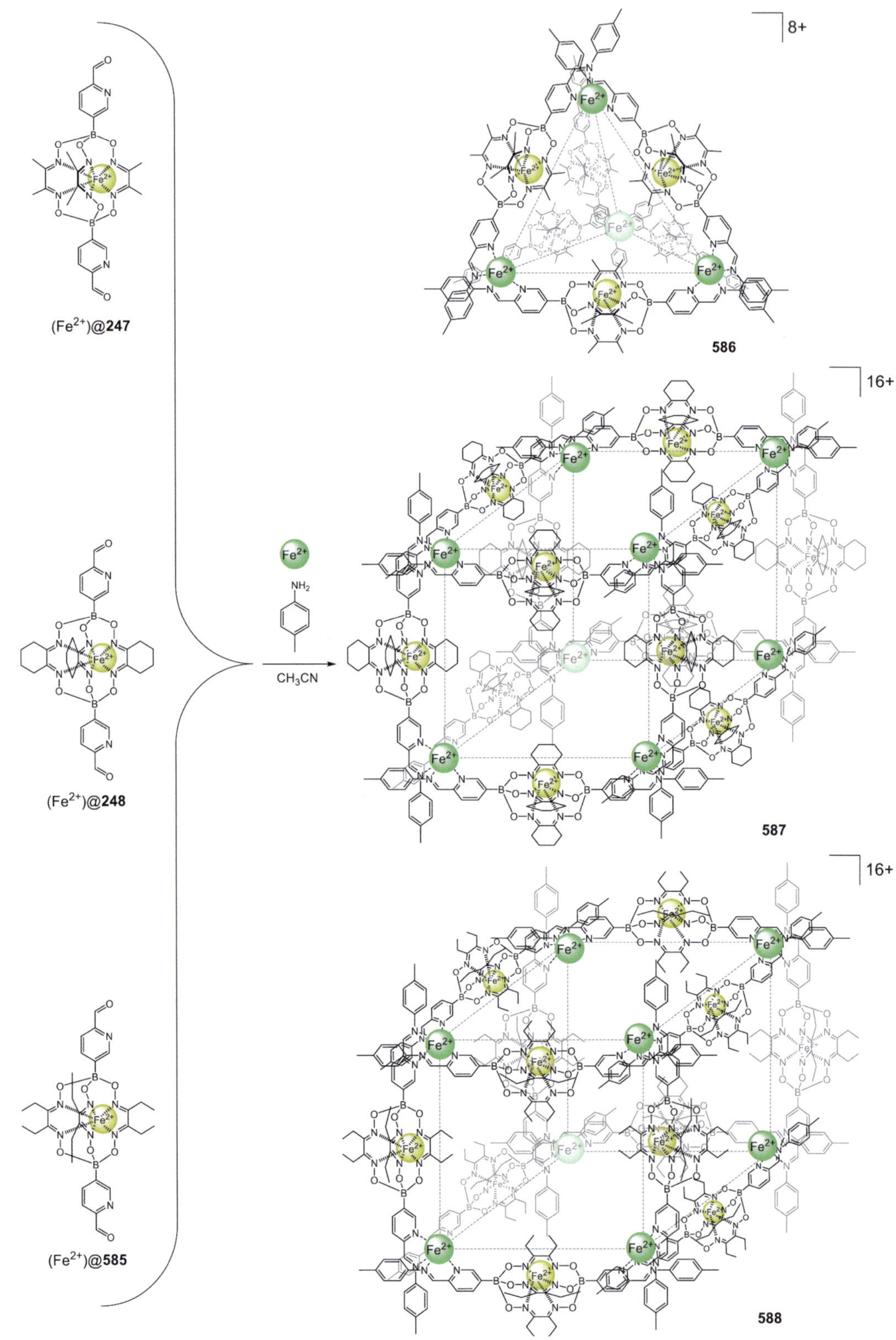

Scheme 2.326 Self-assembly of the coordination cages **586–588**

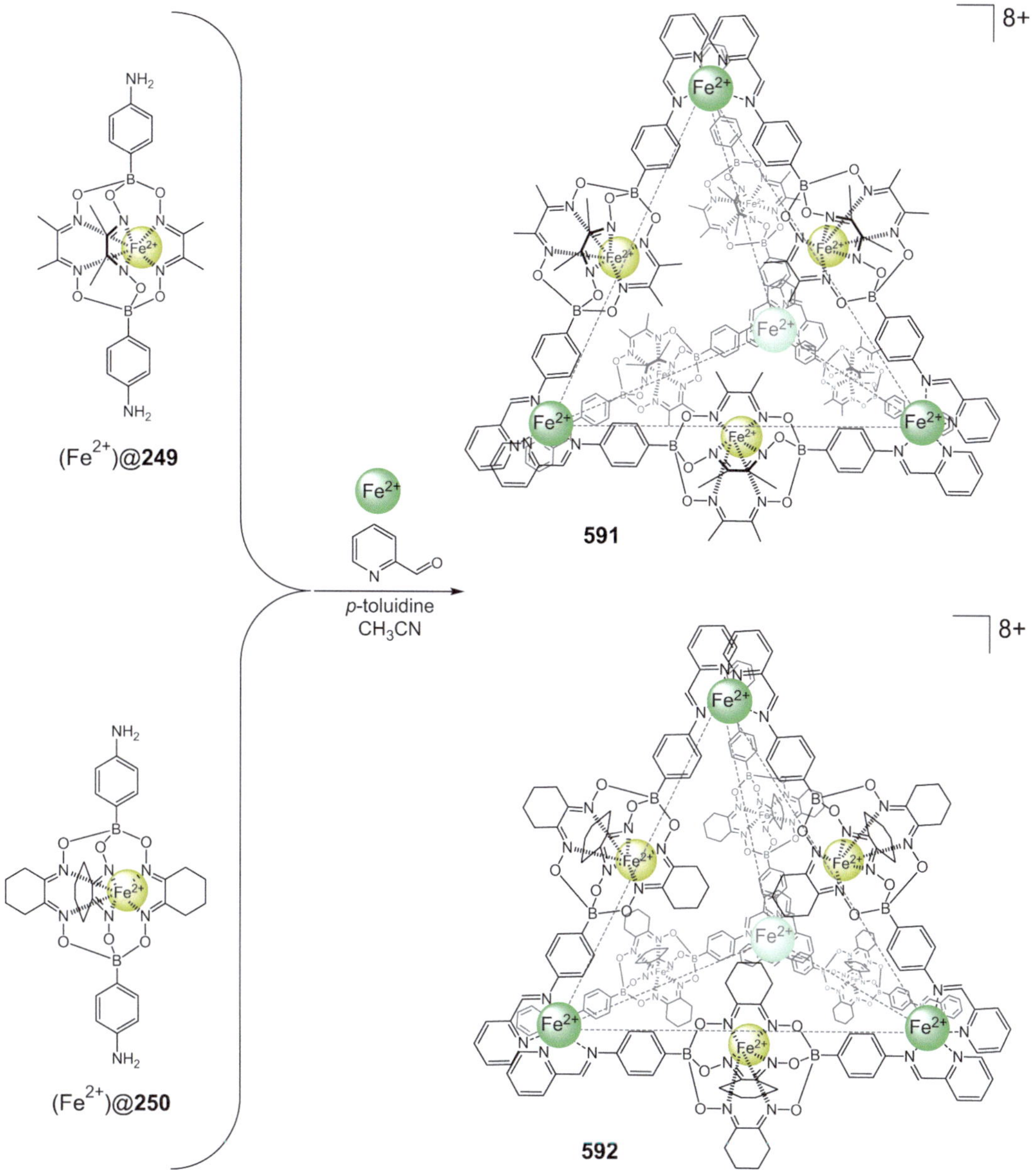

Scheme 2.327 Self-assembly of the coordination cages **591** and **592**

reaction of the same zinc(II) salt with the monoclathrochelate iron(II) tris-dioximate ligand (Fe^{2+})@**212** in DMA by Scheme 2.318 gave 2D-layered coordination polymer *57*. Coordination-driven self-assembly of the tritopic clathrochelate ligand $(Zn^{2+})_2$@**577** by Scheme 2.319 afforded MOF *58*. That of the ditopic macrobicyclic complexes $(Co^{2+})_2$@**575** and $(Co^{2+})_2$@**577–579** with $ZrCl_4$ in DMF by Scheme 2.320 has been used [90] for preparation of the corresponding zirconium(IV)-containing MOFs *59–62* in good yields.

Coordination-driven self-assembly of the corresponding chiral *Cr*$_7$*Ni*-metallomacrocycles, which contain an octagon of metal sites with a pentadeprotonated *N*-ethyl-*D*-glucamine forming the bridges between their Cr sites *via* deprotonated alkoxide side arms at the center,

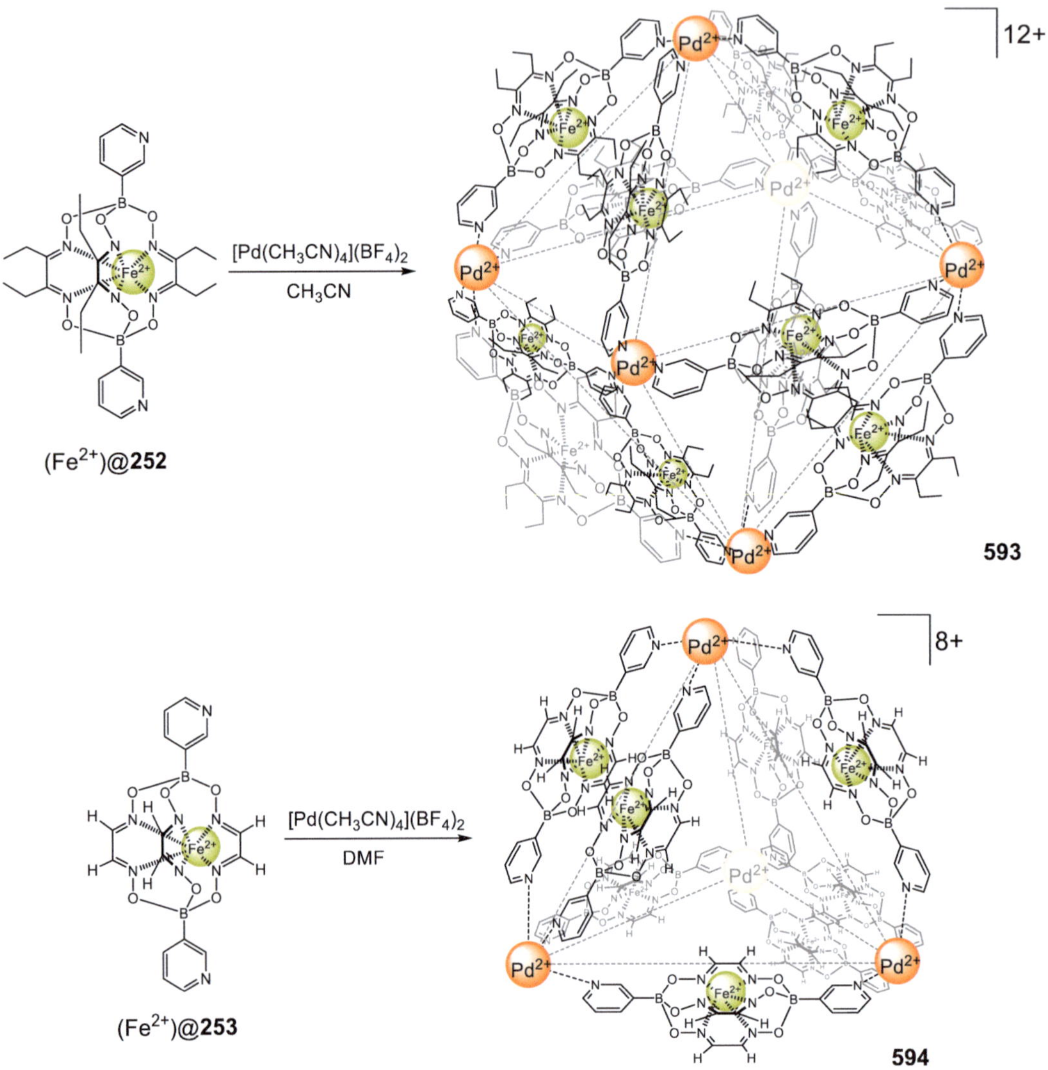

Scheme 2.328 Self-assembly of the coordination cages **593** and **594**

with the 4-pyridyl-terminated clathrochelate ligands (Fe^{2+})@**242**, (Fe^{2+})@**580**, and (Fe^{2+})@**581** by Scheme 2.321 has been used [226] for preparation of the supramolecular clathrochelate-based nanosized dimers *63*, *64*, and *65* with intramolecular Ni...Ni separations of 1.89, 2.71, and 3.07 nm and longest Cr...Cr separations between their rings of 2.24, 3.24, and 3.55 nm, respectively.

Coordination-driven self-assembly of the caging ligand **549** (see Sect. 2.3.1) with $Cu(ClO_4)_2 \cdot 6H_2O$ at pH 6 gave the crystals of the hexanuclear copper(I) complex *66* shown in Scheme 2.322 [227].

Supramolecular self-assembly of sodium *para*-sulfonatocalix[4]arene as a host with cobalt(III) dihydroxyclathrochelate (Co^{3+})@**582** as a guest in their molar ratio 1:3 at pH 4–5 in the presence of an excess of $Pr(O_3SCF_3)$ afforded their 1:1 host–guest complex (Scheme 2.323) [228].

Coordination-driven self-assembly of the clathrochelate ligand synthons (Fe^{2+})@**258** and (Fe^{2+})@**251** with palladium(II) ions by Scheme

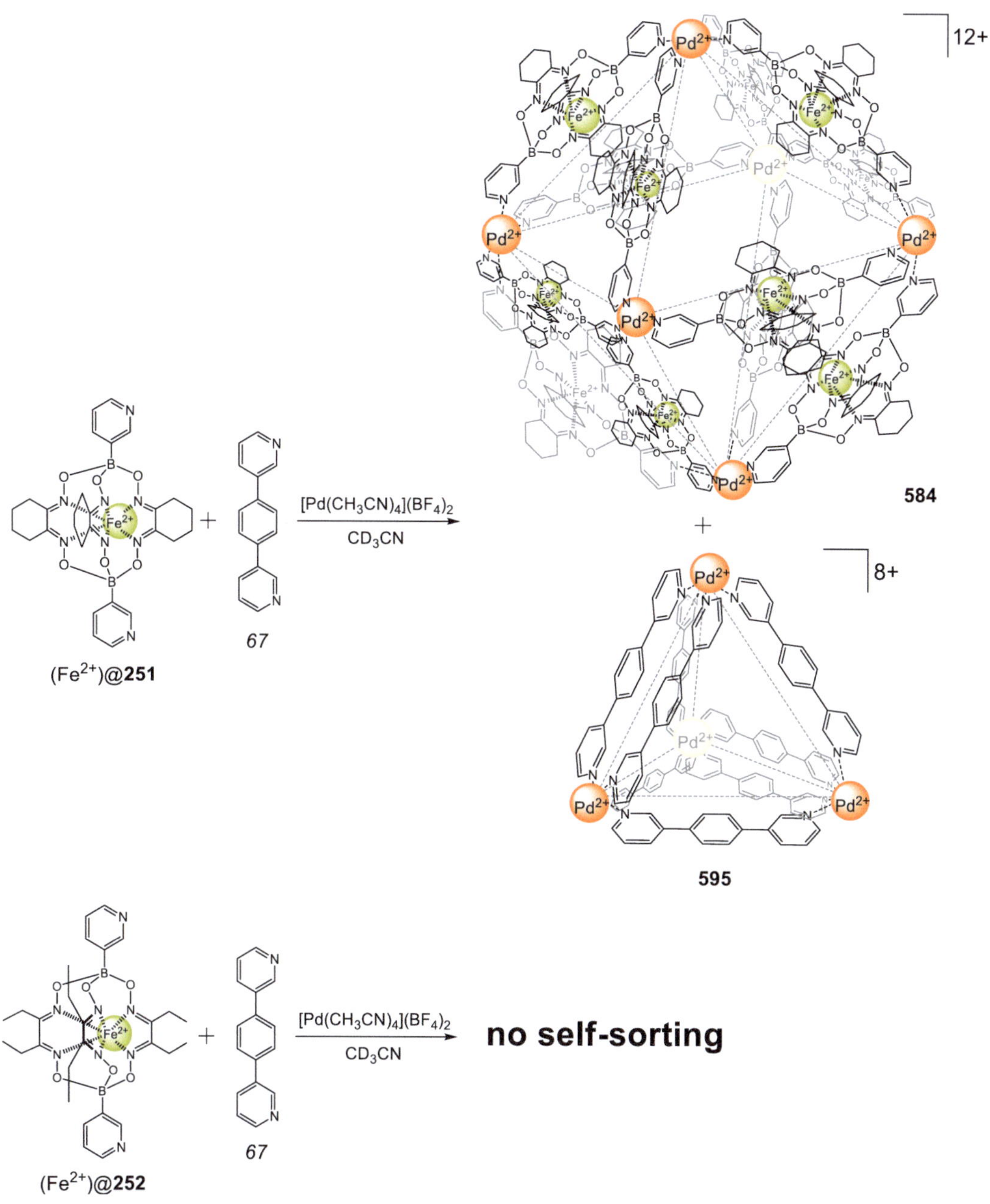

Scheme 2.329 Self-assembly of the coordination cages **584** and **595**

2.324 has been used [91] for preparation of the octahedral coordination cages **583** and **584**.

The possibility to control the geometry and the composition of clathrochelate-based coordination cages *via* the aspect ratio of their macrobicyclic ligand synthons has been demonstrated [89] for a series of the iron(II)- and palladium(II)-cross-linked capsules, using the functionalized clathrochelate complexes as rod-like metalloligand synthons with variable aspect ratios. The boron-capped tris-dioximate ligand synthons are described to be ideally suited

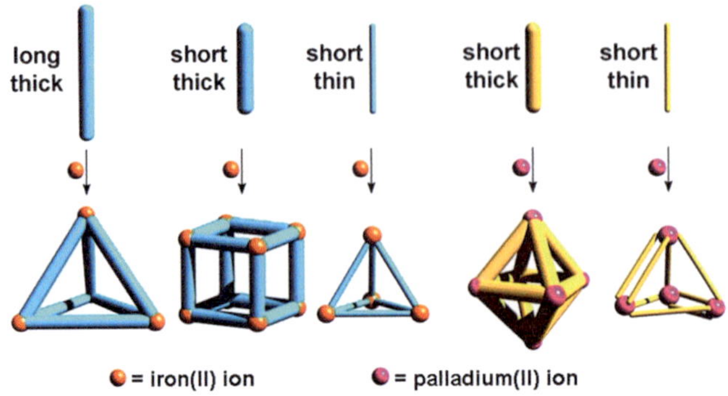

Scheme 2.330 Formation of various types of the coordination cages [89]

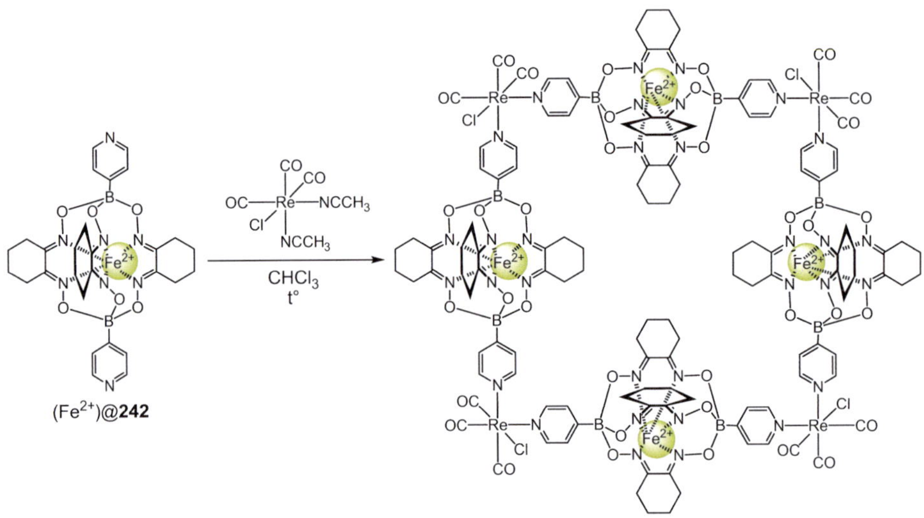

Scheme 2.331 Self-assembly of the clathrochelate-based metallomacrocycle **596**

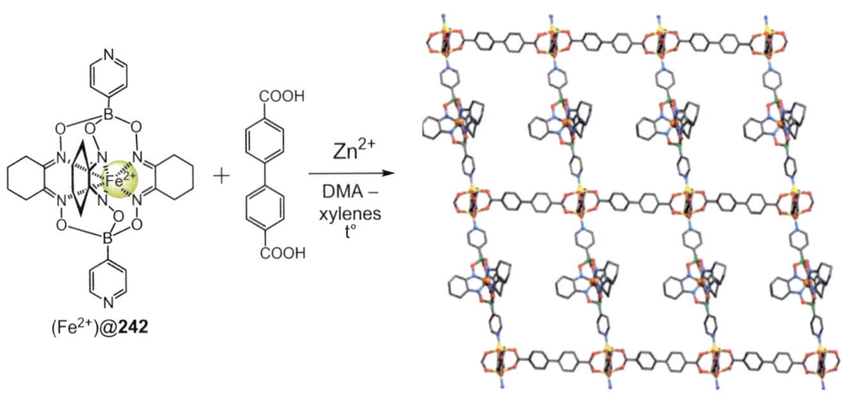

Scheme 2.332 Self-assembly of the clathrochelate-based coordination polymer *68*

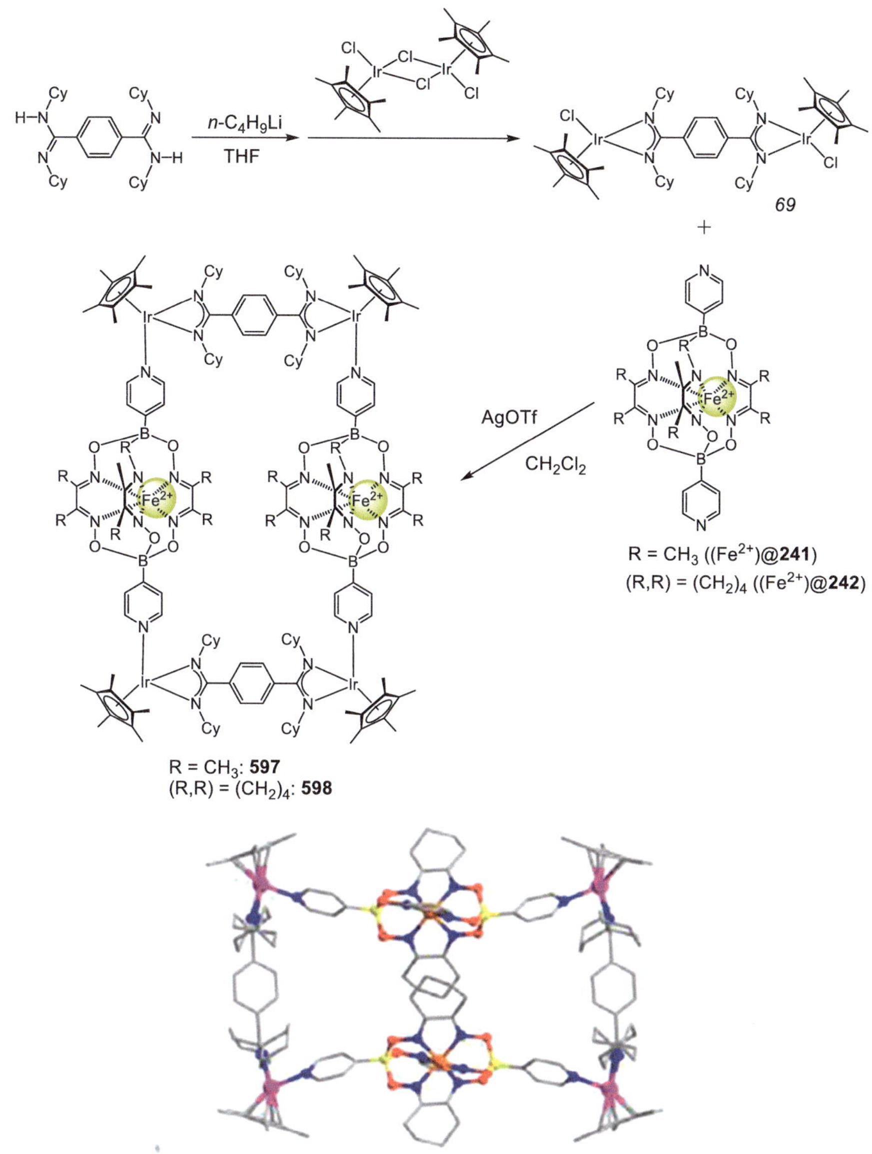

Scheme 2.333 Self-assembly of the clathrochelate-based metallomacrocycles **597** and **598**

to study the influence of the aspect ratio on the self-assembly behavior. Their lengths can be modulated *via* variation of the capping boron-containing fragments, while the lateral size of these ligand synthons can be varied by choosing the appropriate ribbed substituents in α-dioximate chelate fragments. These clathrochelate ligands, possessing a distorted TP–TAP geometries of their FeN_6-coordination polyhedra, have the molecular B…B C_3symmetry pseudoaxis. Thus, they have been regarded [89] as intrinsically three-dimensional metalloligands with a rigid rod geometry. The clathrochelate ligand (Fe^{2+})@**251** (Scheme 2.325) is known to form the Pd_6L_{12}-octahedral coordination cages upon its coordination-driven self-assembly with palladium(II) ions (see above). In order to probe if the assembly process can be influenced by the aspect ratio of the clathrochelate ligands, the cage complexes

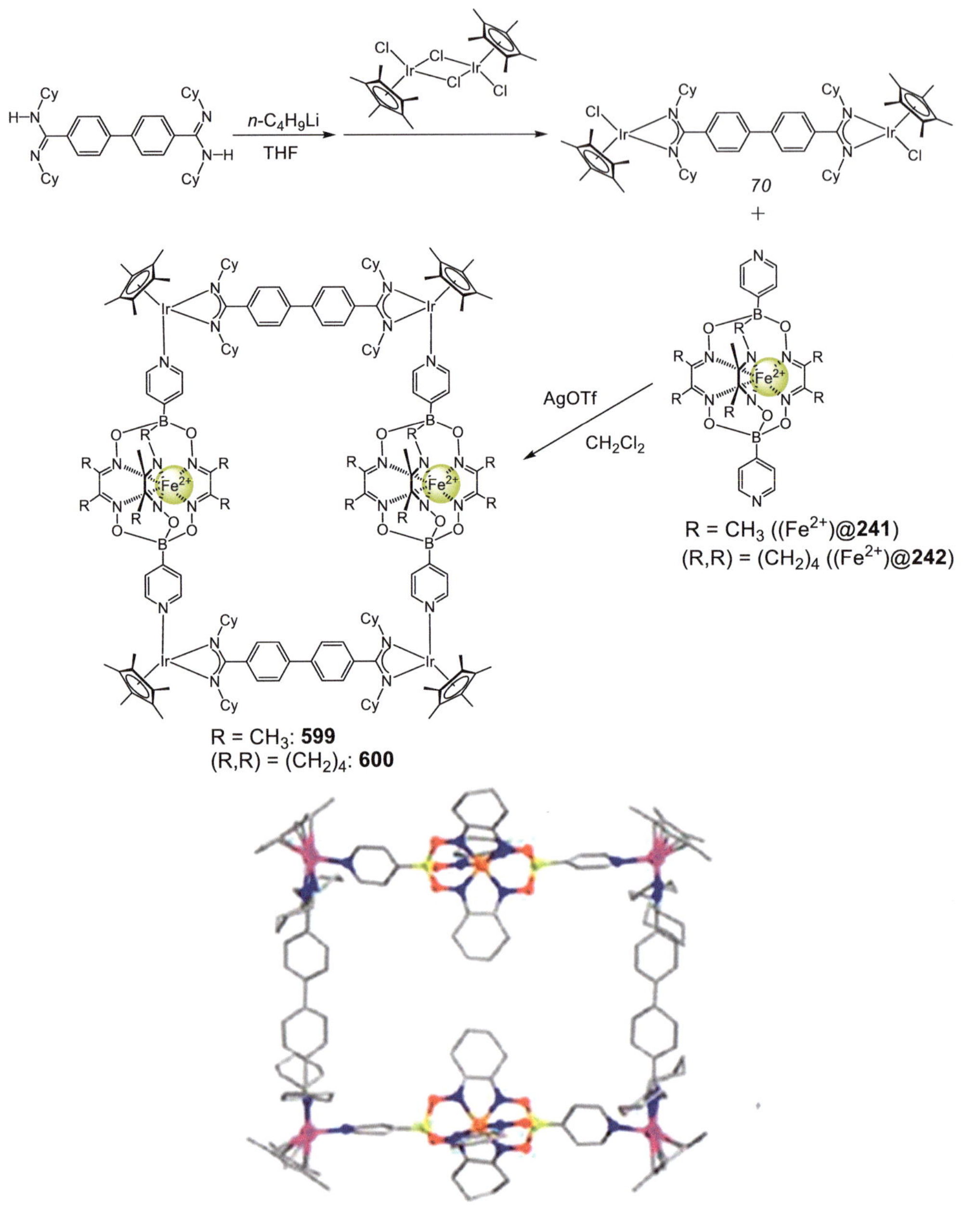

Scheme 2.334 Self-assembly of the clathrochelate-based metallomacrocycles **599** and **600**

(Fe^{2+})@**252** and (Fe^{2+})@**253** (see section "Template Synthesis and Condensation Reactions") have been tested [89] as rigid macrobicyclic ligands. Clathrochelate (Fe^{2+})@**252** is slightly larger than the tris-nioximate (Fe^{2+})@**251** because its six ethyl-ribbed substituents require more space than three cyclohexane substituents of (Fe^{2+})@**251**, when the tris-glyoximate clathrochelate (Fe^{2+})@**253** is the thinnest (thus showing a largest aspect ratio) of the above three macrobicyclic ligands (Scheme 2.325).

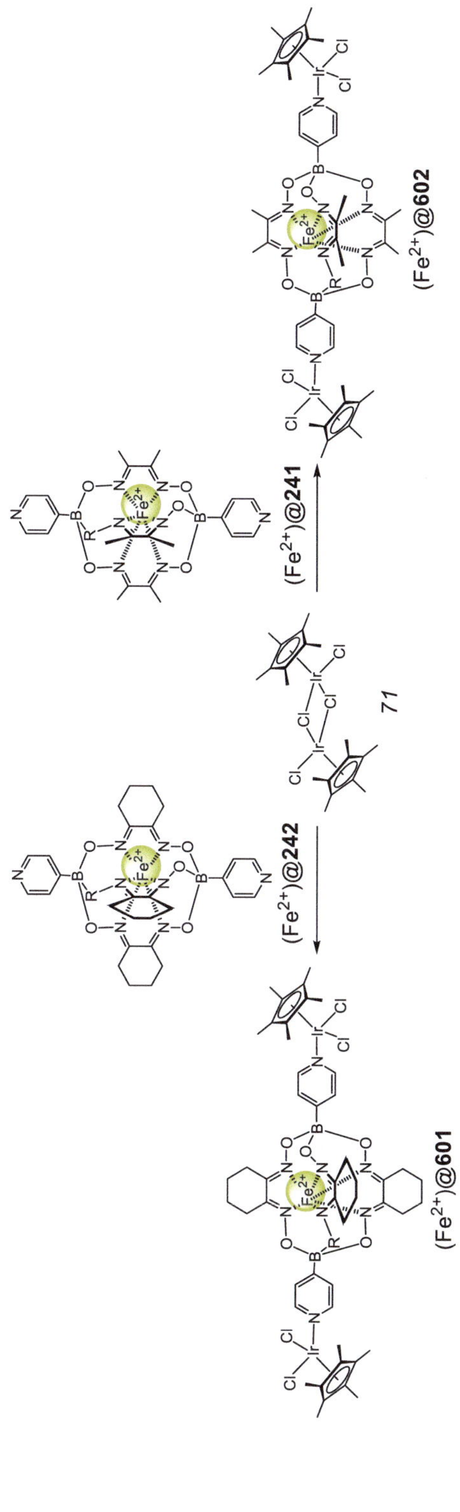

Scheme 2.335 Self-assembly of the polynuclear clathrochelate-based complexes (Fe^{2+})@**601** and (Fe^{2+})@**602**

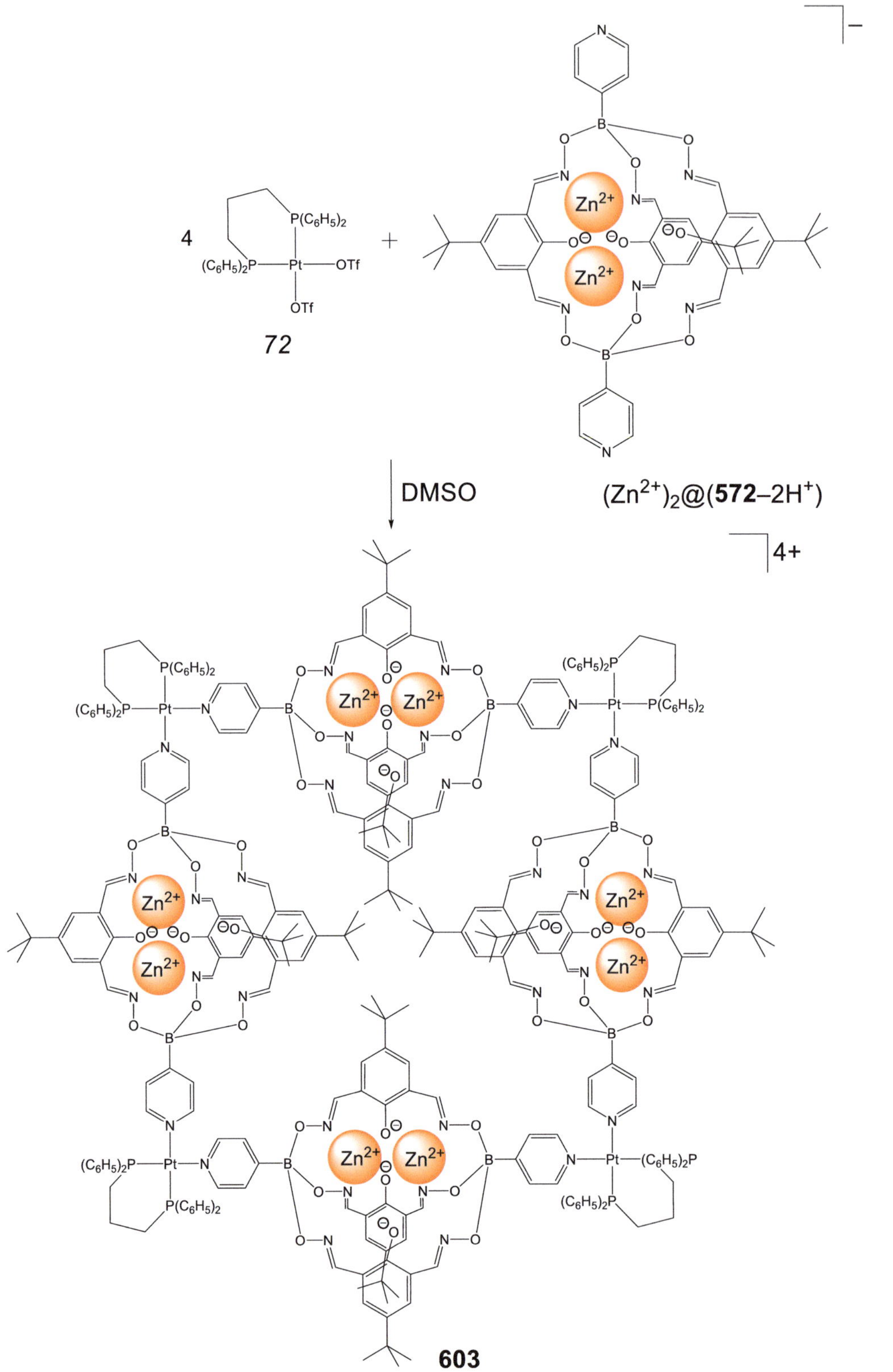

Scheme 2.336 Self-assembly of the clathrochelate-based metallomacrocycle **603**

Scheme 2.337 Self-assembly of the polynuclear clathrochelate-based systems *73* and *74*

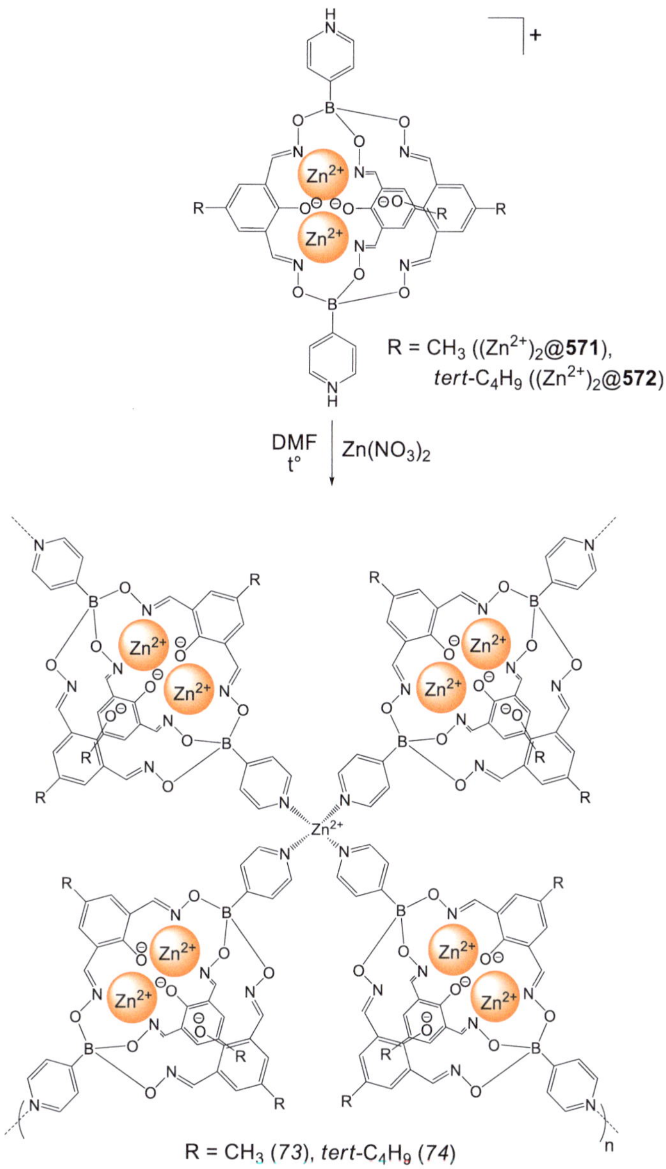

The coordination cages **586–588** have been prepared [89] in high yields using self-assembly by Scheme 2.326 of the clathrochelates (Fe^{2+})@**247**, (Fe^{2+})@**248**, or (Fe^{2+})@**585** with 4-toluidine and $Fe(OTf)_2$ or $Fe(NTf_2)_2$. As follows from high-resolution ESI mass spectrometry data, the cage **586** is a tetrahedral Fe_4L_6-coordination capsule, while the cages **587** and **588** have Fe_8L_{12}-frameworks, possessing an approximately cubic framework with eight iron(II) centers at its vertices.

The lateral sizes of the aniline-terminated iron(II) clathrochelates (Fe^{2+})@**249** and (Fe^{2+})@**250** are described [89] to be the same as those for the above synthons (Fe^{2+})@**247** and (Fe^{2+})@**248**, but their condensation with 2-formylpyridine resulted in the macrobicyclic complexes (Fe^{2+})@**589** and (Fe^{2+})@**590** with an increased distance between their two donor *N,N′*-chelating sites (Scheme 2.325). As a result, the aspect ratio of these clathrochelate ligands is increased. The reactions of (Fe^{2+})@**249** and

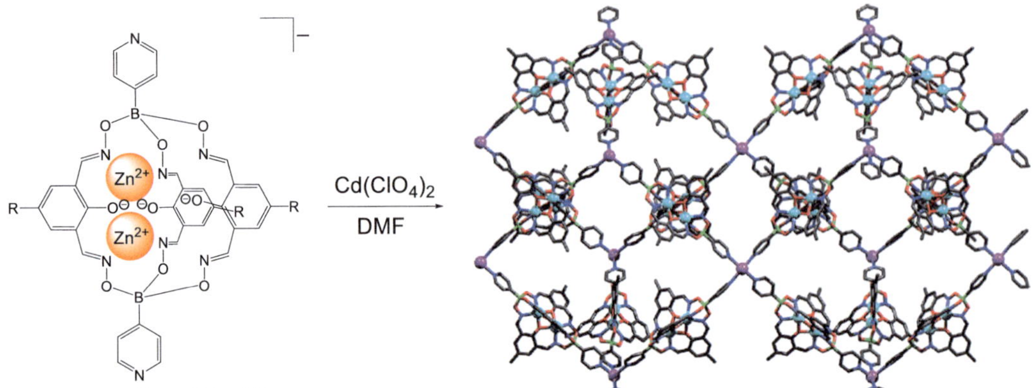

R = CH_3 $((Zn^{2+})_2$@(**571**–2H$^+$)),
tert-C_4H_9 $((Zn^{2+})_2$@(**572**–2H$^+$))

R = CH_3 (*75*), *tert*-C_4H_9 (*76*)

Scheme 2.338 Preparation of the clathrochelate-based coordination polymers *75* and *76*

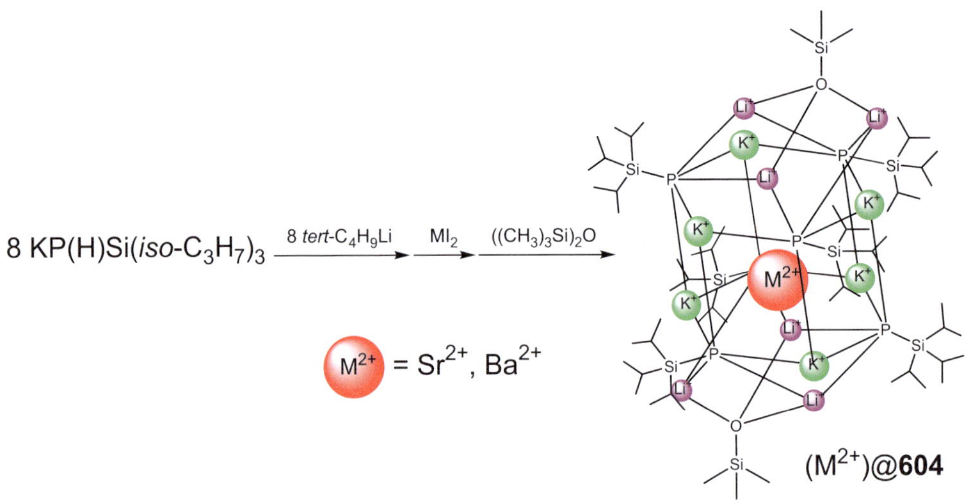

(M^{2+})@**604**

Scheme 2.339 Synthesis of the strontium and barium(II) cage complexes (M^{2+})@**604**

(Fe^{2+})@**250** with 2-formylpyridine and Fe $(OTf)_2$ or $Fe(NTf_2)_2$, allowing to obtain by Scheme 2.327 the Fe_4L_6-coordination cages **591** and **592**, showed several sets of ^{1}H NMR signals, thus suggesting the presence of their diastereoisomers. The formation of such Fe_4L_6-cages with both these aniline-terminated clathrochelates is reported [89] to be consistent with the inference that the aspect ratio and not the absolute width of the macrobicyclic ligand is a decisive factor in controlling of the geometry.

Coordination-driven self-assembly of the 3-pyridyl-terminated tris-diethylglyoximate clathrochelate (Fe^{2+})@**252** with $[Pd(CH_3CN)_4]$ $(BF_4)_2$ by Scheme 2.328 gave an octahedral cage **593**, while its tris-glyoximate analog (Fe^{2+})@**253** formed a tetrahedral coordination cage **594** [89]. The above results (i.e., octahedral vs. tetrahedral geometry of their frameworks) are described to evidence that it is possible to control the coordination cages *via* the aspect ratio of a macrobicyclic ligand: these cages **593** and **594** are based on the clathrochelate ligands

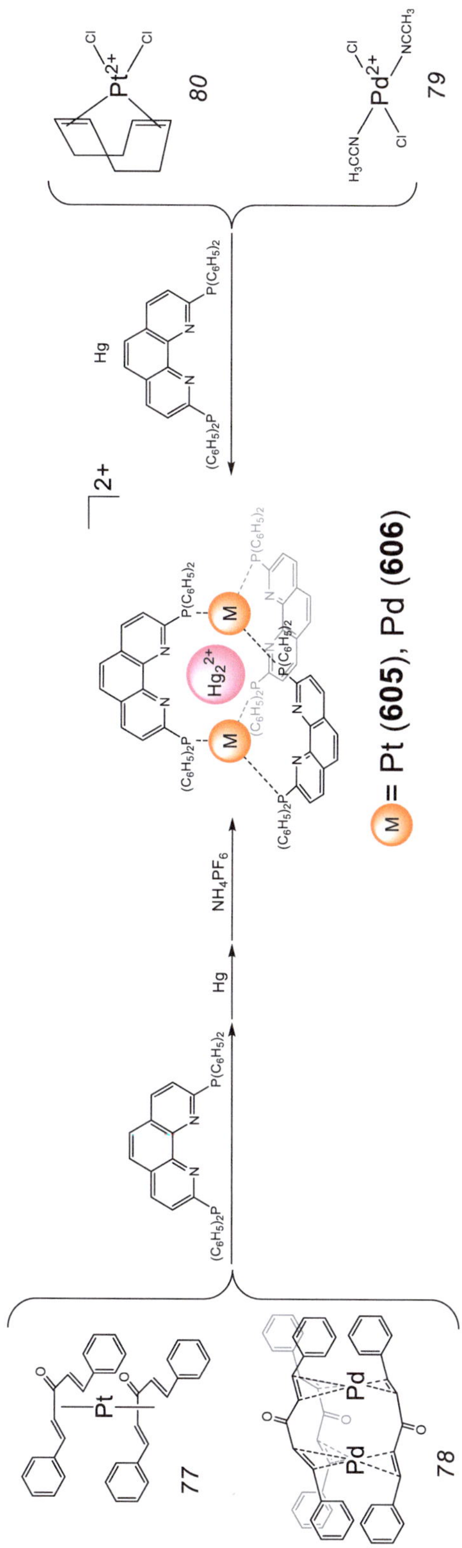

Scheme 2.340 Synthesis of the Hg_2^{2+} cage complexes of the tris-phenanthrolinate ligands **605** and **606**

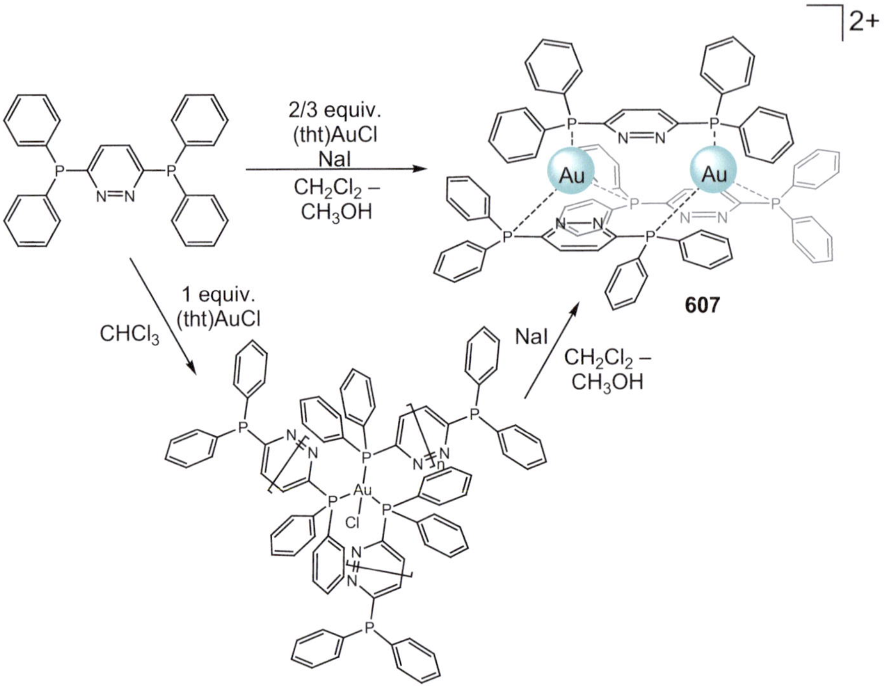

Scheme 2.341 Self-assembly of the gold-capped macrobicyclic compound **607**

(Fe^{2+})@**252** and (Fe^{2+})@**253** having exactly the same length but just a different width.

The mixed-ligand self-sorting experiments by Scheme 2.329 have been also performed [89]. In the presence of palladium(II) ions, ^{1}H NMR spectrum of the mixture of the clathrochelate ligand (Fe^{2+})@**251** and its polyaromatic analog *67* exclusively showed two distinct sets of peaks, corresponding to their homoleptic assemblies **584** and **595** due to so-called "narcissistic self-sorting," In contrast, ^{1}H NMR spectrum of the mixture of the clathrochelate ligand (Fe^{2+})@**252** and the above ligand *67* in the presence of the same metal(II) ions showed the multiple peaks, indicating that a library of the corresponding heteroleptic products had formed; their formation has been confirmed also by ESI MS method. Such different self-sorting behavior of **584** and **593** is explained [89] by their different thermodynamic stability.

So, the cubic Fe_8L_{12}-coordination cages have been obtained [89] from a clathrochelate ligand synthon with an intermediate aspect ratio. By increasing the length or by decreasing the width of such macrobicyclic ligand synthon, the self-assembly process resulted in a formation of the tetrahedral Fe_4L_6-cages. In the case of Pd^{II}-based coordination cages, a clathrochelate ligand synthon with a large aspect ratio gave an entropically favored tetrahedral Pd_4L_8-capsule, whereas an octahedral Pd_6L_{12}-coordination cage has been formed with a macrobicyclic ligand synthon of the same length but with an increased width. The above results can be illustrated by Scheme 2.330. The aspect ratio has been also used [89] to control the composition of dynamic mixtures of the above palladium(II)-capped cages. Thus, the aspect ratio of a rigid rod-type ligand synthon is reported [89] to be a very important parameter for metal-based self-assembly reactions.

Coordination of the 4-pyridylboroncapped iron(II) tris-nioximate (Fe^{2+})@**242** as a clathrochelate ligand synthon toward the

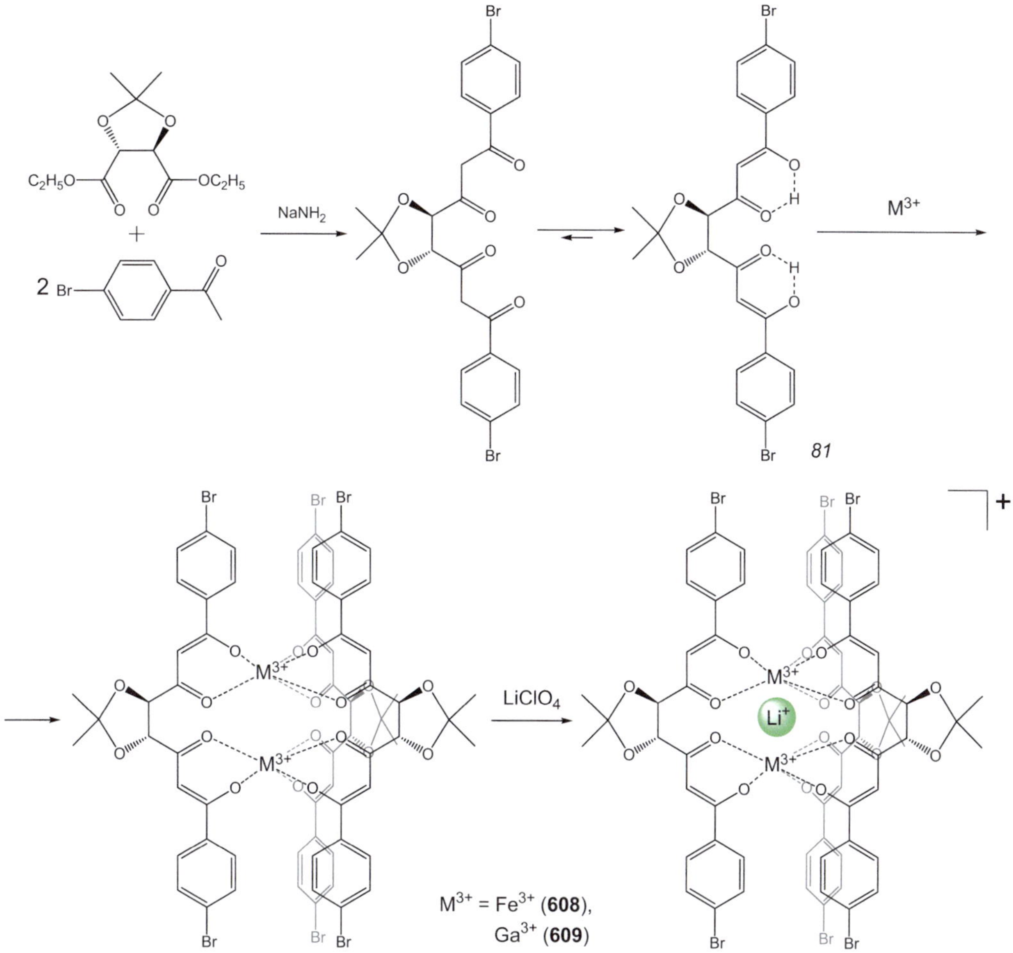

Scheme 2.342 Preparation of the macrobicyclic ligands **608** and **609**

metallocomplex *fac*-$Re(CO)_3(CH_3CN)_2Cl$ by Scheme 2.331 gave the molecular square **596** in a relatively high yield [92]. Heating in a sealed vial the mixture of $Zn(NO_3)_2 \cdot 6H_2O$ and 4,4-′-biphenyldicarboxylic acid with the same clathrochelate ligand in DMA–xylene medium allowed us to obtain a mixed-ligand 2D coordination polymer *68* shown in Scheme 2.332 in a moderate yield.

Self-assembly of Cp*Ir-containing bis-amidinate precursors *69* and *70* with 4-pyridyl-terminated iron(II) tris-dimethylglyoximate and tris-nioximate macrobicyclic ligand synthons (Fe^{2+})@**241** and (Fe^{2+})@**242** by Schemes 2.333 and 2.334 has been used [229] for preparation of the clathrochelate-based Ir,Fe-heterometallic metallomacrocycles **597–600** in high yields. 1:1 stoichiometric reactions of the metallocomplex *71* with the same ligand synthons by Scheme 2.335 at room temperature gave the corresponding Ir_2Fe heterotrinuclear complexes (Fe^{2+})@**601** and (Fe^{2+})@**602** in almost quantitative yields [229].

Coordination-driven self-assembly of the homodinuclear clathrochelate ligand synthon $(Zn^{2+})_2$@(**572**–$2H^+$) with the metallocomplex *72* afforded the molecular square **603** shown in Scheme 2.336 as the single product of this process [225]. The reactions of the macrobicyclic ligand synthons $(Zn^{2+})_2$@**571** and $(Zn^{2+})_2$@**572** with zinc and cadmium(II) ions by Scheme 2.337 have been used for

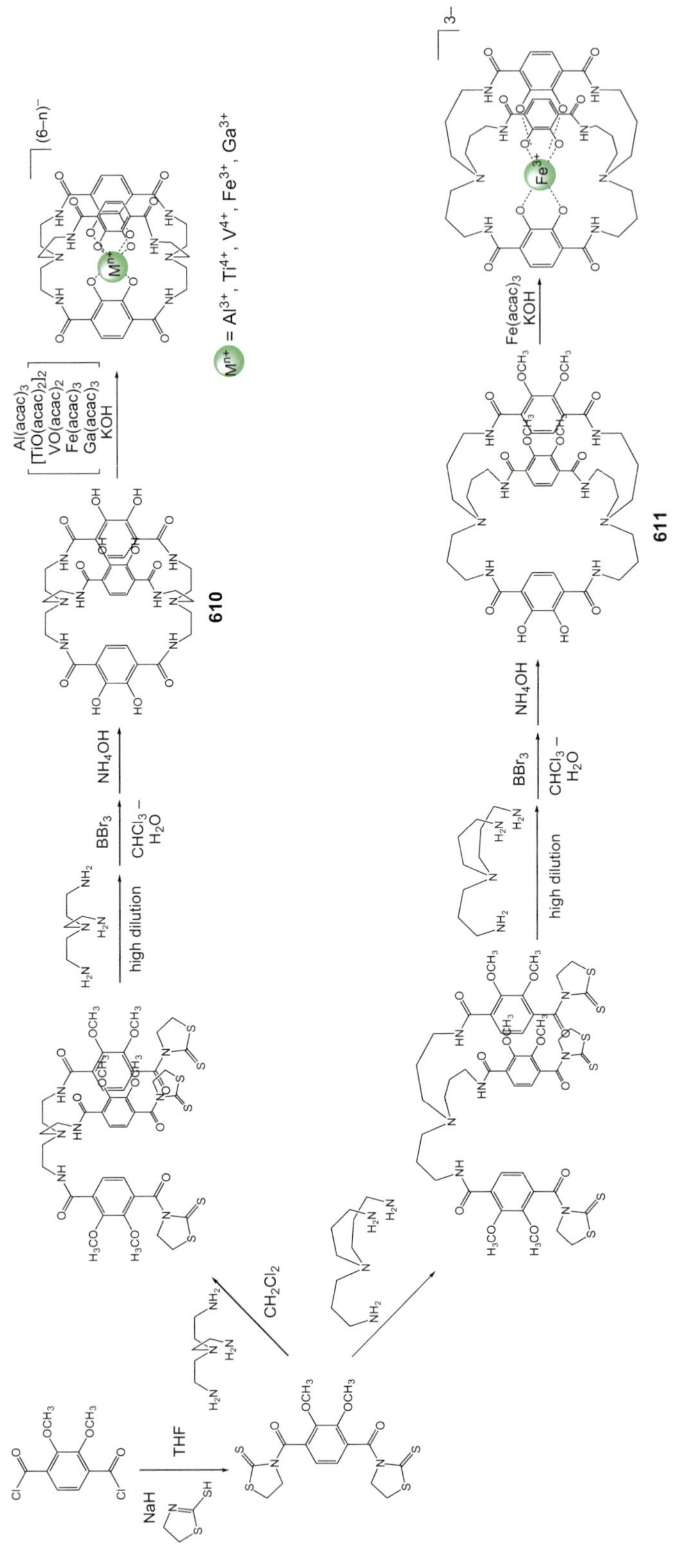

Scheme 2.343 Synthesis of the tris-pyrocatecholate macrobicyclic ligand **611** and its iron(III) cage complex

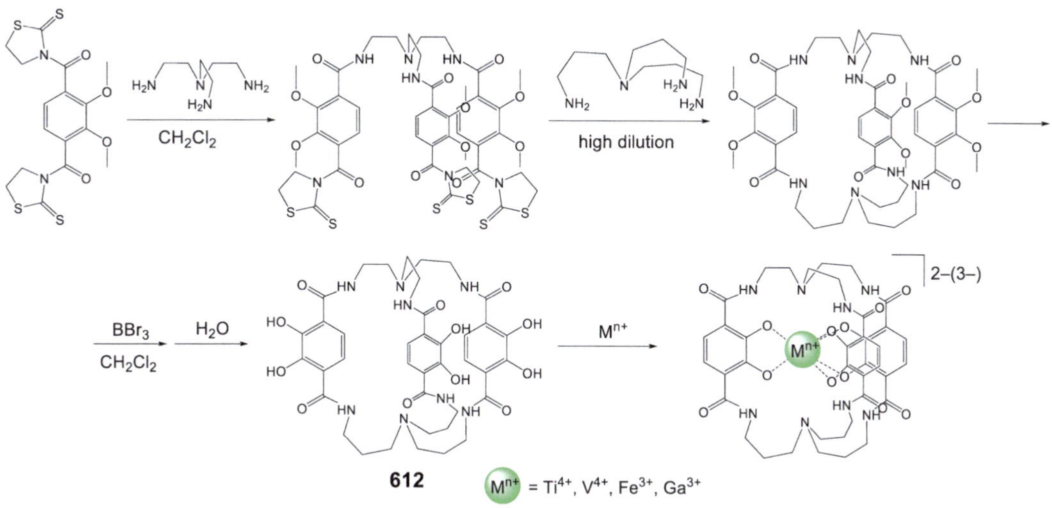

Scheme 2.344 Preparation of the tris-pyrocatecholate macrobicyclic ligand **612** and its cage complexes

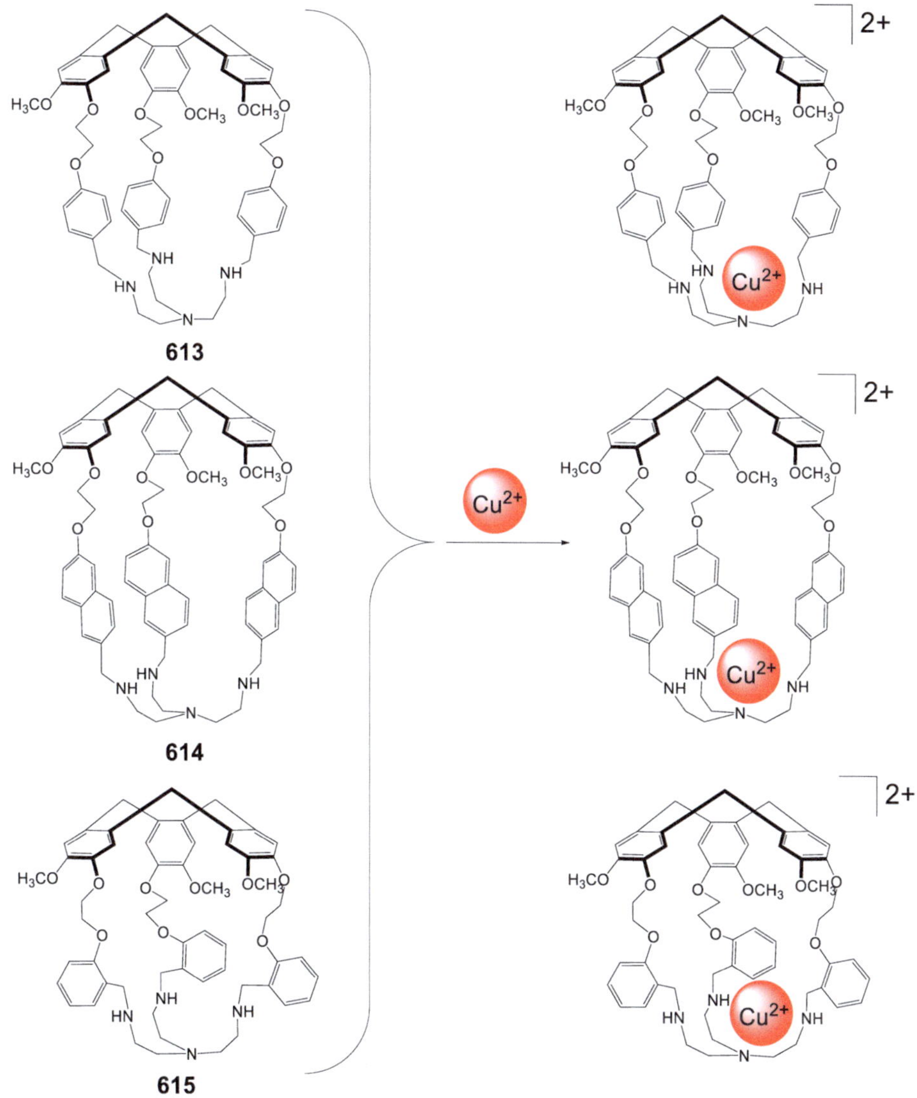

Scheme 2.345 Encapsulation of copper(II) ion by a series of the cryptophane ligands

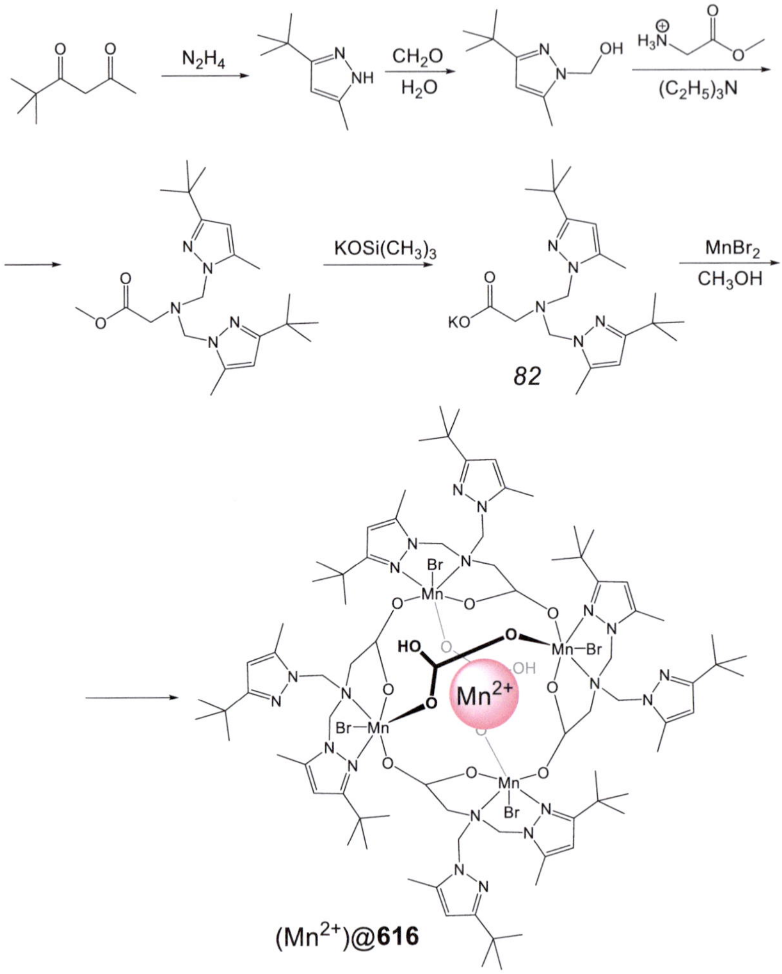

Scheme 2.346 Template self-assembly of the clathrochelate (Mn^{2+})@**616**

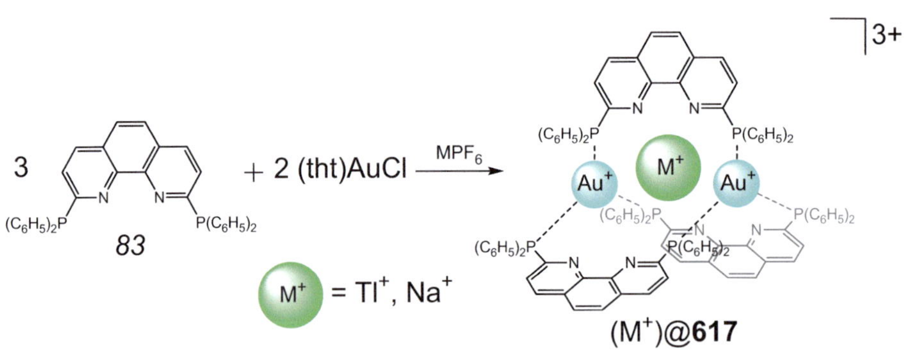

Scheme 2.347 Template synthesis of the gold-capped thallium and sodium(I) cage complexes of **617**

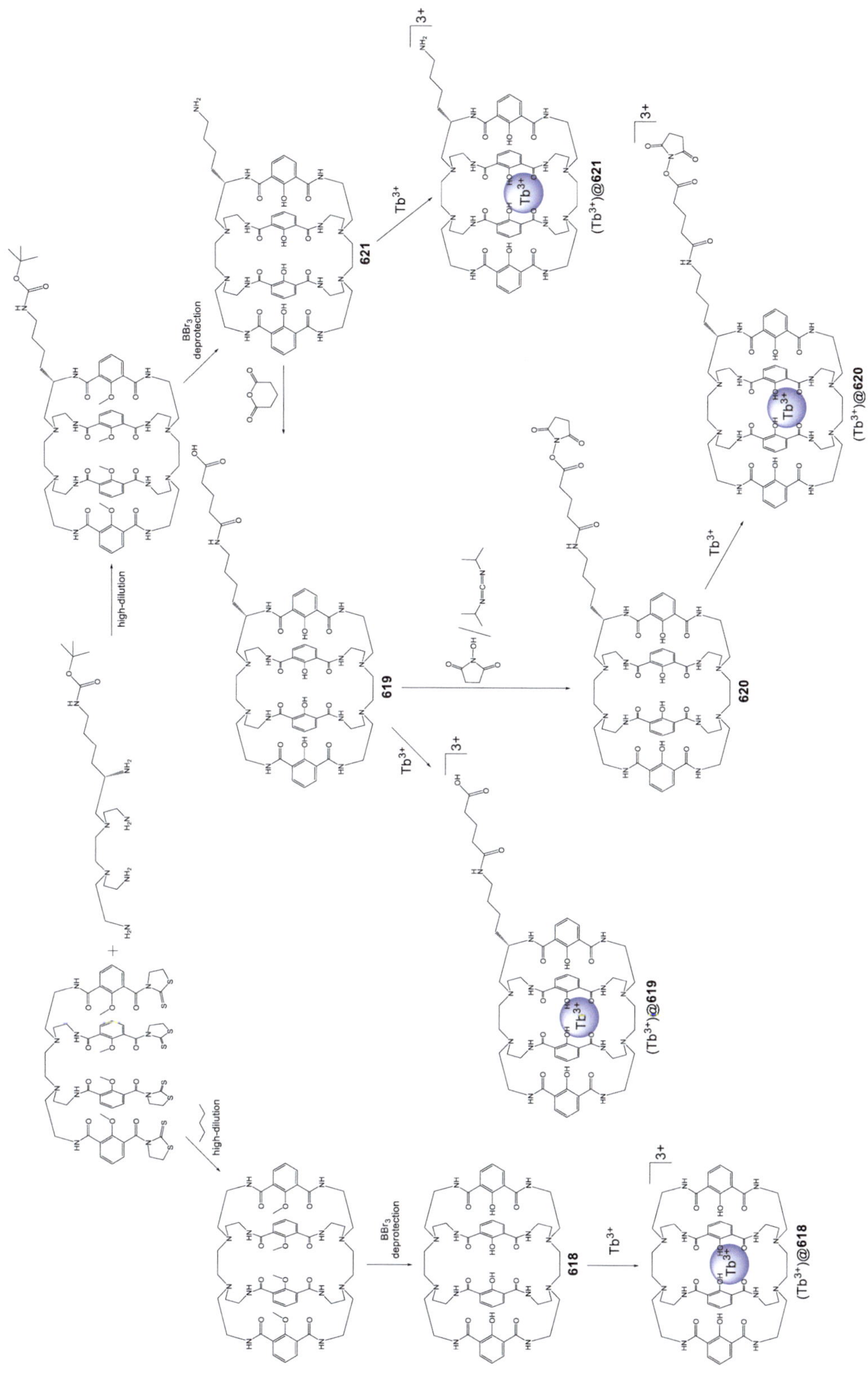

Scheme 2.348 Synthesis of the macrotricyclic ligands **620** and **621**, and their terbium(III) cage complexes

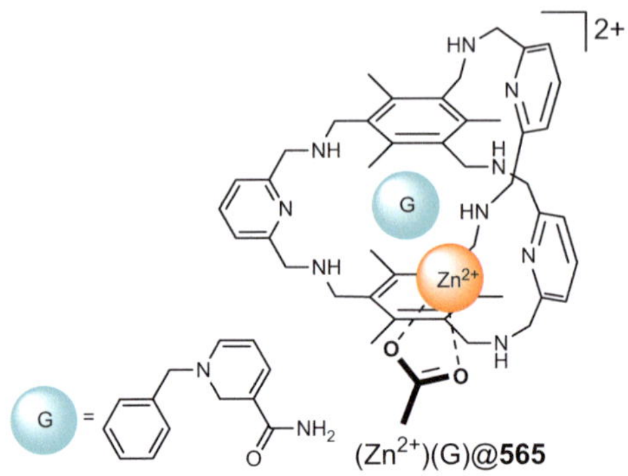

Scheme 2.349 Chemical drawing of zinc(II) complex of the ligand **565**

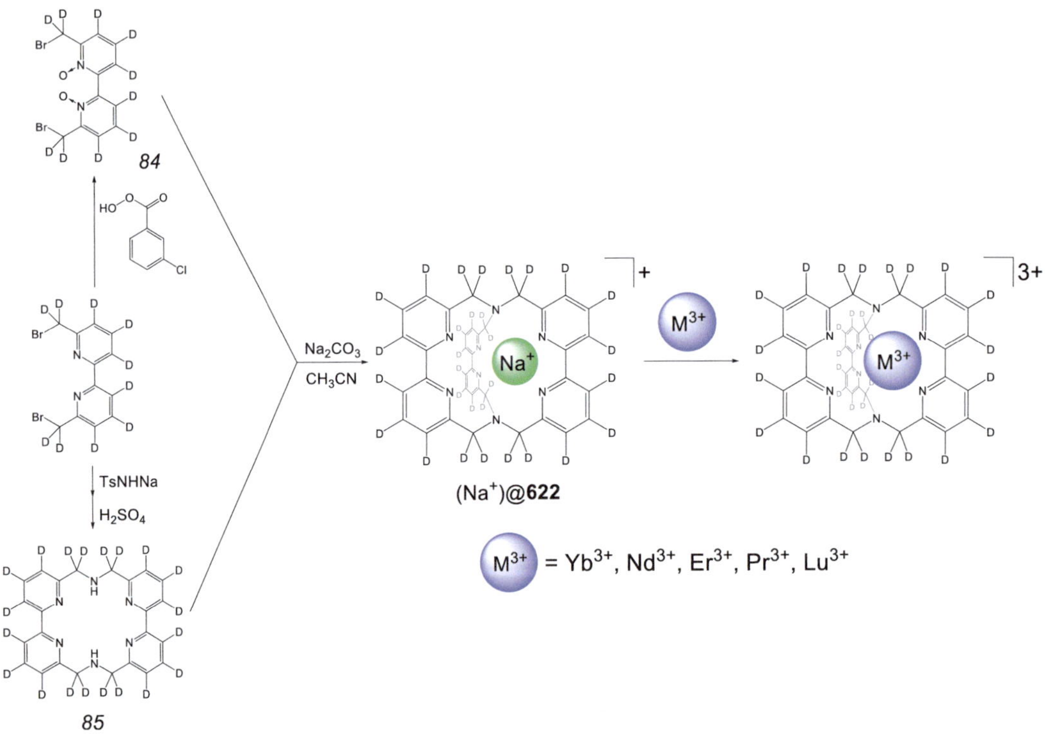

Scheme 2.350 Preparation of potassium(I) complex of the caging ligand **622** and its lanthanide(III) clathrochelates

preparation of the clathrochelate-based coordination polymers *73* and *74* with 2D square-grid-type layer structures (see Sect. 3.2). Overall, these networks are neutral, because the anionic clathrochelate ligand synthons compensate the positive charge of zinc(II) dications. Coordination of the zinc(II) clathrochelate synthons $(Zn^{2+})_2$@(**571**–$2H^+$) and $(Zn^{2+})_2$@(**572**–$2H^+$) toward cadmium(II) ions by Scheme 2.338 under mild reaction conditions has been

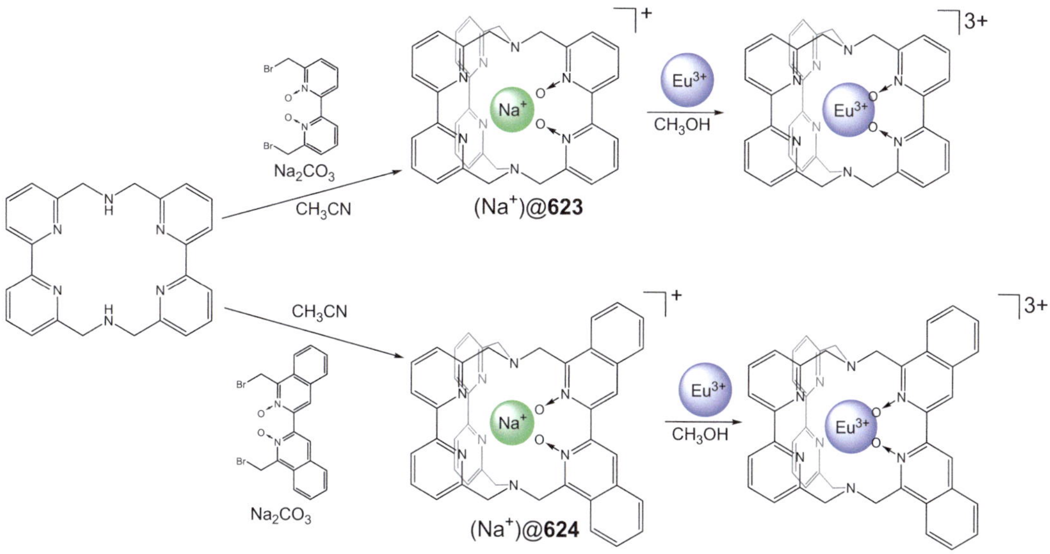

Scheme 2.351 Synthesis of potassium(I) complexes of the macrobicyclic ligands **623** and **624**, and their europium(III) clathrochelates

used [225] for the synthesis of the coordination polymers *75* and *76* in high yields.

2.5 Other Types of Cage Metal Complexes

Lithiation of potassium triisopropylsilylphosphanide by Scheme 2.339, giving the corresponding phosphanediide, and its metathesis reaction with SrI_2 or BaI_2 in the presence of HMDSO have been used [230] for preparation of the strontium and barium(II) cage complexes of the encapsulating ligand **604**, respectively. A lower yield of the strontium(II) clathrochelate is explained [230] by size mismatching between Sr^{2+} ion and an octahedral cavity of **604**; this also caused a higher reactivity of the strontium(II) cage complex (Sr^{2+})@**604** as compared with its barium-encapsulating analog.

Two synthetic approaches shown in Scheme 2.340 have been evaluated [231] for preparation of the Hg_2^{2+}-encapsulating cage complexes of the Pt_2L_3- and Pd_2L_3-coordination capsules **605** and **606**. First of these approaches used the metallocomplexes precursors *77* or *78*, while the second synthetic pathway is based on the coordination-driven self-assembly of the more simple precursors *79* or *80* in the presence of freshly distilled metallic mercury [231].

Their gold-capped analog **607** has been synthesized [232] using various synthetic approaches shown in Scheme 2.341. They included (1) the reaction of the corresponding 1D coordination polymer with an excess of NaI in dichloromethane–methanol mixture, (2) the treatment of the analogous 2D coordination polymer with an excess of NaI in the same medium, and (3) the self-assembly of 3,6-bis (diphenylphosphino)pyridazine with the tetrahydrothiophene solvato-complex (tht)AuCl in their molar ratio 3:2 in the presence of an excess of NaI.

Coordination of a chiral tetraketone ligand synthon *81*, initially prepared using Claisen-type condensation of 4-bromoacetophenone with the diethyl ester of acetone ketal of *L*-tartaric acid, toward the gallium or iron(III) ions by Scheme 2.342 has been used [233] for preparation of the

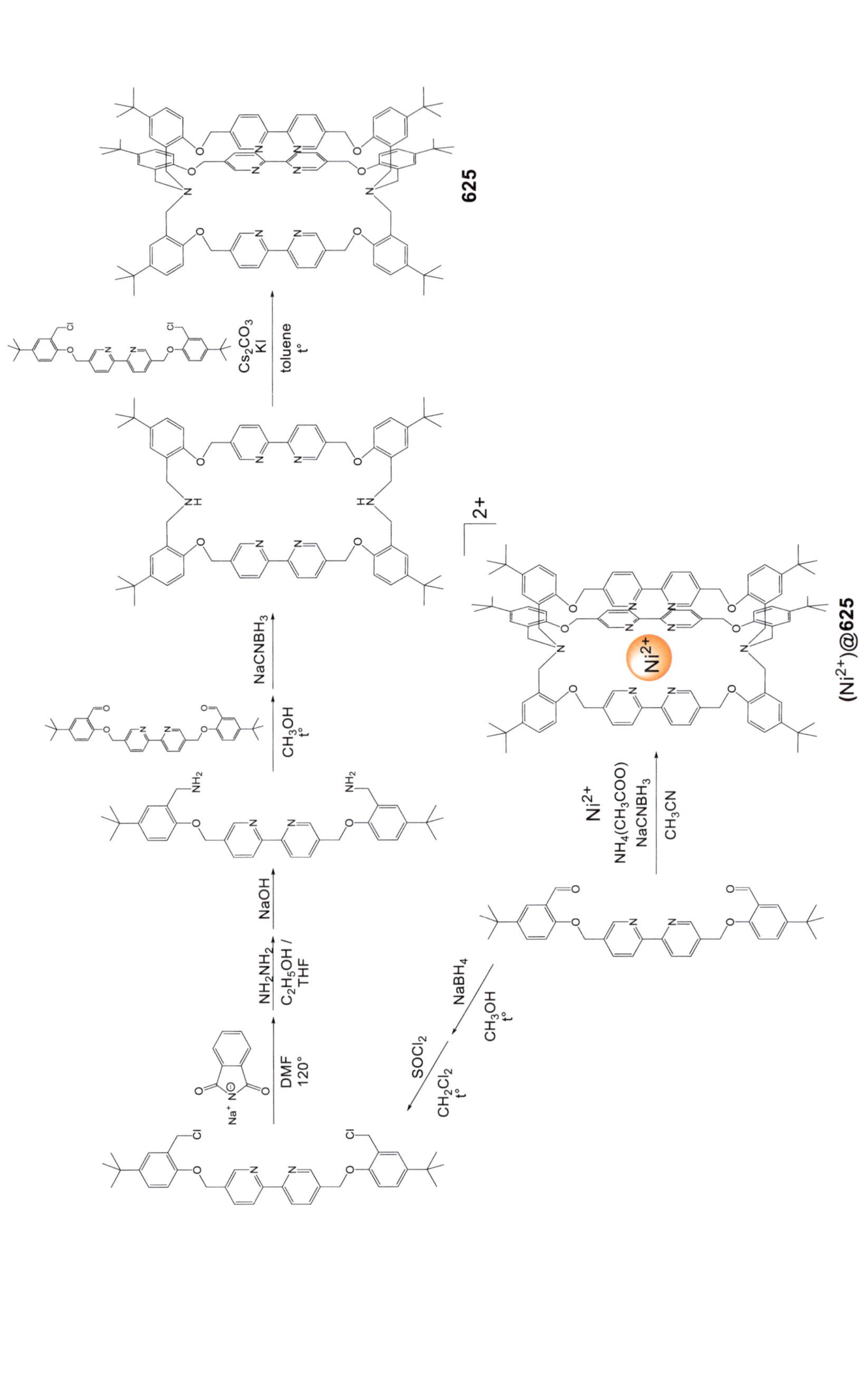

Scheme 2.352 Preparation of the tris-bipyridine macrobicyclic ligand **625** and its nickel(II) cage complex

Scheme 2.353
Coordination-driven self-assembly of the macrobicycle **626**

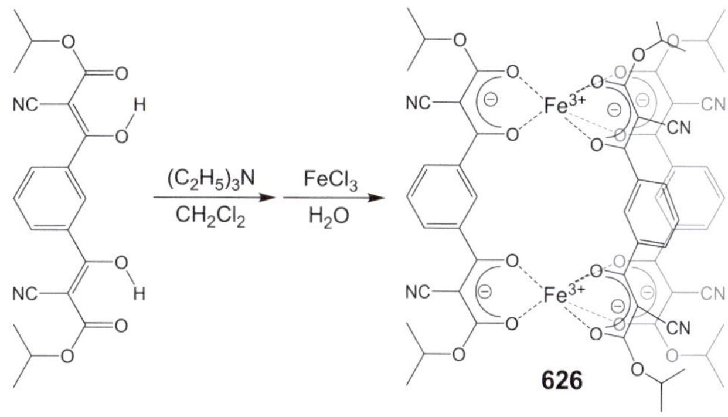

Scheme 2.354
Coordination-driven self-assembly of the clathrochelate (K^+)@**627**

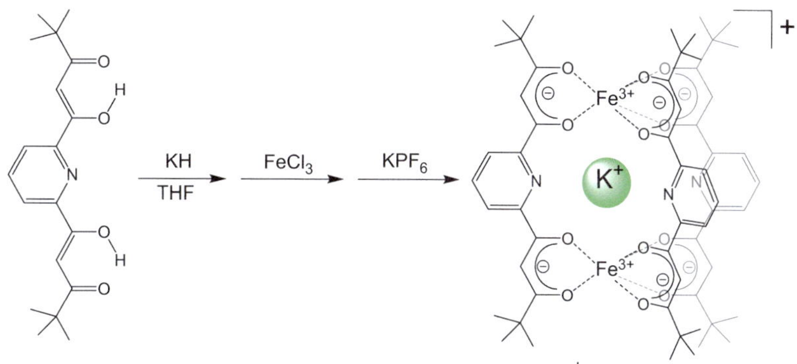

(K^+)@**627**

M_2L_3-helicate macrobicyclic ligands **608** and **609**. They are reported to encapsulate one lithium (I) cation, thus forming the corresponding 1:1 cage complexes.

The macrobicyclic ligand **610** and its extended analog **611** have been prepared [234] using the improved direct synthetic procedures shown in Scheme 2.343. These ligands encapsulate a series of the oxophilic metal(III) and metal(IV) cations by the ligand exchange reactions of their acetylacetonate precursors, thus giving the corresponding 1:1 cage complexes.

A nonsymmetric macrobicyclic analog of **610** and **611**, the caging ligand **612** with non-equivalent capping groups, has been prepared [235] by Scheme 2.344; it forms 1:1 clathrochelates with an encapsulated vanadium or titanium(IV) ion as well as those with a caged iron or gallium(III) trication.

1:1 copper(II) cage complexes of the hemicryptophane ligands **613**, **614**, and **615** with various cavity sizes and shapes have been synthesized [236] by Scheme 2.345.

1:1 manganese(II) cage complex of the M_4L_{16}-coordination capsule **616** has been self-assembled [237] starting from its designed bipyrazole glycinate ligand synthon *82* using one-pot synthetic procedure shown in Scheme 2.346.

Template reaction of a 1,1-phenanthroline-based diphenylphosphine ligand synthon *83* with a tetrahydrothiophene gold(I) solvato-complex by Scheme 2.347 in the presence of the corresponding metal(I) ion as a matrix has been used [238] for the synthesis of 1:1 thallium and sodium(I) cage complexes of the macrobicyclic $Au^I{}_2L_3$-coordination capsule **617**.

A series of the octadentate macrotricyclic 2-hydroxyisophthalamide-based ligands has

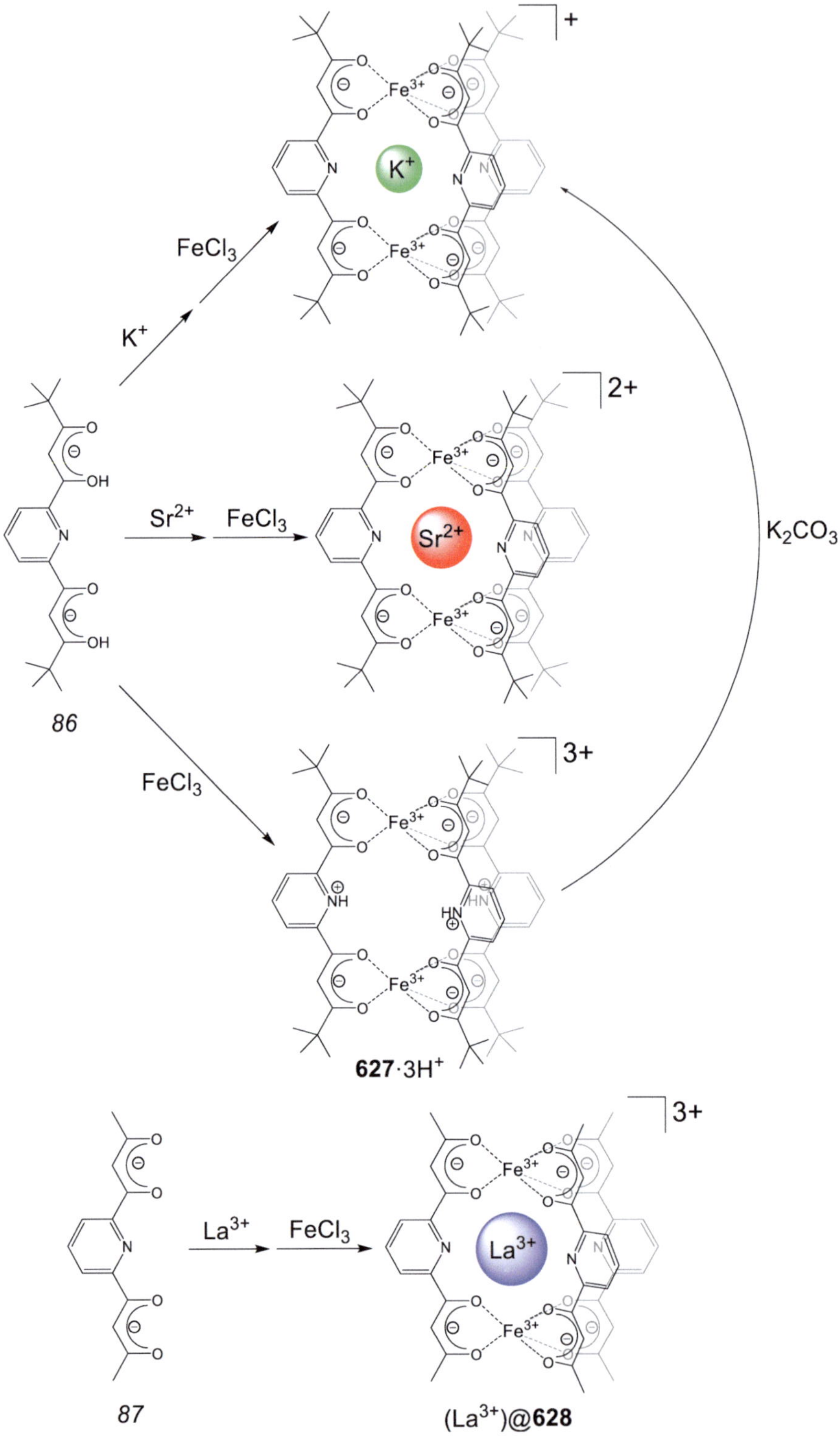

Scheme 2.355 Self-assembly of the macrobicyclic ligand **627**, its cage complexes and the clathrochelate (La^{3+})@**628**

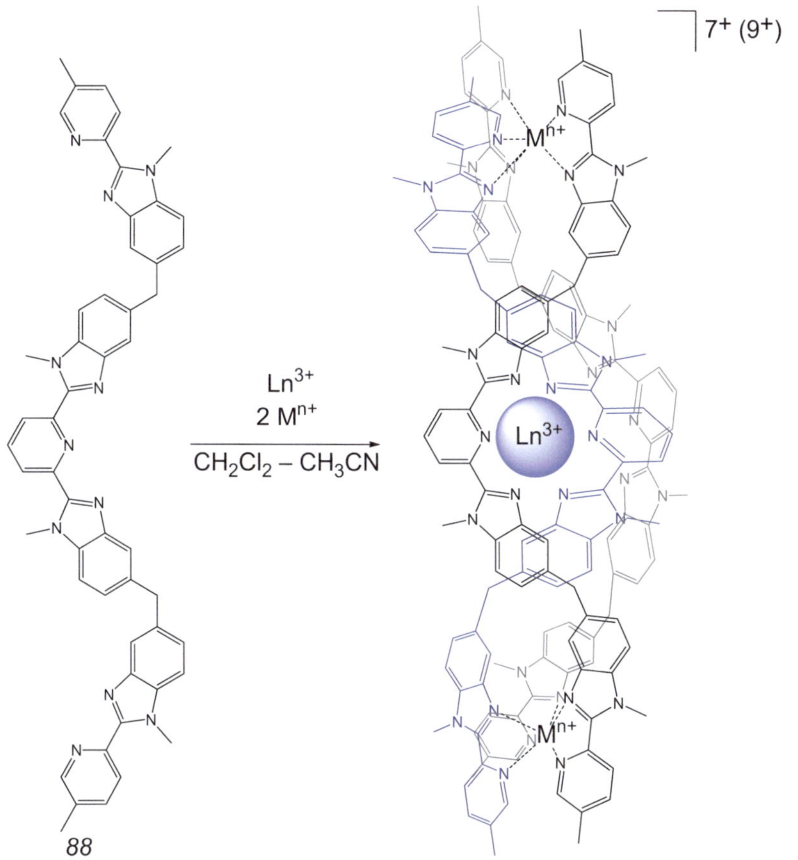

88

M^{n+} = Cu^{2+} (**629**), Cr^{3+} (**630**)

Ln^{3+} = La^{3+}, Eu^{3+}, Gd^{3+}, Tb^{3+}, Lu^{3+} ((M^{3+})@**629**); Eu^{3+}, Tb^{3+} ((M^{3+})@**630**)

Scheme 2.356 Self-assembly of lanthanide(III) cage complexes of the lihands **629** and **630**

been synthesized [239] by Scheme 2.348. These caging ligands encapsulate a terbium(III) cation, thus giving its 1:1 cage complexes.

Preparation of 1:1 zinc(II) cage complex of the macropolycyclic ligand **565** (Scheme 2.349) with co-encapsulated 1-benzylnicotinamide guest molecule is reported [240].

The [D_{30}]-deuterated sodium(I) cage complex of the encapsulating ligand **622** has been straightforwardly synthesized [241, 242] starting from the ligand synthons *84* and *85* shown in Scheme 2.350. All further transformations of this deuterated clathrochelate have been performed without deuterated solvents and/or reagents, still allowing its very high final deuteration level of 98%D. The corresponding deuterated lanthanide(III) cage complexes of **622** and their nondeuterated lutetium(III)-encapsulating analog have been prepared [242] by the exchange reactions of an encapsulated metal ion of the corresponding deuterated and nondeuterated sodium(I)-containing clathrochelate precursors, respectively.

1:1 sodium(I) and europium(III) cage complexes of the 2,2′-bipyridine 1,1′-dioxide- and 3,3′-biisoquinoline 2,2′-dioxide-based macrobicyclic ligands **623** and **624** have been synthesized [243] by Scheme 2.351.

The extended tris-bipyridine N_8O_6-macrobicyclic ligand **625** has been prepared

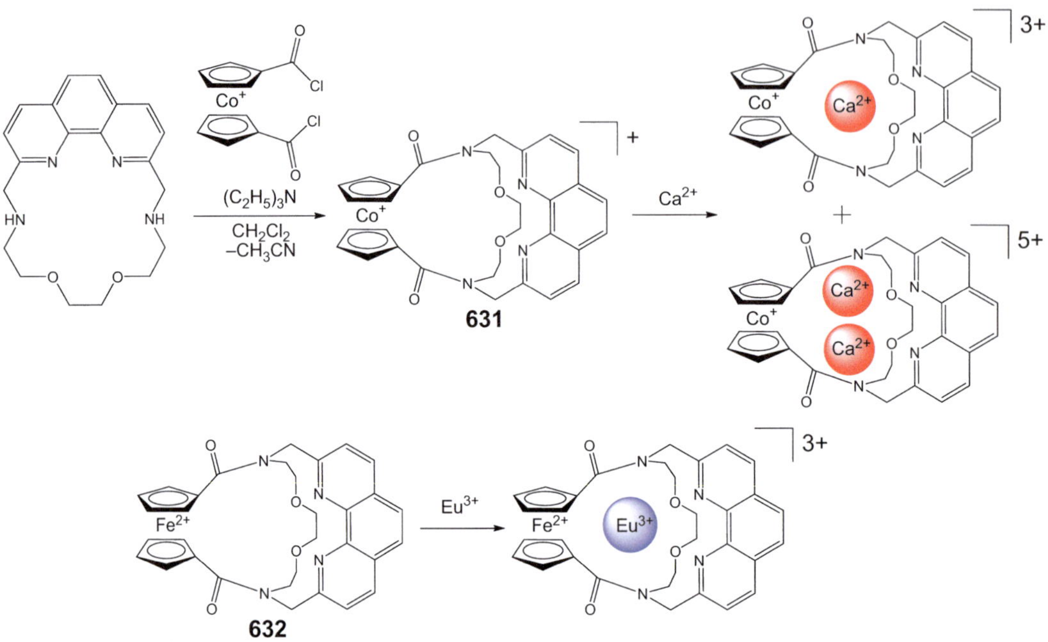

Scheme 2.357 Formation of cage complexes of the cobaltocene- and ferrocene-based ligands **631** and **632**

[244] using two synthetic approaches shown in Scheme 2.352. First of them is based on the classical organic reactions, while the second pathway used template condensation on a nickel(II) ion as a matrix. The cage framework of **625** possesses an inherent helicity after the formation of its 1:1 nickel(II) cage complex (Ni^{2+})@**625** [244].

The Fe_2L_3-coordination capsules **626** and **627** have been self-assembled [245] by Schemes 2.353 and 2.354; the latter macrobicyclic compound encapsulates one potassium(I) ion, thus forming its 1:1 cage complex.

1:1 potassium(I), strontium(II), and lanthanum(III) cage complexes of the pyridine-bis(-β-diketonate) $Fe^{III}{}_2L_3$-coordination capsules **627** and **628** have been prepared [246] by Scheme 2.355 using two synthetic approaches: (1) by direct template self-assembly of the corresponding ligand synthon (*86* or *87*) with iron(III) ions on an appropriate metal cation as a matrix, and (2) by reaction of a protonated form of the initially obtained metal-free Fe_2L_3-capsule **627** with K_2CO_3.

Self-assembly of a heptadentate polyheterocyclic ligand synthon *88* with zinc or chromium(II) ions and lanthanide(III) cations by Scheme 2.356 has been used [247] for the synthesis of 1:1 lanthanium, europium, gadolinium, and lutetium(III) cage complexes of the triple-stranded $M^{II}{}_2L_3$-coordination capsules **629** and **630**. In the case of an attempted cross-linking with chromium(II) ions, the latter underwent a rapid oxidation, thus leading to the chromium (III)-capped coordination capsule **630**.

The cobaltocenium-based redox-active caging ligand **631**, containing the phenanthroline and crown ether ribbed fragments, has been prepared [248] by Scheme 2.357. This ligand encapsulates calcium(II) cation(s), thus forming its 1:1 and 1:2 cage complexes or their mixture (depending on a metal ion-to-ligand molar ratio). Its ferrocene-based analog **632** binds one europium(III) cation, giving the 1:1 cage complex also shown in Scheme 2.357 [248].

The macrobicyclic hexalactam **633** has been synthesized [249] through Stetter cyclization in high dilution conditions using two synthetic approaches shown in Scheme 2.358: (1) by condensation of 2,3-dimethoxyterephthaloyl dichloride with 1,3,5-benzenetriyl-tris-methaneamine and (2) by reaction of a tris-acid chloride ligand synthon *89* and 1,3,5-benzenetriyl-tris-

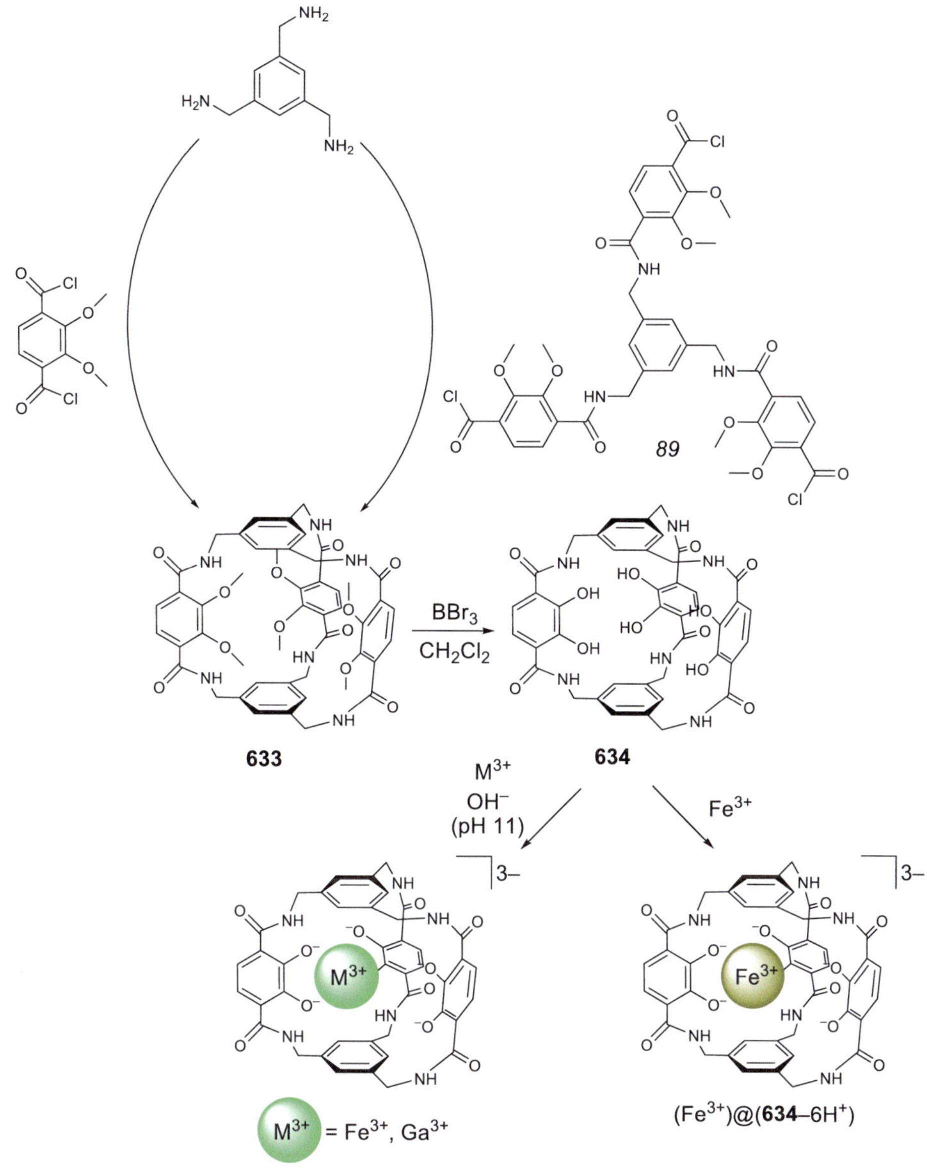

Scheme 2.358 Preparation of the macrobicycles **633** and **634**, and cage complexes of the latter ligand

methaneamine. Cleavage of the methoxyl groups of **633** by BBr_3 allowed obtaining the hexahydroxyl macrobicyclic ligand **634**, a hexaprotonated form of which forms the corresponding 1:1 iron(III) cage complex [249].

Two different synthetic strategies [i.e., (1) the cyclization under high dilution conditions and (2) the stepwise procedure] have been developed [250] for the synthesis of the same tris-catecholate macrobicyclic ligand **634**; they gave the target compound in very low and moderate yields, respectively. A hexadeprotonated form of these ligand binds iron or gallium(III) cations, thus giving their 1:1 cage complexes (M^{3+})@(**634**–$6H^+$), one of which is identical to that described [249]. The extended tris-catecholate macrobicyclic ligands **635** and **636** (Scheme 2.359) also efficiently encapsulate an iron(III) ion, thus giving the corresponding 1:1 clathrochelates [251].

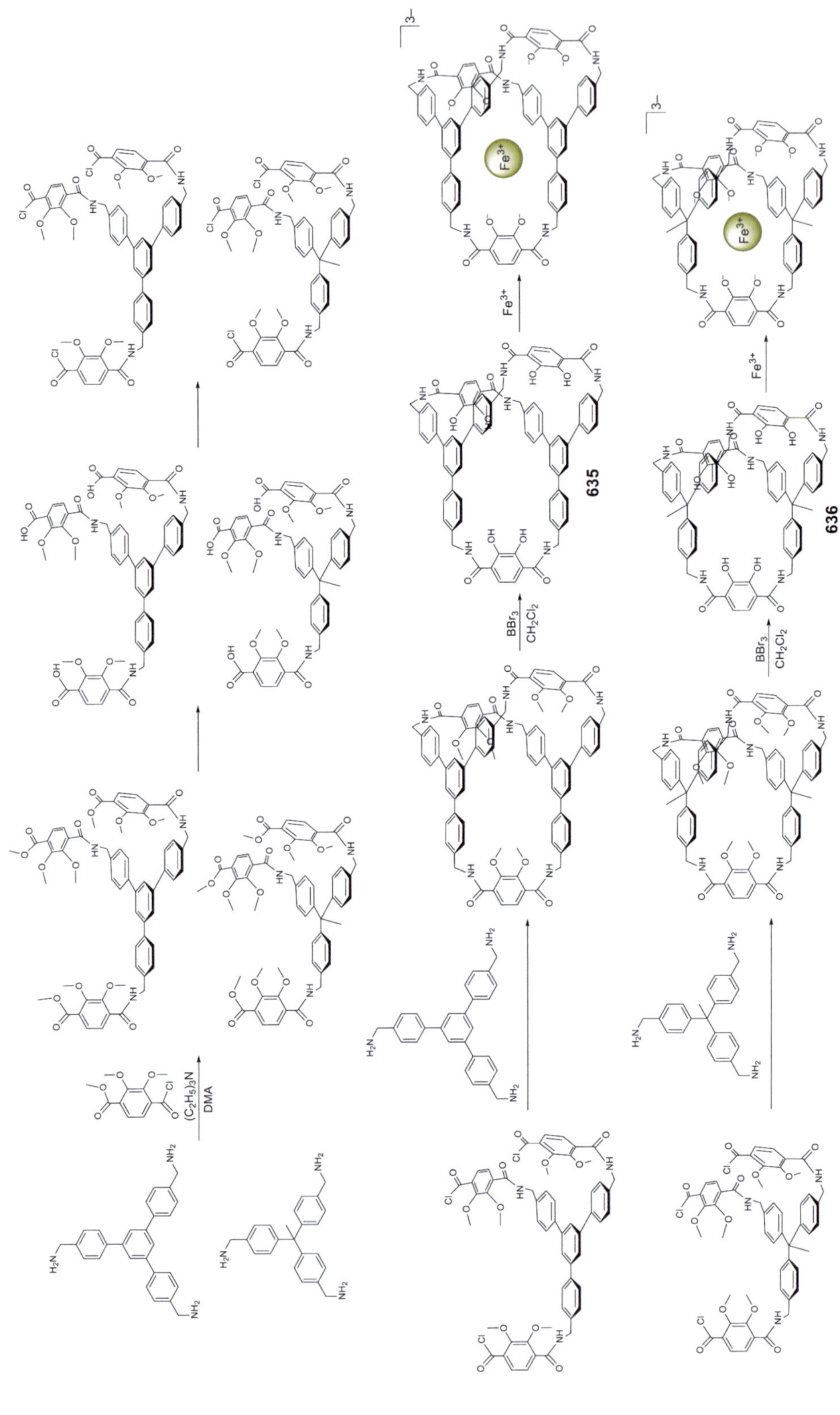

Scheme 2.359 Synthesis of the macrobicyclic tris-pyrocatecholate ligands **635** and **636**, and their iron(III) complexes

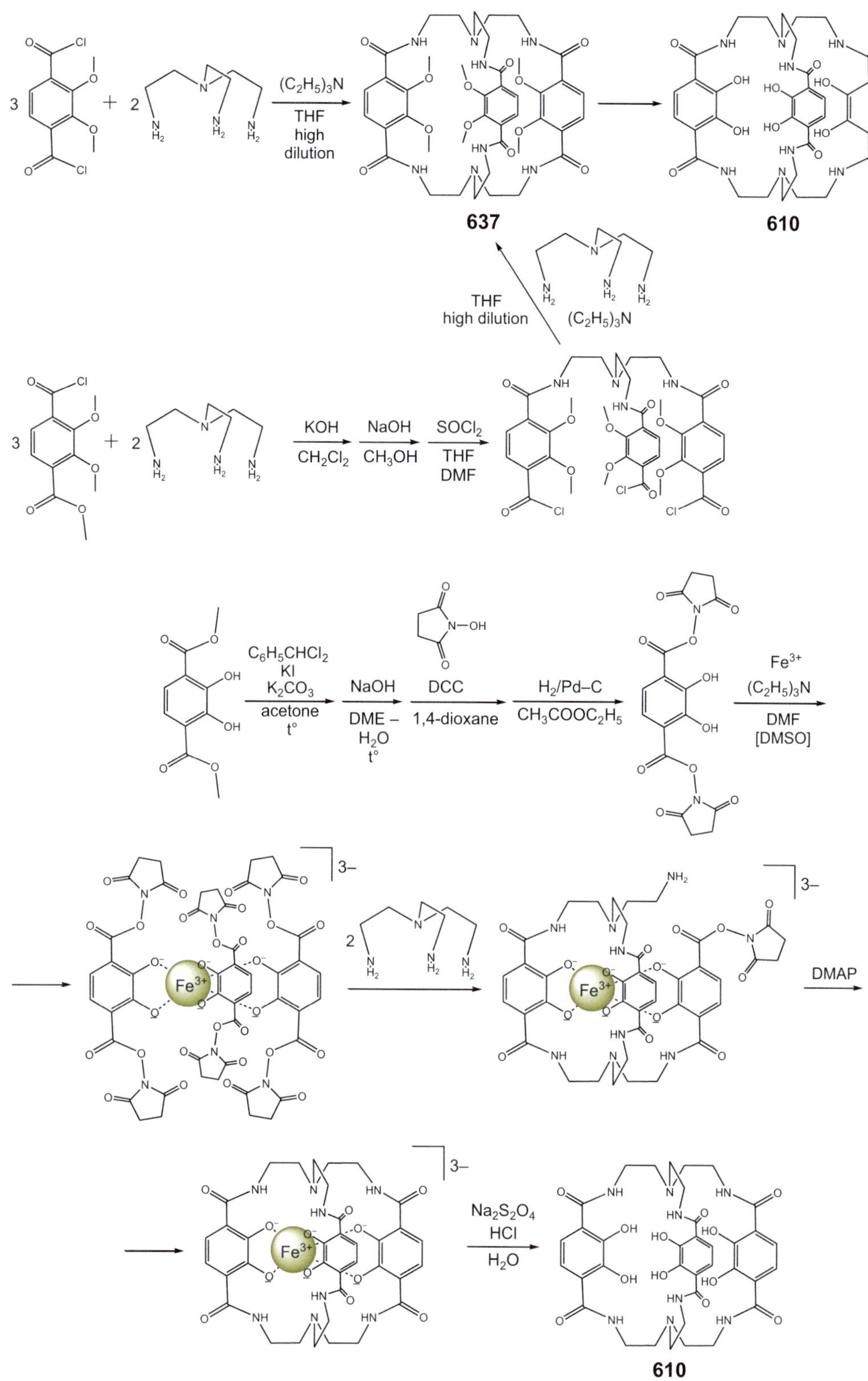

Scheme 2.360 Two synthetic pathways to a metal-free macrobicycle **610**

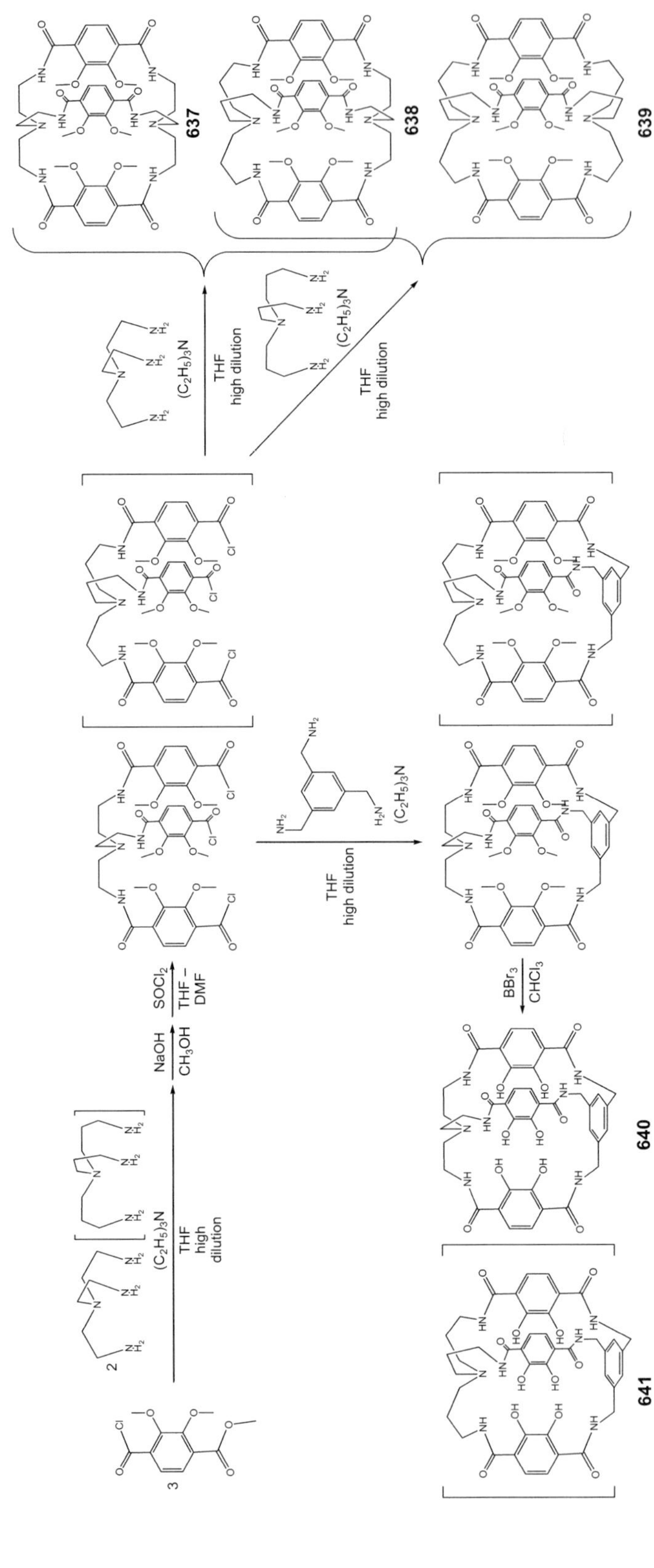

Scheme 2.361 Preparation of a series of the pyrocatecholate-based macrobicyclic compounds

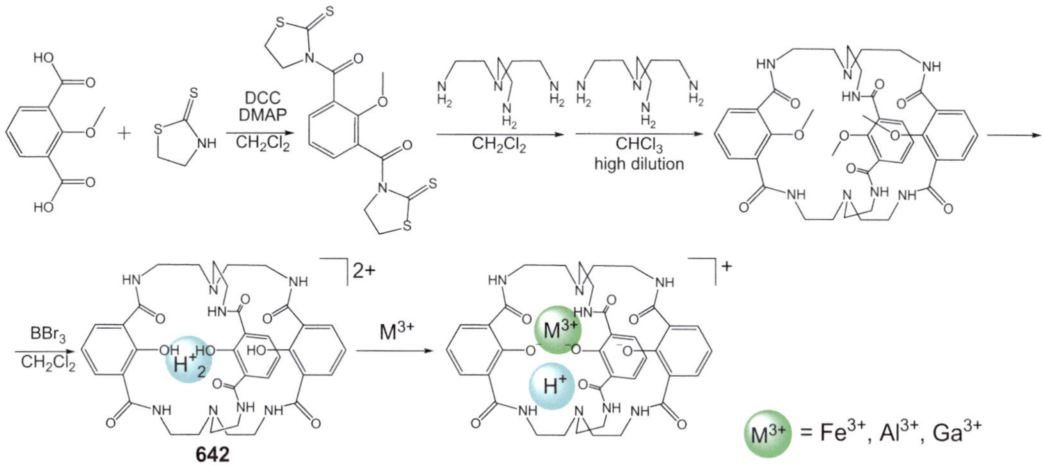

Scheme 2.362 Synthesis of the macrobicyclic ligand **642** and its cage complexes

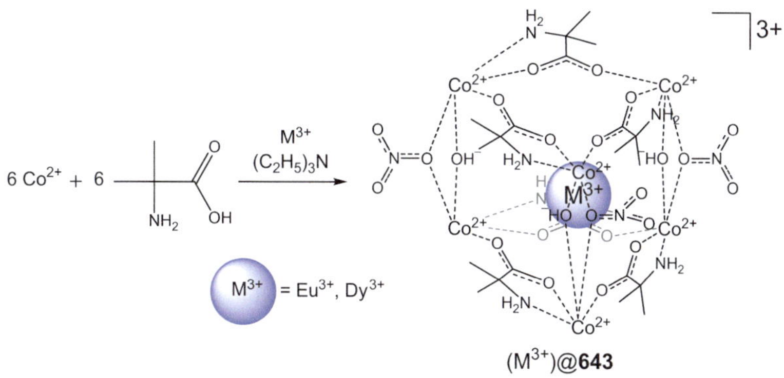

Scheme 2.363 Self-assembly of the europium and dysprosium(III) clathrochelates (M^{3+})@**643**

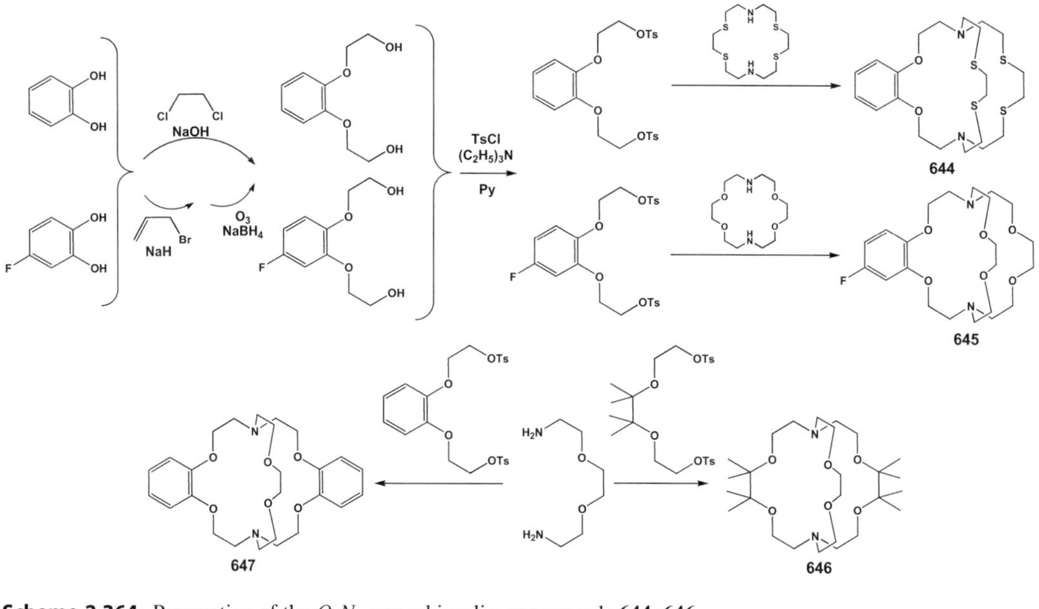

Scheme 2.364 Preparation of the O_6N_2-macrobicyclic compounds **644–646**

The methyl-protected tris-catecholate macrobicycle **637** has been synthesized [252] under high dilution conditions using two synthetic pathways shown in Scheme 2.360. Removal of its methyl protecting groups with BBr_3 allowed us to obtain the target hexahydroxyl-containing encapsulating ligand **610** in a very low total yield [252]. So, template procedure using an iron(II) ion as a matrix (also shown in Scheme 2.360) has been elaborated [252]. As a result, its 1:1 iron(III) cage complex has been prepared in a high yield; the demetallation of (Fe^{3+})@**610** gave a metal-free macrobicycle **610** in a moderate total yield.

A series of the tris-catecholate caging ligands, the potent macrobicyclic iron(III)-sequestering agents, has been synthesized [253] by Scheme 2.361. Coupling of two tripodal subunits using a triple amide bond-forming reaction in the presence of triethylamine under high dilution conditions is a key stage of their synthesis. The obtained methyl-protected macrobicyclic compounds have been deprotected with BBr_3, thus giving their hexahydroxyl-containing derivatives **637–641**. These tris-catecholate macrobicyclic ligands efficiently bind the iron (III) ion, forming its 1:1 cage complexes [253].

The tris-salicylate macrobicyclic ligand **642** has been synthesized [254] using a stepwise synthetic procedure shown in Scheme 2.362. It encapsulates the gallium or iron(III) cation, thus giving the corresponding 1:1 cage complexes.

Unusual 1:1 europium and dysprosium(III) cage complexes of the Co_6L_6-coordination capsule **643** have been self-assembled [255] by template procedure shown in Scheme 2.363.

The macrobicyclic ligands **644–647**, designed for stabilization of unusual oxidation state of an encapsulate metal ion (see Sect. 5.4), have been prepared [256] by Scheme 2.364 using the appropriate reactive intermediates.

References

1. Voloshin YZ, Kostromina NA, Krämer R (2002) Clathrochelates: synthesis, structure and properties. Elsevier, Amsterdam
2. Varzatskii OA, Palchik AV, Starikova ZA, Antipin MY, Lebed EG, Bubnov YN (2006) Mono- and trichloride clathrochelate iron(II) chloroglyoximates and their functionalization: the effect of the substituents in the clathrochelate framework on the reactivity of the chlorine-containing fragments in nucleophilic substitution reactions. Inorg Chim Acta 359:553–569
3. Belov AS, Zelinskii GE, Lebed EG, Vologzhanina AV, Voloshin YZ (2016) Novel N_2-, S_2- and O_2-alicyclic iron(ii) di- and tetrachloroclathrochelates and supramolecular architecture of their crystals. Paper presented at the 8th International symposium "Design and synthesis of supramolecular architectures", Kazan, 25–29 Apr 2016
4. Cai H, Fissekis J, Conti PS (2009) Synthesis of a novel bifunctional chelator AmBaSar based on sarcophagine for peptide conjugation and ^{64}Cu radiolabelling. Dalton Trans 5395–5400
5. Tuntulani T, Thavornyutikarn P, Poompradub S, Jaiboon N, Ruangpornvisuti V, Chaichit N, Asfari Z, Vicens J (2002) Synthesis of tripodal aza crown ether calix[4]arenes and their supramolecular chemistry with transition-, alkali metal ions and anions. Tetrahedron 58:10277–10285
6. Darbost U, Zeng X, Rager MN, Giorgi M, Jabin I, Reinaud O (2004) X-ray and solution structures of the first Zn funnel complex based on a calix[6]aza-cryptand. Eur J Inorg Chem 2004:4371–4374
7. Darbost U, Rager MN, Petit S, Jabin I, Reinaud O (2005) Polarizing a hydrophobic cavity for the efficient binding of organic guests: the case of calix[6]tren, a highly efficient and versatile receptor for neutral or cationic species. J Am Chem Soc 127:8517–8525
8. Izzet G, Douziech B, Prangé T, Tomas A, Jabin I, Le Mest Y, Reinaud O (2005) Calix[6]tren and copper (II): a third generation of funnel complexes on the way to redox calix-zymes. Proc Natl Acad Sci 102:6831–6836
9. Izzet G, Zeng X, Akdas H, Marrot J, Reinaud O (2007) Drastic effects of the second coordination sphere on neutral *vs.* anionic guest binding to a biomimetic Cu(II) center embedded in a calix[6] aza-cryptand. Chem Commun 810–812
10. Bazzicalupi C, Bellusci A, Bencini A, Berni E, Bianchi A, Ciattini S, Giorgi C, Valtancoli B (2002) A new dipyridine-containing cryptand for both proton and Cu(II) encapsulation. A solution and solid state study. J Chem Soc Dalton Trans 2151–2157
11. Fekner T, Gallucci J, Chan MK (2004) Synthesis and metalation of a chiral, pyridine-strapped, cyclic bis (benzimidazole) ligand. Org Lett 6:989–992
12. Gil JM, David S, Reiff AL, Hegedus LS (2005) Synthesis and complexation of "back-to-back" capped bis-dioxocyclams. Inorg Chem 44:5858–5865
13. Vila-Nova SP, Pereira GAL, Albuquerque RQ, Mathis G, Bazin H, Autiero H, de Sá GF, Alves S

Jr (2004) Study of the luminescence of Eu(III), Tb (III) and Gd(III) cryptates containing Py$(CO_2Et)_2$ as ligands. J Lumin 109:173–179
14. Kon N, Takemura H, Otsuka K, Tanoue K, Nakashima S, Yasutake M, Tani K, Kimoto J, Shinmyozu T, Inazu T (2000) Synthesis of macrocyclic cage compounds by diamine–dihalide one-step coupling reaction. J Organomet Chem 65:3708–3715
15. Takemura H, Kon N, Tani K, Takehara K, Kimoto J, Shinmyozu T, Inazu T (1997) Stable Cu^+, Ag^+ complexes of aza-bridged macrocyclic molecules: structure and chemical properties. Perkin Trans 1:239–246
16. Dapporto P, Formica M, Fusi V, Giorgi L, Micheloni M, Pontellini R, Paoli P, Rossi P (2001) Ligational properties of two new phenolic aza cages towards proton and alkali metal ions – a theoretical and an experimental approach. Eur J Inorg Chem 2001:1763–1774
17. Bencini A, Bianchi A, Chimichi S, Ciampolini M, Dapporto P, Garcia-Espana E, Micheloni M, Nardi N, Paoli P, Valtancoli B (1991) Lithium binder in aqueous solution. Synthesis and characterization of the new cage 4,10,15-trimethyl-1,4,7,10,15-pentaazabicyclo[5.5.5] heptadecane (L). Protonation and lithium complex formation. Crystal structures of [HL]$[BPh_4]$ and [LiL] $[BPh_4]$. Inorg Chem 30:3687–3691
18. Bencini A, Bianchi A, Dapporto P, Fusi V, Garcia-Espana E, Micheloni M, Paoletti P, Paoli P, Rodriguez A, Valtancoli B (1993) Thermodynamic, kinetic, and structural study of the ligational properties of the macrobicyclic aza-ligand 4,7,10,17,23-pentamethyl-1,4,7,10,13,17,23-heptaazabicyclo[11.7.5]pentacosane (L1) and of its macrocyclic precursor 1,4,7,13-tetramethyl-1,4,7,10,13,16-hexaazacyclooctadecane (L2). Crystal structure of $[Zn(L1)(H_2O)](BPh_4)_2$. Inorg Chem 32:2753–2760
19. Fuchs A, Kaifer E, Wadepohl H, Himmel HJ (2012) Cryptate complexes with the potential for CO_2 activation: quantum chemical predictions and synthetic efforts. Eur J Inorg Chem 2012:4020–4028
20. Takemura H, Shinmyozu T, Inazu T (1988) Highly symmetrical cage compounds, in which all six, eight or ten lone pairs on nitrogen atoms at bridgehead or intervening pyridine rings are oriented toward the central cavity. Tetrahedron Lett 29:1789–1792
21. Takemura H, Iwanaga T, Shinmyozu T (2006) C–F···Rb+ interaction in a fluorinated cage compound complex. Tetrahedron Lett 47:8989–8991
22. Bianchi A, Ciampolini M, Micheloni M, Nardi N, Valtancoli B, Mangani S, Garcia-Espana E, Ramirez JA (1989) The small cage 12,17-dimethyl-5-oxa-1,9,12,17-tetra-azabicyclo[7.5.5]nonadecane (L): its synthesis, characterization, and 'proton sponge' behaviour. The crystal structure of the dipicrate salt $[H_2(L)](picrate)_2$. Perkin Trans 2:1131–1137
23. Bencini A, Bianchi A, Borselli A, Chimichi S, Ciampolini M, Dapporto P, Micheloni M, Nardi N, Paoli P, Valtancoli B (1990) Selective lithium encapsulation in aqueous solution by the new cage 4,10-dimethyl-1,4,7,10,15-pentaazabicyclo[5.5.5] heptadecane (L). Synthesis, characterization, and structural aspects. Crystal structures of $[LiL](ClO_4)$ and $[CuL]Br_2{\cdot}3H_2O$. Inorg Chem 29:3282–3286
24. Bazzicalupi C, Bencini A, Ciattini S, Denat F, Desogere P, Goze C, Matera I, Valtancoli B (2010) Proton and metal binding by cyclen-based highly rigid cryptands. Dalton Trans 39:11643–11653
25. Mukhopadhyay P, Bharadwaj PK, Savitha G, Krishnan A, Das PK (2002) A new class of three dimensional D-π-A trigonal cryptand derivatives for second-order nonlinear optics. J Mater Chem 12:2237–2244
26. Mukhopadhyay P, Bharadwaj PK, Krishnan A, Das PK (2002) Synthesis and characterization of mono- and bis-D-π-A cryptand derivatives for second-order nonlinear optics and its modulation with different metal ion inputs. J Mater Chem 12:2786–2791
27. Bag B, Bharadwaj PK (2004) Attachment of an electron-withdrawing fluorophore to a cryptand for modulation of fluorescence signaling. Inorg Chem 43:4626–4630
28. Bernier N, Carvalho S, Li F, Delgado R, Félix V (2009) Anion recognition by a macrobicycle based on a tetraoxadiaza macrocycle and an isophthalamide head unit. J Organomet Chem 74:4819–4827
29. Ragunathan KG, Bharadwaj PK (1992) Template synthesis of a cryptand with hetero-ditopic receptor sites. Tetrahedron Lett 33:7581–7584
30. Mukhopadhyay P, Sarkar B, Bharadwaj PK, Nättinen K, Rissanen K (2003) Metal binding characteristics of a laterally nonsymmetric aza cryptand upon functionalization with a π-acceptor group. Inorg Chem 42:4955–4960
31. Bag B, Bharadwaj PK (2005) Fluorescence enhancement of a signaling system in the simultaneous presence of transition and alkali metal ions: a potential AND logic gate. Chem Commun 513–515
32. Sadhu KK, Bharadwaj PK (2008) Translocation of copper within the cavity of cryptands: reversible fluorescence signaling. Chem Commun 4180–4182
33. Gregor T, Weise CF, Lozan V, Kersting B (2007) Straightforward synthesis of novel biphenyl-based macrobicyclic azathioethers. Synthesis 3706–3712
34. Takemura H, Nakamichi H, Nogita R, Iwanaga T, Yasutake M, Shinmyozu T (2003) A comparison of coordination ability of hetero atoms: a Li^+ and Na^+ selective pyridinophane-based cryptand. Tetrahedron Lett 44:5087–5089
35. Sadhu KK, Chatterjee S, Sen S, Bharadwaj PK (2010) Role of spacer in single- or two-step FRET: studies in the presence of two connected cryptands with properly chosen fluorophores. Dalton Trans 39:4146–4154
36. Knight JC, Prabaharan R, Ward BD, Amoroso AJ, Edwards PG, Kariuki BM (2010) A facile one-pot synthesis of a new cryptand via a Pd(II)-catalysed carbonylation reaction. Dalton Trans 39:10031–10033

37. Sadhu KK, Sen S, Bharadwaj PK (2011) Cryptand derived fluorescence signaling systems for sensing Hg(II) ion: a comparative study. Dalton Trans 40:726–734
38. Chand DK, Bharadwaj PK (1996) Synthesis of a heteroditopic cryptand capable of imposing a distorted coordination geometry onto Cu(II): crystal structures of the cryptand (L), [Cu(L)(CN)](picrate), and [Cu(L)(NCS)](picrate) and spectroscopic studies of the Cu(II) complexes. Inorg Chem 35:380–3387
39. Chand DK, Bharadwaj PK (1997) A cobalt (II) cryptate of a heteroditopic cryptand L as an efficient oxygenation catalyst of organic substrates using molecular oxygen and 2-methylpropanal. Inorg Chem 36:5658–5660
40. Esteban-Gómez D, Enríquez-Pérez T, Ferreirós-Martínez R, Mato-Iglesias M, Platas-Iglesias C, de Blas A, Rodríguez-Blas T (2010) Lead (II) complexes of lateral macrobicyclic receptors that incorporate a crown moiety and a pyridine head unit. Eur J Inorg Chem 2010:5027–5034
41. Atkinson IM, Lindoy LF, Matthews OA, Meehan GV, Sobolev AN, White AH (1994) New macrocyclic ligands. V. Synthetic, crystallographic and computational studies of a new N_2O_6-heteroatom cage. Aust J Chem 47:1155–1162
42. Adam KR, Atkinson IM, Kim J, Lindoy LF, Matthews OA, Meehan GV, Raciti F, Skelton BW, Svenstrup N, White AH (2001) Macrocyclic ligand design. A synthetic, solvent extraction, computational and NMR study of the effect of cryptand flexibility on sodium ion affinity. J Chem Soc Dalton Trans 2388–2397
43. Suh MP, Lee J, Han MY, Yoon TS (1997) Template synthesis, properties, and crystal structure of a trigonal bipyramidal cobalt(II) octaazamacrobicyclic complex. Inorg Chem 36:5651–5654
44. Mouaziz H, Montembault V, Soutif JC, Brosse JC (2002) Synthesis of new cryptands starting from *trans* (diprotected and disubstituted) 1,4,7,10-tetraazacyclododecane. Synth Commun 32:3763–3766
45. Coyle J, Drew MGB, Harding CJ, Nelson J, Town RM (1997) Expanded-cap sepulchrates: a new series of mononuclear transition-metal ion cryptates. Dalton Trans 1123–1126
46. Bing-guang Z, Dong G, Chun-ying D, Ping C, Qing-Jin M (2003) A novel pyramidal multiple alkali metal cluster $[K_3(H_2O)]^{3+}$ stabilized within a capsule containing 16-nitrogen donors. Chem Commun 1863–1837
47. Walker GW, Geue RJ, Haller KJ, Rae AD, Sargeson AM (2003) New synthetic routes to hexa-aza cages using cobalt(III) tris(1,2-diamine) templates. Dalton Trans 279–281
48. Walker GW, Geue RJ, Sargeson AM, Behm CA (2003) Surface-active cobalt cage complexes: synthesis, surface chemistry, biological activity, and redox properties. Dalton Trans 2992–3001
49. Brown KN, Geue RJ, Hambley TW, Hockless DCR, Rae AD, Sargeson AM (2003) Specificity in template syntheses of hexaaza-macrobicyclic cages: [Pt(Me_5-tricosatrieneN$_6$)]$^{4+}$ and [Pt(Me_5-tricosaneN$_6$)]$^{4+}$. Org Biomol Chem 1:1598–1608
50. Bygott AMT, Geue RJ, Ralph SF, Sargeson AM, Willis AC (2007) Octahedral and trigonal prismatic structure preferences in a bicyclic hexaamine cage for zinc(II), cadmium(II) and mercury(II) ions. Dalton Trans 4778–4787
51. Bernhardt PV, Dyahningtyas TE, Harrowfield JM, Kim JY, Kim Y, Rukmini E (2003) Chiral resolution of hexaamine cobalt(III) cages: substituent effects on chiral discrimination. Aust J Chem 56:1187–1191
52. Sharrad CA, Lüthi SR, Gahan LR (2003) Embracing ligands. A synthetic strategy towards new nitrogen–thioether multidentate ligands and characterization of the cobalt(III) complexes. Dalton Trans 3693–3703
53. Sharrad CA, Gahan LR (2003) Characterization of $[Fe(AMN_3S_3sarH)]^{3+}$—a rigorously low-spin iron (II) complex. Polyhedron 22:45–51
54. Bell CA, Bernhardt PV, Gahan LR, Martínez M, Monteiro MJ, Rodríguez C, Sharrad CA (2010) Copper(II) complexes of a hexadentate mixed-donor N_3S_3 macrobicyclic cage: facile rearrangements and interconversions. Chem Eur J 16:3166–3175
55. Bernhardt PV, Font H, Gallego C, Martínez M, Rodríguez C (2012) Reversible rearrangements of Cu(II) cage complexes: solvent and anion influences. Inorg Chem 51:12372–12379
56. Geue RJ, Höhn A, Ralph SF, Sargeson AM, Willis AC (1994) A new template encapsulation strategy for larger cavity metal ion cages. J Chem Soc Chem Commun 1513–1515
57. Bernhardt PV, Bramley R, Geue RJ, Ralph SF, Sargeson AM (2007) An expanded cavity hexaamine cage for copper(II). Dalton Trans 1244–1249
58. Qin CJ, James L, Chartres JD, Alcock LJ, Davis KJ, Willis AC, Sargeson AM, Bernhardt PV, Ralph SF (2011) An unusually flexible expanded hexaamine cage and its Cu^{II} complexes: variable coordination modes and incomplete encapsulation. Inorg Chem 50:9131–9140
59. Bernhardt PV, Harrowfield JM, Kim Y, Lee YH, Park YC, Sujandi S (2007) A functionalised macrobicycle complex available for surface immobilisation and protein grafting. Bull Kor Chem Soc 28:589–595
60. Kurata H, Nakaminami H, Matsumoto K, Kawase T, Oda M (2001) Synthesis of novel cage molecules bicapped with tris(2-thienyl)methanes. Chem Commun 529–530
61. Kurata H, Haruki K, Nakaminami H, Kawase T, Oda M (2003) Novel monocation, dication, and dianion of a cage molecule, trithienylmethanophane. Chem Lett 32:422–423
62. Alberto R, Angst D, Ortner K, Abram U, Schubiger PA, Kaden TA (2007) Syntheses of a series of S_6

thioether cages and their coordination chemistry with Ag^+. New J Chem 31:409–417
63. Cai H, Li Z, Huang CW, Park R, Shahinian AH, Conti PS (2010) An improved synthesis and biological evaluation of a new cage-like bifunctional chelator, 4-((8-amino-3,6,10,13,16,19-hexaazabicyclo[6.6.6]icosane-1-ylamino)methyl) benzoic acid, for 64Cu radiopharmaceuticals. Nucl Med Biol 37:57–65
64. Liu S, Li Z, Yap LP, Huang CW, Park R, Conti PS (2011) Efficient preparation and biological evaluation of a novel multivalency bifunctional chelator for ^{64}Cu radiopharmaceuticals. Chem Eur J 17:10222–10225
65. Ma MT, Karas JA, White JM, Scanlon D, Donnelly PS (2009) A new bifunctional chelator for copper radiopharmaceuticals: a cage amine ligand with a carboxylate functional group for conjugation to peptides. Chem Commun 3237–3239
66. Donnelly PS, Harrowfield JM (2002) Synthesis with coordinated ligands: biomolecule attachment to cage amines. J Chem Soc Dalton Trans 906–913
67. Dittrich B, Harrowfield JM, Koutsantonis GA, Nealon GL, Skelton BW (2010) Long tailed cage amines: synthesis, metal complexation, and structure. Dalton Trans 39:3433–3448
68. Koutsantonis GA, Nealon GL, Buckley CE, Paskevicius M, Douce L, Harrowfield JM, McDowall AW (2007) Wormlike Micelles from a Cage Amine Metallosurfactant. Langmuir 23:11986–11990
69. Harrowfield JM, Koutsantonis GA, Kraatz HB, Nealon GL, Orlowski GA, Skelton BW, White AH (2007) Cages on surfaces: Thiol functionalisation of Co^{III} sarcophagine complexes. Eur J Inorg Chem 2007:263–278
70. Sargeson AM, Lay PA (2009) Dependence of the properties of cobalt(III) cage complex as a function of the derivatization of amine substituents. Aust J Chem 62:1280–1290
71. Harrowfield JM, Koutsantonis GA, Lengkeek NA, Skelton BW, White AH (2010) Structural and electrochemical studies of Co(III) cage amine complexes with pendent thienylmethylamino groups. Inorg Chem 49:3152–3161
72. Harrowfield JM, Koutsantonis GA, Nealon GL, Skelton BW, White AH (2005) Amphiphile structures in the solid state: complex cations with lipophilic substituents. Eur J Inorg Chem 2005:2384–2392
73. Koutsantonis GA, Lee J, Lengkeek NA, Nealon GL, Skelton BW, White AH (2010) Synthesis and characterisation of trinuclear metal complexes derived from carboxymethyl-substituted sarcophagine macrobicyclic cage amines. J Incl Phenom Macrocycl Chem 66:61–66
74. Voloshin YZ, Varzatskii OA, Vorontsov II, Antipin MY, Lebedev AY, Belov AS, Palchik AV (2003) Template synthesis and structures of apically functionalized iron(II) clathrochelates. Russ Chem Bull Int Ed 52:1552–1561
75. Voloshin YZ, Lebed EG, Erdyakov SY, Makarenko IG, Lyssenko KA, Starikova ZA, Gurskii ME, Bubnov YN (2006) Application of the allylboration reaction of terminal acetylenes with allyldihaloboranes for the preparation of capping agents for the synthesis of precursors of polymeric iron(II) clathrochelates. Russ Chem Bull Int Ed 55:1971–1981
76. Voloshin YZ, Makarov IS, Vologzhanina AV, Monakov YB, Islamova RM, Polshin EV, Bubnov YN (2008) Template synthesis and structures of bis-ferrocenylboronate macrobicyclic iron(ii) tris-dioximates. Russ Chem Bull Int Ed 57:1199–1206
77. Voloshin YZ, Varzatskii OA, Belov AS, Lebedev AY, Makarov IS, Gurskii ME, Antipin MY, Starikova ZA, Bubnov YN (2007) Cage iron(II) complexes with apical and ribbed adamantyl substituents: the creation of second (hydrophobic) shell of an encapsulated metal ion. Inorg Chim Acta 360:1543–1554
78. Liu W, Huang W, Pink M, Lee D (2010) Layer-by-layer synthesis of metal-containing conducting polymers: caged metal centers for interlayer charge transport. J Am Chem Soc 132:11844–11846
79. Liu W, Huang W, Chen CH, Pink M, Lee D (2012) Charge injection and transport in metal-containing conducting polymers: spectroelectrochemical mapping of redox activities. Chem Mater 24:3650–3658
80. Belov AS, Dolganov AV, Novikov VV, Vologzhanina AV, Fedin MV, Kuznetsov EV, Bubnov YN, Voloshin YZ (2013) Template synthesis, structure and electropolymerization of the 2-thiopheneboron-capped cobalt(II) clathrochelates. Inorg Chem Commun 29:160–164
81. Dolganov AV, Belov AS, Novikov VV, Vologzhanina AV, Mokhir A, Bubnov YN, Voloshin YZ (2013) Iron *vs* Cobalt clathrochelate electrocatalysts of HER: first example on a cage iron complex. Dalton Trans 42:4373–4376
82. Lebed EG, Belov AS, Dolganov AV, Vologzhanina AV, Szebesczyk A, Gumienna-Kontecka E, Kozlowski H, Bubnov YN, Dubey IY, Voloshin YZ (2013) First clathrochelate iron and cobalt(II) - tris-dioximates with reactive apical substituents. Inorg Chem Commun 30:53–57
83. Zhang J, Ma Z, Zhao H, Ge C, Wang Y, Zhang X (2016) Bis-boron-capped tris(dioxime) cage complexeswith terminal carbonyl groups. Inorg Chem Commun 65:63–67
84. Zelinskii GE, Belov AS, Lebed EG, Vologzhanina AV, Novikov VV, Voloshin YZ (2016) Synthesis, structure and reactivity of iron(II) clathrochelates with terminal formyl (acetal) groups and their further functionalization for target delivery to biological systems. Inorg Chim Acta 440:154–164
85. Voloshin YZ, Lebedev AY, Novikov VV, Dolganov AV, Vologzhanina AV, Lebed EG, Pavlov AA, Starikova ZA, Buzin MI, Bubnov YN (2013) Template synthesis, X-ray structure, spectral and redox properties of the paramagnetic alkylboron-capped cobalt(II) clathrochelates and their diamagnetic iron (II)-containing analogs. Inorg Chim Acta 399:67–78

86. Vologzhanina AV, Belov AS, Novikov VV, Dolganov AV, Romanenko GV, Ovcharenko VI, Korlyukov AA, Buzin MI, Voloshin YZ (2015) Synthesis and temperature-induced structural phase and spin transitions in hexadecylboron-capped cobalt (II) hexachloroclathrochelate and its diamagnetic iron(II)-encapsulating analogue. Inorg Chem 54:5827–5838
87. Lebed EG, Belov AS, Dolganov AV, Vologzhanina AV, Novikov VV, Kuznetsov EV (2013) Template synthesis, structure and properties of 4-pyridinylboron-capped iron(II) clathrochelate precursors for Bubnov diallylation reaction. Inorg Chem Commun 33:57–62
88. Erdyakov SY, Voloshin YZ, Makarenko IG, Lebed EG, Potapova TV, Ignatenko AV, Vologzhanina AV, Gurskii ME, Bubnov YN (2009) New *o*-carboranyl-containing capping agents for *d*-metal tris-dioximates and first bis-C-carboranylboron-capped iron(II) clathrochelates: synthesis and X–ray structure. Inorg Chem Commun 12:135–139
89. Jansze S, Cecot G, Wise MD, Zhurov KO, Ronson TK, Castilla AM, Finelli A, Pattison P, Solari E, Scopelliti R, Zelinskii GE, Vologzhanina AV, Voloshin YZ, Nitschke J, Severin K (2016) Ligand aspect ratio as a decisive factor for the self-assembly of coordination cages. J Am Chem Soc 138:2046–2054
90. Marmier M, Wise MD, Holstein JJ, Pattison P, Schenk K, Solari E, Scopelliti R, Severin K (2016) Carboxylic acid functionalized clathrochelate complexes: large, robust and easy-to-access metalloligands. Inorg Chem 55:4006–4015
91. Wise MD, Holstein JJ, Pattison P, Besnard C, Solari E, Scopelliti R, Bricogne G, Severin K (2015) Large, heterometallic coordination cages based on ditopic metallo-ligands with 3-pyridyl donor groups. Chem Sci 6:1004–1010
92. Wise MD, Ruggi A, Pascu M, Scopelliti R, Severin K (2013) Clathrochelate-based bipyridyl ligands of nanoscale dimensions: easy-to-access building blocks for supramolecular chemistry. Chem Sci 4:1658–1662
93. Voloshin YZ, Varzatskii OA, Belov AS, Starikova ZA, Dolganov AV, Magdesieva TV (2008) New antimony-capped iron(II) and cobalt(III) clathrochelate precursors of the polytopic hybrid cage complexes: synthesis, X-ray structures and electrochemistry. Polyhedron 27:325–334
94. Voloshin YZ, Varzatskii OA, Korobko SV, Antipin MY, Vorontsov II, Lyssenko KA, Kochubey DI, Nikitenko SG, Strizhakova NG (2004) Pathways of directed synthesis of C_2-unsymmetric iron(II) clathrochelates and polyclathrochelates starting from antimony- and germanium-containing precursors. Inorg Chim Acta 357:3187–3204
95. Voloshin YZ, Korobko SV, Dolganov AV, Novikov VV, Vologzhanina AV, Bubnov YN (2011) New types of the germanium-capped clathrochelate iron (II) and cobalt(III) tris-dioximates: the synthesis, structure and electrochemical properties. Inorg Chem Commun 14:1043–1047
96. Belaya IG, Svidlov SV, Dolganov AV, Zelinskii GE, Potapova TV, Vologzhanina AV, Varzatskii OA, Bubnov YN, Voloshin YZ (2013) Apically linked iron(II) α-dioximate and α-oximehydrazonate bis-clathrochelates: synthesis, structure and electrocatalytic properties. Dalton Trans 42:13667–13678
97. Novikov VV, Ananyev IV, Pavlov AA, Fedin MV, Lyssenko KA, Voloshin YZ (2014) Spin crossover anticooperativity induced by weak intermolecular interactions. J Phys Chem Lett 5:496–500
98. Novikov VV, Pavlov AA, Belov AS, Vologzhanina AV, Savitsky A, Voloshin YZ (2014) Transition ion strikes back: large magnetic susceptibility anisotropy in cobalt(II) clathrochelates. J Phys Chem Lett 5:3799–3803
99. Dolganov AV, Belov AS, Novikov VV, Vologzhanina AV, Romanenko GV, Budnikova YG, Zelinskii GE, Buzin MI, Voloshin YZ (2015) First iron and cobalt(II) hexabromoclathrochelates: structural, magnetic, redox and electrocatalytic behavior. Dalton Trans 44:2476–2487
100. Varzatskii OA, Denisenko IN, Volkov SV, Belov AS, Dolganov AV, Vologzhanina AV, Novikov VV, Bubnov YN, Voloshin YZ (2013) Perfluoroarylation of iron(II) di- and hexaiodoclathrochelates: synthesis, X-ray structure and properties of the first cage complexes with inherent pentafluorophenyl substituent(s). Eur J Inorg Chem 2013:3178–3184
101. Zelinskii G, Novikov V, Vologzhanina A, Voloshin Y, Bubnov Y, Mokhir A (2015) Template synthesis of iron and cobalt(II) (per)fluoroclathrochelates with superhydrophobic exterior. Paper presented at the international symposia on metal complexes, Wrocław, 24–28 June 2015
102. Belov AS, Zelinskii GE, Varzatskii OA, Belaya IG, Vologzhanina AV, Dolganov AV, Novikov VV, Voloshin YZ (2015) Molecular design of cage iron (II) and cobalt(II,III) complexes with a second fluorine-enriched superhydrophobic shell. Dalton Trans 44:3773–3784
103. Voloshin YZ, Varzatskii OA, Vorontsov II, Antipin MY (2005) Tuning a metal's oxidation state: the potential of clathrochelate systems. Angew Chem Int Ed 44:3400–3402
104. Burdukov AB, Vershinin MA, Boguslavsky EG, Eltsov IV, Romanenko GV, Belov AS, Voloshin YZ (2013) New hydroxylamine-containing macrobicyclic encapsulating ligand: unexpected double addition of ethyl radicals to the azomethine fragment of a boron-capped iron(II) clathrochelate dioximate. Inorg Chem Commun 36:117–121
105. Voloshin YZ, Varzatskii OA, Novikov VV, Strizhakova NG, Vorontsov II, Vologzhanina AV, Lyssenko KA, Romanenko GV, Fedin MV,

Ovcharenko VI, Bubnov YN (2010) Tris-dioximate cobalt(I, II, and III) clathrochelates: stabilization of different oxidation and spin states of an encapsulated metal ion by ribbed functionalization. Eur J Inorg Chem 2010:5401–5415
106. Voloshin YZ, Varzatskii OA, Palchik AV, Strizhakova NG, Vorontsov II, Antipin MY, Kochubey DI, Novgorodov BN (2003) First trigonal-antiprismatic tris-dichloroglyoximate iron(II) clathrochelate and its reactivity in nucleophilic substitution reactions. New J Chem 27:1148–1155
107. Voloshin YZ, Varzatskii OA, Palchik AV, Vorontsov II, Antipin MY, Lebed EG (2005) Template synthesis and structure of mono- and trisubstituted ribbed-functionalized iron(II) clathrochelates. Inorg Chim Acta 358:131–146
108. Voloshin YZ, Belov AS, Vologzhanina AV, Aleksandrov GG, Dolganov AV, Novikov VV, Varzatskii OA, Bubnov YN (2012) Synthesis, structure, properties and immobilization on a gold surface of the monoribbed-functionalized tris-dioximate cobalt(II) clathrochelates and an electrocatalytic hydrogen production from H^+ ions. Dalton Trans 41:6078–6093
109. Belov AS, Belaya IG, Novikov VV, Starikova ZA, Polshin EV, Dolganov AV, Lebed EG (2013) Structure, spectral and electrochemical properties of the 2,6-di-*tert*-butylphenol-functionalized iron and cobalt(II) clathrochelates and their phenylsulfide analogs. Inorg Chim Acta 394:269–281
110. Voloshin YZ, Varzatskii OA, Belov AS, Vologzhanina AV, Starikova ZA, Dolganov AV, Novikov VV (2009) Synthesis, X-ray structures and properties of the first tris-dioximate cobalt clathrochelates with nonequivalent chelate ribbed fragments. Inorg Chim Acta 362:5144–5150
111. Varzatskii OA, Denisenko IN, Belov AS, Vologzhanina AV, Bubnov YN, Volkov SV, Voloshin YZ (2014) Metal-catalyzed cross-coupling reactions of iron(II) cage complexes: new furyl-containing macrobicyclic scaffold, a reactive halogenoclathrochelate precursor and its ribbed-functionalized derivatives. Inorg Chem Commun 44:134–138
112. Voloshin YZ, Belov AS, Varzatskii OA, Shul'ga SV, Stuzhin PA, Starikova ZA, Lebed EG, Bubnov YN (2012) Synthesis and structure of the first clathrochelate iron(II) tris-dioximates with inherent nitrile substituent(s) and new dehalogenation – reduction reaction at a quasi-aromatic macrobicyclic framework. Dalton Trans 41:921–928
113. Varzatskii OA, Novikov VV, Shulga SV, Belov AS, Vologzhanina AV, Negrutska VV, Dubey IY, Bubnov YN, Voloshin YZ (2014) Copper-promoted reductive homocoupling of quasi-aromatic iron(II) clathrochelates: boosting the inhibitory activity in a transcription assay. Chem Commun 50:3166–3168
114. Voloshin YZ, Varzatskii OA, Vorontsov II, Antipin MY, Lebedev AY, Belov AS, Strizhakova NG (2004) Ribbed-functionalized iron(II) tris-dioximate clathrochelates with pendant fragments of various types: synthetic pathways, structures, and properties. Russ Chem Bull Int Ed 53:92–98
115. Voloshin YZ, Belov AS, Lebedev AY, Varzatskii OA, Antipin MY, Starikova ZA, Kron TE (2004) Synthesis and structure of the first ribbed-functionalized quinoxaline clathrochelate: design of cage complexes for efficient intercalation into DNA structure. Russ Chem Bull Int Ed 53:1218–1222
116. Voloshin YZ, Varzatskii OA, Starikova ZA, Antipin MY, Lebedev AY, Belov AS (2004) Supramolecular organization of the crystals of allylsulfide clathrochelate: inf luence of the nature of solvate molecules. Russ Chem Bull Int Ed 53:1496–1502
117. Burdukov AB, Boguslavsky EG, Reznikov VA, Pervukhina NV, Vershinin MA, Voloshin YZ, Varzatskii OA, Bubnov YN (2005) Cage complexes as a molecular scaffold for assembling of polyfunctional and multicentered systems: Synthesis and structures of the first nitroxide clathrochelates. Russ Chem Bull Int Ed 54:1125–1130
118. Voloshin YZ, Varzatskii OA, Zhizhin KY, Kuznetsov NT, Bubnov YN (2006) Cage complexes as a molecular scaffold for polyfunctional and polytopic systems: synthesis of the first *closo*-borate iron(II) clathrochelate. Russ Chem Bull Int Ed 55:22–25
119. Voloshin YZ, Varzatskii OA, Belov AS, Bubnov YN (2006) Cage complexes as a molecular "scaffold" for assembling of polyfunctional and multicentered systems: synthesis of ribbed-functionalized dopamine-containing iron(II) clathrochelate and its properties as a receptor for boric acid. Russ Chem Bull Int Ed 55:1119–1125
120. Burdukov AB, Vershinin MA, Pervukhina NV, Voloshin YZ, Varzatskii OA (2006) Reactivity of iron(II) dichloride clathrochelate: synthesis and properties of mono-and disubstituted amino clathrochelates. Russ Chem Bull Int Ed 55:1982–1988
121. Vershinin MA, Burdukov AB, Boguslavskii EG, Pervukhina NV, Kuratieva NV, Eltsov IV, Reznikov VA, Varzatskii OA, Voloshin YZ, Bubnov YN (2011) The first monoribbed-functionalized tris-dioximate iron(II) clathrochelate with two inherent NH_2-substituents, its reactivity, acid-base and coordination-chemical properties. Inorg Chim Acta 366:91–97
122. Vershinin MA, Burdukov AB, Pervukhina NV, Eltsov IV, Varzatskii OA, Voloshin YZ (2008) The first example of the alkoxy-containing iron(ii) tris-dioximate clathrochelate: synthesis, structure, and properties. Russ Chem Bull Int Ed 57:1309–1313
123. Voloshin YZ, Varzatskii OA, Kochubey DI, Vorontsov II, Bubnov YN (2009) Synthesis and

structure of monoribbed-functionalized disulfide iron(II) clathrochelates and their coordination as the ligands toward platinum(II) and platinum (IV) ions. Inorg Chim Acta 362:149–158
124. Belov AS, Varzatskii OA, Vologzhanina AV, Viswanathan S, Radecki J, Bubnov YN (2009) Synthesis, structure and electron-mediator properties of the mono- and difunctionalized macrobicyclic iron (II) tris-dioximates with thiol terminated ribbed spacer substituents. Inorg Chim Acta 362:2982–2988
125. Voloshin YZ, Varzatskii OA, Belov AS, Starikova ZA, Suponitsky KY, Novikov VV, Bubnov YN (2008) Interaction of dichloride iron(II) clathrochelate with dimercaptomaleodinitrile: synthesis of the precursor of monoribbed-functionalized phthalocyaninoclathrochelates and the unexpected formation of a new thiophene-containing heterocyclic system in the ribbed chelate fragment of the clathrochelate framework. Inorg Chem 47:2155–2161
126. Voloshin YZ, Varzatskii OA, Starikova ZA, Belov AS, Ignatenko AV, Bubnov YN (2008) Synthesis and structure of the first chlathrochelate ligand with a *cis*-diimidazolyl donor fragment. Inorg Chem Commun 11:988–991
127. Burdukov AB, Vershinin MA, Eltsov IV, Pervukhina NV, Voloshin YZ (2009) The first ribbed functionalized tris-dioximate clathrochelates with inherent hydroxyl substituents: synthesis, spectra and X-ray structure. Inorg Chem Commun 12:919–922
128. Belov AS, Prikhod'ko AI, Novikov VV, Vologzhanina AV, Bubnov YN, Voloshin YZ (2012) First "click" synthesis of the ribbed-functionalized metal clathrochelates: cycloaddition of benzyl azide to propargylamine iron(II) macrobicycle and the unexpected transformations of the resulting cage complex. Eur J Inorg Chem 2012:4507–4514
129. Burdukov AB, Vershinin MA, Pervukhina NV, Boguslvaskii EG, Eltsov IV, Shundrin LA, Selector SL, Shokurov AV, Voloshin YZ (2014) Towards the clathrochelate-based electrochromic materials: the case study of the first iron(II) cage complex with an annelated quinoxaline fragment. Inorg Chem Commun 44:183–187
130. Belov AS, Vologzhanina AV, Novikov VV, Negrutska VV, Dubey IY, Mikhailova ZA, Lebed EG, Voloshin YZ (2014) Synthesis of the first morpholine-containing iron (II) clathrochelates: a new class of efficient functionalized transcription inhibitors. Inorg Chim Acta 421:300–306
131. Petrov PA, Virovets AV, Plyusnin PE, Filatov EY, Eltsov IV, Voloshin YZ, Konchenko SN (2014) Ribbed-monofunctionalized iron(II) clathrochelate with *tert*-butyl sulfide substituents: synthesis, structure, and thermochemical transformations. Russ J Inorg Chem 59:1162–1167
132. Vershinin MA, Burdukov AB, Pervukhina NV, Eltsov IV, Boguslvaskii EG, Voloshin YZ (2015) Iron(II) clathrochelate with an annelated 4,5-dimercapto-[1,3]dithiol-2-thione moiety: synthesis, structure, and redox reactions. Russ J Inorg Chem 60:861–865
133. Vershinin MA, Burdukov AB, Pervukhina NV, Eltsov IV (2015) Structure of the product of 1,8-diazabicyclo[5.4.0]undecane-7-ene substitution for chlorine atoms in dichlorine substituted Fe (II) clathrochelate. J Struct Chem 56:994–997
134. Burdukov AB, Šipoš R, Vershinin MA, Pervukhina NV, Kuratieva NV, Plyusnin PE, Eltsov IV, Voloshin YZ (2015) Two alternative approaches to modification of a cage complex: nucleophilic substitution and electrophilic addition for the synthesis of an iron(II) clathrochelate with an annulated imidazole fragment. J Coord Chem 68:3894–3902
135. Artyushin OI, Odinets IL, Matveeva EV, Vologzhanina AV, Varzatskii OA, Lubimov SE, Voloshin YZ (2014) Clathrochelates meet phosphorous. 1. Thiophosphorylation of an iron(II) dichlathrochelate precursor, synthesis of *N,S*-donor clathrochelate ligands and their palladium (II) complexes as potent catalysts of Suzuki cross-coupling reaction. Dalton Trans 43:9677–9689
136. Artyushin OI, Matveeva EV, Vologzhanina AV, Voloshin YZ (2016) Clathrochelates meet phosphorus. New thio- and phosphorylation reactions of an iron(II) dichloroclathrochelate precursor and preparation of its first phosphorus(III)-containing macrobicyclic derivative. Dalton Trans 45:5328–5333
137. Belaya IG, Zelinskii GE, Belov AS, Varzatskii OA, Novikov VV, Dolganov AV, Kozlowski H, Szyrwiel Ł, Bubnov YN, Voloshin YZ (2012) Synthesis, spectra and properties of the first protono- and ionogenic tris-dioximate iron(II) clathrochelates. Polyhedron 40:32–39
138. Voloshin YZ, Varzatskii OA, Belov AS, Starikova ZA, Strizhakova NG, Dolganov AV, Kochubey DI, Bubnov YN (2010) Synthesis, X-ray structure and redox properties of the macrobicyclic iron(II) N_2- and S_2-containing *vic*-dioximates. Inorg Chim Acta 363:134–146
139. Voloshin YZ, Varzatskii OA, Belov AS, Starikova ZA, Dolganov AV, Novikov VV, Bubnov YN (2011) Synthesis, structural and electrochemical features of alicyclic and aromatic α,α'-N_2- and -S_2-dioximate macrobicyclic cobalt(II,III) and ruthenium(II) tris-complexes. Inorg Chim Acta 370:322–332
140. Belov AS, Vologzhanina AV, Fedorov YV, Kuznetsov EV, Voloshin YZ (2013) Unexpected transformation of mono- to bis-macrobicyclic dimethylglyoximate framework in a chloroform solution: photochemical, MALDI-TOF MS and X-ray diffraction studies. Inorg Chem Commun 35:242–246
141. Belaya IG, Belov AS, Platonov VE, Maksimov AM, Vologzhanina AV, Starikova ZA, Dolganov AV, Novikov VV, Bubnov YN (2012) Formation of the second superhydrophobic shell around an

encapsulated metal ion: synthesis, X-ray structure and electrochemical study of the clathrochelate and bis-clathrochelate iron(II) and cobalt(II, III) dioximates with ribbed perfluoroarylsulfide substituents. Dalton Trans 41:737–746
142. Kuznetsov NT, Belaya Makarenko IG, Dolganov AV, Zelinskii GE, Matveev EY, Zhizhin KY, Voloshin YZ, Bubnov YN (2011) First example of the ribbed-functionalized iron(ii) clathrochelate with six pendante *closo*-borate substituents. Russ Chem Bull Int Ed 60:2518–2521
143. Varzatskii OA, Shul'ga SV, Belov AS, Novikov VV, Dolganov AV, Vologzhanina AV, Voloshin YZ (2014) Copper (I)-and copper (0)-promoted homocoupling and homocoupling–hydrodehalogenation reactions of dihalogenoclathrochelate precursors for C–C conjugated iron (II) bis-cage complexes. Dalton Trans 43:17934–17948
144. Solovieva SE, Rovnova TA, Latipov SK, Voloshin YZ, Antipin IS, Konovalov AI (2013) Synthesis of conjugates of iron(II) tris-dioximates and the dithiol-terminated calyx[4]arenas. Phosphorus Sulfur Silicon Relat Elem 188:503–506
145. Solovieva SE, Tyuftin AA, Muravev AA, Gruner M, Habicher W, Korobko SV, Antipin IS, Konovalov AI, Bubnov YN, Voloshin YZ (2013) A new type of polytopic coordination compound: the synthesis and NMR studies of the first hybrid thiacalix[4]arenoclathrochelates. Polyhedron 50:90–100
146. Vershinin MA, Burdukov AB, Eltsov IV, Reznikov VA, Boguslavsky EG, Voloshin YZ (2011) Unexpected radical substitution of the dichlorine-containing iron(II) clathrochelate with 1,4-dioxane derivatives: novel approach to functionalization of its macrobicyclic framework. Polyhedron 30:1233–1237
147. Vershinin MA, Burdukov AB, Starikova ZA, Novikov VV, Voloshin YZ (2011) Reactions of dichloro-substituted iron(ii) clathrochelate with 1,4-dioxane radical derivatives: synthesis, structure, and spectral characteristics of the dioxane ring opening product in the ribbed fragment of the macrobicyclic ligand. Russ Chem Bull Int Ed 60:2510–2517
148. Burdukov AB, Vershinin MA, Pervukhina NV, Kozlova SG, Eltsov IV, Voloshin YZ (2011) Free-radical reactions of the tris-dioximate clathrochelates: synthesis and X-ray structure of the first cyclohexyl-substituted monoribbed-functionalized macrobicyclic iron(II) complex. Russ Chem Bull Int Ed 60:2504–2509
149. Vershinin MA, Burdukov AB, Pervukhina NV, Eltsov IV, Voloshin YZ (2012) Free-radical reaction of the iron(II) dichloroclathrochelate with tetrahydrofurane radical derivatives: synthesis and structure of the monotetrahydrofuryl-containing cage complex. Macroheterocycles 5:11–16
150. Vershinin MA, Burdukov AB, Pervukhina NV, Eltsov IV, Voloshin YZ (2015) Free-radical reaction of iron(II) dichloroclathrochelate with 1,3-dioxolane radical derivative: synthesis and structure of macrobicyclic tris-dioximate bearing 1,3-dioxolan-2-yl ribbed substituent. Macroheterocycles 8:71–74
151. Vershinin MA, Burdukov AB, Pervukhina NV, Eltsov IV, Voloshin YZ (2013) Stereoselective C-alkylation of an iron(II) dichloroclathrochelate *via* free-radical reactions with alcohols. Inorg Chem Commun 30:159–162
152. Vershinin MA, Burdukov AB, Pervukhina NV, Eltsov IV (2015) Structure of the product of free-radical substitution of chlorine atoms in iron(II) clathrochelate with two ethanol fragments. J Struct Chem 56:379–381
153. Burdukov AB, Vershinin MA, Pervukhina NV, Eltsov IV, Voloshin YZ (2015) Reactivity of coordinated α-dioximate ribbed fragments of the iron(II) clathrochelate framework in the system DMSO–H_2O–Fe(II)–H_2O_2. Curr Inorg Chem 5:71–76
154. Varzatskii OA, Voloshin YZ, Korobko SV, Shulga SV, Krämer R, Belov AS, Vologzhanina AV, Bubnov YN (2009) On a way to new types of the polyfunctional and polytopic systems based on cage metal complexes: cadmium-promoted nucleophilic substitution with low-active nucleophilic agents. Polyhedron 28:3431–3438
155. Varzatskii OA, Voloshin YZ, Stuzhin PA, Shul'ga SV, Volkov SV, Vologzhanina AV, Lebed EG, Bubnov YN (2011) New cadmium-promoted reaction of a C-nucleophile: synthesis and X-ray structure of the first dicyanopyrazine iron(II) clathrochelate. Inorg Chem Commun 14:1504–1507
156. Voloshin YZ, Varzatskii OA, Shul'ga SV, Denisenko IN, Vologzhanina AV, Bubnov YN (2012) Copper(I)-promoted halogen exchange in the iron(II) dichloroclathrochelate. Inorg Chem Commun 17:128–131
157. Varzatskii OA, Denisenko IN, Volkov SV, Dolganov AV, Vologzhanina AV, Bubnov YN, Voloshin YZ (2013) First example of perfluoroalkylation of a quasi-aromatic encapsulating ligand: 2,5-dithiahexane-assisted reaction of the iron(II) diiodoclathrochelate with trifluoromethylcopper(I). Inorg Chem Commun 33:147–150
158. Svidlov SV, Varzatskii OA, Potapova TV, Vologzhanina AV, Bukalov SS, Leites LA, Voloshin YZ, Bubnov YN (2014) C-carboranylation of a quasi-aromatic iron(II) cage complex and its organic aromatic analog by the metal-promoted (catalyzed) cross-coupling reactions. Inorg Chem Commun 43:142–145
159. Voloshin YZ, Makarov IS, Vologzhanina AV, Polshin EV, Monakov YB, Islamova RM, Bubnov YN (2008) Synthesis and structure of the ferrocenylboron-capped clathrochelate iron(II) oximehydrazonates. Russ Chem Bull Int Ed 57:1215–1222
160. Voloshin YZ, Erdyakov SY, Makarenko IG, Lebed' EG, Potapova TV, Svidlov SV, Starikova ZA, Pol'shin EV, Gurskii ME, Bubnov YN (2007) Macrocyclization of the semiclathrochelate *o*-carboranylboronate and *n*-butylboronate iron(II)

oximehydrazonates: synthesis and structure of clathrochelate products and unexpected allosteric effect of the apical substituent. Russ Chem Bull Int Ed 56:1787–1794
161. Voloshin YZ, Makarenko IG, Erdyakov SY, Svidlov SV, Vologzhanina AV, Lebed EG, Ignatenko AV, Gurskii ME, Bubnov YN (2010) Template synthesis, structure and spectra of the semiclathrochelate and clathrochelate 1,4-pentadienylboron-capped iron(II) oximehydrazonates. Inorg Chim Acta 363:395–403
162. Voloshin YZ, Varzatskii OA, Tomilova LG, Breusova MO, Magdesieva TV, Bubnov YN, Krämer R (2007) First hybrid oximehydrazonate phthalocyaninoclathrochelates: the synthesis and properties of lutetium phthalocyanine-capped, cage iron(II) complexes. Polyhedron 26:2733–2740
163. Varzatskii OA, Penkova LV, Kats (Menkach) SV, Dolganov AV, Vologzhanina AV, Pavlov AA, Novikov VV, Bogomyakov AS, Nemykin VN, Voloshin YZ (2014) Chloride ion-aided self-assembly of the pseudo-clathrochelate metal tris-pyrazoloximates. Inorg Chem 53:3062–3071
164. Pavlov AA, Nelyubina YV, Kats SV, Penkova LV, Efimov NN, Dmitrienko AO, Vologzhanina AV, Belov AS, Voloshin YZ, Novikov VV (2016) Polymorphism in a cobalt-based single-ion magnet tuning its barrier to magnetization relaxation. J Phys Chem Lett 7:4111–4116
165. Galaup C, Azéma J, Tisnès P, Picard C, Ramos P, Juanes O, Brunet E, Rodríguez-Ubis JC (2002) Luminescence of Eu^{3+} and Tb^{3+} complexes of two macrobicyclic ligands derived from a tetralactam ring and a chromophoric antenna. Helv Chim Acta 85:1613–1625
166. Porwanski S, Marsura A (2009) Tandem Staudinger–Aza-Wittig templated reaction: De novo synthesis of sugar–ureido cryptands. Eur J Org Chem 2009:2047–2050
167. Beer PD, Cowley AR, Jeffery JC, Paul RL, Wong WWH (2003) Self-assembled xanthate-transition metal polyether macrocycles and cryptands. Polyhedron 22:795–801
168. Takemura H, Kon N, Yasutake M, Kariyazono H, Shinmyozu T, Inazu T (1999) A potassium complex of a fluorine-containing macrocyclic cage compound: interactions between fluorine atoms and metal ions. Angew Chem Int Ed 38:959–961
169. Takemura H, Nakashima S, Kon N, Inazu T (2000) C–F···cation interactions: alkali metal complexes of hexafluoro cage compound. Tetrahedron Lett 41:6105–6109
170. Takemura H, Kon N, Yasutake M, Shinmyozu T (2003) Structurally isomeric two pyridino macrocycles: complexation and structures. Tetrahedron 59:427–431
171. Cross JP, Dadabhoy A, Sammes PG (2004) The sensitivity of the lehn cryptand—europium and terbium (III) complexes to anions compared to a coordinatively saturated systems. J Lumin 110:113–124
172. Burai L, Toth E, Bazin H, Benmelouka M, Jaszberenyi Z, Helm L, Merbach AE (2006) Positively charged Gd^{III} cryptates: slow, associative water exchange. Dalton Trans 629–634
173. Di Pietro C, Guglielmo G, Campagna S, Diotti M, Manfredi A, Quici S (1998) A new tetraoxa-tetraaza macrobicyclic anthracenyl luminescent receptor: control of the receptor-substrate stoichiometry. New J Chem 22:1037–1039
174. Janczak J, Kubiak R (1994) Synthesis and crystal structure of an indium(III) complex with a bicyclic phthalocyanine at 300 K. J Chem Soc Dalton Trans 2539–2543
175. Janczak J, Kubiak R (1995) Synthesis and crystal structure of thallium(III) bicyclic phthalocyanine. Acta Chem Scand 49:871–877
176. Sun X, Johnson DW, Raymond KN, Wong EH (2001) A silver-linked supramolecular cluster encapsulating a cesium cation. Inorg Chem 40:4504–4506
177. Voloshin YZ, Varzatskii OA, Korobko SV, Chernii VY, Volkov SV, Tomachynski LA, Pehn'o VI, Antipin MY, Starikova ZA (2005) Ditopic macropolycyclic complexes: synthesis of hybrid phthalocyaninoclathrochelates. Inorg Chem 44:822–824
178. Dudkin SV, Erickson NR, Vologzhanina AV, Novikov VV, Rhoda HM, Holstrom CD, Zatsikha YV, Yusubov MS, Voloshin YZ, Nemykin VN (2016) Preparation, X-ray structures, spectroscopic, and redox properties of di- and trinuclear iron-zirconium and iron-hafnium porphyrinoclathrochelates. Inorg Chem 55:11867–11882
179. Varzatskii OA, Kats (Menkach) SV, Penkova LV, Volkov SV, Dolganov AV, Vologzhanina AV, Bubnov YN, Voloshin YZ (2013) Cage metal complexes: synthesis, X-ray structure, and spectral and redox behavior of the first hybrid iron(II) clathrochelatoscorpionate and its pyrazoloxime-armed macrocyclic intermediate. Eur J Inorg Chem 2013:1987–1992
180. El Ghachtoulia S, Guillota R, Leibl W, Aukauloo A (2014) Intramolecular photoinduced electron transfer in cobalt(II) chlathrochelate-zinc porphyrin assemblies leading to long-lived Co(I) species. J Porphyrins Phthalocyanines 18:1125–1130
181. Basallote MG, Blanco E, Blázquez M, Fernández-Trujillo MJ, Litrán R, Máñez MÁ, Ramírez del Solar M (2003) Exploring the properties and optical sensing capability of sol−gel materials containing a covalently bonded binucleating cryptand. Chem Mater 15:2025–2032
182. Hu XL, Luo QH, Wang ZL, Wang YN, Zhou ZY (2003) Synthesis, spectroscopy and crystal structure of a new heterodinuclear silver(I)—copper (I) cryptate. Transit Met Chem 28:640–643
183. Jazwinski J, Lehn JM, Lilienbaum D, Ziessel R, Guilhem J, Pascard C (1987) Polyaza macrobicyclic cryptands: synthesis, crystal structures of a

cyclophane type macrobicyclic cryptand and of its dinuclear copper(I) cryptate, and anion binding features. Chem Commun 22:1691–1694
184. Lu Q, McKee V, Nelson J (1994) Redox-linked pyrrole NH deprotonation in a dicopper azacryptate system. Chem Commun 649–651
185. Howarth OW, Nelson J, McKee V (2000) Steric constraint generating large through-space $^1H–^{203,205}$Tl coupling in a dithallium(I) cryptate. Chem Commun 21–22
186. Chen QY, Luo QH, Hu XL, Shen MC, Chen JT (2002) Heterodinuclear cryptates [EuML(dmf)] $(ClO_4)_2$ (M=Ca, Cd, Ni, Zn): tuning the luminescence of europium(III) through the selection of the second metal ion. Chem Eur J 8:3984–3999
187. Brooker S, Ewing JD, Nelson J, Jeffery JC (2002) Heterobinuclear cryptates; cooperative binding generates two different coordination sites within a symmetrical cryptand. Inorg Chim Acta 337:463–466
188. Drew MGB, Farrell D, Morgan GG, McKee V, Nelson J (2000) d^{10} Cations within triple-helical cryptand hosts; a structural and modelling study. Dalton Trans 1513–1519
189. Escuer A, McKee V, Nelson J, Ruiz E, Sanz N, Vicente R (2005) Ferromagnetic interaction in $\mu_{1,3}$-cyanamido-derived copper(II) cryptates. Chem Eur J 11:398–405
190. Shu-Yan Y, Qin-Hui L, Meng-Chang S, Xiao-Yun H (1994) Crystal structure of the first mixed-valence silver(I,III) cryptate. Inorg Chem 38:1251–1252
191. Brooker S, Ewing JD, Ronson TK, Harding CJ, Nelson J, Speed DJ (2003) Redox-adaptable copper hosts. Pyridazine-linked cryptands accommodate copper in a range of redox states. Inorg Chem 42:2764–2773
192. Drew MGB, McDowell D, Nelson J (1988) A new ditopic polyaza macrobicyclic ligand: X-ray crystallographic structure determination. Polyhedron 7:2229–2232
193. Drew MGB, Marrs DJ, Hunter J, Nelson J (1992) Divergent and convergent forms of a new Schiff-base cryptand; X-ray crystallographic and molecular mechanics investigations. J Chem Soc Dalton Trans 11–18
194. Lu Q, Latour JM, Harding CJ, Martin N, Marrs DJ, McKee V, Nelson J (1994) Dicopper cryptates with 1,1 and 1,3 bridging ligands: spectroscopic, magnetic and electrochemical properties. J Chem Soc Dalton Trans 1471–1478
195. Lamarque L, Miranda C, Navarro P, Escarti F, García-España E, Latorre J, Ramirez JA (2000) Dopamine interaction with a polyamine cryptand of 1H-pyrazole in the absence and in the presence of Cu (II) ions. Crystal structure of $[Cu_2(H_{-1}L]$ $(ClO_4)_3{\cdot}2H_2O$. Chem Commun 1337–1338
196. Lamarque L, Navarro P, Miranda C, Arán VJ, Ochoa C, Escartí F, García-España E, Latorre J, Luis SV, Miravet JF (2001) Dopamine interaction in the absence and in the presence of Cu^{2+} ions with macrocyclic and macrobicyclic polyamines containing pyrazole units. Crystal structures of $[Cu_2(L_1)(H_2O)_2](ClO_4)_4$ and $[Cu_2(H_{-1}L_3)]$ $(ClO_4)_3{\cdot}2H_2O$. J Am Chem Soc 123:10560–10570
197. Drew MGB, Harding CJ, Howarth OW, Lu Q, Marrs DJ, Morgan GG, McKee V, Nelson J (1996) Thiophene-linked azacryptand sites for dicopper and disilver; thiophene sulfur as an inert spacer? J Chem Soc Dalton Trans 3021–3030
198. Ma Z, Cao R (2005) An N_8O_6 cryptand: the host for Ag^+ guests and H_2O molecules. J Mol Struct 738:137–142
199. Drew MGB, Howarth OW, Morgan GG, Nelson J (1994) Proton transfer in a phenol-based compartmental cryptate; solution and solid-state structural studies. J Chem Soc Dalton Trans 3149–3158
200. Drew MGB, Howarth OW, Harding CJ, Martin N, Nelson J (1995) Kinetically stable lanthanide cryptates. J Chem Soc Chem Commun 903–905
201. Drew MGB, Harding CJ, McKee V, Morgan GG, Nelson J (1995) Geometric control of manganese redox state. J Chem Soc Chem Commun 1035–1038
202. Platas C, Avecilla F, de Blas A, Geraldes CFGC, Rodríguez-Blas T, Adams H, Mahía J (1999) ^1H NMR in solution and solid state structural study of lanthanide(III) cryptates. Inorg Chem 38:3190–3199
203. Platas C, Avecilla F, de Blas A, Rodríguez-Blas T, Geraldes CFGC, Tóth E, Merbach AE, Bünzli JCG (2000) Mono- and bimetallic lanthanide(III) phenolic cryptates obtained by template reaction: solid state structure, photophysical properties and relaxivity. J Chem Soc Dalton Trans 611–618
204. Avecilla F, Bastida R, de Blas A, Fenton DE, Macías A, Rodríguez A, Rodríguez-Blas T, García-Granda S, Corzo-Suárez R (1997) Metal template synthesis of lanthanide cryptates. Crystal structure of a dysprosium cryptate. J Chem Soc Dalton Trans 409–414
205. Rodríguez-Cortiñas R, Avecilla F, Platas-Iglesias C, Imbert D, Bünzli JCG, de Blas A, Rodríguez-Blas T (2002) Structural and photophysical properties of heterobimetallic 4f-Zn iminophenolate cryptates. Inorg Chem 41:5336–5349
206. Deeney FA, Harding CJ, Morgan GG, McKee V, Nelson J, Teat SJ, Clegg W (1998) Response to steric constraint in azacryptate and related complexes of iron-(II) and -(III). J Chem Soc Dalton Trans 1837–1844
207. McKee V, Dorrity MRJ, Malone JF, Marrs D, Nelson J (1992) Novel group 1 cation cryptates: X-ray and ^{23}Na NMR studies. J Chem Soc Chem Commun 383–386

208. Anaud-Neu F, Fuangswasdi S, Maubert B, Nelson J, McKee V (2000) Binding properties of octaamino-cryptands. Inorg Chem 39:573–579
209. Motekaitis RJ, Martell AE, Lehn JM, Watanabe E (1982) Bis(2,2′,2″-triaminotriethylamine) cryptates of cobalt(II), nickel(II), copper(II), and zinc(II). Protonation constants, formation constants, and hydroxo bridging. Inorg Chem 21:4253–4257
210. Lehn J-M, Pine SH, Watanabe E, Willard AK (1977) Binuclear cryptates. Synthesis and binuclear cation inclusion complexes of bis-tren macrobicyclic ligands. J Am Chem Soc 99:6766–6768
211. Drew MGB, Hunter J, Marrs DJ, Nelson J, Harding C (1992) Cascade complexes of an octaaza cryptand: co-ordinated azide with linear M-NNN-M geometry. Dalton Trans 3235–3242
212. Escuer A, Harding CJ, Dussart Y, Nelson J, McKee V, Vicente R (1999) Constrained ferromagnetic coupling in dinuclear $\mu_{1,3}$ -azido nickel(II) cryptate compounds. Crystal structure and magnetic behaviour of $[Ni_2(L1)(N_3)(H_2O)][CF_3SO_3]_3 \cdot 2H_2O \cdot EtOH$ {L1 = N $[(CH_2)_2NHCH_2(C_6H_4\text{-}m)CH_2NH(CH_2)_2]_3N$}. J Chem Soc Dalton Trans 223–228
213. Fabbrizzi L, Faravelli I (1998) A fluorescent cage for anion sensing in aqueous solution. J Chem Soc Chem Commun 971–972
214. Amendola V, Bastianello E, Fabbrizzi L, Mangano C, Pallavicini P, Perotti A, Lanfredi AM, Ugozzoli F (2000) Halide-ion encapsulation by a fl exible dicopper(II) bis-tren cryptate. Angew Chem Int Ed 39:2917–2920
215. Amendola V, Fabbrizzi L, Mangano C, Pallavicini P, Zema M (2002) A di-copper(II) bis-tren cage with thiophene spacers as receptor for anions in aqueous solution. Inorg Chim Acta 337:70–74
216. Maurizot V, Yoshizawa M, Kawano M, Fujita M (2006) Control of molecular interactions by the hollow of coordination cages. Dalton Trans 2750–2756
217. Farrell D, Harding CJ, McKee V, Nelson J (2006) Effect of methylation on the coordination of copper by small azacryptands; the role of geometrically constrained hydrogen bonding in stabilizing terminally coordinated oxygen species. Dalton Trans 3204–3211
218. Bond AD, Derossi S, Harding CJ, McInnes EJL, McKee V, McKenzie CJ, Nelson J, Wolowska J (2005) Cascade complexation: a single cyano bridge links a pair of Cu(II) cations. Dalton Trans 2403–2409
219. Christinat N, Scopelliti R, Severin K (2008) Multicomponent assembly of boronic acid based macrocycles and cages. Angew Chem Int Ed 47:1848–1852
220. Alliger GE, Müller P, Cummins CC, Nocera DG (2010) Cofacial dicobalt complex of a binucleating hexacarboxamide cryptand ligand. Inorg Chem 49:3697–3699
221. Alliger GE, Müller P, Do LH, Cummins CC, Nocera DG (2011) Family of cofacial bimetallic complexes of a hexaanionic carboxamide cryptand. Inorg Chem 50:4107–4115
222. Ambrosi G, Dapporto P, Formica M, Fusi V, Giorgi L, Guerrir A, Micheloni M, Paoli P, Pontellini R, Rossi P (2006) Synthesis of a large amino-phenolic cage. Synthesis, crystal structures, and acid–base and coordination behavior toward cations and anions. Inorg Chem 45:304–314
223. Funahashi Y, Shirota T, Kinoshita W, Toyama T, Nagata K, Nishikawa T, Fukui K, Inomata T, Ozawa T, Masuda H (2012) Synthesis of biomimetic models of multinuclear copper sites for O_2-activation. Paper presented at the 40 international conference on coordination chemistry, Valencia Conference Centre, Valencia, 9–13 Sept 2012
224. Khanra S, Weyhermüller T, Bill E, Chaudhuri P (2006) Deliberate synthesis for magnetostructural study of linear tetranuclear complexes $B^{III}Mn^{II}Mn^{II}B^{III}$, $Mn^{III}Mn^{II}Mn^{II}Mn^{III}$, $Mn^{IV}Mn^{II}Mn^{II}Mn^{IV}$, $Fe^{III}Mn^{II}Mn^{II}Fe^{III}$, and $Cr^{III}Mn^{II}Mn^{II}Cr^{III}$. Influence of terminal ions on the exchange coupling. Inorg Chem 45:5911–5923
225. Pascu M, Marmier M, Schouwey C, Scopelliti R, Holstein JJ, Bricogne G, Severin K (2014) Anionic bipyridyl ligands for applications in metallasupramolecular chemistry. Chem Eur J 20:5592–5600
226. Ardavan A, Bowen AM, Fernandez A, Fielding AJ, Kaminski D, Moro F, Muryn CA, Wise MD, Ruggi A, McInnes EJL, Severin K, Timco GA, Timmel CR, Tuna F, Whitehead GFS, Winpenny REP (2015) Engineering coherent interactions in molecular nanomagnet dimers. npj Quant Inform. doi:10.1038/npjqi.2015.12
227. Pitarch-Jarque J, Belda R, Lloret F, Ferrando-Soria J, Navarro P, Lopera A, García-España E (2015) "3 + 1 = 6 + 2" in Cu(II) coordination chemistry of 1*H*-pyrazole aza cryptands. Dalton Trans 44:3378–3383
228. Smith CB, Barbour LJ, Makha M, Raston CL, Sobolev AN (2006) Lanthanide-induced helical arrays of [{Co(III) sepulchrate}@{*p*-sulfonatocalix[4]arene}] supermolecules. Chem Commun 2006 (9):950–952
229. Zhang YY, Lin YJ, Jin GX (2014) Nano-sized heterometallic macrocycles based on 4-pyridinylboron-capped iron(II) clathrochelates: syntheses, structures and properties. Chem Commun 50:2327–2329
230. Westerhausen M, Weinrich S, Kramer G, Piotrowski H (2002) Cage compounds of heterotrimetallic mixed phosphanediide/silanolate aggregates with an alkaline earth metal-centered phosphorus octahedron. Inorg Chem 41:7072–7076
231. Catalano VJ, Malwitz MA, Noll BC (2002) Pd(0) and Pt(0) metallocryptands encapsulating a spinning mercurous dimer. Inorg Chem 41:6555–6559
232. Catalano VJ, Malwitz MA, Horner SJ, Vasquez J (2003) Three- and four-coordinate gold(I) complexes of 3,6-bis(diphenylphosphino)pyridazine: monomers, polymers, and a metallocryptand cage. Inorg Chem 42:2141–2148

233. Albrecht M, Schmid S, deGroot M, Weis P, Fröhlich R (2003) Self-assembly of an unpolar enantiomerically pure helicate-type metalla-cryptand. Chem Commun 2526–2527
234. Karpishin TB, Stack TDP, Raymond KN (1993) Octahedral vs trigonal prismatic geometry in a series of catechol macrobicyclic ligand-metal complexes. J Am Chem Soc 115:182–192
235. Albrecht M, Franklin SJ, Raymond KN (1994) Macrobicyclic tris(catecholate ligand) complexes: spectroscopy, electrochemistry, and the structure of $K_2[(H_2\text{-bicappedTRENCAM})MoO_2]$. Inorg Chem 33:5785–5793
236. Perraud O, Tommasino J-B, Robert V, Albela B, Khrouz L, Bonneviot L, Dutasta J-P, Martinez A (2013) Hemicryptophane-assisted electron transfer: a structural and electronic study. Dalton Trans 42:1530–1535
237. Gätjens J, Kampf JW, Pecoraro VL (2009) Preparation of a new 16-MC-4 structure type that captures Mn(II) in the central cavity. Inorg Chim Acta 362:878–886
238. Catalano VJ, Bennett BL, Kar HM, Noll BC (1999) Synthesis and characterization of trigonal gold (I) cage complexes: luminescent metallocryptates encapsulating Tl(I) and Na^+ Ions. J Am Chem Soc 121:10235–10236
239. Xu J, Corneillie TM, Moore EG, Law GL, Butlin NG, Raymond KN (2011) Octadentate cages of Tb (III) 2-hydroxyisophthalamides: a new standard for luminescent lanthanide labels. J Am Chem Soc 133:19900–19910
240. Murase M, Inomata T, Ozawa T, Masaoka S, Funahashi Y, Masuda H (2012) Structure and reactivity of a cage-type Zn(II) complex as a functional model of alcohol dehydrogenase. Paper presented at the 40 international conference on coordination chemistry, Valencia Conference Centre, Valencia, 9–13 Sept 2012
241. Bischof C, Wahsner J, Scholten J, Trosien S, Seitz M (2010) Quantification of C–H quenching in near-IR luminescent ytterbium and neodymium cryptates. J Am Chem Soc 132:14334–14355
242. Doffek C, Alzakhem N, Molon M, Seitz M (2012) Rigid, perdeuterated lanthanoid cryptates: extraordinarily bright near-IR luminophores. Inorg Chem 51:4539–4545
243. Lehn JM, Roth CO (1991) Synthesis and properties of sodium and europium(III) cryptates incorporating the 2,2′-bBipyridine 1,1′-dioxide and 3,3′-biisoquinoline 2,2′-dioxide units. Helv Chim Acta 74:572–578
244. Perkins DF, Lindoy LF, Meehan GV, Turner P (2004) Inherent helicity in an extended tris-bipyridyl molecular cage. Chem Commun 152–153
245. Saalfrank RW, Dresel A, Seitz V, Trummer S, Hampel F, Teichert M, Stalke D, Stadler C, Daub J, Schünemann V, Trautwein AX (1997) Topologic equivalents of coronands, cryptands and their inclusion complexes: synthesis, structure and properties of (2)-metallacryptands and (2)-metallacryptates. Chem Eur 3:2058–2062
246. Saalfrank RW, Seitz V, Caulder DL, Raymond KN, Teichert M, Stalke D (1998) Self-assembly of {2}-metallacryptands and {2}-metallacryptates. Eur J Inorg Chem 1998:1313–1317
247. Cantuel M, Gumy F, Bünzli JCG, Piguet C (2006) Encapsulation of labile trivalent lanthanides into a homobimetallic chromium(III)-containing triple-stranded helicate. Synthesis, characterization, and divergent intramolecular energy transfers. Dalton Trans 2647–2660
248. Hall CD, Djedovic N (2002) The synthesis and complexation of a cobaltocenium-based redox-active cryptand containing the phenanthroline unit. J Organomet Chem 648:8–13
249. Kiggen W, Vögtle F (1984) Functionalized, oligocyclic large cavities—a novel siderophore. Angew Chem Int Ed 23:712–713
250. Kiggen W, Vögtle F, Franken S, Puff H (1986) Large oligocyclic cavities for strong cation complexation. Tetrahedron 42:1859–1872
251. Stutte P, Kiggen W, Vögtle F (1987) Large molecular cavities bearing siderophore type functions. Tetrahedron 43:2065–2074
252. McMurry TJ, Rodgers SJ, Raymond KN (1987) Ferric ion sequestering agents. 16. Template and stepwise syntheses of a macrobicyclic catechoylamide ferric ion sequestering agent. J Am Chem Soc 109:3451–3453
253. McMurry TJ, Hosseini MW, Garrett TM, Hahn FE, Reyes ZE, Raymond KN (1987) Ferric ion sequestering agents. 17. Macrobicyclic iron(III) sequestering agents. J Am Chem Soc 109:7196–7198
254. Cohen SM, Petoud S, Raymond KN (1999) A novel salicylate-based macrobicycle with a "split personality". Inorg Chem 38:4522–4529
255. Orfanoudaki M, Tamiolakis I, Siczek M, Lis T, Armatas GS, Pergantis SA, Milios CJ (2011) Unique trigonal prism encapsulated Ln complexes: a $[Co^{II}{}_6Eu]$ and a $[Co^{II}{}_6Dy]$ cage. Dalton Trans 40:4793–4796
256. Gamage NDH, Mei Y, Garcia J, Allen MJ (2010) Oxidatively stable, aqueous europium(II) complexes through steric and electronic manipulation of cryptand coordination chemistry. Angew Chem Int Ed 49:8923–8925

Spacial and Electronic Structure of Cage Metal Complexes 3

Molecular structures of cage metal complexes in the solid state have been mainly elucidated by single crystal X-ray diffraction experiments and, in several cases, by EXAFS method; those in solutions have been obtained mainly using various NMR techniques.

Most of the clathrochelate metal tris-dioximates are intracomplexes, and their crystals are formed by weak intermolecular contacts, such as van der Waals interactions. In some cases, the 3D shape of these complexes allows for an intramolecular non-covalent bonding. In the crystals of their ionic-associate clathrochelate analogs, strong electrostatic (Coulombic) interactions have been observed (especially, in the case of polycationic polyamine macrobicyclic species).

The review [1], published in 2004, highlighted main trends in the structural chemistry of α-dioximate, oximehydrazonate, and dihydrazonate clathrochelates, such as the determination of the crystal and molecular structures of various types of cage complexes, their symmetry and functionality as well as the supramolecular organization and phase transitions in their crystals, the structure–spectra relationships, and the use of X-ray diffraction data for explanation of the kinetic and thermodynamic parameters of their synthesis and decomposition.

3.1 Mononuclear Polyamine Cage Complexes and Their Analogs

In the molecule of 1:1 zinc(II) cage complex of the calix[6]azaclathrochelate ligand **3** (Fig. 3.1), the encapsulated metal ion coordinates four nitrogen atoms of its *tren*-based capping fragment and acetonitrile solvent molecule as well, thus forming a slightly tetragonally distorted TBP ZnN_5-coordination polyhedron [2]. The average Zn–N distance in its trigonal base is approximately 2.07 Å, while that for the cross-linking tertiary nitrogen atom in the apical position of **3** is equal to 2.197 Å.

The zinc(II) cage complex of the macrobicyclic ligand **3** is described [3] to co-encapsulate also the solvate ethanol molecule, thus forming the corresponding heteroguest 1:1:1 compound that is stabilized by the C–H...π interactions and strong hydrogen bonding with the oxygen atom of one of its calixarene fragments. Its caged metal ion coordinates four amine nitrogen atoms of **3** and this solvate molecule as well, thus giving the ZnN_4O-coordination polyhedron. According to the solution NMR data, the molecule (Zn^{2+})@**3** is achiral, and this asymmetry is caused by coordination of **3** toward Zn^{2+} ion; it is effectively transmitted

Y. Voloshin et al., *Cage Metal Complexes*, DOI 10.1007/978-3-319-56420-3_3

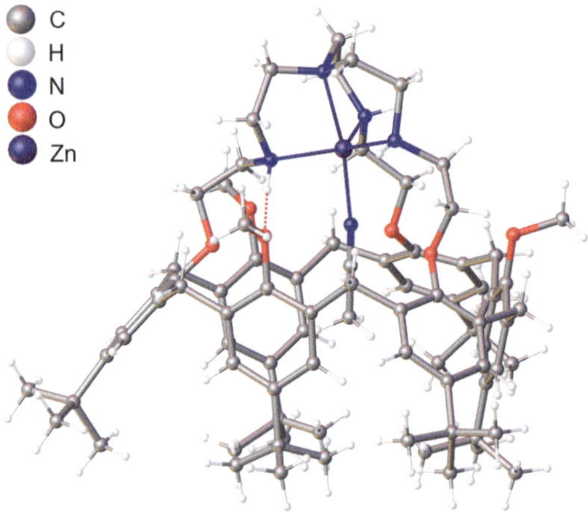

Fig. 3.1 X-ray structure of the clathrochelate dication (Zn^{2+})@**3** [2]

to co-encapsulated ethanol molecule. The use of more bulky guest molecules, such as imidazol, benzylamine, and dodecylamine, as well as the protonation of **3**, resulted in demetallation of the parent zinc(II) cage complex with extrusion of its encapsulated metal ion. The following deprotonation of this caging ligand with strong organic bases caused an opposite process of re-encapsulation of a zinc(II) cation [3].

Single crystal X-ray diffraction study of the monoprotonated form of a metal-free N_5-macrobicycle **36** has been performed [4]. Five nitrogen atoms of its framework, forming a slightly distorted TBP polyhedron, are in an *endo*-configuration, thus suggesting the encapsulation of H^+ ion. Three methyl substituents at the basal nitrogen atoms adopted a *head*-to-*tail*-configuration, determining the molecular C_3 pseudosymmetry of a cage framework. The N...N distance between its bridgehead nitrogen atoms (3.03 Å) is described [4] to be the shortest among those reported in literature to date for the protonated azacages (3.70–4.70 Å). In its 1:1 lithium(I) cage complex, the encapsulated metal ion has an almost regular TBP LiN_5-coordination polyhedron with Li–N distances in the range 2.01–2.08 Å [4].

The X-ray structure of 1:1 zinc(II) cage complex of the N_5-macrobicyclic ligand **37** is described [5]. The encapsulated pentacoordinate zinc(II) ion binds four nitrogen atoms of this encapsulating ligand and one solvate water molecule as well, thus forming a distorted TBP ZnN_4O-coordination polyhedron. In this polyhedron, two nitrogen and one oxygen atoms of **37** and its caged Zn^{2+} ion are located in its equatorial plane, while the bridgehead nitrogen atoms of this ligand occupy the apical positions. The Zn–N distances for these axially coordinated atoms are substantially greater than those with equatorial nitrogen donors [5].

Single crystal X-ray diffraction study of 1:1 copper(II) cage complex (Cu^{2+})@**3** has been performed [6]. Its encapsulated copper(II) ion coordinates four nitrogen atoms of this calix[6]-tren ligand and one solvate ethanol molecule, forming a TBP CuN_4O-coordination polyhedron. One of the secondary amine nitrogen atoms of **3** adopted a configuration different from those for two other nitrogen atoms, thus breaking a C_{3v} symmetry of this clathrochelate dication, having the molecular C_3 pseudoaxis [6].

The X-ray structure of the diprotonated form of a metal-free N_4O-macrobicycle **40** is described [7]. Its methylated tertiary nitrogen atoms are protonated, and the hydrogen atoms of these protonated tripodal amine fragments form hydrogen bonds with the apical oxygen atom of **40**. All its four nitrogen atoms are coplanar, thus forming an *endo*-conformation of the basal plane of a slightly distorted square pyramidal N_4O-cage framework; all of the carbon atoms of its ethylene moieties are below their mean plane [7].

Comparison of the X-ray structures of 1:1 copper(II) cage complexes of the N_4O-macrobicyclic ligands **42** and **43** has been performed [8]. In the clathrochelate dication (Cu^{2+})@**42**, its pentacoordinate encapsulated copper(II) ion binds four nitrogen atoms and one oxygen of a caging ligand. The geometry of CuN_4O-ccordination polyhedron is intermediate between a TBP polyhedron, in which the apical positions are occupied by two tertiary nitrogen atoms, while the remaining secondary nitrogen atoms and the above oxygen atom define its equatorial plane, and a square pyramidal polyhedron, which the basal plane is defined by all four nitrogen atoms and the apical position is occupied by this oxygen atom. The encapsulated copper(II) ion of (Cu^{2+})@**42** is deviated by approximately 0.41 Å above the mean plane of these nitrogen atoms. In the copper(II) cage complex of **43**, its caged metal ion coordinates four nitrogen and one oxygen atoms of the bridging fragments of this encapsulating ligand. CuN_4O-coordination polyhedron possesses a square pyramidal geometry: its basal plane is defined by the above four nitrogen atoms, while one oxygen atom is in its apical position; the caged copper(II) ion of (Cu^{2+})@**43** is above this basal plane by 0.342 Å [8].

The X-ray structures of 1:1 lithium(I) and copper(II) cage complexes of the N_5-macrobicyclic ligand **41** have been compared [9]. The encapsulated copper(II) ion coordinates its five nitrogen atoms, thus forming a TBP CuN_5-coordination polyhedron, that is slightly more distorted than LiN_5-coordination polyhedron of an encapsulated lithium(I) ion in the clathrochelate (Li^+)@**41**. The conformation of **41** is less strained in its copper(II) cage complex than that in the above lithium(I) compound [9].

X-ray diffraction data of [10] suggests that in 1:1 cobalt(II) cage complex of the octaazamacrobicycle **86**, this ligand plays the role of a N_5-pentadentate ligand. As a result, its encapsulated cobalt(II) ion has a TBP coordination polyhedron, formed by two capping tertiary and three secondary amine nitrogen atoms of **86**. This cation is located by approximately 0.37 Å out of the trigonal N_3-plane of the above coordination polyhedron toward one of its two capping nitrogen atoms. The clathrochelate dication (Co^{2+})@**86**, adopting a three-blade propeller shape, has the molecular C_3 pseudoaxis lying through its encapsulated metal ion [10].

As follows from single crystal X-ray diffraction data of [11] for 1:1 cadmium(II) complex of the encapsulating ligand **60**, its caged metal ion coordinates all eight donor atoms of this ligand (Fig. 3.2). The molecule (Cd^{2+})@**60** is substantially distorted as compared to that of a metal-free macrobicycle **60**. In particular, the N...N separation between two bridgehead nitrogen atoms of (Cd^{2+})@**60** is approximately 4.98 Å, while that in the molecule **60** is close to 6.25 Å; the encapsulated cadmium(II) ion preferentially occupies the apical N_4-fragment of its cavity [11].

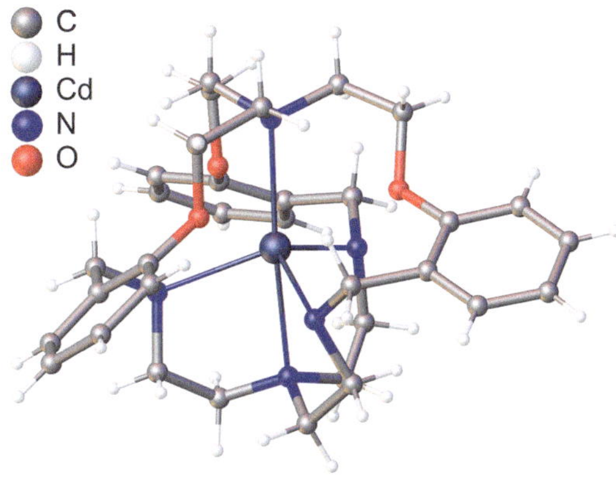

Fig. 3.2 X-ray structure of the clathrochelate (Cd^{2+})@**60** [11]

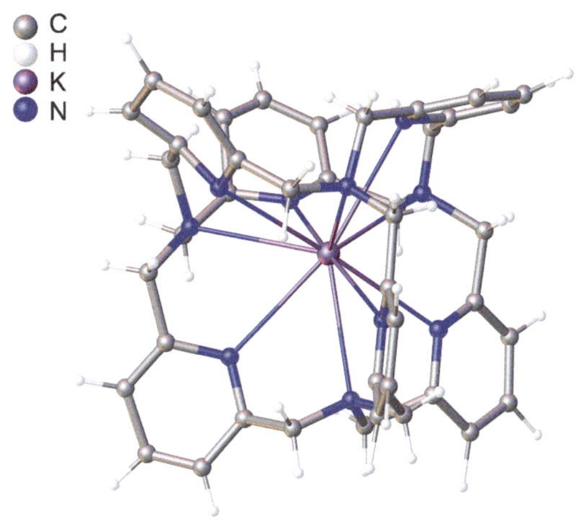

Fig. 3.3 X-ray structure of the clathrochelate cation (K^+)@**33** [12]

In the X-ray structure of 1:1 potassium(I) cage complex of the encapsulating ligand **33** [12], its four bridgehead nitrogen atoms form a slightly distorted tetrahedron (Fig. 3.3). The encapsulated potassium(I) ion occupies the center of **33**, six nitrogen atoms of which are coordinated toward this ion, thus forming an almost octahedral KN_6-coordination polyhedron with K–N distances in the range 3.07–3.26 Å [12].

The X-ray structures of a protonated form of the azaphenolic macrobicyclic compound **34** and 1:1 sodium(I) cage complex of its extended analog **35** are reported [13]. Two acidic hydrogen atoms of the above protonated form of **34** are hydrogen-bonded to the bridgehead nitrogen atoms of its N_4O-macrobicyclic framework and also form the hydrogen bonds with its oxygen atom. The nitrogen atoms point toward the phenolic group of **34**, while two other nitrogen atoms are on the opposite side of the N_4-mean plane with their attached methyl groups in equatorial and axial positions with respect to this plane. In (Na^+)@**35** (Fig. 3.4), the encapsulated metal ion coordinates five nitrogen atoms of its macrobicyclic ligand and the oxygen atom of phenol group as well, thus resulting in a strongly distorted pyramidal NaN_5O-coordination polyhedron with N_5-pentagonal base and the apex occupied by the above oxygen atom; the Na–N distances in this polyhedron fall in the range 2.455–2.731 Å. The clathrochelate cation (Na^+)@**35** has a pseudosymmetry plane that passes through its phenolic moiety and the bond between two opposite carbon atoms as well as through the encapsulated sodium(I) ion [13].

Single crystal X-ray diffraction data of [14] showed that the macrobicyclic ligand **91** in its 1:1 cage complex with $[K_3(H_2O)]^{3+}$ ion is slightly distorted from the expected T_d symmetry and 16 amine nitrogen atoms of this ligand are oriented inward the cavity of **91**, encapsulating the above tricationic guest.

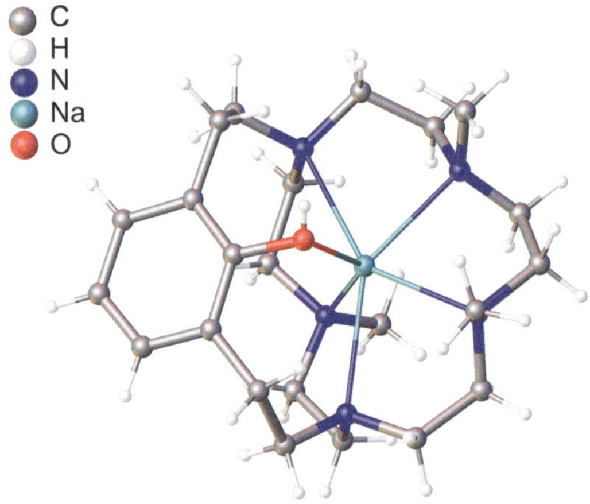

Fig. 3.4 X-ray structure of the clathrochelate cation (Na^+)@**35** [13]

As follows from single crystal X-ray diffraction data of [15], the encapsulated strontium or barium(II) cation of their 1:1 cage complexes with a mixed phosphanediide–silanolate clathrochelate ligand **604** coordinates six donor phosphine groups of the phosphanediide ribbed fragments of this ligand, thus giving the corresponding MP_6-coordination polyhedron. Six P_3-faces of the SrP_6- and BaP_6-octahedral polyhedra are capped by potassium(I) cations, and their P–P edges of two opposite P_3-faces are bridged by lithium(I) ions. The latter ions also coordinate the oxygen atom of the corresponding $(CH_3)_3SiO$ donor groups, thus forming their pyramidal LiP_2O-coordination polyhedra. K_6P_6-framework of **604** possesses the geometry of a distorted hexagonal prism in both these cage complexes; this distortion is explained [15] by electrostatic attractions between their donor phosphorus atoms and an encapsulated metal(II) ion.

The X-ray structure of 1:1 platinum(IV) cage complex of the triimine macrobicyclic ligand **96** is reported [16]. All its imine groups are attached to one capping tripodal fragment of **96** with a chiral amine site (ΛSSS, ΔRRR) and a chiral methine carbon site (ΛRRR, ΔSSS) on each ligand's strand. Its PtN_6-coordination polyhedron has a slightly distorted octahedral geometry with two types of the Pt–N bonds caused by the presence of the imine and the amine donor groups (the average Pt–N distances are equal to 2.050 Å and to 2.072 Å, respectively). In its polyamine derivative (Pt^{4+})@**97**, the average distance of this type is increased up to 2.10 Å. All its three six-membered chelate rings adopted a somewhat flattened *chair*-conformation. For a Λ configuration about its encapsulated platinum(IV) metallocenter, all the nitrogen atoms have an *S*-configuration and the methyl groups on the strap methine carbon atoms of **96** are all oriented equatorially, thus giving an overall molecular C_3 symmetry of this clathrochelate cation [16].

In the X-rayed crystal of the cage complex (Co^{3+})@(**99** · $2H^+$) [17], its macrobicyclic cation adopted a D_3lel_3-configuration, in which the five-membered chelate cycles are parallel to the molecular C_3 axis, while the apical capping groups of **99** are related by C_2 symmetry axis.

The clathrochelate cation (Co^{3+})@**98** also contains the C_2 symmetry axis; its CoN_6-coordination polyhedron possesses a distorted TAP geometry [17]. A slightly distorted octahedral 1:1 cobalt(III) cage complex of the expanded N_4S_2-macrobicyclic ligand **100** also adopted a lel_3-conformation of its ribbed chelate fragments [18].

As follows from single crystal X-ray diffraction data of [19], each of the two capping gallium (III) ions of a metal-free Ga_2L_3-coordination capsule **609** coordinates its three ligand synthons in such a manner that these metallocenters have the almost octahedral GaO_6-coordination polyhedra. The cavity of **609** is reported [19] to be enough for encapsulation of Li^+ ion, but it is too small for caging of more bulky sodium and potassium(I) cations.

Detailed analysis of the X-ray structures of the macrobicyclic tris-catecholate ligand **610** and its 1:1 aluminum, iron, gallium(III), titanium, and vanadium(IV) cage complexes, as well as that of an iron(III) clathrochelate, the derivative of the extended covalent capsule **611**, has been performed [20] to study the stabilization of a TP geometry in these macrobicyclic compounds. Although the crystal field stabilization energy is reported to play no role in any of them, their stereochemistries were found to be intermediate between a pseudo-TP and a pseudo-O_h. The geometries of the clathrochelate anions $[(Ti^{4+})$@**610**$]^{2-}$, $[(V^{4+})$@**610**$]^{2-}$ and $[(Fe^{3+})$@**610**$]^{3-}$ are very close to a TP, whereas their cage analogs $[(Al^{3+})$@**610**$]^{3-}$ and $[(Ga^{3+})$@**610**$]^{3-}$ possess the pseudooctahedral geometries. The spectral data of [20] evidenced that π-bonding within *d*-transition metal clathrochelates stabilizes a TP geometry within the constraints of their macrobicyclic ligand **610**. A metal-free macrobicyclic compound $H_6 \cdot$ **610** also possesses a TP geometry, and its framework is described [20] to be more strained in the above pseudooctahedral cage complexes.

Single crystal X-ray diffraction studies of 1:1 sodium and thallium(I) cage complexes of the diphosphine-based Au_2L_3-coordination capsule **617** have been performed [21]. Both these clathrochelate dications contain two apical gold(I) ions, capping their D_3-symmetric cage frameworks. In the molecule of a thallium(I) clathrochelate, these metallocenters are reported to strongly interact with its encapsulated metal ion at the distances of 2.9171 and of 2.9109 Å. Such strong Au…Tl interactions caused the deviations of planarity of their trigonal AuP_3-coordination polyhedra. The position of an encapsulated metal ion in the above 1:1 sodium(I) cage complex has been modeled [21] as three partially occupied sites, each one is slightly shifted from the center of **617** in the direction of its phenanthroline fragment. The intramolecular Au…Au separations for these thallium and sodium(I) clathrochelates are equal to 5.821 Å and 5.609 Å, respectively, and their macrobicyclic frameworks have a helical structure with its rotational C_3 axis lying through their capping gold(I) ions [21].

The supramolecular complex (Co^{3+})@**582**·(*para*-sulfonatocalix[4]arene) (Fig. 3.5) has been structurally characterized [22] by single crystal X-ray diffraction. Its clathrochelate guest is "perched" with respect to the "basket" of the calixarene host that adopted its *pinched cone*-conformation. This "basket" is too small to completely accommodate the above cobalt(III) clathrochelate, which is oriented with its ethylene group directed inward. The apical hydroxyl groups of this macrobicyclic guest are pointed

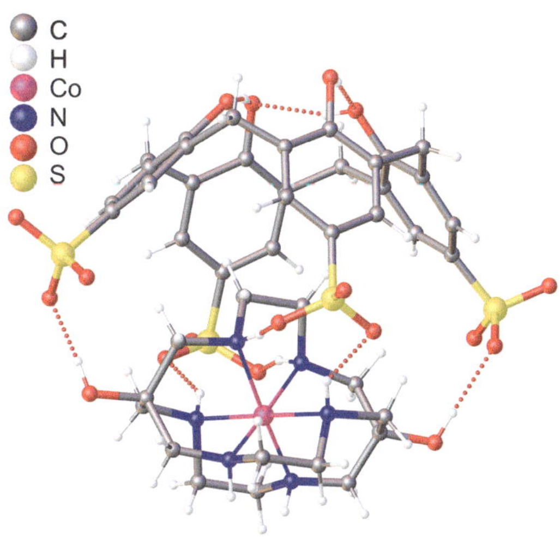

Fig. 3.5 Supramolecular complex between the clathrochelate trication (Co^{3+})@**582**· and *para*-sulfonatocalix[4]arene [22]

directly toward two opposite sulfonate groups of a calixarene host, held by hydrogen bonding. The clathrochelate trication (Co^{3+})@**582** is equipropably disordered on its Δ- and Λ-enantiomers [22].

The X-ray structure of the iron(II) N_3S_3-clathrochelate (Fe^{2+})@**101** is reported [23]. The apical amino group of its encapsulating macrobicyclic ligand adopting a lel_3-conformation is protonated. The Fe–N and Fe–S distances in this clathrochelate (*av.* 2.043 Å and 2.224 Å, respectively) are characteristic of the low-spin iron(II) complexes [23].

As follows from single crystal X-ray diffraction data of [24], the caged metal ion in the -gallium(III)-encapsulating macrobicyclic tetracation (Ga^{3+})@**99** · H^+ coordinates six amine donor atoms of **99**, thus giving GaN_6-coordination polyhedron; it possesses the geometry of a trigonally distorted octahedron. An unit cell of its X-rayed crystal contains both the enantiomers of this cage complex. The encapsulating ligand **99** adopted a lel_2ob-conformation with two parallel and one oblique orientation of the chelate C–C bonds with respect to its molecular C_3 pseudoaxis [24].

The X-ray structures of four copper(II) cage complexes of the macrobicyclic homologous polyamine ligands **125** · H^+, **126** · H^+, **127**, and **128** shown in Scheme 3.1 have been compared [25]. In the clathrochelate dication (Cu^{2+})@**128**, the encapsulated metal ion is located on the rotation C_2 axis of **128** and coordinates its six secondary amino groups, thus forming CuN_6-coordination polyhedron with an intermediate TP–TAP geometry. Both the apical amino groups at its cage framework, adopting a lel_3-conformation, are acylated. The monoacylated complex (Cu^{2+})@**125** · H^+ also adopts a TP-distorted pseudooctahedral CuN_6-polyhedron. Tetragonal elongation of a cage framework of this clathrochelate resulted from the Jahn–Teller distortion of electronic configuration d^9 of the hexacoordinate copper(II) ion, causing non-equivalence of the Cu–N bonds; the encapsulating ligand **125** · H^+ has a common lel_3-conformation. The copper(II) clathrochelate derivative of the diacylated macrobicyclic ligand **127** [25] is an intracomplex with two

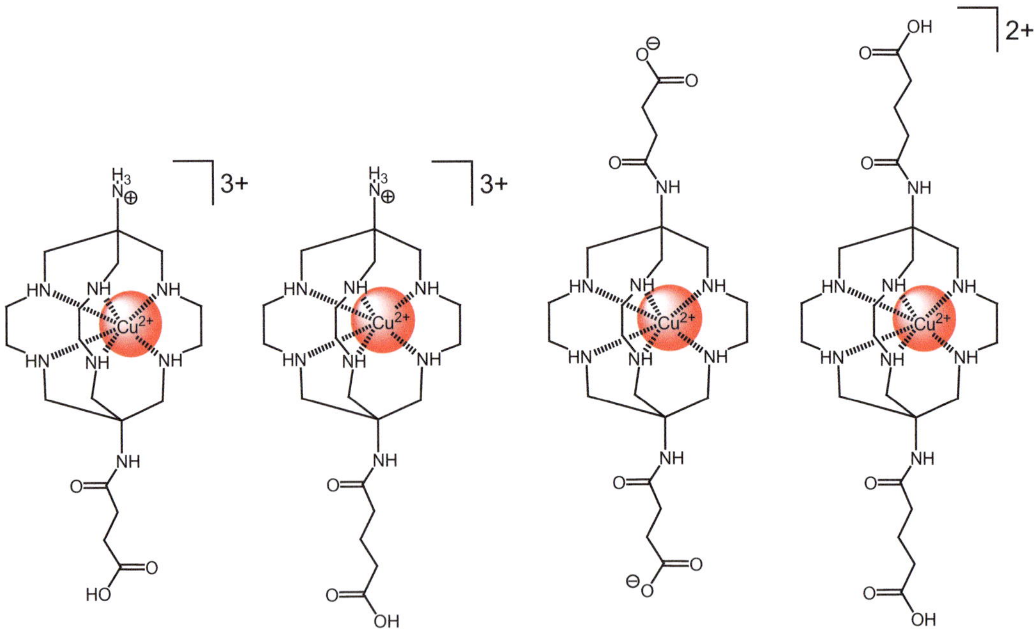

(Cu^{2+})@**125**·H^+ (Cu^{2+})@**126**·H^+ (Cu^{2+})@**127** (Cu^{2+})@**128**

Scheme 3.1 Chemical drawings of a series of the carboxyl-terminated polyamine copper(II) clathrochelates

deprotonated terminal carboxylic groups of its encapsulating ligand; both the *Δ*- and *Λ*-isomers of this clathrochelate are presented in a unit cell of its X-rayed single crystal. Substantial Jahn–Teller tetragonal distortion of a cage framework of (Cu^{2+})@**127** has been also observed [25].

The X-ray structures of 1:1 penta- and hexacoordinate copper(II) cage complexes of a flexible and expanded hexaamine macrobicyclic ligand **102** have been studied [26]. In the former case, the encapsulated copper(II) ion coordinates six amine donor groups of **102** in such a manner that one pair of its *trans*-Cu–N bonds is substantially longer than other four bonds of this type. Such axial elongation of a clathrochelate molecule is explained [26] by its Jahn–Teller structural distortion. In the pentacoordinate copper (II) cage complex of **102**, the encapsulated copper(II) ion binds five of the six donor amine groups of this macrobicyclic ligand, while non-coordinated secondary amino group forms an intramolecular hydrogen bond. Its CuN_5-coordination polyhedron possesses the geometry intermediate between a square pyramid and a trigonal bipyramid. In the copper(II) cage complex of a monoprotonated form of this ligand (Cu^{2+})@**102** · H^+, its encapsulated metal ion also coordinates five secondary amine nitrogen atoms of **102**, while the remaining amino group is protonated and, thus, do not form the coordination bond with this ion. The geometry of such CuN_5-coordination polyhedron is described [26] to be close to an axially elongated square pyramid; such elongation is caused by its Jahn–Teller distortion, characteristic of electronic configuration d^9.

Single crystal X-ray diffraction study of 1:1 potassium(I) cage complex of the fluorinated encapsulating ligand **517** has been performed [27]. Its six fluorine atoms are coordinated toward the encapsulated metal ion, thus forming a distorted octahedral KF_6-coordination polyhedron in which four equatorial fluorine atoms and this ion are placed in an almost one plane. The distances between the caged K^+ cation and the fluorine atoms of **517** are in the range 2.56–2.92 Å, while those between this ion and the nitrogen atoms of the caging ligand are much greater (3.28–3.74 Å). On the other hand, the bridgehead nitrogen atoms of **517** are reported [27] to be non-coordinated toward such encapsulated metal ion.

The X-ray structures of a diprotonated form of the caging ligand **38** and its lithium(I) 1:1 cage complex are described [28]. Coordination polyhedron of an encapsulated lithium(I) ion possesses the geometry of a TBP (Fig. 3.6) with Li–N distances in the range 1.983–2.025 Å (*av.* 2.004 Å) and with Li–O bond lengths from 1.926 to 1.945 Å (*av.* 1.936 Å).

According to the X-ray diffraction data of [29], the (*Δ*,*Λ*)-*fac*-Fe_2L_3-coordination capsule **627** in its 1:1 potassium(I) cage complex (Fig. 3.7) has the molecular C_{3h} symmetry. Its encapsulated metal ion coordinates six oxygen and three pyridine nitrogen atoms of **627**, two capping iron(III)-containing fragments of which are in the opposite configurations, thus forming the almost octahedral FeO_6-coordination polyhedra. 1:1 lithium(I) cage complex of the same encapsulating ligand **627** is reported [29] to be achiral.

The X-ray structure of 1:1 strontium(II) cage complex of the coordination capsule **627** is described [30]. The encapsulated strontium(II) ion is slightly shifted from the molecular

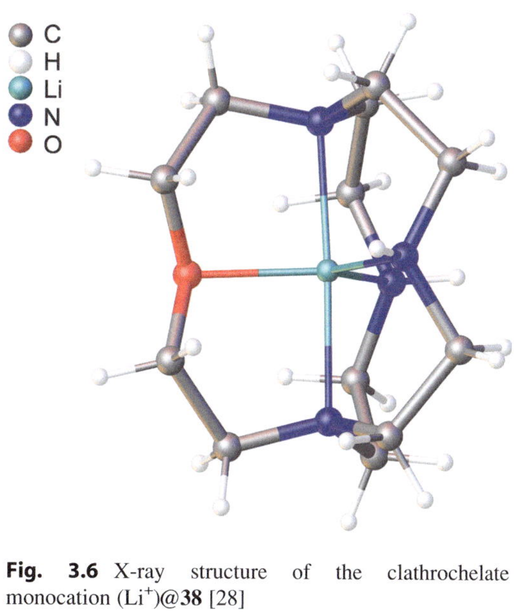

Fig. 3.6 X-ray structure of the clathrochelate monocation (Li^+)@**38** [28]

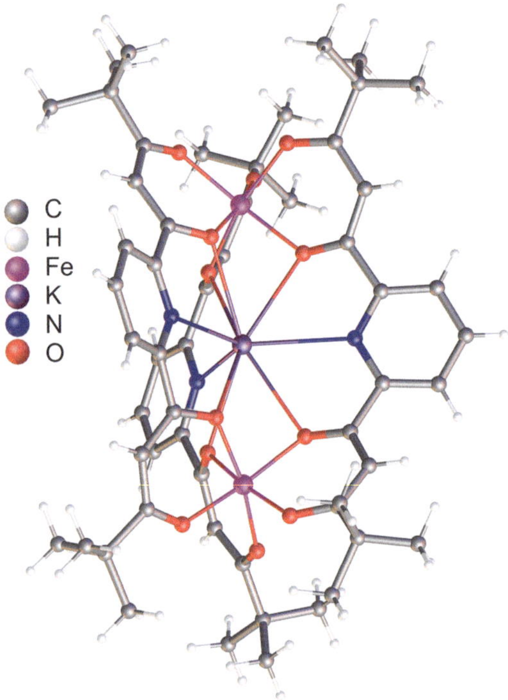

Fig. 3.7 X-ray structure of the clathrochelate cation (K^+)@**627** [29]

Fe…Fe C_3 pseudoaxis, connecting the capping iron ions; the intramolecular Fe…Fe separation between them is approximately 7.15 Å. The above strontium(II) cation coordinates six oxygen and three nitrogen donor atoms of **627** and one molecule of water as well. The ligand synthons are almost octahedrally coordinated toward the apical iron metallocenters in a helical fashion, and in a chiral form of the racemic dication (Sr^{2+})@**627**, both these centers have the identical coordination environments. Therefore, this clathrochelate is either a (*Δ*,*Δ*)-*fac*- or (*Λ*,*Λ*)-*fac*-cage complex, and its crystal contains a homochiral coordination capsule. The X-rayed crystal of a chiral racemic lanthanum(III) cage complex of the encapsulating ligand **628** contains in its asymmetric unit cell two slightly different, independent, and homochiral cage trications (*Δ*,*Δ*)- or (*Λ*,*Λ*)-(Ln^{3+})@**628** of the same chirality. Each of the two capping iron(III) ions coordinates six oxygen atoms of three ligand synthons, thus forming a slightly distorted pseudooctahedral FeO_6-coordination polyhedron. These apical cross-linking ions are located on the molecular C_3 pseudoaxis of its D_{3h}-symmetric clathrochelate framework with intramolecular Fe…Fe separation of 7.19 Å. The encapsulated undecacoordinate lanthanum(III) cation binds six internal oxygen atoms of three ligand synthons, thus forming a pseudo-TP O_6-coordination polyhedron. Three nitrogen atoms of these synthons and the solvate THF and water molecules are also coordinated toward the caged Ln^{3+} ion equatorially in a near perfect plane. The intramolecular Ln…Fe separation is approximately 3.60 Å, and this encapsulated metal ion lies almost on the molecular Fe…Fe axis [30]. The lanthanum(III) clathrochelate (Ln^{3+})@**628** gives a racemic mixture of its homochiral (*Δ*,*Δ*)-*fac*- and (*Λ*,*Λ*)-*fac*-stereoisomers. In the X-rayed crystal of a metal-free coordination capsule **627**, the pyridine nitrogen atoms of its three ligand synthons are protonated. The intramolecular Fe…Fe separation in this macrobicyclic trication (approximately 6.42 Å) is much smaller than those in the above cage complexes of **627**. So, the encapsulation of a metal cation caused the expansion of its macrobicyclic framework and, therefore, an elongation of the Fe…Fe distances in it. In a chiral racemic macrobicyclic trication **627** · $3H^+$, each of the capping iron(III) metallocenters coordinates six oxygen atoms of three ribbed fragments of **627** and, as a result, this coordination capsule is either (*Δ*,*Δ*)-*fac*- or (*Λ*,*Λ*)-*fac*-dinuclear triple helicate; their racemic mixture underwent a self-selectively sorting by chirality, thus generating its enantiomerically pure crystals [30].

As follows from single crystal X-ray diffraction data of [31], the cobalt(III) clathrochelate trication (Co^{3+})@**648** adopts an optically active *lel*$_3$-[Δ(λλλ)] conformation in its hexahydrated salt with dihydrogentriphosphate counterions. The geometry of this macrobicyclic trication allowed a formation of the hydrogen-bonded cobalt(III) clathrochelate-based supramolecular assemblies in its X-rayed crystal: the pairs of N–H groups belonging to a macrobicyclic cation are bridged by hydrogen-bonding anions or water molecules [31]. Six amine groups of a caging ligand are pointed, in pairs, toward three different positions around this clathrochelate cation, adopting a *lel*$_3$-conformation.

The X-ray structure of the cobalt(III) clathrochelate (Co^{3+})@**648**$)_2(SO_4)_3 \cdot 10H_2O$ is described [32]. Its encapsulated cobalt(III) ion coordinates six nitrogen atoms of the secondary amino groups of **648**, thus giving a slightly distorted pseudooctahedral CoN_6-coordination polyhedron with Co–N distances in the range 1.978–1.993 Å. The clathrochelate trications in its crystal are surrounded by sulfate anions and solvate water molecules as well, thus being involved in an extensive intermolecular hydrogen bonding to form 3D supramolecular network as shown in Fig. 3.8 [32].

An orientational disorder of one nitrate monocation in the X-rayed crystal of cage complex Λ-((Co^{3+})@**648**$)(NO_3)_3$ of the same encapsulating ligand, resulted from the symmetry requirements, is described [33]. Single crystal axial CD spectra of this clathrochelate trication (Co^{3+})@**648** has been measured [34] in the temperature range 9–295 K. The rotational strength $R(^1T_1(E) \leftarrow {}^1A_1)$ for a Λ-conformer of this cage complex decreases from approximately 1×10^{-40} c.g.s. units at 295 K to approximately zero value at 80 K. Solid-state diffuse reflectance CD spectra of Λ-(−)(S)- and Λ-(+)(R)-isomers of this cage complex are described [35]; CD data obtained have been thoroughly analyzed and compared with its solution, nujol mull, and KBr pellet UV-vis spectra. The partial molar volumes and refractions of an aqueous solution of (Co^{3+})@**648** have been determined [36].

The detailed analysis of the X-ray structures of three cobalt(III) cage complexes with solvate water molecules, the derivatives of the macrobicyclic trications (Co^{3+})@**649** and (Co^{3+})@**650**, has been performed [37]. The observed hydrogen bonding of these clathrochelate cations with chloride counterions in one of these cage complexes (Fig. 3.9) or that

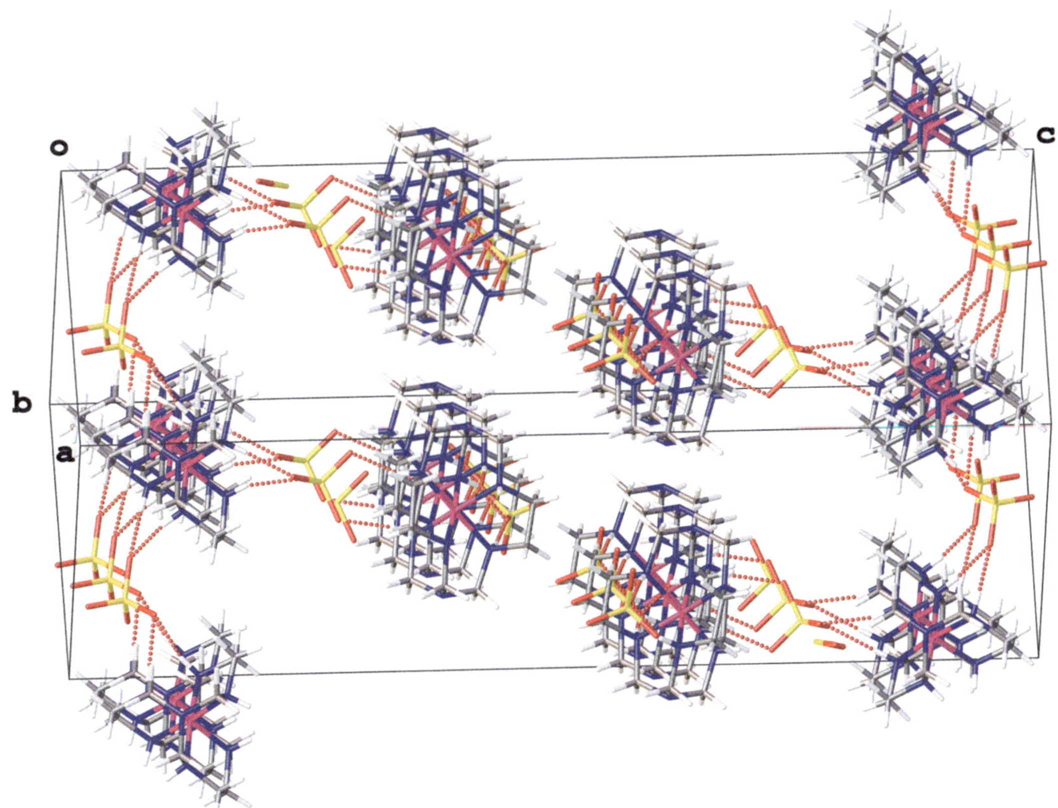

Fig. 3.8 Crystal packing of $((Co^{3+})$@**648**$)_2(SO_4)_3 \cdot 10H_2O$ [32]

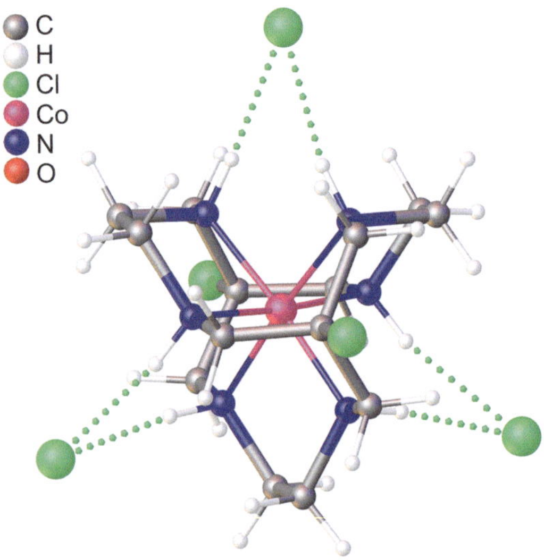

Fig. 3.9 Equatorial hydrogen bonding between the chloride counterions (*green circle*) and the clathrochelate trication (Co^{3+})@**649**, depicted with *dashed line* [37]

with solvate water molecules caused the retention of a *lel*$_3$-conformation of their encapsulating ligands [37].

The X-ray structures of two diamagnetic cobalt (III) and one paramagnetic copper(II) polyamine clathrochelates (Co^{2+})@**180**, (Co^{2+})@**181** · $2H^+$, and (Cu^{2+})@**183** (Scheme 3.2) with short (C_4) and long (C_{13}) paraffin apical substituents are described [38]. The aggregation of such lipophilic substituents at their cage frameworks caused a formation of the clathrochelate layers in their crystals (see, for example, crystal packing of (Cu^{2+})@**183** shown in Fig. 3.10).

The X-ray structure of the polyamine clathrochelate trication Δ-*lel*$_3$-(Co^{3+})@**92** (Fig. 3.11) is reported [39]. Its caged cobalt(III) ion coordinates six secondary amine nitrogen donor atoms of the encapsulating ligand **92**, thus forming an almost octahedral CoN_6-coordination

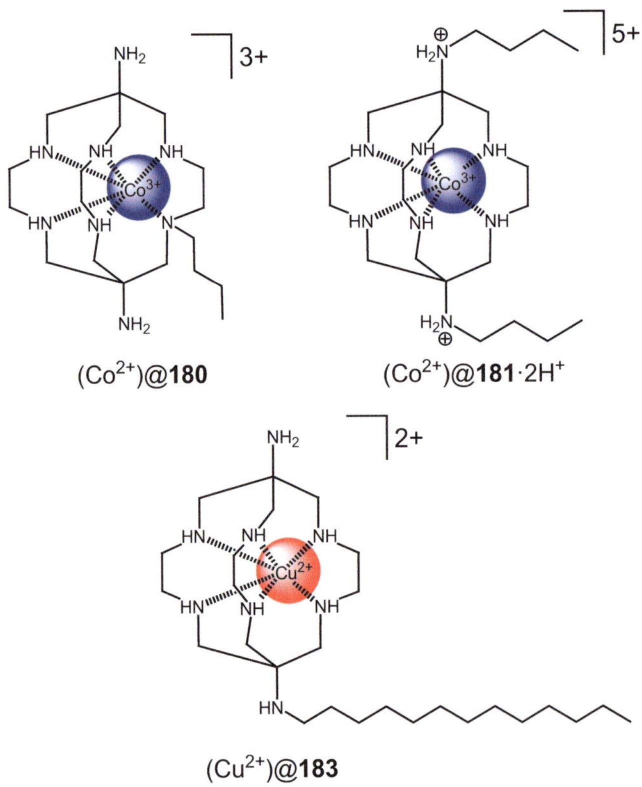

Scheme 3.2 Chemical drawings of a series of the polyamine metal(II) clathrochelates

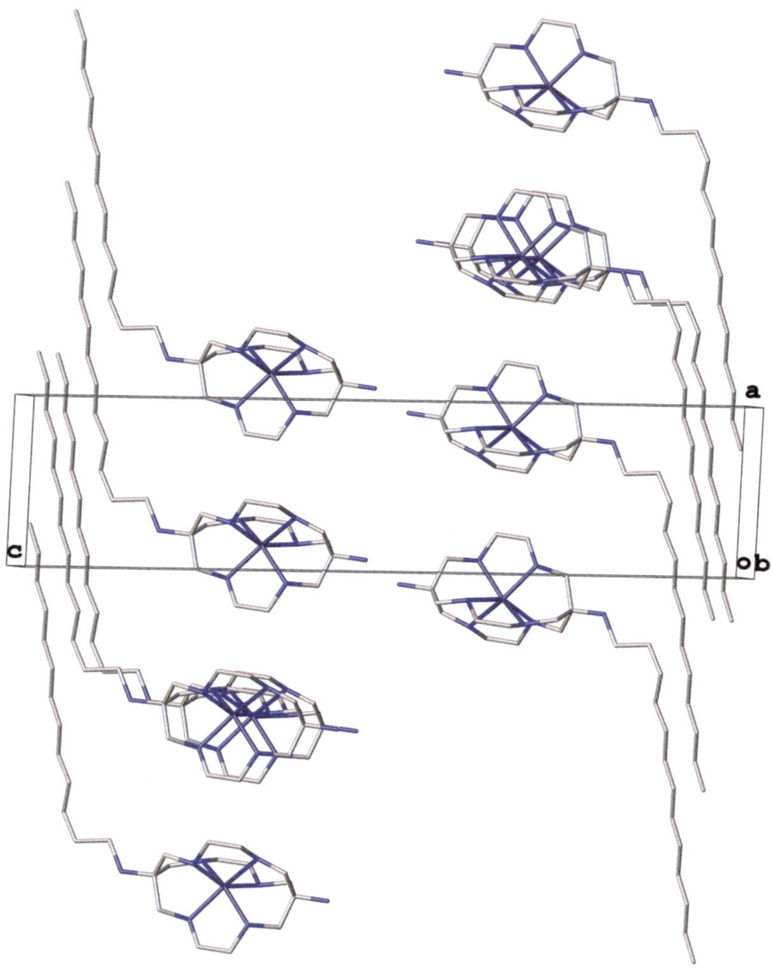

Fig. 3.10 Crystal packing of the copper(II) cage complex (Cu^{2+})@**183** [38]

polyhedron. This macrobicyclic cation retains an absolute Δ configuration of its encapsulated metallocenter and the same (*S*)-configuration about each coordinated secondary nitrogen atom. The C–C bonds of the chelate five-membered cycles of (Co^{3+})@**92** are parallel to its molecular C_3 pseudoaxis, thus suggesting a lel_3-conformation, fixed by rigid (*R*,*R*)-*trans*-1,2-cyclohexanediamine ribbed substituents in a Δ stereoisomer. All these cyclohexane fragments have the same *chair*-conformation, whereas six-membered chelate cycles in the apical fragments of a macrobicyclic framework of (Co^{3+})@**92** are described [39] to be conformationally more flexible, thus allowing to adopt its right- or left-handed helical arrangement with respect to the above molecular C_3 pseudoaxis. Both the tripodal capping fragments of the encapsulating ligand **92** adopt their slight - right-handed conformations; so, this 1:1 cage complex possesses a general Δ-D_3-lel_3-configuration [39].

As follows from the X-ray diffraction data of [40], the central copper(II) ion in its centrosymmetric trinuclear $Co^{III}Cu^{II}Co^{III}$ complex with two cobalt(III)-encapsulating clathrochelate ligands (Co^{3+})@**185**–$2H^+$ (Fig. 3.12) is effectively sandwiched between these macrobicycles, which are coordinated toward it in a facial mode through their iminodiacetate apical substituents

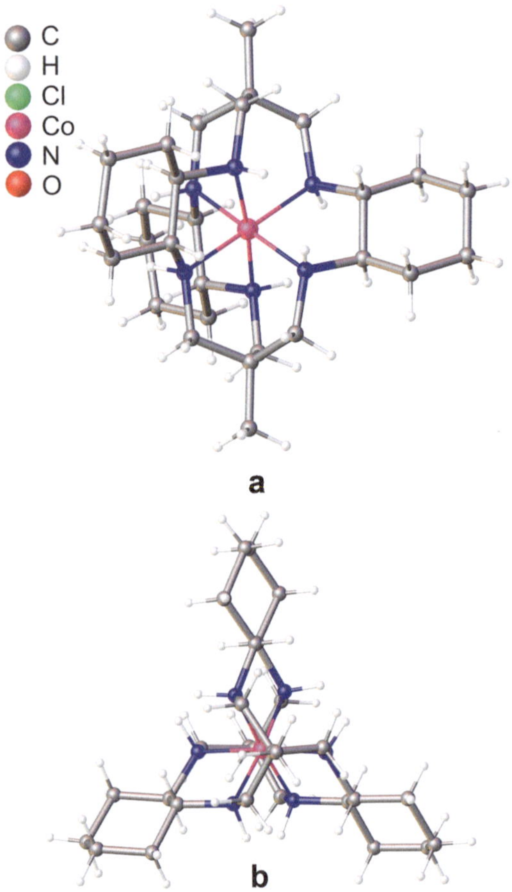

Fig. 3.11 Side (**a**) and top (**b**) views of the clathrochelate trication Δ-*lel*$_3$-(Co^{3+})@**92** [39]

acting as the tridentate ligands. In the Jahn–-Teller distorted pseudooctahedral coordination polyhedron of this cation, the Cu–N distances are close to 2.35 Å, while the Cu–O distances are approximately 2.08 and 1.91 Å; two clathrochelate ligands with uncommon ob_3-conformations have an opposite chirality. The X-rayed crystal of its trinuclear $Co^{III}Ni^{II}Co^{III}$ analog also contains the centrosymmetric nickel (II) complex with the same macrobicyclic ligands; the central metal(II) cation is situated on its crystallographic inversion center [40]. The pseudooctahedral NiO_4N_2-coordination polyhedron is formed by four oxygen atoms of two terminal carboxyl groups of these macrobicyclic ligands and by two tertiary nitrogen atoms of their apical substituents as well [40].

The X-ray structures of 1:1 cadium and mercury(II) cage complexes of an extended hexaamine macrobicyclic ligand **97** and their analogs have been analyzed [41]. Their MN_6-coordination polyhedra possess a TP geometry, which is unusual for the labile polyamine macrobicyclic ligands. The diprotonated form of a metal-free macrobicycle **97** also adopts the same TP geometry, while its 1:1 zinc(II) cage complex has an almost octahedral ZnN_6-coordination polyhedron. The above coordination geometries (*i.e.*, O_h and TP) are described [41]

Fig. 3.12 X-ray structure of 1:2 copper(II) complex of the cobalt(III)-encapsulating clathrochelate ligand (Co^{3+})@(**185**–$2H^+$) [40]

to be largely determined by best fit between the metal ion size and the appropriate cavity of a given conformation of the corresponding macrobicyclic ligand.

Single crystal X-ray structural study of 1:1 rubidium(I) cage complex of the fluorinated ligand **27** (Fig. 3.13) has been performed [42]. The observed average Rb^{+}...F distance of 2.803 Å proves a presence of the attractive interactions between the encapsulated rubidium (I) cation and this caging ligand.

The X-ray diffraction data of [43] for 1:1 manganese(II) cage complex of the octaamine ligand **90** evidenced that its encapsulated metal ion is symmetrically placed within a slightly elliptical cavity of **90** adopting an overall *lel*$_3$-conformation.

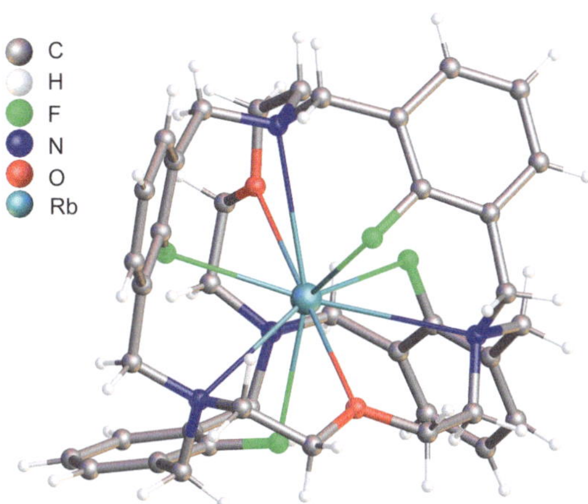

Fig. 3.13 X-ray structure of the clathrochelate cation (Rb^{+})@**27** [42]

DFT theoretical calculations have been performed [44] to evaluate the molecular structures and the formation energies of 1:1 alkali and alkaline earth metal cage complexes of the encapsulating ligands **90**, **648**, and **651** (Scheme 3.3) as well as to predict the selectivity of such encapsulation. The ligands **648** and **651** have been deduced to favor the binding of Li^{+} and Mg^{2+} ions, while **90** prefers the selective encapsulation of K^{+} and Ba^{2+} cations.

The X-ray structures of 1:1 gallium and iron(III) cage complexes of the tris-salicylate macrobicyclic ligand **642** have been discussed [45]. These clathrochelates have an unusual conformational distortion of their cage frameworks, breaking the molecular C_2 axis of its parent metal-free macrobicycle. Two tripodal *tren*-based capping fragments of these frameworks possess the inverted conformations relative to one another. One of them is "*puckered in*", thus pointing the amide protons to outside of a framework, while the second capping fragment with amide protons pointing into the cavity of **642** is "*puckered out*." The latter conformation of this tripodal fragment stabilizes the above cage complexes by formation of the intramolecular hydrogen bonds. The tertiary nitrogen atoms of the capping fragment possessing a "*puckered in*" conformation are protonated. The distortion angles φ of these iron(III)- and gallium(III)-encapsulating

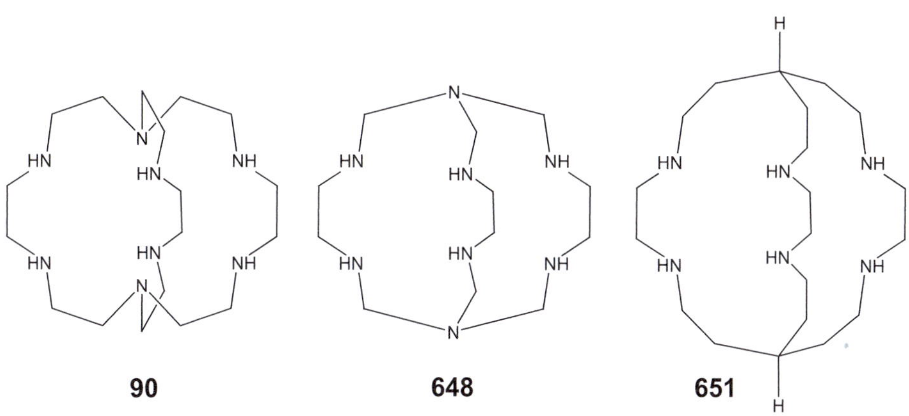

90 **648** **651**

Scheme 3.3 Chemical drawings of a series of the polyamine macrobicycles clathrochelate molecules are very similar (51° and 55°, respectively) [45].

Single crystal X-ray diffraction study of 1:1 europium and dysprosium(III) cage complexes of the $Co^{II}{}_6L_6$-coordination capsule **643** has been performed [46]. A cage framework of the europium(III)-encapsulating clathrochelate cation consists of a TP Co_6-core in which all its hexacoordinate cobalt(II) metallocenters have the slightly distorted octahedral coordination polyhedra, while the encapsulated europium(III) ion has a tricapped TP coordination polyhedron. The intramolecular Co...lanthanide separations in these clathrochelates are approximately 3.55 Å [46].

As follows from X-ray diffraction data of [47] for the clathrochelate cation (Pb^{2+})@**78**, its encapsulated lead(II) ion is asymmetrically placed within the cavity of **78**. This hexacoordinate cation binds one pyridyl and two secondary amine nitrogen atoms as well as two oxygen atoms of crown ether moiety and one nitrogen atom of the coordinated isothiocyanate counterion, thus forming its PbN_4O_2-coordination polyhedron [47].

The X-ray diffraction data of [48] evidenced that the molecule of a metal-free N_2O_6-macrobicycle **80** adopts a propeller-shaped chiral configuration; the N...N distance between its bridgehead nitrogen atoms is approximately 5.44 Å. Six oxygen atoms of a D_{3h}-pseudosymmetric cage framework of **80** form an achiral slightly distorted TP core with the distortion angle φ of 2.7°. The lone pairs of the above bridgehead nitrogen atoms are oriented outside of its cavity, thus giving an *exo–exo*-configuration of **80**. This cavity has a spheroidal geometry that is suitable for encapsulation of the appropriate metal ions [48].

The X-ray structure of 1:1 sodium(I) cage complex of the N_2O_6-macrobicyclic ligand **83** is reported [49]. The clathrochelate cation (Na^+)@**83** has an overall molecular C_3 pseudosymmetry; its bridgehead nitrogen atoms adopted the *endo–endo*-configurations and are weakly coordinated toward the encapsulated sodium(I) ion with Ag–N distances in the range 2.85–2.95 Å; all six oxygen atoms of **83** are also coordinated toward this cation [49].

3.2 Mononuclear Polyazomethine Clathrochelates and Their Derivatives

The X-ray structures of 1:1 lanthanum, europium, and terbium(III) cage complexes of the tris-bipyridine macrobicyclic ligands **519** and **520** have been compared [50]. Their encapsulated lanthanide cations coordinate eight nitrogen atoms of these caging ligands, thus forming the MN_8-coordination polyhedra. The ligands **519** and **520** have a shallow bowl geometry with two bipyridines spread widely open, while the third one is tilted to one side toward a bipyridine fragment of the former type (Fig. 3.14), thus allowing an additional coordination of two monodentate ligands, such as chloride counterions or water molecules [50]. The experimental luminescent study of 1:1 sodium(I), europium, and gadolinium(III) cage complexes of the macrobicyclic tris-bipyridinate ligands **519, 520,** and **652–655** shown in Scheme 3.4 and the theoretical DFT and TD-DFT calculations of the effects of the substituents at a clathrochelate framework on their UV-vis spectra have been performed [51].

The nature of solvate molecules is reported [52] to affect the supramolecular organization of the crystals of the bis-allylsulfide clathrochelate

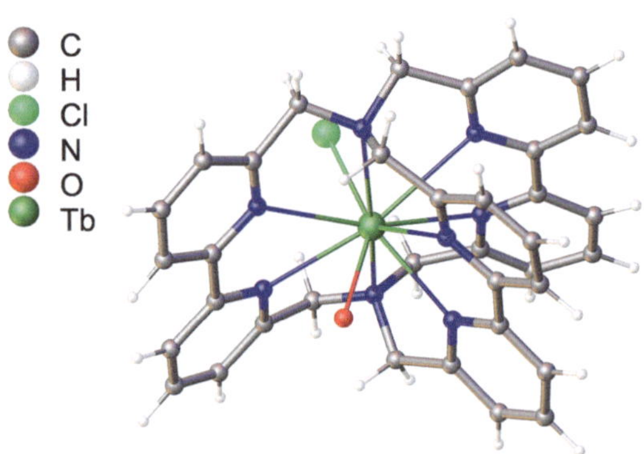

Fig. 3.14 Top view of 1:1 terbium(III) cage complex of the encapsulating ligand **519** [50]

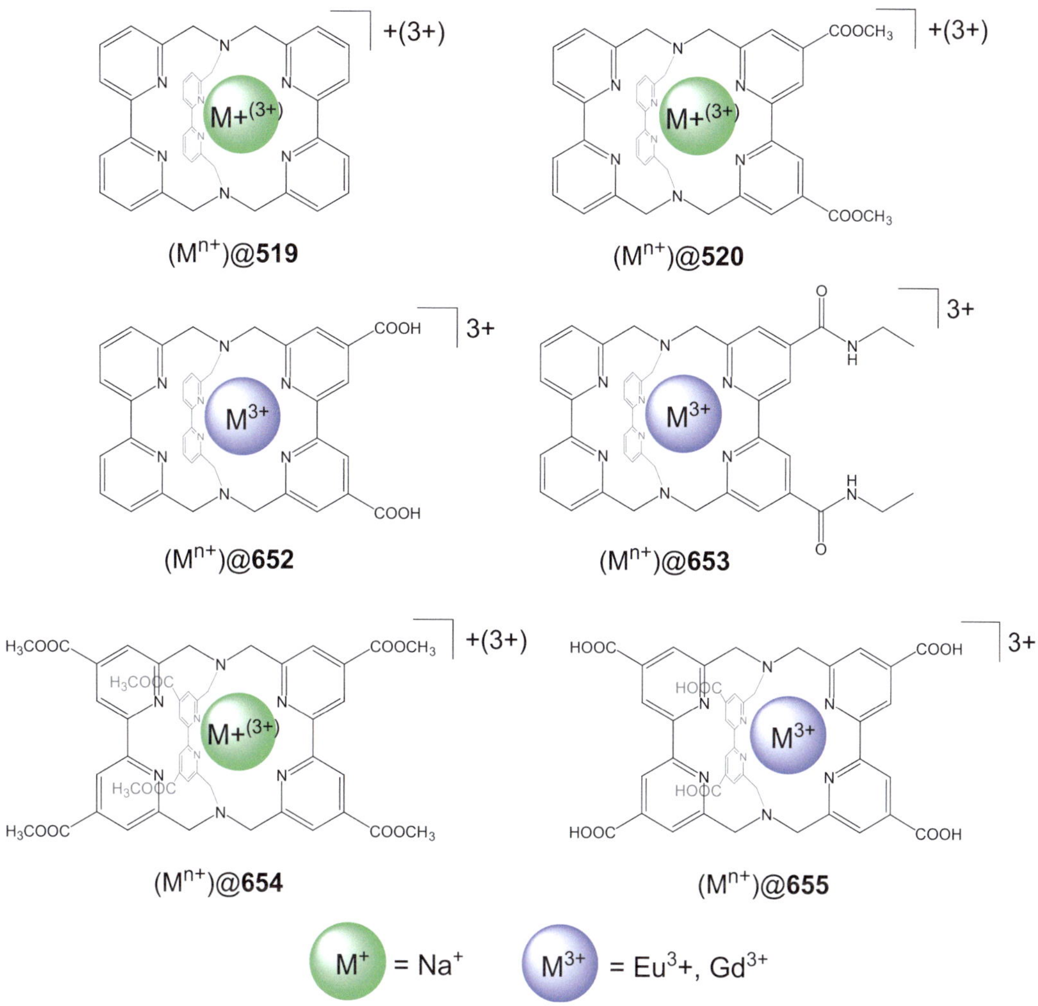

Scheme 3.4 Chemical drawings of a series of the tris-bipyridinate metal clathrochelates

(Fe^{2+})@**317**. The crystals of this cage complex with polar low donor chloroform solvate molecules (**A**), with polar and donor acetonitrile solvate molecules (**B**), and with nonpolar *n*-hexane (**C**) have been studied by single crystal X-ray diffraction. The nature of these solvate molecules [52] does not affect the molecular structure of this clathrochelate. The crystal structures of **A** and **B** have similar packing of the molecules of (Fe^{2+})@**317**, but are substantially different from that of **C** (Fig. 3.15). In all these crystals, the solvate molecules are located in the centrosymmetric cavities formed by these three-dimensional macrobicyclic molecules and are disordered about the corresponding center of symmetry. The clathrochelate entities of **C** are shifted along the crystallographic *c* axis. These cage molecules are reported to have the different orientations of their molecular C_3 pseudoaxes with respect to the crystallographic *a*, *b*, and *c* axes. In the crystals **A** and **B**, these pseudoaxes [52] are oriented perpendicular to the crystallographic axis *a* and are inclined to *b* and *c* by approximately 45°. In contrast, the C_3 pseudoaxes in the crystal **C** are perpendicular to the axes *a* and *b* and parallel to the axis *c*. The centrosymmetric cavities in **A–C**, having an ellipsoid shape, are formed by phenyl groups of α-benzildioximate chelate fragments and allylsulfide substituents at a cage framework.

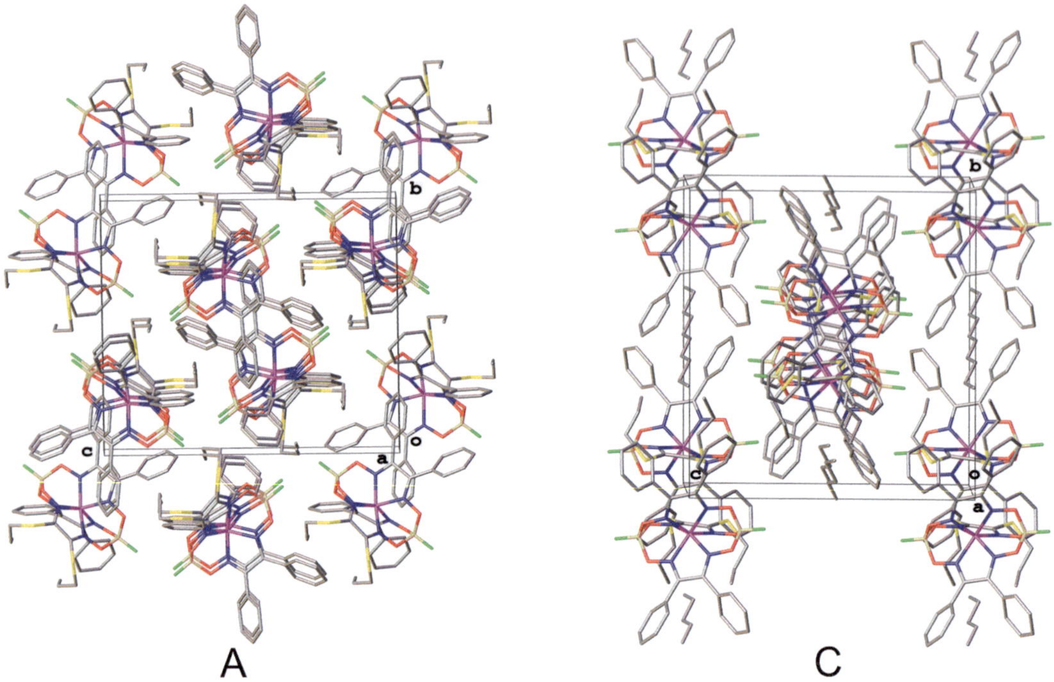

Fig. 3.15 Crystal packing of the solvates **A** and **C** [52]

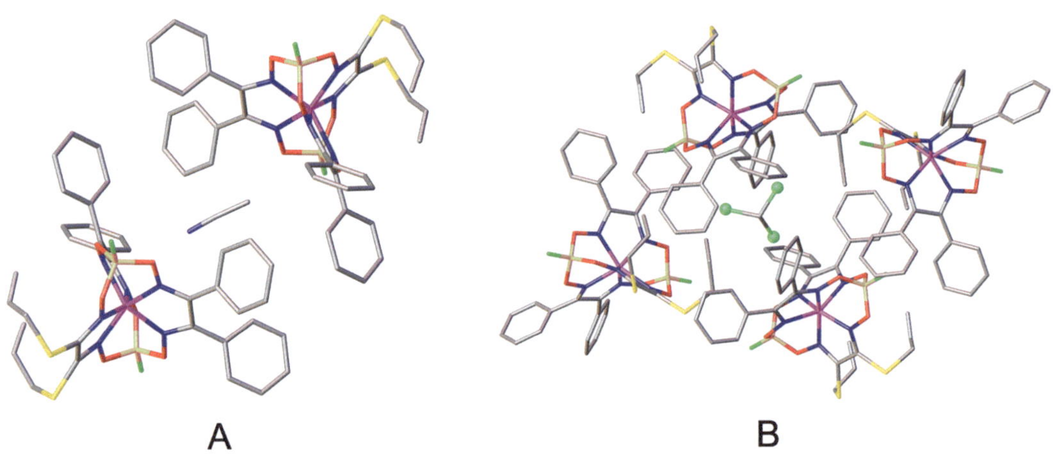

Fig. 3.16 Location of the solvent molecules in the cavities of the crystals **A** and **B** [52]

Their schematic representation with the included solvate molecules, denoted by circles, is shown in Figs. 3.16 and 3.17; the supramolecular interactions determining their sizes are also shown in these figures. In the X-rayed crystals **A** and **B**, the intermolecular weak specific dipole–dipole Cl...C and N...C interactions have been observed [52] but not found in the crystal **C**. In **A** the centrosymmetrical cavity is formed by four phenyl and two allyl groups of the two clathrochelate molecules (Fig. 3.16). In **B** it is formed by six phenyl and two allyl substituents of the two macrobicyclic molecules. Unlike **A**, the long axis of this ellipsoidal cavity is determined by carbon atoms of the allylsulfide substituents of (Fe^{2+})@**317**, while two other

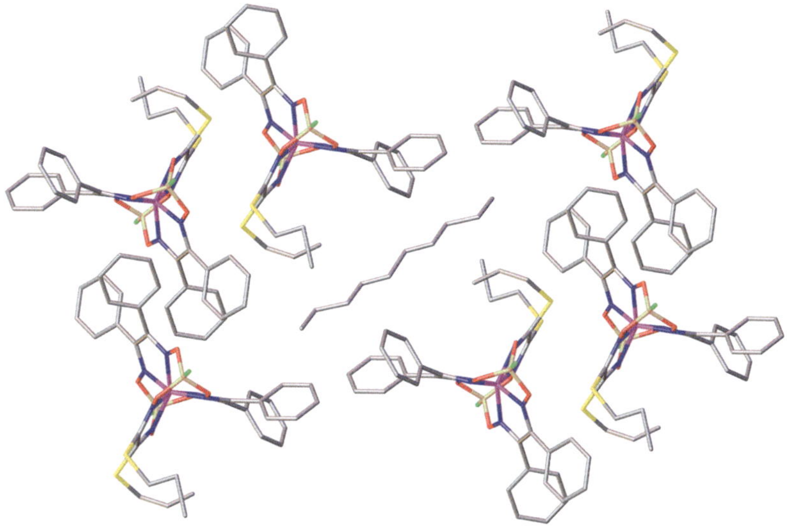

Fig. 3.17 Location of *n*-hexane molecules in the cavities of the crystal **C** [52]

axes of this ellipsoid are determined by phenyl substituents of α-benzildioximate chelate fragments of these macrobicyclic molecules. Its cavity volume is approximately 37.6 Å^3, while that of acetonitrile molecule is close to 22 Å^3. As a result, this solvate molecule is disordered over four positions. In the crystal **C**, the vacant cavity with an estimated volume of 75.8 Å^3 is more elongated, as it is formed by six clathrochelate molecules (i.e., carbon atoms of their twelve phenyl and four allylsulfide substituents, Fig. 3.17). No shortened contacts between the solvate and clathrochelate molecules were observed in this crystal, while its solvate hexane molecule is also disordered about the center of symmetry. The comparative analysis of the volumes of the above solvate molecules and the elliptical cavities in the crystals **A–C** [52] showed that the above cavities in all the X-rayed crystals of (Fe^{2+})@**317** are suitable for inclusion of these solvent molecules, allowing for their disordering about the center of symmetry. Their crystal packings are determined not only by the volume and shape of the solvate molecules, but also by their polarity: chlorine and nitrogen atoms of these molecules were found to be involved in the specific dipole–dipole host–guest interactions. Therefore, such weak intermolecular interactions between the clathrochelate molecules and the solvate guest molecules affect the supramolecular architecture, which is determined mainly by polarity rather than by the geometry of a given solvate molecule [52].

^{57}Fe Mössbauer and EXAFS data have been used [53] for deducing the molecular structures of polyazomethine iron(II) mono- and bis-clathrochelates (Scheme 3.5) with nonequivalent capping groups. An intermediate TP–TAP geometry of their FeN_6-coordination polyhedra has been established using the magnitudes of QS in their ^{57}Fe Mössbauer spectra; the deduced φ values are collected in Table 3.1. While the diboron-capped clathrochelate (Fe^{2+})$_2$@**273** has an almost TP coordination polyhedron, those of the antimony-capped iron(II) tris-nioximate semiclathrochelate and the SnB-, SbB-, and SbGe-capped mono- and bis-clathrochelates are intermediate between a TP and a TAP; the geometry of an oximehydrazonate iron(II) - bis-clathrochelate approaches a TAP [53].

EXAFS data and MM calculations have been used [53] to obtain RDA surrounding of the scattering nucleus (*i.e.*, to determine the coordination number and the type of these atoms as well as the distances from the scattering atoms to the atoms of their nearest environment); the coordination polyhedron of an encapsulated

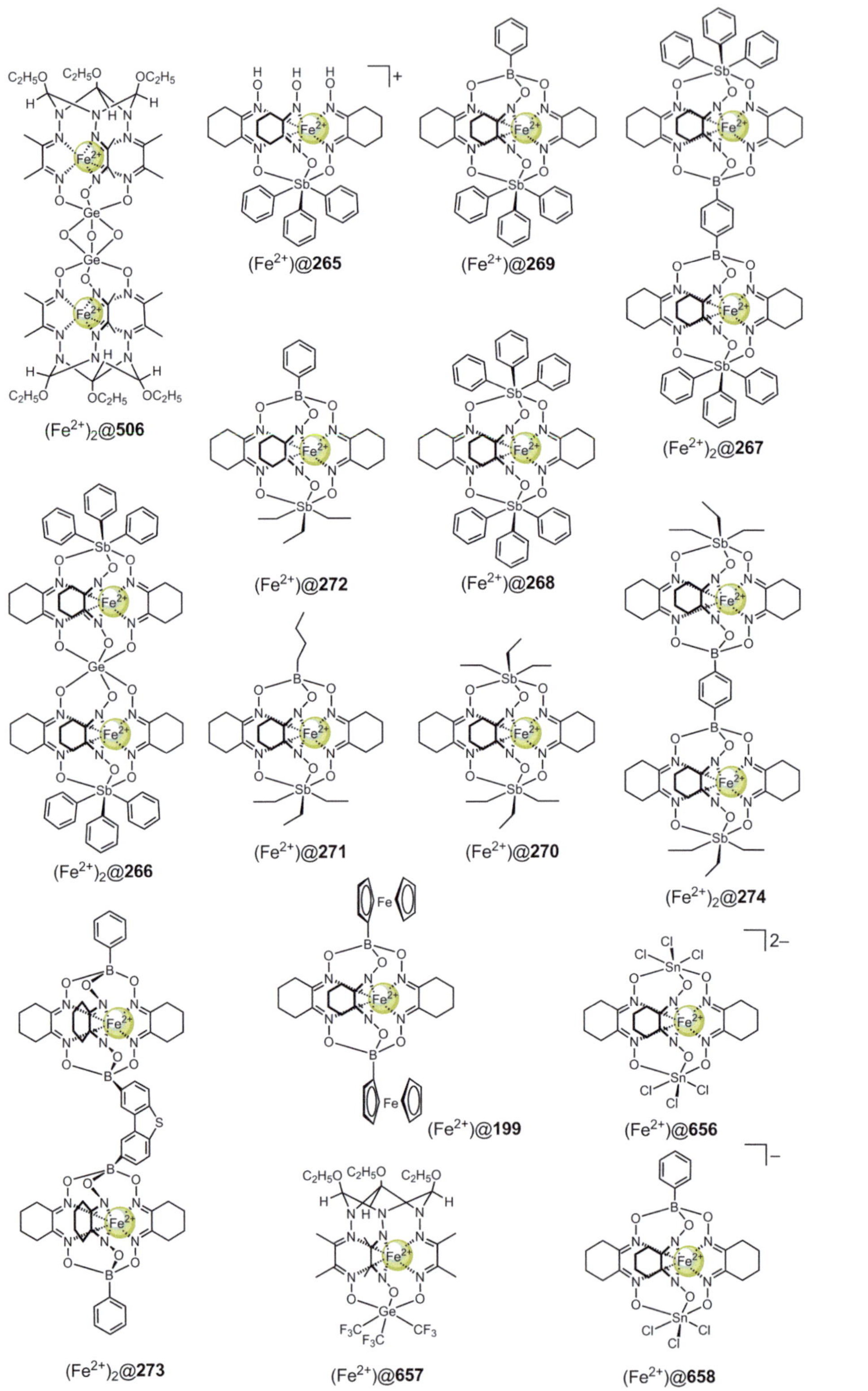

Scheme 3.5 Chemical drawings of a series of the iron(II) mono- and bis-clathrochelates

Table 3.1 Parameters of ^{57}Fe Mössbauer spectra (mm s^{-1}) and the distortion angle φ values (°) for the iron(II) mono- and bis-clathrochelates and their analogs [53]

Compound	IS	QS (deducted sign of QS)	φ Experimental (deduced)
$(Fe^{2+})_2$@**506**	0.40	(–)0.42	(50–55)
(Fe^{2+}) · **265**	0.39	(–)0.24	(35–40)
(Fe^{2+})@**269**	0.32	(+)0.27	(25–30)
(Fe^{2+})@**267**	0.30	(+)0.27	(25–30)
$(Fe^{2+})_2$@**266**	0.32	(–)0.19	(32–37)
(Fe^{2+})@**272**	0.32	(+)0.21	(25–30)
(Fe^{2+})@**274**	0.33	(+)0.26	(25–30)
$(Fe^{2+})_2$@**273**	0.33	(+)0.66	(12–17)
(Fe^{2+})@**271**	0.32	0.00	(27–32)
(Fe^{2+})@**270**	0.35	(–)0.25	35.7[a]
(Fe^{2+})@**656**	0.37	(–)0.18	37.5[a]
$(Fe^{2+})_2$@**268**	0.33	(–)0.22	(35–40)
(Fe^{2+})@**657**	0.40	(–)0.47	(50–55)
(Fe^{2+})@**199**	0.31	(+)0.72	9.5[a]
(Fe^{2+})@**658**	0.33	0.00	(27–32)

[a]X-ray data

iron(II) ion has been retained, and the model structures for each of the complexes under study have been calculated using known X-ray diffraction data for a series of the analogous iron (II) clathrochelates. The intramolecular distances between the scattering atoms and those of their nearest environment, calculated from EXAFS data, are collected in Tables 3.2 and 3.3. For the first coordination sphere of an encapsulated iron (II) ion, the coordination numbers were found [53] to be close to six, and the nature of a caging ligand determines the coordination numbers of its next coordination spheres. The use of two partially correlating EXAFS experiments for the same molecule is described [53] to substantially increase a reliability of the results obtained. A good agreement between the experimental EXAFS and the calculated MM data has been observed for the distances up to 4.5 Å. Not only the experimental radial distributions but also the calculated angular characteristics were found [53] to be in a good agreement with ^{57}Fe Mössbauer and single crystal X-ray diffraction data. The above conclusions on the geometry of the FeN_6-coordination polyhedra in these cage complexes have been confirmed by single crystal X-ray diffraction data of [53] for the clathrochelate (Fe^{2+})@**270**. The six-membered carbocycles in its α-dioximate chelate-ribbed fragments adopted a *half-chair*-conformation with β-carbon atoms displaced by approximately 0.36 Å to the opposite sides of the mean plane of other their carbon atoms. The average intramolecular Fe…Sb separation is equal to 3.76 Å, while the Sb…Sb distance is approximately 7.52 Å. So, the intramolecular Sb…Fe…Sb chain is almost linear [53]. A slightly distorted octahedral SbO_3C_3-coordination polyhedron is formed by three facially bonded oxygen atoms and three facially bonded carbon atoms of ethyl substituents. The geometry of FeN_6-coordination polyhedron of an encapsulated iron(II) ion of (Fe^{2+})@**270** is closer to a TAP than to a TP (φ = 35.7°). The average Fe–N distance (1.914 Å) is characteristic of the low-spin macrobicyclic iron(II) tris-dioximates, while an increase in φ led to a decrease in the height h of this intermediate TP–TAP coordination polyhedron up to 2.26 Å [53].

The X-ray structures of the *n*-butylboron-capped cobalt(II) hexachloroclathrochelate precursor (Co^{2+})@**293** and its reduced cobalt(I)-encapsulating derivative have been compared [54]. Passing from a cobalt(II) clathrochelate to a macrobicyclic cobalt(I) anion caused the insignificant changes of the average Co–N distances

Table 3.2 Calculated average distances from an encapsulated iron(II) ion and EXAFS data (Å) for the antimony- and germanium-capped cage complexes [53]

	Fe→N		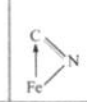		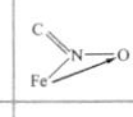		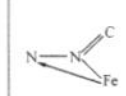		Fe→Q		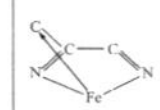			Other distances		
	cn = 6		cn = 6		cn = 6(3)		cn = 3		cn = 1(2)		cn = 6			cn = 1(2)		
Compound	a	b (c)[d]	a	b (c)[d]	a	b (c)[d]	a	b (c)[d]	a	b (c)[d]	a	b	c	a	b	c
$(Fe^{2+})\cdot$ **265**	1.91	1.94	2.75	2.84	2.89	3.02			3.71	3.72	4.20	4.44	4.30			
(Fe^{2+})@**269**	1.91	1.92	2.76	2.89	2.90	3.18			3.77	3.75	4.22	4.39	4.34	3.07	2.97 (Fe → B)	3.26
$(Fe^{2+})_2$@**267**	1.91	1.91	2.76	2.88	2.88	3.08			3.71	3.73	4.22	4.42	4.33	3.08	3.10 (Fe → B)	3.28
$(Fe^{2+})_2$@**266**																
Sb:	1.91	1.91	2.76	2.83	2.90	3.03			3.70	3.75	4.22	4.37	4.29			
Ge:									3.54	3.51						
(Fe^{2+})@**268**	1.91	1.90	2.76	2.81	2.89	3.02			3.75	3.71	4.21	4.37	4.27			
(Fe^{2+})@**270**	1.91	1.92	2.76	2.81	2.89	3.03			3.67	3.76	4.22	4.14	4.27			
(Fe^{2+})@**272**	1.91	1.92	2.77	2.80	2.91	2.97			3.73	3.78	4.22	4.54	4.26	3.09	3.19 (Fe → B)	3.19
$(Fe^{2+})_2$@**274**	1.91	1.92	2.77	2.81	2.89	2.96			3.70	3.79	4.22	4.38	4.29	3.08	3.24 (Fe → B)	3.16
(Fe^{2+})@**271**	1.91	1.95	2.77	2.90	2.93	3.00			3.76	3.72	4.24	4.11	4.35	3.13	3.13 (Fe → B)	3.19
$(Fe^{2+})_2$@**506**	1.91	1.93	2.77	2.86	2.91	2.98	2.92	3.21	3.53	3.43	4.22	4.43	4.29	2.77	2.76 (Ge → Ge)	2.76
(Fe^{2+})@**659**	1.91	1.93	2.75	2.86	2.88	3.04	2.91	3.30	3.63	3.70	4.20	4.46	4.31			
(Fe^{2+})@**660**	1.91	1.93	2.77	2.88	2.92	3.25	2.91	3.06	3.75	3.72	4.21	4.44	4.33	3.33	3.50 (Fe → C_3)	3.32
(Fe^{2+})@**661**	1.91	1.93	2.75	2.86	2.88	3.01	2.91	3.29	3.56	3.50	4.20	4.43	4.30			
(Fe^{2+})@**662**	1.91	1.93	2.77	2.84	2.92	2.98	2.92	3.21	3.62	3.54	4.22	4.45	4.29	3.33	3.39 (Fe → C_3)	3.34
(Fe^{2+})@**657**	1.91	1.94	2.77	2.84	2.91	2.97	2.91	3.22	3.58	3.48	4.22	4.44	4.28	3.30	3.41 (Fe → C_3)	3.36
$(Fe^{2+})_2$@**663**	1.91	1.93	2.74	2.74	2.90	2.83	2.90	3.05	3.56	3.48	4.21	4.26	4.18	3.67	3.64 (Fe → Fe)	3.51
(Fe^{2+})@**664**	1.91	1.92	2.76	2.77	2.90	2.93			3.57	3.55	4.22	4.36	4.23			

a: First *HyperChem* approximation
b: EXAFS data
c: Second *HyperChem* approximation
[d]This parameter was fixed in the second *HyperChem* calculation

Table 3.3 Calculated average distances from the capping atoms Q and EXAFS data (Å) for the antimony- and germanium-capped iron(II) cage complexes [53]

	Q → O		Q → C		N—O→Q			Q—C→C(F)			Q → Fe			C←(N—O)—Q		
	cn = 3		cn = 3		cn = 3(6)			cn = 6(9)			cn = 1(2)			cn = 3		
Compound	a	b (c)[d]	a	b (c)[d]	a	b	c	a	b	c	a	b	c	a	b	c
(Fe^{2+}) · **265**	2.03	2.16	2.15	2.13	2.98	3.00	3.03	3.12	3.15	3.09	3.71	3.73	3.72	4.01	4.12	4.10
(Fe^{2+})@**269**	2.03	2.10	2.15	2.18	3.03	3.05	3.12	3.10	3.22	3.13	3.77	3.80	3.75	4.13	4.10	4.30
$(Fe^{2+})_2$@**267**	2.02	2.13	2.15	2.14	3.02	3.11	3.12	3.11	3.07	3.10	3.71	3.70	3.73	4.11	4.19	4.27
(Fe^{2+})@**266***Sb:*	2.03	2.14	2.15	2.15	3.00	3.08	3.06	3.10	3.12	3.10	3.70	3.76	3.75	4.12	4.24	4.16
Ge:	1.85	1.92			2.81	2.82	2.80				3.54	3.49	3.51	3.90	4.45	3.9–4.46
(Fe^{2+})@**268**	2.03	2.14	2.15	2.17	2.99	3.19	3.02	3.12	3.12	3.11	3.75	3.73	3.71	4.04	4.11	4.09
(Fe^{2+})@**270**	2.02	2.14	2.19	2.15	2.98	3.11	3.02	3.12	3.22	3.08	3.67	3.73	3.73	4.06	4.03	4.08
(Fe^{2+})@**272**	2.02	2.15	2.18	2.13	3.01	3.13	3.10	3.10	3.08	3.10	3.73	3.82	3.78	4.12	4.13	4.23
$(Fe^{2+})_2$@**274**	2.02	2.16	2.18	2.16	2.99	3.14	3.10	3.10	3.12	3.11	3.70	3.74	3.79	4.10	4.15	4.25
(Fe^{2+})@**271**	2.02	2.16	2.18	2.16	3.02	3.09	3.12	3.10	3.18	3.06	3.76	3.75	3.72	4.15	4.27	4.29
$(Fe^{2+})_2$@**506**	1.84 1.90	1.87 1.95			2.81	2.82	2.81	4.10	4.21	4.03	3.53	3.47	3.43	3.91	3.83	3.93
(Fe^{2+})@**659**	2.03	2.12	2.15	2.16	2.95	3.03	3.00	3.11	3.16	3.11	3.63	3.70	3.70	4.00	3.98	4.08
(Fe^{2+})@**660**	2.02	2.16	2.00	2.14	3.02	3.36	3.06	3.02	3.10	3.10	3.75	3.80	3.72	4.13	4.11	4.17
(Fe^{2+})@**661**	1.87	1.99	2.03	2.09	2.83	2.90	2.86	2.91	2.95	2.96	3.56	3.52	3.50	3.92	3.94	3.97
(Fe^{2+})@**662**	1.85	1.96	2.02	2.08	2.86	3.05	2.85	2.83	2.91	2.95	3.62	3.60	3.54	3.95	3.92	3.96
(Fe^{2+})@**657**	1.85	1.85	2.02	2.07	2.85	3.30	2.77	2.84	2.90	2.95	3.58	3.53	3.48	3.97	4.25	3.91
$(Fe^{2+})_2$@**663**	1.85	1.87	2.03	2.06	2.85	2.83	2.79	2.84	2.93	2.87	3.56	3.49	3.48	3.93	3.95	3.95
(Fe^{2+})@**664**	1.86	1.97	2.03	2.09	2.84	2.88	2.86	2.84	2.96	2.96	3.57	3.57	3.55	3.94	3.96	3.96

a: First *HyperChem* approximation
b: EXAFS data
c: Second *HyperChem* approximation
[d]This parameter was fixed in the second *HyperChem* calculation

(1.984 Å vs. 2.003 Å, respectively) due to a contribution of Co(I)-to-cage CT ($\pi d \rightarrow \pi^{*} cage$ back donation) in the latter clathrochelate. Their CoN_6-coordination polyhedra possess a TP geometry with the distortion angle φ of 7.8 and of 1.3°, respectively. Such almost TP geometry is explained [54] by high Shannon radius of Co^{2+} ion that poorly fits the cavity size of **293**. Further rotational–translational elongation of a quasiaromatic rigid clathrochelate framework allowed one to encapsulate a cobalt(I) ion with even larger Shannon radius. The caged cobalt(II) ion was found [54] to be substantially (by approximately 0.3 Å) shifted from the geometrical center of this macrobicyclic framework in the direction of one of the three chelate ribbed fragments due to the structural Jahn–Teller distortion of its low-spin electronic configuration d^7.

The electron density study from X-ray diffraction of the crystal (Fe^{2+})@**193** · $CHCl_3$ (see Scheme 3.6 for the chemical drawing of this clathrochelate) has been performed [55]. The

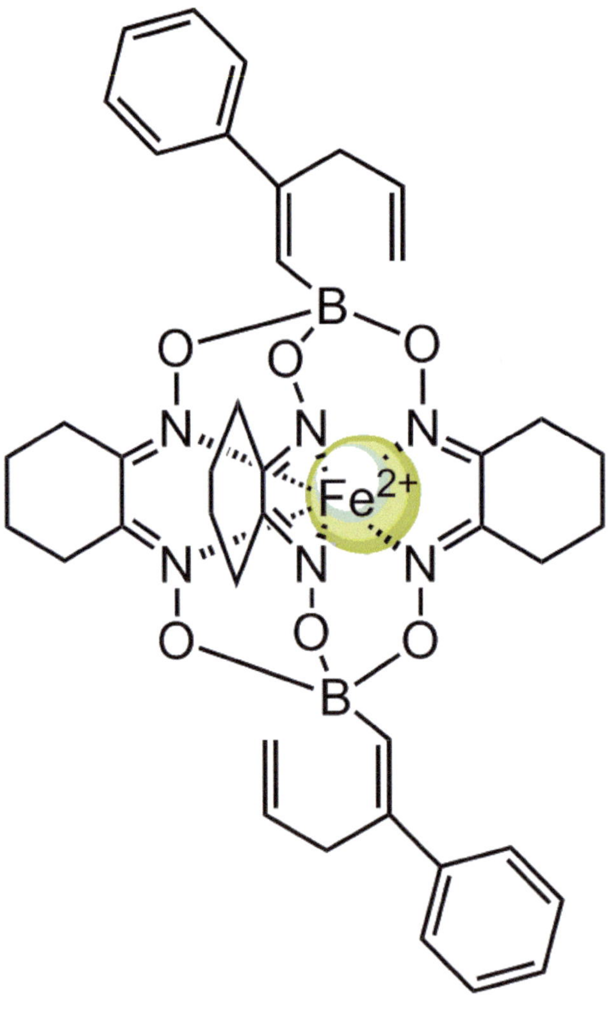

(Fe^{2+})@**193**

Scheme 3.6 Chemical drawing of the molecule (Fe^{2+})@**193**

analysis of DED distribution showed that its anisotropy is observed in the vicinity of an encapsulated iron(II) ion, while the DED maxima occur in the plane passing through the middles of the chelate C–C bonds in a macrobicyclic ligand (Fig. 3.18) and along the B...Fe...B molecular C_3 pseudoaxis as well (Fig. 3.19). The clathrochelate framework of this cage molecule has been characterized by the expected features of DED distribution: its maxima are localized on the chemical bonds and near lone electron pairs of the nitrogen and oxygen atoms of this framework. Ellipticity of the chelate C–C bond (Fig. 3.18) evidenced a substantial contribution of the π-component. Another important feature of these cross-sections is a substantial polarization of the electron pairs of nitrogen atoms and a continuous distribution connecting the electron pairs of nitrogen atoms and the DED maximum corresponding to the *d*-orbitals of the encapsulated iron(II) ion [55]. Similar character of DED distribution and the presence of π-component of the chelate C–C bond suggested an electron delocalization in the five-membered chelate metallocycles of this polyazomethine cage framework (Fig. 3.19).

Topological analysis of $\rho(r)$ in terms of Bader theory of critical points (CPs) [55] showed the presence of CPs (3, −1) in the region of all the expected interactions, including the Fe–N bonds. In addition to CPs (3, −1), a number of CPs (3, +1) corresponding to closure of three saturated and three aromatic six-membered cycles, three five-membered chelate metallocycles, and six six-membered FeN_2O_2B-chelate cycles in the apical fragments have been also found in the molecule (Fe^{2+})@**193**. Two polyhedral CPs (3, +3) were found along its molecular B...Fe...B axis [55]. Charges of the atoms have been determined by integration of atomic basins. The obtained charge values evidenced that main negative charge in its cage framework is localized on the oxygen and nitrogen atoms of a caging ligand, and the positive charge is localized on the capping boron atoms and on the carbon atoms of five-membered chelate metallocycle. Although the formal oxidation state of an encapsulated iron ion is +2, its charge was found [55] to be approximately +0.63 *e*.

Single crystal X-ray structure of the monoribbed-functionalized iron (II) clathrochelate (Fe^{2+})@**665** with two bulky *tert*-butyl amine substituents in a *vic*-position of the same chelate ribbed fragment is reported [56]. The geometry of its FeN_6-coordination polyhedron is almost intermediate between a TP and a TAP with the average distortion angle φ of 27.6° and the height h equal to 2.30 Å. The *tert*-

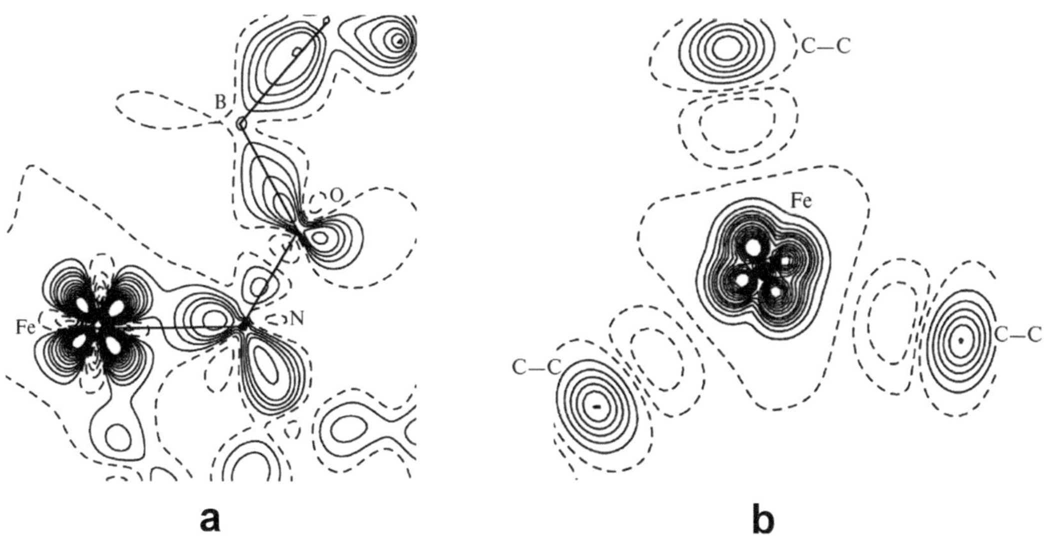

Fig. 3.18 Cross-section of DED in the plane passing through the Fe, N, and B atoms of the molecule (Fe^{2+})@**193** (**a**) and that through the middles of its chelate C–C bonds (**b**). The contours are drawn with their increment of 0.1 e Å^{-3} and the negative isolines are shown with *dashed line* [55]

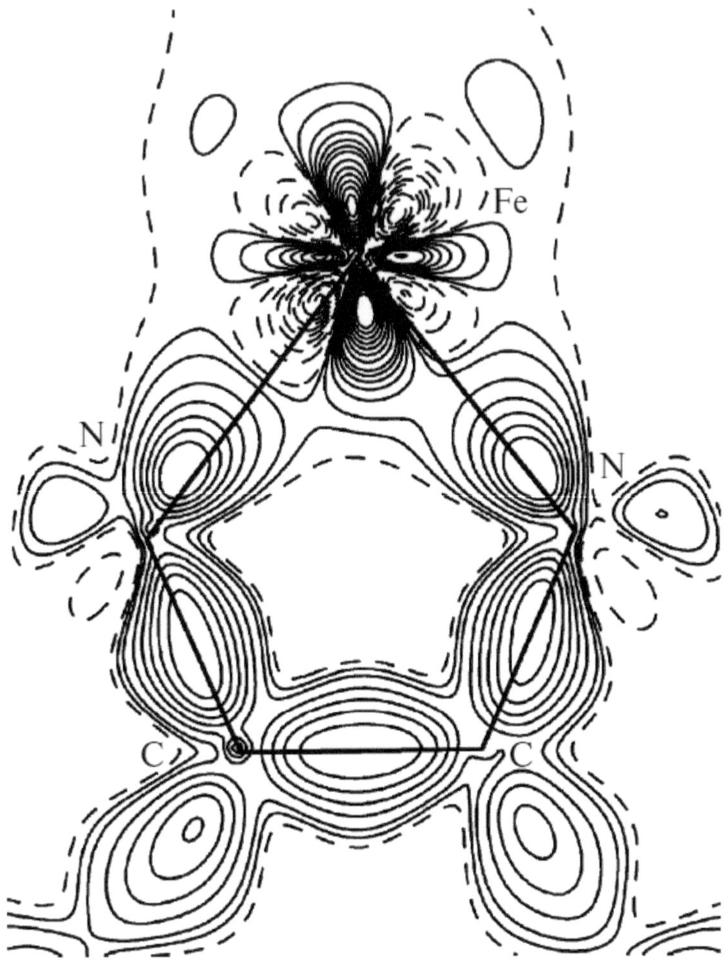

Fig. 3.19 DED distribution in a five-membered chelate metallocyclic fragment of the molecule (Fe^{2+})@**193**. The contours are drawn with their increment of 0.1 e Å^{-3}, and the negative isolines are shown with *dashed line* [55]

butylamine ribbed substituents at a clathrochelate framework of (Fe^{2+})@**665** have the different orientations relative to it: one of them is virtually coplanar with the corresponding amine-containing oxime fragment (the dihedral angle is equal to 17.5°), whereas the second substituent of this type is almost perpendicular to the corresponding donor fragment with the dihedral angle of 86.3°. The former *tert*-butylamine group forms the short C...O contact with neighboring oxygen atom of a caging ligand at the distance of approximately 3.05 Å [56].

The X-ray structures of three ferrocenylboron-capped iron(II) tris-dioximates (Fe^{2+})@**198**, (Fe^{2+})@**199**, and (Fe^{2+})@**200** have been compared [57]. The crystal (Fe^{2+})@**198** · $CHCl_3$ contains two types of the crystallographically non-equivalent clathrochelate molecules ((Fe^{2+})@**198**A and (Fe^{2+})@**198**B). The planes of their Cp fragments, being in a hindered conformation, were found to be almost parallel in all the cage molecules, while the apical ferrocenyl substituents are oriented covectorially to each other in the direction of the B...Fe...B molecular C_3 pseudoaxis. The TP–TAP distortion angles φ of their FeN_6-coordination polyhedra varied from 9.5 to 24.9°, while the heights h of these polyhedra fall in the range 2.31–2.38 Å. The small value of φ in the tris-nioximate molecule (Fe^{2+})@**199** (approximately 9.5°) is explained [57] by relative structural rigidity of its alicyclic substituents: an increase in h to 2.38 Å is caused by sterical clashes between the Cp fragments of the ferrocenylboronate apical groups and the neighboring cyclohexane ribbed substituents of the same caging ligand. In the case of the methyl (**198**A and **198**B) or cyclooctane (**200**) ribbed substituents, the absence of such clashes allowed one a substantial distortion of the corresponding TP–TAP coordination polyhedra, and, therefore, an increase in φ for the clathrochelates (Fe^{2+})@**198**A, (Fe^{2+})@**198**B, and (Fe^{2+})@**200** as compared with (Fe^{2+})@**199**. In the molecules of the former macrobicyclic complexes both the apical ferrocenyl substituents are located between the planes formed by their two ribbed fragments. As a result, h values for the former cage complexes did not exceed 2.34 Å. On the other hand, in the molecule (Fe^{2+})@**199,** only one of these apical substituents is located between the "petals" of its chelate fragments [57].

The molecules (Fe^{2+})@**199** in the X-rayed crystal of this cage complex form the clathrochelate layers shown in Fig. 3.20*via* the C–H…π(Cp) contacts of their methyl ribbed

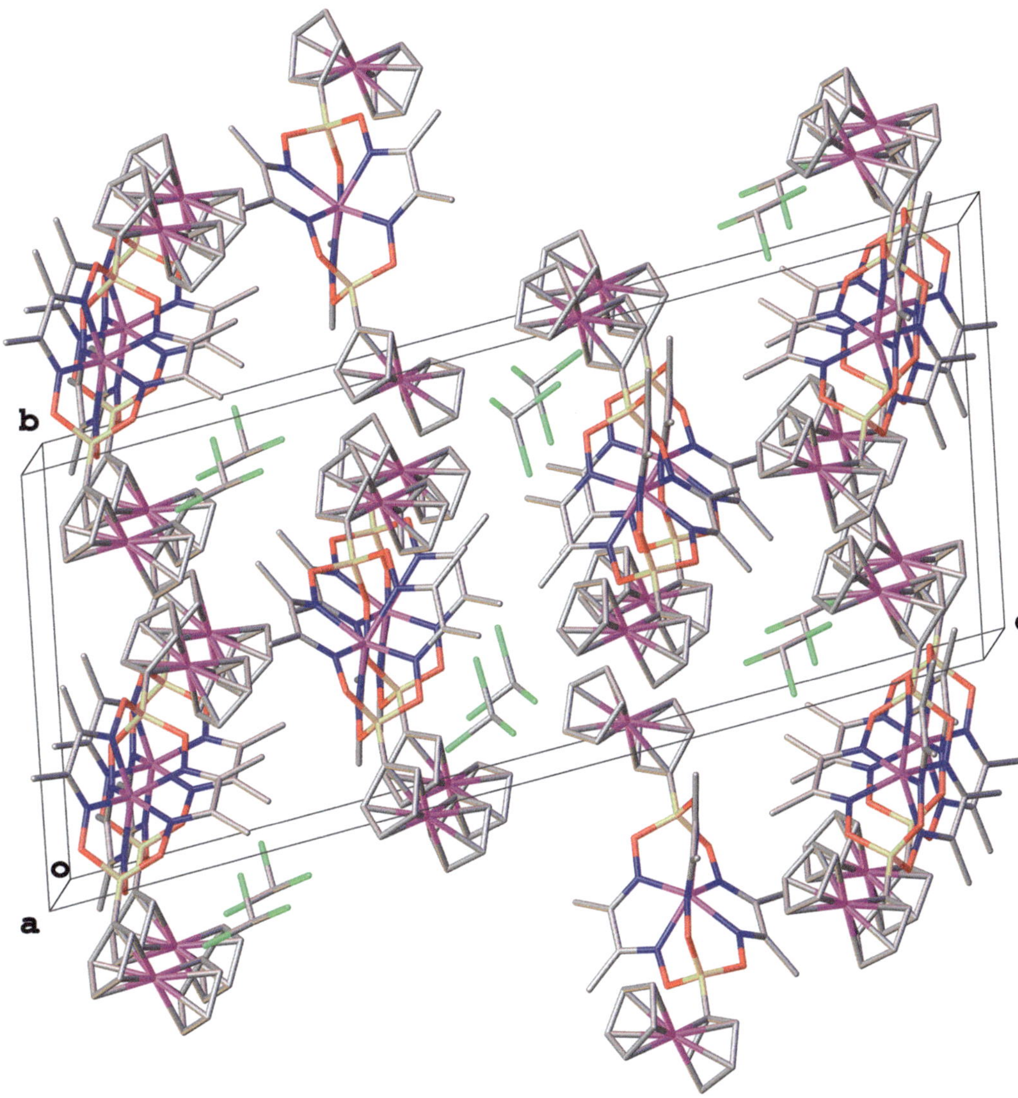

Fig. 3.20 Crystal packing of (Fe^{2+})@**198** · $CHCl_3$ [57]

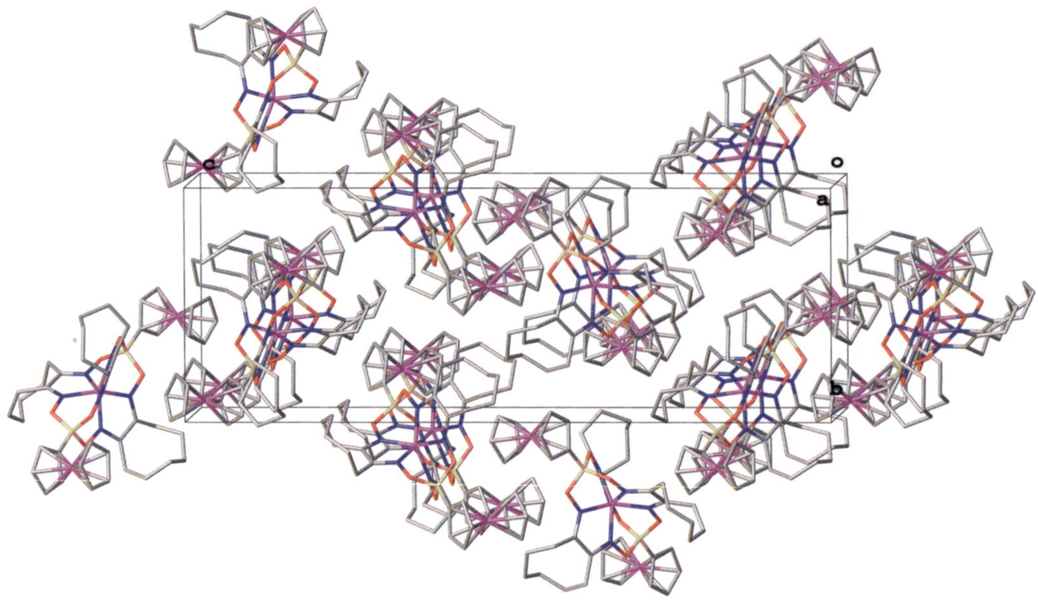

Fig. 3.21 Crystal packing of (Fe^{2+})@**200** [57]

substituents and through the Cp–H. . .O interactions as well. In the crystal (Fe^{2+})@**198** · $CHCl_3$, each of these layers contains the clathrochelate molecules of the same crystallographic type (**198**A or **198**B) with their B. . .Fe. . .B molecular C_3 pseudoaxes lying in the planes of these layers. The same intermolecular C–H. . .Cp contacts, formed by ferrocenyl substituents and cyclooctane carbocycles, as well as the Cp–H. . .O interactions have been also observed in the crystal (Fe^{2+})@**200** (Fig. 3.21). The structural lability of the cyclooctane ribbed substituents of **200** allowed the molecules of its iron(II) cage complex to arrange according to the "bump-into-hollow" principle through the attractive H. . .H interactions between such substituents of the neighboring macrobicyclic molecules of (Fe^{2+})@**200** [57].

The X-rayed crystal of the clathrochelate (Fe^{2+})@**327** [58] contains no cavities suitable for inclusion of the solvent molecules. The shortened contacts between the hydrogen atoms of the phenyl substituent of its α-benzildioximate chelate fragments and the fluorine atoms of the cross-linking O_3BF groups of the neighboring clathrochelate entities, shown in Fig. 3.22, have been observed; the latter contacts have been considered [58] as weak C–H. . .F hydrogen bonds.

The solvate water molecule in the crystal (Fe^{2+})@**337** · $2(CH_3)_2CO$ · $0.5H_2O$ [59] forms the hydrogen bonds with both the heterocyclic substituents at a cage framework (Fig. 3.23). Such hydrogen bonds are rather weak, but they governed the mutual orientation of these imidazolyl substituents. The solvate acetone molecules were found to occupy the vacant cavities of this crystal close to the apical fluorine atom of (Fe^{2+})@**337**. Such solvate molecule forms the weak contacts both with the apical fragment and with the imidazolyl substituent of (Fe^{2+})@**337** [59].

The X-ray structures of triethylammonium and tetra-*n*-butylammonium salts of the clathrochelate anion (Fe^{2+})@**338** (Fig. 3.24) are described [60]. Their FeN_6-coordination polyhedra with heights h of approximately 2.31 Å possess a distorted TP–TAP geometry with the average distortion angles φ of 25.3° and of 26.8°, respectively; the Fe–N distances fall in the range 1.892–1.944 Å, and the greatest Fe–N bond lengths have been observed for their OClGm chelate cycles. Triethylammonium counterion forms a short hydrogen bond with

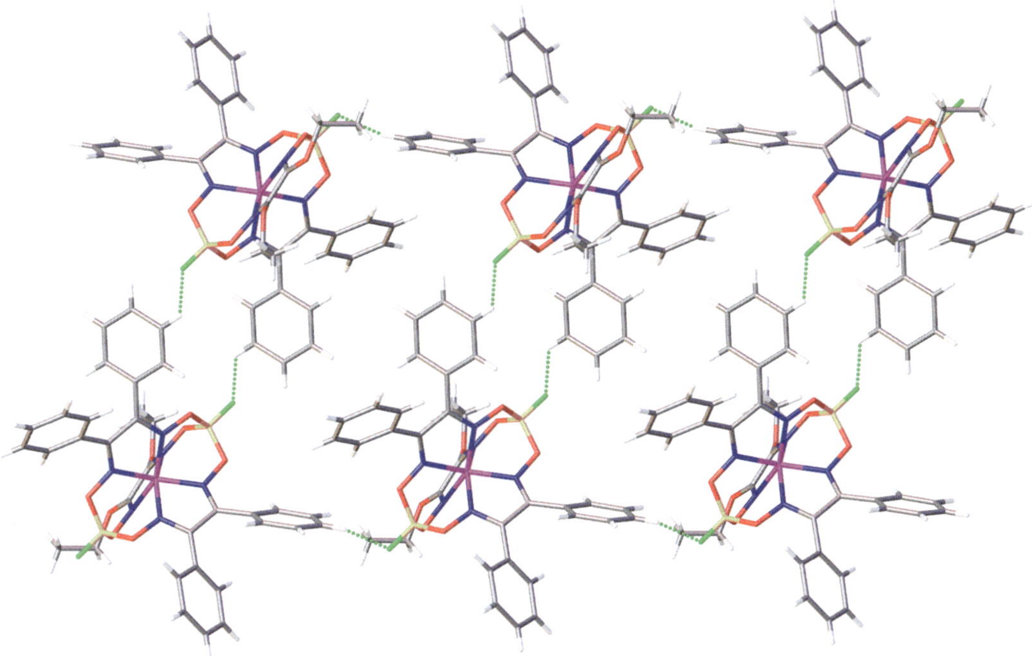

Fig. 3.22 Intermolecular C–H...F interactions in the crystal (Fe^{2+})@**327**; only hydrogen atoms involved in such contacts are shown [58]

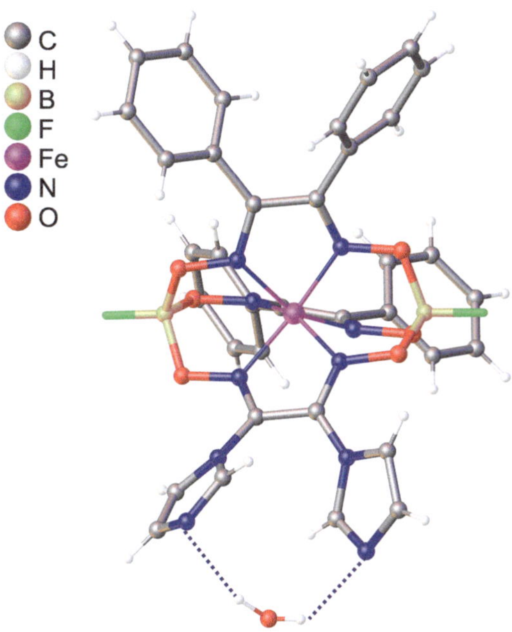

Fig. 3.23 Mutual orientation of the imidazolyl substituents in a *vic*-position of the same chelate fragment of (Fe^{2+})@**337** and their hydrogen bonding with solvate water molecule [59]

the oxygen atom of the deprotonated hydroxyl group of this macrobicyclic anion (Fig. 3.24). The shortened C–O distance in the OClGm ribbed fragment, approaching to the value characteristic of the double C=O bond, is most intriguing structural feature of (Fe^{2+})@**338**: such C–O bond has an essential double character, that is explained [60] by strong delocalization of the negative charge from its inherent oxygen atom into a quasiaromatic cage framework, thus causing a domination of the resonance structure **b** (Scheme 3.6). Delocalization of this negative charge on the nitrogen atom of the corresponding donor group is described [60] to be stabilized by positively charged encapsulated iron(II) ion. Predominance of the resonance structure **b** also caused an elongation of the C=N bond in the OC=N fragment up to 1.322 and to 1.332 Å for triethylammonium and tetra-*n*-butylammonium salts of (Fe^{2+})@**338**, respectively, as compared with the same bonds in their chlorooximate donor groups (1.290 Å and 1.302 Å, respectively).

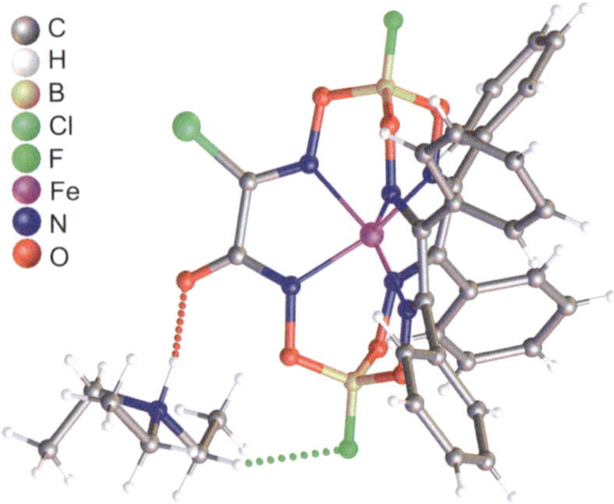

Fig. 3.24 X-ray structure of the clathrochelate anion and triethylammonium cation in the crystal $((C_2H_5)_3NH) \cdot (Fe^{2+})$@**338** [60]

Comparison of the X-ray structures of the monoribbed-functionalized iron(II) clathrochelates (Fe^{2+})@**466**, (Fe^{2+})@**471**, and (Fe^{2+})@**469** has been performed [61]. The distortion angles φ of their TP–TAP FeN_6-coordination polyhedra vary from 25.3 to 26.9°, while the bite (chelate) angles α fall in the range 78.2–79.0°. As a result, their heights h persist (approximately 2.31 Å). At the same time, the nonbonding interactions of the functionalizing ribbed substituents at a cage framework were found [61] to affect the molecular geometry of these clathrochelates. Sterical clashes between the bulky *n*-butyl and crown ether substituents in a *vic*-position of the same ribbed fragment in the molecules (Fe^{2+})@**471** and (Fe^{2+})@**469** resulted in an *anti*-orientation of these functionalizing groups, whereas in the molecule (Fe^{2+})@**466**, its *syn*-oriented *para*-$C_6H_5(COOCH_3)$ aromatic substituents, situated rather close to each other, were found to be an almost parallel because of stacking interactions between them [61]. These interactions also caused a substantial distortion of the corresponding functionalized ribbed fragment, whereas those in the molecules (Fe^{2+})@**471** and (Fe^{2+})@**469** are less distorted. The strong hydrogen bonding between two macrobicyclic entities (Fe^{2+})@**466** has been observed [61] at the distance of approximately 2.92 Å; this caused a formation of the corresponding clathrochelate dimers in its crystal (Fig. 3.25).

The X-ray structure of the dicyanopyrazine-functionalized iron(II) clathrochelate tris-dioximate (Fe^{2+})@**474** is discussed [62]. Its encapsulated iron(II) ion is slightly shifted from the center of a cage framework in the direction of the functionalized chelate fragment and the Fe–N distances for this fragment are smallest in a clathrochelate framework; its TP–TAP distortion is equal to 25.8°. The ribbed pyrazine substituent at this framework forms the intramolecular F. . .π contact with the fluorine atom of a neighboring clathrochelate molecule at the distance of approximately 1.47 Å (Fig. 3.26); the short intermolecular Cl. . .N contacts with $r_i = 3.17$ Å have been also observed [62] in its X-rayed crystal.

The detailed analysis of the X-ray structure of the monoribbed-functionalized tris-dioximate iron (II) clathrochelate (Fe^{2+})@**325** with two inherent amine substituents (Fig. 3.27a) has been performed [63]. The Fe–N distances in its molecule are substantially nonequivalent: they vary from 1.897 to 1.937 Å with the longest those for the amine-substituted donor groups. The average distortion angle φ of its TP–TAP coordination polyhedron is equal to 26.3°, while the height h is approximately 2.30 Å. The above functionalized donor oxime groups are flattened, thus indicating their substantial amide character [63]. FeN_6-coordination polyhedron of its derivatives (Fe^{2+})@**494** (Fig. 3.27b), resulted from an electrophilic addition of benzil to the above iron(II) diaminoclathrochelate, is very similar to that of the diaminomacrobicyclic precursor (Fe^{2+})@**325**. The annulated ribbed fragment of (Fe^{2+})@**494** contains two phenyl substituents at one carbon atom, while its second carbon atom forms a double C=O bond. Main bond lengths in this annulated heterocyclic fragment are characteristic of the π-conjugated systems, and its conjugation is broken at the sp^3-hybridizied carbon atom, thus resulting in the deviation from a planarity. The rigidity of a clathrochelate framework of (Fe^{2+})@**494** caused a flattening of the inherent ribbed piperazinone cyclic fragment of this annulated ribbed substituent [63].

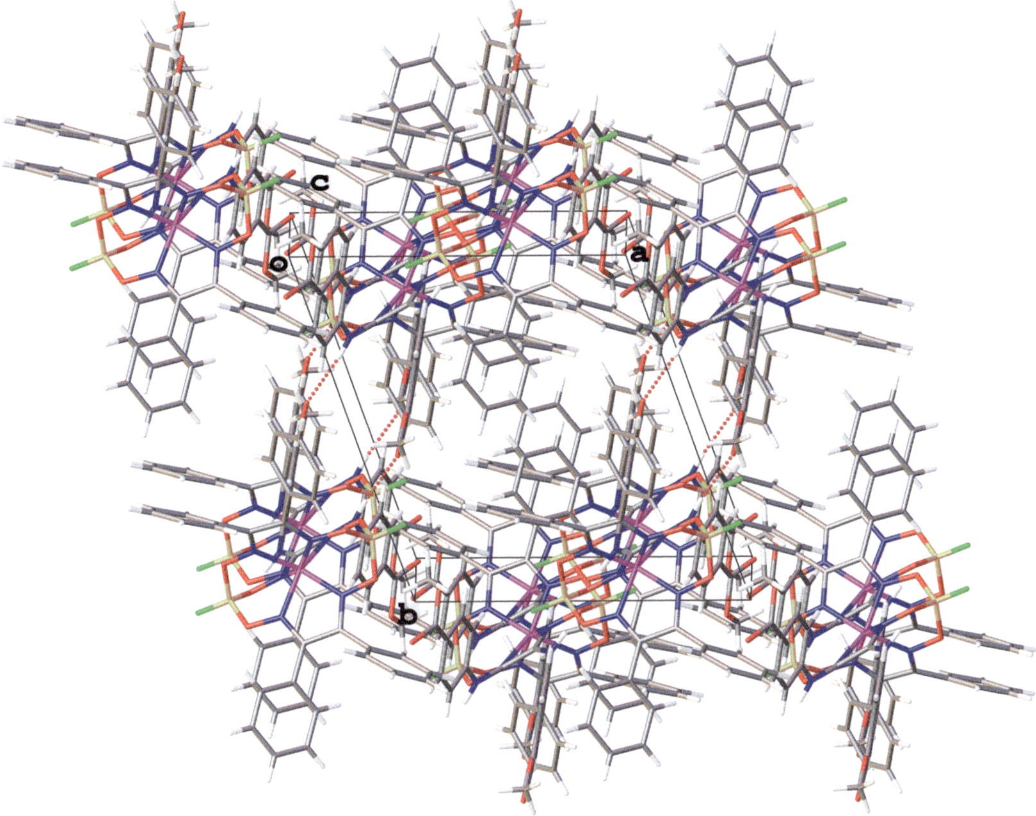

Fig. 3.25 Crystal packing of (Fe^{2+})@**466** [61]

The X-rayed crystal of the 1,4-dioxane-2-yl-substituted iron(II) tris-dioximate (Fe^{2+})@**455** [64] contains two independent clathrochelate molecules (types **A** and **B**) differing in a conformation of their functionalizing ribbed substituent and in a slight turn of the phenyl substituents in one of the α-benzildioximatate chelate fragments of their macrobicyclic ligands as well. The average distortion angels φ in the TP–TAP FeN_6-coordination polyhedra are equal to 24.5° and to 26.2°, respectively, while the average chelate angles α are almost the same (approximately 78.60°). In the above crystal, these clathrochelate molecules formed the chains shown in Fig. 3.28*via* the intermolecular Cl…N contacts of different types [64].

The detailed analysis of the electronic and molecular structures of the boron-capped metal tris-dioximates in their different oxidation and spin states has been performed [65]. The structural limitations of these clathrochelate ligands are caused by rigidity of the polyazomethine quasiaromatic frameworks governed by the energetics of their covalent bonds and bond angles. These macrobicyclic frameworks generally possess a D_3 symmetry with the molecular C_3 pseudoaxis lying through the capping boron atoms and with three C_2 pseudoaxes lying through the centers of the chelate C–C bonds. The changes of the torsion angles in the B–O–N=C and N=C–C=N moieties allowed a cage framework to undergo a rotational–translation elongation (shortening) around this pseudoaxis. The donor–acceptor interactions of an encapsulated metal ion with six donor nitrogen atoms of three π-conjugated α-dioximate fragments of a clathrochelate ligand also affect these structures [65]. Because the

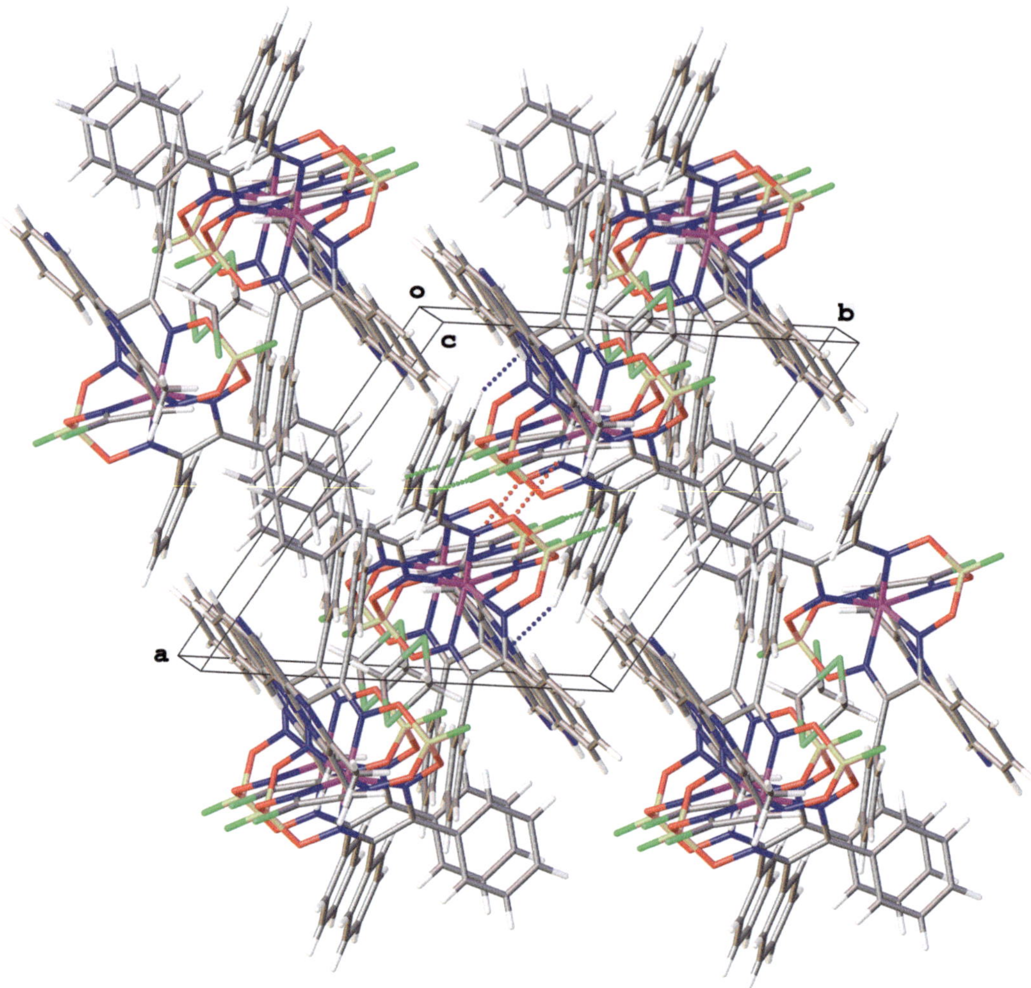

Fig. 3.26 Crystal packing of (Fe^{2+})@**474** · 0.5 CH_2Cl_2. The intermolecular F...π and Cl...N contacts are depicted with *dashed line* [62]

ribbed chelate cycles of MN_6-coordination polyhedron are formed by relatively rigid N=C–C=N fragments, the bite (chelate) angles α persist, while the heights h of these polyhedra undergo a variation with the distortion angles φ. On the one hand, a decrease in φ leads to an increase in h and permits one to encapsulate the relatively large metal ions (i.e., Co^+, Co^{2+}, and Ru^{2+} cations). In particular, in the cage complexes (Co^{2+})@**293**, (Co^{2+})@**295**, (Co^+)@**294**, and (Co^+)@**293** with encapsulated Co^{2+} cation or Co^+ ion, the angles φ are close to 0°. On the other hand, in the case of a given encapsulated metal ion (and, therefore, the same M–N distances) passing from the small capping boron atoms to the larger tin(IV)-, germanium(IV)-, and antimony(V) cross-linking ions, one can achieve a TAP geometry, which is energetically preferable for a given electronic configuration of an encapsulated metal ion. These geometrical changes are described [65] as the above elongation (shortening). In the case of the Co^{3+} and Fe^{2+} ions, when the size of a metal ion is smaller or close to that of the cavity of a hypothetical metal-free clathrochelate ligand, an increase in φ leads to a decrease in h. As a result, a macrobicyclic framework is able not only to adapt itself to a metal ion size, but also to use one more factor, such as an energetical preference

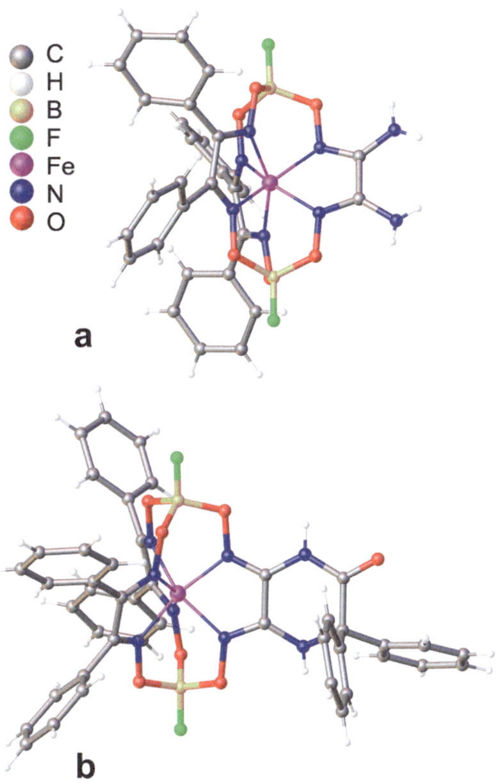

Fig. 3.27 X-ray structures of the clathrochelates (Fe^{2+})@**325** (**a**) and (Fe^{2+})@**494** (**b**) [63]

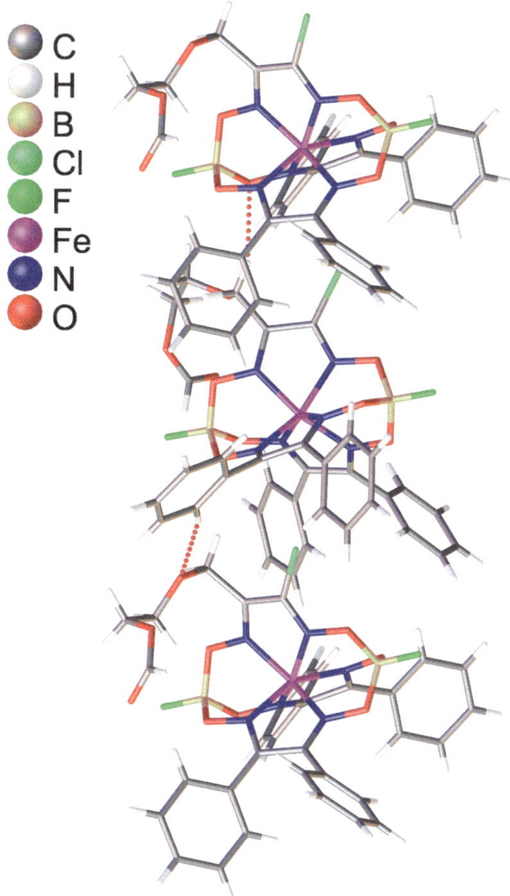

Fig. 3.28 Formation of a clathrochelate chain of the crystal (Fe^{2+})@**455**; the shortened contacts between its cage entities are depicted with *dashed line* [64]

of a TAP geometry over a TP one for the metal cations with electronic configurations d^6 (Fe^{2+}, Ru^{2+}, and Co^{3+}) and d^7 (Co^{2+}).

A series of the boron-capped tris-dioximates with encapsulated cobalt ion in its different oxidation ((Co(I)/Co(II) and Co(II)/Co(III) pairs) and spin (1/2 and 3/2 for Co^{2+} ion) states have been analyzed [65] to evaluate the relationship between the electronic configuration and spin state of this ion and the molecular geometry of its cage complexes. The X-ray structures of the cobalt, iron, and ruthenium(II) complexes of the same tris-dichloroglyoximate macrobicyclic ligand with electronic configurations $3d^7$, $3d^6$, and $4d^6$, respectively, have been also compared [65]. In the clathrochelates (Co^{2+})@**293**, (Co^{2+})@**295**, and (Co^{2+})@**312**, their encapsulated cobalt ion is in the same oxidation state 2+, but its spin states are different. In the crystals of (Co^{2+})@**293** and (Co^{2+})@**312** at 100 K, this ion is in the low-spin state ($s = 1/2$) as follows from the average Co–N distances, and the substantial variation of the Co–N distances (approximately 0.16 Å) resulted from the shift of an encapsulated metal ion in the direction of one of the three chelate dioximate fragments of their cage frameworks: the average Co–N distances for this fragment in both these molecules are equal to 1.89 Å, whereas for two other chelate cycles the Co–N bond lengths are in the range 2.04–2.05 Å. This fact is explained [65] by the structural Jahn–Teller distortion of their electronic configuration d^7. In order to estimate the possible role of disorder in these two cage molecules, the peculiarities of their thermal motion parameters have been analyzed: the corresponding ADPs have been used [65] to estimate an ordering of the molecular species in a crystal. These parameters for encapsulated cobalt(II) ion in the molecules (Co^{2+})@**293** and (Co^{2+})@**312** at 100 K evidenced their substantional anisotropy, and the maximal displacement has been observed along the line perpendicular to the C_2 axis; the longest Co–N bonds corresponded to a superposition of their two Jahn–Teller distorted structures. In the complexes (Co^{2+})@**295** · 3 C_6H_6 and (Co^{2+})@**295**· 3 $C_6H_5CH_3$ at the same temperature, the encapsulated cobalt(II) ion is in its high-spin state with $s = 3/2$. In these isostructural crystals, it occupies the center of CoN_6-coordination polyhedron, and the Co–N distances are markedly greater as those in (Co^{2+})@**293** and (Co^{2+})@**312**. No disorder has been observed [65] in the case of (Co^{2+})@**295**, thus suggesting almost isotropic thermal motion parameters of its Co^{2+} ion. The aromatic solvate molecules of the above crystals are situated between the ribbed fragments of (Co^{2+})@**295**, forming the shortened contacts with its chelate C–C bonds (Fig. 3.29); the distances between π-system of **295** and these solvate molecules are rather small, thus allowing their stacking interactions. These interactions significantly affected the φ values for the molecules (Co^{2+})@**295** in these crystals (approximately 0° and 7.8°, respectively).

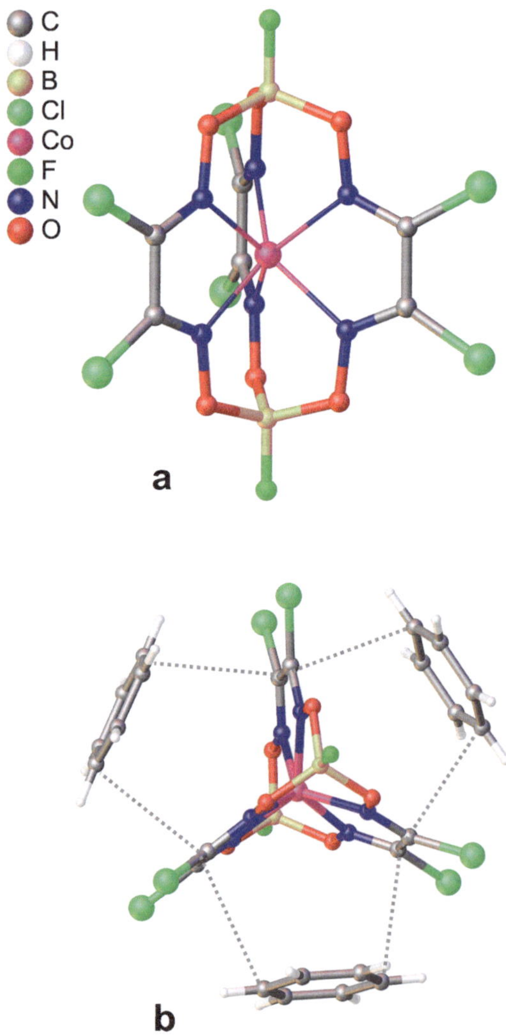

Fig. 3.29 X-ray structures of the clathrochelate (Co^{2+})@**295** (**a**) and its solvate with benzene (**b**) [65]

According to the magnetometry data of [65], the spin state of an encapsulated cobalt(II) ion in (Co^{2+})@**293** is temperature-dependent (see Chap. 4) and, therefore, the multitemperature X-ray diffraction study of its single crystal in the temperature range 30–295 K has been performed. Above 60 K, this ion demonstrated the substantial and almost constant ADPs (0.036–0.042 Å^2). Its X-rayed single crystal underwent the structural phase transition at approximately 40–50 K from a base-centered cell (space group $C2/c$) to a primitive one ($P2_1/c$). The lowering of the site symmetry of a cobalt ion from C_2 to C_1 upon this phase transition led to the ordering of the crystal structure and caused both the changes of the ADPs for this encapsulated ion and the variation of the Co–N distances. At 30 K, these ADPs are in the range 0.008–0.010 Å^2, and the greatest distances of this type have been observed [65] only for one of the three chelate α-dioximate fragments, while other bonds of this type are markedly shorter. The ordering of the crystal (Co^{2+})@**293** at 30 K is evidenced that only one of its Jahn–Teller distorted structure is most energetically preferable. The discrimination of one of the three equivalent structures is explained [65] by the crystal packing effects. The crystal packing of (Co^{2+})@**293** at 30 K (Fig. 3.30) and that above 60 K (Fig. 3.31) is slightly different. From 30 to 60 K, each molecule of (Co^{2+})@**293** formed two pairs of the intermolecular Cl. . .O contacts. As a result, the formation of the infinite supramolecular layers has been observed. Both the "infinite layers" and the clathrochelate dimers were not found in this crystal above 60 K. So, the above phase transition is explained [65] by formation of the additional Cl. . .O contacts at low temperatures.

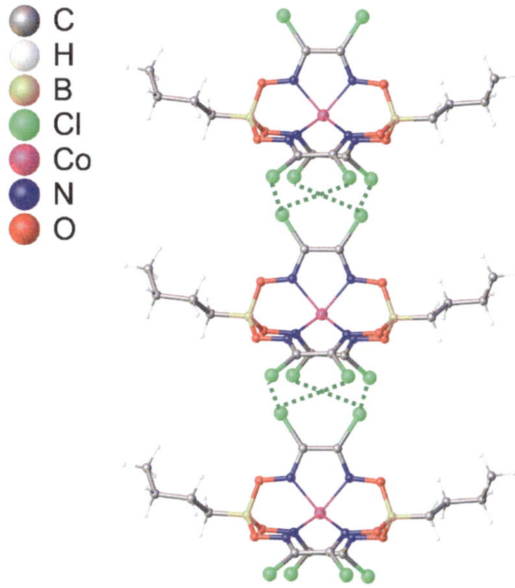

Fig. 3.31 Crystal packing of (Co^{2+})@**293** above 60 K [65]

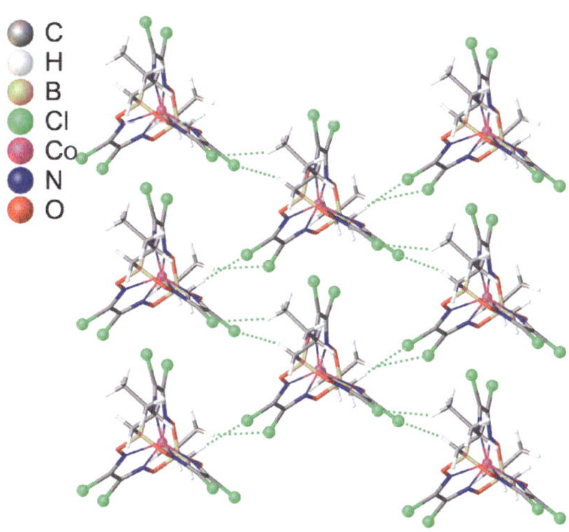

Fig. 3.30 Crystal packing of (Co^{2+})@**293** at 30 K [65]

A systematic increase in the average N–O and C–Cl distances has been observed passing from the cobalt(II) hexachloroclathrochelate (Co^{2+})@**293** to the cobalt(I) cage complexes (Co^{+})@**294** and (Co^{+})@**293**. As a result, the average N=C–C=N torsion angle decreases from 4.0° for the molecule (Co^{2+})@**293** to 3.1 and 1.0° for the complexes (Co^{+})@**294** and (Co^{+})@**293** with decrease in a TAP distortion of the coordination polyhedron, possessing its D_3 symmetry that approaches a higher D_{3h} symmetry in the latter cases. Indeed, the distortion angle φ of the cobalt(II) clathrochelate (Co^{2+})@**293** is equal to 7.8°, whereas those for its cobalt(I)-encapsulating analogs (Co^{+})@**294** and (Co^{+})@**293** are close to zero ($\varphi = 1.3°$). In contrast, their h values persist. So, the variation of the corresponding CoN_6-coordination polyhedron is described [65] not only as its rotational–translational elongation (shortening) along the molecular C_3 pseudoaxis, but also as an expansion (compression) along its C_2 symmetry axes.

Single crystal X-ray data of [65] suggest that the capping cobalt(II) ion of the trinuclear $Co^{III}Co^{II}Co^{III}$ molecule (Co^{3+})$_2$@**425** is located in its inversion center, while the encapsulated cobalt(III) ions and their nearest environments are in the independent part of this molecule. The Co(III)–N distances are substantially shorter than the average Co(II)–N distance (1.971 Å) in its cobalt(II)-encapsulating analog (Co^{2+})@**312**. Due to non-equivalence of the capping fragments of (Co^{3+})$_2$@**425**, the coordination polyhedra of its encapsulated cobalt(III) ions possess a distorted truncated pyramidal geometry that approaches to a TAP ($\varphi = 40.2°$), whereas the geometry of (Co^{2+})@**312** is close to a TP ($\varphi = 14.5°$). As a result, the h value for (Co^{3+})$_2$@**425** is substantially smaller than that for (Co^{2+})@**312** [65].

The X-ray structures of the alicyclic and aromatic S_2-dioximate macrobicyclic cobalt and ruthenium(II) tris-dioximates (Co^{2+})@**404**, (Co^{2+})@**405** (two types **A** and **B**), and (Ru^{2+})@**405** have been compared [66] with those of their iron(II)-encapsulating analogs [67]. The CoN_6- and RuN_6-coordination polyhedra possess the geometry of a slightly distorted TP with the average distortion angles φ of 17.0°, 5.2°, and 6.3°, respectively. The pronounced discrepancies have been observed [67] between main geometrical characteristics of the above cobalt (II) complexes, on the one hand, and those of the ruthenium(II) clathrochelate, on the other hand. The average Co–N and Ru–N distances (1.96 Å and 1.97 Å) are close, although the Co–N bond lengths vary in a rather wide range: the encapsulated cobalt(II) ion is shifted from the centers of the cavities of their macrobicyclic ligands in the direction of one of the three chelate α-dioximate fragments by 0.14, 0.16, and 0.21 Å for (Co^{2+})@**404** and (Co^{2+})@**405** (types **A** and **B**), respectively, while the encapsulated Ru^{2+} ion in the molecule (Ru^{2+})@**405** occupies the center of its cavity. The average Co–N distances for the above ribbed fragment in both the molecules (Co^{2+})@**404** and (Co^{2+})@**405** are equal to 1.88 Å, whereas the average Co–N bond lengths for two other chelate cycles are approximately 1.92 and 2.10 Å. This fact is explained [66] by the Jahn–Teller structural distortion of these low-spin cobalt(II) complexes. The heights h of their TP coordination polyhedra persist, whereas the distortion angles φ are affected by the nature of the ribbed substituents: the molecules (Co^{2+})@**405** and (Ru^{2+})@**405** have bulky dithia-containing aromatic ribbed substituents, and the optimization of intramolecular sterical clashes between them led to decrease in a TAP distortion as compared with that for the alicyclic molecule (Co^{2+})@**404** having the conformationally labile α,α'-S_2-cyclohexane ribbed substituents. The bond delocalization in the case of the highly conjugated macrobicyclic frameworks of these cage complexes is described [67] to affect also the exocyclic C–S bonds in the ribbed N=C–S fragments.

The geometry of the FeN_6-coordination polyhedra of the iron(II)-encapsulating analogs (Fe^{2+})@**405** and (Fe^{2+})@**405** · 0.3 CH_2Cl_2 [67] is similar to those of the CoN_6- and RuN_6-polyhedra, but the former ones are more

TAP-distorted; therefore, their heights *h* are markedly smaller because of the smaller Fe–N distances. The encapsulated iron(II) ion, like the Ru^{2+} cation and unlike the Co^{2+} ion, occupies the centers of the corresponding FeN_6-coordination polyhedra [67]. The bite angles α in the above iron(II) clathrochelates are close to each other and are less acute than those in the case of the analogous cobalt and ruthenium(II) complexes. So, all the cage metal complexes formed by the same macrobicyclic ligand have similar sterical clashes, and main geometrical parameters of their coordination polyhedra vary with Shannon radius of an encapsulated metal ion. The bulky and rigid ribbed substituents caused a decrease in a TP–TAP distortion of MN_6-coordination polyhedron and an increase in its height *h*. The same structural changes of MN_6-coordination polyhedron have been observed [66] for an encapsulated metal ion having a smaller size (Shannon radius). In the crystals of the *ortho*-phenylenedisulfide clathrochelates (Co^{2+})@**405**, (Ru^{2+})@**405,** and (Fe^{2+})@**405**, their ribbed chelate fragments have different conformations. One of the three *ortho*-phenylenedisulfide fragments of the cobalt- and ruthenium-encapsulating clathrochelates is flattened, whereas two others are bent. In the molecule (Fe^{2+})@**405**, two of its three *ortho*-phenylenedisulfide fragments are relatively flattened, whereas the third moiety of this type is markedly bent. In the crystals of the above complexes (M^{2+})@**405**, weak stacking interactions between the flattened and parallel *ortho*-phenylenedisulfide ribbed fragments has been observed [66]. Such interactions involved only their C_2S_2–C_6H_4 moieties (Figs. 3.32 and 3.33), whereas the markedly bent *ortho*-phenylenedisulfide fragments of the above cage molecules do not form any stacking interactions [66].

The X-ray structures of the heterodihalogenoclathrochelate (Fe^{2+})@**475** and its homodihalogenoclathrochelate analog (Fe^{2+})@**309** and (Fe^{2+})@**302** have been compared [68]. The distortion angles φ of their intermediate TP–TAP FeN_6-coordination polyhedra are approximately 24.7°, 23.6°, and 24.9°, respectively.

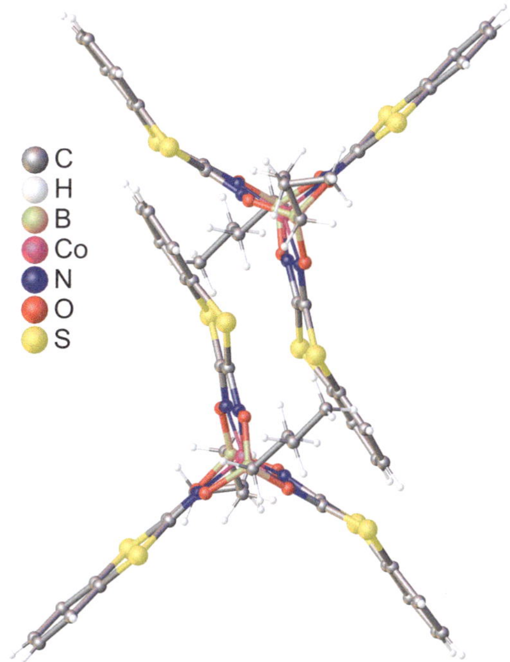

Fig. 3.32 Overlapping between the aromatic ribbed fragments of the clathrochelate molecules (Co^{2+})@**405** [66]

The encapsulated iron(II) ion in the molecules (Fe^{2+})@**475** and (Fe^{2+})@**309** is located in the geometrical center of these polyhedra, whereas in the molecule (Fe^{2+})@**302** this ion is slightly shifted toward its two α-benzildioximate chelate fragments. The infinite chains of the clathrochelate molecules in the crystal (Fe^{2+})@**475** · CH_2Cl_2 are formed through the intermolecular C–H…O contacts at the distance of approximately 2.42 Å (Fig. 3.34). Each solvate dichloromethane molecule is involved in I…Cl halogen bonding (r_i = 3.32 Å) with one of the macrobicyclic entities as well as in Cl…H and C–H…π interactions with other clathrochelate molecule at the distances of approximately 2.89 Å. In the crystal (Fe^{2+})@**475** · 2 C_6H_6, its cage entities form the centrosymmetric dimers (Fig. 3.35) *via* the I(1)…F(1) halogen bonds at the distance of approximately 2.94 Å; the electron lone pair of the above fluorine atom is pointed in the direction of the vacant σ*-orbital that is localized on the I–C bond of a macrobicyclic

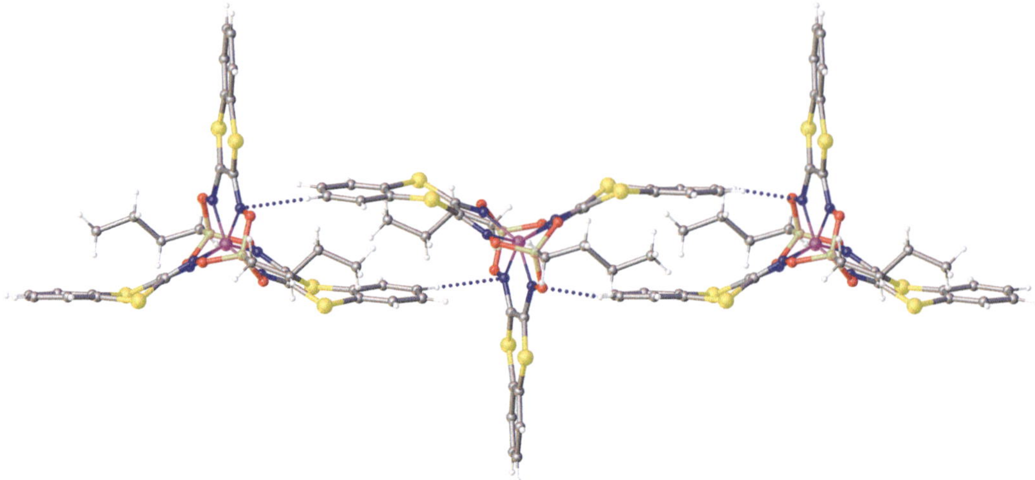

Fig. 3.33 Overlapping between the aromatic ribbed fragments of the clathrochelate molecules (Fe^{2+})@**405** [67]

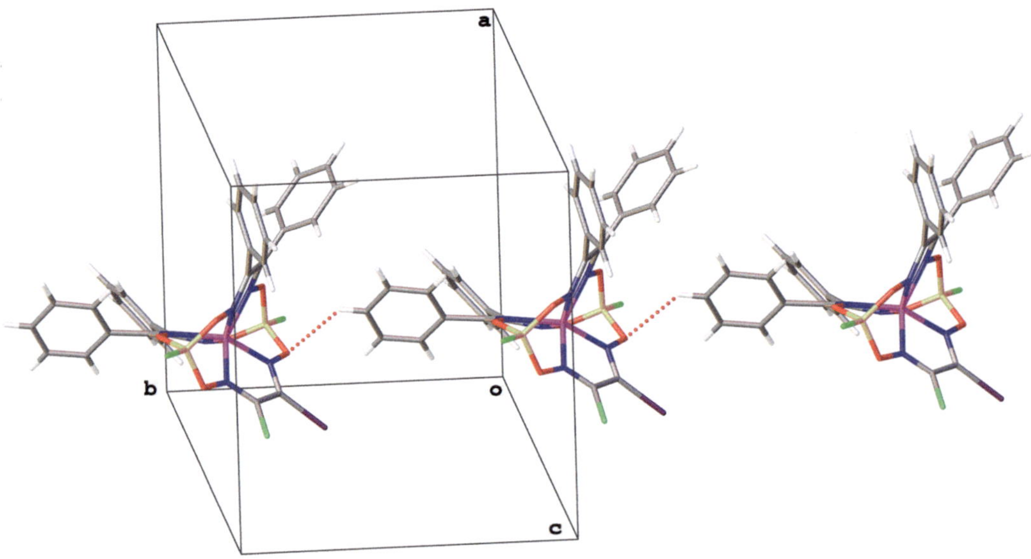

Fig. 3.34 Crystal packing of (Fe^{2+})@**475** · CH_2Cl_2 [68]

ligand of the neighboring clathrochelate molecule [69].

The X-rayed crystal of the mononitrile iron(II) clathrochelate (Fe^{2+})@**480** [69] contains in its asymmetric unit two types (**A** and **B**) of the cage molecules and five solvate benzene molecules as well; in one of the former macrobicyclic entities (type **A**), its single nitrile substituent is disordered. The FeN_6-coordination polyhedra of **A** and **B** with the angles φ and α equal to 24.1, 22.6° and 79.2, 78.4°, respectively, possess the geometry intermediate between a TP and a TAP. The heights h of these TP–TAP polyhedra are the same (2.33 Å). The Fe–N distances for the functionalized chelate fragments of (Fe^{2+})@**480**A and (Fe^{2+})@**480**B (*av*. 1.87 Å) are markedly smaller than those for their α-benzildioximate ribbed moieties (*av*.

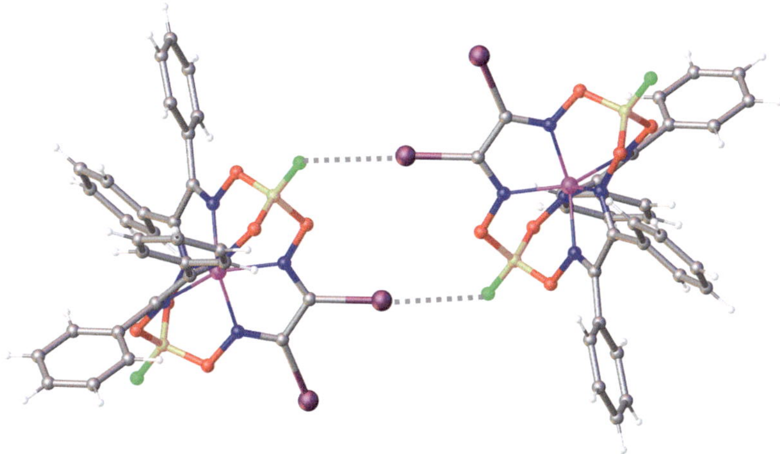

Fig. 3.35 Formation of a clathrochelate dimer in the crystal (Fe^{2+})@**309** · 2 C_6H_6; the F…I halogen bonds are shown with *dashed line* [69]

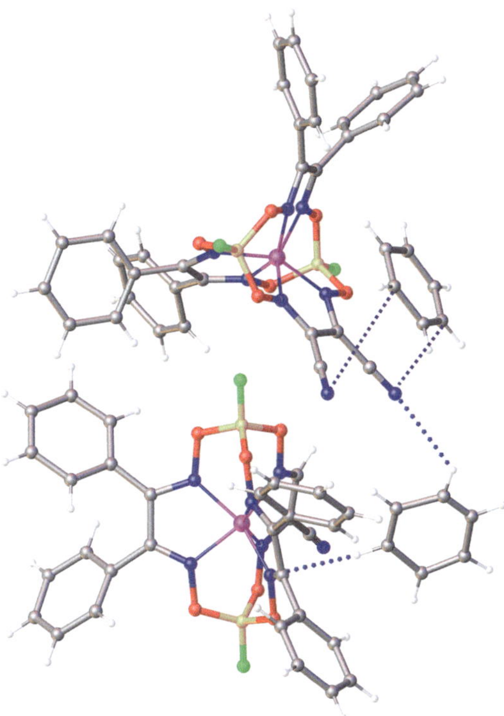

Fig. 3.36 Formation of a clathrochelate dimer by the molecules of (Fe^{2+})@**480**A; their intermolecular contacts are depicted with *dashed line* [69]

1.92 Å) [69]. The intermolecular interactions of the molecule (Fe^{2+})@**480**A in the crystal (Fe^{2+})@**480** · 2.5 C_6H_6 (Fig. 3.36) are different from those for the macrobicycle (Fe^{2+})@**480**B (Fig. 3.37): the former complex forms a series of the shortened intermolecular contacts at the distances less than 3.5 Å. In particular, its chelate fragment N=C(H)–C(CN)=N is involved in stacking interaction with one of the solvate benzene molecules (the intermolecular C…C distance is approximately 3.27 Å), in two shortened N…C contacts (r_i = 3.39 and 3.37 Å) with the second molecule of this type as well as in two N…C contacts with phenyl substituent of the chelate α-benzildioximate fragment of the neighboring molecule (Fe^{2+})@**480**B (r_i = 3.25 and 3.26 Å), and in intermolecular C…F interaction with the apical fluorine atom of this clathrochelate (r_i = 3.02 Å). The intermolecular interactions of the complex (Fe^{2+})@**480**B in this crystal are less pronounced: its functionalized ribbed fragment is involved only in the above intermolecular interactions with the neighboring molecule of (Fe^{2+})@**480**A [69].

The X-ray structure of the hybrid carboranyloclathrochelate (Fe^{2+})@**483** is described [70]. Presence of the bulky ribbed carboranyl substituent in its molecule caused an additional rotation–contraction distortion of a rigid TP–TAP macrobicyclic framework around the molecular B…Fe…B C_3 pseudoaxis. Such inherent substituent at a cage framework forms the strong intramolecular C(carborane)–H…O

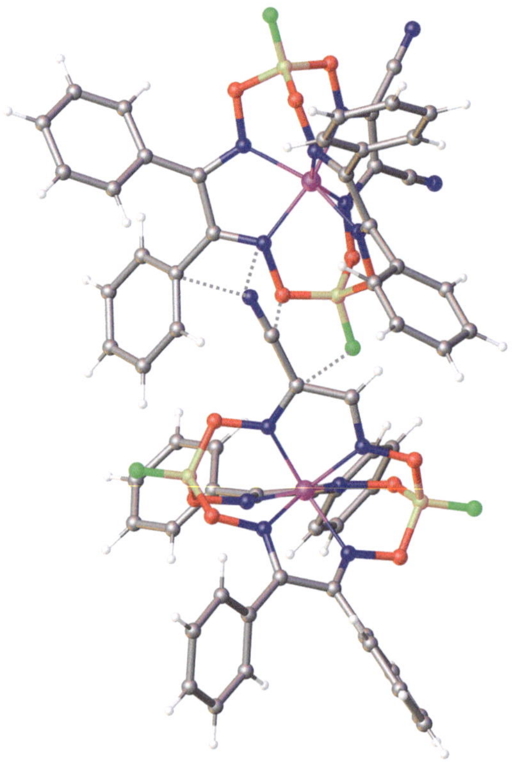

Fig. 3.37 Formation of a clathrochelate dimer by the molecules of (Fe^{2+})@**480**B; their intermolecular contacts are depicted with *dashed line* [69]

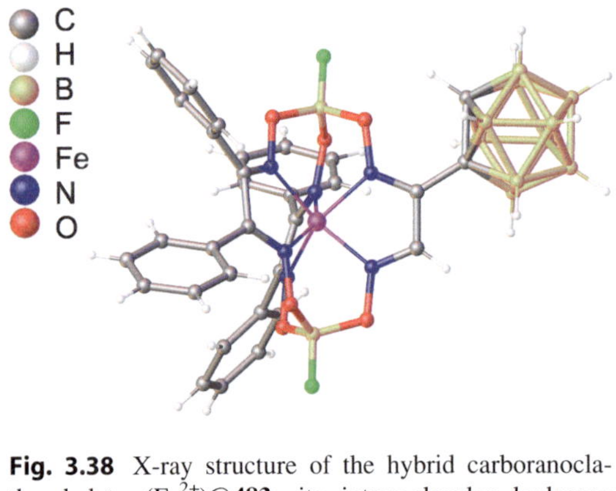

Fig. 3.38 X-ray structure of the hybrid carboranoclathrochelate (Fe^{2+})@**483**; its intramolecular hydrogen bond is depicted with *dashed line* [70]

hydrogen bond at the distance of approximately 2.77 Å (Fig. 3.38) with a formation of the six-membered cyclic moiety [70].

The X-ray structures of four C–C conjugated iron(II) bis-clathrochelates $(Fe^{2+})_2$@**441**, $(Fe^{2+})_2$@**446**, $(Fe^{2+})_2$@**484,** and $(Fe^{2+})_2$@**485** and their dibromomonoclathrochelate precursor (Fe^{2+})@**310** (Scheme 3.7) are reported [71, 72]; the example of molecular structure of the iron(II) dimethine bis-clathrochelate $(Fe^{2+})_2$@**485** is shown in Fig. 3.39. The average Fe–N distances in these cage complexes fall in the narrow range 1.90–1.92 Å, characteristic of the low-spin iron(II) complexes with *N*-donor ligands, while their FeN_6-polyhedra possess a distorted TP–TAP geometry with the distortion angles φ from 24.3 to 27.8°. The encapsulated iron(II) ion is situated almost in the centers of these polyhedra with heights h in the range 2.24–2.32 Å and the bite angles α of approximately 78°. The chelate C–C bond lengths in their functionalized ribbed fragments (1.47–1.49 Å) are greater than those in the α-benzildioximate chelate moieties (1.43–1.45 Å), while the C–C distances between their two conjugated clathrochelate entities (1.46–1.47 Å) are close to the average value for the single $C(sp^2)$–$C(sp^2)$ bond (1.45 Å). The torsion C=C–C=C angle ψ between these C–C conjugated macrobicyclic fragments decreases from the bis-clathrochelates $(Fe^{2+})_2$@**485** (160°) and $(Fe^{2+})_2$@**441** (122°) to their iodo- and piperidinyl-containing analogs $(Fe^{2+})_2$@**484** (88°) and $(Fe^{2+})_2$@**446** (53°). So, the latter bis-clathrochelate with most bulky ribbed substituents showed the lowest ψ value. This fact is explained [71] by sterical clashes between these substituents in their *chair*-conformations with nitrogen lone pairs directed toward each other. Overlapping of one of the two macrobicyclic fragments of the bis-clathrochelates $(Fe^{2+})_2$@**485** and $(Fe^{2+})_2$@**446** makes the C_3 pseudoaxes of their cage entities to be almost perpendicular [71].

Crystal packing of these rigid polyazomethine iron(II) bis-clathrochelates allowed the formation of vacant cavities suitable for inclusion of the solvate molecules. Indeed, their X-rayed crystals contain up to six solvent guests, forming

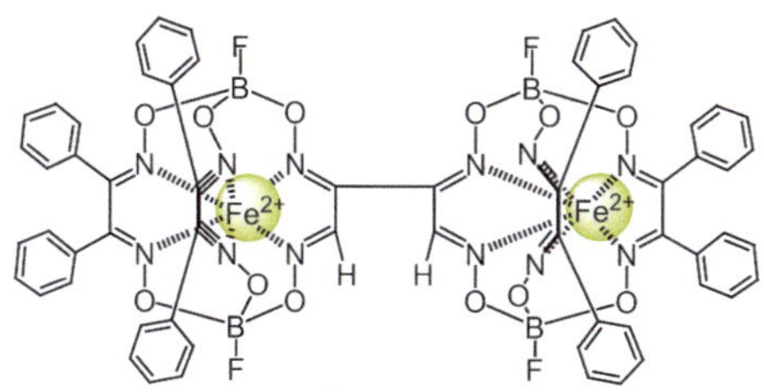

$(Fe^{2+})_2$@**485**

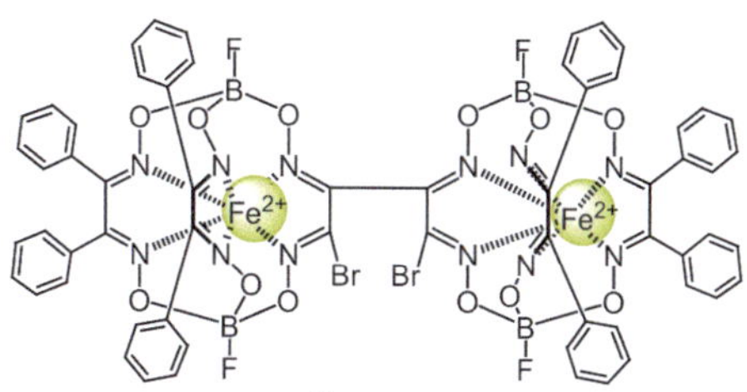

$(Fe^{2+})_2$@**441**

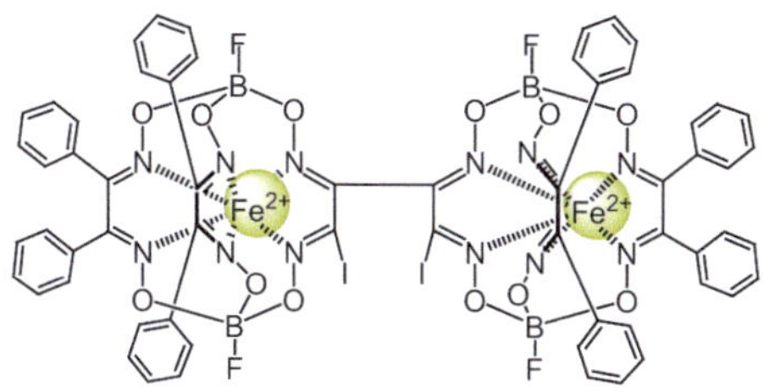

$(Fe^{2+})_2$@**484**

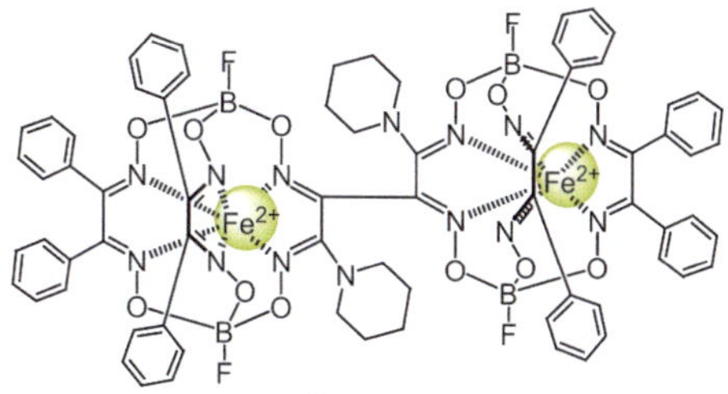

$(Fe^{2+})_2$@**446**

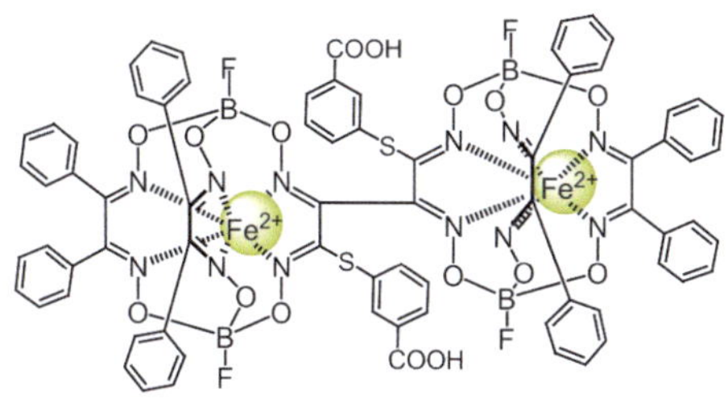

$(Fe^{2+})_2$@**445**

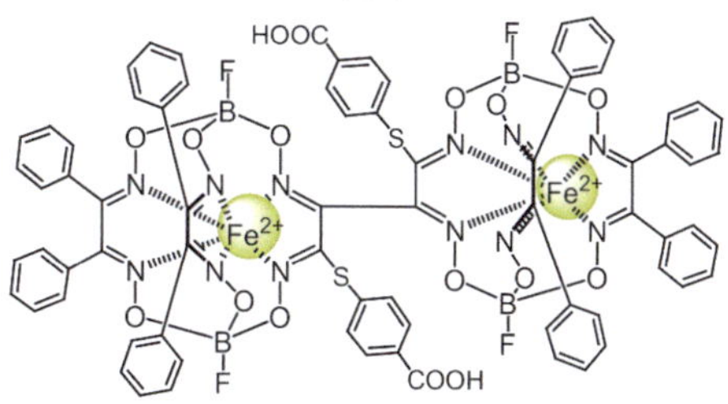

$(Fe^{2+})_2$@**444**

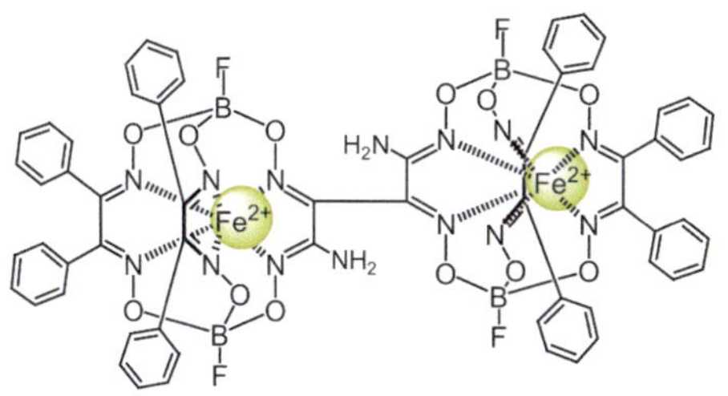

$(Fe^{2+})_2$@**443**

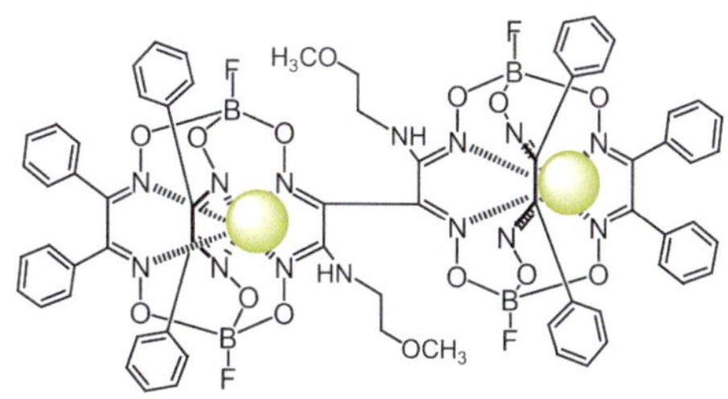

$(Fe^{2+})_2$@**442**

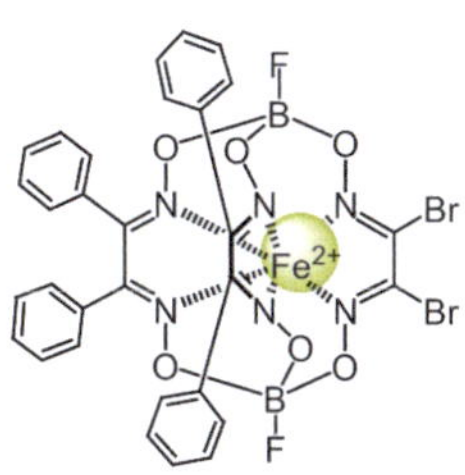

(Fe^{2+})@**310**

Scheme 3.7 Chemical drawings of a series of the iron(II) bis-clathrochelates and their dibromoclathrochelate precursor

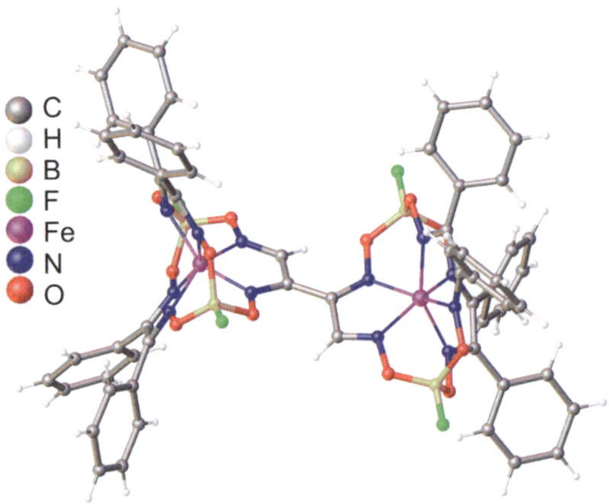

Fig. 3.39 X-ray structure of the bis-clathrochelate $(Fe^{2+})_2$@**485** [71]

the systems of intermolecular van der Waals π. . .π, C–H. . .π and C–H. . .X interactions with the bis-clathrochelate molecules; the supramolecular architecture of these crystals have been thoroughly analyzed [72] using the Hirshfeld surfaces and their 2D projections. NMR method has been used for characterization of the conformational behavior of the above bis-cage compounds in their solutions. For the bis-clathrochelates $(Fe^{2+})_2$@**442**, $(Fe^{2+})_2$@**443**, $(Fe^{2+})_2$@**444**, $Fe^{2+})_2$@**445**, and $(Fe^{2+})_2$@**446** (Scheme 3.7), four types of the ribbed phenyl groups have been observed in their ^{13}C NMR spectra. This number of signals corresponds to the symmetry of a system in which there is no additional mirror plane, as determined by the torsion angle ψ close to 0 or 180° (Scheme 3.8); the level of such induced non-equivalence depends on its value in a nonlinear way. In contrast, the spectrum of the dimethine bis-clathrochelate $(Fe^{2+})_2$@**485** contains half as much signals in the aromatic region of its ^{13}C NMR spectrum; this result is explained [72] by higher symmetry of the molecule $(Fe^{2+})_2$@**485** with ψ close to 0 or 180°, thus being in good agreement with the experimental X-ray value in its crystal (approximately 161°; a residual non-equivalence is small). This conformation is possible in the case of the above dimethine bis-cage complex, because the large ribbed substituents in the adjacent positions not only prohibit a free rotation around the single C–C bond between two macrobicyclic entities but also prevent the conformations with ψ close to 0 or 180° to be populated due to the intramolecular sterical clashes [72].

The X-ray structures of two types (**A** and **B**) of the dithiopheneboron-capped cobalt (II) tris-α-benzildioximate (Co^{2+})@**207** and that of its reduced cobalt(I)-encapsulating analog (Co^{+})@**207** are described [73]. The molecules (Co^{2+})@**207**A and (Co^{2+})@**207**B possess a distorted TP geometry with two Co–N distances which are substantially different from those for other four Co–N bonds due to the structural Jahn–Teller distortion of electronic configuration d^7. The cobalt(I)-encapsulating clathrochelate monoanion (Co^{+})@**207** (Fig. 3.40) also possesses a TP geometry, but with narrowly distributed Co–N distances (*av.* 1.994 Å), which are substantially increased as compared with those for its cobalt (II)-containing clathrochelate precursor (*av.* 1.962 Å). This result is explained [73] by higher electronic population of the antibonding MO in the clathrochelate anion (Co^{+})@**207**, while other geometrical parameters of a cage framework persist upon such $Co^{2+/+}$ reduction.

Structural peculiarities of a series of the homologous iron(II)-encapsulating clathrochelates (Fe^{2+})@**666**, (Fe^{2+})@**667**, and (Fe^{2+})@**296** with one ribbed glyoximate, methylglyoximate, or dimethylglyoximate chelate fragment, the molecules of which are also formed by two α-benzildioximate ligand synthons (Scheme 3.9), have been analyzed [74]. An unit cell of the crystal (Fe^{2+})@**666** · 1.5 $CHCl_3$ contains two types (**A** and **B**) of its clathrochelate molecules. The observed higher TP distortion of FeN_6-coordination polyhedron of the molecule (Fe^{2+})@**666**B as compared with that of (Fe^{2+})@**666**A is explained [74] by difference in their intermolecular interactions in this crystal. The C–C distances in the glyoximate and

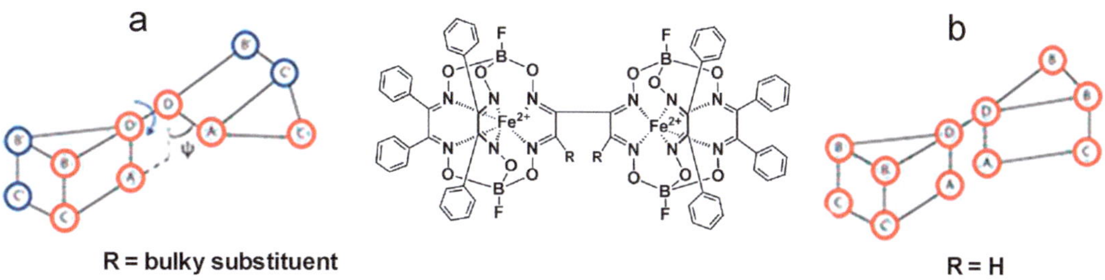

Scheme 3.8 Lowering in a symmetry of the C–C-conjugated iron(II) bis-clathrochelates, induced by nonzero value of the torsion angle ψ (**a**) and the corresponding symmetric geometry with ψ = 0° (**b**) [72]

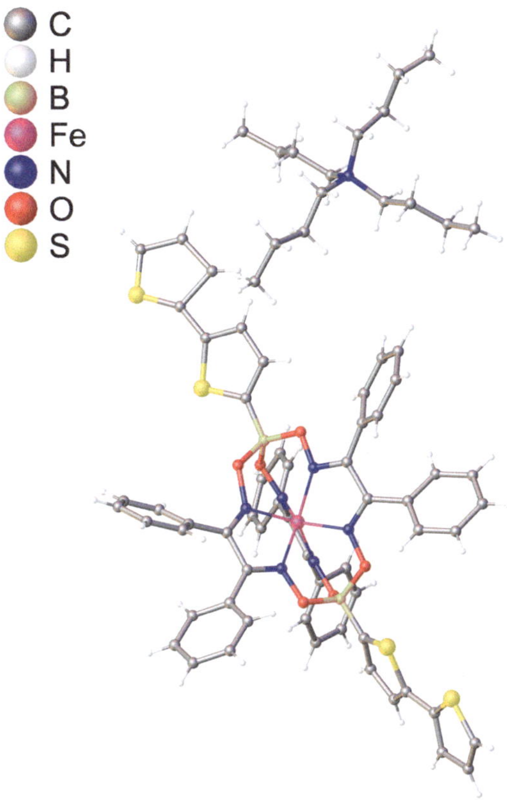

Fig. 3.40 X-ray structure of the cobalt(I) clathrochelate $((n\text{-}C_4H_9)_4N^+)[(Co^+)@\mathbf{207}]^-$ [73]

methylglyoximate ribbed fragments of the molecules $(Fe^{2+})@\mathbf{666}$ and $(Fe^{2+})@\mathbf{667}$ (*av.* 1.414 Å) are substantially shortened as compared with those in the α-benzildioximate and dimethylglyoximate chelate cycles (approximately 1.45 Å). The crystals $(Fe^{2+})@\mathbf{666} \cdot 1.5\ CHCl_3$ and $(Fe^{2+})@\mathbf{296} \cdot CH_2Cl_2$ have the laminated structures, the chains of the clathrochelate molecules of which are differently oriented relative to their crystallographic axes. In the crystals $(Fe^{2+})@\mathbf{667} \cdot CH_2Cl_2$ and $(Fe^{2+})@\mathbf{296} \cdot 0.5\ C_6H_6 \cdot 0.5\ iso\text{-}C_8H_{18}$, their solvate molecules occupy the vacant cavities between the clathrochelate dimers of $(Fe^{2+})@\mathbf{667}$ (Fig. 3.41) and between the macrobicyclic molecules of $(Fe^{2+})@\mathbf{296}$ (Fig. 3.42) [74].

The X-ray structures of a series of the alkylboron-capped iron and cobalt (II) clathrochelates are reported [75]. Their molecular geometries are governed by both the nature of an encapsulated metal ion and that of the ribbed substituents. Due to the rigidity of the chelate N=C–C=N fragments, their bite angles persist, whereas the heights *h* of the MN_6-coordination polyhedra were found to be affected by distortion of a clathrochelate framework around the molecular C_3 pseudoaxis. The geometry of the FeN_6-coordination polyhedra of the clathrochelates $(Fe^{2+})@\mathbf{230}$ and $(Fe^{2+})@\mathbf{234}$ is intermediate between a TP and a TAP ($\varphi = 24.6$ and 13.5°); their heights *h* are equal to 2.33 Å and 2.39 Å, respectively. The Shannon radius of the low-spin cobalt(II) ion is appreciably greater than that of the low-spin iron(II) ion and this caused the rotational–translation elongation of a cage framework and increasing in *h* values for the analogous cobalt(II) complexes [75]. Such distortion angle φ is lower in the tris-glyoximate molecule $(Co^{2+})@\mathbf{234}$ with the geometry approaching to a TP, whereas the sterical clashes between the bulky aromatic ribbed substituents

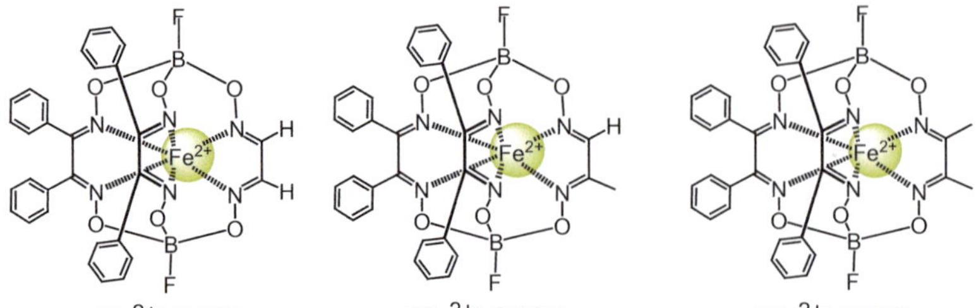

Scheme 3.9 Chemical drawings of a series of the fluoroboron-capped iron(II) tris-dioximates

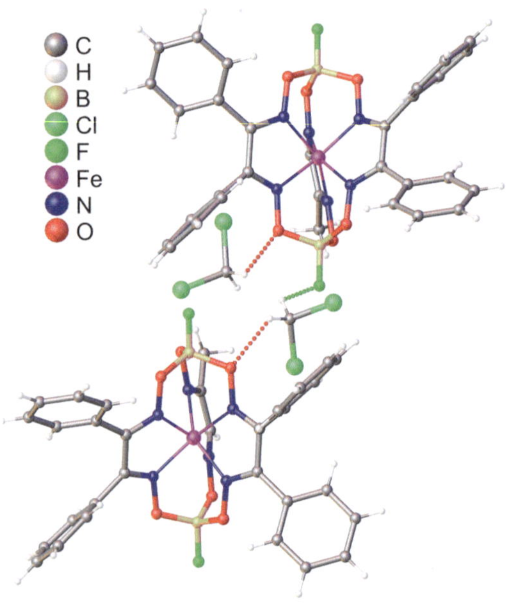

Fig. 3.41 Formation of a clathrochelate dimer in the crystal (Fe^{2+})@**667** · CH_2Cl_2 [74]

caused a substantial distortion of TP–TAP coordination polyhedron of the tris–αbenzildioximate complex (Co^{2+})@**230** ($\varphi = 13.1°$). The caged cobalt(II) ion in these clathrochelates is shifted by approximately 0.17 Å from the centers of their cage frameworks, causing a substantial alternation of the Co–N distances due to the Jahn–Teller distortion of its electronic configuration d^7. The ADPs of an encapsulated cobalt(II) ion at 100 K evidenced that in the molecules (Co^{2+})@**234** and (Co^{2+})@**230** only one of the three possible Jahn–Teller distorted structures is energetically preferable due to the effects of their crystal packing. In particular, the intermolecular C–H. . .π and C–H. . .H–C contacts and two stackings formed by phenyl ribbed substituents at a cage framework of (Co^{2+})@**230** have been observed [75] in its X-rayed crystal, thus causing a formation of the infinite *zigzag* chains shown in Fig. 3.43. These chains are additionally stabilized by the C–H. . .N contacts between the donor nitrogen atoms and the hydrogen atoms of phenyl substituents, forming the above stackings. Two independent macrobicyclic molecules (Co^{2+})@**234**A and (Co^{2+})@**234**B in the crystal (Co^{2+})@**234** are described [75] to form the clathrochelate dimers shown in Fig. 3.44*via* the formation of the intermolecular C. . .O contacts at the distances from 3.1 to 3.3 Å; the C_3 pseudoaxes of their clathrochelate entities are almost perpendicular. The crystal packing of (Fe^{2+})@**234** is similar to that of (Co^{2+})@**234**, but its two macrobicyclic molecules forming a clathrochelate dimer are slightly shifted with respect to each other. So, such intermolecular C. . .O contacts in these so-called "perpendicular dimers" of (Fe^{2+})@**234** are weak. In the crystal (Fe^{2+})@**230** · C_7H_{16}, the intermolecular C–H. . .π, C–H. . .O, and C–H. . .H–C interactions involve only its clathrochelate molecules: they form closest packing shown in Fig. 3.45, whereas the solvate *n*-heptane molecules are situated in the vacant cavities between them [75].

Combined DSC and X-ray diffraction study of multiple phase transitions in the crystals of the hexadecylboron-capped aliphatic and aromatic

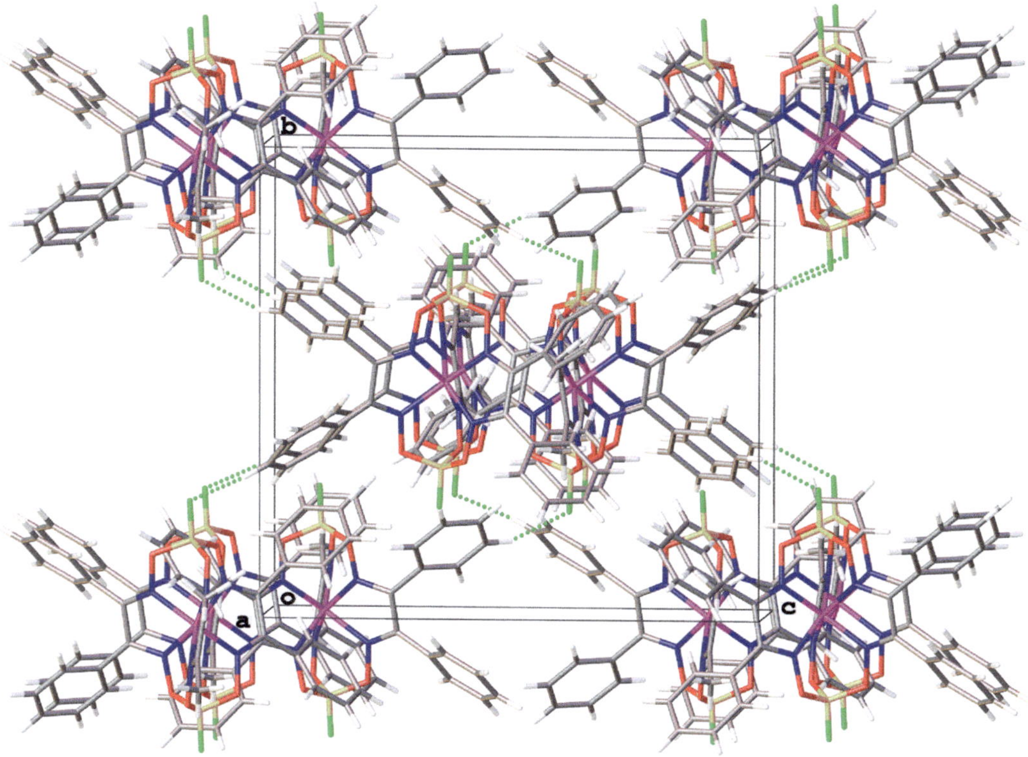

Fig. 3.42 Crystal packing of the complex (Fe^{2+})@**667** · 0.5 C_6H_6 · 0.5 *iso*-C_8H_{18} [74]

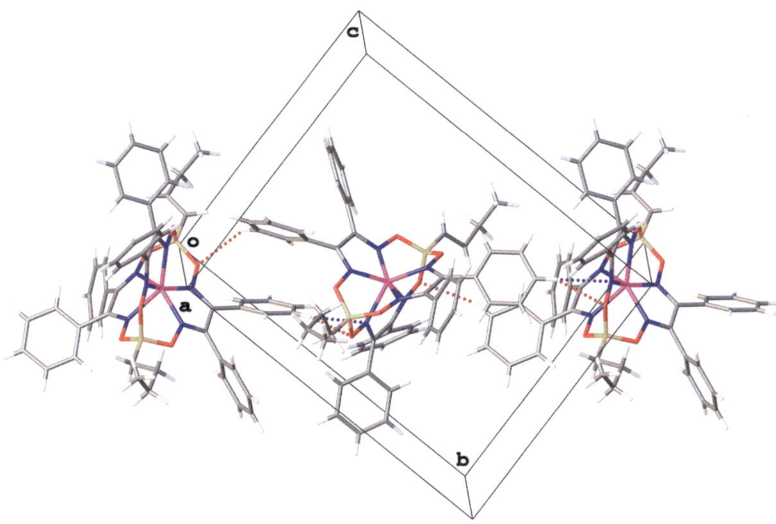

Fig. 3.43 Crystal packing of the complex (Co^{2+})@**230** [75]

iron and cobalt(II) tris-dioximate clathrochelates (Fe^{2+})@**238**, (Co^{2+})@**237**, (Co^{2+})@**236**, and (Co^{2+})@**235** has been also performed [75]; the former aliphatic clathrochelates were found to undergo the structural phase transitions, whereas their aromatic analog did not do this.

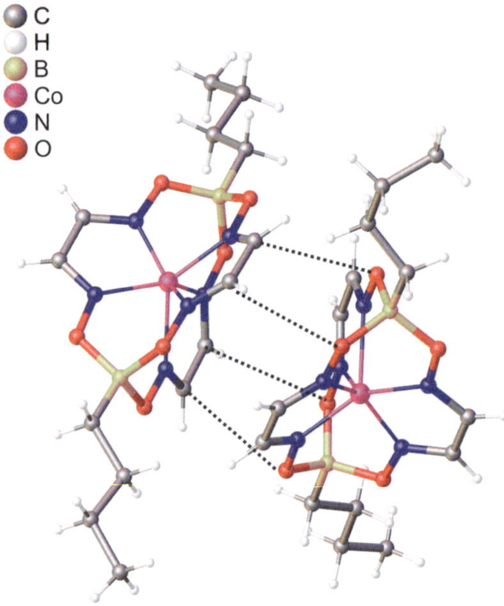

Fig. 3.44 Formation of a clathrochelate dimer in the crystal (Co^{2+})@**234** [75]

Combined XANES and X-ray photoelectron study of the cobalt(I, II, and III) cage tris-dioximates shown in Scheme 3.10 has been performed [76]. The obtained Co K-edge spectra evidenced a lowering of the Co*1s* ionization potentials from cobalt(III) cage complexes to their cobalt(II)-containing analogs and then to the cobalt(I) clathrochelates, while high symmetry of the CoN_6-coordination polyhedra of an encapsulated cobalt(I) ion caused the absence of pre-edge structure and specific peculiarities in the spectra of its cage complexes. Their Co*2p* core-level spectra contained very weak *shake-up* satellites, thus suggesting a hybridization of the cobalt-localized *3d*-orbitals and the valent orbitals of their encapsulating ligands. The binding energy $Co2p_{3/2}$ increases with the formal oxidation state of an encapsulated cobalt ion(I, II or III) from 780.5–780.8 and 780.9–781.2 to 781.8–782.2 eV, respectively. The obtained Cl*2p*, C*1s*, N*1s*, O*1s*, B*1s*, and Co*2p* core-level

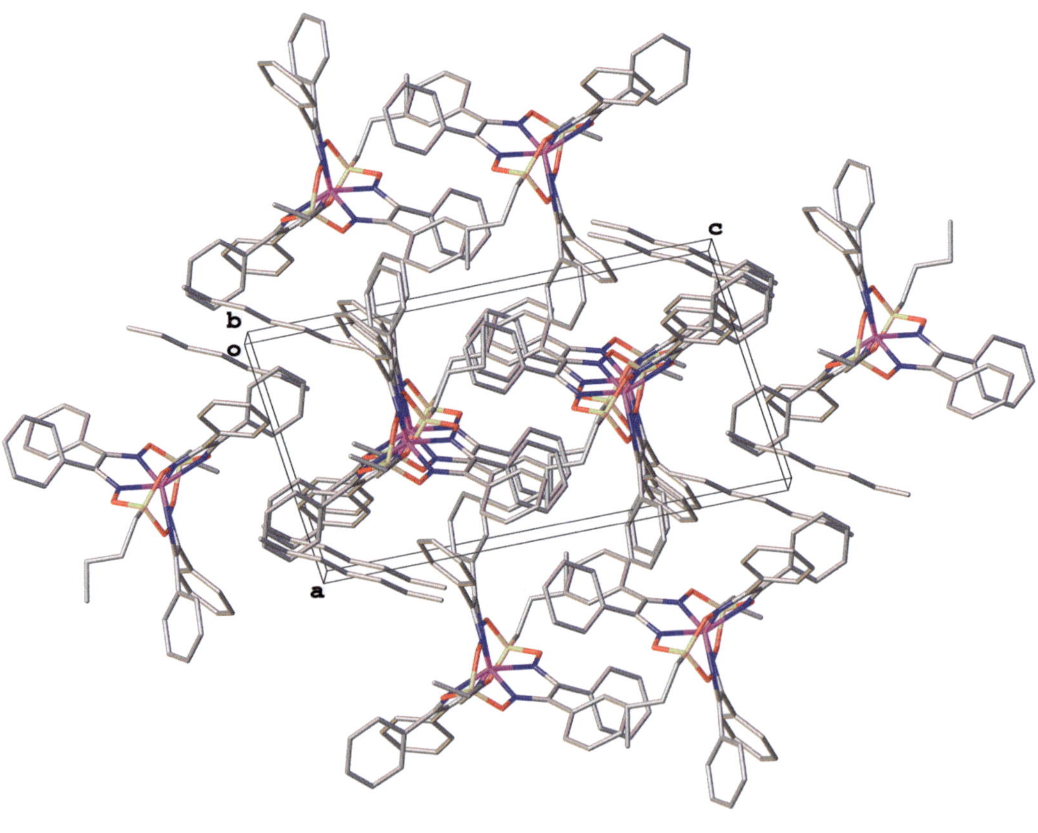

Fig. 3.45 Crystal packing of the complex (Fe^{2+})@**230** [75]

Scheme 3.10 Chemical drawings of a series of the cobalt(I,II and III) clathrochelates

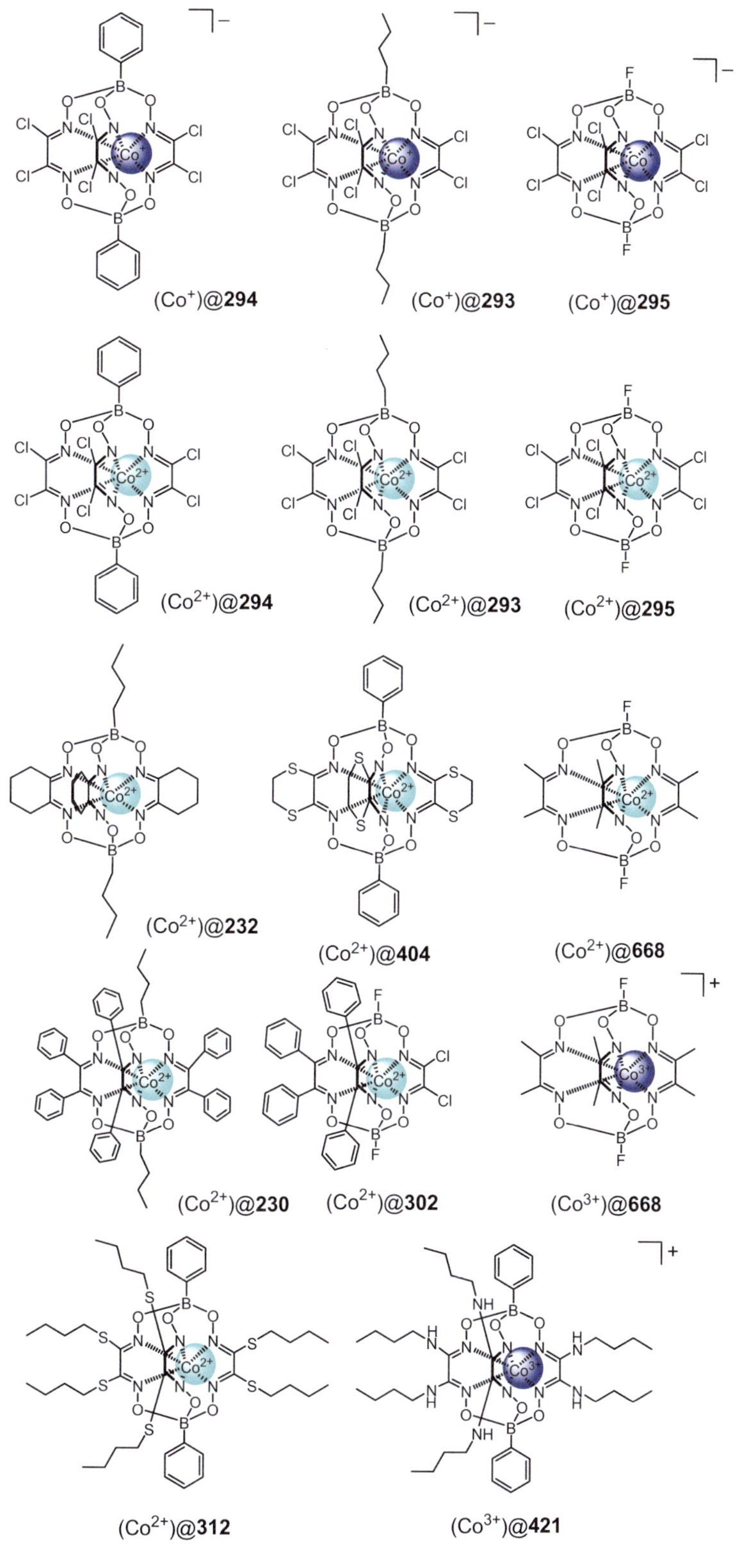

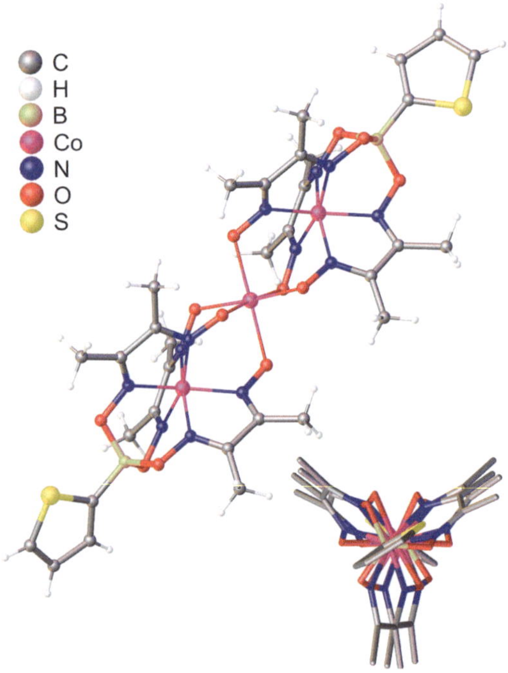

Fig. 3.46 X-ray structure of the $Co^{III}Co^{II}Co^{III}$ bis-clathrochelate $(Co^{3+})_2$@**426** (*top*) and its projection along the molecular C_3 pseudoaxis (*bottom*) [77]

spectra and XANES data have been used [76] to evaluate the effects of both the electronic and molecular structures of their highly conjugated polyazomethine macrobicyclic ligands on metal-localized redox processes of these cage complexes.

The X-ray structures of the tris-dimethylglyoximate cobalt(II) monoclathrochelate (Co^{2+})@**208** and the bis-clathrochelate product of its transformation, the $Co^{III}Co^{II}Co^{III}$ cage complex $(Co^{3+})_2$@**426** (Fig. 3.46), have been compared [77]. The N–O distances in their donor oxime groups depend on the nature of cross-linking groups in a cage framework: the average N–O bond lengths are approximately 1.38 and 1.31 Å for the boron- and cobalt-capped fragments, respectively. The average Co–N distances are substantially affected by oxidation state of an encapsulated cobalt ion, being equal to 1.97 and to 1.91 Å for the above cobalt(II) clathrochelate and its bis-clathrochelate cobalt(III)-encapsulating derivative, respectively. The encapsulated cobalt(II) ion in the molecule (Co^{2+})@**208** is shifted from the center of its CoN_6-coordination polyhedron by approximately 0.2 Å due to the Jahn–Teller distortion of its electronic configuration d^7, whereas the caged cobalt(III) ions in the complex $(Co^{3+})_2$@**426** occupy the centers of its two monoclathrochelate entities. The geometries of their $Co^{II}N_6$- and $Co^{III}N_6$-coordination polyhedra are also substantially different. A rigidity of the encapsulating macrobicyclic tris-dioximate ligand **208** did not allow adopting an energetically preferable TAP geometry due to high Shannon radius of an encapsulated low-spin cobalt(II) ion and the molecule (Co^{2+})@**208** possesses an almost TP geometry with the distortion angle φ of approximately 4°; the height h of its $Co^{II}N_6$-coordination polyhedron is equal to 2.45 Å. The encapsulation of a substantially smaller cobalt(III) ion caused the rotational–translational shortening of $Co^{III}N_6$-coordination polyhedron ($h = 2.24$ Å) with the φ angle of approximately 40°. The capping cobalt(II) ion in the trinuclear $Co^{III}Co^{II}Co^{III}$ bis-clathrochelate $(Co^{3+})_2$@**426** occupies the center of its pseudooctahedral $Co^{II}O_6$-coordination polyhedron. The above structural Jahn–Teller distortion of (Co^{2+})@**208** is additionally stabilized in its crystal by intra- or intermolecular contacts of this cage molecule. Its dimethylglyoximate ribbed chelate fragments are nonequivalent due to their C–H…N interactions at the distance of approximately 2.68 Å, assembling such macrobicycles into the infinite chains shown in Fig. 3.47; the encapsulated cobalt(II) ion of (Co^{2+})@**208** is shifted in the direction of the above interacting chelate fragment. Although this metal ion is almost completely isolated from the external factors by a three-dimensional rigid macrobicyclic ligand, the rather short Co…H contacts at the distance of 3.11 Å have been noted [77].

The multitemperature X-ray diffraction and DSC data for the cobalt(II) hexabromoclathrochelate (Co^{2+})@**287** and its iron(II)-containing analog (Fe^{2+})@**287** have been used

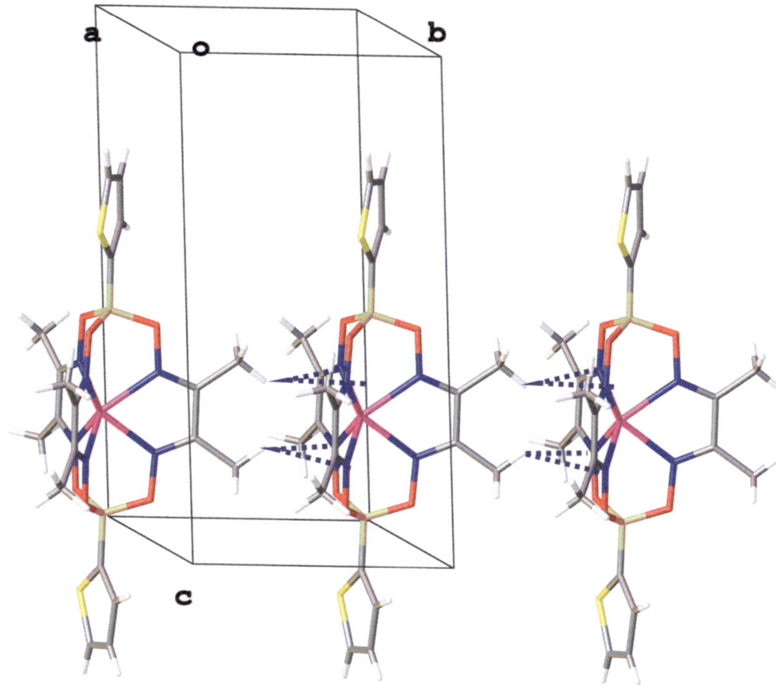

Fig. 3.47 Infinite chain of the clathrochelate molecules (Co^{2+})@**208**, formed *via* the C–H. . .N interactions (depicted with *dashed line*) [77]

[78] to evaluate the relationship between the geometrical characteristics and SCO behavior of the above cobalt(II) cage complex. The variations of the radii, electronic structures, and spin states of their encapsulated metal ions caused the substantial differences in the geometry of the corresponding MN_6-coordination polyhedra. The Fe–N distances were found to be in the range 1.89–1.94 Å, and the encapsulated iron(II) ion is situated almost in the center of its TP–TAP coordination polyhedron, having the distortion angles φ of 20.9° and of 22.1° at 250 K and 220 K, respectively. On the other hand, the average Co–N distances are longer, thus causing the rotational–translational elongation of a quasiaromatic macrobicyclic framework. The φ values for the molecule (Co^{2+})@**287** increase from 3.7 to 10.7° with lowering in temperature due to a depopulation of its high-spin state. Main differences between high- and low-spin states of this cobalt (II) cage complex are the alternation of its Co–N distances and the change in shift of a caged cobalt(II) ion from the center of CoN_6-coordination polyhedron. This shift is high for the Jahn–Teller distorted low-spin cobalt(II) ion at low temperatures, while with increase in temperature, the Co–N distances became more isotropic due to a gradual population of its symmetric high-spin state. On the other hand, the substantial anisotropy in this polyhedron was found even at 290 K, thus being in agreement with the incomplete nature of such SCO, determined [78] by SQUID magnetometry. The observed phase transition in the crystal (Co^{2+})@**287** at 273K is explained by movement of its apical alkyl substituents. DSC study of the fine-crystalline sample of this cage complex allowed to observe the reversible structural phase transition close to 273 K: cooling of this sample from 300 to 203 K showed the exothermic peak at 274 K, while a consequence heating resulted in the endothermic peak at 281 K. DSC curve for its iron(II)-encapsulating analog (Fe^{2+})@**287** is characterized by the similar T_{CR} values for its exothermic and endothermic effects at approximately 281 K and 287 K, respectively. All the above temperature effects on DSC curves for

both these cage complexes of **287** were assigned [78] to the conformation changes of its *n*-butyl apical substituents.

The DSC, SQUID magnetometry, and X-ray diffraction experiments have been performed [79] to study the temperature-induced structural phase and spin transitions in the hexadecylboron-capped iron and cobalt(II) hexachloroclathrochelates (Fe^{2+})@**240** and (Co^{2+})@**240**. DSC experiment for (Fe^{2+})@**240** revealed only one phase transition before melting at 173 K, while the DSC curve for (Co^{2+})@**240** showed two crystal-to-crystal transitions in the temperature range 150–420 K. The heating of the fine-crystalline sample of (Co^{2+})@**240** caused the phase transitions at 250 and 273 K, exhibiting hysteresis. The type of these crystal-to-crystal phase transitions has been assigned [79] using the multitemperature single crystal X-ray diffraction experiments. The above cage complexes were found to be isostructural above 40 K. The heights *h* of their MN_6-polyhedra are affected by their distortion around the molecular C_3 pseudoaxis and, therefore, by the nature of an encapsulated metal ion: if FeN_6-coordination polyhedron possesses the geometry intermediate between a TP and a TAP ($\varphi = 17.5°$ and $h = 2.38$ Å at 240 K), a cage framework of the molecule (Co^{2+})@**240** underwent a translational–rotational elongation, leading to increase in *h* for its almost TP CoN_6-polyhedron up to approximately 2.50 Å. Single crystal X-ray diffraction data for (Co^{2+})@**240** showed a number of the geometrical parameters of its coordination polyhedron, evidencing the temperature SCO, such as the highest (in the temperature range under study) shift of an encapsulated cobalt(II) ion from the geometrical center of this polyhedron at 40 K and the biggest φ value (3.1°) as well. These parameters decreases with temperature up to 0.026–0.035 Å and 2.3°, respectively, while the average Co–N distance, that is the shortest at 40 K, increases with temperature. The encapsulated cobalt ion has a significant anisotropy of its thermal ellipsoid at low temperatures, and this ellipsoid is elongated in the direction perpendicular to the molecular C_2 axis; this fact is explained [79] by superposition of its two positions. The apical long-chain alkyl substituents of (Fe^{2+})@**240** are ordered in the temperature range 168–300 K; the thermal ellipsoids of their carbon atoms underwent only the thermal expansion and the crystal-to-crystal phase transition, detected using DSC method for its fine-crystalline sample at 173 K, is assigned [79] to the movement of these apical substituents. DSC curve upon cooling of the fine-crystalline sample of (Co^{2+})@**240** showed two phase transitions at 273 and 250 K. Comparison of its molecular structures at various temperatures revealed the larger shifts of carbon and hydrogen atoms of the long-chain alkyl substituents in the case of the cobalt(II) cage complex (Co^{2+})@**240** as compared with its iron(II)-encapsulating analog. After cooling, single crystal of this cobalt(II) clathrochelate has been heated again; an increase in temperature from 231 to 260 K has been accompanied with lower atomic shifts in these apical substituents and the further heating above 280 K has not been accompanied with any disorder; this fact is in good agreement with DSC data. The π…Cl supramolecular interactions in the crystal (Co^{2+})@**240** shown in Fig. 3.48 substantially affect its SCO behavior; the reverse drop on the χT(T) curve of its fine-crystalline sample (see Chap. 4) is explained [79] by differences in the temperature behavior of the long-range and short-range cooperativity effects in this crystal.

The X-ray structures of the clathrochelates (Fe^{2+})@**219** (for its two polymorphs), (Fe^{2+})@**226**, and (Fe^{2+})@**220** have been analyzed [80]. Their molecular packings followed Kitaigorodsky *aufbau* principle with the molecular "bumps" (in this case, the ribbed substituents and the apical aromatic groups), situated within the intramolecular cavities formed by two ribbed α-dioximate chelate fragments of these cage molecules. The monoclinic and triclinic polymorphs of (Fe^{2+})@**219** have been X-ray structurally characterized [80]: the conformations of this macrobicyclic molecule in them are very close (Fig. 3.49a). Two of the three vacant cavities between the ribbed α-dioximate chelate fragments of (Fe^{2+})@**219** in its monoclinic and triclinic polymorps as well as those in the crystals

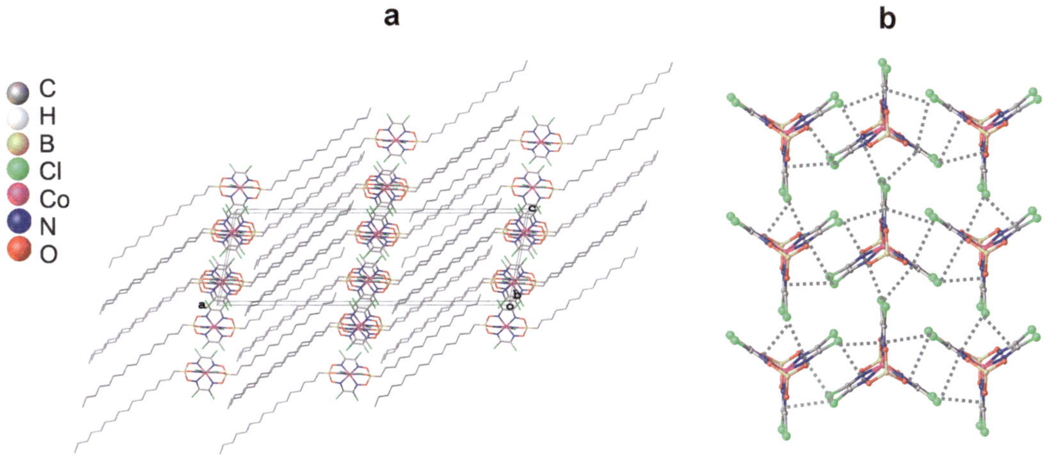

Fig. 3.48 Crystal packing of (Co^{2+})@**240** (**a**) and halogen bonding between its macrobicyclic molecules (**b**) [79]

of both the macrobicyclic tris-nioximates (Fe^{2+})@**226** and (Fe^{2+})@**220** are occupied by apical aromatic substituents of two neighboring clathrochelate molecules. In contrast, third cavity of this type is occupied by hydrogen atoms of one (for both the above macrobicyclic tris-nioximates) or two (for the polymorphs of (Fe^{2+})@**219**) aliphatic ribbed substituent(s) at neighboring cage framework(s). Such cavities in the cases of (Fe^{2+})@**219** are occupied by methyl ribbed substituents belonging to two neighboring macrobicyclic molecules, thus forming the trimeric clathrochelate supramolecular assemblies as shown in Fig. 3.49b, c. At the same time, despite the very similar conformations of these macrobicyclic molecules, the positions of their apical aromatic substituents in the above assemblies are different in the crystals of the above polymorphs of (Fe^{2+})@**219** [80].

The X-ray structures of the hexaiodoclathrochelate (Fe^{2+})@**288**, its triribbed-functionalized pentafluorophenyl-substituted derivative (Fe^{2+})@**289,** and the monoribbed-functionalized cage complex (Fe^{2+})@**479** are reported [81]. The average Fe–N distance in (Fe^{2+})@**288** (1.89 Å) is slightly shortened as compared with that characteristic of the boron-capped iron(II) tris-dioximates (1.90–1.92 Å). If an encapsulated iron(II) ion in all these

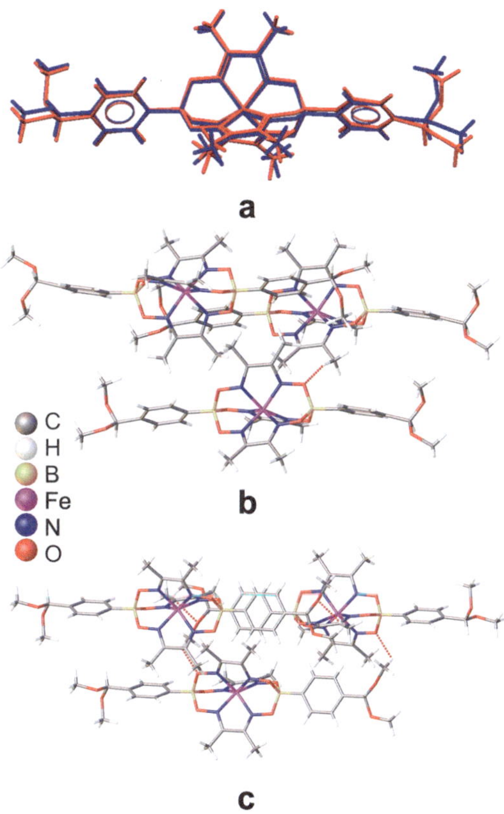

Fig. 3.49 Comparison of the X-ray structures of (Fe^{2+})@**219** in its monoclinic (shown in *red*) and triclinic (shown in *blue*) polymorphs (**a**) and the fragments of their crystal packings (**b** and **c**, respectively) [80]

macrobicyclic molecules is situated almost in the centers of their TP–TAP distorted FeN_6-coordination polyhedra, the φ value for (Fe^{2+})@**288**, on the one hand, and those for (Fe^{2+})@**479** and (Fe^{2+})@**289**, on the other hand, are substantially different: the former cage complex possesses an almost TP geometry ($\varphi = 4.5°$), whereas that of the latter clathrochelates is intermediate between a TP and a TAP with φ of approximately 25°. Such translational–rotational shortening of the coordination polyhedra caused a decrease of their heights h from 2.40 to 2.33–2.35 Å, respectively. In general, passing from the phenyl ribbed substituents to their perfluorinated analogs resulted in a rotational expansion of the corresponding macrobicyclic framework with six inherent aromatic substituents [81].

The detailed analysis of the molecular and crystal X-ray structures of the cage iron (II) complexes with apical and ribbed adamantyl substituents has been performed [82]. These clathrochelate molecules usually possess a "whirligig" shape: the capping boron atoms occupy their apical positions, whereas the ribbed chelate fragments form the central belt (toroid). It can be assumed that due to the absence of polar terminal substituents, van der Waals interactions play a key role in their crystal packing: the above shape of these cage molecules is hardly suitable for the tightest packing of a space and, therefore, one should expect the appearance of cavities in their crystals. Among them, only the crystals of the hexaadamantylsulfide cage complex (Fe^{2+})@**382** (Fig. 3.50) did not contain any

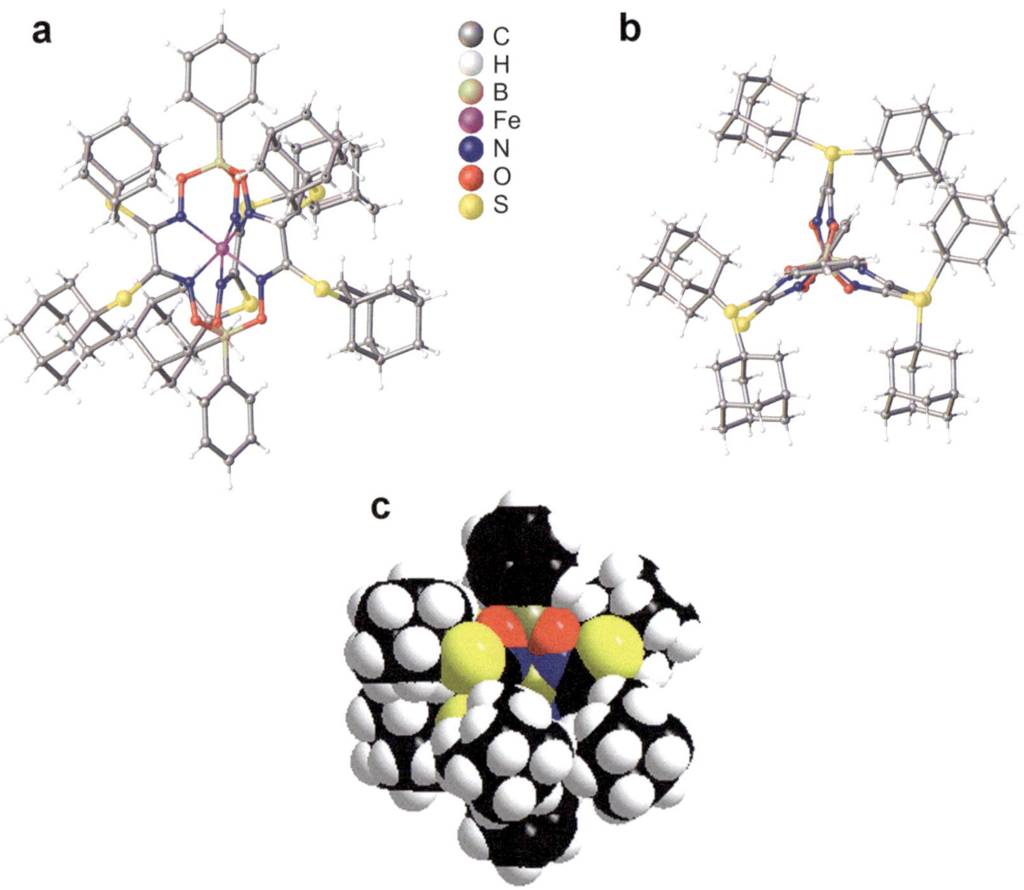

Fig. 3.50 General views of the molecule (Fe^{2+})@**382** in the different projections (**a** and **b**) and its spacefilling representation (**c**): formation of second hydrophobic shell around the encapsulated iron(II) ion [82]

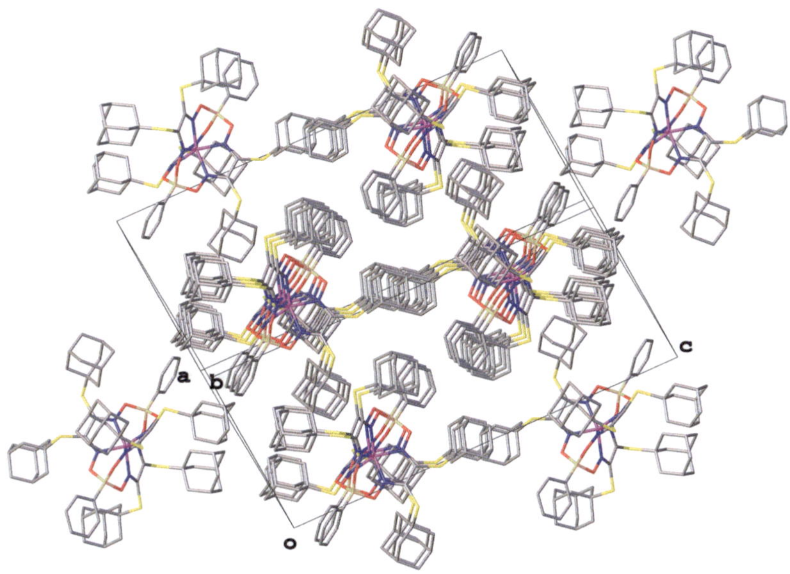

Fig. 3.51 Crystal packing of (Fe^{2+})@**382** [82]

solvate molecules. The molecule of (Fe^{2+})@**382** (Fig. 3.50a, b) is symmetric, but in fact its shape is far from a "whirligig" since both its phenyl and two of the six adamantyl substituents are located in one belt (Fig. 3.50c). The clathrochelate molecules of (Fe^{2+})@**382** in its X-rayed crystal are oriented perpendicular each other, thus forming the layers (Fig. 3.51) *via* the weak S...S interactions between them (Fig. 3.52) [82].

Cage complex (Fe^{2+})@**203** crystallizes with one hexane solvate molecule; its macrobicyclic framework possesses a "whirligig" shape. The packing of its X-rayed crystal (Fig. 3.53) is mainly governed by van der Waals interactions; all the clathrochelate molecules (Fe^{2+})@**203** are oriented perpendicular to each other. Three types of their mutual arrangement have been observed [82]. First of them is formed by those oriented by two ribbed chelate fragments with respect to each other; in Fig. 3.53, this refers to the macrobicyclic molecules related by symmetry centers in the tops of an unit cell. The clathrochelate molecules of the second type are related by other symmetry centers and occupy the ribs of this cell. Such cage molecules are oriented by only one α-dioximate chelate fragment of their macrobicyclic ligand. Clathrochelate molecules of the third type also occupy the ribs of an unit cell and are oriented by their adamantyl substituents. The cavities suitable for inclusion of solvate hexane molecule are only between the second and third types of the above clathrochelate molecules and formed by the chelate ribbed substituents of the above cages of the third type and by the adamantyl apical substituents as well [82].

The X-rayed crystal of the clathrochelate (Fe^{2+})@**204** having a "whirligig" shape (Fig. 3.54) also contains the hexane solvate molecules. This clathrochelate and its macrobicyclic homolog (Fe^{2+})@**203** have the different sizes of their ribbed substituents (eight- and six-membered alicycles, respectively). In the molecule (Fe^{2+})@**203**, these substituents are flattened, whereas in the clathrochelate (Fe^{2+})@**204** they are strongly bent (the average bend angle is approximately 70°) and in two of these three bent alicycles their twisted fragments are directed toward each other. So, the molecule (Fe^{2+})@**204** is more bulky and, therefore, has a less dense packing than (Fe^{2+})@**203**. This caused the difference in symmetry of their crystals and the different pattern of these crystal packings as well. All the macrobicyclic molecules in the crystal

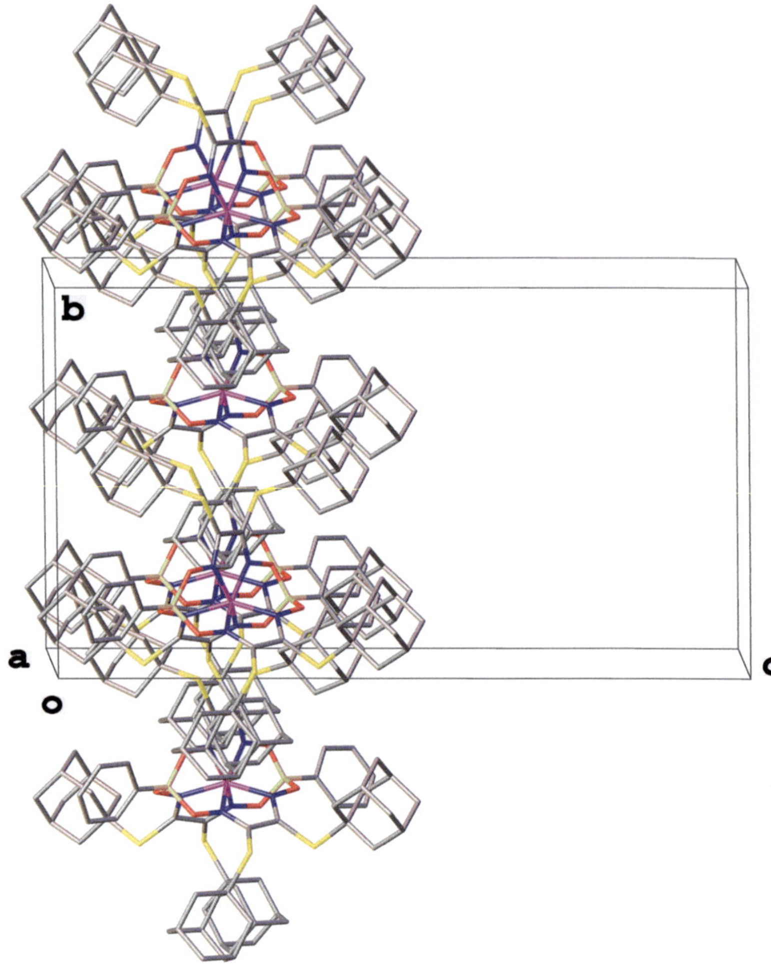

Fig. 3.52 Formation of a clathrochelate layer in the crystal $(Fe^{2+})@\mathbf{382}$ [82]

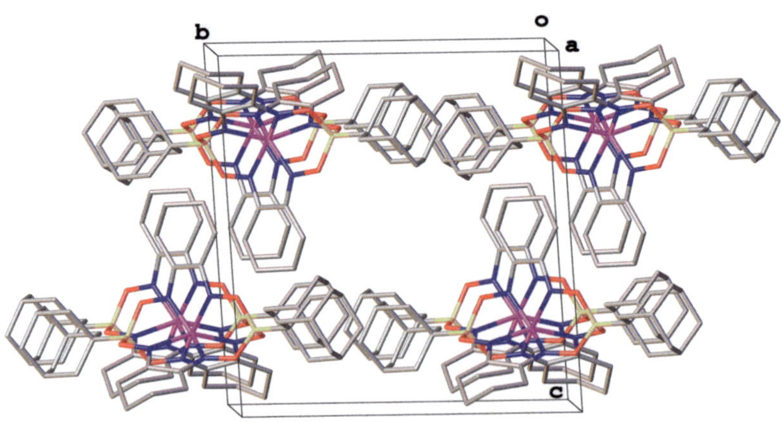

Fig. 3.53 Crystal packing of $(Fe^{2+})@\mathbf{203} \cdot C_6H_{12}$ [82]

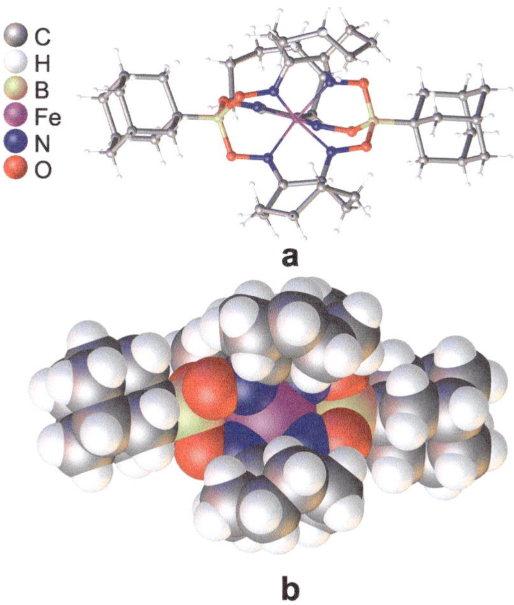

Fig. 3.54 X-ray structure of (Fe^{2+})@**204** with atoms shown as *spheres* (**a**) and its spacefilling representation (**b**): formation of the second hydrophobic shell around the encapsulated iron(II) ion [82]

(Fe^{2+})@**204** (Fig. 3.55) are oriented perpendicular to each other and their molecular C_3 pseudoaxes are parallel. The size of its central cavities is too small for inclusion of a hexane molecule and only the cavities formed by clathrochelate molecules that are oriented to each other by their two dioximate chelate fragments (i.e., the macrobicyclic entities occupying the ribs of an unit cell, Fig. 3.55) include such solvent guest molecules [82].

The crystal (Fe^{2+})@**202** contains 1.5 solvate *iso*-octane molecule per one clathrochelate entity [82]. The shape of its cage molecule is close to a "whirligig" (Fig. 3.56) with substantially higher toroid (approximately 6.8 Å) as compared with those for the molecules (Fe^{2+})@**203** and (Fe^{2+})@**204** (approximately 3.0 Å and 3.6 Å, respectively). The molecules (Fe^{2+})@**202** in this crystal are oriented crosswise (Fig. 3.57a); their mutual orientation is also shown in Fig. 3.57b.

Thus, the crystal packings of the above adamantyl-functionalized iron(II) clathrochelates are determined mainly by van der Waals interactions between their hydrophobic peripheral groups and by whole shape of the macrobicyclic molecules as well as by the size and lability of the apical and ribbed substituents at a cage framework [82].

The X-ray structures of the *n*-butylboron-capped clathrochelate and bis-clathrochelate iron(II) and cobalt(II, III) dioximates with ribbed perfluoroarylsulfide substituents (Fe^{2+})@**429**, (Co^{2+})@**427** (Fig. 3.58), (Co^{2+})@**428** (Fig. 3.59), (Co^{2+})@**429** (Fig. 3.60), and $(Co^{3+})_2$@**430** (Fig. 3.61) have been compared [83]. Their geometry is affected by both the nature of an encapsulated metal ion and that of the ribbed perfluoroarylsulfide substituents at a cage framework. The chelate angles α in these molecules persist, and the rigidity of their chelate moieties N=C–C=N caused the distortion of these cage molecules around the molecular B…B C_3 pseudoaxis. Only bis-capping $Co^{II}O_6$-fragment of the $Co^{III}Co^{II}Co^{III}$bis-clathrochelate $(Co^{3+})_2$@**430** possesses a TAP geometry, whereas all the $Co^{II}N_6$-coordination polyhedra of (Co^{2+})@**427**, (Co^{2+})@**428**, and (Co^{2+})@**429** possess an almost TP geometry with the average distortion angles φ in the range 0.7–9.5°. The geometry of the coordination polyhedra of the encapsulated iron(II) and cobalt(III) ions in the molecules (Fe^{2+})@**429** and $(Co^{3+})_2$@**430** is intermediate between a TP and a TAP (φ = 25.2 and 40.3°, respectively); their heights h (2.4 and 2.3 Å, respectively) are smaller than those for the above cobalt(II) clathrochelates. The average M–N distances were found [83] to be affected by the size of an encapsulated metal ion. The wide range of the Co–N distances for the above cobalt(II) clathrochelates and the substantial (approximately 0.2 Å) shifts of their encapsulated metal(II) ions from the centers of the corresponding cage frameworks are explained [83] by the Jahn–Teller distortion of low-spin electronic configuration d^7, whereas the encapsulated low-spin iron(II) and cobalt (III) ions with electronic configuration d^6 are located in the centers of their TP–TAP coordination polyhedra.

All the X-rayed crystals of the cobalt(II) cage complexes of this type contain the similar infinite

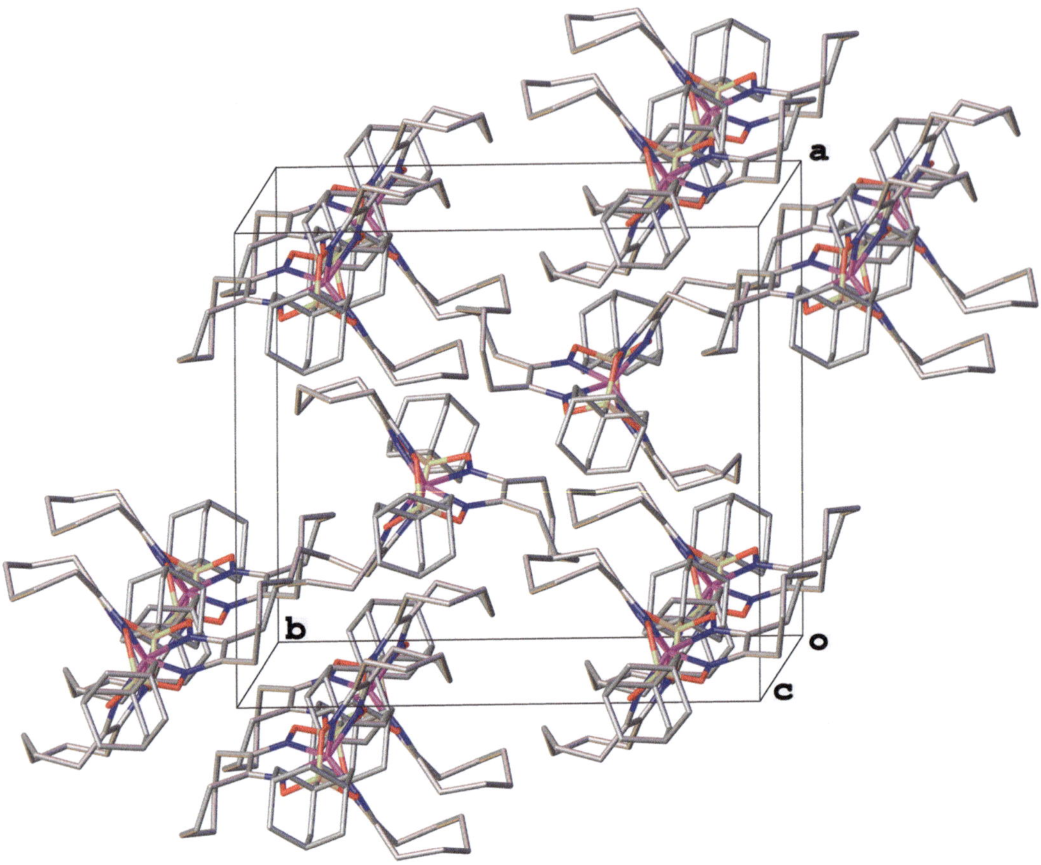

Fig. 3.55 Crystal packing of (Fe^{2+})@**204** · C_6H_{14} [82]

chains shown in Fig. 3.62, formed by intermolecular S. . .S contacts between the neighboring clathrochelate entities at the distances from 3.8 to 4.0 Å [83].

The same structural effects have been observed [84] for a series of their pentafluorophenylboron-capped analogs, the iron (II) hexachloroclathrochelate (Fe^{2+})@**292**, and the iron and cobalt(II) (per)fluoroarylsulfide macrobicyclic and bis-macrobicyclic complexes (Fe^{2+})@**431**, (Fe^{2+})@**432** (Fig. 3.63), (Co^{2+})@**432**, and $(Co^{3+})_2$@**434** (Fig. 3.64). The rotational–translational elongation (shortening) along the molecular C_3 pseudoaxis, accompanied by the changes in the geometry of the corresponding MN_6-coordination polyhedra from a TP to a TAP, allowed an encapsulation of the Fe^{2+}, Co^{2+}, and Co^{3+} ions. The type of an encapsulated metal ion and its oxidation state affected the corresponding M–N bond lengths: the average Fe–N distances (approximately 1.92 Å) are smaller than the Co(II)–N bond lengths (approximately 1.98 Å); passing from the cobalt(II) clathrochelate (Co^{2+})@**432** to its bis-clathrochelate analog $(Co^{3+})_2$@**434** with encapsulated cobalt(III) ions has been accompanied by shortening of the average Co–N bonds to approximately 1.92 Å [84]. For the molecule (Co^{2+})@**432**, the Jahn–Teller structural distortion has been observed as an alternation of the Co–N distances: they vary within 0.23 Å in its $Co^{II}N_6$-polyhedron and within 0.06 Å in the $Fe^{II}N_6$- and $Co^{III}N_6$-polyhedra of the corresponding iron(II) and cobalt(III) cage complexes. Despite these variations, their bite angles α fall in the narrow range 79.6–81.4°. The pentafluoroboron-capped (per)fluorophenylsulfide iron(II) and cobalt(III) mono- and

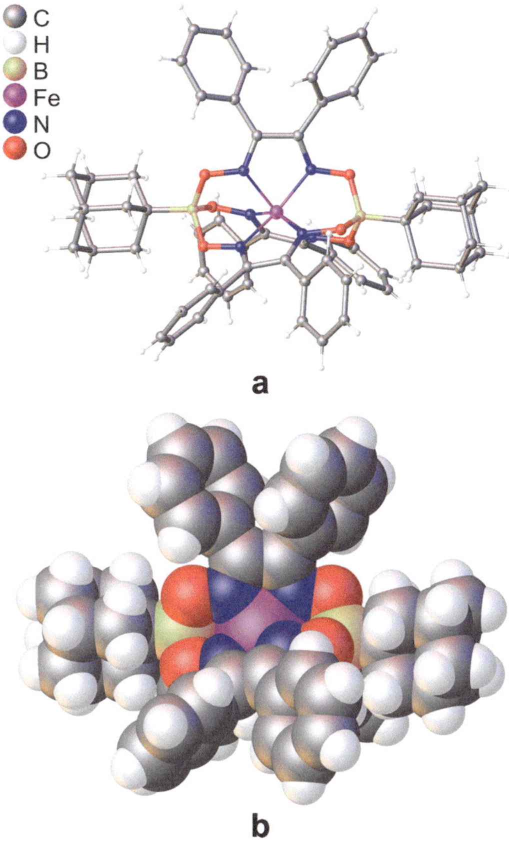

Fig. 3.56 X-ray structure of (Fe^{2+})@**202** with atoms shown as *spheres* (**a**) and its spacefilling representation (**b**): formation of the second hydrophobic shell around the encapsulated iron(II) ion [82]

bis-clathrochelates (Fe^{2+})@**431**, (Fe^{2+})@**432,** and $(Co^{3+})_2$@**434** adopt the geometry intermediate between a TP and a TAP with the φ angles of 21.7°, 26.5°, and 39.9°, respectively. At the same time, the analogous low-spin cobalt(II) cage complex possesses much less distorted, an almost TP $Co^{II}N_6$-coordination polyhedron with φ equal to 13.6°; an almost TP geometry of FeN_6-polyhedron has been also observed in the hexachloroclathrochelate (Fe^{2+})@**292**. An increase in φ from 0° to 30° has been accompanied by decrease in the height *h* of the MN_6-coordination polyhedra from 2.33–2.38 Å for the above iron(II) cage complexes to 2.45–2.51 Å for their cobalt(II)-encapsulating analogs [84]. Like in the case of their *n*-butylboron-capped cage complexes of this type (see above), the crystals of these pentafluoroboron-capped (per)fluorosulfide mono- and bis-clathrochelates contain the S…S bonded clathrochelate chains with r_i(S…S) of approximately 4.0 Å. In the crystal of the iron(II) hexachloroclathrochelate (Fe^{2+})@**292**, two dichloroglyoximate ribbed fragments of its macrobicyclic ligand form the Cl…Cl and Cl…F halogen bonds [84].

The X-ray structures of the tris-perfluoro-α-benzildioximate cobalt(II) clathrochelates (Co^{2+})@**291**, (Co^{2+})@**289**, and (Co^{2+})@**290** (Fig. 3.65) and that of their iron(II)-encapsulating analog (Fe^{2+})@**289** [81] have been compared [85]. Main bond lengths in the cage frameworks of these clathrochelates as well as the chelate N=C–C=N angles are similar for all these clathrochelates, while the translational–rotational elongation (shortening) along the molecular C_3 pseudoaxis, accompanied by changes in the geometry of MN_6-coordination polyhedra from a TP to a TAP allowed encapsulation of both the Fe^{2+} and Co^{2+} ions. CoN_6-coordination polyhedron of a caged cobalt(II) ion adopts an almost TP geometry for the clathrcohelate (Co^{2+})@**290** (φ = 1.4°), while those for the cage complexes (Co^{2+})@**291** and (Co^{2+})@**289** are intermediate between a TP and a TAP (their distortion angles φ are in the range 10–14°). That for the iron(II) complex (Fe^{2+})@**289** with the same macrobicyclic ligand is equal to 25.4° [81]. Such translational–rotational shortening of FeN_6-coordination polyhedron also caused a decrease in its height *h* up to 2.33 Å as compared with the CoN_6-polyhedra (2.44–2.46 Å). The degree of a TP–TAP distortion is affected by both the nature of the ribbed substituents at a cage framework and that of an encapsulated metal ion. Moreover, these distortions of the cage molecules are affected not only by sterical clashes in their *vic*-substituted α-dioximate chelate fragments but also by the electromeric (inductive and mesomeric) effects of the above ribbed substituents at a quasiaromatic polyazomethine macrobicyclic framework. If such inductive effects persist, those mesomeric effects depend on a conjugation between the quasiaromatic system of a cage framework

Fig. 3.57 Packing of the crystal (Fe^{2+})@**202** (**a**) and the mutual orientation of its clathrochelate molecules (**b**) [82]

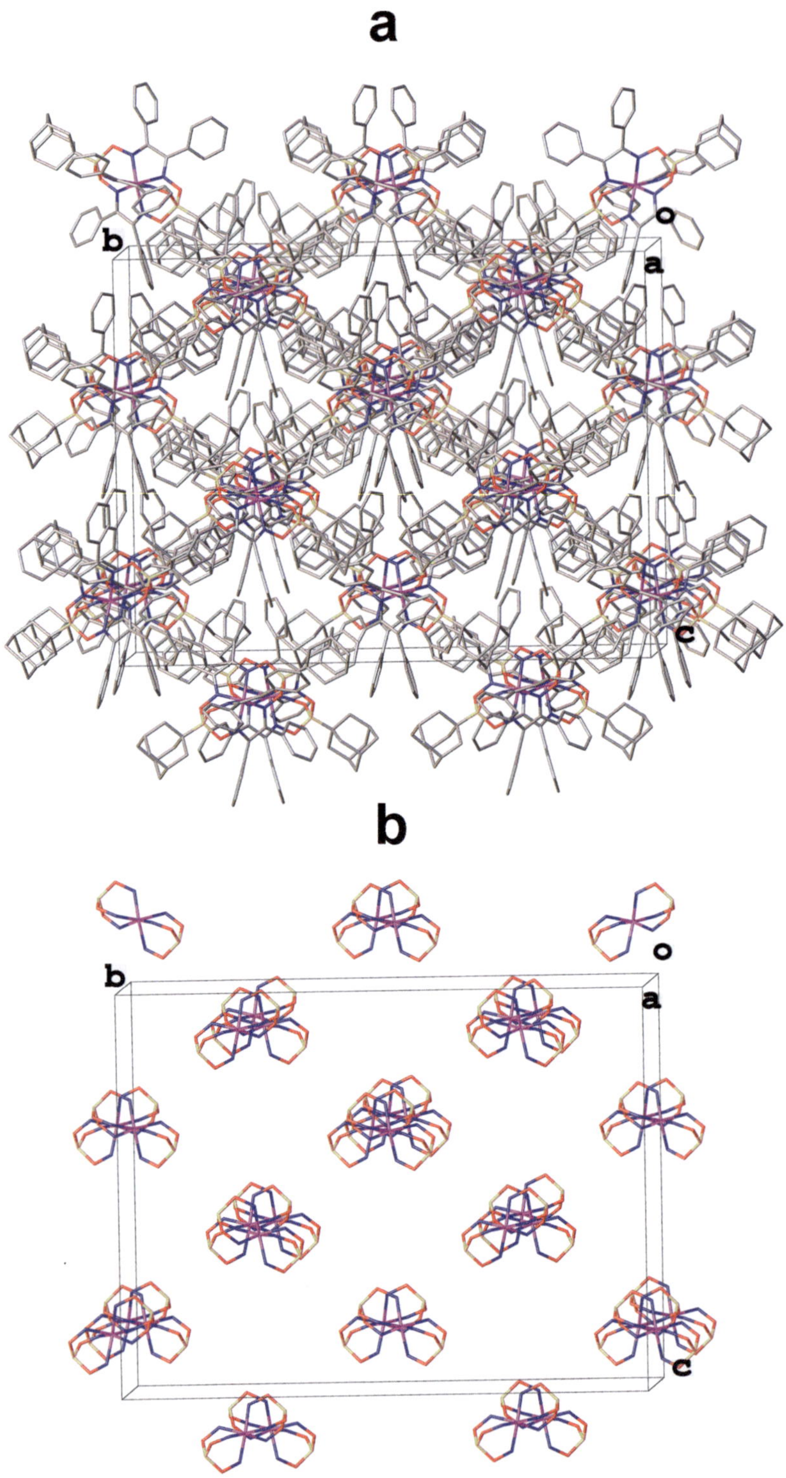

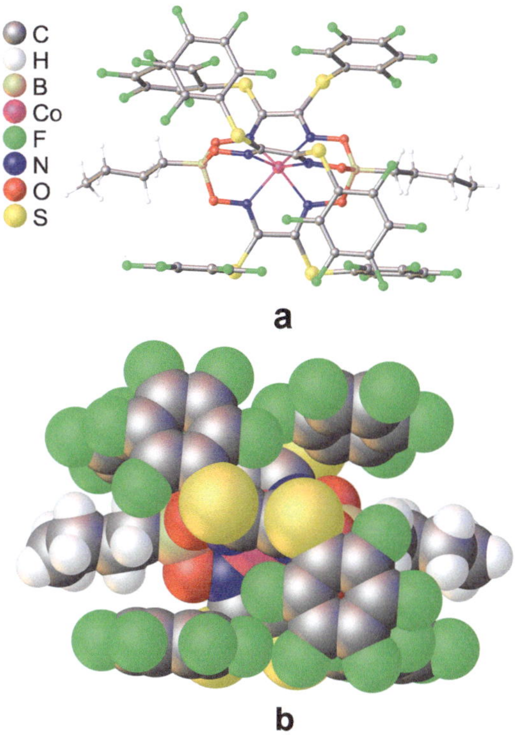

Fig. 3.58 X-ray structure of (Co^{2+})@**427** with atoms shown as *spheres* (**a**) and its spacefilling representation (**b**): formation of the second superhydrophobic shell around the encapsulated cobalt(II) ion [83]

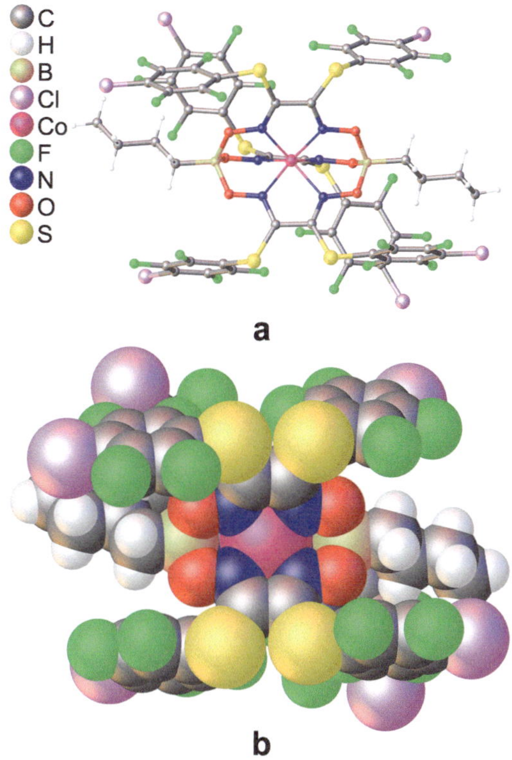

Fig. 3.59 X-ray structure of (Co^{2+})@**428** with atoms shown as *spheres* (**a**) and its spacefilling representation (**b**): formation of the second superhydrophobic shell around the encapsulated cobalt(II) ion [83]

and the δ- and π-systems of a ribbed substituent at this framework (first of all, on the planarity of such chelate fragments, e.g., the dihedral angles between the mean plane of π-conjugated chelate cycle and that of the inherent aromatic substituent). Sterical clashes between bulky pentafluorophenyl ribbed substituents in *vic*-substituted α-dioximate fragments caused a substantial rotation of these substituents relative to the mean planes in such chelate cycles: the average dihedral angles between them are equal to 58.9, 55.7, and 58.3° for the molecules (Fe^{2+})@**289**, (Co^{2+})@**289**, and (Co^{2+})@**291**, respectively. The Co–N distances in the above cobalt(II) (per)fluoroclathrochelates are alternated due to their Jahn–Teller structural distortion: these distances for an encapsulated low-spin cobalt(II) ion vary by 0.21 Å, and this cation is substantially shifted from the centers of the CoN_6-coordination polyhedra, whereas in the molecule of (Fe^{2+})@**289** its encapsulated low-spin iron(II) ion is situated almost in the center of FeN_6-coordination polyhedron [81].

Unexpected double addition of ethyl radicals to the azomethine fragment of a mixed dimethylglyoximate-α-benzildioximate iron(II) clathrochelate (Fe^{2+})@**296** has been confirmed [86] both in solution and in the solid state using NMR and single crystal X-ray diffraction experiments, respectively. The obtained NMR and X-ray structural data perfectly match each other: the resulting hydroxylamine-containing macrobicyclic encapsulating ligand **462** (Fig. 3.66) contains two ethyl groups attached to both the nitrogen and carbon atoms of the azomethine fragment of the same methylglyoximate donor group of the initial macrobicyclic ligand **296**; such addition gave two chiral centers at these atoms [86]. The cage

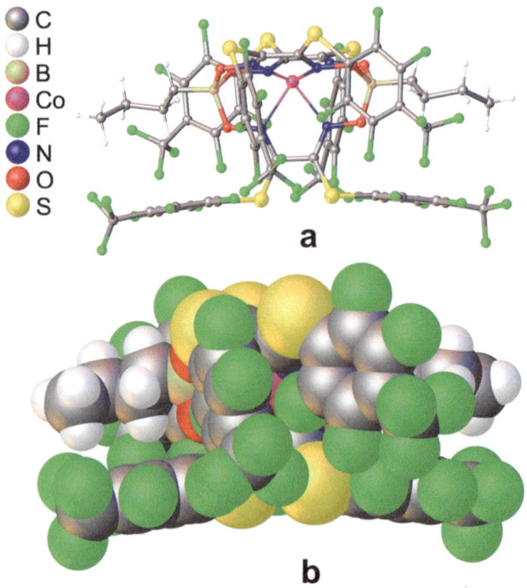

Fig. 3.60 X-ray structure of (Co^{2+})@**429** with atoms shown as *spheres* (**a**) and its spacefilling representation (**b**): formation of the second superhydrophobic shell around the encapsulated cobalt(II) ion [83]

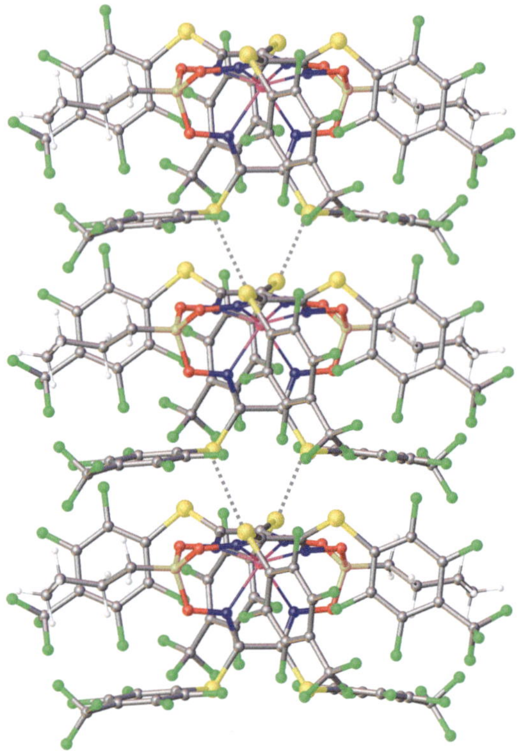

Fig. 3.62 Formation of 1D clathrochelate chain in the crystal (Co^{2+})@**429** [83]

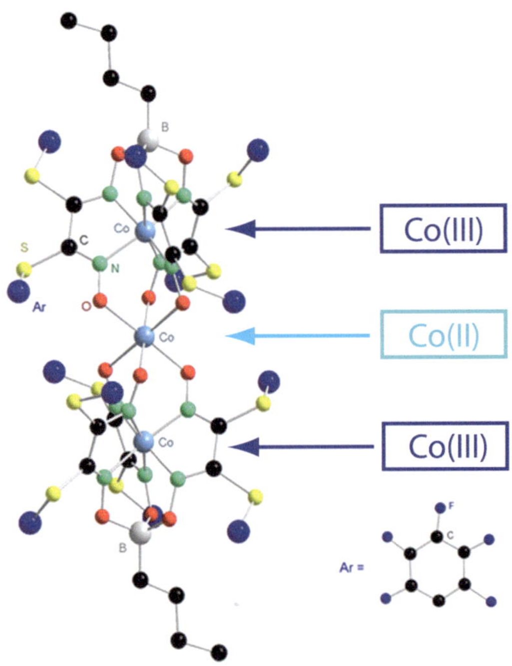

Fig. 3.61 X-ray structure of the trinuclear $Co^{III}Co^{II}Co^{III}$ bis-clathrochelate $(Co^{3+})_2$@**430** with atoms shown as *spheres* [83]

complex (Fe^{2+})@**462** crystallizes in the space group $P2_1/n$ and, therefore, its X-rayed crystal is achiral. A cage framework of (Fe^{2+})@**462** is a substantially TAP-distorted: if the Fe–N distances in its α-benzildioximate chelate moieties fall within the narrow range 1.88–1.90 Å, and that for the oxime nitrogen atom of its functionalized ribbed fragment is approximately 1.93 Å, the Fe–N distance for the hydroxylamine-containing donor group of this clathrochelate molecule is equal to 2.04 Å. The N–O bond lengths in the α-benzildioximate chelate fragments and the methyl-containing oxime donor group of (Fe^{2+})@**462** are approximately 1.39 Å and 1.40 Å, respectively, while that in its hydroxylamine donor group is substantially higher (approximately 1.44 Å). The oxime carbon atoms in this cage complex are trigonal, while its quaternary carbon atom was found [86] to be substantially tetrahedral. The tetrahedral environment that included an encapsulated

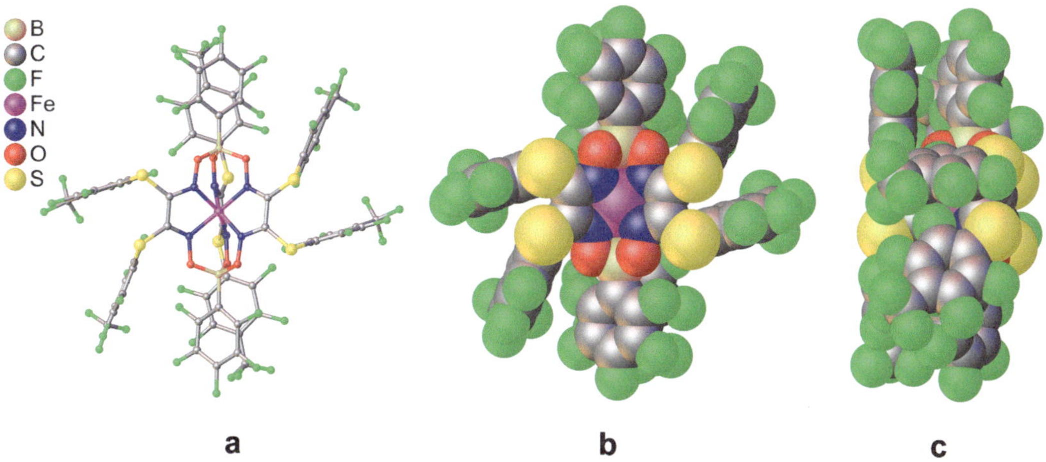

Fig. 3.63 X-ray structure of (Fe^{2+})@**432** with atoms shown as *spheres* (**a**) and its spacefilling representations [face (**b**) and side (**c**) views]: formation of the second superhydrophobic shell around the encapsulated cobalt(II) ion [84]

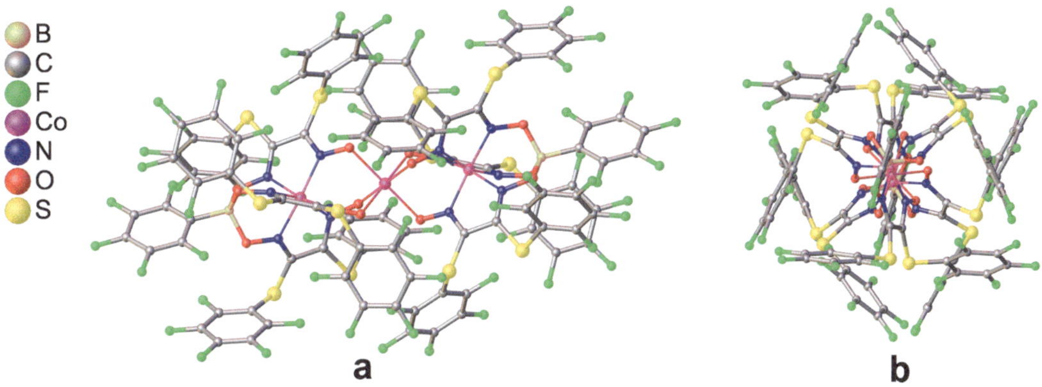

Fig. 3.64 Side (**a**) and top (**b**) views of the trinuclear $Co^{III}Co^{II}Co^{III}$ bis-clathrochelate $(Co^{3+})_2$@**434** with atoms shown as *spheres* [84]

iron(II) ion has been also observed for the hydroxylamine nitrogen donor atom of (Fe^{2+})@**296**. So, the above transformation of one of the oxime groups to the hydroxylamine donor fragment resulted in an elongation of the corresponding N–O bond and in the formation of the tetrahedral environments of the corresponding carbon and nitrogen atoms as well [86].

The X-ray structures of the macrobicyclic complexes (Fe^{2+})@**459** and (Fe^{2+})@**460** are described [87]. The molecule (Fe^{2+})@**459** has a chiral center at the ethyl carbon atom attached to its clathrochelate framework, and this cage complex crystallizes in the space group $P2_1/n$. So, its X-rayed single crystal contains both of the enantiomers of (Fe^{2+})@**459**. The FeN_6-coordination polyhedra of (Fe^{2+})@**459** and (Fe^{2+})@**460** with heights h of approximately 2.32 Å possess an intermediate TP–TAP geometry with the average distortion angles φ of 23.4° and of 25.4°, respectively. The intramolecular O…Cl distance with $r_i = 2.85$ Å is described [87] to evidence the formation of the corresponding hydrogen bond in the molecule (Fe^{2+})@**460** (Fig. 3.67).

The X-ray structure of the hybrid phthalocyaninoclathrochelate, a mixed hafnium, boron-capped iron(II) cage complex (Fe^{2+})@**527**, is reported [88]. Its FeN_6-coordination polyhedron

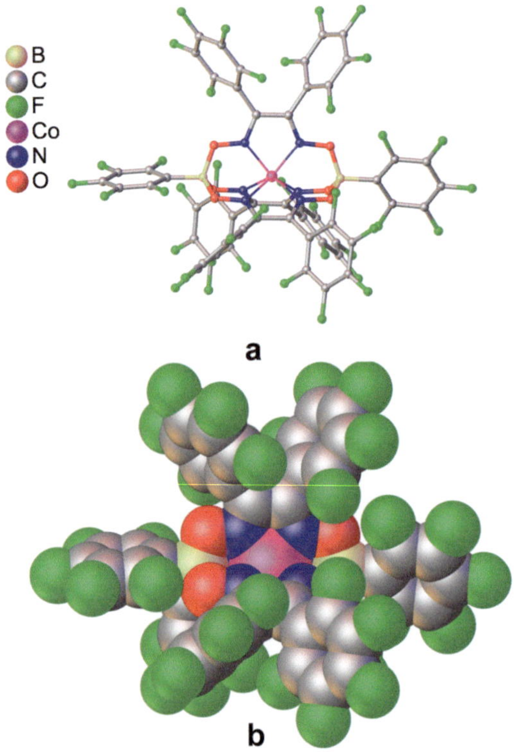

Fig. 3.65 X-ray structure of the molecule (Co^{2+})@**290** with atoms shown as *spheres* (**a**) and its spacefilling representation (**b**): formation of the second superhydrophobic shell

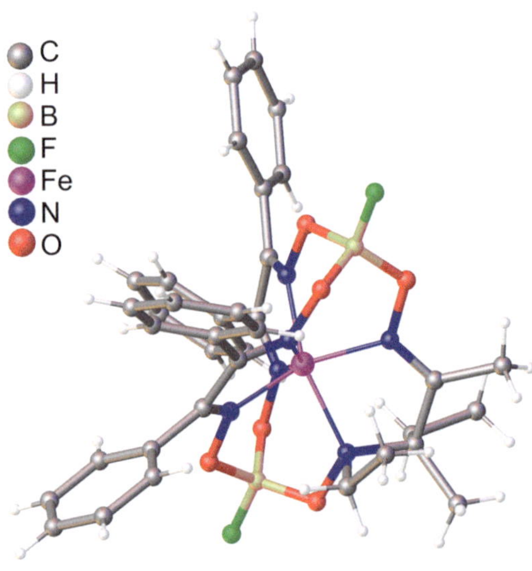

Fig. 3.66 X-ray structure of the hydroxylamine-containing clathrochelate (Fe^{2+})@**462** [86]

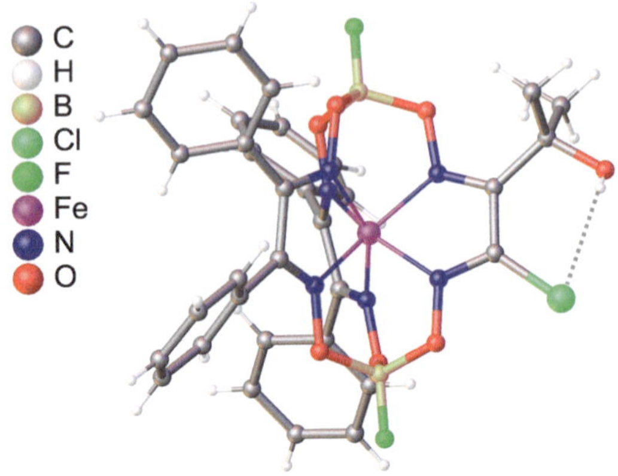

Fig. 3.67 X-ray structure of the clathrochelate (Fe^{2+})@**460** and the intramolecular hydrogen bond in its molecule, depicted with *dashed line* [87]

possesses a truncated trigonal-pyramidal geometry (due to non-equivalence of two capping fragments of a clathrochelate framework) with the distortion angle φ of approximately 30°. An apparent stacking between the "base-to-base"--oriented neighboring macrobicyclic molecules is characteristic feature of the X-rayed crystal of (Fe^{2+})@**527** (Fig. 3.68). π-systems of two coplanar "overlapping" pyrrol fragments of their phthalocyanine macrocyclic bases interacted at the distance of 3.35 Å. Four coordinated nitrogen atoms of these macrocyclic ligands are located in one plane, whereas the N_4O_3-heptacoordinate hafnium(IV) ion is above their N_4-mean plane by approximately 1.1 Å. Each of the phthalocyanine macrocyclic fragments of (Fe^{2+})@**527** was found [88] to be not planar: the dihedral angle between its planar fragments with a bend along the N. . .N(opposite) line is equal to 10.3°.

A comparative study of the X-ray structures of the hybrid porphyrinoclathrochelates (Fe^{2+})@**530**, (Fe^{2+})@**532**, and (Fe^{2+})@**533** (Fig. 3.69) has been performed [89]. The FeN_6-polyhedra of the complexes (Fe^{2+})@**532** and (Fe^{2+})@**533** with equivalent capping groups possess an intermediate TP–TAP geometry with the distortion angles φ of approximately 33°. The Fe–N distances for their zirconium- and hafnium-capped tripodal fragments are greater than those for the boron-cross-linked apical group of (Fe^{2+})@**530** (the average values are

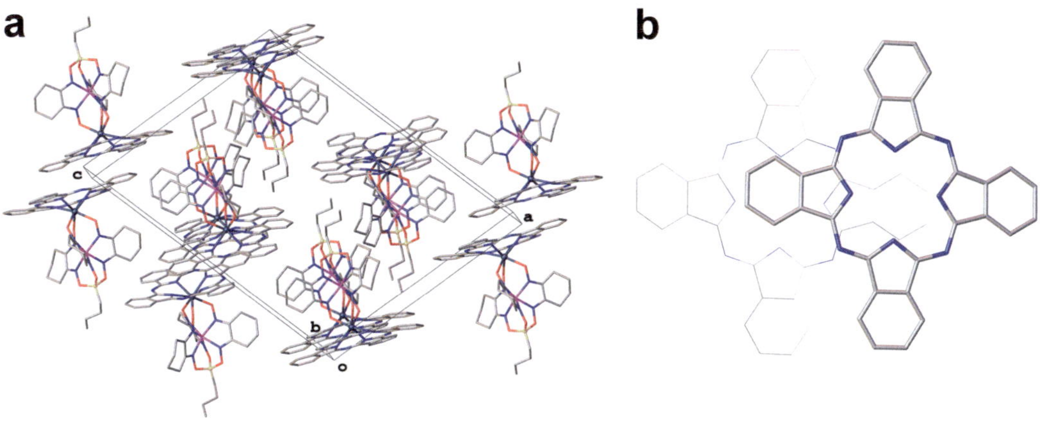

Fig. 3.68 Crystal packing of (Fe^{2+})@**527** · $3C_6H_6$ (**a**) and stacking interactions between the neighboring phthalocyanine fragments of two molecules of (Fe^{2+})@**527** in its clathrochelate dimer (**b**) [88]

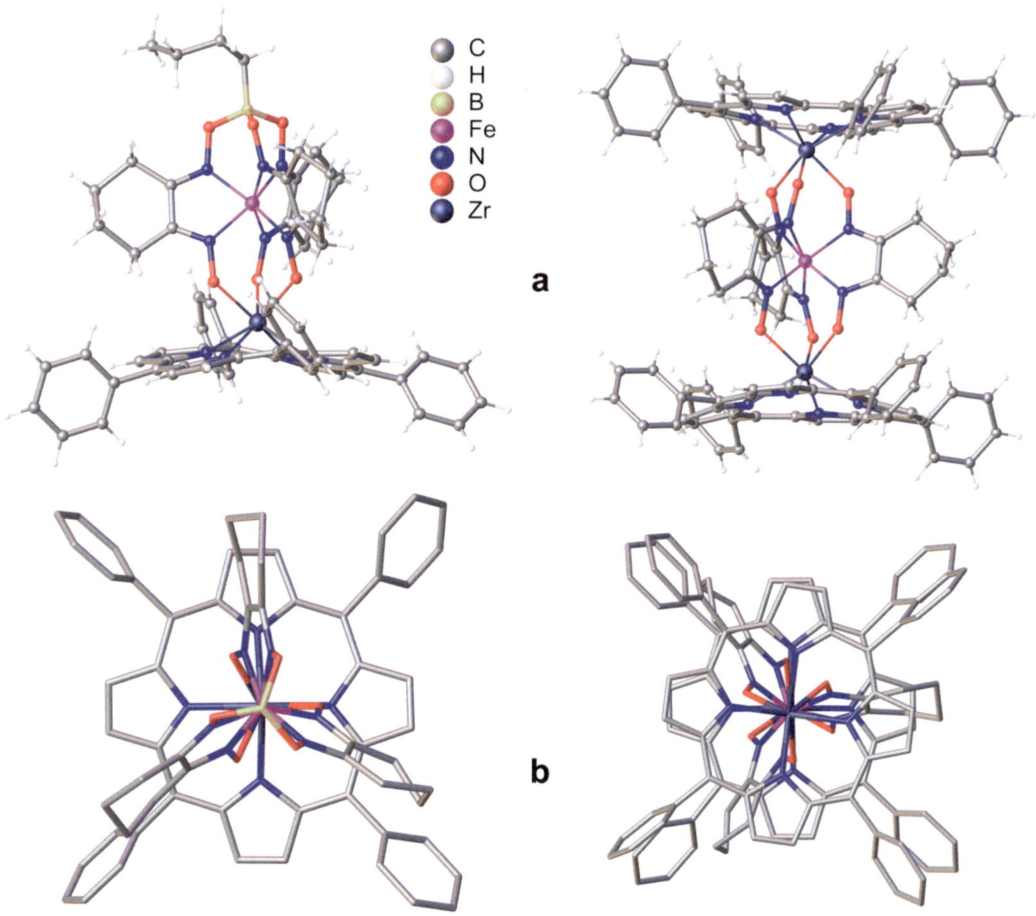

Fig. 3.69 *Side* (**a**) and *top* (**b**) views of the molecules (Fe^{2+})@**530** (on left) and (Fe^{2+})@**532** (on right) [89]

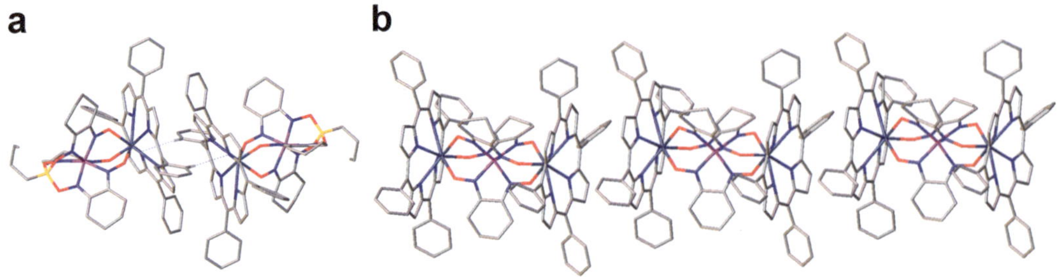

Fig. 3.70 Formation of a clathrochelate dimer in the crystal (Fe^{2+})@**530** · 0.5 C_6H_{14} (**a**) and the intermolecular stacking and C–H...N interactions in the crystal (Fe^{2+})@**532** · C_8H_{18} · 0.25 C_7H_8 (**b**) [89]

1.92 Å and 1.90 Å, respectively). As a result, the encapsulated iron(II) ion in the latter boron, zirconium-capped macrobicyclic molecule is slightly shifted from the center of its FeN_6-polyhedron along the molecular M...Fe...B C_3 pseudoxis in the direction of the apical boron atom. Coordination polyhedron of (Fe^{2+})@**530** possesses the geometry of a distorted truncated trigonal pyramid with φ of approximately 30° and h equal to 2.30 Å. The above φ and h values are intermediate between those for its TP diboron- and TAP dizirconium(dihafnium)-capped iron (II)-encapsulating analogs with equivalent capping groups. The apical heptacoordinate zirconium and hafnium(IV) ions in the molecules (Fe^{2+})@**530**, (Fe^{2+})@**532**, and (Fe^{2+})@**533** coordinate the oxygen atoms of three donor oxime fragments of their three ribbed chelate α-dioximate fragments and four nitrogen atoms of the corresponding porphyrin macrocycle. These nitrogen atoms are located in almost one plane, whereas a central metal(IV) ion is deviated from this plane at the distance of approximately 1.1 Å. These macrocycles are slightly bended along their N...N (opposite) line with the corresponding dihedral angles from 6.4 to 16.9°. Despite this bent and the presence of four bulky phenyl substituents, the X-rayed crystal (Fe^{2+})@**530** · 0.5 C_6H_{14} contains the clathrochelate dimers formed *via* the intermolecular C–H...N and stacking interactions (Fig. 3.70). The infinite stacks are also observed in the crystals (Fe^{2+})@**532** · C_8H_{18} · 0.25 $C_6H_5CH_3$ and (Fe^{2+})@**533** · 0.33 C_8H_{18} (Fig. 3.70), and their solvate molecules are situated between these stacks [89].

The electronic and molecular structures of the hybrid hafnium-, zirconium(IV)-, and lutetium (III)-capped iron(II) phthalocyaninoclathrochelates (Fe^{2+})@**527**, (Fe^{2+})@**526**, and (Fe^{2+})@**529** have been studied [90] using the single crystal X-ray diffraction, UV-vis, MCD, and (spectro)electrochemical experimental and the theoretical DFT and TD-DFT calculations as well. UV-vis and MCD data and the above theoretical predictions evidenced that their electronic structure can be described as a superposition of the weakly interacting metal phthalocyanine and clathrochelate fragments, while the spectroelectrochemical data and DFT calculations suggested that the HOMO in all these hybrid complexes is localized on their phthalocyanine fragment. As follows from single crystal X-ray diffraction data, FeN_6-coordination polyhedron of the zirconium(IV)-capped phthalocyaninoclathrochelate (Fe^{2+})@**526** (Fig. 3.71) possesses the geometry of a distortion truncated trigonal pyramide, approaching to an intermediate TP–TAP geometry with φ of 30.2°; its height h is equal to 2.30 Å. The Fe–N distances for the zirconium(IV)-capped tripodal fragment of (Fe^{2+})@**526** (*av.* 1.92 Å) are greater than those for its boron-capped fragment (approximately 1.90 Å). The zirconium(IV)-containing phthalocyanine terminal fragment is slightly bended and its heptacoordinate metallocenter is located by approximately 1 Å above the phthalocyanine macroring; it coordinates four nitrogen atoms of this macroheterocycle and three oxygen atoms of the donor oxime groups of three ribbed chelate α-dioximate fragments, thus giving its ZrN_4O_3-

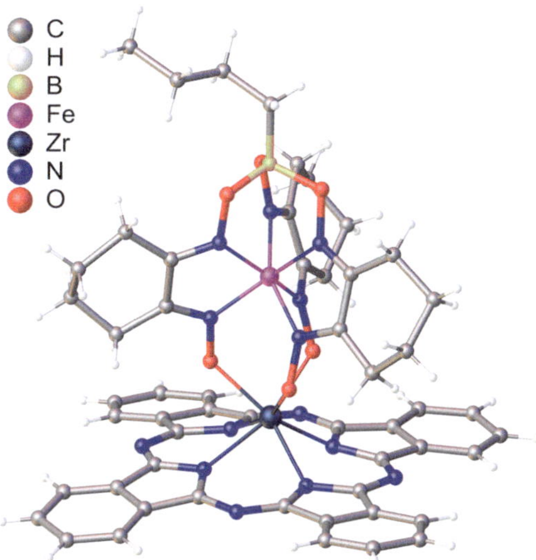

Fig. 3.71 X-ray structure of the hybrid phthalocyanino-clathrochelate (Fe^{2+})@**526** [90]

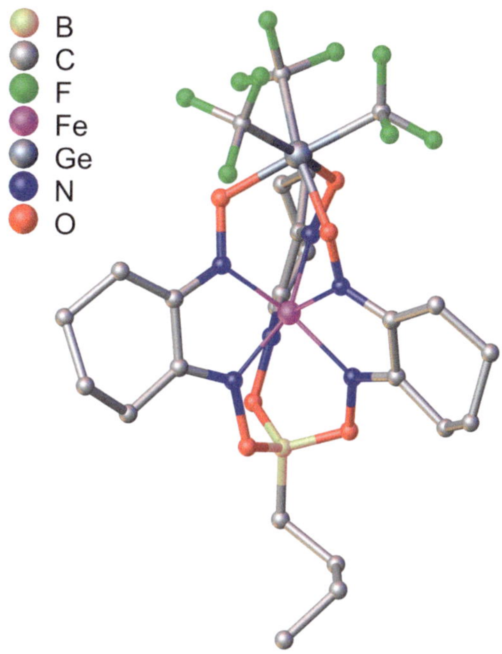

Fig. 3.72 X-ray structure of the macrobicyclic anion (Fe^{2+})@**275** [91]

coordination polyhedron. The intermolecular partial stacking interactions between the phthalocyanine terminal fragments of the neighboring hybrid cage molecules, forming a clathrochelate dimer with the closest contacts between its pyrrole units of approximately 3.5 Å, have been observed in the X-rayed crystal of (Fe^{2+})@**526** [90].

The X-ray structure of the boron, germanium-capped clathrochelate anion (Fe^{2+})@**275** (Fig. 3.72) is reported [91]. Its FeN_6-coordination polyhedron possesses the geometry of a truncated trigonal pyramid that is close to intermediate between a TP and a TAP with φ of 27.6°. Such geometry of this coordination polyhedron of (Fe^{2+})@**275** is explained [91] by non-equivalence of its two capping fragments: most of the diboron-capped iron (II) clathrochelates with known X-ray structures possess a TP geometry, whereas that approaches a TAP in the case of their macrobicyclic analogs, which are capped with bigger Lewis-acidic ions, such as a hexacoordinate Ge^{4+} cation. The encapsulated iron(II) ion is situated almost in the middle between the mean planes of the N_3-bases of this coordination polyhedron: the distances from it to these planes are equal to 1.19 and 1.14 Å for the boron- and germanium-capped tripodal fragments, respectively. So, this caged iron(II) ion is only slightly shifted from the geometrical center of a clathrochelate framework of (Fe^{2+})@**275** in its direction of the germanium(IV)-containing apical fragment.

The X-ray structures of the clathrochelates (Fe^{2+})@**215** and (Fe^{2+})@**212** with terminal carboxylic groups are described [92]. Their encapsulated low-spin iron(II) ions have the characteristic distorted TP–TAP coordination polyhedra in which the Fe–N distances fall in the narrow ranges 1.908–1.916 and 1.904–1.931, respectively. Two carbon atoms of the terminal carboxylic groups in the latter cage complex are approximately 17.92 Å apart. The X-rayed crystal of the polymeric derivative of the clathrochelate ligand (Fe^{2+})@**212**, the coordination polymer *57*, contains 2D sheets in which its doubly deprotonated macrobicyclic molecules play a role of the bridges between the zinc(II)-containing paddlewheel cores (Fig. 3.73); these zinc(II) ions also apically coordinate DMA molecules. Each layer of *57* is identical in its composition and orientation, but rather

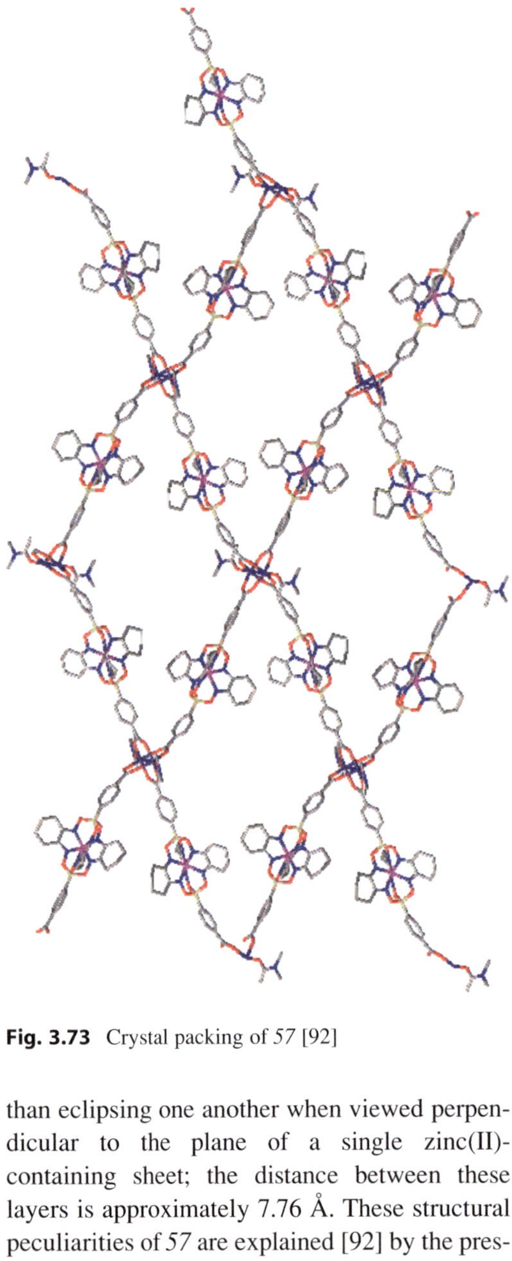

Fig. 3.73 Crystal packing of *57* [92]

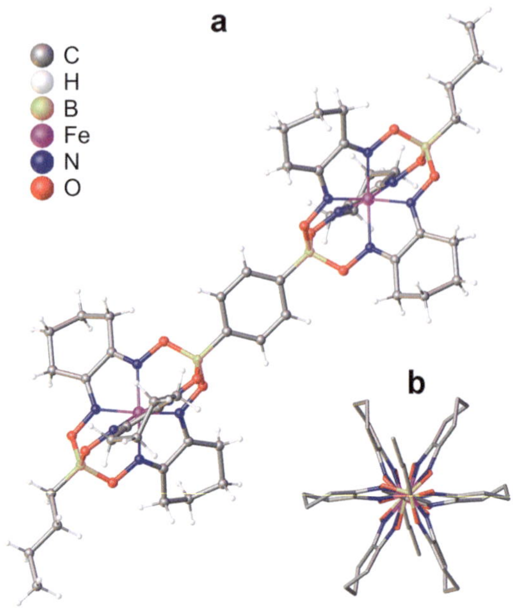

Fig. 3.74 *Side* (**a**) and *top* (**b**) views of the bis-clathrochelate molecule $(Fe^{2+})_2$@**277** [93]

than eclipsing one another when viewed perpendicular to the plane of a single zinc(II)-containing sheet; the distance between these layers is approximately 7.76 Å. These structural peculiarities of *57* are explained [92] by the presence of the above apically coordinated DMA molecules and by bulkiness of the cyclohexane ribbed substituents at a cage framework as well. 1D channels in the X-rayed crystal of *57* are filled with disordered solvent guest molecules, permeating the material: their removal is reported [92] to cause a total collapsing of its coordination network structure, thus giving a nonporous material.

In the X-ray structure of the iron(II) -bis-clathrochelate $(Fe^{2+})_2$@**277** (Fig. 3.74a), the Fe–N distances fall in a narrow range close to 1.90 Å [93]. Both its FeN_6-coordination polyhedra possess the geometry intermediate between a TP and a TAP with the average distortion angles φ of approximately 20.4°. The rigidity of a phenylene spacer fragment of the molecule $(Fe^{2+})_2$@**277** caused an appearance of its rotational B...Fe...B...B...Fe...B molecular C_3 pseudoaxis lying through both the macrobicyclic entities of a bis-clathrochelate framework. These macrobicyclic entities have a staggered conformation of their ribbed cyclohexane chelate fragments (Fig. 3.74b) [93].

The quantum chemical calculations of the -boron-capped iron(II) tris-dioximates (Fe^{2+})@**669**, (Fe^{2+})@**670**, (Fe^{2+})@**671**, (Fe^{2+})@**672**, and (Fe^{2+})@**673** shown in Scheme 3.11 have been performed [94]. Their apical substituents that are reported only slightly affect the main CT effects; these computations revealed also the strong electrostatic character of bonding within their capping tetrahedral BO_4 fragments.

The molecular and electronic structures of the cobalt(I, II and III) cage complexes of the -fluoroboron-capped tris-dimethylglyoximate ligand **668** have been theoretically analyzed [95]. The performed DFT and TD-DFT

Scheme 3.11 Chemical drawing of a series of the boron-capped iron(II) tris-dioximates

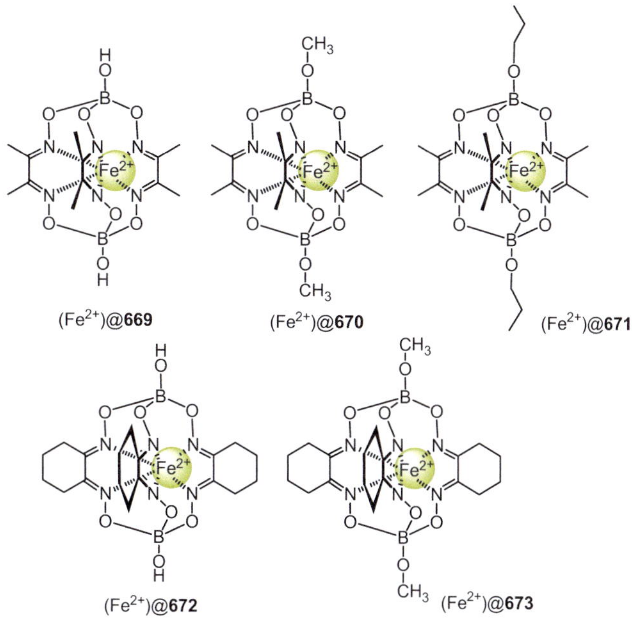

calculations confirmed the known experimental data for the above cobalt(II) and cobalt(III) clathrochelates. The reversibility of the $Co^{2+/+}$ reduction of an encapsulated cobalt ion has been evidenced [95] by the presence of the isosbestic points on both the direct reduction and the backward reoxidation waves. The reduced cobalt(I) clathrochelate monoanion is a high-spin paramagnetic complex with $s = 1$, having a highly symmetric cage framework with the molecular C_3 pseudoaxis. Substantial mixing of $d\pi$-type orbitals of this caged ion with the imine-type orbitals of **668** is reported [95] to be responsible for the NIR spectrum of its cobalt(I) cage complex; one or even two electrons can be transferred from this encapsulated ion to the orbitals of **668** (i.e., a metal-to-ligand back donation). As a result, an excess of electronic density is localized on the imine carbon atoms of its quasiaromatic cage framework as evidenced by occupation number of the corresponding orbitals of **668** [95].

As follows from single crystal X-ray diffraction data of [96], the cavities of the clathrochelate-based octahedral coordination capsules **583** and **584** are almost completely occupied by solvent guest molecules and NO_3^- or BF_4^- anions as well. These anions are located in their interior binding pockets close to the vertex palladium(II) ions. The cavity volumes of these coordination cages are approximately 5200 Å^3, but they are reported [96] to be mostly occupied by the ribbed substituents of the macrobicyclic ligand syntones. As a result, these cages have the central cavities with a hydrophobic internal environment formed by methyl or cyclohexane ribbed substituents of the clathrochelate synthons; their void volumes have been deduced [96] to be approximately 1400/1800 Å^3, 1300 Å^3, and 1100 Å^3, respectively.

Single crystal X-ray structures of a series of clathrochelate-based coordination cages **591** (Fig. 3.75a), **592** (Fig. 3.76a), **593** (Fig. 3.75b), and **594** (Fig. 3.76b) and those in solution, obtained using various NMR techniques, are described [97]. The dodecanuclear coordination cages **591** and **592** were found to crystallize as a racemic mixture of their $\Delta\Delta\Delta\Lambda$ and $\Lambda\Lambda\Lambda\Delta$ isomers. The former cage with a cavity volume of approximately 480 Å^3 has the molecular C_3 pseudosymmetry axis, while the latter coordination capsule with such volume of 377 Å^3 possesses its crystallographic C_3 symmetry. The

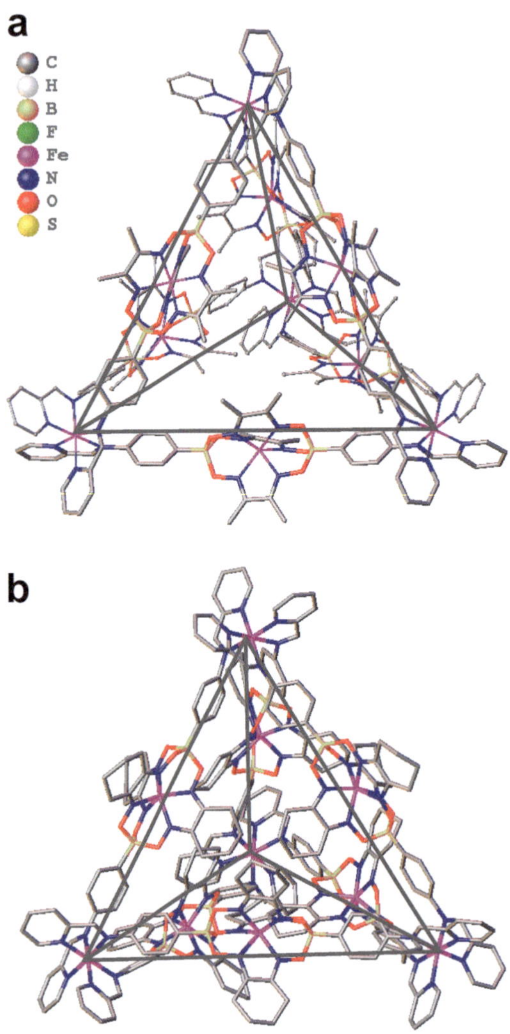

Fig. 3.75 X-ray structures of the coordination cages **591** (**a**) and **592** (**b**) [97]

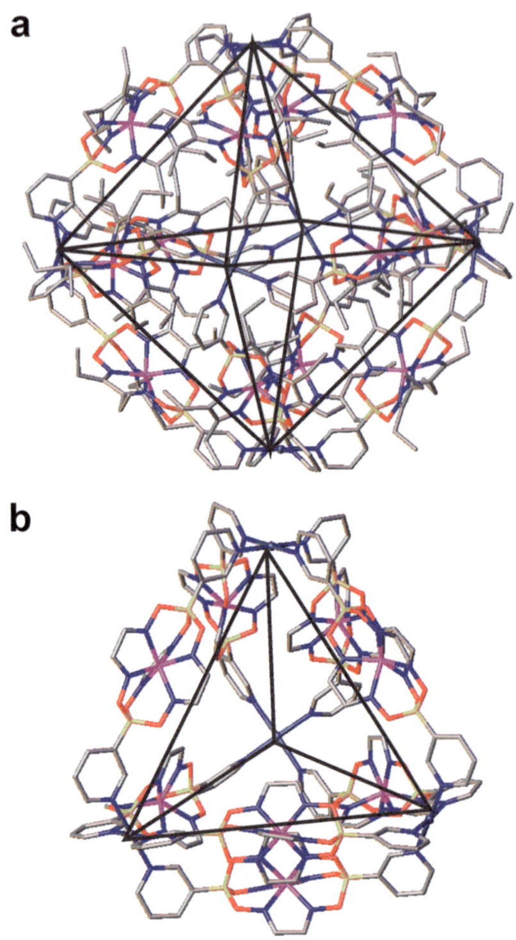

Fig. 3.76 X-ray structures of the coordination cages **593** (**a**) and **594** (**b**) [97]

average distances between their four FeN_6-vertices are approximately 20 Å, while the average Pd...Pd distances within the triangular faces of the coordination cage **593** are close to 16 Å. At the same time, the solution NMR data of [97] revealed a reduced symmetry of the cage **594** as compared with that of the coordination capsule **593**: instead of one set of signals for the bridging clathrochelate ligand synthons in the latter case, two sets of signals with equal intensity have been observed for **594**. This evidenced that the above symmetry reduction has caused no restricted rotational freedom of these bridging synthons, but the presence of two kinds of such macrobicyclic synthons, which are magnetically distinct. DOSY NMR spectrum of **594** in its DMSO-d_6 solution confirmed the presence of only one type of assembly with a diffusion constant of 7.64×10^{-7} cm^2 s^{-1}, being larger than that for **593** in the same solvent (6.14×10^{-7} cm^2 s^{-1}), thus suggesting that **594** has a smaller size. As follows from high-resolution mass spectrometry data, it has a Pd_4L_8-framework with two doubly bridged and four singly bridged Pd...Pd edges. Such geometry of **594** has been confirmed [97] by single crystal X-ray diffraction experiment (Fig. 3.76b). Four capping palladium(II) ions form a slightly distorted tetrahedron with the

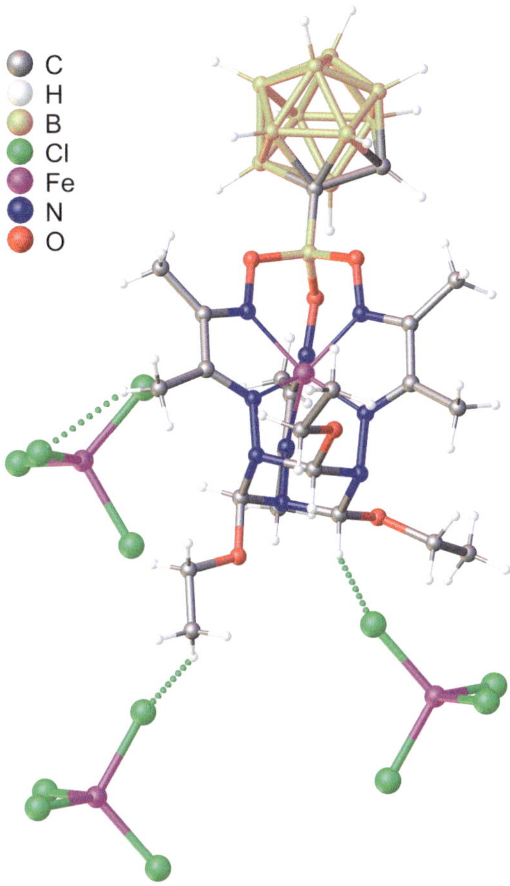

Fig. 3.77 X-ray structure of the *ortho*-carboranyl-capped clathrochelate $((Fe^{2+})@\mathbf{497})^+ \cdot (FeCl_4^-)$ [98]

Pd. . .Pd distances for the doubly bridged edges being slightly smaller than those for the singly bridged connections (14.7 Å and 15.5 Å, respectively). The distance between two encapsulated iron(II) ions in the doubly bridged fragment of **594** is only 5.7 Å, thus suggesting a tight packing of these two clathrochelate ligand synthons [97].

The X-ray structures of the *ortho*-carboranyl- and *n*-butylboron-capped iron(II) oximehydrazonate cations $[(Fe^{2+})@\mathbf{497}]^+$ and $[(Fe^{2+})@\mathbf{498}]^+$ (Figs. 3.77 and 3.78) have been compared [98]. Main geometrical parameters of their TP–TAP distorted cage frameworks are characteristic of the boron-capped clathrochelate tris-oximehydrazonates. All the ethoxy groups and the semiclathrochelate fragment of $(Fe^{2+})@\mathbf{498}$ are located on the same side of the

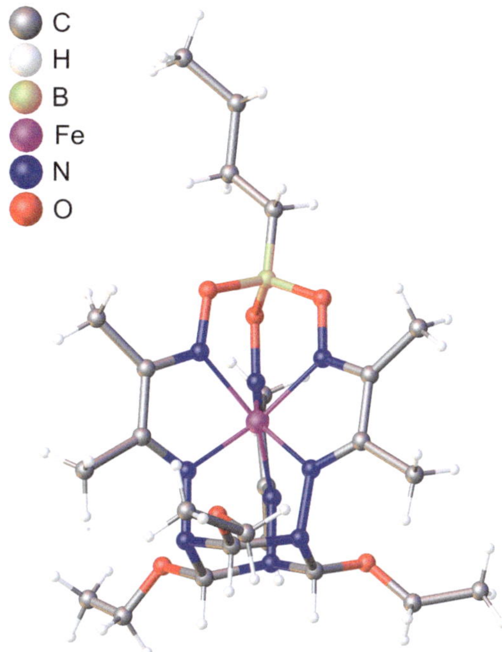

Fig. 3.78 X-ray structure of the *n*-butylboron-capped oximehydrazonate clathrochelate cation $((Fe^{2+})@\mathbf{498})^+$ with atoms shown as spheres [98]

capping 1,3,5-triazacyclohexane ring; so, this macropolycyclic molecule has a common *syn,syn,syn*-conformation. On the other hand, the clathrochelate cation $(Fe^{2+})@\mathbf{497}$ has an unusual *syn,syn,anti*-orientation of the ethoxyl groups in its cross-linking 1,3,5-triazacyclohexane fragment [98].

The X-rayed crystal $((Fe^{2+})@\mathbf{497})^+\cdot (FeCl_4^-)$ is stabilized by the Cl. . .H–C hydrogen bonds between its counteranion and an *anti*-oriented ethoxyl substituent at a clathrochelate framework of this cage complex (Fig. 3.77). The formation of the oximehydrazonate clathrochelate cation $((Fe^{2+})@\mathbf{497})^+$ with such unusual *syn,syn,anti*-orientation of its pendant ethoxyl substituents is explained [98] by the presence of the bulky carboranyl apical substituent in **497**: one hydrogen atom of this polyhedral substituent, being located between the oximehydrazonate chelate fragments of this encapsulating ligand, caused the sterical clashes for one of the ethoxyl groups on the first stage of macrobicyclization of the corresponding carboranylboron-capped

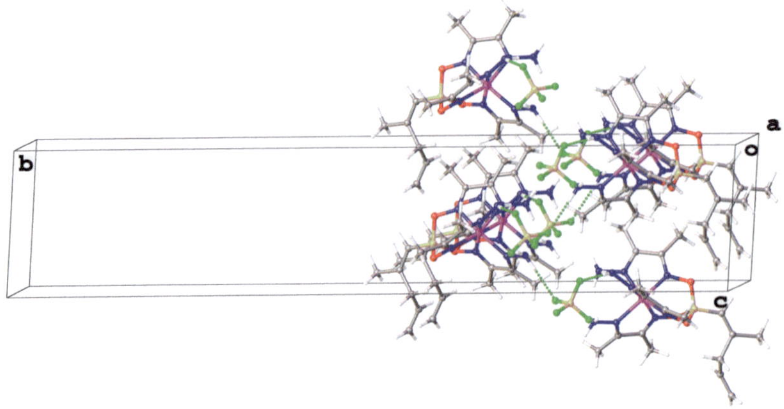

Fig. 3.79 Crystal packing of the semiclathrochelate *31*·(BF_4) [99]

semiclathrochelate precursor with TOF, thus resulting in its *anti*-orientation.

The X-ray structures of the 1,4-pentadienylboron-capped semiclathrochelate and clathrochelate iron(II) oximehydrazonate cations *31* and $((Fe^{2+})@$**500**$)^+$ are described [99]. The Fe–N distances in the macrobicycle $(Fe^{2+})@$**500** are smaller for its boron-capped tripodal fragment than those for the 1,3,5-triazacyclohexane cross-linking moiety. The height *h* of an intermediate TP–TAP coordination polyhedron of the semiclathrochelate *31* is smaller than that of its macrobicyclic derivative $(Fe^{2+})@$**500** (2.24 Å and 2.35 Å, respectively). This fact is explained [99] by sterical clashes caused by macrobicyclization of the above semiclathrochelate precursor, resulting in a TP distortion of its FeN_6-coordination polyhedron around the molecular C_3 pseudoaxis (the angles φ are equal to 36.1° and 20.8°, respectively). So, such macrobicyclization of *31* caused the rotational–translation elongation of this FeN_6-coordination polyhedron. All the ethoxyl substituents of the macropolycyclic cation $(Fe^{2+})@$**500** and its semiclathrochelate fragment have the same orientations with respect to the 1,3,5-triazacyclohexane cross-linking group, thus adopting a common *syn,syn,syn*-configuration.

The ionic associates in the crystal *31* · (BF_4^-) (Fig. 3.79) are formed by the corresponding Coulombic interactions as well as by two strong N–H…F hydrogen bonds [99]. Each of its anions forms such bonds with three semiclathrochelate cations *31*, and each of these cations is bound with four BF_4^- anions. As a result, the infinite H-bonded layers have been observed in this crystal; these layers are connected with each other by weak C–H…O interactions. In the crystal $((Fe^{2+})@$**500**$)^+$ · (BF_4^-) · C_6H_6 (Fig. 3.80), only the Coulombic and weak intermolecular interactions were found [99]. Its BF_4^- anion is surrounded by one solvate benzene molecule and three clathrochelate cations $(Fe^{2+})@$**500**. One of them and BF_4^- anion are hydrogen-bonded with the C…F and H…F distances of 3.667, 3.336 Å, and 2.687, 2.422 Å, respectively, while two other macrobicyclic cations are involved in the dipole–dipole C–H…F interactions through their methyl substituents [99].

The ferrocenylboron-capped oximehydrazonate semiclathrochelate cation *25* (Fig. 3.81) and its clathrochelate derivative $(Fe^{2+})@$**495** (Fig. 3.82) have been X-ray structurally characterized [100]. The apical cyclopentadienyl fragments of these complexes are almost parallel; they adopt a hindered conformation with deviation angles of 2.2° and of 8.8°, respectively. The distortion angles φ of their TP–TAP FeN_6-coordination polyhedra are substantially different (37.4° and 22.7°, respectively), while the average Fe–N distances and the bite angles α persist. As a result, the height *h* of such polyhedron for *25* is substantially lower (2.23 Å) than that of its

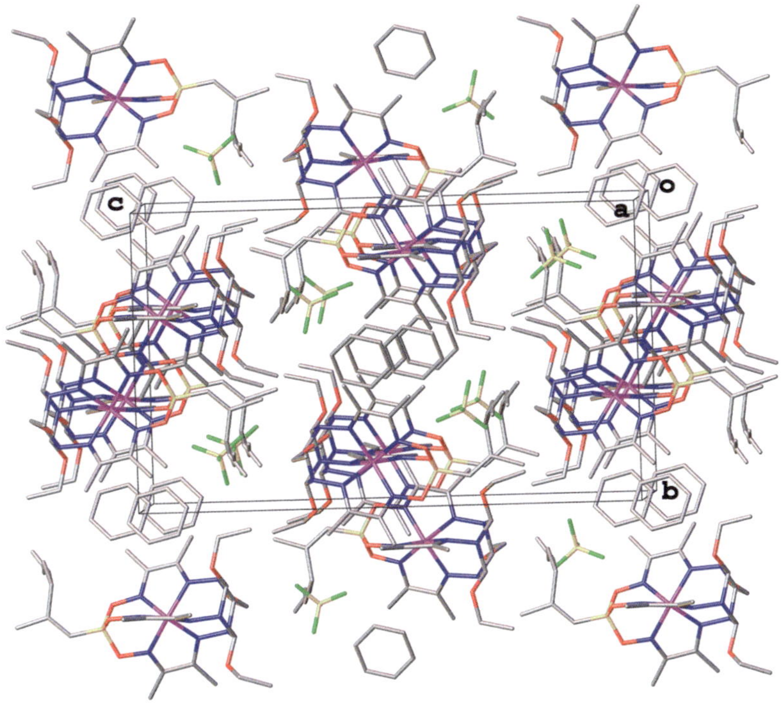

Fig. 3.80 Crystal packing of the clathrochelate $((Fe^{2+})@\mathbf{500})\cdot(BF_4)\cdot C_6H_6$ [99]

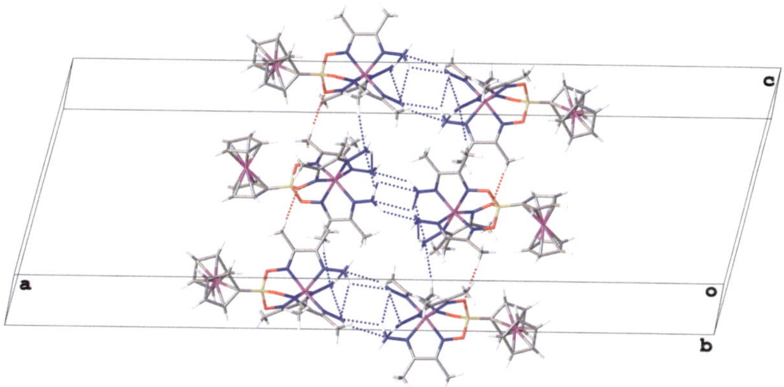

Fig. 3.81 Packing of the semiclathrochelate cations *25*; the N–H...N and C–H...O hydrogen bonds are depicted with *dashed line* [100]

clathrochelate derivative (2.34 Å). So, FeN_6-coordination polyhedron underwent a rotational–translational elongation in the course of this macrobicyclization, causing the additional sterical clashes in a rigid quasiaromatic clathrochelate framework of (Fe^{2+})@**495** and, therefore, leading to a more TP geometry. All the ethoxyl substituents and the semiclathrochelate fragment of this macrobicyclic complex have the same orientations relative to the capping 1,3,5-triazacyclohexane moiety, thus suggesting a *syn,syn,syn*-conformation of **495** [100].

In the X-rayed crystal of the semiclathrochelate cation *25*, the hydrogen atoms of its hydrazonate NH_2-groups are involved in the system of intramolecular and intermolecular N–H...N hydrogen bonds and those with counteranions as well. These complex

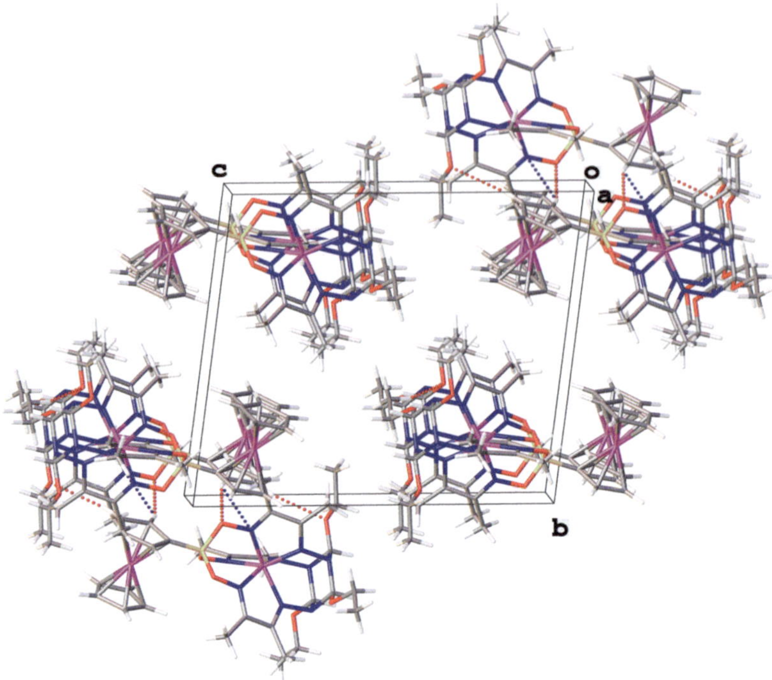

Fig. 3.82 Packing of the clathrochelate cations $[(Fe^{2+})@\mathbf{495}]^+$; the Cp–H. . .O hydrogen bonds are depicted with *dashed line* [100]

cations give the corresponding dimers shown in Fig. 3.81 by intermolecular N–H. . .N hydrogen bonding: two supramolecular bonds of this type have been observed in this crystal. These dimers form the chains *via* the weak C–H. . .O contacts between the ribbed methyl substituents of the oximehydrazonate ligand synthons and their oxime donor fragments. The X-rayed crystal of its clathrochelate derivative $(Fe^{2+})@$**495** do not contain such hydrogen bonds, and the macrobicyclic entities have been arranged in this crystal (Fig. 3.82) according to the "bump-into-hollow" principle; these entities also form the clathrochelate dimers with opposite orientations of their macrobicyclic frameworks *via* the Cp–H. . .O attractive interactions [100].

The X-ray structures of a hybrid iron(II) clathrochelatoscorpionate, the dioximatopyrazoloximate cage complex $(Fe^{2+})@$**534**, and its pyrazoloxime-armed macrocyclic intermediate *46* have been compared [101]. Despite the semiclathrochelate character of the former complex, the average B–O, O–N, C=N, and C–C bond lengths in the encapsulating macrobicyclic ligand **534** and in its pyrazol-armed macrocyclic analog were found to be characteristic of the boron-capped iron(II) tris-dioximates; the Fe–N distances in the clathrochelate $(Fe^{2+})@$**534** and those for the α-benzildioximate chelate fragments of *46* (1.88–1.94 Å) are almost the same. On the other hand, the Fe–N distances for five-membered pyrazoloximate chelate cycle of *46* are substantially greater (approximately 1.97 Å). The FeN_6-coordination polyhedra of these complexes possess an intermediate TP–TAP geometry; that of $(Fe^{2+})@$**534** is slightly elongated (its height h is equal to 2.31 Å and the average distortion angle φ is approximately 24.6°) as compared with FeN_6-coordination *polyhedron of the molecule *46* (h = 2.28 Å and $\varphi = 29.6°$). The oxygen atom of the solvate ethanol molecule in the X-rayed crystal of *46* · C_2H_5OH forms two hydrogen bonds, thus giving a pseudobridging fragment and, therefore, a pseudomacrobicyclic structure of this pyrazoloxime-armed complex (Fig. 3.83). Its central iron(II) ion is shifted from the center of FeN_6-coordination polyhedron in the direction

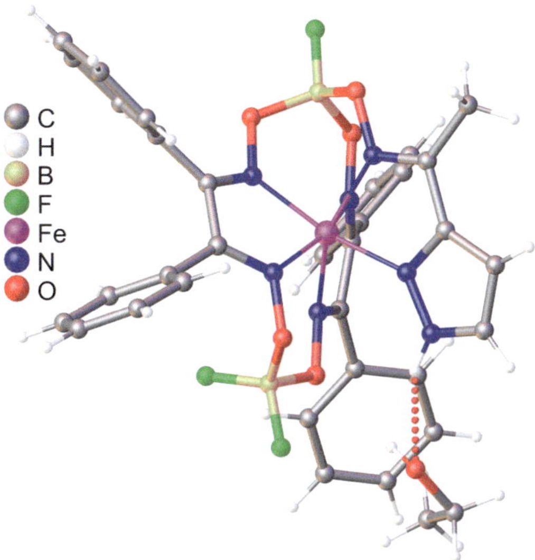

Fig. 3.83 X-ray structure of the macrocyclic complex *46*·C_2H_5OH showing its hydrogen-bonded solvate ethanol molecule; the corresponding hydrogen bonds are depicted with *dashed line* [101]

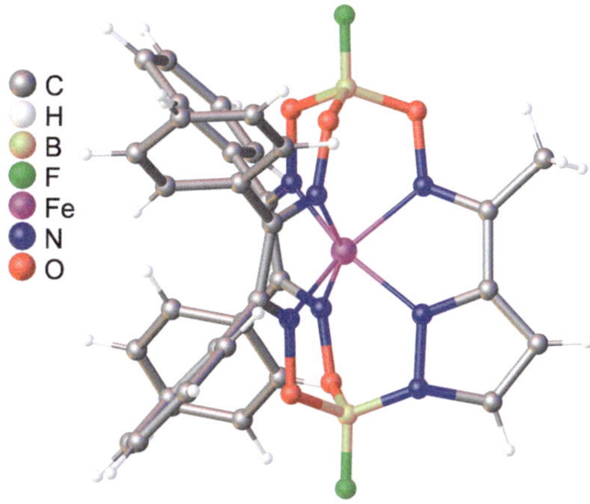

Fig. 3.84 X-ray structure of the clathrochelato-scorpionate (Fe^{2+})@**534** [101]

of the capping O_3BF fragment, whereas the encapsulated iron(II) ion in the molecule (Fe^{2+})@**534** (Fig. 3.84) is located in the center of its cage framework [101].

As follows from single crystal X-ray diffraction data of [102] for the pseudoclathrochelate metal(II) tris-pyrazoloximates (M^{2+})@**509**, the B–O, O–N, N=C, and C–C distances in their pseudomacrobicyclic ligand **509** are characteristic of both their boron-capped tris-dioximate clathrochelate analogs and the above hybrid iron(II) clathrochelatoscorpionate (Fe^{2+})@**534**. At the same time, the absence of one rigid apical group and the lability of their chloride-containing pseudocapping fragment allowed an elongation of the tripodal pseudocapped tris-pyrazolyl fragment of **509** as compared with its boron-containing tripodal cap. The MN_6-coordination polyhedra of these complexes possess a truncated trigonal pyramidal, that is close to only slightly distorted quasi-TP geometry with the distortion angles φ between its nonequivalent N_3-bases of approximately 0°. The average N...N distances in their boron-capped tripodal fragment vary from 2.77 to 2.84 Å, thus being substantially lower than those in their μ_3-Cl^--pseudocapped tris-pyrazolyl fragment (3.13–3.40 Å). The M–N distances fall in a narrow range for all these pseudoclathrochelates, while the pseudoencapsulated metal ions are situated in the geometrical centers of their pseudomacrobicyclic frameworks [102]. The above distances for *3d*-transition metal pseudoclathrochelates are characteristic of their high-spin complexes increasing in a row: cobalt (II) (*av.* 2.13 Å) < iron(II) (*av.* 2.18 Å) < manganese(II) (*av.* 2.24 Å). The pseudomacrobicyclic frameworks underwent an elongation–shortening along the B...M...Cl molecular C_3 pseudoaxis, thus allowing to bind a series of high-spin *3d*-metal ions. On the other hand, these frameworks do not undergo a rotational distortion around this axis: their φ angles did not exceed 4°. The above elongation of the tripodal tris-pyrazoloximate framework allowed its pseudocapping (cross-linking) with μ_3-Cl^- anion through the formation of three N–H...Cl hydrogen bonds, stabilizing a pseudomacrobicyclic structure (see, for example, Fig. 3.85). So, the geometry of the resulting pseudoclathrochelate ionic associates is the same for all these pseudocage complexes. Fourth lone electron pair of this chloride counterion allowed it to be involved in the intermolecular hydrogen bonding; the C–H...Cl-bonded dimers (Fig. 3.86) have been observed [102] in all their X-rayed crystals.

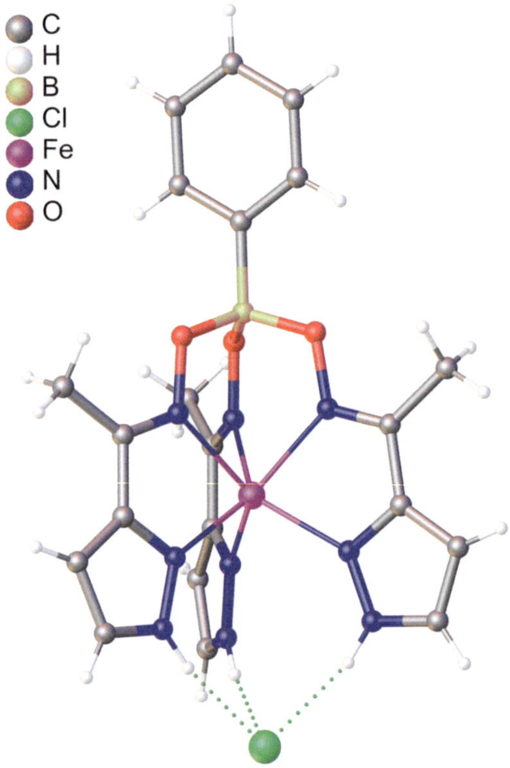

Fig. 3.85 X-ray structure of the quasiclathrochelate complex (Fe^{2+})@**509** [102]

The structures and the formation energies of 1:1 alkali and alkaline earth cage complexes of the macrobicyclic tris-bipyridinate and tris-phenantrolinate ligands **519** and **674** (Scheme 3.12) have been theoretically predicted [103] using quantum chemical calculations; the results obtained have been used to evaluate their cationic selectivity. Both these caging ligands are reported to strongly bind Ca^{2+} and Sr^{2+} ions; **519** also prefers an encapsulation of K^+ over other alkali cations, while **674** favors Na^+ and K^+ ions [103].

The X-ray structure of 1:1 copper(II) cage complex of the dipyridine-containing encapsulating ligand **5** (Fig. 3.87) is described [104]. Its CuN_5-coordination polyhedron possesses the geometry of a distorted square pyramid; the aromatic and benzylic nitrogen atoms form the basal plane of this polyhedron, while the methylated nitrogen atom occupies its apical position. The caged metal ion of (Cu^{2+})@**5** lies

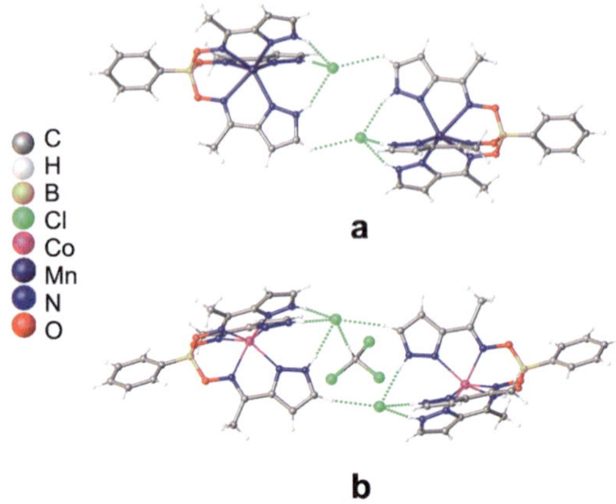

Fig. 3.86 Formation of C–H...Cl-bonded dimers and other supramolecular interactions in the crystals (Mn^{2+})@**509** · C_2H_5OH (**a**) and (Co^{2+})@**509** · $CHCl_3$ (**b**); the corresponding hydrogen bonds are depicted with *dashed line* [102].

at approximately 0.077 Å above this basal plane [104].

As follows from single crystal X-ray diffraction data of [105], in 1:1 copper(II) cage complex of the pyridine-strapped, cyclic bis(benzimidazole) caging ligand **6** (Fig. 3.88), CuN_5O-coordination polyhedron of its hexacoordinate encapsulated metal ion possesses a pseudooctahedral geometry. Four internal nitrogen donor atoms of **6** are coordinated in an equatorial fashion toward this ion, while two remaining donor oxygen and nitrogen atoms are provided by the solvate methanol molecule and the strap-incorporated pyridine group, respectively. The Cu–O and Cu–N distances are equal to 1.98 and to 1.99 Å and, therefore, the encapsulated copper (II) ion lies above the equatorial N_4-mean plane by approximately 0.31 Å in the direction of its axially bound methanol molecule. Two sets of the Cu–N bonds have been observed in the clathrochelate molecule (Cu^{2+})@**6,** and this fact is explained [105] by the structural Jahn–Teller distortion of electronic configuration d^9 of its encapsulated copper(II) ion.

The X-ray structures of the tris-bipyridinate N_8O_6-caging ligand **625** and its 1:1 nickel(II) cage complex are described [106]. Both the helical enantiomers of **625** are present in its X-rayed centrosymmetric crystal, and in both these cases

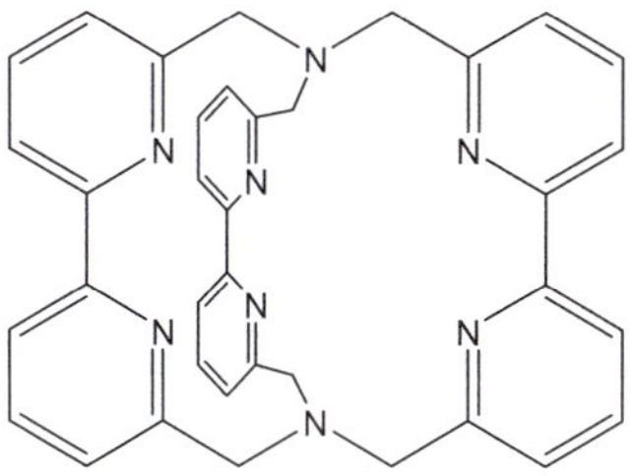

519

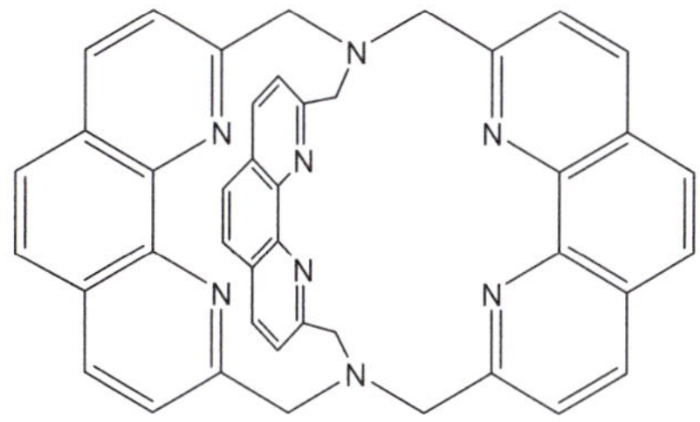

674

Scheme 3.12 Chemical drawings of the macrobicyclic compounds **519** and **674**

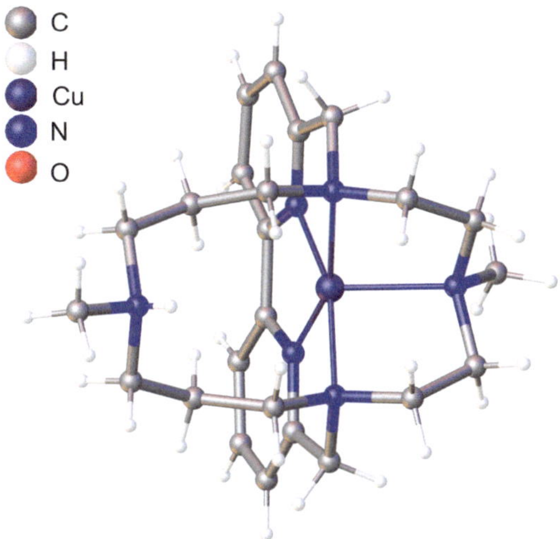

Fig. 3.87 X-ray structure of the clathrochelate (Cu^{2+})@**5** [104]

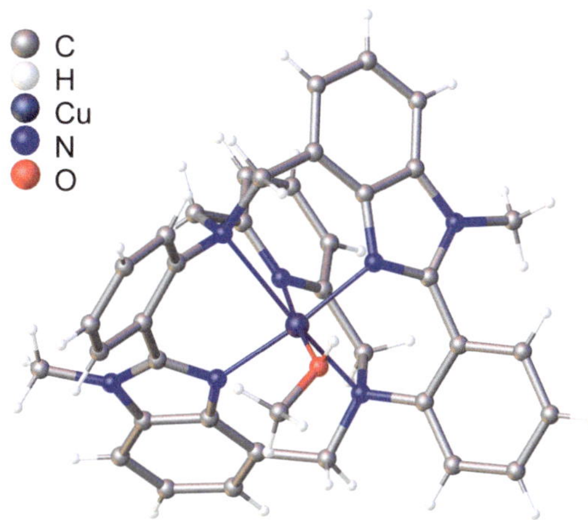

Fig. 3.88 X-ray structure of the clathrochelate dication (Cu^{2+})@**6** [105]

a bipyridyl ribbed moiety of one enantiomer of **625** is slotted into the "groove" formed by two analogous fragments of its complementary enantiomer. The X-rayed crystal of this macrobicyclic complex contains its two crystallographically independent molecules (types **A** and **B**), each located on the molecular C_3 pseudoaxis, forming the clathrochelate dimers *via* the groove interactions. The N. . .N distances between the bridgehead nitrogen atoms of **625** and of its nickel(II) complex are approximately 13.27 and 13.45 (for type **A**), 12.57 Å (for type **B**), respectively. NiN_6-coordination polyhedron of an encapsulated nickel(II) ion of the clathrochelate (Ni^{2+})@**625** possesses a slightly distorted pseudooctahedral geometry; its molecule adopts an *exo*-configuration of each of its bridgehead nitrogen lone pairs, and this encapsulated metal ion promotes a triple-helical distortion of the caging ligand **625** that is elongated in its above cage complex by approximately 22 Å [106].

The X-ray structures of the pyridinophane N_4O_2-caging ligand **66** and its 1:1 lithium(I) cage complex has been compared [107]. An encapsulation of the lithium(I) cation by **66** caused the small movement of an ethylenedioxy moiety of this encapsulating ligand, whereas the positions of its nitrogen atoms persist. The encapsulated lithium (I) ion in its clathrochelate complex (Li^{+})@**66** coordinates all the donor atoms of a caging ligand, thus forming LiN_4O_2-coordination polyhedron [107].

The encapsulated indium(III) ion in its 1:1 cage complex with the macrobicyclic bis-phthalocyanine macrobicyclic ligand **522** [108] coordinates six nitrogen donor atoms of this ligand (Fig. 3.89), thus forming a distorted TP–TAP InN_6-coordination polyhedron with the In–N distances in the range 2.203–2.267 Å. A cage framework of **522** consists of the phthalocyaninate platform of four isoindole units, which is bridged by two additional isoindole rings. So, it is described [108] as two phthalocyaninate anions having one common half. Two sp^3-hybridized bridgehead carbon atoms of this macrobicyclic ligand belong to two opposite isoindole rings of the above phthalocyaninate moieties, and the dihedral angles between two neighboring planes of these phthalocyaninate moieties are equal to 105.9, 120.0, and 134.1°. The mutual arrangement of the clathrochelate molecules (In^{3+})@**522** in an unit cell of its X-rayed crystal is shown in Fig. 3.90. The shortest intermolecular distance between their two neighboring phthalocyaninate moieties is equal to 3.25 Å, thus suggesting the weak stacking interactions between them. These clathrochelate molecules adopt a propeller-like shape, and the ribbed fragments of the adjacent cage molecules were found [108] to be interpenetrated.

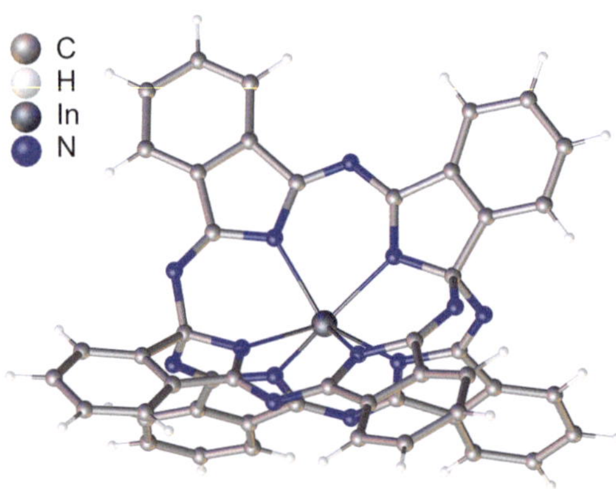

Fig. 3.89 X-ray structure of the clathrochelate (In^{3+})@**522** [108]

The X-ray structures of Hg_2^{2+}-encapsulating cage complexes of the platinum(0)- and palladium(0)-capped D_3-symmetric coordination capsules **605** and **606** have been compared [109]. The encapsulated dimeric dication Hg_2^{2+} coordinates the phenanthroline groups of two phosphorous-containing ribbed fragments of the above macrobicyclic ligands, and their molecular C_3 pseudoaxes are oriented nearly perpendicularly to the Hg...Hg axis of this caged dication; the Hg...Hg separations are equal to 2.688 Å and

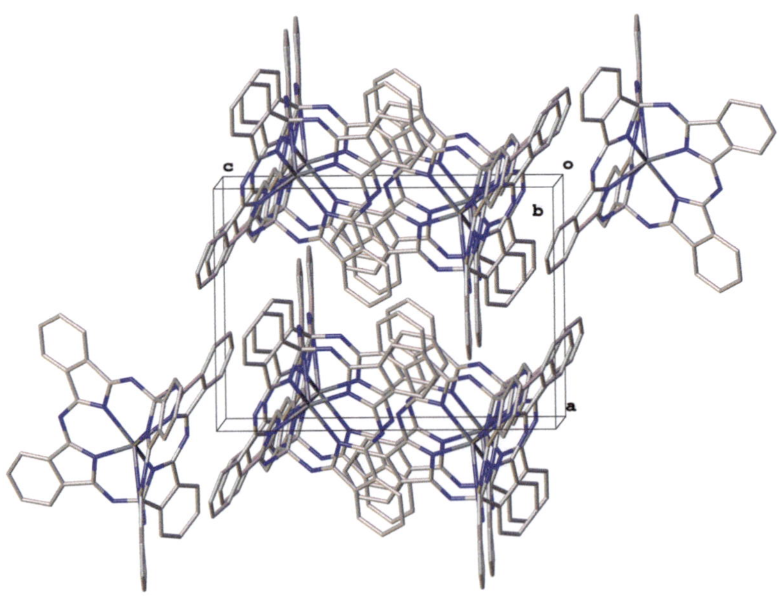

Fig. 3.90 An unit cell of the crystal (In^{3+})@**522** [108]

2.736 Å, respectively. These encapsulated dimeric guests strongly and nearly symmetrically interact with the capping metallocenters of **605** and **606**, having the trigonal coordination polyhedra; the average distances Hg_2^{2+}...Pt and Hg_2^{+}...Pd are equal to 2.77 and 2.81 Å. Coordination polyhedra of their encapsulated Hg_2^{2+} dications possess a distorted trigonal geometry, and the observed short Hg...Hg separations are explained [109] by twisting of two apical capping PtP_3 or PdP_3 fragments.

3.3 Polynuclear Cage Complexes and Their Derivatives

Detailed analysis of the X-ray structures of the cage complexes, the derivatives of the Schiff-base encapsulating ligands **540** and **554–557** and their non-macrocyclic analogs (Scheme 3.13), has been performed [110]. Their molecular geometries are affected by both the oxidation and spin states of an encapsulated metal ion as well as by matching of its size, depending on the above two parameters, to that of the cavity of a clathrochelate ligand.

The X-ray structure of the heterodinuclear silver(I), copper(I) 1:1:1 cage complex of the imine macrobicyclic ligand **537** with intramolecular Ag...Cu separation of 7.422 Å is described [111] to evidence no interaction between its two co-encapsulated metal ions. The caged silver (I) ion coordinates one bridgehead and three imine nitrogen atoms of **537**, thus forming its tetrahedral AgN_4-coordination polyhedron, the geometry of which is slightly distorted from a trigonal pyramid. The encapsulated copper (II) ion binds two imine and one bridgehead

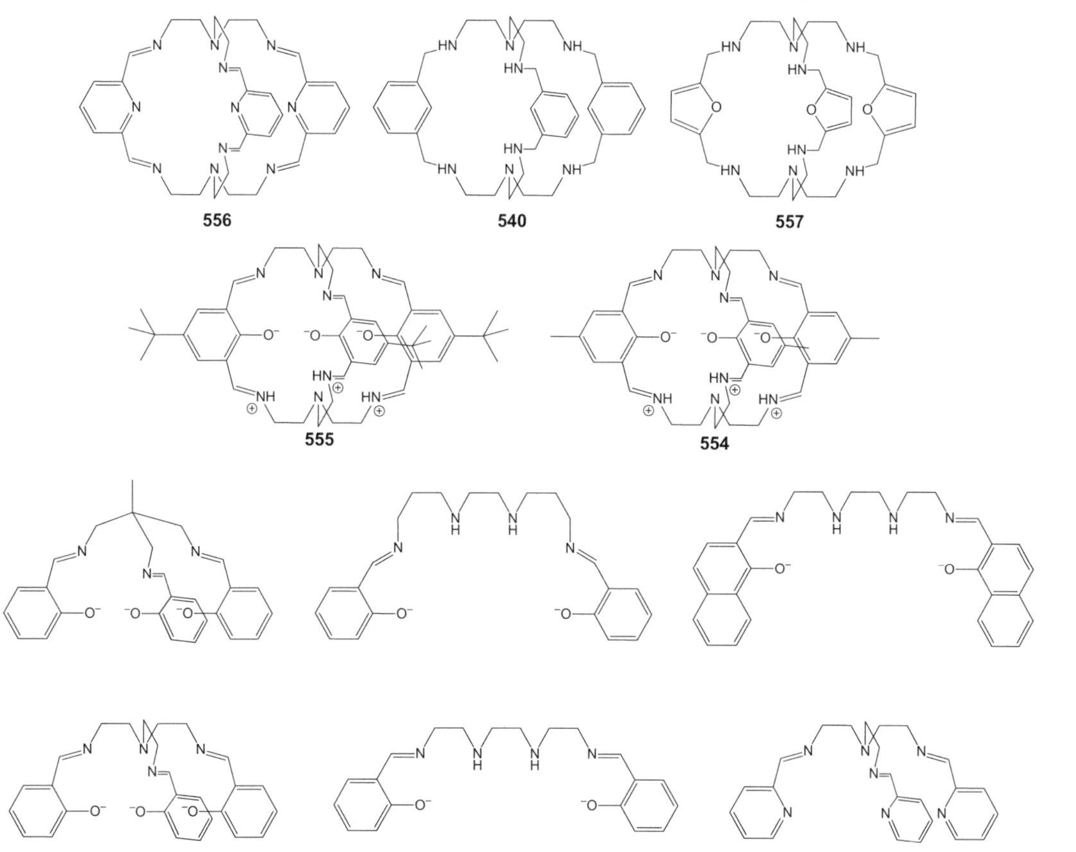

Scheme 3.13 Chemical drawing of a series of the binucleating macrobicyclic ligands and their open-chain analogs

nitrogen atoms of **537**; this ion is located by approximately 0.21 Å below their N_3-mean plane; its CuN_3-coordination polyhedron possesses a distorted planar triangular geometry [111].

The X-ray structures of the Schiff-base caging ligand **538** and its homodinuclear 1:2 copper (I) cage complex have been compared [112]. This hexaimine ligand has an elongated ellipsoidal shape and its molecular C_3 pseudoaxis lies through the bridgehead nitrogen atoms of **538**. In the complex $(Cu^+)_2$@**538**, each of the two co-encapsulated copper(I) ions is located in the cavity of this ligand close to its end and coordinates four nitrogen atoms of a tripodal aminetriimine capping fragment of **538**. The intramolecular Cu…Cu separation between these copper(I) cations is approximately 11.07 Å. Their CuN_4-coordination polyhedra possess an unusual tetrahedral geometry with this metal cation lying by 0.2 Å outside of its N_4-coordination environment toward the geometrical center of a caging ligand [112]. The N…N separations between the bridgehead nitrogen atoms of a metal-free compound **538** and its copper(I) cage complex are almost the same (15.5 Å). On the other hand, an encapsulation of the copper(I) ions by **538** induced the substantial changes in its conformation: the bridgehead N–C bonds are staggered in a metal-free macrobicycle **538** having the molecular symmetry close to D_3, but they are eclipsed in its copper(I) cage complex in which a cage framework lost its C_3 symmetry axis [112].

As follows from X-ray diffraction data of [113], 1:2 homodinuclear silver(I) cage complex of the Schiff-base macrobicyclic ligand **537** is disordered in its X-rayed crystal. The fully ordered encapsulated silver(I) ion has an almost trigonal-pyramidal AgN_4-coordination polyhedron, formed by three imine and one tertiary amine nitrogen atoms in the trigonal plane and in the apical position, respectively. The Ag–N (apical) distance (2.457 Å) is substantially greater than three Ag–N(imine) distances (*av.* 2.29 Å). The major occupancy site of the second encapsulated silver(I) ion lies at end of the cavity of **537,** and this ion coordinates its two imine and one bridgehead amine nitrogen atoms as well as the oxygen atom of perchlorate counterion, thus forming AgN_3O-coordination polyhedron [113].

The X-ray structure of 1:2 homodinuclear copper(II) cage complex of the triple-deprotonated form of its tris-pyrrole macrobicyclic ligand **539** is described [114]. One of the co-encapsulated copper(II) ions has CuN_4-coordination polyhedron with the geometry intermediate between a tetrahedron and a planar square, while CuN_5-coordination polyhedron of the second caged cation possesses an almost square pyramidal geometry; the intramolecular Cu…Cu separation in the molecule $(Cu^+)_2$@**539** is equal to 3.006 Å [114].

As follows from X-ray diffraction data of [115], in 1:2 homodinuclear thallium(I) cage complex of the hexaamine macrobicyclic ligand **540** with the intramolecular Tl…Tl separation 4.376 Å each of its co-encapsulated thallium (I) ions coordinates one bridgehead and three secondary amine nitrogen atoms of this ligand and is displaced from the N_3-plane of the above secondary amine groups by approximately 1.1 Å toward the geometrical center of **540**. This macrobicyclic ligand has its fully extended conformation without helicity, thus allowing accommodation of these two large spherical guest cations [115].

Single crystal X-ray diffraction data of [116] evidenced that a metal-free coordination capsule **607** is capped by two iodine atoms as well as by two gold(I) ions, which are bridged by three 3,6-bis(diphenylphosphino)pyridazine ligand syntones, the nitrogen atoms in the pyridazine rings of which are oriented slightly away from the cavity of a cage framework. The AuP_3I-coordination polyhedra of the cross-linking gold (I) ions possess a distorted tetrahedral geometry and are located at the distances of approximately 0.53 Å from the center of its cavity; the intramolecular Au…Au separation in the molecule **607** is equal to 7.373 Å [116].

The X-ray structure of 1:2 homodinuclear silver(I) cage complex of the octaamine ligand **90** with intramolecular Ag…Ag separation of 2.814 Å is reported [117]. Each of its encapsulated silver(I) ions coordinates one bridgehead and three amine nitrogen donor atoms, a pair of these atoms from one ribbed fragment of **90** and one nitrogen atom from its other moiety of this type.

1:1:1 heterodinuclear praseodymium, ytterbium, and lutetium(III) cage complexes of the iminophenolate macrobicyclic ligand **554** with co-encapsulated zinc(II) ion and their homodinuclear 1:2 zinc(II)-encapsulating analog $(Zn^{2+})_2$@**554** have been X-ray structurally characterized [118]. Their caging ligand is helically wrapped around two co-encapsulated metal ions, thus giving the molecular C_3 pseudoaxis. The intramolecular lanthanide...Zn separations in the above heteronuclear cage complexes fall in the range 3.270–3.325 Å, while the intramolecular Zn...Zn distance in the corresponding homodinuclear zinc(II) clathrochelate is equal to 3.104 Å. Three five-membered chelate fragments of **554** adopt the $(\lambda\lambda\delta)_5$ or $(\delta\delta\lambda)_5$ conformations, while three pseudochelate rings formed by its coordination toward an encapsulated zinc(II) ion adopted the $(\lambda'\lambda'\lambda')_5$ or $(\delta'\delta'\delta')_5$ configurations. The conformations of the above clathrochelate cations in the solid state were found to be either Λ-$(\delta\delta\lambda)_5(\delta'\delta'\delta')_5$ or its enantiomeric form Δ-$(\lambda\lambda\delta)_5(\lambda'\lambda'\lambda')_5$. As follows from ^{1}H NMR data, these cage frameworks persist as time-averaged C_3 symmetry in solution and their conformations are described [118] as Λ-$(\delta\delta\delta)_5(\delta'\delta'\delta')_5$ or Δ-$(\lambda\lambda\lambda)_5(\lambda'\lambda'\lambda')_5$.

The mononuclear dysprosium(III) complex of an anionic form of the phenolic macrobicyclic ligand **554** is reported [119] to show an asymmetric positioning of the encapsulated metal ion into the cavity of this caging ligand. The dysprosium(III) cation coordinates its seven donor atoms as well as two oxygen atoms of a bidentate nitrate counterion. In the molecule of 1:2 homodinuclear dysprosium(III) cage complex of **554**, the co-encapsulated metal ions were found to be held into its cavity at a very short distance from each other: the intramolecular Dy...Dy separation is approximately 3.484 Å. Each of these co-caged nonacoordinate dysprosium(III) ions binds one bridgehead and three azomethine nitrogen atoms of **554**, its three μ_2-phenolate oxygen atoms, and two oxygen atoms of a bidentate nitrate counterion to form their DyN_4O_5-coordination polyhedra, possessing a distorted monocapped dodecahedral geometry. The anionic macrobicyclic ligand **554**–3H$^+$ adopted a *SSS-endo–endo*-conformation, and its cavity is substantially more elliptical in this homodinuclear cage complex than that in the above mononuclear clathrochelate [119].

The molecular structures of the paramagnetic mononuclear lanthanide(III) cage complexes of the potentially binucleating encapsulating ligand **554** in their aqueous solutions have been studied [120] using the ^{1}H NMR LIS and LIR methods with particular attention to the changes induced by the lanthanide ion contraction. The experimental Ln–N distances in these clathrochelates progressively decrease along the lanthanide series both in the solid state and in solution. Their solution structures are reported [120] to be in excellent agreement with the known single crystal X-ray structures of 1:1 neodymium and yttrium(III) complexes of **554**. Despite the observed "breaks" in the contact–pseudocontact shift separation plots of the ^{1}H LIS values near the middle of the lanthanide series, none of the calculated structural parameters showed the substantial difference between the cage complexes of light and heavy lanthanide ions; only a slight decrease of the metal-to-proton distances was found [120] for the latter clathrochelates.

The X-ray structures of 1:2 homodinuclear lanthanide(III) and heterodinuclear 1:1:1 lanthanide(III)–*3d*-metal(II) cage complexes of a triple-deprotonated form of the imine macrobicyclic ligand **554** shown in Scheme 3.14 have been compared [121]. Each of the co-encapsulated lanthanide(III) ions of the corresponding homodinuclear clathrochelate coordinates three phenolate oxygen atoms of **554**, three imine nitrogen atoms, and one bridgehead nitrogen atom as well as one nitrate counterion in a bidentate fashion, thus forming its LnN_4O_5-coordination polyhedron. In the heterodinuclear cage complexes with co-encapsulated copper or zinc(II) ions, the coordination polyhedron of their caged lantanide(III) ion is almost the same, except that nitrate counterion is coordinated in a monodentate fashion.

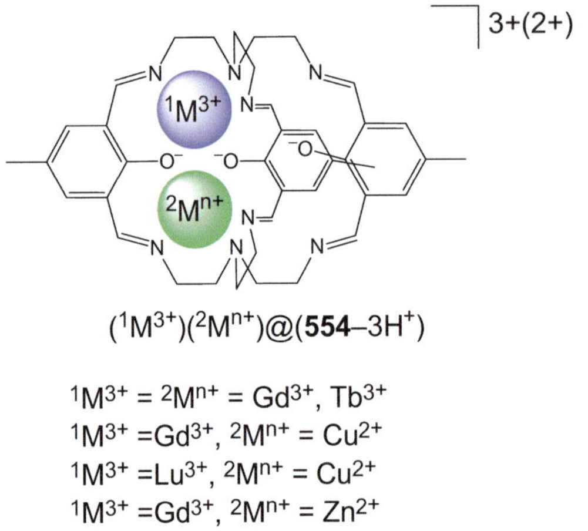

Scheme 3.14 Chemical drawing of 1:1:1 heterodinuclear cage complexes of the macrobicyclic ligand **554**–3H$^+$

As a result, this lanthanide ion is octacoordinate and its *LnN$_4$O$_4$*-coordination polyhedron possesses a distorted dodecahedral geometry. The co-encapsulated hexacoordinate metal (II) ion binds three μ_2-phenolate oxygen atoms and other three imine nitrogen atoms, thus giving its slightly distorted octahedral *MN$_3$O$_3$*-coordination polyhedron, two almost parallel triangular faces of which are defined by three phenolate oxygen atoms and by three imine nitrogen atoms. The anionic macrobicyclic ligand **554**–3H$^+$ adopts a *SSS-endo-endo*-conformation; its imine nitrogen atoms are pointed at the same side of the aromatic ring in three ribbed fragments, and the lone pairs of the bridgehead nitrogen atoms are directed toward its cavity. The rotational distortion of the macrobicyclic trianion [**554**–3H$^+$]$^{3-}$ around the molecular C_3 pseudoaxis resulted in the triple-helical structure of a cage framework, thus inducing its structural chirality. The TAP cavity of this framework is defined by six azomethine nitrogen atoms. Its N_3-faces are connected by three diazomethine ribbed fragments, generating two possible helical structures which correspond to two different (Λ or Δ) optical isomers of this caging ligand and thus suggesting its either left-handed (Λ) or right-handed (Δ) structural chirality about the molecular C_3 axis [121].

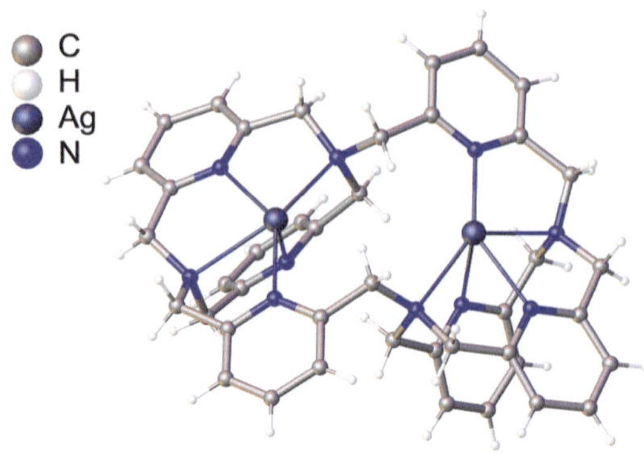

Fig. 3.91 X-ray structure of the clathrochelate (Ag$^+$)$_2$@**518** [122]

The X-ray structure of homodinuclear 1:2 silver(I) cage complex of the encapsulating ligand **518** [122] is shown in Fig. 3.91. Each of its two co-encapsulated Ag$^+$ ions coordinates three pyridine and two bridgehead nitrogen atoms of this ligand, thus forming a distorted pyramidal *AgN$_3$O$_2$*-coordination polyhedron. Each pyridinophane unit of **518** has a *syn*-configuration, while the methylene fragment of this unit adopts a *boat–boat*-conformation. The Ag–N (bridgehead) distance (*av.* 2.52 Å) is greater than the Ag–N (pyridine) bond length (*av.* 2.36 Å); the intramolecular Ag…Ag separation was found [122] to be approximately 4.84 Å, thus suggesting no interaction between these co-encapsulated metal ions.

As follows from single crystal X-ray diffraction data of [123], in the molecule of 1:1:1 heteronuclear europium(III), nickel(II) cage complex of the triple-deprotonated form of macrobicyclic ligand **541** (Fig. 3.92), the octacoordinate encapsulated europium(III) ion occupies one end of its cavity and *EuN$_4$O$_4$*-coordination polyhedron is formed by one bridgehead nitrogen atom, three imine nitrogen atoms, and three phenolate μ_2-oxygen atoms of 541, and one oxygen atom of DMF solvate molecule as well. The opposite end of this cavity is occupied by the nickel(II) ion having a distorted octahedral *NiN$_3$O$_3$*-coordination polyhedron, formed by three μ_2-phenolate oxygen atoms and three imine nitrogen atoms of **541**. The observed intramolecular Eu…Ni separation of 3.2204 Å

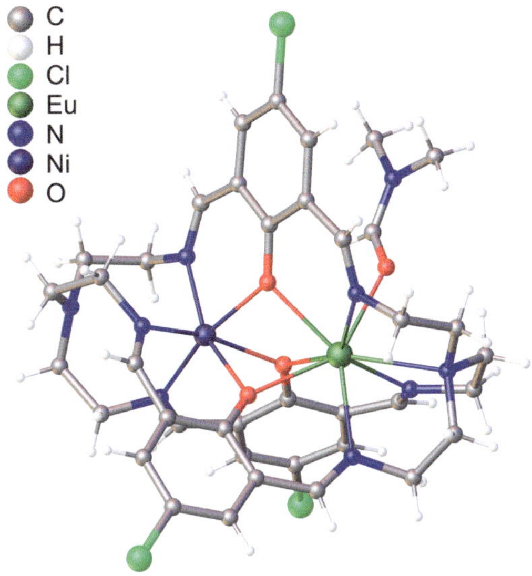

Fig. 3.92 X-ray structure of the clathrochelate dication $(Eu^{3+})(Ni^{2+})@(\mathbf{541}–3H^+)^{3-}$ [123]

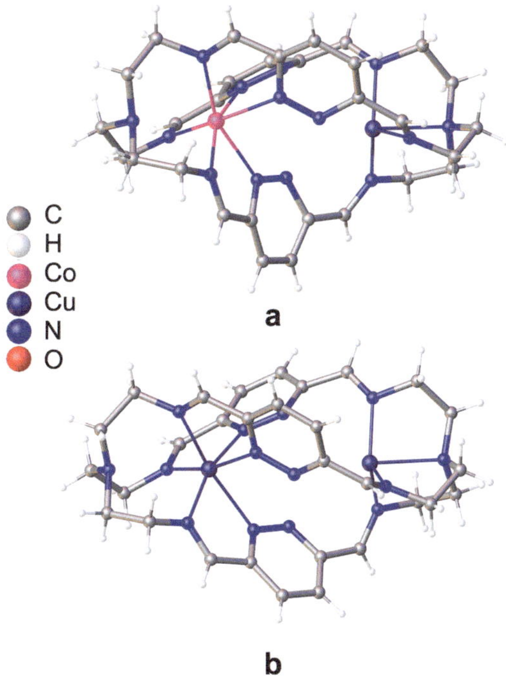

Fig. 3.93 X-ray structures of the clathrochelate trications $(Co^{2+})(Cu^+)@\mathbf{545}$ (**a**) and $(Cu^{2+})(Cu^+)@\mathbf{545}$ (**b**) [124]

evidenced a weak interaction between these co-encapsulated metal ions [123].

The X-ray structures of three heterodinuclear 1:1:1 cage complexes of the encapsulating ligand **545** have been compared [124]. In the case of tetrafluoroborate and perchlorate salts of heterodinuclear cobalt(II), copper (I) clathrochelate trication (Fig. 3.93a), its co-encapsulated cobalt(II) and copper(I) ions are end-for-end disordered within the cavity of **545**, three ribbed fragments of which are also disordered. The caged cobalt(II) ion has a distorted octahedral CoN_6-coordination polyhedron, formed by three pyridazine and three imine nitrogen atoms. The co-encapsulated copper(I) ion has a distorted trigonal-pyramidal CuN_4-coordination polyhedron, formed by three imine nitrogen donor atoms and one bridgehead nitrogen atom of **545**. In order to provide the binding site for this copper(I) ion, it adopted a conformation with mutually *cis*-imine groups. In the analogous heterodinuclear copper(II), copper(I) clathrochelate trication (Fig. 3.93b), its two co-encapsulated metal ions possess the substantially different geometries: the caged copper(I) ion has a distorted trigonal-pyramidal CuN_4-coordination polyhedron, while the encapsulated copper (II) ion has CuN_6-polyhedron with a Jahn–Teller distorted TP–TAP geometry with the distortion angle φ of 45.4°. The encapsulating ligand **545** in the latter cage complex also adopted its conformation with *cis*-imine groups [124].

The X-ray structures of 1:2 copper(I) cage complexes of the Schiff-base binucleating caging ligands **543** and **544** and that of the heterodinuclear copper(I), silver(I) clathrochelate complex of **544** are described [125]. The intramolecular Cu…Cu separation in the clathrochelate dication $(Cu^+)_2@\mathbf{543}$ (Fig. 3.94a) is approximately 4.23 Å, and the observed substantial deviation from the planarity in the bis-iminophenylene moiety of **543** is explained by sterical clashes in the central fragment of its cage framework. In the copper(I) cage complex of the furane-based macrobicyclic ligand **544** (Fig. 3.94b), the intramolecular Cu…Cu separation is substantially greater (4.54 Å), while the intramolecular Cu…Ag separation in the heterodinuclear cage complex of **544** (Fig. 3.94c) is intermediate

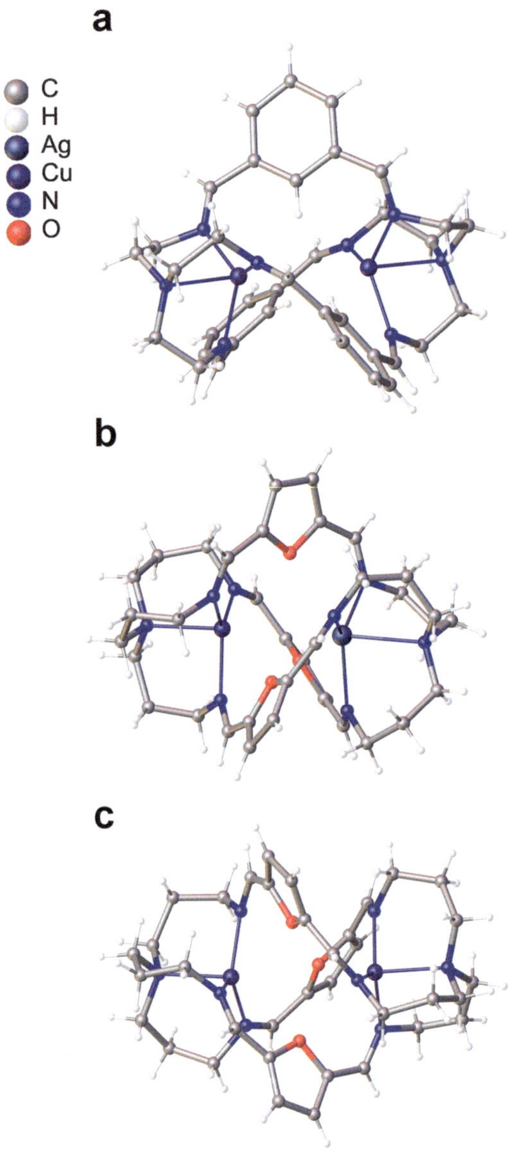

Fig. 3.94 X-ray structures of the homodinuclear clathrochelate dications $(Cu^{+})_2$@**543** (**a**) and $(Cu^{+})_2$@**544** (**b**) and that of their heterodinuclear analog $(Cu^{+})(Ag^{+})$@**544** (**c**) [125]

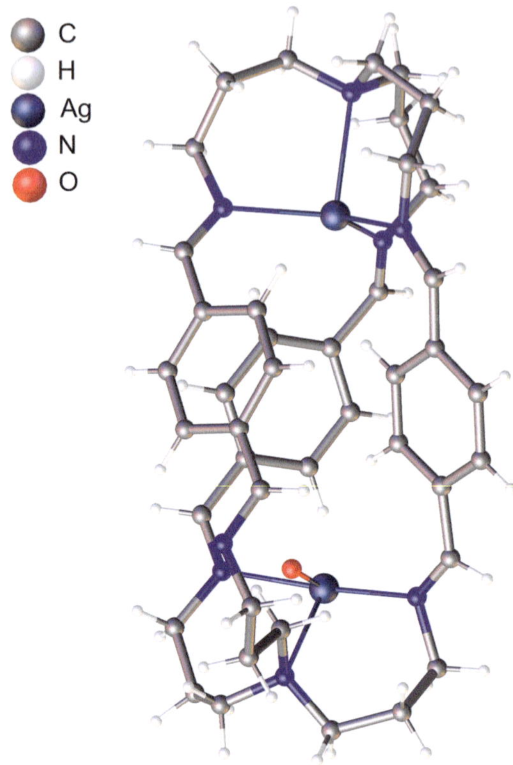

Fig. 3.95 X-ray structure of a mixed-valence silver(I), silver(III) cage complex of **537** [126]

between those of its homodinuclear copper(I)- and silver(I)-encapsulating clathrochelate analogs [125].

Single crystal X-ray diffraction study of a mixed-valence Ag^{I},Ag^{III}-encapsulating 1:1:1 cage complex of the macrobicyclic ligand **537** (Fig. 3.95) has been performed [126]. The coordination polyhedra of its co-encapsulated metal ions are substantially different. The caged silver(I) cation binds one bridgehead and three imine nitrogen atoms, thus forming a distorted tetrahedral $Ag^{I}N_4$-coordination polyhedron. The Ag^{I}–N distances fall in the range 2.29–2.46 Å, and a silver(I) ion is deviated by approximately 0.19 Å above the N_3-mean plane of the above polyhedron. On the other hand, the encapsulated silver(III) ion coordinates two imine nitrogen atoms and one oxygen atom, thus giving its trigonal and planar N_2O-coordination environment. The Ag^{III}–N distances are substantially (by approximately 0.1 Å) smaller than those for the above co-encapsulated silver(I) cation; the intramolecular $Ag^{I}...Ag^{III}$ separation between them is approximately 7.41 Å [126].

The detailed analysis of the X-ray structures of the copper(II) mononuclear, 1:2 and 1:1:1 $Cu^{I}Cu^{I}$-, $Cu^{I}Cu^{II}$-, and $Cu^{II}Cu^{II}$-homo- and heterodinuclear cage complexes of the encapsulating ligands **545** and **546** has been performed [127]. In the macrobicyclic dication $(Cu^+)_2$@**545**, its two co-encapsulated copper(I) ions having the trigonal-pyramidal CuN_4-coordination polyhedra occupy two apical ends of **545**, formed by three ribbed imine and one capping amine nitrogen atoms. In a mixed-valence complex $(Cu^+)(Cu^{2+})$@**545**, one nitrogen atom of each pyrazine fragment of **545** is coordinated toward the encapsulated copper(II) cation. In the case of 1:1 copper(II) cage complex of **545**, its three polymorphs (two of which are isomorphic) have been characterized [127] by single crystal X-ray diffraction. In one of them, the encapsulated copper(II) ion has a distorted TBP CuN_6-coordination polyhedron, formed by three imine, one bridgehead amine, and one pyridazine nitrogen atoms of this macrobicyclic ligand. In its other X-rayed polymorph, CuN_6-coordination polyhedron of an encapsulated copper(II) ion is formed by three imine and three pyridazine nitrogen atoms, thus possessing a distorted pseudooctahedral geometry. Two co-encapsulated copper(II) ions of 1:2 homodinuclear cage complex $(Cu^{2+})_2$@**546** have the different coordination polyhedra: first of them coordinates one apical tertiary, two secondary amine, and two pyridazine nitrogen atoms of **546**, whereas CuN_5-coordination polyhedron of its co-caged copper(II) ion is formed by one bridgehead, one pyridazine, and three amine nitrogen donor atoms of this caging ligand. Both these coordination polyhedra have the geometry of a distorted square pyramid [127].

The X-ray structure of 1:2 homodinuclear silver(I) cage complex of the macrobicyclic ligand **33** with intramolecular Ag...Ag separation of 2.78 Å is described [128]. Its macrobicyclic framework is substantially distorted, thus producing enough space for caging of two Ag^+ ions. Each of these ions coordinates one bridgehead and two pyridine nitrogen atoms of **33**, thus giving its pyramidal AgN_3-coordination polyhedron [128].

As follows from single crystal X-ray diffraction data of [129] for homodinuclear copper(I) cage complex of the hexaimine macrobicyclic ligand **548**, this clathrochelate dication has the molecular C_3 pseudoaxis lying through both the encapsulated copper ions and the bridgehead nitrogen atoms of **548** as well. The N...N distance between these bridgehead atoms (8.89 Å) are shortened by approximately 1.5 Å as compared with a parent metal-free macrobicycle. This fact is explained [129] by adoption of a *cis,cis*-conformation in the bis-iminofuran ribbed fragments of **548** in its above cage complex, while a metal-free macrobicycle adopts a *trans, trans*-conformation. The intramolecular Cu...Cu separation in the above copper(II) clathrochelate is approximately 4.2 Å, and each of the co-encapsulated copper(I) ions coordinates four nitrogen atoms of **548**, thus forming its trigonal-pyramidal CuN_4-coordination polyhedron [130].

The X-ray structure of the homodinuclear copper(II) cage complex of a monodeprotonated form of the tris-pyrazolate macrobicyclic ligand **549** is reported [131, 132]. Each of the two co-encapsulated copper(II) ions has a square pyramidal CoN_6-coordination polyhedron, the base of which is formed by two secondary nitrogen atoms of the bridging fragment of **549** and by two nitrogen atoms of its two different pyrazolate moieties, acting as exobidentate ligand syntones; the axial positions are occupied by bridgehead nitrogen atoms. The intramolecular Cu...Cu separation in this 1:2 host–guest complex is approximately 3.96 Å [132].

As follows from the X-ray diffraction data of [133] for the macrobicyclic dication $(Ag^+)_2$@**550**, the two silver(I) ions are co-encapsulated within the cavity of **550** at its two ends. Each of these heptacoordinate caged silver(I) cations has a capped octahedral AgN_4S_3-coordination polyhedron with one bridgehead nitrogen atom in its apical position, three imine nitrogen atoms in the capped face of this polyhedron, and three sulfur atoms in its uncapped face. The Ag–N (bridgehead) distances are substantially greater than the Ag–N(imine) ones, whereas the contacts between the caged silver(I) ions and the thiophene sulfur atoms of **550** are described [133] to be long. The intramolecular Ag...Ag separation in this cage complex is approximately 4.65 Å, while the N...N distance between its bridgehead nitrogen atoms is close to 9.69 Å. In the homodinuclear silver(I) clathrochelate dication formed by its amine derivative **552**, each of the tetracoordinate encapsulated metal ions forms three Ag–N(imine) bonds (2.27–2.36 Å) and that with one bridgehead nitrogen atom (approximately 2.49 Å). The intramolecular Ag...Ag and N(bridgehead)...N (bridgehead) separations in this clathrochelate dication are equal to 5.004 Å and to 9.962 Å, respectively. The greater helicity of a macrobicyclic framework of $(Ag^+)_2$@**552** as compared with that of $(Ag^+)_2$@**550** was found [133]. An unsymmetric unit of the X-rayed crystal of the homodinuclear copper(I) cage complex $(Cu^+)_2$@**550** contains half of each of the two independent macrobicyclic cations with intermolecular Cu...Cu separations of 4.733 and of 4.826 Å. Each of the co-encapsulated copper ions coordinates three imine nitrogen atoms and is displaced by approximately 0.5 Å from their N_3-plane toward the center of **550**; these caged ions and the bridgehead nitrogen atoms of **550** are weakly interacting [133].

The encapsulated metal(II) ions in all their isostructural cage complexes of the hexacarboxamide ligand **563** are reported [134] to occupy its apical *tren* pockets, thus forming the distorted trigonal-pyramidal $M^{II}N_4$-coordination polyhedra with a cofacial orientation; the dipropoxyphenoxyl substituents of this encapsulating ligand are splayed out to the periphery of their molecules. The $M^{II}...M^{II}$ separation is small in its binuclear manganese (II) cage complex (approximately 6.08 Å), and these co-encapsulated ions are shifted from the mean plane of the equatorial donor nitrogen atoms of **563**. For the analogous zinc (II) clathrochelate, this separation is approximately 6.42 Å, and its metallocenters are relaxed into the above tripodal *tren*-based binding pockets. This fact is explained [134] by difference in a size of these encapsulated metal ions. The homodinuclear iron(II) clathrochelate $(Fe^{2+})_2$@**563** is reported [134] to undergo both the electrochemical and chemical oxidations, thus giving its high-spin diiron(III) co-encapsulating derivative (as follows from its ^{57}Fe Mössbauer spectrum). Re-reduction of this iron(III) compound with cobaltocene gave the initial homodinuclear iron(II) cage complex, which has been identified by NMR spectroscopy; these electrochemical and chemical redox reactions were found [134] to be reversible.

The X-ray structure of 1:2 homodinuclear silver(I) cage complex of the *meso*-N_8O_6-caging ligand **553** is reported [135]. This macrobicyclic ligand adopts an *endo–endo*-conformation; the distance between its two bridgehead nitrogen atoms is approximately 19.6 Å, and the molecular C_3 pseudoaxis lies through these apical atoms. Each of the two co-encapsulated silver(I) ions is located at the N_4-end of **553** and coordinates three secondary amine and one bridgehead nitrogen atoms, thus giving an irregular AgN_4-coordination polyhedron with the average Ag–N (secondary amine) and Ag–N(bridgehead) distances of 2.382 Å and of 2.473 Å, respectively. The intramolecular Ag...Ag separation between these caged ions is equal to 14.666 Å [135].

The X-rayed crystal of 1:2 homodinuclear silver(I) cage complexes of the hexaimine macrobicyclic ligand **542** contains the clathrochelate dications, two co-encapsulated silver(I) ions of which are located at the distance of approximately 6.06 Å apart [136]. Each of these ions coordinates the bridgehead nitrogen and three azomethine nitrogen atoms, thus forming its AgN_4-coordination polyhedron [136].

Single crystal X-ray diffraction study of a metal-free macrobicycle **542** is reported [130]. Its X-rayed crystal contains the discrete macrobicyclic entities, forming the hydrogen bonds with outside-located solvate water molecules. The conformation of this macrobicycle is substantially different from that in its above homodinuclear 1:2 silver(I) cage complex in which the lone pairs of the imine nitrogen atoms of **542** are pointed toward these encapsulated metal ions. In contrast, the lone pairs of a metal-free macrobicycle **542** are pointed outwards of its cavity and toward the above hydrogen-bonded water molecules as well [130].

In the molecule of 1:2 homodinuclear manganese(II) cage complex of a triple-deprotonated form of the boron-capped macrobicyclic ligand

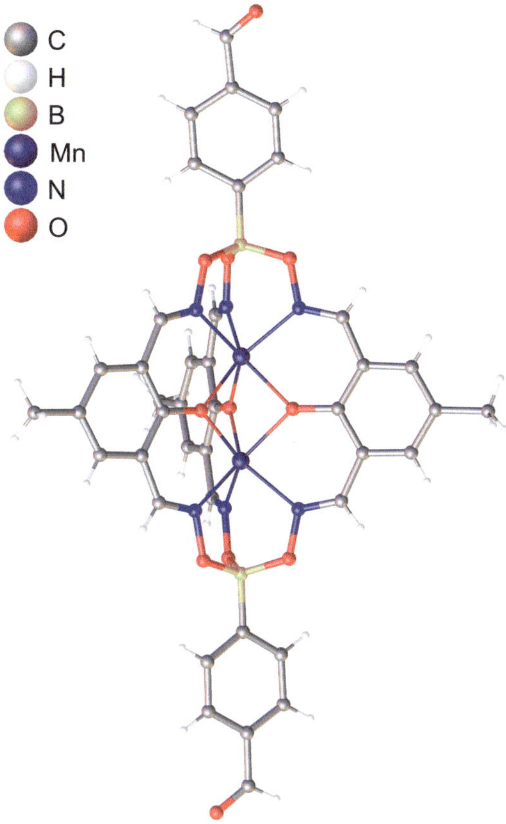

Fig. 3.96 X-ray structure of 1:2 homodinuclear manganese(II) cage complex $(Mn^{2+})_2$@**573** [137]

573 (Fig. 3.96) with terminal carbonyl groups [137], each of its co-encapsulated hexacoordinate manganese(II) ions has a slightly distorted TP MnN_3O_3-coordination polyhedron (φ = 1.86 and 1.52°) with three μ_2-bridging phenoxyl oxygen atoms and a face-shared TBP Mn_2O_3-core. The intramolecular Mn…Mn separation in the complex $(Mn^{2+})_2$@**573** is equal to 2.898 Å, while the B…B distance in this clathrochelate monoanion having the molecular C_3 pseudoaxis is approximately 9.13 Å. The observed average Mn–O and Mn–N distances (approximately 2.13 Å and 2.17 Å, respectively) are described [137] to be characteristic of the high-spin manganese(II) complexes with electronic configuration d^5. In 1:2 cobalt (II) clathrochelate complex $(Co^{2+})_2$@**573**, the Co–O and Co–N distances fall in the ranges 2.077–2.124 Å and 2.081–2.103 Å, respectively, while the distortion angles φ of their almost TP CoN_3O_3-coordination polyhedra are equal to 2.60 and to 6.87°. The intramolecular Co…Co and B…B separations in the molecule $(Co^{2+})_2$@**573** are approximately 2.96 Å and 9.14 Å, respectively [137].

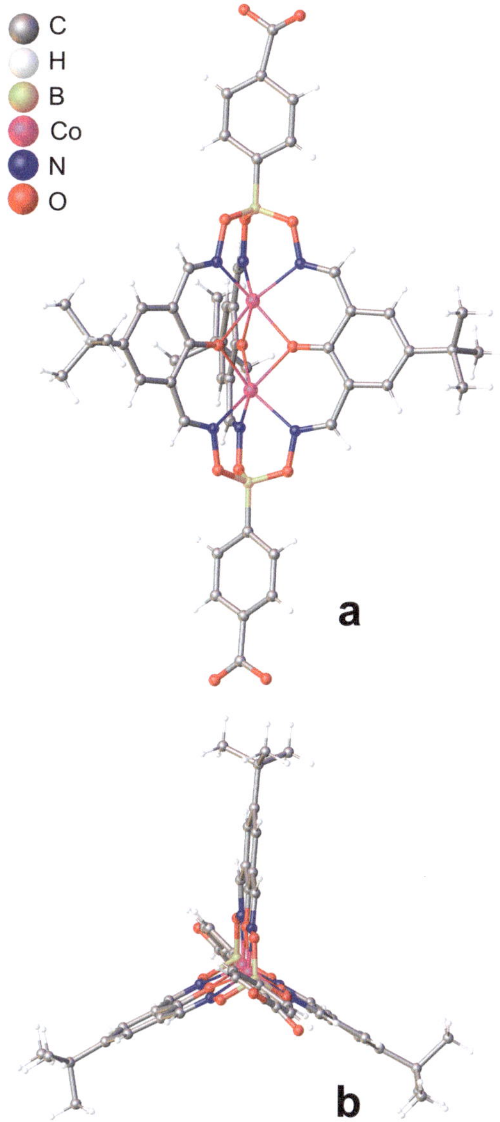

Fig. 3.97 *Side* (**a**) and *top* (**b**) views of the homodinuclear cobalt(II) clathrochelate $(Co^{2+})_2$@**575** [92]

Single crystal X-ray diffraction experiments have been used [92] to structurally characterize the homodinuclear cobalt(II) clathrochelate $(Co^{2+})_2$@**575** with terminal carboxylic groups (Fig. 3.97) and MOFs based on this cage

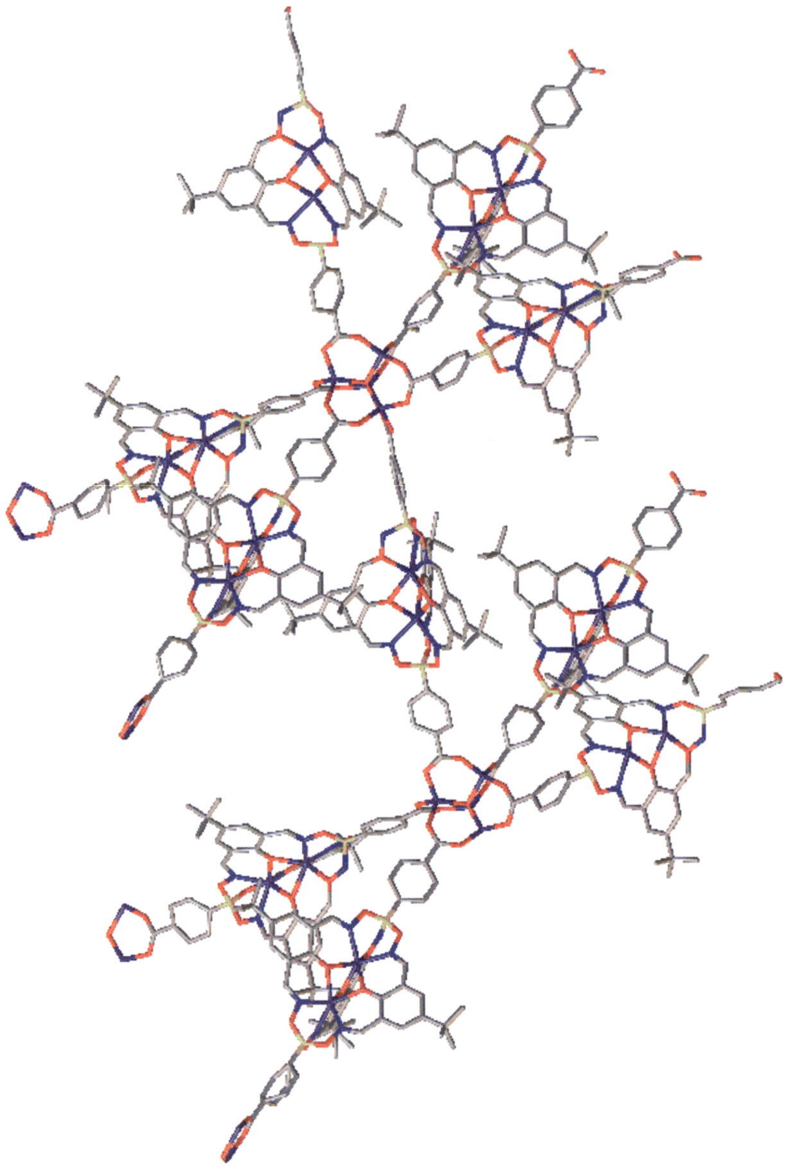

Fig. 3.98 Crystal packing of MOF *54* [92]

complex as a macrobicyclic ligand. Each of its two co-encapsulated cobalt(II) ions has a distorted TP CoN_3O_3-coordination polyhedron. Two carbon atoms of the terminal carboxylic groups of $(Co^{2+})_2$@**575** are 21 Å apart from each other. A framework of its complex with zinc (II) ions, MOF *54*, contains these clathrochelate ligands $(Co^{2+})_2$@**575**, coordinated toward Zn_4O clusters by their carboxylate groups (Fig. 3.98). These clusters are composed of four tetrahedral zinc(II) cations and their central μ_4-oxygen atom. If the X-rayed single crystal of MOFs *55* and *55a* (Figs. 3.99 and 3.100) has almost the same crystal packings like MOF *54* (the observed small differences are explained [92] by different solvents used during their synthesis), MOF *54a*

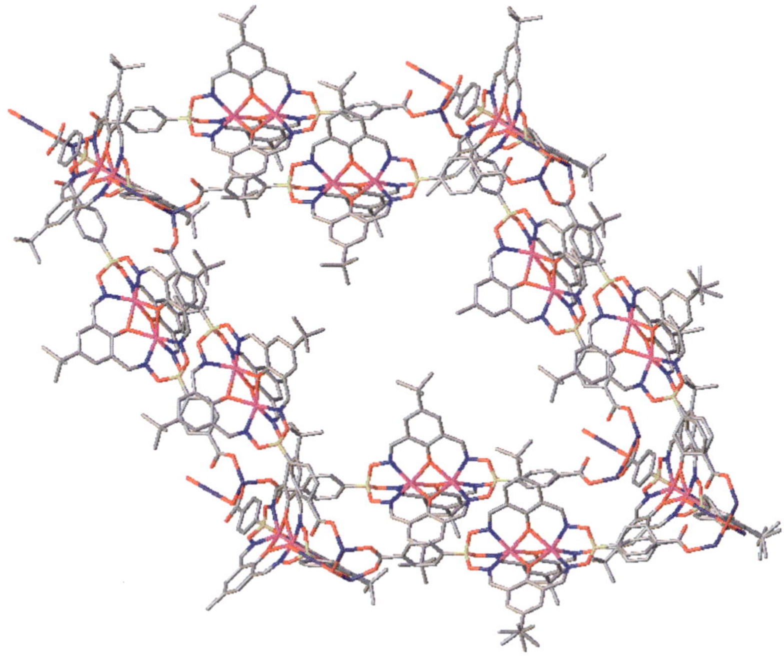

Fig. 3.99 Crystal packing of MOF *55* [92]

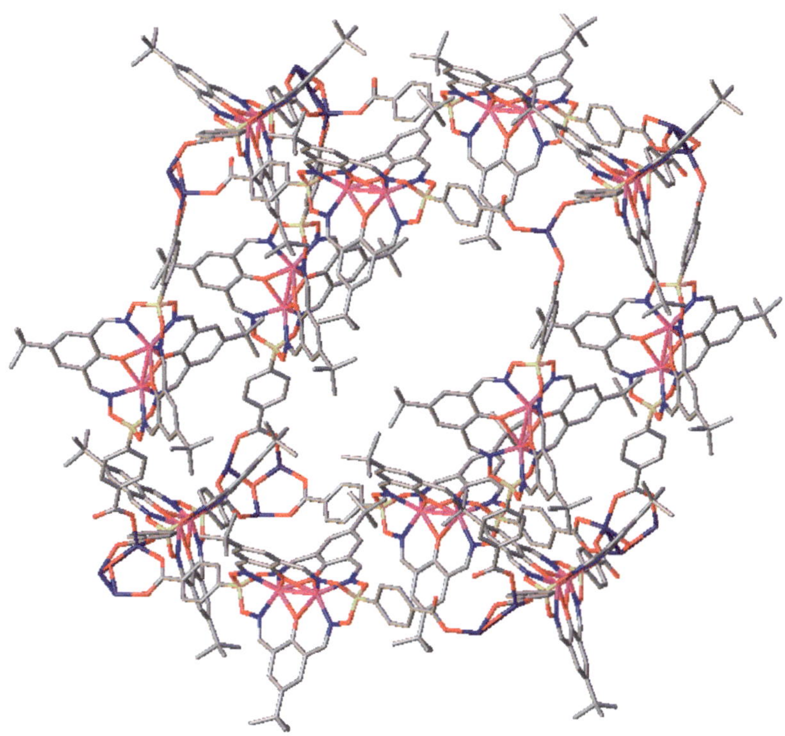

Fig. 3.100 Crystal packing of MOF *55a* [92]

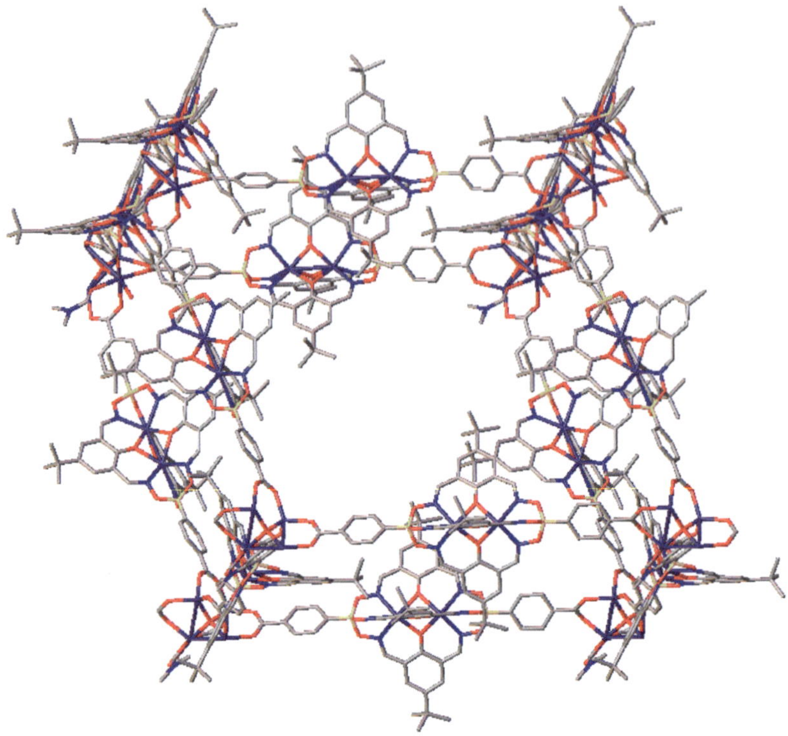

Fig. 3.101 Crystal packing of MOF *54a* [92]

reveals the different behavior: its three-dimensional framework is composed of the same clathrochelate ligands $(Co^{2+})_2$@**575**, coordinated toward Zn_4O clusters (Fig. 3.101) in which their three zinc(II) cations adopt the tetrahedral coordination polyhedra, whereas fourth cation of this type is hexacoordinate and has an octahedral coordination polyhedron. Each of the former tetracoordinate zinc(II) ions coordinates three carboxylate oxygen atoms from the different clathrochelate ligands $(Co^{2+})_2$@**575** and this central μ_4-oxygen atom as well, while the latter hexacoordinate zinc(II) ion, adopting an unusual octahedral geometry, coordinates the above μ_4-oxygen central atom, three carboxylate oxygen atoms, and two oxygen atoms of one DMF and one water solvate molecules. The presence of one octahedral Zn^{2+} ion within the Zn_4O cluster of MOF *54a* is reported [92] to substantially affect its crystal structure, containing one pair of the clathrochelate ligands $(Co^{2+})_2$@**575**, the axis of which are tilted with respect to each other; the angles between their opposite pairs are equal to 180, 180, and 135°. This resulted in the distortion of this *zigzag*-shaped crystal structure (Fig. 3.101). The separations between the central μ_4-oxygen atoms of adjacent Zn_4O cores are approximately 27.47 and 27.89 Å. As follows from the X-ray diffraction data [92] for MOF *56*, its crystal structure is very similar to those of *54a*, *55*, and *55a*. Passing from more bulky *tert*-butyl side chains to the bromide substituents did not affect the twofold network interpenetration in the X-ray crystal of *56*. The channels in this crystal are occupied by the tetraethylammonium cations [92]. The X-rayed crystal of MOF *58* (Fig. 3.102) contains a three-dimensional network in which all the clathrochelate ligands $(Co^{2+})_2$@**575** are coordinated toward Zn_4O cluster, which is formed by three tetrahedral zinc(II) ions and one octahedral ion of this type that binds the central μ_4-oxygen atom, three oxygen atoms from the deprotonated carboxyl groups of **575**, and two DMSO

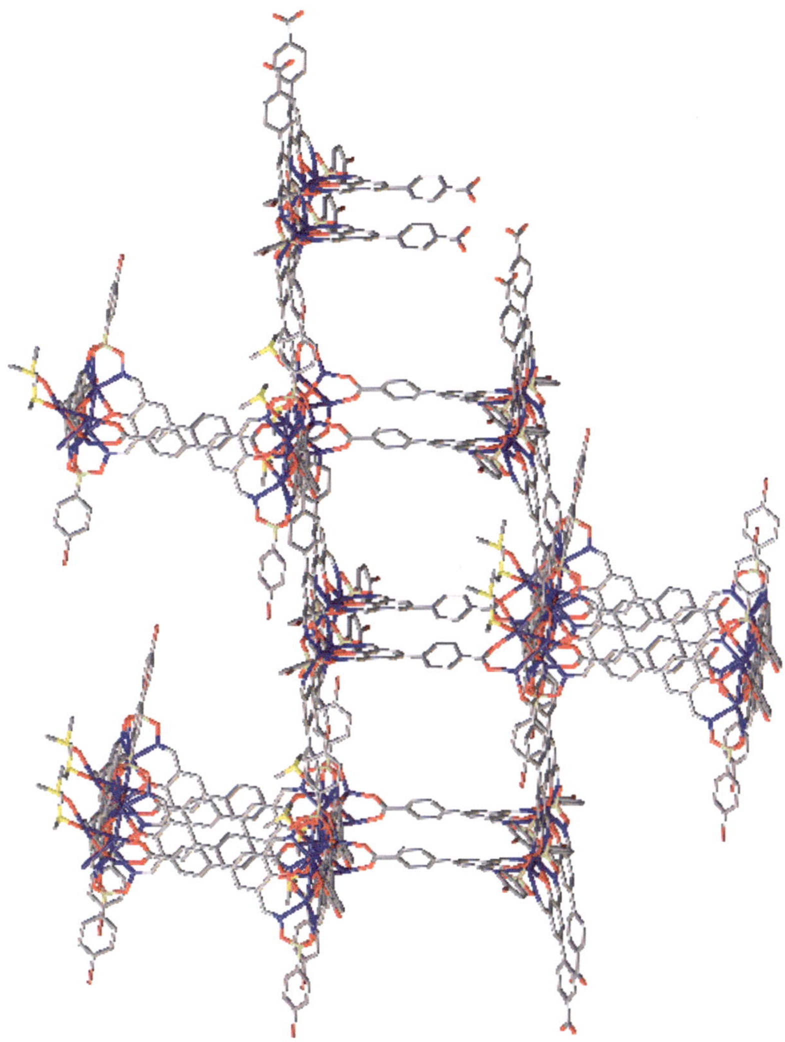

Fig. 3.102 Crystal packing of MOF *58* [92]

molecules as well. The network of MOF *58* was found [92] to be twofold interpenetrated and anionic, thus also containing the included tetraethylammonium counterions.

References

1. Voloshin YZ, Antipin MY (2004) Structural chemistry of complexes with an encapsulated metal ion. α-Dioximate, oximehydrazonate, and dihydrazonate clathrochelates. Russ Chem Bull Int Ed 53:2097–2108
2. Darbost U, Zeng X, Rager MN, Giorgi M, Jabin I, Reinaud O (2004) X-ray and solution structures of the first Zn funnel complex based on a calix[6]azacryptand. Eur J Inorg Chem 2004:4371–4374
3. Darbost U, Rager MN, Petit S, Jabin I, Reinaud O (2005) Polarizing a hydrophobic cavity for the efficient binding of organic guests: the case of calix[6]tren, a highly efficient and versatile receptor for neutral or cationic species. J Am Chem Soc 127:8517–8525
4. Bencini A, Bianchi A, Chimichi S, Ciampolini M, Dapporto P, Garcia-Espana E, Micheloni M, Nardi N, Paoli P, Valtancoli B (1991) Lithium binder in aqueous solution. Synthesis and characterization of the new cage 4,10,15-trimethyl-1,4,7,10,15-pentaazabicyclo[5.5.5]heptadecane (L). Protonation and lithium complex formation. Crystal structures of [HL][BPh_4] and [LiL][BPh_4]. Inorg Chem 30:3687–3691

5. Bencini A, Bianchi A, Dapporto P, Fusi V, Garcia-Espana E, Micheloni M, Paoletti P, Paoli P, Rodriguez A, Valtancoli B (1993) Thermodynamic, kinetic, and structural study of the ligational properties of the macrobicyclic aza-ligand 4,7,10,17,23-pentamethyl-1,4,7,10,13,17,23-heptaazabicyclo[11.7.5]pentacosane (L1) and of its macrocyclic precursor 1,4,7,13-tetramethyl-1,4,7,10,13,16-hexaazacyclooctadecane (L2). Crystal structure of $[Zn(L1)(H_2O)](BPh_4)_2$. Inorg Chem 32:2753–2760
6. Izzet G, Douziech B, Prangé T, Tomas A, Jabin I, Le Mest Y, Reinaud O (2005) Calix[6]tren and copper (II): a third generation of funnel complexes on the way to redox calix-zymes. Proc Natl Acad Sci 102:6831–6836
7. Bianchi A, Ciampolini M, Micheloni M, Nardi N, Valtancoli B, Mangani S, Garcia-Espana E, Ramirez JA (1989) The small cage 12,17-dimethyl-5-oxa-1,9,12,17-tetra-azabicyclo[7.5.5]nonadecane (L): its synthesis, characterization, and 'proton sponge' behaviour. The crystal structure of the dipicrate salt $[H_2(L)](picrate)_2$. Perkin Trans 2:1131–1137
8. Bazzicalupi C, Bencini A, Ciattini S, Denat F, Desogere P, Goze C, Matera I, Valtancoli B (2010) Proton and metal binding by cyclen-based highly rigid cryptands. Dalton Trans 39:11643–11653
9. Bencini A, Bianchi A, Borselli A, Chimichi S, Ciampolini M, Dapporto P, Micheloni M, Nardi N, Paoli P, Valtancoli B (1990) Selective lithium encapsulation in aqueous solution by the new cage 4,10-dimethyl-1,4,7,10,15-pentaazabicyclo[5.5.5] heptadecane (L). Synthesis, characterization, and structural aspects. Crystal structures of $[LiL](ClO_4)$ and $[CuL]Br_2{\cdot}3H_2O$. Inorg Chem 29:3282–3286
10. Suh MP, Lee J, Han MY, Yoon TS (1997) Template synthesis, properties, and crystal structure of a trigonal bipyramidal cobalt(II) octaazamacrobicyclic complex. Inorg Chem 36:5651–5654
11. Bag B, Bharadwaj PK (2004) Attachment of an electron-withdrawing fluorophore to a cryptand for modulation of fluorescence signaling. Inorg Chem 43:4626–4630
12. Kon N, Takemura H, Otsuka K, Tanoue K, Nakashima S, Yasutake M, Tani K, Kimoto J, Shinmyozu T, Inazu T (2000) Synthesis of macrocyclic cage compounds by diamine–dihalide one-step coupling reaction. J Organomet Chem 65:3708–3715
13. Dapporto P, Formica M, Fusi V, Giorgi L, Micheloni M, Pontellini R, Paoli P, Rossi P (2001) Ligational properties of two new phenolic aza cages towards proton and alkali metal ions – a theoretical and an experimental approach. Eur J Inorg Chem 2001:1763–1774
14. Bing-guang Z, Dong G, Chun-ying D, Ping C, Qing-Jin M (2003) A novel pyramidal multiple alkali metal cluster $[K_3(H_2O)]^{3+}$ stabilized within a capsule containing 16-nitrogen donors. Chem Commun 1863–1837
15. Westerhausen M, Weinrich S, Kramer G, Piotrowski H (2002) Cage compounds of heterotrimetallic mixed phosphanediide/silanolate aggregates with an alkaline earth metal-centered phosphorus octahedron. Inorg Chem 41:7072–7076
16. Brown KN, Geue RJ, Hambley TW, Hockless DCR, Rae AD, Sargeson AM (2003) Specificity in template syntheses of hexaaza-macrobicyclic cages: $[Pt(Me_5\text{-tricosatrieneN}_6)]^{4+}$ and $[Pt(Me_5\text{-tricosaneN}_6)]^{4+}$. Org Biomol Chem 1:1598–1608
17. Bernhardt PV, Dyahningtyas TE, Harrowfield JM, Kim JY, Kim Y, Rukmini E (2003) Chiral resolution of hexaamine cobalt(III) cages: substituent effects on chiral discrimination. Aust J Chem 56:1187–1191
18. Sharrad CA, Lüthi SR, Gahan LR (2003) Embracing ligands. A synthetic strategy towards new nitrogen–thioether multidentate ligands and characterization of the cobalt(III) complexes. Dalton Trans 3693–3703
19. Albrecht M, Schmid S, deGroot M, Weis P, Fröhlich R (2003) Self-assembly of an unpolar enantiomerically pure helicate-type metalla-cryptand. Chem Commun 2526–2527
20. Karpishin TB, Stack TDP, Raymond KN (1993) Octahedral vs. trigonal prismatic geometry in a series of catechol macrobicyclic ligand-metal complexes. J Am Chem Soc 115:182–192
21. Catalano VJ, Bennett BL, Kar HM, Noll BC (1999) Synthesis and characterization of trigonal gold (I) cage complexes: Luminescent metallocryptates encapsulating Tl(I) and Na^+ Ions. J Am Chem Soc 121:10235–10236
22. Smith CB, Barbour LJ, Makha M, Raston CL, Sobolev AN (2006) Lanthanide-induced helical arrays of [{Co(III) sepulchrate}@{*p*-sulfonatocalix [4]arene}] supermolecules. Chem Commun 2006 (9):950–952
23. Sharrad CA, Gahan LR (2003) Characterization of $[Fe(AMN_3S_3sarH)]^{3+}$—a rigorously low-spin iron (II) complex. Polyhedron 22:45–51
24. Ma MT, Neels OC, Denoyer D, Roselt P, Karas JA, Scanlon DB, White JM, Hicks RJ, Donnelly PS (2011) Gallium-68 complex of a macrobicyclic cage amine chelator tethered to two integrin-targeting peptides for diagnostic tumor imaging. Bioconjug Chem 22:2093–2103
25. Ma MT, Cooper MS, Paul RL, Shaw KP, Karas JA, Scanlon DB, White JM, Blower PJ, Donnelly PS (2011) Macrobicyclic cage amine ligands for copper radiopharmaceuticals: a single bivalent cage amine containing two Lys^3-bombesin targeting peptides. Inorg Chem 50:6701–6710
26. Qin CJ, James L, Chartres JD, Alcock LJ, Davis KJ, Willis AC, Sargeson AM, Bernhardt PV, Ralph SF

(2011) An unusually flexible expanded hexaamine cage and its Cu^{II} complexes: variable coordination modes and incomplete encapsulation. Inorg Chem 50:9131–9140
27. Takemura H, Kon N, Yasutake M, Kariyazono H, Shinmyozu T, Inazu T (1999) A potassium complex of a fluorine-containing macrocyclic cage compound: interactions between fluorine atoms and metal ions. Angew Chem Int Ed 38:959–961
28. Fuchs A, Kaifer E, Wadepohl H, Himmel HJ (2012) Cryptate complexes with the potential for CO_2 activation: quantum chemical predictions and synthetic efforts. Eur J Inorg Chem 2012:4020–4028
29. Saalfrank RW, Dresel A, Seitz V, Trummer S, Hampel F, Teichert M, Stalke D, Stadler C, Daub J, Schünemann V, Trautwein AX (1997) Topologic equivalents of coronands, cryptands and their inclusion complexes: synthesis, structure and properties of (2)-metallacryptands and (2)-metallacryptates. Chem Eur J:2058–2062
30. Saalfrank RW, Seitz V, Caulder DL, Raymond KN, Teichert M, Stalke D (1998) Self-assembly of {2}-metallacryptands and {2}-metallacryptates. Eur J Inorg Chem 1998:1313–1317
31. Bacchi A, Ferranti F, Pelizzi G (1996) Conformation and packing of cobalt(III) sepulchrate dihydrogentriphosphate hexahydrate. Acta Cryst C52:1624–1627
32. Ribeiro S, Cunha-Silva L, Balula SS, Gago S (2014) Cobalt(III) sepulchrate complexes: application as sustainable oxidative catalysts. New J Chem 38:2500–2507
33. Schönleber A, van Smaalen S, Larsen FK (2010) Orientational disorder in Λ-cobalt(III) sepulchrate trinitrate. Acta Cryst C66:m107–m109
34. Dubicki L, Ferguson J, Geue RJ, Sargeson AM (1980) Circular dichroism of cobalt(III) complexes: dependence on the outer sphere configuration. Chem Phys Lett 74:393–397
35. Bilotti I, Biscarini P, Castiglioni E, Ferranti F, Kuroda R (2012) Reflectance circular dichroism of solid-state chiral coordination compounds. Chirality 14:750–756
36. Fucaloro A, Zanella A, Widjaja S, Widjaja J (2005) Partial molar volumes and refractions of cobalt(III) complexes, Part 1: Homologous series of hexaaminecobalt(III) complexes. J Solut Chem 34:1357–1370
37. Ling I, Sobolev AN, Hashim R, Harrowfield JM (2014) Stereochemistry of cage amine complexes – probing the ligand conformational flexibility with hydrogen bonds. CrystEngComm 16:11058–11063
38. Harrowfield JM, Koutsantonis GA, Nealon GL, Skelton BW, White AH (2005) Amphiphile structures in the solid state: complex cations with lipophilic substituents. Eur J Inorg Chem 2005:2384–2392
39. Walker GW, Geue RJ, Haller KJ, Rae AD, Sargeson AM (2003) New synthetic routes to hexa-aza cages using cobalt(III) tris(1,2-diamine) templates. Dalton Trans 279–281
40. Koutsantonis GA, Lee J, Lengkeek NA, Nealon GL, Skelton BW, White AH (2010) Synthesis and characterisation of trinuclear metal complexes derived from carboxymethyl-substituted sarcophagine macrobicyclic cage amines. J Incl Phenom Macrocycl Chem 66:61–66
41. Bygott AMT, Geue RJ, Ralph SF, Sargeson AM, Willis AC (2007) Octahedral and trigonal prismatic structure preferences in a bicyclic hexaamine cage for zinc(II), cadmium(II) and mercury(II) ions. Dalton Trans 4778–4787
42. Takemura H, Iwanaga T, Shinmyozu T (2006) C–F···Rb+ interaction in a fluorinated cage compound complex. Tetrahedron Lett 47:8989–8991
43. Coyle J, Drew MGB, Harding CJ, Nelson J, Town RM (1997) Expanded-cap sepulchrates: a new series of mononuclear transition-metal ion cryptates. Dalton Trans 1123–1126
44. Puchta R, Meier R, van Eldik R (2007) Host–guest complexes of bicyclic hexaamine cryptands – prediction of ion selectivity by quantum chemical calculations. III. Aust J Chem 60:889–897
45. Cohen SM, Petoud S, Raymond KN (1999) A novel salicylate-based macrobicycle with a "split personality". Inorg Chem 38:4522–4529
46. Orfanoudaki M, Tamiolakis I, Siczek M, Lis T, Armatas GS, Pergantis SA, Milios CJ (2011) Unique trigonal prism encapsulated Ln complexes: a $[Co^{II}{}_6Eu]$ and a $[Co^{II}{}_6Dy]$ cage. Dalton Trans 40:4793–4796
47. Esteban-Gómez D, Enríquez-Pérez T, Ferreirós-Martínez R, Mato-Iglesias M, Platas-Iglesias C, de Blas A, Rodríguez-Blas T (2010) Lead (II) complexes of lateral macrobicyclic receptors that incorporate a crown moiety and a pyridine head unit. Eur J Inorg Chem 2010:5027–5034
48. Atkinson IM, Lindoy LF, Matthews OA, Meehan GV, Sobolev AN, White AH (1994) New macrocyclic ligands. V. Synthetic, crystallographic and computational studies of a new N_2O_6-heteroatom cage. Aust J Chem 47:1155–1162
49. Adam KR, Atkinson IM, Kim J, Lindoy LF, Matthews OA, Meehan GV, Raciti F, Skelton BW, Svenstrup N, White AH (2001) Macrocyclic ligand design. A synthetic, solvent extraction, computational and NMR study of the effect of cryptand flexibility on sodium ion affinity. J Chem Soc Dalton Trans 2388–2397
50. Bkouche-Waksman I, Guilhem J, Pascard C, Alpha B, Deschenaux R, Lehn JM (1991) Crystal structures of the lanthanum(III), europium(III), and terbium(III) cryptates of tris(bipyridine) macrobicyclic ligands. Helv Chim Acta 74:1163–1170

51. Dominique G, Hervé B, Jean-Marc B, Marion B, Gérard M (2007) Luminescent Eu(III) and Gd(III) trisbipyridine cryptates: Experimental and theoretical study of the substituent effects. Chem Phys Chem 8:480–488
52. Voloshin YZ, Varzatskii OA, Starikova ZA, Antipin MY, Lebedev AY, Belov AS (2004) Supramolecular organization of the crystals of allylsulfide clathrochelate: influence of the nature of solvate molecules. Russ Chem Bull Int Ed 53:1496–1502
53. Voloshin YZ, Varzatskii OA, Korobko SV, Antipin MY, Vorontsov II, Lyssenko KA, Kochubey DI, Nikitenko SG, Strizhakova NG (2004) Pathways of directed synthesis of C_2-unsymmetric iron (II) clathrochelates and polyclathrochelates starting from antimony- and germanium-containing precursors. Inorg Chim Acta 357:3187–3204
54. Voloshin YZ, Varzatskii OA, Vorontsov II, Antipin MY (2005) Tuning a metal's oxidation state: the potential of clathrochelate systems. Angew Chem Int Ed 44:3400–3402
55. Voloshin YZ, Lebed EG, Erdyakov SY, Makarenko IG, Lyssenko KA, Starikova ZA, Gurskii ME, Bubnov YN (2006) Application of the allylboration reaction of terminal acetylenes with allyldihaloboranes for the preparation of capping agents for the synthesis of precursors of polymeric iron (II) clathrochelates. Russ Chem Bull Int Ed 55:1971–1981
56. Burdukov AB, Pervukhina NV, Vershinin MA, Voloshin YZ (2007) Molecular structure of di-*tert*-butylamine-containing the monoedge-functionalized clathrochelate iron(II) *tris*-dioximate. J Struct Chem 48:358–361
57. Voloshin YZ, Makarov IS, Vologzhanina AV, Monakov YB, Islamova RM, Polshin EV, Bubnov YN (2008) Template synthesis and structures of bis-ferrocenylboronate macrobicyclic iron(ii) tris-dioximates. Russ Chem Bull Int Ed 57:1199–1206
58. Vershinin MA, Burdukov AB, Pervukhina NV, Eltsov IV, Varzatskii OA, Voloshin YZ (2008) The first example of the alkoxy-containing iron(ii) tris-dioximate clathrochelate: synthesis, structure, and properties. Russ Chem Bull Int Ed 57:1309–1313
59. Voloshin YZ, Varzatskii OA, Starikova ZA, Belov AS, Ignatenko AV, Bubnov YN (2008) Synthesis and structure of the first chlathrochelate ligand with a *cis*-diimidazolyl donor fragment. Inorg Chem Commun 11:988–991
60. Burdukov AB, Vershinin MA, Eltsov IV, Pervukhina NV, Voloshin YZ (2009) The first ribbed functionalized tris-dioximate clathrochelates with inherent hydroxyl substituents: Synthesis, spectra and X-ray structure. Inorg Chem Commun 12:919–922
61. Varzatskii OA, Voloshin YZ, Korobko SV, Shulga SV, Krämer R, Belov AS, Vologzhanina AV, Bubnov YN (2009) On a way to new types of the polyfunctional and polytopic systems based on cage metal complexes: cadmium-promoted nucleophilic substitution with low-active nucleophilic agents. Polyhedron 28:3431–3438
62. Varzatskii OA, Voloshin YZ, Stuzhin PA, Shul'ga SV, Volkov SV, Vologzhanina AV, Lebed EG, Bubnov YN (2011) New cadmium-promoted reaction of a C-nucleophile: synthesis and X-ray structure of the first dicyanopyrazine iron(II) clathrochelate. Inorg Chem Commun 14:1504–1507
63. Vershinin MA, Burdukov AB, Boguslavskii EG, Pervukhina NV, Kuratieva NV, Eltsov IV, Reznikov VA, Varzatskii OA, Voloshin YZ, Bubnov YN (2011) The first monoribbed-functionalized tris-dioximate iron(II) clathrochelate with two inherent NH_2-substituents, its reactivity, acid-base and coordination-chemical properties. Inorg Chim Acta 366:91–97
64. Vershinin MA, Burdukov AB, Starikova ZA, Novikov VV, Voloshin YZ (2011) Reactions of dichloro-substituted iron(ii) clathrochelate with 1,4-dioxane radical derivatives: synthesis, structure, and spectral characteristics of the dioxane ring opening product in the ribbed fragment of the macrobicyclic ligand. Russ Chem Bull Int Ed 60:2510–2517
65. Voloshin YZ, Varzatskii OA, Novikov VV, Strizhakova NG, Vorontsov II, Vologzhanina AV, Lyssenko KA, Romanenko GV, Fedin MV, Ovcharenko VI, Bubnov YN (2010) Tris-dioximate cobalt(I, II, and III) clathrochelates: stabilization of different oxidation and spin states of an encapsulated metal ion by ribbed functionalization. Eur J Inorg Chem 2010:5401–5415
66. Voloshin YZ, Varzatskii OA, Belov AS, Starikova ZA, Dolganov AV, Novikov VV, Bubnov YN (2011) Synthesis, structural and electrochemical features of alicyclic and aromatic α,α'-N_2- and -S_2-dioximate macrobicyclic cobalt(II,III) and ruthenium(II) tris-complexes. Inorg Chim Acta 370:322–332
67. Voloshin YZ, Varzatskii OA, Belov AS, Starikova ZA, Strizhakova NG, Dolganov AV, Kochubey DI, Bubnov YN (2010) Synthesis, X-ray structure and redox properties of the macrobicyclic iron(II) N_2- and S_2-containing *vic*-dioximates. Inorg Chim Acta 363:134–146
68. Voloshin YZ, Varzatskii OA, Shul'ga SV, Denisenko IN, Vologzhanina AV, Bubnov YN (2012) Copper(I)-promoted halogen exchange in the iron(II) dichloroclathrochelate. Inorg Chem Commun 17:128–131
69. Voloshin YZ, Belov AS, Varzatskii OA, Shul'ga SV, Stuzhin PA, Starikova ZA, Lebed EG, Bubnov YN (2012) Synthesis and structure of the first

clathrochelate iron(II) tris-dioximates with inherent nitrile substituent(s) and new dehalogenation – reduction reaction at a quasi-aromatic macrobicyclic framework. Dalton Trans 41:921–928
70. Svidlov SV, Varzatskii OA, Potapova TV, Vologzhanina AV, Bukalov SS, Leites LA, Voloshin YZ, Bubnov YN (2014) C-carboranylation of a quasi-aromatic iron(II) cage complex and its organic aromatic analog by the metal-promoted (catalyzed) cross-coupling reactions. Inorg Chem Commun 43:142–145
71. Varzatskii OA, Shul'ga SV, Belov AS, Novikov VV, Dolganov AV, Vologzhanina AV, Voloshin YZ (2014) Copper (I)-and copper (0)-promoted homocoupling and homocoupling–hydrodehalogenation reactions of dihalogenoclathrochelate precursors for C–C conjugated iron (II) bis-cage complexes. Dalton Trans 43:17934–17948
72. Varzatskii OA, Novikov VV, Shulga SV, Belov AS, Vologzhanina AV, Negrutska VV, Dubey IY, Bubnov YN, Voloshin YZ (2014) Copper-promoted reductive homocoupling of quasi-aromatic iron (II) clathrochelates: boosting the inhibitory activity in a transcription assay. Chem Commun 50:3166–3168
73. Liu W, Huang W, Chen CH, Pink M, Lee D (2012) Charge injection and transport in metal-containing conducting polymers: spectroelectrochemical mapping of redox activities. Chem Mater 24:3650–3658
74. Voloshin YZ, Belov AS, Starikova ZA, Vologzhanina AV (2013) Structural peculiarities of a homologous series of iron(II) cage complexes with ribbed glyoximate, methylglyoximate, and dimethylglyoximate chelate fragments. Russ Chem Bull Int Ed 62:1858–1865
75. Voloshin YZ, Lebedev AY, Novikov VV, Dolganov AV, Vologzhanina AV, Lebed EG, Pavlov AA, Starikova ZA, Buzin MI, Bubnov YN (2013) Template synthesis, X-ray structure, spectral and redox properties of the paramagnetic alkylboron-capped cobalt(II) clathrochelates and their diamagnetic iron (II)-containing analogs. Inorg Chim Acta 399:67–78
76. Kochubey DI, Kaichev VV, Saraev AA, Tomyn SV, Belov AS, Voloshin YZ (2013) Combined XANES and XPS study of the electrocatalytically active cobalt(I) cage complexes and the clathrochelate cobalt(II)- and cobalt(III)-containing precursors and analogs. J Phys Chem C 117:2753–2759
77. Belov AS, Vologzhanina AV, Fedorov YV, Kuznetsov EV, Voloshin YZ (2013) Unexpected transformation of mono- to bis-macrobicyclic dimethylglyoximate framework in a chloroform solution: photochemical, MALDI-TOF MS and X-ray diffraction studies. Inorg Chem Commun 35:242–246
78. Dolganov AV, Belov AS, Novikov VV, Vologzhanina AV, Romanenko GV, Budnikova YG, Zelinskii GE, Buzin MI, Voloshin YZ (2015) First iron and cobalt(II) hexabromoclathrochelates: structural, magnetic, redox and electrocatalytic behavior. Dalton Trans 44:2476–2487
79. Vologzhanina AV, Belov AS, Novikov VV, Dolganov AV, Romanenko GV, Ovcharenko VI, Korlyukov AA, Buzin MI, Voloshin YZ (2015) Synthesis and temperature-induced structural phase and spin transitions in hexadecylboron-capped cobalt (II) hexachloroclathrochelate and its diamagnetic iron(II)-encapsulating analogue. Inorg Chem 54:5827–5838
80. Zelinskii GE, Belov AS, Lebed EG, Vologzhanina AV, Novikov VV, Voloshin YZ (2016) Synthesis, structure and reactivity of iron(II) clathrochelates with terminal formyl (acetal) groups and their further functionalization for target delivery to biological systems. Inorg Chim Acta 440:154–164
81. Varzatskii OA, Denisenko IN, Volkov SV, Belov AS, Dolganov AV, Vologzhanina AV, Novikov VV, Bubnov YN, Voloshin YZ (2013) Perfluoroarylation of iron(II) di- and hexaiodoclathrochelates: synthesis, X-ray structure and properties of the first cage complexes with inherent pentafluorophenyl substituent(s). Eur J Inorg Chem 2013:3178–3184
82. Voloshin YZ, Varzatskii OA, Belov AS, Lebedev AY, Makarov IS, Gurskii ME, Antipin MY, Starikova ZA, Bubnov YN (2007) Cage iron (II) complexes with apical and ribbed adamantyl substituents: the creation of second (hydrophobic) shell of an encapsulated metal ion. Inorg Chim Acta 360:1543–1554
83. Belaya IG, Belov AS, Platonov VE, Maksimov AM, Vologzhanina AV, Starikova ZA, Dolganov AV, Novikov VV, Bubnov YN (2012) Formation of the second superhydrophobic shell around an encapsulated metal ion: synthesis, X-ray structure and electrochemical study of the clathrochelate and bis-clathrochelate iron(II) and cobalt(II, III) dioximates with ribbed perfluoroarylsulfide substituents. Dalton Trans 41:737–746
84. Belov AS, Zelinskii GE, Varzatskii OA, Belaya IG, Vologzhanina AV, Dolganov AV, Novikov VV, Voloshin YZ (2015) Molecular design of cage iron (II) and cobalt(II,III) complexes with a second fluorine-enriched superhydrophobic shell. Dalton Trans 44:3773–3784
85. Zelinskii G, Novikov V, Vologzhanina A, Voloshin Y, Bubnov Y, Mokhir A (2015) Template synthesis of iron and cobalt(II) (per)fluoroclathrochelates with superhydrophobic exterior. Paper presented at the international symposia on metal complexes, Wrocław, 24–28 June 2015

86. Burdukov AB, Vershinin MA, Boguslavsky EG, Eltsov IV, Romanenko GV, Belov AS, Voloshin YZ (2013) New hydroxylamine-containing macrobicyclic encapsulating ligand: unexpected double addition of ethyl radicals to the azomethine fragment of a boron-capped iron(II) clathrochelate dioximate. Inorg Chem Commun 36:117–121
87. Vershinin MA, Burdukov AB, Pervukhina NV, Eltsov IV, Voloshin YZ (2013) Stereoselective C-alkylation of an iron(II) dichloroclathrochelate *via* free-radical reactions with alcohols. Inorg Chem Commun 30:159–162
88. Voloshin YZ, Varzatskii OA, Korobko SV, Chernii VY, Volkov SV, Tomachynski LA, Pehn'o VI, Antipin MY, Starikova ZA (2005) Ditopic macropolycyclic complexes: synthesis of hybrid phthalocyaninoclathrochelates. Inorg Chem 44:822–824
89. Dudkin SV, Erickson NR, Vologzhanina AV, Novikov VV, Rhoda HM, Holstrom CD, Zatsikha YV, Yusubov MS, Voloshin YZ, Nemykin VN (2016) Preparation, X-ray structures, spectroscopic, and redox properties of di- and trinuclear iron-zirconium and iron-hafnium porphyrinoclathrochelates. Inorg Chem 55:11867–11882
90. Sabin JR, Varzatskii OA, Voloshin YZ, Starikova ZA, Novikov VV, Nemykin VN (2012) Insight into the electronic structure, optical properties, and redox behavior of the hybrid phthalocyaninoclathrochelates from experimental and density functional theory approaches. Inorg Chem 51:8362–8372
91. Voloshin YZ, Korobko SV, Dolganov AV, Novikov VV, Vologzhanina AV, Bubnov YN (2011) New types of the germanium-capped clathrochelate iron (II) and cobalt(III) tris-dioximates: the synthesis, structure and electrochemical properties. Inorg Chem Commun 14:1043–1047
92. Marmier M, Wise MD, Holstein JJ, Pattison P, Schenk K, Solari E, Scopelliti R, Severin K (2016) Carboxylic acid functionalized clathrochelate complexes: large, robust and easy-to-access metalloligands. Inorg Chem 55:4006–4015
93. Belaya IG, Svidlov SV, Dolganov AV, Zelinskii GE, Potapova TV, Vologzhanina AV, Varzatskii OA, Bubnov YN, Voloshin YZ (2013) Apically linked iron(II) α-dioximate and α-oximehydrazonate bis-clathrochelates: synthesis, structure and electrocatalytic properties. Dalton Trans 42:13667–13678
94. Krámos B, Kovács A (2008) Metal–ligand interactions in Fe(II)-dioxime complexes. J Mol Struct THEOCHEM 867:1–4
95. Nguyen MTD, Charlot MF, Aukauloo A (2011) Structural, electronic, and theoretical description of a series of cobalt clathrochelate complexes in the Co (III), Co(II) and Co(I) oxidation states. J Phys Chem A 115:911–922
96. Wise MD, Holstein JJ, Pattison P, Besnard C, Solari E, Scopelliti R, Bricogne G, Severin K (2015) Large, heterometallic coordination cages based on ditopic metallo-ligands with 3-pyridyl donor groups. Chem Sci 6:1004–1010
97. Jansze S, Cecot G, Wise MD, Zhurov KO, Ronson TK, Castilla AM, Finelli A, Pattison P, Solari E, Scopelliti R, Zelinskii GE, Vologzhanina AV, Voloshin YZ, Nitschke J, Severin K (2016) Ligand aspect ratio as a decisive factor for the self-assembly of coordination cages. J Am Chem Soc 138:2046–2054
98. Voloshin YZ, Erdyakov SY, Makarenko IG, Lebed' EG, Potapova TV, Svidlov SV, Starikova ZA, Pol'shin EV, Gurskii ME, Bubnov YN (2007) Macrocyclization of the semiclathrochelate *o*-carboranylboronate and *n*-butylboronate iron (II) oximehydrazonates: synthesis and structure of clathrochelate products and unexpected allosteric effect of the apical substituent. Russ Chem Bull Int Ed 56:1787–1794
99. Voloshin YZ, Makarenko IG, Erdyakov SY, Svidlov SV, Vologzhanina AV, Lebed EG, Ignatenko AV, Gurskii ME, Bubnov YN (2010) Template synthesis, structure and spectra of the semiclathrochelate and clathrochelate 1,4-pentadienylboron-capped iron (II) oximehydrazonates. Inorg Chim Acta 363:395–403
100. Voloshin YZ, Makarov IS, Vologzhanina AV, Polshin EV, Monakov YB, Islamova RM, Bubnov YN (2008) Synthesis and structure of the ferrocenylboron-capped clathrochelate iron (II) oximehydrazonates. Russ Chem Bull Int Ed 57:1215–1222
101. Varzatskii OA, Kats (Menkach) SV, Penkova LV, Volkov SV, Dolganov AV, Vologzhanina AV, Bubnov YN, Voloshin YZ (2013) Cage metal complexes: synthesis, X-ray structure, and spectral and redox behavior of the first hybrid iron (II) clathrochelatoscorpionate and its pyrazoloxime-armed macrocyclic intermediate. Eur J Inorg Chem 2013:1987–1992
102. Varzatskii OA, Penkova LV, Kats (Menkach) SV, Dolganov AV, Vologzhanina AV, Pavlov AA, Novikov VV, Bogomyakov AS, Nemykin VN, Voloshin YZ (2014) Chloride ion-aided self-assembly of the pseudo-clathrochelate metal tris-pyrazoloximates. Inorg Chem 53:3062–3071
103. Puchta R, van Eldik R (2007) Host–guest complexes of oligopyridine cryptands: prediction of ion selectivity by quantum chemical calculations. Eur J Inorg Chem 2007:1120–1127
104. Bazzicalupi C, Bellusci A, Bencini A, Berni E, Bianchi A, Ciattini S, Giorgi C, Valtancoli B (2002) A new dipyridine-containing cryptand for both proton and Cu(II) encapsulation. A solution and solid state study. J Chem Soc Dalton Trans 2151–2157
105. Fekner T, Gallucci J, Chan MK (2004) Synthesis and metalation of a chiral, pyridine-strapped, cyclic bis (benzimidazole) ligand. Org Lett 6:989–992

106. Perkins DF, Lindoy LF, Meehan GV, Turner P (2004) Inherent helicity in an extended tris-bipyridyl molecular cage. Chem Commun 152–153
107. Takemura H, Nakamichi H, Nogita R, Iwanaga T, Yasutake M, Shinmyozu T (2003) A comparison of coordination ability of hetero atoms: a Li^+ and Na^+ selective pyridinophane-based cryptand. Tetrahedron Lett 44:5087–5089
108. Janczak J, Kubiak R (1994) Synthesis and crystal structure of an indium(III) complex with a bicyclic phthalocyanine at 300 K. J Chem Soc Dalton Trans 2539–2543
109. Catalano VJ, Malwitz MA, Noll BC (2002) Pd(0) and Pt(0) metallocryptands encapsulating a spinning mercurous dimer. Inorg Chem 41:6555–6559
110. Deeney FA, Harding CJ, Morgan GG, McKee V, Nelson J, Teat SJ, Clegg W (1998) Response to steric constraint in azacryptate and related complexes of iron-(II) and -(III). J Chem Soc Dalton Trans 1837–1844
111. Hu XL, Luo QH, Wang ZL, Wang YN, Zhou ZY (2003) Synthesis, spectroscopy and crystal structure of a new heterodinuclear silver(I)—copper (I) cryptate. Transit Met Chem 28:640–643
112. Jazwinski J, Lehn JM, Lilienbaum D, Ziessel R, Guilhem J, Pascard C (1987) Polyaza macrobicyclic cryptands: synthesis, crystal structures of a cyclophane type macrobicyclic cryptand and of its dinuclear copper(I) cryptate, and anion binding features. Chem Commun 22:1691–1694
113. Morgan G, McKee V, Nelson J (1994) Disorder in a disilver azacryptate structure. Inorg Chem 33:4427–4428
114. Lu Q, McKee V, Nelson J (1994) Redox-linked pyrrole NH deprotonation in a dicopper azacryptate system. Chem Commun 649–651
115. Howarth OW, Nelson J, McKee V (2000) Steric constraint generating large through-space ^{1}H–$^{203,205}Tl$ coupling in a dithallium(I) cryptate. Chem Commun 21–22
116. Catalano VJ, Malwitz MA, Horner SJ, Vasquez J (2003) Three- and four-coordinate gold(I) complexes of 3,6-bis(diphenylphosphino)pyridazine: monomers, polymers, and a metallocryptand cage. Inorg Chem 42:2141–2148
117. McKee V, Nelson J, Speed DJ, Town RM (2001) Conformational and coordination plasticity in silver (I) cryptates. Dalton Trans 3641–3646
118. Rodríguez-Cortiñas R, Avecilla F, Platas-Iglesias C, Imbert D, Bünzli JCG, de Blas A, Rodríguez-Blas T (2002) Structural and photophysical properties of heterobimetallic 4f-Zn iminophenolate cryptates. Inorg Chem 41:5336–5349
119. Platas C, Avecilla F, de Blas A, Rodríguez-Blas T, Geraldes CFGC, Tóth E, Merbach AE, Bünzli JCG (2000) Mono- and bimetallic lanthanide(III) phenolic cryptates obtained by template reaction: solid state structure, photophysical properties and relaxivity. J Chem Soc Dalton Trans 611–618
120. Platas C, Avecilla F, de Blas A, Geraldes CFGC, Rodríguez-Blas T, Adams H, Mahía J (1999) ^{1}H NMR in solution and solid state structural study of lanthanide(III) cryptates. Inorg Chem 38:3190–3199
121. Avecila F, Platas-Iglesias C, Rodríguez-Cortiñas R, Guillemot G, Bünzli JCG, Brondino CD, Geraldes CFGC, de Blas A, Rodríguez-Blas T (2002) Structural characterisation, EPR and magnetic properties of f–f and f–d lanthanide(III) phenolic cryptates. Dalton Trans 4658–4665
122. Takemura H, Kon N, Yasutake M, Shinmyozu T (2003) Structurally isomeric two pyridino macrocycles: complexation and structures. Tetrahedron 59:427–431
123. Chen QY, Luo QH, Hu XL, Shen MC, Chen JT (2002) Heterodinuclear cryptates [EuML(dmf)] $(ClO_4)_2$ (M=Ca, Cd, Ni, Zn): tuning the luminescence of europium(III) through the selection of the second metal ion. Chem Eur J 8:3984–3999
124. Brooker S, Ewing JD, Nelson J, Jeffery JC (2002) Heterobinuclear cryptates; cooperative binding generates two different coordination sites within a symmetrical cryptand. Inorg Chim Acta 337:463–466
125. Drew MGB, Farrell D, Morgan GG, McKee V, Nelson J (2000) d^{10} Cations within triple-helical cryptand hosts; a structural and modelling study. Dalton Trans 1513–1519
126. Shu-Yan Y, Qin-Hui L, Meng-Chang S, Xiao-Yun H (1994) Crystal structure of the first mixed-valence silver(I,III) cryptate. Inorg Chem 38:1251–1252
127. Brooker S, Ewing JD, Ronson TK, Harding CJ, Nelson J, Speed DJ (2003) Redox-adaptable copper hosts. Pyridazine-linked cryptands accommodate copper in a range of redox states. Inorg Chem 42:2764–2773
128. Takemura H, Kon N, Tani K, Takehara K, Kimoto J, Shinmyozu T, Inazu T (1997) Stable Cu^+, Ag^+ complexes of aza-bridged macrocyclic molecules: structure and chemical properties. Perkin Trans 1:239–246
129. Lu Q, Latour JM, Harding CJ, Martin N, Marrs DJ, McKee V, Nelson J (1994) Dicopper cryptates with 1,1 and 1,3 bridging ligands: spectroscopic, magnetic and electrochemical properties. J Chem Soc Dalton Trans 1471–1478
130. Drew MGB, Marrs DJ, Hunter J, Nelson J (1992) Divergent and convergent forms of a new Schiff-base cryptand; X-ray crystallographic and molecular mechanics investigations. J Chem Soc Dalton Trans 11–18
131. Lamarque L, Miranda C, Navarro P, Escarti F, García-España E, Latorre J, Ramirez JA (2000) Dopamine interaction with a polyamine cryptand of 1H-pyrazole in the absence and in the presence of Cu (II) ions. Crystal structure of $[Cu_2(H_{-1}L](ClO_4)_3{\cdot}2H_2O$. Chem Commun 1337–1338
132. Lamarque L, Navarro P, Miranda C, Arán VJ, Ochoa C, Escartí F, García-España E, Latorre J,

Luis SV, Miravet JF (2001) Dopamine interaction in the absence and in the presence of Cu^{2+} ions with macrocyclic and macrobicyclic polyamines containing pyrazole units. Crystal structures of $[Cu_2(L_1)(H_2O)_2](ClO_4)_4$ and $[Cu_2(H_{-1}L_3)](ClO_4)_3 \cdot 2H_2O$. J Am Chem Soc 123:10560–10570
133. Drew MGB, Harding CJ, Howarth OW, Lu Q, Marrs DJ, Morgan GG, McKee V, Nelson J (1996) Thiophene-linked azacryptand sites for dicopper and disilver; thiophene sulfur as an inert spacer? J Chem Soc Dalton Trans 3021–3030
134. Alliger GE, Müller P, Do LH, Cummins CC, Nocera DG (2011) Family of cofacial bimetallic complexes of a hexaanionic carboxamide cryptand. Inorg Chem 50:4107–4115
135. Ma Z, Cao R (2005) An N_8O_6 cryptand: the host for Ag^+ guests and H_2O molecules. J Mol Struct 738:137–142
136. Drew MGB, McDowell D, Nelson J (1988) A new ditopic polyaza macrobicyclic ligand: X-ray crystallographic structure determination. Polyhedron 7:2229–2232
137. Zhang J, Ma Z, Zhao H, Ge C, Wang Y, Zhang X (2016) Bis-boron-capped tris(dioxime) cage complexes with terminal carbonyl groups. Inorg Chem Commun 65:63–67

4 Magnetic Properties, EPR, and Paramagnetic NMR Spectra of Cage Metal Complexes

Magnetometry, EPR, and NMR methods have been widely used to study electronic structure of paramagnetic transition *d*- and *f*-metal cage complexes, their magnetic characteristics, and SCO and SMM behavior of the (pseudo) clathrochelates as well as to deduce the geometry of coordination polyhedra of encapsulated metal ion(s).

Both the high- and low-spin states of the polyamine cobalt(II) clathrochelate (Co^{2+})@**675** with permethylated apical amine substituents (Scheme 4.1) have been characterized [1] by EPR method. For the sample of this cage complex, obtained by chemical reduction of its cobalt (III)-encapsulating precursor, the well-defined EPR spectrum at 4 K was found to be characteristic of the high-spin cobalt(II) complexes. When this clathrochelate has been prepared by radiation-induced electron addition to the same macrobicyclic precursor at 77 K, the low-spin complex (Co^{2+})@**675** has been initially formed, which underwent a conversion into its high-spin state on annealing to approximately 115 K [1].

EPR spectra of 1:1 copper(II) cage complexes of the encapsulating hemicryptophane ligands **613–615** have been recorded [2]. A pseudo-axial EPR signal (g_1 = 2.00–2.02, g_2 = 2.13–2.14, and g_3 = 2.20–2.22) with a resolved hyperfine structure ($A_1 = 57–68 \times 10^{-4}$, $A_2 = 40–52 \times 10^{-4}$, and $A_3 = 108–127 \times 10^{-4}$ cm^{-1}) has been observed in all these spectra; the above *g*-values are characteristic of the copper(II) complexes with a slightly distorted TBP $Cu^{II}N_4$-coordination polyhedron. The unpaired electron of a caged Cu^{2+} ion is described [2] to predominantly occupy its d_{Z^2}-orbital lying along the molecular C_3 pseudoaxis, and the confinement imposed by a hemicryptophane caging ligand promoted the angular distortions around this encapsulated metal ion.

SQUID magnetometry data for the *n*-butylboron-capped cobalt(II) hexachloroclathrochelate (Co^{2+})@**293** (μ_{eff} = 1.55 M.B.) have been used [3] for evidence of the low-spin state ($s = 1/2$) of an encapsulated metal ion, whereas its reduced derivative, the cobalt(I) cage complex (Co^{+})@**293**, was found to be in high-spin state (μ_{eff} = 2.85 MB, $s = 1$). X-band EPR spectral parameters for the former cobalt (II) clathrochelate ($g_x = 2.252$, $g_y = 2.070$, and $g_z = 1.984$ and $A_{xx} = 125 \times 10^{-4}$, $A_{yy} = 9 \times 10^{-4}$, and $A_{zz} = 50 \times 10^{-4}$ cm^{-1} at 10 K) suggest a strong anisotropy of the *g*-factor and of the hyperfine coupling constant as well. This spectrum is described [3] in terms of a spin doublet $s = 1/2$ interacting with a single ^{59}Co nucleus ($I = 7/2$), characteristic of the low-spin complexes with electronic configuration d^7. On the other hand, EPR spectrum of the above paramagnetic cobalt(I) cage complex with electronic configuration d^8 has been not observed [3] within the temperature range down to that of liquid helium; this fact is explained by a long spin-lattice relaxation time, by the

Y. Voloshin et al., *Cage Metal Complexes*, DOI 10.1007/978-3-319-56420-3_4

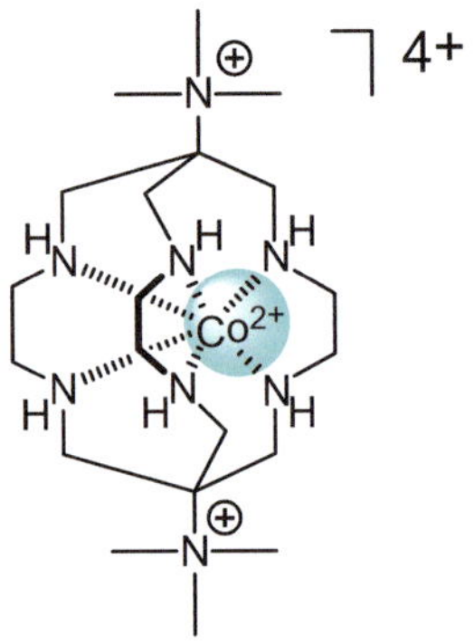

Scheme 4.1 Chemical drawing of the clathrochelate (Co^{2+})@**675**

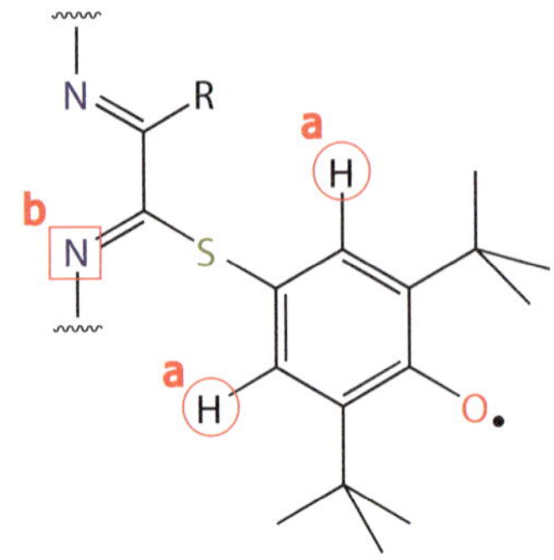

Fig. 4.1 Chemical drawing of the phenolsulfide-substituted ribbed α-dioximate fragment of a clathrochelate framework and the atoms causing the hyperfine splittings in EPR spectra [5]

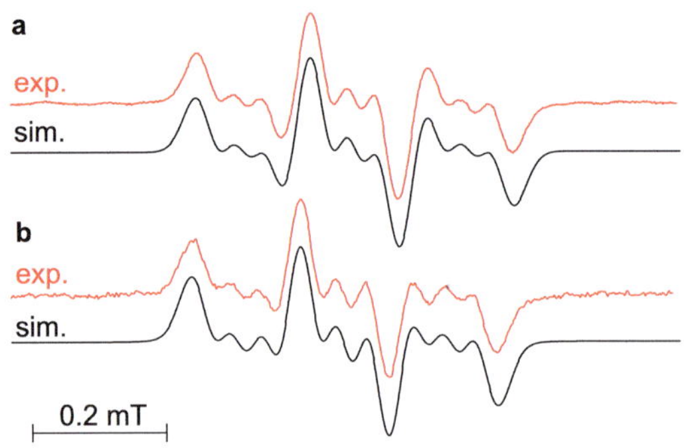

Fig. 4.2 Experimental and simulated EPR spectra of the oxidized radical derivatives of the monophenolsulfide clathrochelates (Fe^{2+})@**400** (**a**) and (Fe^{2+})@**399** (**b**) [5]

magnitude and the sign of spin–orbital interaction, as well as by the ligand field symmetry.

Multifrequency EPR spectroscopy at 9.5 and 275.7 GHz has been used [4] to study an electronic structure of the mononuclear manganese (II) tris-pyrazoloximate pseudoclathrochelate (Mn^{2+})@**509**. The observed virtually axial zero-field splitting with $D = -9.7$ GHz (-0.32 cm^{-1}) is described [4] to be the largest value reported at the moment for the manganese(II) complexes with MnN_6-coordination polyhedra.

Chemical oxidation of the phenol-functionalized iron(II) clathrochelate (Fe^{2+})@**306**, (Fe^{2+})@**396**, (Fe^{2+})@**397**, (Fe^{2+})@**398**, (Fe^{2+})@**399**, and (Fe^{2+})@**400** and their cobalt(II)-encapsulating analog (Co^{2+})@**397** has been studied [5] using EPR method. EPR spectra of the monoradical macrobicyclic complexes, generated in situ by oxidation of their above clathrochelate precursors with PbO_2, have been simulated taking into account the hyperfine coupling with protons of the phenoxyl fragment (Fig. 4.1a) as well as with the adjacent nitrogen atom of the same functionalized α-dioximate chelate cycle (Fig. 4.1b). On the other hand, such interactions with protons of *tert*-butyl groups have not been observed in these spectra. The oxidized clathrochelate monoradicals are stable for at least 2–3 days, whereas the formation of their biradical clathrochelate derivatives has not been detected [5]. The experimental and simulated EPR spectra of the oxidized radical derivatives of the cage complexes (Fe^{2+})@**399** and (Fe^{2+})@**400** are shown in Fig. 4.2. Both these spectra contain nine poorly resolved lines caused by hyperfine interaction of an unpaired electron with two equivalent *meta*-protons of the phenoxyl fragment and one nitrogen atom of a clathrochelate framework, whereas an interaction with the second distal nitrogen atom of this functionalized chelate fragment has not been observed due to its remoteness from the radical center. Passing from the methine hydrogen atom in the molecule (Fe^{2+})@**399** to the methyl substituent in the molecule (Fe^{2+})@**400** caused no significant changes in a delocalization of this unpaired electron. The nature of apical substituents at a clathrochelate framework (i.e., passing from phenyl to *n*-butyl groups) did not affect the line shape of EPR spectra for the above clathrochelate radicals; their isotropic hyperfine tensor values also persist [5].

EPR spectrum of the diphenoxyl clathrochelate radical, resulting from an oxidation of the complex (Fe^{2+})@**398**, is very similar to that of its mono- and hexafunctionalized analogs (Fe^{2+})@**399**, (Fe^{2+})@**400**, and (Fe^{2+})@**396**, (Fe^{2+})@**397**. An increase in the isotropic values of hyperfine splitting constant with nitrogen atom a_N in a row, (Fe^{2+})@**400**, (Fe^{2+})@**399** < (Fe^{2+})@**398** < (Fe^{2+})@**396**, (Fe^{2+})@**397**, is explained [5] by increasing in a delocalization of the unpaired electron over a highly conjugated quasiaromatic clathrochelate framework with attached phenol-terminated substituents. Passing from the iron (II) clathrochelate (Fe^{2+})@**397** to its cobalt(II)-containing analog (Co^{2+})@**397** resulted in a complication of the EPR spectrum due to the hyperfine interactions of the formed radical species with an encapsulated cobalt(II) ion ($I = 7/2$). The spectrum of (Co^{2+})@**397** at 20 K is characteristic of the low-spin cobalt(II) complexes; no hyperfine splittings due to an interaction of the unpaired electron with six donor nitrogen atoms have been observed [5] in this spectrum due to the large linewidth of its components. EPR spectrum of an oxidized radical derivative of the clathrochelate ((Fe^{2+})@**306**) was found [5] to contain only one broadened line due to the inherent character of its single phenoxyl substituent; the hyperfine coupling constant with nitrogen atom a_N is equal to 0.338 mT.

Spin state of the paramagnetic tris-pyrazoloximate 3*d*-metal pseudoclathrochelates (M^{2+})@)**509** ($M^{2+} = Co^{2+}$, Fe^{2+}, and Mn^{2+}) has been studied [6] using SQUID magnetometry, NMR, and EPR methods. All these complexes are in high-spin state with $s = 3/2$, 2, and 5/2, respectively, in the temperature range 5–300 K. High-spin state of the encapsulated iron(II) ion in its pseudoclathrochelate (Fe^{2+})@**509** has been also confirmed [6] using ^{57}Fe Mössbauer experiments.

The paramagnetic shifts in NMR spectra of these paramagnetic pseudoclathrochelates, the signal assignments of which were performed [6] with the assistance of DFT calculations, have been governed by very different mechanisms. The cobalt(II) complex (Co^{2+})@**509** demonstrated very large values of the paramagnetic shifts of both the chelate and the capping fragments' nuclei, thus suggesting the substantial pseudocontact contribution. In the case of its iron(II)-containing analog (Fe^{2+})@**509**, large paramagnetic shifts have been observed for the signals of the chelate fragments' nuclei, whereas those of the phenyl substituents were affected only slightly. This suggests the prevalence of the contact contribution into the value of the paramagnetic shift with a very small pseudocontact input. For the manganese (II) complex (Mn^{2+})@**509**, no pseudocontact interaction has been expected [6], and the observed shifts in its spectra are explained by direct delocalization of the unpaired electron over a polyazomethine pseudomacrobicyclic ligand; the paramagnetic broadening did not allow to observe the signals of the carbon nuclei close to its central metal ion. So, the magnetic properties of these paramagnetic 3*d*-metal tris-pyrazoloximate pseudoclathrochelates were found [6] to be strongly affected by the nature of this ion.

Combined EPR and magnetometry study of a series of the tris-dioximate cobalt(I) and cobalt (II) hexachloroclathrochelates has been performed [7]; the temperature plots of the effective magnetic moments μ_{eff} of the boron-capped cobalt(II) complexes with different apical substituents are shown in Fig. 4.3. Such plot for the fluoroboron-capped clathrochelate (Co^{2+})@**295** substantially differs from those for other cobalt (II) hexachloroclathrochelates under study: μ_{eff} for the former compound (3.87 M.B. at 4 K) is close to spin-only value for the high-spin cobalt (II) complexes, thus suggesting that its high-spin state is preferably populated even at very low temperatures, whereas those for the *n*-butylboron- and phenylboron-capped cage complexes (Co^{2+})@**293** and (Co^{2+})@**294** (Fig. 4.3) suggest the gradual and incomplete spin doublet–quartet transitions. At the low temperatures, the μ_{eff} values for both these clathrochelates were found [7] to be characteristic of the systems with $s = 1/2$, while an increase in temperature caused the gradual increase in μ_{eff} due to the $1/2 \rightarrow 3/2$ spin transition that is not

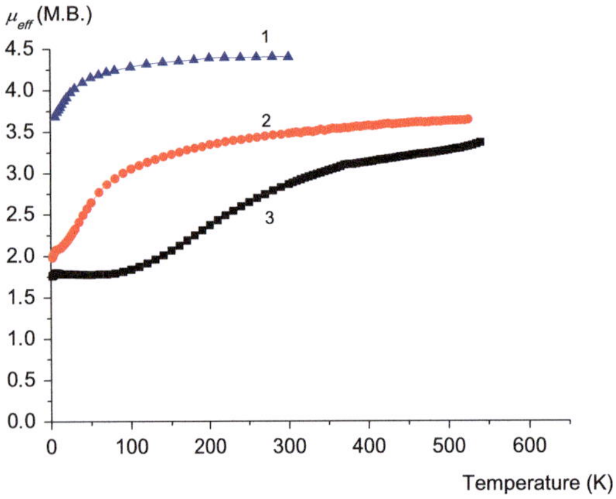

Fig. 4.3 Temperature plots of μ_{eff} for the fine-crystalline samples of the clathrochelates (Co^{2+})@**295** (*filled triangle*, 1), (Co^{2+})@**294** (*filled circle*, 2), and (Co^{2+})@**293** (*filled square*, 3) [7]

complete even at 300 K. The μ_{eff} values for their cobalt(I)-encapsulating clathrochelate derivatives are temperature independent above 30 K being characteristic of the high-spin electronic configuration d^8 with $s = 1$. On the other hand, the cobalt(II) clathrochelates with six sulfide-ribbed substituents did not undergo temperature SCO and are the low-spin compounds with $s = 1/2$ in the temperature range 4–400 K [7].

The low-temperature glassy EPR spectra for most of the above cobalt(II) clathrochelates are characteristic of the low-spin cobalt (II) complexes with slightly rhombic *g*-tensors. The observed well-resolved eight-line splittings in the downfield region of these spectra are explained [7] by the hyperfine interactions with ^{59}Co nucleus ($I = 7/2$), while the additional splittings resulted from interactions with donor nitrogen atoms of their caging ligands. The five-line splitting pattern, which has been observed [7] in the high-field region of the spectrum for the fluroboron-capped complex (Co^{2+})@**293**, is resulted from an interaction of the unpaired electron with only two nitrogen atoms of this type due to the structural Jahn–Teller distortion of its CoN_6-coordination polyhedron, causing the shift of an encapsulated cobalt(II) ion from the center of this polyhedron in the direction of one of the three ribbed chelate fragments (see Sect. 3.2). X-band EPR spectra, characteristic of high-spin states of these cobalt(II) cage complexes, cannot be measured [7]. Indeed, their high-spin states become substantially populated only above 100 K, while a very fast relaxation of their EPR signals makes it impossible to observe the spectra above 30 K. On the other hand, according to the magnetochemical data, a high-spin state is significantly populated for two of these clathrochelates below 30 K, and the absence of their EPR signals is explained [7] by considerable broadening of these signals because of the extremely fast relaxation occurring even at low temperatures. The above high-spin cobalt (I) clathrochelates are reported [7] to be also EPR silent in X-band due to long spin-lattice relaxation time as well as because of the magnitude and the sign of the spin–orbital interaction and the ligand field symmetry; no EPR transitions have been detected even using EPR submillimeter spectrometry in the range 90–560 GHz at 4 K [7].

EPR spectrum of the mononitroxyl-containing clathrochelate (Fe^{2+})@**318** [8] contains the triplet signal of its radical substituent with $g = 2.0067$ and $a_N = 15.8$ G. The spectrum of its difunctionalized analog (Fe^{2+})@**319** has a more complex character: the intensity of its central line is substantially higher as compared with that in the spectrum of (Fe^{2+})@**318** and the additional broadened lines appeared. This EPR spectral pattern is described [8] to be characteristic of the nitroxide biradicals with their exchange interaction energy close to the hyperfine coupling constant.

SQUID magnetometry method has been used [9] to evidence the low-spin state of the cobalt (II) hexasulfide clathrochelates (Co^{2+})@**405** and (Co^{2+})@**404**: the observed μ_{eff} values are close to pure spin-only value for the paramagnetic complexes with $s = 1/2$ and the Curie–Weiss law is held rather well from 4 to 200 K, thus indicating that only doublet state of these clathrochelate molecules is populated. Their low-spin character has also been confirmed [9] using the glassy EPR spectra of these cobalt (II) clathrochelates at 30 K, the parameters of which are characteristic of the Jahn–Teller-distorted low-spin cobalt(II) complexes. The g- and hyperfine tensors are rhombic and these spectra contained the well-resolved eight-line splittings in the downfield region caused by the hyperfine interactions with ^{59}Co nucleus ($I = 7/2$) as well as the additional small splittings resulted from such interactions with donor nitrogen atoms of a caging ligand [9].

Their hexafluorosulfide macrobicyclic analogs (Co^{2+})@**427**, (Co^{2+})@**428**, and (Co^{2+})@**429** [10] have almost the same EPR spectral characteristics, and their identity evidenced a similar level of the spin density delocalization over the ribbed substituents for all these cobalt (II) complexes; their low-spin state has been also confirmed by SQUID magnetometry experiments.

Magnetic properties of the alkylboron-capped aliphatic and aromatic cobalt(II) clathrochelates have been studied [11] using EPR, SQUID magnetometry, and NMR methods. The low-temperature glassy EPR spectra of these cobalt clathrochelates are characteristic of the low-spin Jahn–Teller-distorted cobalt(II) complexes (see above). The nature of their ribbed substituents only slightly affected the magnetic tensor's values. In all the spectra, except those for the α-benzildioximate cobalt(II) clathrochelates, the additional splittings assigned to the hyperfine interactions with donor nitrogen nuclei of a caging ligand have been observed. As follows from the SQUID magnetometry data of [11], their magnetic moments μ_{eff} are close to the spin-only value with $s = 1/2$ in the temperature range 5–400 K.

The paramagnetic shift values, obtained as the difference between the chemical shifts in NMR spectra of the paramagnetic cobalt(II)- and diamagnetic iron(II)-encapsulating clathrochelate analogs, have been also analyzed [11]. An encapsulated low-spin cobalt(II) ion can cause both the contact and pseudocontact paramagnetic shifts due to the direct spin delocalization and the dipolar interaction, respectively. These cobalt(II) cage complexes in their solutions undergo the dynamic Jahn–Teller exchange between their three possible distorted molecular structures at room temperature, whereas one Jahn–Teller-distorted configuration is more stabilized in the solid state. The above exchange in their solutions led to a partial averaging of both the Fermi and dipolar contributions to the paramagnetic shifts and, thus, complicated their assignment; so, only few qualitative conclusions are drawn [11]. The relatively low anisotropy of g-tensor is described to imply the similarly moderate values of the pseudocontact shifts, and the Jahn–Teller averaging is expected to decrease the dipolar contribution. The paramagnetic shifts of the apical substituents in these cage complexes have the pseudocontact nature due to their remoteness from an encapsulated paramagnetic metallocenter. No substantial differences between the cobalt(II) clathrochelates with different ribbed substituents have been observed [11] except that for the macrobicyclic tris-α-benzildioximates, which demonstrated much lower values of the paramagnetic shifts. This fact is explained by delocalization of a spin density on their aromatic ribbed substituents causing an increase in the ligand-centered pseudocontact shifts with simultaneous decrease in the point-dipole contribution. The above results are in good agreement with the absence of the nitrogen hyperfine coupling in EPR spectra: the spin density localized on the donor nitrogen atoms of their caging ligands is effectively reduced by its delocalization on six inherent aromatic substituents at a quasiaromatic cage framework. The possibility of such direct spin delocalization was confirmed [11] by their nuclei's chemical shifts shown in Fig. 4.4. The glyoximate, dimethylglyoximate, and α-benzildioximate cobalt(II) clathrochelates showed an alternation of the direction of such shifts, thus suggesting the presence of a spin density in π-system of a macrobicyclic framework due to the significant overlap of SOMO of an encapsulated cobalt(II) ion with π-orbitals of its caging ligand. The observed dependence of a chemical shift on the distance of a given nucleus from the paramagnetic metallocenter evidenced also some σ-contributions. So, the paramagnetic shifts of nuclei of their ribbed chelate substituents have the complex character: they are caused by π-spin delocalization, spin polarization σ-contribution, and non-averaged part of

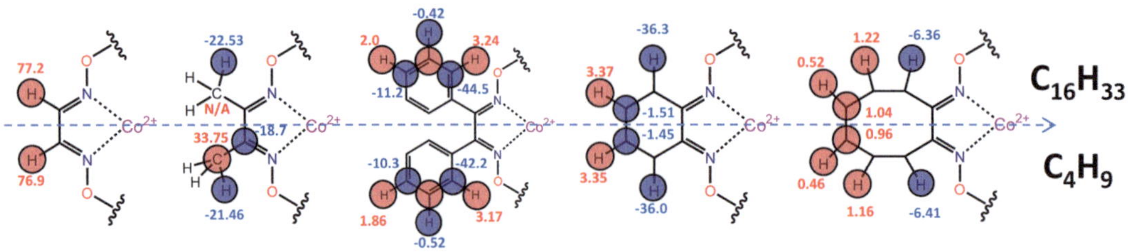

Fig. 4.4 Low-field (in *violet*) and high-field (in *red*) paramagnetic NMR shifts (in ppm) of the ribbed fragments' nuclei of the alkylboron-capped cobalt(II) clathrochelates [11]

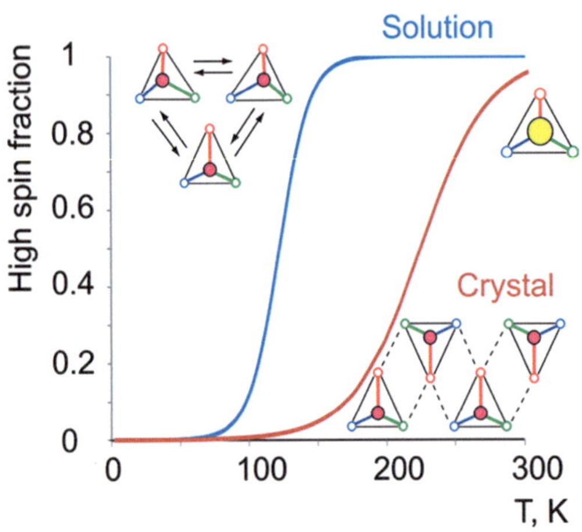

Fig. 4.5 Temperature SCO for the clathrochelate (Co^{2+})@**285** in solution and in its single crystal [12]

the pseudocontact interactions [11]. The abundance of these interactions is explained [11] by magnetic non-equivalence of all three α-dioximate chelate fragments in the Jahn–Teller-distorted cage molecules and the dynamic exchange between three possible configurations mixing the above contributions together.

Detailed study of the magnetic properties of the methylboron-capped cobalt(II) hexachloroclathrochelate (Co^{2+})@**285** both in its solution and in the solid state has been performed [12] using the NMR, EPR, and SQUID magnetometry experiments. As follows from the data of Evans method, the dissolution of (Co^{2+})@**285**, which exhibits the incomplete and gradual SCO in the solid state (Fig. 4.5), caused a population of its high-spin state above 200 K. This method could not be used at low temperatures, and the above SCO has been studied [12] down to 10 K using EPR spectroscopy by measuring the relative integral intensity of EPR signal of the low-spin state of this cage complex. As follows from EPR data, SCO in solution is complete and substantially more abrupt than that in the fine-crystalline sample of (Co^{2+})@**285**; in the latter case, it depends on the size of its crystallites. The multitemperature X-ray diffraction experiments have been used [12] to study SCO in the single crystal of this clathrochelate. The Co–N distances in its TP molecule revealed a substantial temperature-dependent anisotropy with low-temperature structural Jahn–Teller distortion. An increase in temperature caused more uniform distribution of these bond lengths, but some anisotropy persists even at room temperature due to the residual population of low-spin state of this cobalt(II) cage complex. The rigidity of its encapsulating ligand does not favor a *trans*-like Jahn–Teller distortion, while a less common *cis*-distortion has been observed [12]. In the latter case, the geometry of a macrobicyclic ligand does not change significantly, and the degeneracy is lifted by the shift of an encapsulated cobalt (II) ion from the center of the ligand's cavity toward two of the three ribbed chelate fragments. ADPs of this caged metal ion were found to become significantly anisotropic at the temperatures higher than 100 K, when its high-spin state is populated. As this anisotropy is caused by dynamic superposition of two cobalt ion's positions corresponding to its different spin states, the analysis of ADPs allowed to observe SCO in the single crystal of (Co^{2+})@**285**. The comparison of SCO curves for its solution and X-rayed crystal (Fig. 4.5) evidenced the stabilization of low-spin state of this clathrochelate by the crystal lattice: its single crystal had the highest temperature of SCO, while in the diluted solution of (Co^{2+})@**285**, this temperature is the lowest and such SCO is most abrupt, thus suggesting an anticooperative behavior. These anticooperative effects are assigned [12] to some intermolecular interactions that effectively stabilize the low-spin state of (Co^{2+})@**285**. The loss of anisotropy in the case of this cage complex has been assigned to the dynamic averaging between its three equivalent Jahn–Teller-distorted structures. Such Jahn–Teller distortion is static in the crystal and is dynamic in solution, and the above SCO behavior is explained [12] as follows. The crystal lattice increases its energy barrier by stabilizing the corresponding Jahn–Teller-distorted configuration. The dissolution of (Co^{2+})@**285** removes the corresponding intermolecular interactions, and all its three Jahn–Teller-distorted structures have the same energy. The low-spin–high-spin energy difference for this macrobicyclic intracomplex is small enough for its high-spin state to become populated at relatively low temperatures, and the specific intermolecular weak π...Cl interactions in its crystal (Fig. 4.6) are reported [12] to be responsible for stabilization of the low-spin state of (Co^{2+})@**285**. Disrupting these interactions by either including solvent molecules into the crystal lattice or by using bulky apical substituents preventing the close packing of the clathrochelate molecules resulted in the gradual and complete SCO without an additional stabilization of the low-spin state of the corresponding cage complex. Although the average energy of these π...Cl interactions, estimated [12] from X-ray diffraction data, is 1.2 kcal mol^{-1} only, this weak stabilization is quite significant in comparison with the enthalpy of spin transition in solution (1.7 kcal mol^{-1}). This fact has been used [12] for explanation of the above changes in the SCO curves for the clathrochelate (Co^{2+})@**285**. So, even weak intermolecular interactions can cause a pronounced anticooperativity of SCO, being more gradual in the solid state than in solution.

The fine-crystalline samples of the cobalt (II) clathrochelates (Co^{2+})@**240**, (Co^{2+})@**286**, and (Co^{2+})@**295** (Scheme 4.2) are reported [13] to undergo temperature-induced SCO. Given the D_3-symmetry of these macrobicycles, orientation of their main axis of the magnetic susceptibility tensor has been chosen along the molecular B...B axis. As the conformational dynamics of (Co^{2+})@**286** is limited to the rotation around the B–C bond ensuring the axiality of the magnetic anisotropy tensor and the rotation of the CF_3 groups, the values of NMR paramagnetic shifts of its ^1H and ^{19}F nuclei have been directly used [13] to obtain the magnetic susceptibility anisotropy value of $\Delta\chi_{ax} = 12.6 \times 10^{-32}$ m^3. The small number of nuclei in the capping fragments of the clathrochelate (Co^{2+})@**286** and the nonzero contact contribution reduced the precision of $\Delta\chi_{ax}$ determination. In contrast, the paramagnetic shift of all the nuclei of the hexadecyl long-chain aliphatic substituents of the complex (Co^{2+})@**240** as compared with its diamagnetic analog (Fe^{2+})@**240** has been observed (Fig. 4.7a). All these shifts have been considered [13] as pseudocontact, correlating with those calculated from molecular geometry for a given value of $\Delta\chi_{ax}$. The aliphatic substituents

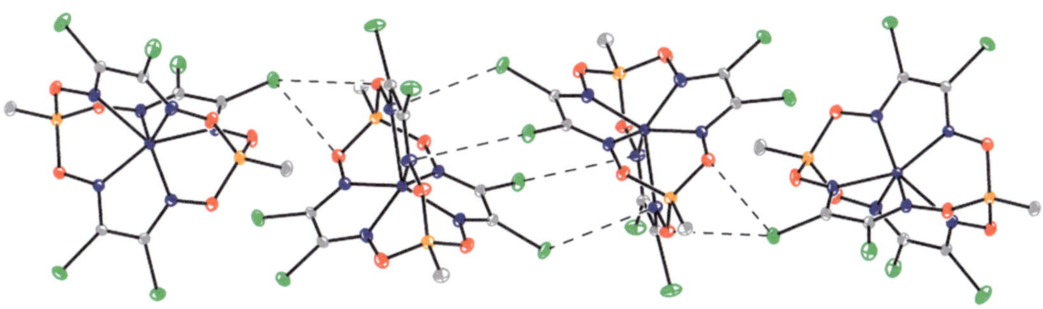

Fig. 4.6 Packing of the crystal (Co^{2+})@**285**; intermolecular π...Cl interactions are depicted with *dashed line* [12]

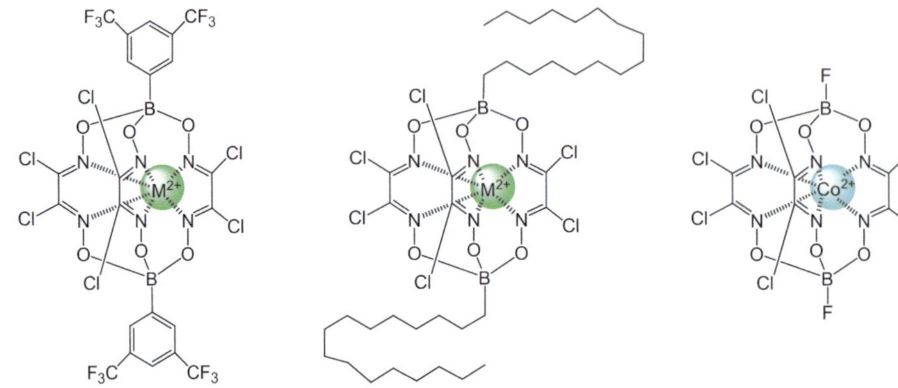

$M^{2+} = Co^{2+}$, (Co^{2+})@**286**
$M^{2+} = Fe^{2+}$, (Fe^{2+})@**286**

$M^{2+} = Co^{2+}$, (Co^{2+})@**240**
$M^{2+} = Fe^{2+}$, (Fe^{2+})@**240**

(Co^{2+})@**295**

Scheme 4.2 Chemical drawing of a series of the metal(II) hexachloroclathrochelates

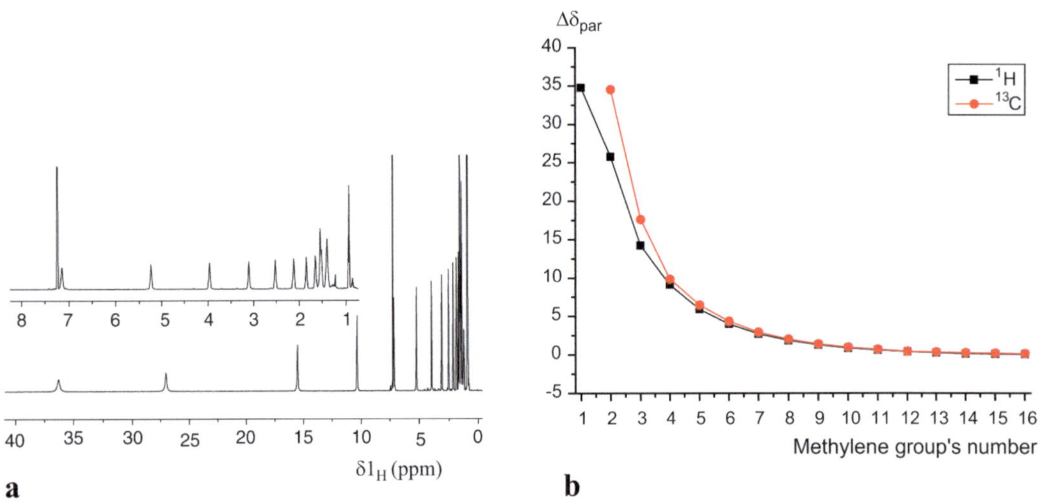

Fig. 4.7 ^{1}H NMR spectrum of the clathrochelate (Co^{2+})@**240** (**a**) and the paramagnetic contributions to the chemical shifts in its ^{1}H (*filled square*) and ^{13}C (*filled red circle*) NMR spectra versus the number of a given methylene group of its hexadecyl substituent (**b**)

of (Co^{2+})@**240** and (Fe^{2+})@**240** are highly flexible, and their conformational mobility substantially affects the above pseudocontact shifts. The fit of paramagnetic shifts to the molecular structure of (Co^{2+})@**240** resulted in a poor agreement between their experimental and calculated values, and simplified model that assumes highly limited number of conformations for each C–C bond has been used. Such accounting [13] for the above conformational mobility substantial improved compliance between calculated and experimental pseudocontact shifts; very small effect of the apical groups on the magnetic properties of an encapsulated metal ion has been also observed.

The difference between the magnetic anisotropy of the cobalt(II) cage complexes (Co^{2+})@**286** and (Co^{2+})@**240** can be also caused by the breakdown of a point-dipole approximation or underestimation of the contact shifts in the case of (Co^{2+})@**286**, and the maps characterizing the probability density distribution of the aliphatic chain have been evaluated [13]. They showed that the magnetic anisotropy of such cobalt(II) cage complexes is high enough to allow not only the study of their static structure but also to deduce a highly complicated dynamic behavior of aliphatic chains. These clathrochelates are reported [13] to have the largest reported at the moment $\Delta\chi_{ax}$ values for known cobalt(II) complexes with the possible source of this high magnetic anisotropy being a large negative ZFS value. In general, the Van Vleck equation may be employed to determine the components of g-tensor and ZFS by fitting temperature dependence of the magnetic susceptibility, but in the case of these cobalt (II) clathrochelates this approach could not be applied because their low-temperature magnetic behavior is masked by SCO. EPR spectroscopy is also described [12] as not a good method because the signals of a high-spin cobalt(II) ion are very difficult to detect in the presence of its predominantly populated low-spin state, owing to the sign and the value of ZFS energy as well as due to short electronic spin-lattice relaxation time. So, the isostructural fluoroboron-capped cobalt(II) cage complex (Co^{2+})@**295** that is in its high-spin state at all the temperatures from 4 to 300 K has been studied [13] by EPR and SQUID magnetometry methods. High-spin complexes with a very large negative ZFS value are usually EPR silent. This fact is explained by the following effects: (1) the inter-Kramers transitions cannot be induced by typical microwave frequencies, (2) the transition between $m_s = \pm 3/2$ levels of the lowest Kramers doublet is formally forbidden, and (3) $m_s = \pm 1/2$ doublet is thermally depopulated at low temperatures required for the detection of the fast-relaxing high-spin cobalt(II) ions. Indeed, the sample of (Co^{2+})@**295** was found [13] to be EPR silent in the X-band (9.5 GHz) within a wide temperature range, while in the Q-band (34 GHz), a very broad EPR signal assigned to the $m_s = \pm 1/2$ transition has been detected. The magnetic field position of this signal (~700 mT at 34.1 GHz) corresponds to a system with total spin $s = 3/2$: it has a large negative ZFS with the perpendicular component of g-tensor of approximately 1.9 and the parallel component being invisible owing to a very large linewidth. The temperature plot of the EPR spectral intensity was found to be also in good agreement with the large negative value of ZFS. The signals observed above 30 K became more pronounced with increase in the temperature up to 100 K, implying, together with the decrease in the EPR intensity at higher temperatures, a slow population of a higher-lying Kramers doublet followed by a gradual decrease in the population difference between its components. Temperature plot of the EPR intensity has been used [13] to estimate the ZFS value of $D \approx -40\ \text{cm}^{-1}$ that is in reasonable agreement with its value of $-63\ \text{cm}^{-1}$ obtained from the analysis of the temperature-dependent magnetic susceptibility for the fine-crystalline sample of (Co^{2+})@**295**. These D values have been applied [13] as initial parameters for the analysis of the temperature-dependent paramagnetic shifts for the cage complexes (Co^{2+})@**286** and (Co^{2+})@**240**.

Detailed study of the magnetic behavior of the above hexadecylboron-capped cobalt (II) hexachloroclathrochelate (Co^{2+})@**240** in solution and in the solid state has been performed [14] using solution NMR, frozen solution EPR, and SQUID magnetometry methods. Although the apical hexadecyl substituents of its diamagnetic iron(II)-encapsulating analog (Fe^{2+})@**240** showed a heavily overlapped set of the signals in its ^{1}H NMR spectrum, the paramagnetic nature of an encapsulated cobalt(II) ion in the cobalt (II) cage complex (Co^{2+})@**240** with the same macrobicyclic ligand caused very large downfield paramagnetic shifts of all the nuclei of its apical hexadecyl substituents (Fig. 4.7a). As a result, its ^{1}H NMR spectrum contains 16 well-resolved signals. Their paramagnetic shifts are described [14] to be pseudocontact in nature, thus depending on the intramolecular distances: the nuclei which are closer to the encapsulated paramagnetic high-spin metallocenter have the larger chemical shifts (Fig. 4.7b), whereas the paramagnetic shifts for known low-spin cobalt (II) clathrochelates have significantly lower magnitude and an opposite sign. Decrease in temperature caused a steady increase in paramagnetic shifts of the signals in NMR spectra of (Co^{2+})@**240** according to the Curie law; even

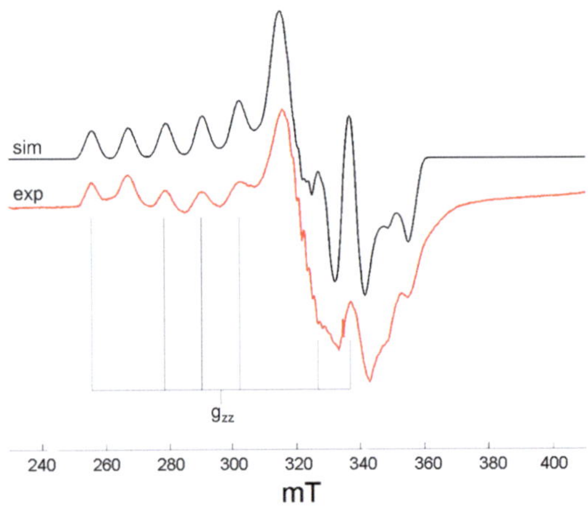

Fig. 4.8 Experimental and simulated X-band EPR spectra of the frozen toluene solution of (Co^{2+})@**240** at 30 K [14]

at 200 K, the population of its low-spin state has not been observed. Because further lowering of the temperature in the solution NMR experiments could not be performed, EPR method has been used [14] to study the spin state of (Co^{2+})@**240** at low temperatures. X-band EPR spectrum of its frozen toluene solution at 30 K (Fig. 4.8) is characteristic of the low-spin cobalt(II) clathrochelates (see above). Additional splitting with the signal ratio 1:2:3:2:1, assigned to the superhyperfine interaction of a cobalt nucleus with two ^{14}N nuclei ($I_N = 1$) of one of the three ribbed chelate fragments of (Co^{2+})@**240**, has been observed in the second-derivative EPR spectrum. The performed DFT calculations of the spin density distribution in its cage molecule confirmed the substantial delocalization of its unpaired electron over two nitrogen atoms of the α-dioximate-ribbed fragment of a macrobicyclic framework having the largest Co–N bond lengths. Such lowering of the molecular symmetry of (Co^{2+})@**240** is explained [14] by the structural Jahn–Teller distortion of electronic configuration d^7, thus confirming its low-spin state. The intensity of EPR signal of this cage complex is substantially decreased at higher temperatures due to both the loss of spin polarization and the depopulation of its low-spin state caused by SCO 1/2 ↔ 3/2. The signals of high-spin state of (Co^{2+})@**240** have not been observed at any temperature because high-spin cobalt(II) clathrochelates are known to be EPR silent in X-band due to a very large negative zero-field splitting energy and a fast transverse relaxation (see above). So, the magnetic properties of this hexadecylboron-capped macrobicyclic cobalt(II) complex in solution are identical to those of its methylboron-capped analog (Co^{2+})@**285** (see above): this evidenced only small effect of the apical substituents at a cage framework on a spin state of these cobalt (II) hexachloroclathrochelates. In contrast, those affected their magnetic properties in the solid state [14]. As follows from SQUID magnetometry data (Fig. 4.9), the hexadecylboron-capped cobalt(II) clathrochelate (Co^{2+})@**240** is the low-spin complex below 50K (μ_{eff} = 1.83 MB) and a gradual increase in μ_{eff} has been observed at higher temperatures. Such SCO is not complete even at 340 K, thus suggesting the anticooperative behavior for clathrochelate (Co^{2+})@**240** in its single crystal similar to that for its methylboron-capped analog (Co^{2+})@**285** (see above). An unusual feature of the SQUID curve for the former hexadecyl-terminated cage complex is the reproducible abrupt drop of its magnetic properties at 230 K (Fig. 4.9) [14].

The similar magnetic behavior has been observed [15] for the *n*-butylboron-capped cobalt (II) hexabromoclathrochelate (Co^{2+})@**287**. The SQUID magnetometry data (Fig. 4.10) evidenced that this complex is in its low-spin state below 100 K; a gradual increase in μ_{eff} has been observed at higher temperatures, and such SCO is not complete even at 400 K due to substantial stabilization of this state by intermolecular interactions in its crystals. At very low temperatures, the μ_{eff} value undergoes abruptly dropping below the spin-only value due to the intermolecular antiferromagnetic interactions in them. The sudden decrease in μ_{eff} (i.e., the reverse spin transition) at approximately 273 K is assigned [15] to the structural phase transition in these crystals. On the other hand, the *n*-butyl apical substituents of (Co^{2+})@**287** are relatively short as compared with hexadecyl substituents of its analog (Co^{2+})@**240**. As a result, an overall effect of the above structural phase transition for

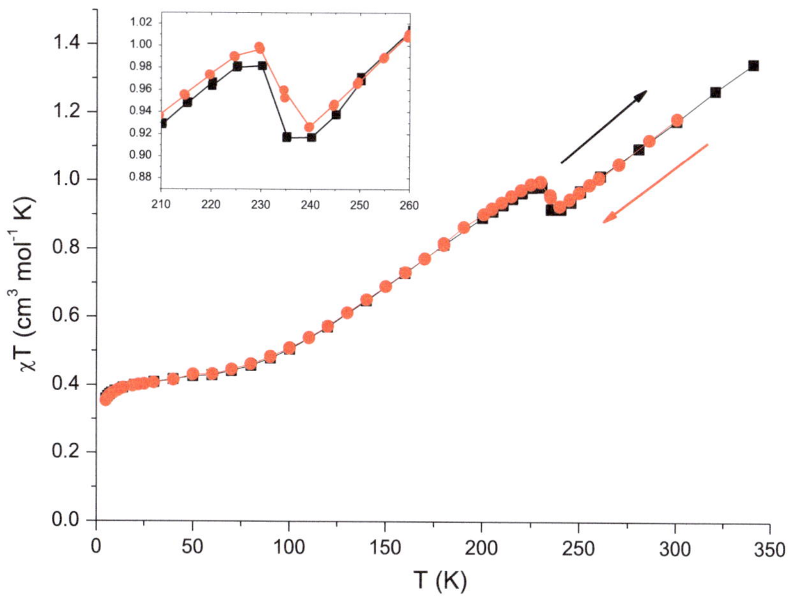

Fig. 4.9 SQUID magnetometry data for the hexadecylboron-capped cobalt(II) hexachloroclathrochelate (Co^{2+})@**240** on its cooling (*filled red circle*) and its heating (*filled square*). Inset: the "reverse" fragment of this curve around 230 K [14]

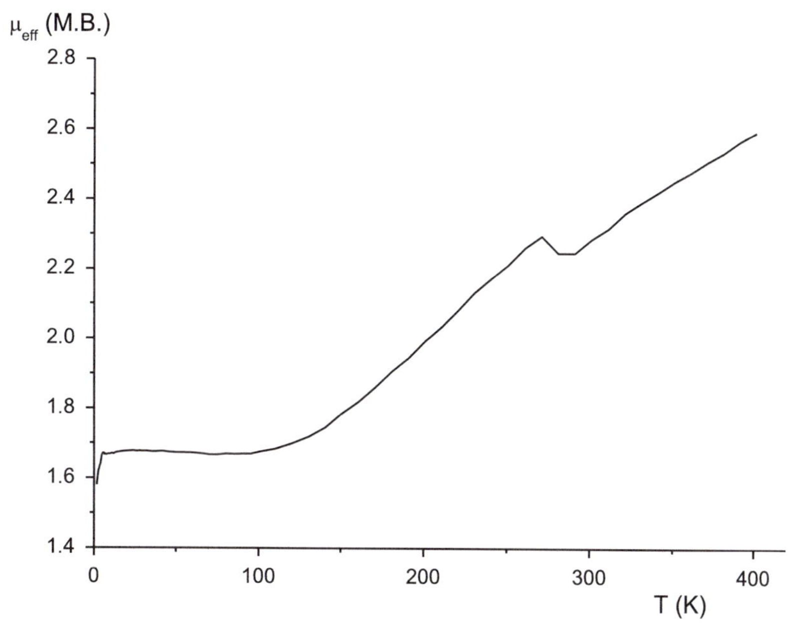

Fig. 4.10 Temperature plot of μ_{eff} for the fine-crystalline sample of (Co^{2+})@**287** [15]

(Co^{2+})@**287** is substantially smaller than that in the case of the latter clathrochelate (see Fig. 4.9) and no hysteresis has been observed [15]. An assignment of the above magnetic anomaly to structural reorganization of the apical substituents of (Co^{2+})@**287** has been confirmed by single crystal X-ray diffraction and DSC experiments.

The pseudoclathrochelate cobalt(II) tris-pyrazoloximate (Co^{2+})@**509** (Fig. 4.11a) demonstrated more than twofold increase in

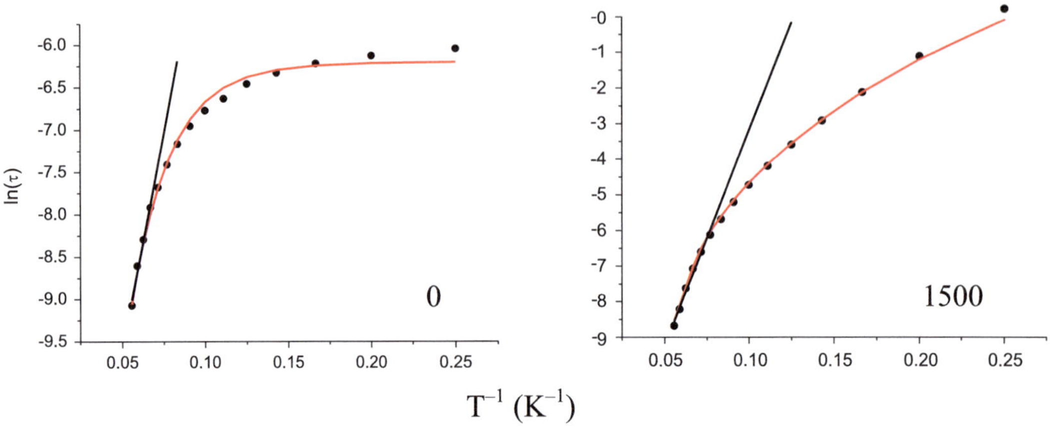

Fig. 4.11 SMM behavior in the cobalt(II) pseudoclathrochelate (Co^{2+})@**509** [16]

the values of pseudocontact shifts in NMR spectra as compared with its boron-capped hexachloroclathrochelate analogs with an encapsulated cobalt(II) ion (see above). So, (Co^{2+})@**509** has been tested [16] as a potent SMM. Variable-temperature magnetic susceptibility measurements evidenced that its pseudoencapsulated cobalt(II) ion is in high-spin state ($s = 3/2$) with a significant orbital contribution to the net magnetic moment of this complex; its significant magnetic anisotropy has been also observed and the following magnetic parameters were found: $D = -82$ cm^{-1}, E/D = 0.003, $g_{\|} = 2.9$, $g_{\perp} = 2.2$. An acceptable fit has been obtained [16] only with the negative value of D and the axial g-tensor. As the large D value prevents the population of the excited states at low temperatures, the corresponding isofield magnetization curves (Fig. 4.11) are reported [16] to be almost superimposable. No magnetic hysteresis has been observed in the variable-field magnetization data at 2 K, thus suggesting a quantum tunneling of magnetization. The above pseudoclathrochelate (Co^{2+})@**509** was found [16] to be EPR silent at X- and Q-band frequencies in the temperature range 4–150 K, thus confirming the large negative value of D. Its value as well as that of g-tensor has been calculated [16] also from NMR paramagnetic shifts, giving the following parameters: $D = -109$ cm^{-1}, $g_{\|} = 2.9$, $g_{\perp} = 2.2$, which are close to those obtained from the above magnetometry experiments.

SMM behavior in this complex with very high Orbach relaxation barrier of 152 cm^{-1} is explained [16] by high magnetic anisotropy, caused by a TP geometry of CoN_6-coordination polyhedron with orbitally degenerated ground state. Its hexadecylboron-capped analog (Co^{2+})@**510** is reported [17] to form two polymorphs (α and β), and both these polymorphs with almost the same pseudocage frameworks showed a SMM behavior (Fig. 4.12), but even such small structural differences [16] caused the change an effective relaxation barrier by approximately 70 cm^{-1}. This result is explained [17] by high orbital contribution to the orbital moment that is not quenched by the crystal field of such TP pseudocaging ligand.

Radical nature of the reaction of the iron (II) dichloroclathrochelate (Fe^{2+})@**302** with 1,4-dioxane derivatives and the structure of its radical intermediates have been evidenced [18] using EPR experiments. This quasiaromatic cage complex was found to effectively compete with O_2 and PBN for these radical derivatives. The radical character of the above process has been confirmed using the experiment with *tert*-C_4H_9OOH as a radical initiator: the reflux of (Fe^{2+})@**302** in freshly purified 1,4-dioxane under inert atmosphere in the presence of this initiator gave the macrobicyclic products of radical substitution, the mono- and difunctionalized clathrochelates (Fe^{2+})@**453** and (Fe^{2+})@**452**. Its key intermediate (Fe^{2+})@**454** shown in Scheme 4.3 has been

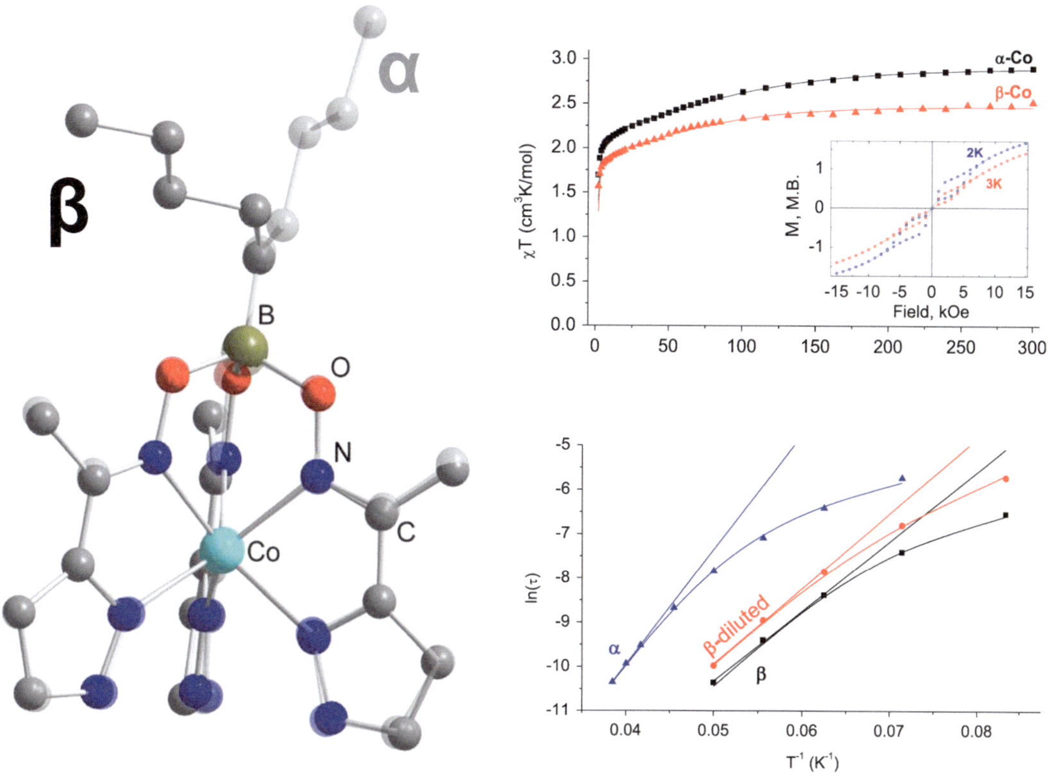

Fig. 4.12 Impact of crystal polymorphism of (Co^{2+})@**510** of its SMM properties [17]

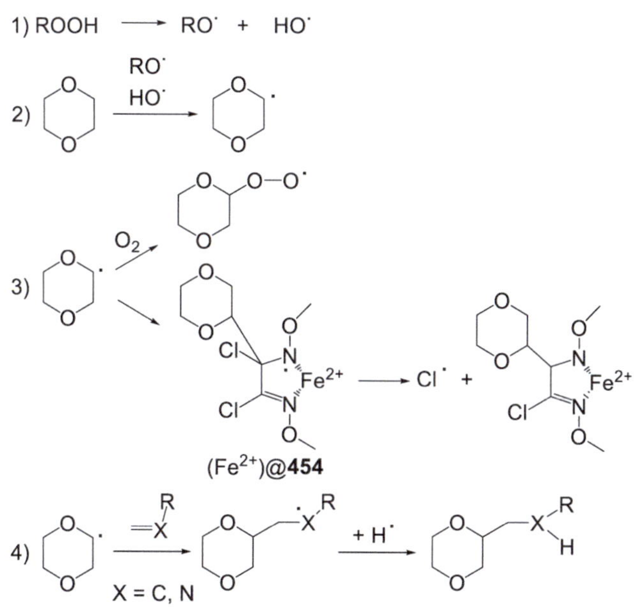

Scheme 4.3 Formation of the radical derivatives of 1.4-dioxane and their interaction with the dichloroclathrochelate (Fe^{2+})@**454**

experimentally detected [18] using EPR method: *tert*-C_4H_9OOH as a radical initiator has been added to the 1,4-dioxane solution of (Fe^{2+})@**302**, and the reaction mixture has been deoxygenated and then heated. The obtained solution exhibited a weak broad triplet EPR signal with $g = 2.0035$, $a_N = 16.3$ G, and the average line width of 9 G. The observed Lande factor value evidenced that its spin density is substantially localized on an atom, lighter than oxygen, and the hyperfine splitting was found [18] to be characteristic of the nitrogen nucleus. These EPR data are in good agreement with the proposed chemical structure of the radical clathrochelate intermediate (Fe^{2+})@**454**. The same method has been used [19] to identify the clathrochelate spin adduct (Fe^{2+})@**462**: the saturated benzene solution of its precursor (Fe^{2+})@**296** has been treated by THF solution of $(C_2H_5)_3B$ and the triplet EPR signal with $g = 2.0032$ and $a_N = 16.2$ G appeared. These parameters are similar to those for the above spin adduct of the dichloroclathrochelate (Fe^{2+})@**302** with 1,4-dioxane-2-yl radical [19].

Magnetic behavior of the homodinuclear gadolinium(III) cage complex of the macrobicyclic ligand **554** and its Gd^{III},Cu^{II}-encapsulating heterodinuclear analog [20] evidenced a substantial magnetic interaction between their two co-encapsulated metal ions. Such magnetic interaction is very large antiferromagnetic for cage complex $(Gd^{3+})_2$@**554**, whereas it is relatively weakly ferromagnetic for the above heteronuclear clathrochelate. This fact is explained [20] by strong bending of the bridging fragments in the molecule (Gd^{3+})(Cu^{2+})@**554**. The above intracavity magnetic interactions have been also confirmed [20] using the EPR experiments.

The homodinuclear manganese(II) cage complexes of the boron-, manganese(III,IV)-, and iron(III)-capped macrobicyclic ligands **566**, **568**, **569**, and **570** have been characterized [21] by magnetic susceptibility measurements in the temperature range 2.0–290 K at different field strengths. Their magnetic behavior evidenced a weak antiferromagnetic coupling between two co-encapsulated manganese(II) ions in the central tris-phenoxo-bridged dimanganese(II) core of these clathrochelates. In the case of the magnetically active capping metal ions, weak ferromagnetic or antiferromagnetic couplings between them and the encapsulated metallocenters have been observed [21]; the charge on the above cross-linking ions was found to substantially affect the strength of the exchange coupling in the central dimanganese(II) core.

Formation of 1:1 europium(III) cage complex of the imine Schiff-base macrobicyclic ligand **554** [22] has been confirmed in [23] using EPR experiments in the range 3.9–300 K.

The pulsed EPR spectroscopy, which measured a coherence time for each *Cr_7Ni*-building block of the clathrochelate-based nanomagnet dimers *63*, *64*, and *65* (see Sect. 2.4), and the DEER spectroscopy, which allowed us to obtain the two-qubit gate time, have been used [24] to study the characteristics of these polynuclear complexes as potent logic gates. The non-optimized coherence time was found to be approximately 600 ns in each case. For the complex *65* with the longest link, the two-qubit gate time is approximately 550 ns, which is too long, as it approaches the coherence time. For the polynuclear complex *63* with the shortest macrobicyclic linker, the two-qubit gate time is equal to 157 ns, i.e., substantially shorter than the coherence time. So the gate time (from too short and too long to just right) can be controlled by using the clathrochelate linkers of an appropriate length [24].

References

1. Symons MCR, Taiwo T, Sargeson AM, Ali MM, Tabl AS (1996) EPR spectra for high- and low-spin Co^{II} encapsulated complexes. Inorg Chim Acta 241:5–8
2. Perraud O, Tommasino J-B, Robert V, Albela B, Khrouz L, Bonneviot L, Dutasta J-P, Martinez A (2013) Hemicryptophane-assisted electron transfer: a structural and electronic study. Dalton Trans 42:1530–1535
3. Voloshin YZ, Varzatskii OA, Vorontsov II, Antipin MY (2005) Tuning a metal's oxidation state: the potential of clathrochelate systems. Angew Chem Int Ed 44:3400–3402
4. Azarkh M, Penkova LV, Kats SV, Varzatskii OA, Voloshin YZ, Groenen EJJ (2014) A mononuclear Mn(II) pseudoclathrochelate complex studied by multi-frequency electron-paramagnetic-resonance spectroscopy. J Phys Chem Lett 5:886–889

5. Belov AS, Belaya IG, Novikov VV, Starikova ZA, Polshin EV, Dolganov AV, Lebed EG (2013) Structure, spectral and electrochemical properties of the 2,6-di-*tert*-butylphenol-functionalized iron and cobalt(II) clathrochelates and their phenylsulfide analogs. Inorg Chim Acta 394:269–281
6. Varzatskii OA, Penkova LV, Kats (Menkach) SV, Dolganov AV, Vologzhanina AV, Pavlov AA, Novikov VV, Bogomyakov AS, Nemykin VN, Voloshin YZ (2014) Chloride ion-aided self-assembly of the pseudo-clathrochelate metal tris-pyrazoloximates. Inorg Chem 53:3062–3071
7. Voloshin YZ, Varzatskii OA, Novikov VV, Strizhakova NG, Vorontsov II, Vologzhanina AV, Lyssenko KA, Romanenko GV, Fedin MV, Ovcharenko VI, Bubnov YN (2010) Tris-dioximate cobalt(I, II, and III) clathrochelates: stabilization of different oxidation and spin states of an encapsulated metal ion by ribbed functionalization. Eur J Inorg Chem 2010:5401–5415
8. Burdukov AB, Boguslavsky EG, Reznikov VA, Pervukhina NV, Vershinin MA, Voloshin YZ, Varzatskii OA, Bubnov YN (2005) Cage complexes as a molecular scaffold for assembling of polyfunctional and multicentered systems: synthesis and structures of the first nitroxide clathrochelates. Russ Chem Bull Int Ed 54:1125–1130
9. Voloshin YZ, Varzatskii OA, Belov AS, Starikova ZA, Dolganov AV, Novikov VV, Bubnov YN (2011) Synthesis, structural and electrochemical features of alicyclic and aromatic α,α'-N_2- and -S_2-dioximate macrobicyclic cobalt(II,III) and ruthenium (II) tris-complexes. Inorg Chim Acta 370:322–332
10. Belaya IG, Belov AS, Platonov VE, Maksimov AM, Vologzhanina AV, Starikova ZA, Dolganov AV, Novikov VV, Bubnov YN (2012) Formation of the second superhydrophobic shell around an encapsulated metal ion: synthesis, X-ray structure and electrochemical study of the clathrochelate and bis-clathrochelate iron(II) and cobalt(II, III) dioximates with ribbed perfluoroarylsulfide substituents. Dalton Trans 41:737–746
11. Voloshin YZ, Lebedev AY, Novikov VV, Dolganov AV, Vologzhanina AV, Lebed EG, Pavlov AA, Starikova ZA, Buzin MI, Bubnov YN (2013) Template synthesis, X-ray structure, spectral and redox properties of the paramagnetic alkylboron-capped cobalt(II) clathrochelates and their diamagnetic iron (II)-containing analogs. Inorg Chim Acta 399:67–78
12. Novikov VV, Ananyev IV, Pavlov AA, Fedin MV, Lyssenko KA, Voloshin YZ (2014) Spin crossover anticooperativity induced by weak intermolecular interactions. J Phys Chem Lett 5:496–500
13. Novikov VV, Pavlov AA, Belov AS, Vologzhanina AV, Savitsky A, Voloshin YZ (2014) Transition ion strikes back: large magnetic susceptibility anisotropy in cobalt(II) clathrochelates. J Phys Chem Lett 5:3799–3803
14. Vologzhanina AV, Belov AS, Novikov VV, Dolganov AV, Romanenko GV, Ovcharenko VI, Korlyukov AA, Buzin MI, Voloshin YZ (2015) Synthesis and temperature-induced structural phase and spin transitions in hexadecylboron-capped cobalt (II) hexachloroclathrochelate and its diamagnetic iron(-II)-encapsulating analogue. Inorg Chem 54:5827–5838
15. Dolganov AV, Belov AS, Novikov VV, Vologzhanina AV, Romanenko GV, Budnikova YG, Zelinskii GE, Buzin MI, Voloshin YZ (2015) First iron and cobalt (II) hexabromoclathrochelates: structural, magnetic, redox and electrocatalytic behavior. Dalton Trans 44:2476–2487
16. Novikov VV, Pavlov AA, Nelyubina YV, Boulon ME, Varzatskii OA, Voloshin YZ, Winpenny REP (2015) A trigonal prismatic mononuclear cobalt (II) complex showing single-molecule magnet behavior. J Am Chem Soc 137:9792–9795
17. Pavlov AA, Nelyubina YV, Kats SV, Penkova LV, Efimov NN, Dmitrienko AO, Vologzhanina AV, Belov AS, Voloshin YZ, Novikov VV (2016) Polymorphism in a cobalt-based single-ion magnet tuning its barrier to magnetization relaxation. J Phys Chem Lett 7:4111–4116
18. Vershinin MA, Burdukov AB, Eltsov IV, Reznikov VA, Boguslavsky EG, Voloshin YZ (2011) Unexpected radical substitution of the dichlorine-containing iron(II) clathrochelate with 1,4-dioxane derivatives: novel approach to functionalization of its macrobicyclic framework. Polyhedron 30:1233–1237
19. Burdukov AB, Vershinin MA, Boguslavsky EG, Eltsov IV, Romanenko GV, Belov AS, Voloshin YZ (2013) New hydroxylamine-containing macrobicyclic encapsulating ligand: unexpected double addition of ethyl radicals to the azomethine fragment of a boron-capped iron(II) clathrochelate dioximate. Inorg Chem Commun 36:117–121
20. Avecila F, Platas-Iglesias C, Rodríguez-Cortiñas R, Guillemot G, Bünzli JCG, Brondino CD, Geraldes CFGC, de Blas A, Rodríguez-Blas T (2002) Structural characterisation, EPR and magnetic properties of f–f and f–d lanthanide(III) phenolic cryptates. J Chem Soc Dalton Trans 4658–4665
21. Khanra S, Weyhermüller T, Bill E, Chaudhuri P (2006) Deliberate synthesis for magnetostructural study of linear tetranuclear complexes $B^{III}Mn^{II}Mn^{II}B^{III}$, $Mn^{III}Mn^{II}Mn^{II}Mn^{III}$, $Mn^{IV}Mn^{II}Mn^{II}Mn^{IV}$, $Fe^{III}Mn^{II}Mn^{II}Fe^{III}$, and $Cr^{III}Mn^{II}Mn^{II}Cr^{III}$. Influence of terminal ions on the exchange coupling. Inorg Chem 45:5911–5923
22. Platas C, Avecilla F, de Blas A, Geraldes CFGC, Rodríguez-Blas T, Adams H, Mahía J (1999) ^{1}H NMR in solution and solid state structural study of lanthanide(III) cryptates. Inorg Chem 38:3190–3199
23. Leniec G, Typek J, Kaczmarek SM (2008) Magnetic properties of a new Er(III) macrobicyclic complex studied by EPR. Appl Magn Res 35:197–203
24. McInnes EJL, Timco GA, Whitehead GFS, Winpenny REP (2015) Heterometallic rings: their physics and use as supramolecular building blocks. Angew Chem Int Ed 54:14244–14269

5 Practical Applications of Cage Metal Complexes

Main fields of potent practical applications of the metal clathrochelates and the corresponding clathrochelate-containing derivatives and nanosystems summarized in this chapter are based on their unprecedented chemical robustness, unusual redox and magnetic properties of an encapsulation metal ion, and three-dimensional shape of these cage molecules as well as on special selectivity of their self-assembly reactions.

The following main trends and the applications of transition metal clathrochelates in medicine and biochemistry, such as encapsulation of radioactive metal ion for radiodiagnostics and radiotherapy (Scheme 5.1), clathrochelate-based imaging agents for MRI applications, antidotes and prolonged pharmaceuticals, boron-enriched targets for ^{10}B-NCT treatment, antihelminthic and antiparasitic detergents, clathrochelate-based antioxidants, membrane transport of an encapsulated metal ion, the interaction of cage metal complexes with nucleic acids, and the potential of their self-assembly reactions in immunology and molecular biology for recognition of antibodies, antigens, and DNA sites as well as the design of clathrochelate-based HIV inhibitors for the AIDS drug therapy are highlighted in the review [1], published in 2007.

For the first time, the concept of so-called "topological drugs" has been introduced [2]; main recent results in this field are highlighted in the review [3]. The biological activity of regular pharmaceuticals, such as antibiotics, based on low-weight organic molecules is in most cases governed by their interactions with biological targets *via* either covalent or supramolecular bonding. The selectivity and specificity of these interactions determine the potential of such chemical compounds in drug therapy (e.g., in the cases of viral or neurodegenerative diseases). The modern trends in drug development lead to discovery of the new biological targets, including the hidden allosteric sites (in this case, the binding of regulator compound as a guest in a special allosteric center as a host that is removed from the active site of the enzyme causes a change in its catalytic properties) and the interfaces of macromolecular interactions. A concept of "allosteric drugs" implies an existence of such hidden sites in nearly all existing proteins, offering the potential opportunity to develop their inhibitors working as shown in Fig. 5.1a [3]. These sites are typically not known and their choice depends on the ability to create the artificial molecules (xenobiotics, i.e., the chemical compounds that are foreign to a living organism) that are topologically complementary to them. Another modern approach to drug therapy is targeting of protein–protein interactions of biological macromolecules, and relatively small but bulky inhibitor molecules of suitable size and shape can be used for modulating such interactions as it shown in Fig. 5.1b [3].

Such new targets demand for the guest molecules with the strictly controlled geometry

Y. Voloshin et al., *Cage Metal Complexes*, DOI 10.1007/978-3-319-56420-3_5

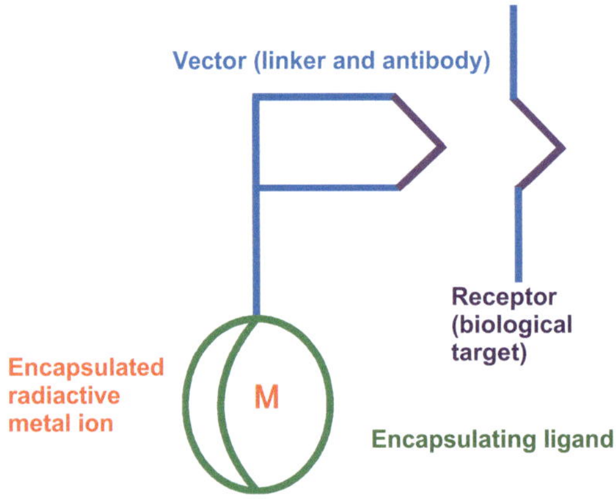

Scheme 5.1 Encapsulation of a metal ion and its target delivery [1]

and rather extensive surface [2]. A concept of so-called topological drugs suggests the use of rigid three-dimensional molecules of inorganic, organic, and hybrid compounds (as guests) that form multicentered supramolecular interactions in vacant cavities of protein macromolecules and their complexes (as hosts): in the case of polyhedral and macrobicyclic compounds (Scheme 5.2) as molecular scaffolds for design of biologically active molecules: their three-dimensional geometrically determinated cores are prospective for specific supramolecular interactions with big (giant) protein molecules. The molecules of cage metal compounds as well as those of the fullerene- and metallacarborane-based inhibitors are large and bulky enough to ensure the strong multicentered supramolecular binding by van der Waals interactions. The development of such topological drugs based on the relatively large and bulky inorganic, organic, and hybrid molecules is modern and prospective pathway to new antitumor and antiviral drug candidates [3]; probably the drug resistance is less likely to develop in this case. Such xenobiotics, which have neither the natural analogs nor the structural similarity to the biological molecules, are of particular interest. An example of the similar polyhedral xenobiotics is carbocyclic adamantane derivatives shown in Scheme 5.3, which are widely used in drug therapy of the human diseases, such as Parkinson's and viral, dermatological and psychical diseases. Their activity is usually attributed to the presence of a bulky carbocyclic polyhedron allowing them to be included in the lipid bilayers of biological membranes and to directly interact with biological targets, such as hydrophobic parts of the proteins (in particular, their receptors). Thus, therapeutic effects of these diamondoids are based, first of all, on the geometric complementarity of their molecules to such receptors [4, 5].

Mono- and polyclathrochelates are topological analogs of the above polyhedral compounds with similar geometrical parameters. The applicability of these chemically robust cage complexes is also determined by their synthetical availability and structural diversity. The apical and ribbed functionalizations by Scheme 5.4 can be easily performed by the convenient, well-known organic reactions using commercially available chemical reagents. The possibility to "build" the structure of an inhibitor from a single center in three (for the metal monoclathrochelates) or six (for their bis-clathrochelate analogs) different directions using up to 8 and up to 14 substituents (Scheme 5.2), respectively, combined with a wide range of functionalizing groups available for such modification, is a way for the design of cage-based inhibitors whose shapes closely match the macromolecular surfaces of a very complex topology (e.g., the above protein–protein and protein–DNA interaction interfaces). So, their geometrically determinated rigid cage frameworks with a given location of the pharmacophore substituents are suitable three-dimensional molecular scaffolds for the design of topological drugs; their chemical reactivity and physical and physicochemical properties (solubility, redox characteristics, etc.) can be varied as well. As a result, the specific abiotic interactions of such cage molecules with biological macromolecules and their macromolecular complexes are possible, and these supramolecular interactions may change both the structure and the functions of the biological systems. Their geometrically determinated rigid macrobicyclic ligands with biomimetic substituents binding to the host biological macromolecules and their complexes seem to be prospective for the drug design and may be considered as a cage scaffold for the

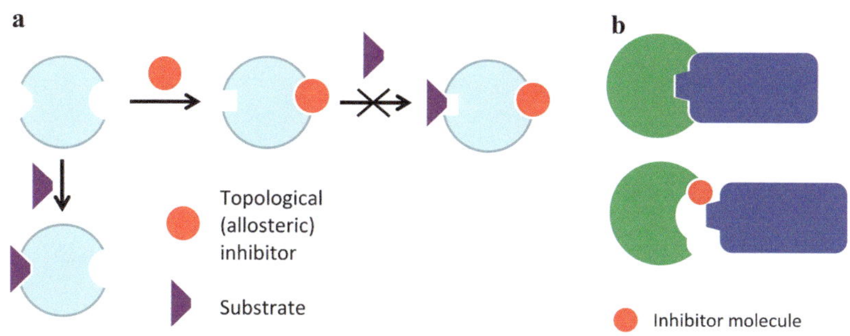

Fig. 5.1 Schematic representation of topological (allosteric) (**a**) and interface macromolecular protein–protein (**b**) inhibitions [3]

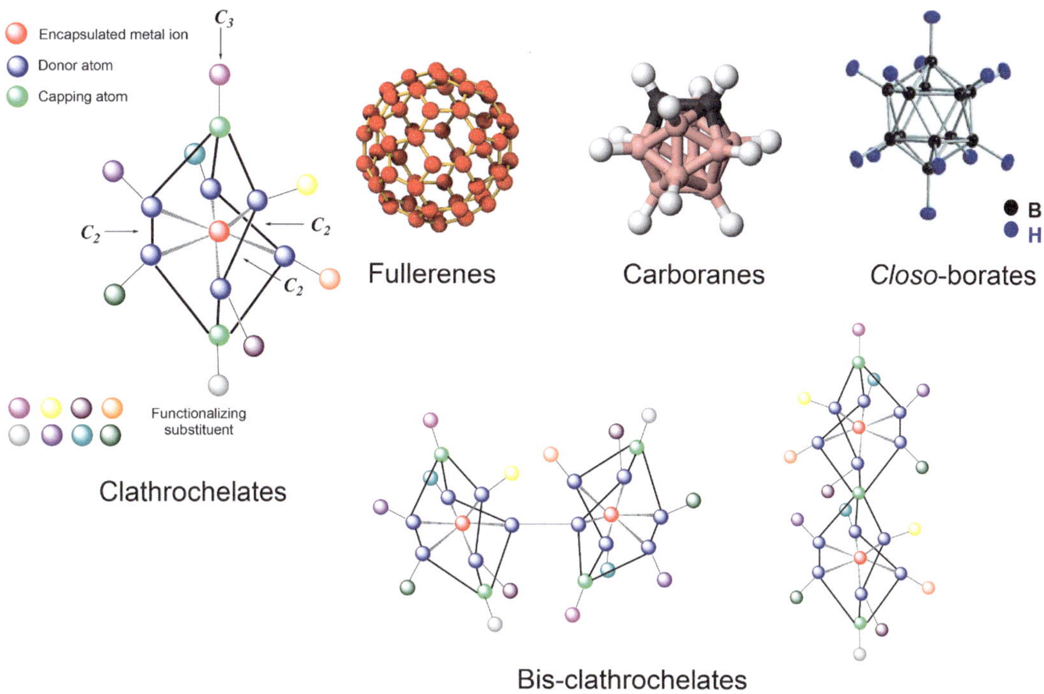

Scheme 5.2 Main types of polyhedral and mono- and bis-clathrochelate compounds

direct "bottom-up" synthesis of biological regulators based on the "geometrical recognition" and "topological invasion" principles. Thus, transition metal clathrochelates are reported [2, 3] to be prospective for the design of topological drugs owing to the availability of these three-dimensional macrobicycles, obtained using the efficient synthetic approaches for large-scale synthesis of the apically and ribbed-functionalized cage complexes of different symmetry, chemical structure, and reactivity with given physical and chemical properties.

The design and straightforward apical functionalization of the polyamine caging ligands for efficient binding and target delivery of an encapsulated radioactive 64-copper(II) ion to a

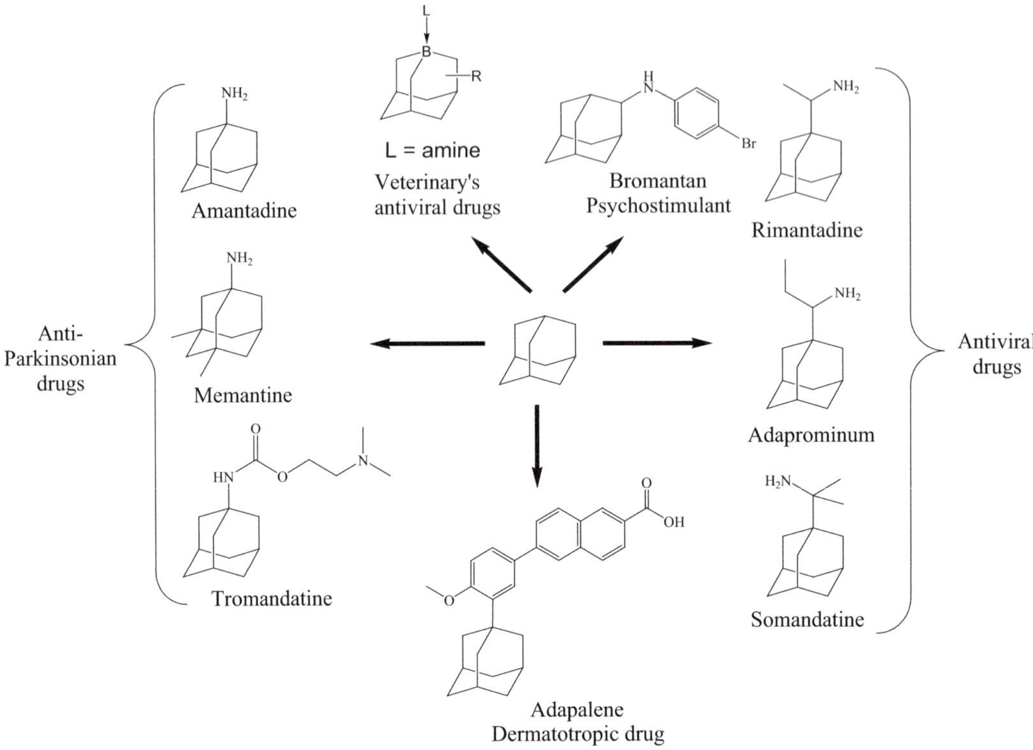

Scheme 5.3 Adamantane-based drugs

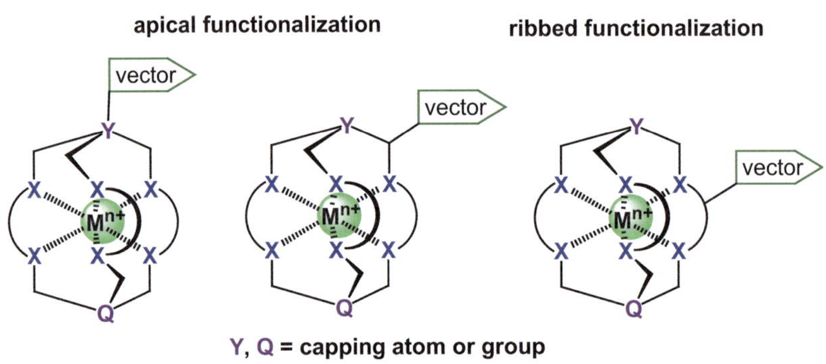

Scheme 5.4 Functionalization of the clathrochelate ligands

given biosystem is highlighted [6]. The potential of cobalt clathrochelate-based electrocatalysts for hydrogen production in proton exchange membrane of the water electrolyzers is discussed [7, 8]. The use of the metal clathrochelates in controlled radical polymerization of vinyl monomers is very recently highlighted in the review [9].

5.1 Biochemical and Biomedical Aspects of Cage Metal Complexes

5.1.1 Supramolecular Interactions of Clathrochelates with Biological Systems and Their Bioactivity

The clathrochelate trication (Co^{3+})@**648** is reported [10] to be ideal for the study of the electrostatic, covalent, and hydrogen bonding interactions of the small molecules with nucleic acid fragments, while the use of mononucleotide is allowed to determine the fundamental interactions in the absence of complexity of a nucleic acid. The interactions of chiral forms of (Co^{3+})@**648** with guanosine 5′-monophosphate have been studied both in the solid state and in diluted aqueous solutions using FT-IR and ^{1}H, ^{31}P NMR methods. This mononucleotide showed no bonding preference for enantiomers of the above cobalt(III) cage complex [10].

Antibiotic lasalocid A (*90*) with known ionophoric activity has been used [11] for the extraction and membrane transport of a series of the polyamine cobalt(III) cage complexes shown in Scheme 5.5. These processes are reported to be structure- and enantioselective, thus resulting in the partial resolution of their enantiomers. In the majority of cases, the aqueous receiving phases were enriched in their Δ isomers at the end of each run, whereas the aqueous source phases, which were initially racemic, mainly contained the Λ isomers of these clathrochelates [11].

Successful use of the clathrochelate complexes in biochemistry, (radio)diagnostics, and (radio)therapy is strongly affected by their chemical stability under physiological conditions as well as by their supramolecular interactions with biomolecules as well as with macromolecular complexes formed by them. In particular, the interactions of the iron(II) mono- and bis-clathrochelates (Fe^{2+})@**387**, (Fe^{2+})@**388**, and $(Fe^{2+})_2$@**445** (Scheme 5.6) with BSA,

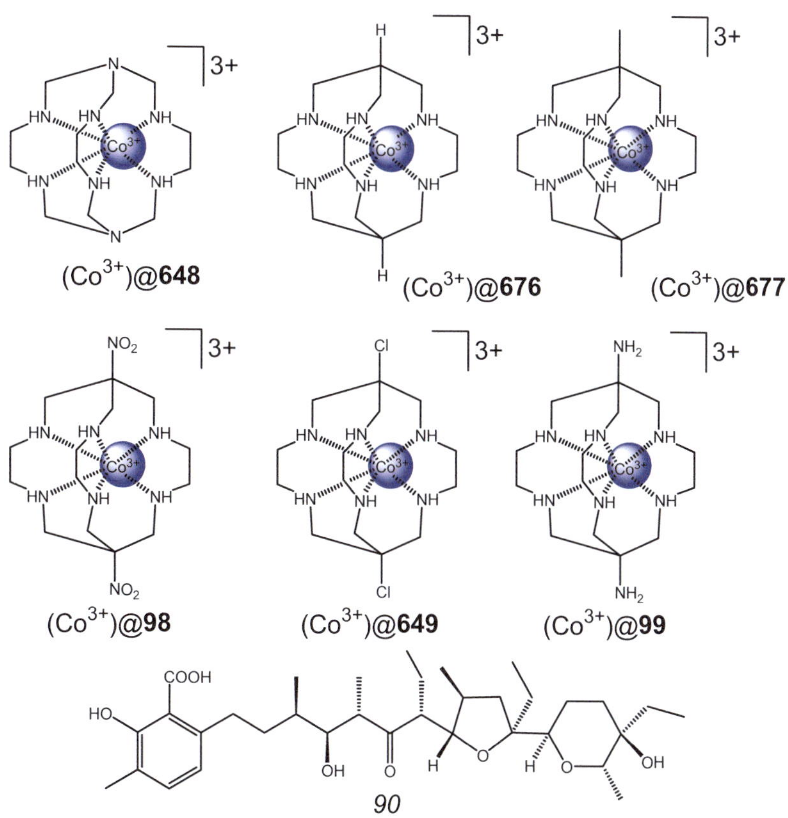

Scheme 5.5 Chemical drawings of a series of polyamine cobalt(III) clathrochelates and antibiotic lasalocid A

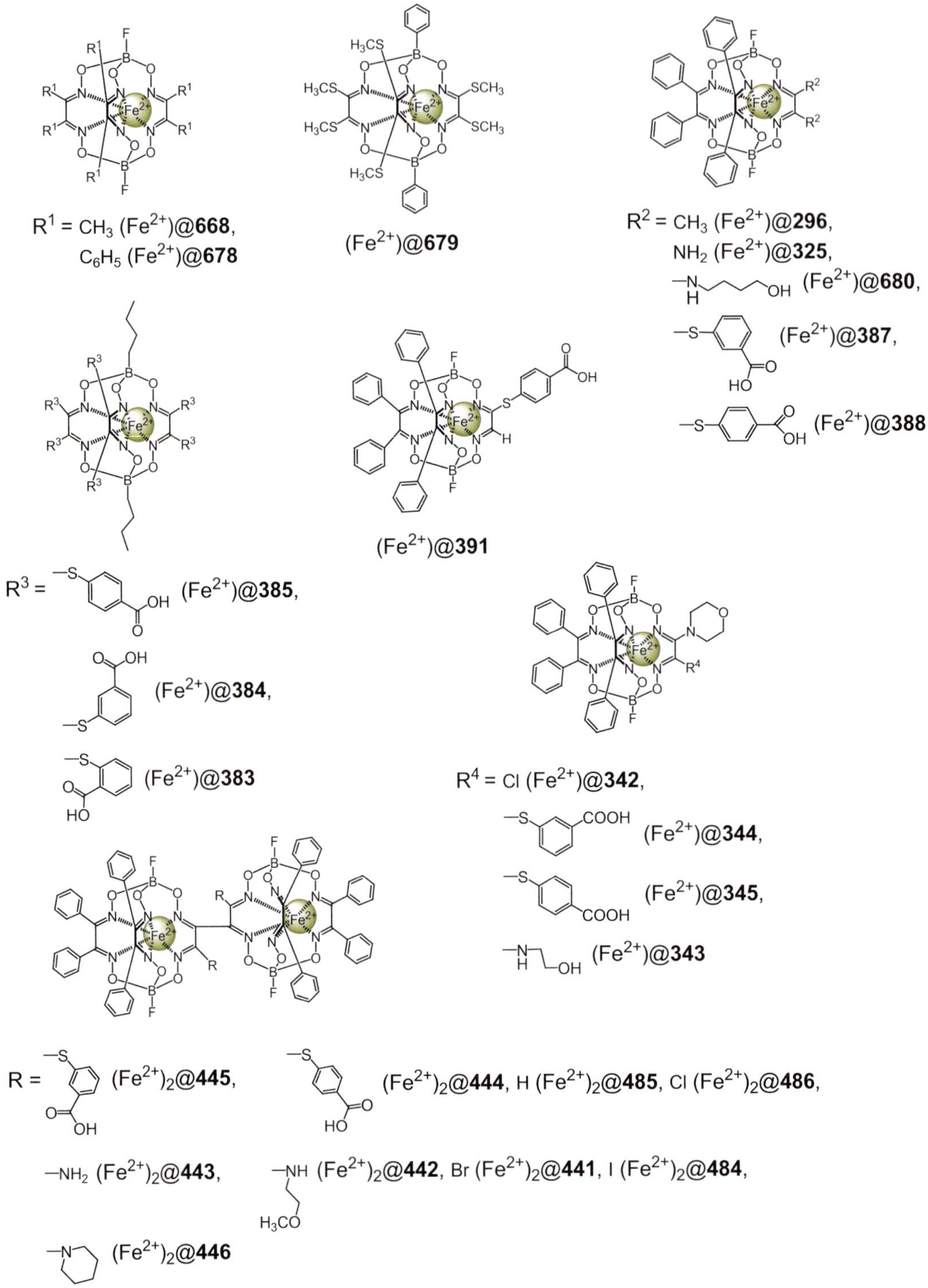

Scheme 5.6 Iron(II) mono- and bis-clathrochelates tested as potent biological effectors

β-lactoglobulin, lysozyme, and insulin have been studied [12] by steady-state and time-resolved fluorescent spectroscopies. The strong changes in fluorescent characteristics in the presence of such cage complexes have been observed [12] only in the case of the transport protein BSA: their binding to BSA led to the substantial quenching of protein fluorescence and to decrease in its excited-state lifetime, thus evidencing the BSA–clathrochelate supramolecular binding. So, the above cage complexes are versatile guests to form strong assemblies with blood transport proteins, the serum albumins, but discriminating other globular proteins [12].

The stereospecific interactions of a chiral polyamine cobalt(III) clathrochelate (Co^{3+})@**648** with guanosine 5′-monophosphate have been characterized [10] in the solid state and in aqueous solutions using FT-IR and multinuclear NMR spectra and by MM calculations as well: the formation of their diastereomeric supramolecular assemblies through hydrogen bonding (without bonding preference for the enantiomers of this cage complex) is evidenced [10]. Light scattering electron microscopy and UV spectra have been used [13] to study the structural effects of the same cobalt(III) complex on DNA condensation. The macrobicyclic trication (Co^{3+})@**648** effectively precipitates this nucleic acid at high concentrations being substantially more efficient condensation agent than its non-macrocyclic analogs. This suggests that although electrostatic interactions are the main driving force for binding of multivalent complex cations to DNA in solution, DNA condensation also depends on the structure of the condensing agent used; the above supramolecular recognition and consequent DNA condensation are reported [13] to be affected by changes in its size and chemical constitution, as well as by its surface terminal groups.

Interaction of the zinc(II) diaminoclathrochelate (Zn^{2+})@**99** with BSA and HSA has been studied [14] under the simulative physiological conditions using various experimental techniques and molecular docking calculations as well. This cage complex effectively quenches the intrinsic fluorescence of these transport proteins *via* a static quenching process. The fluorescence data have also been used for determination of binding sites and binding constants. The above experiments and calculations allowed the authors of [14] to evaluate the nature of this supramolecular self-assembly that proceeds *via* hydrogen bonding and van der Waals interactions. The distance between HAS (or BSA) as a donor and this zinc(II) cage complex as an acceptor of 12.8 Å and of 12.7 Å, respectively, has been deduced [14] using FRET experiments.

The same experimental and theoretical approaches have been used [15] for in vitro study of the supramolecular interactions of its manganese(II)-encapsulating analog (Mn^{2+})@**99** with the same proteins. FT-IR data evidenced that protein conformations and microenvironments have been changed in the presence of this cage complex, which also has strongly affected their intrinsic fluorescence by static quenching. Hydrogen bonding and weak van der Waals interactions are described [16] to play a key role in the formation of these clathrochelate–protein supramolecular assemblies; the parent metal-free macrobicyclic compound also exhibited the good binding properties to HSA and BSA, characterized by relatively high binding constants. FT-IR data of [16] evidenced that these multicentered interactions only slightly change the secondary protein structures of BSA and HSA.

Functionalization of one of the three ribbed fragments of the iron(II) cage complex (Fe^{2+})@**316** with annelated quinoxaline fragment has been proposed [17] for its efficient intercalation into DNA structure. The intensively colored clathrochelate (Fe^{2+})@**489** with annulated highly conjugated π-system, which is suitable for intercalation into DNA structure, has been proposed [18] for the design of prospective molecular UV-vis probes for biochemical studies. The boron-capped iron(II) clathrochelates (Fe^{2+})@**188** and (Fe^{2+})@**191** with apical phenylethynyl and biphenyl substituents, exhibiting an intrinsic luminescence and also allowing their intercalation into DNA structure, are described [19] as prospective molecular luminescent bioprobes; their analog (Fe^{2+})@**190** with apical uracil groups can form the strong supramolecular assemblies with nucleic acids *via* their biospecific pairing interactions.

Supramolecular bindings of the polyamine ruthenium(II) complex of the pyridyl-terminated cobalt(III)-encapsulating clathrochelate ligand (Co^{3+})@**681** (Scheme 5.7) with calf thymus DNA and with lipopolysaccharide from *Escherichia coli C* have been studied [20] using UV-vis spectrophotometry; both these bindings are reported to increase the rate of autooxidation of this complex.

The hydrazonate iron(II) cage complexes (Fe^{2+})@**280** and (Fe^{2+})@**281** with vector apical isoniazid substituents (Scheme 5.8) have been proposed [21] for target delivery of these clathrochelates to biological systems.

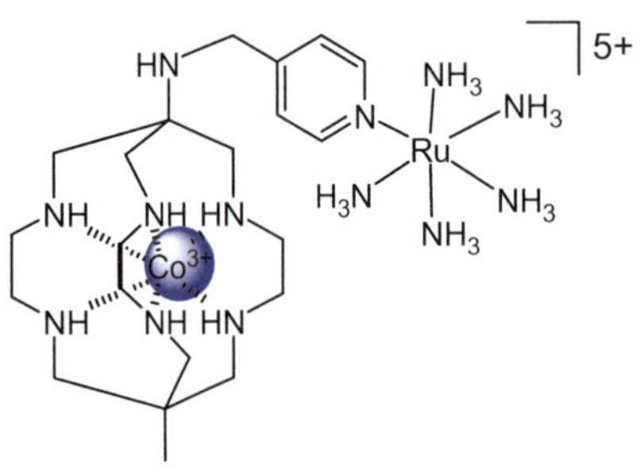

$Ru(NH_3)_5 \cdot (Co^{3+})$@**681**

Scheme 5.7 Chemical drawing of the complex $Ru(NH_3)_5 \cdot (Co^{3+})$@**681**

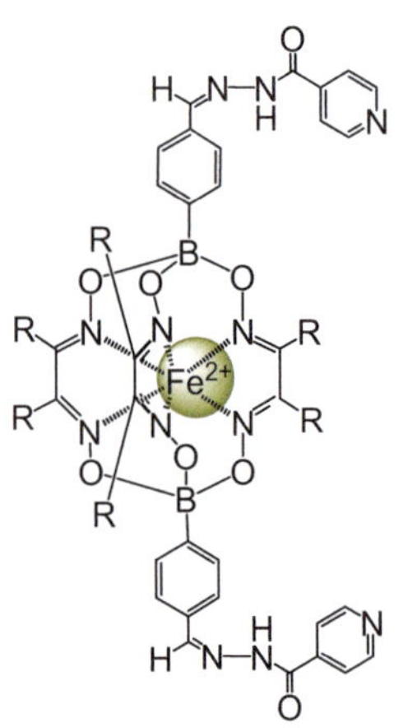

(Fe^{2+})@**280** : R = CH_3
(Fe^{2+})@**281** : R,R = $(CH_2)_4$

Scheme 5.8 Chemical drawings of the hydrazonate iron(II) clathrochelates

5.1.2 Antihelminthic and Antiparasitic Agents and Other Types of Prodrugs

A comparative study of the toxicity of the cobalt(II) bis(salicylideniminato-3-propyl) methylamine and the cobalt(III) clathrochelate (Co^{3+})@**648** on *Escherichia coli* DH5α growth has been performed [22]. The toxicity of the former cobalt(II) complex was found to be larger than that of the latter cobalt(III) cage complex. On the other hand, the SEM data suggested that both these compounds have the same toxic mechanism on *E. coli* DH5α, which has been attributed [22] to destroying the cell walls of these bacteria. The accumulation of intracellular cobalt by *E. coli* DH5α, resulting from its interaction with such cobalt complexes, has been confirmed [22] using ICP technique.

The cobalt(III) cage complexes (Co^{3+})@**93**, (Co^{3+})@**94**, and (Co^{3+})@**95** (Scheme 5.9) with long-chain paraffin apical substituents are reported [23] to be surface active: they reduce the surface tension of water to the levels, approaching those of organic solvents. The in vitro testings showed that these macrobicyclic surfactants are lethal at their millimolar levels to

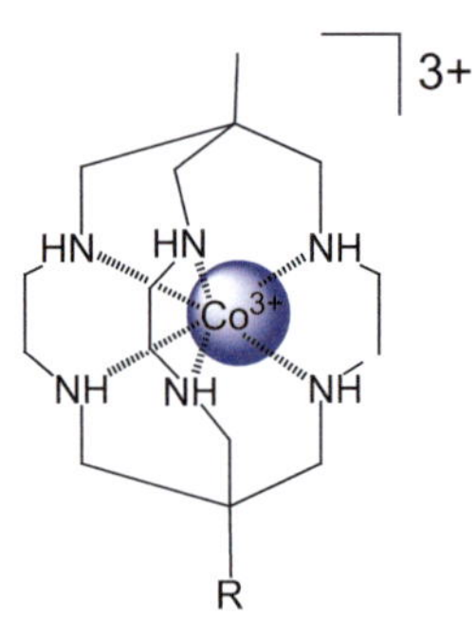

R = n-C_8H_{17} ((Co^{3+})@**93**)
n-$C_{12}H_{25}$ ((Co^{3+})@**94**)
n-$C_{16}H_{33}$ ((Co^{3+})@**95**)

Scheme 5.9 Chemical drawing of paraffin-functionalized cobalt(II) clathrochelates the tapeworm *Hymenolepis diminuta* and to the parasitic eukaryote *Tritrichomonas foetus* as well. Their biological activity is explained [23] by insertion of the paraffin tail at a cage framework into the exterior membrane of these organisms and by consequent incorporation of the triple positively charged macrobicyclic head, thus perturbing the normal membrane potential; this leads to disintegration of the membranes and to death of these organisms [23].

The in vitro and in vivo anti-*Trypanosoma cruzi* activity of the tris-pyrazole macrobicyclic polyamine **549** and its macrocyclic analogs has been tested [24] on acute and chronic phases of Chagas disease; their activity against both these phases was found. All the tested compounds are more active against the parasite and less toxic against Vero cells than the reference drug, benznidazole, and **549** is reported to be most active among them, especially in the chronic phase of this disease. Moreover, these macrocyclic and macrobicyclic compounds are almost inactive against human SOD but active against the Fe-SOD of the above parasite [24] that is important from the point of view of their potent medical applications.

The in vitro leishmanicidal activity and cytotoxicity of the same macrobicyclic polyamine has been assayed [25] on *Leishmania infantum* and *Leishmania braziliensis* species. The cage compound **549** is more active and less toxic than the reference drug, glucantime, and both the infection rates and the ultrastructural alterations evidenced the highly leishmanicidal and induced extensive parasite cell damage by this macrobicycle. Its in vitro leishmanicidal activity is assigned [25] to inhibition of the Fe-SOD of the above parasites; the redox stress derived from such inhibition has been used for explanation of the observed catabolic changes.

The pyridine-spaced octaamine caging ligand **558**, showing the high selectivity of encapsulation of a copper(II) ion as compared with zinc(II) cation, has been proposed [26] as potent prodrug for the treatment of the copper-excess pathologies.

The clathrochelate-based polytopic systems with terminal bulky phenol groups are reported [1] to be prospective molecular platforms for the design of efficient metal-containing antioxidants shown in Scheme 5.10. The effect of an encapsulated metal ion on the characteristics of their radical reactions is explained [1] by superposition of two main factors: (1) a steric factor (i.e., an increase in the structural rigidity of a macrobicyclic molecule and the sterical clashes in its reactive radical centers) and (2) an electronic factor (i.e., a decrease in the spin density on its radical center due to delocalization of the unpaired electron on a quasiaromatic cage framework). The inhibitory activity of the clathrochelate antioxidants may be even higher within the cells due to inclusion of their three-dimensional cage molecules with lipophilic peripheral groups into the lipid bilayer of a cellular membrane as shown in Scheme 5.11 [1].

5.1.3 Clathrochelate-Based Bioprobes, Biomimetic and Bioactive Cage Metal Compounds

The trinuclear copper(I) and copper(II) cage complexes of the encapsulating ligand **565** (Scheme 5.12) have been proposed [27] as potent biomimetic models of the O_2-reduction site of multicopper oxidases.

1:1:1 heteroguest zinc(II) cage complex (Zn^{2+})@**565** (see Scheme 2.348) with co-encapsulated 1-benzylnicotinamide guest molecule is reported [28] to be the functional model of alcohol dehydrogenase.

1:1 terbium(III) cage complexes of the macrotricyclic ligands **618**, **619**, and **621** have been elaborated [29] as prospective luminescent labels in for TR-FRET-based bioassays, exhibiting exceptionally high quantum yields, large extinction coefficients, and long luminescence lifetimes as well. These properties persist in their diluted solution, and the above clathrochelates have an excellent long-term stability, even in the presence of such competing ligands, as EDTA or phosphate buffer. The photophysical properties of the bis-capped terbium(III) complex also persist in protein bioconjugates, such as the streptavidin conjugate. So, these cage complexes have been proposed

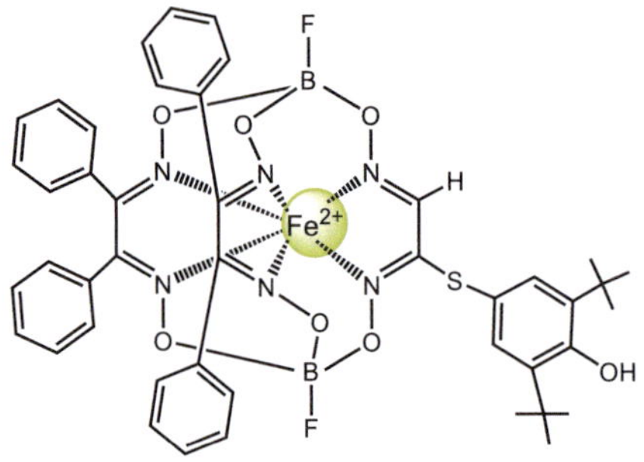

(Fe^{2+})@**399**

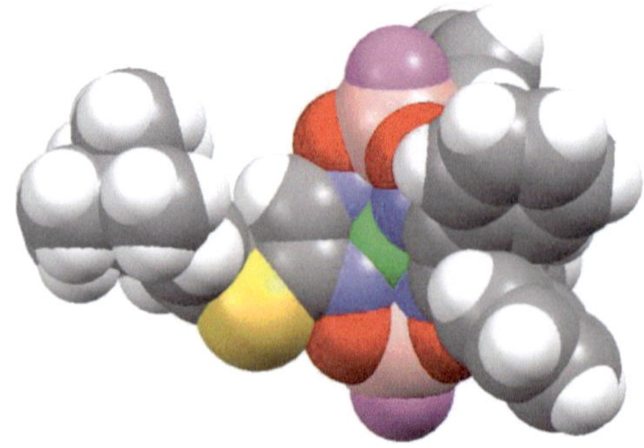

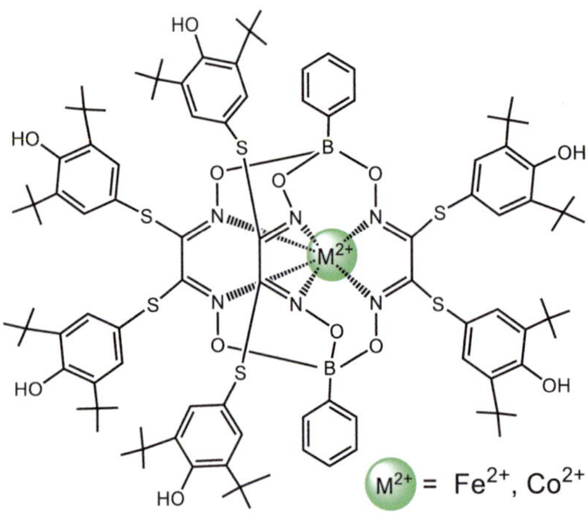

(M^{2+})@**397**

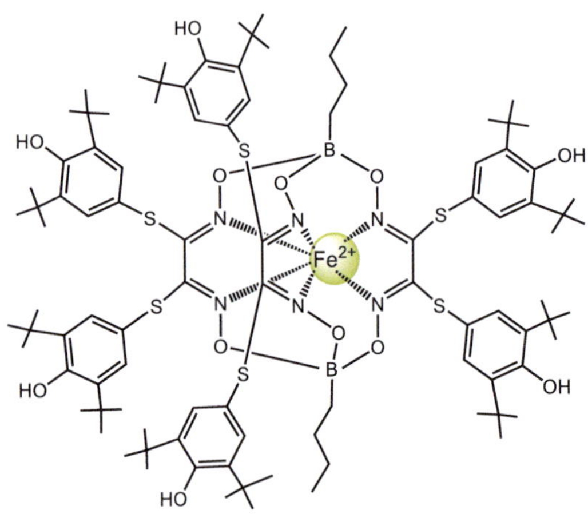

(Fe^{2+})@**396**

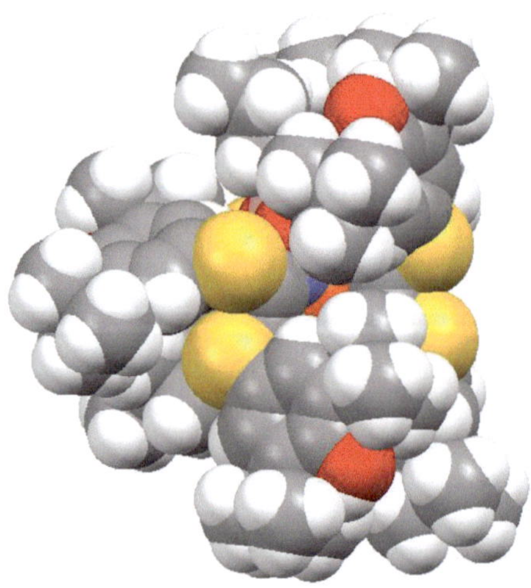

Scheme 5.10 Chemical drawings of the phenol-terminated iron and cobalt(II) clathrochelates and their space-filling representations [1]

[29] as potent long-lifetime lanthanide reporters for their use in high-throughput screening and for diagnostic applications using homogeneous TR-FRET assays as well.

The macrobicyclic ligands shown in Scheme 5.13 are described [30] to encapsulate the europium(III) ion, thus giving its 1:1 cage complexes with improved photophysical properties,

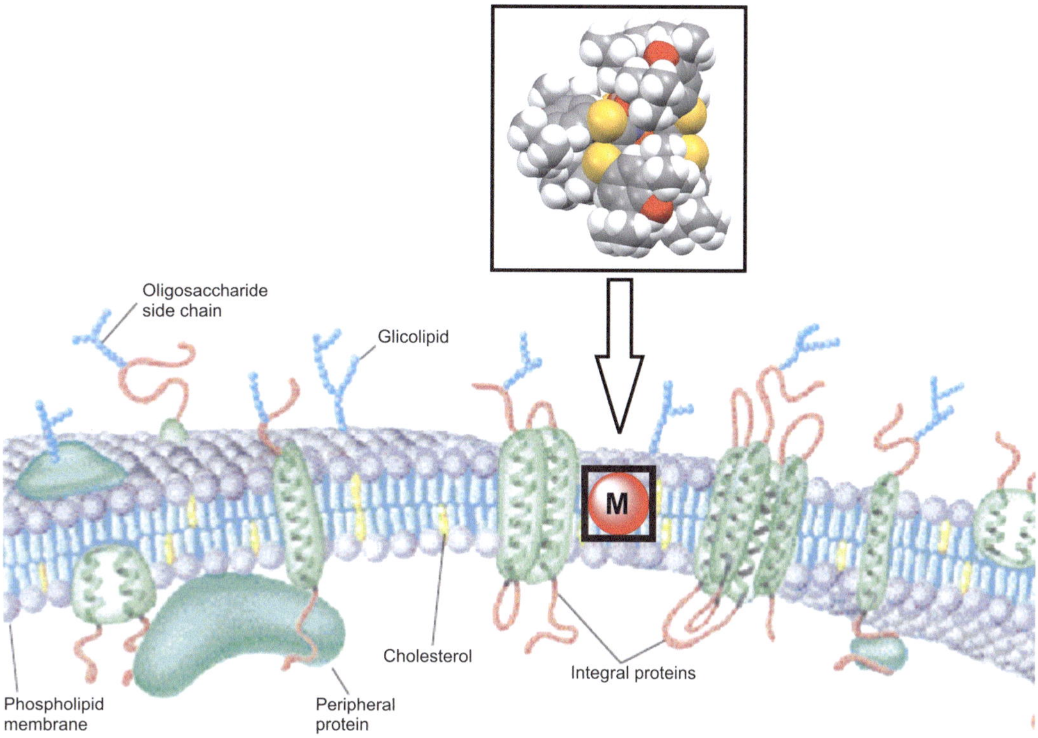

Scheme 5.11 Schematic representation of inclusion of the phenol-terminated iron(II) clathrochelate into a lipid bilayer of a cellular membrane [1]

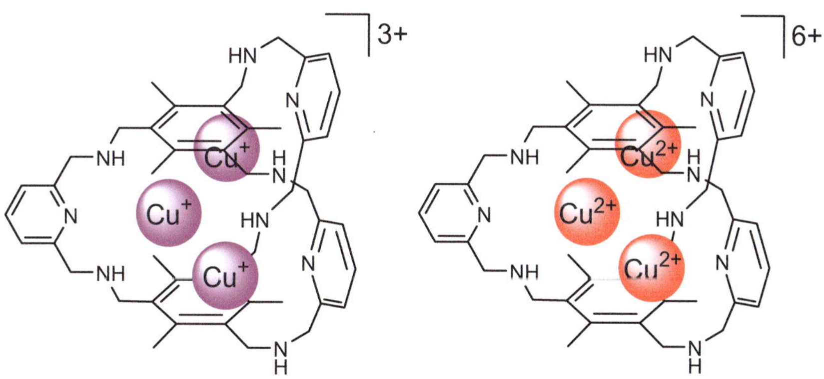

Scheme 5.12 Chemical drawings of the trinuclear copper complexes of **565**

which are suitable for the design of the corresponding clathrochelate-based bioassays.

Self-assembly of cage metal complexes has been proposed [1] for ligase mimics and as covalent linkers (Scheme 5.14) in immunology and molecular biology. Both these applications are based on the fact that the formation of the -boron-capped tris-dioximate iron(II) clathrochelates in many cases easily proceeds under mild, nearly physiological, conditions and can be described as the three-component self-assembly of the corresponding α-dioxime with an appropriate boron-containing Lewis acid on an iron(II) ion as a matrix, giving the thermodynamically stable and kinetically very inert clathrochelate intracomplexes. The single and

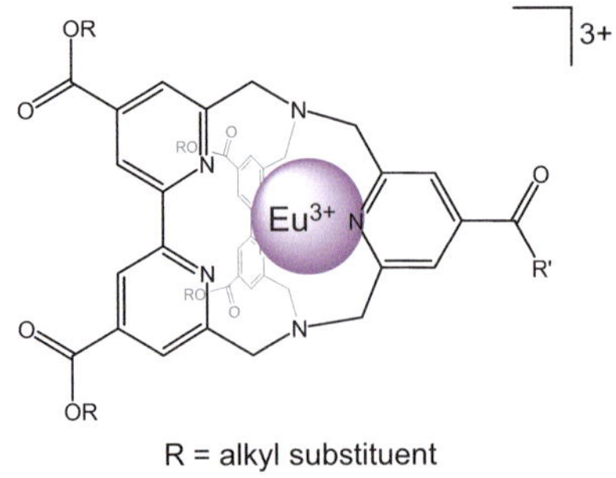

Scheme 5.13 General chemical drawing of the europium(III) clathrochelates designed for bioassays

robust clathrochelate product is formed in the reaction mixture, and it is almost unreactive toward further transformations under mild (and, in some cases, under vigorous) reaction conditions. This unique self-assembly process, which allows obtaining a cage complex with low polarity from its nontoxic abiogenic polar ligand synthons, is described [1] to have no analogs in the natural biological systems. Therefore, it can be used for the design of linker systems that form under specific conditions

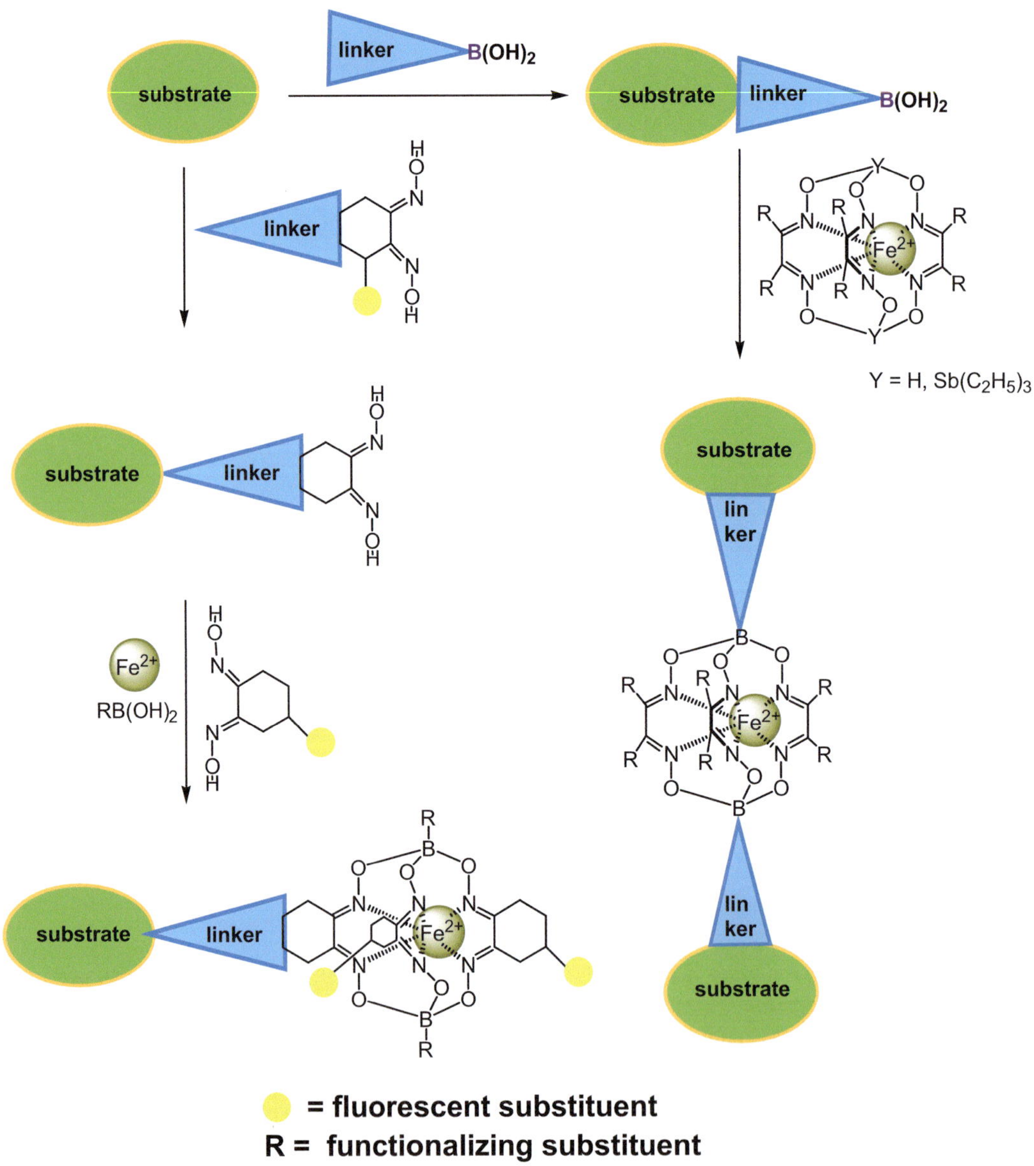

Scheme 5.14 Ligase mimics with clathrochelate linkers [1]

without overlapping with biogenic processes. Intensive visible coloration of the clathrochelate products and their well-defined redox characteristics, such as one-electron metal-centered oxidation waves in a wide range of the potentials, have been proposed [1] for UV-vis and electrochemical detections of these linking reactions. These can be used for binding the substrates initially modified either by trigonal boron-containing groups or by α-dioximate fragments and their mixture when other components (i.e., iron(II) ions, the corresponding boronic acid, or α-dioxime) are present in solution. These systems are suitable for covalent immobilization and affinity binding of macromolecules, antibodies, and enzymes as well as for linking of the substrates with active sites of enzymes and with fluorescent probes as well [1].

The promising applications of the clathrochelate linkers for ligases mimics and for binding of the nucleotide sequences on a single-stranded DNA template are also described [1]. The complementary interaction of oligonucleotide sequences, functionalized with reactive sites (boronic acid residues, Scheme 5.15) or α-dioxime fragments (Scheme 5.16), and a single-stranded DNA together has been proposed [1] for highly selective binding of the oligonucleotide sequences by formation of the corresponding clathrochelate linker. Such binding can be controlled by the presence or absence of the third component (either iron(II) ions, or α-dioxime, or Lewis–acidic boron compound) and is an irreversible process affected by the concentrations of these reactants. The reaction conditions under which this self-assembly proceed only in the case of a complementary interaction of the above functionalized oligonucleotides with a single-stranded nucleic acid can be thus evaluated. The above ligand synthons may be also functionalized with fluorescent substituents; a fluorescent output on such self-assembly reaction is described [1] to be suitable for the design of the corresponding clathrochelate-based fluorescent bioprobes.

The design of novel isotope-coded mass tags for quantitation approaches (first of all, those for biological systems), based on the lanthanide tris-bipyridinate clathrochelates shown in Scheme 5.17, is described [31]. The selective and flexible incorporation of the stable isotopes ^{2}H, ^{13}C, and ^{15}N in these cage complexes has been used for generation of manifold mass differences, and, at the same time, several samples could be quantified relative to each other (multiplexed analysis). An encapsulation of the lanthanide ion has been used [31] for the absolute quantification using ICP-MS measurements in which the relatively high mass differences can be thus achieved, and, therefore, the substances with high molecular weights, such as peptides or proteins, can be analyzed.

The efficient and versatile method for conjugation of α-dioximes to *N*- and *C*-terminal PNA sites (Scheme 5.18) using the tris-dioximate iron(II) clathrochelates as initial α-dioxime-containing precursors has been developed [32]. Interaction of the ribbed-functionalized iron(II) clathrochelates (Fe^{2+})@**682** and (Fe^{2+})@**368** containing an amine-terminated spacer ribbed substituent (Schemes 5.19 and 5.20) with PNAs, which are the analogs of DNA with *N*-(2-aminoethyl)glycine fragment instead of the sugar–phosphate backbone, has been studied [32]. PNA's advantages over the natural nucleic acids are their higher chemical stability and specificity in oligonucleotide binding. So, they are widely used for analysis of the nucleic acids and in control of gene expression in vitro and, in some cases, also in vivo. Conjugation of the metal complexes to the PNA terminal sites allowed one to change their optical and electronic properties, chemically immobilize them on a given surface, control the binding properties of PNA to DNA, as well as perform the metal ion-catalyzed reactions on DNA templates [32]. Synthetic pathways and procedures for preparation of the α-dioxime–PNA conjugates with modified *C*- or *N*-terminus and the multi-stranded complexes of these conjugates with metal ions, thus giving the stable *d*-metal bis- and tris-α-dioximates (Scheme 5.19), have been developed [32]. Conjugation of an α-dioxime with PNA requires a protection of its oxime groups due to occurrence of their side acylation

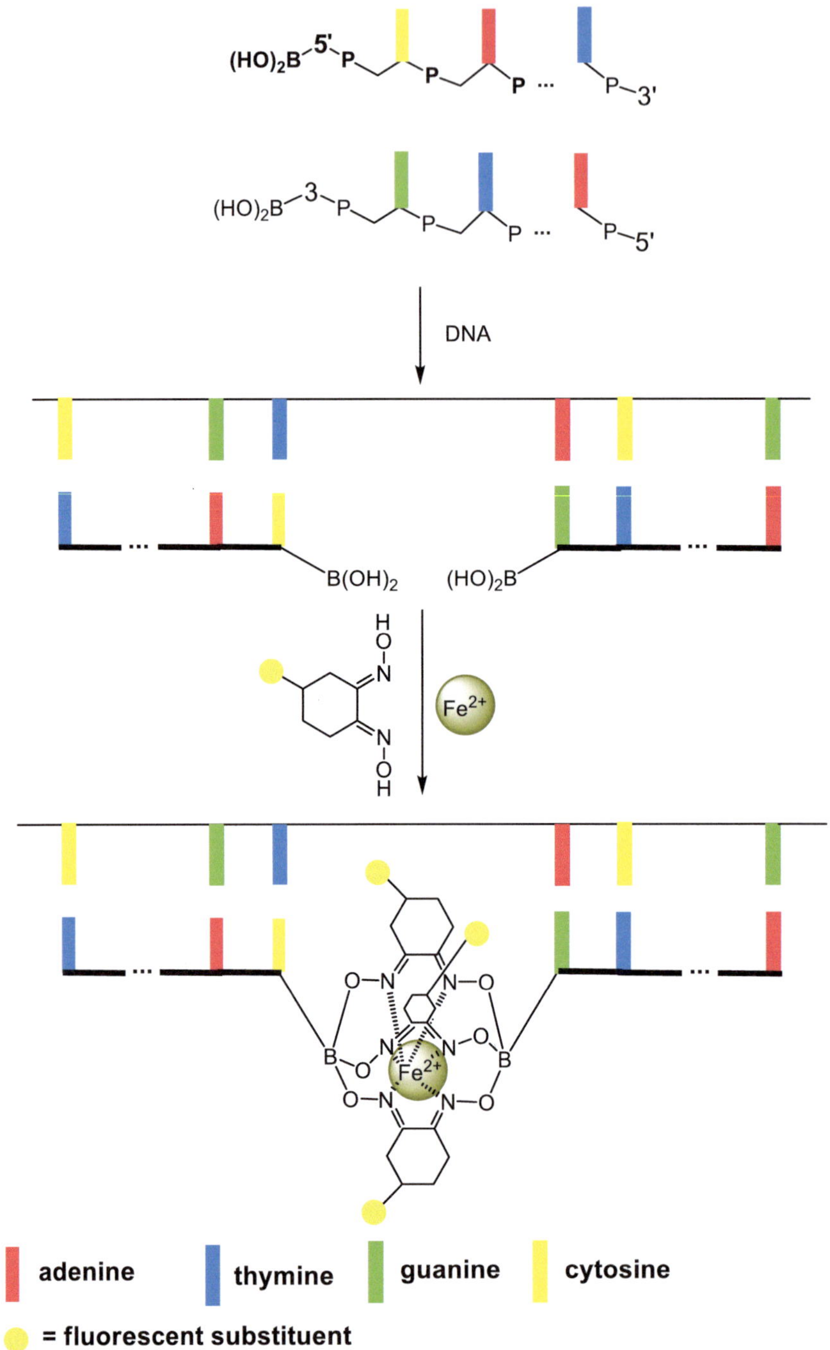

Scheme 5.15 Formation of a clathrochelate linker by oligonucleotides with boronic acid residues [1]

reaction in the PNA synthesis [in particular, in the case of monomer coupling or acylation with acetic anhydride on its final stage (capping)]. Therefore, α-dioximes have been protected [32] *via* formation of their iron(II) clathrochelates. These cage complexes play a role of bifunctional protecting agents: their donor nitrogen atoms are protected by coordination toward a caged metal ion, while the oxime oxygen atoms are protected by their covalent binding with cross-linking tetrahedral

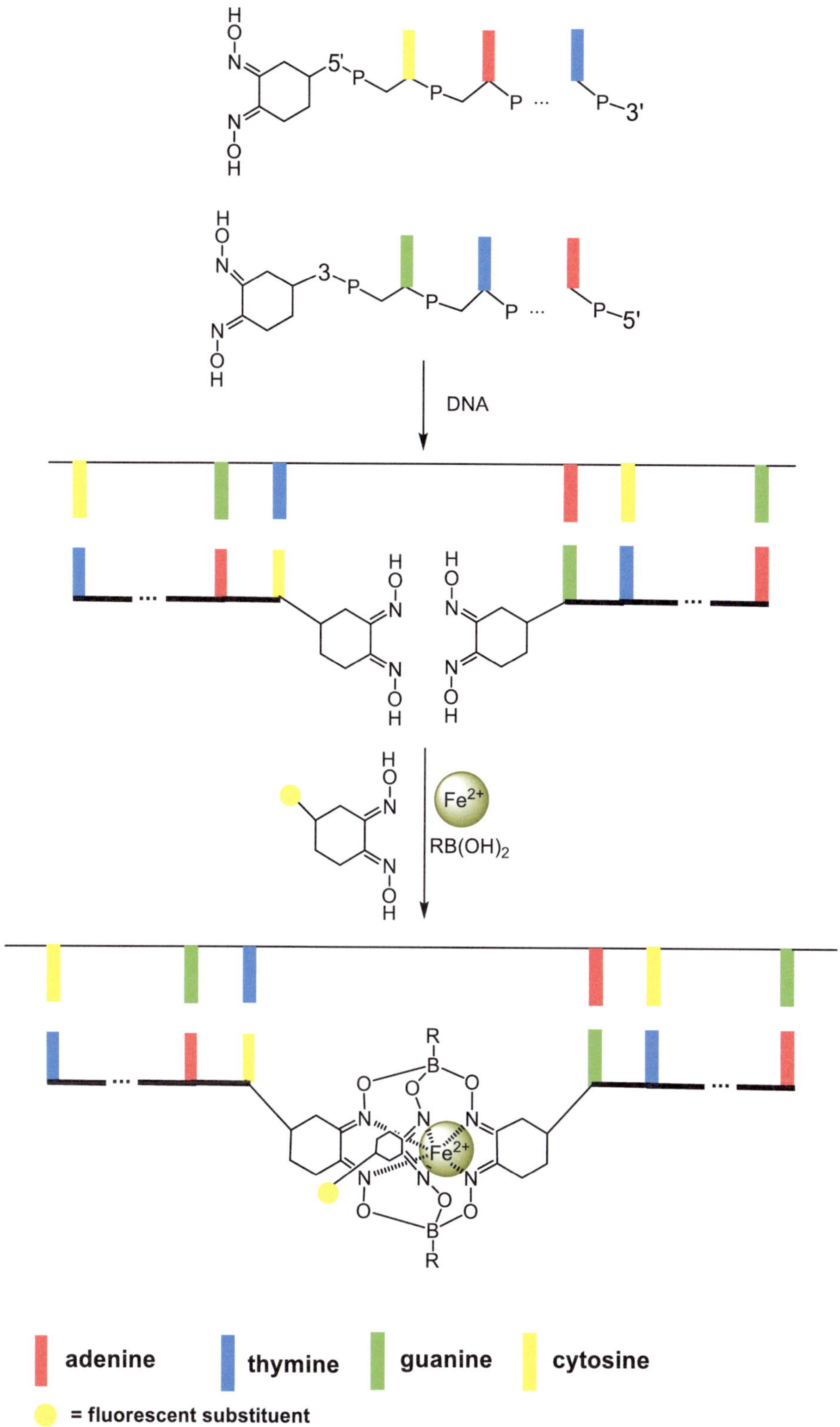

Scheme 5.16 Formation of a clathrochelate linker by oligonucleotides with α-dioxime fragments [1]

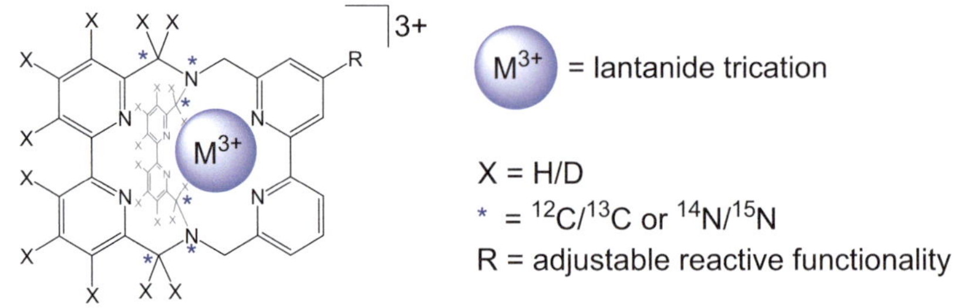

Scheme 5.17 General chemical drawing of the tris-bipyridinate lanthanide clathrochelates, the potent isotope-coded mass tags [31]

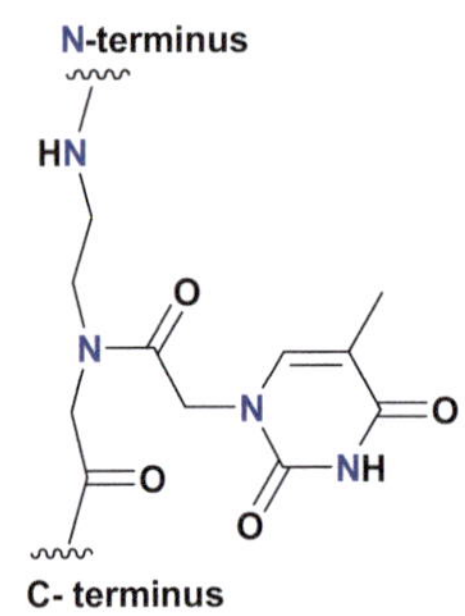

Scheme 5.18 *N*- and *C*-terminal sites of PNA

boron atom. The linker fragment of these cage complexes with terminal amino group has been used [32] for their covalent binding to PNA. Clathrochelates (Fe^{2+})@**682** and (Fe^{2+})@**368** underwent the complete destruction of their cage frameworks in trifluoroacetic acid—*meta*-cresol mixture, thus giving the corresponding α-dioximes by Scheme 5.19. So, the α-dioximate fragments of the above clathrochelate–PNA conjugates have been deprotected without their side reactions during the standard PNA cleavage–deprotection procedure. Moreover, these cage complexes are reported [32] to be stable in the presence of piperidine under the conditions of cleavage of the fluorenylmethoxycarbonyl *N*-protecting group as well as in acetic anhydride–lutidine mixture that is typically used on the capping step of the PNA synthesis. Their attachment to the *N*-terminal PNA sites by Scheme 5.20, the final PNA cleavage from a polymeric matrix, and deprotection have been used [32] for preparation of the target α-dioxime–PNA conjugates in high yields. Such multi-stranded PNAs, self-assembled in the presence of the complementary oligonucleotides, have been proposed [32] as prospective building blocks for preparation of the corresponding biological nanostructures and assemblies.

The first example of an efficient transcription inhibition by the iron(II) clathrochelates in a model in vitro system based on T7 RNA polymerase (T7 RNAP) is reported [33, 34]; this small single-subunit polymerase has been often used as a convenient and reliable model in the search for transcription inhibitors. The *in vitro* testing of the iron(II) clathrochelates (Fe^{2+})@**668**, (Fe^{2+})@**678**, (Fe^{2+})@**679**, (Fe^{2+})@**296**, (Fe^{2+})@**325**, (Fe^{2+})@**680**, (Fe^{2+})@**387**, and (Fe^{2+})@**388** (see Scheme 5.6) in the transcription assay revealed the structure- and concentration-dependent inhibition of the transcription by most of these xenobiotics (Fig. 5.2); two monoribbed-functionalized iron(II) clathrochelates (Fe^{2+})@**387** and (Fe^{2+})@**388**, the derivatives of *meta*- and *para*-mercaptobenzoic acids, were found to be highly efficient inhibitors of the T7 RNAP, suppressing the transcription in the submicromolar range (Fig. 5.3).

The heterofunctionalized transcription inhibitors, based on the morpholine-containing iron(II) clathrochelates (Fe^{2+})@**342**, (Fe^{2+})@**343**, (Fe^{2+})@**344**, and (Fe^{2+})@**345** (Scheme 5.6) with good ADMET properties, such as sufficient water solubility, membrane permeability, and low toxicity, are reported [35] to be also the efficient structure- and concentration-dependent inhibitors in a transcription assay based on T7 RNAP in their low- and submicromolar concentration range.

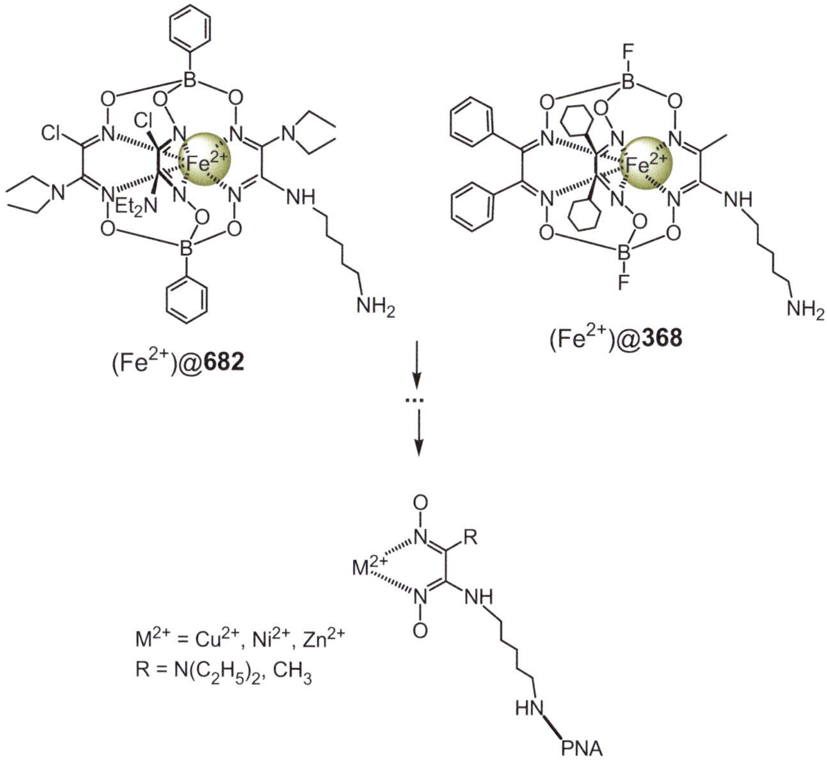

Scheme 5.19 Amine-terminated iron(II) clathrochelates used for functionalization of PNA [32]

The C–C-conjugated iron(II) bis-clathrochelates (see Sect. 2.4) have been designed [36] as potent, highly efficient transcription inhibitors of T7 RNAP, forming the quaternary supramolecular complexes (see below). In the case of such purely supramolecular interactions, very close fit of the structure of a rigid metal complex to the macromolecule's binding site is the top-priority task. Tighter binding can be obtained by expanding the surface of a metal complex while keeping the complementarity of its shape to that of a biomolecular target. For example, the above inhibitory activity can be increased by joining of two identical rigid clathrochelate fragments by a flexible linker. On the other hand, incorporation of such flexible linker into the inhibitor's structure can cause a significant increase in a conformational flexibility, thus decreasing the binding selectivity. So, an alternative approach, which employs close and relatively rigid joining of identical molecular scaffolds having the target bioactivity, to expand the binding surface of the inhibitor with a relatively low increase in the conformational flexibility, has been elaborated [36]: the boosting of inhibitory activity of the iron(II) cage complexes has been achieved by direct joining of two macrobicyclic fragments by a single C–C bond. These C–C conjugated bis-clathrochelate compounds shown in Scheme 5.6 (see Sect. 5.1.1) are prone to the limited torsion mobility, allowing them to accommodate themselves to the shape of a binding site while keeping the number of conformational states low to ensure higher selectivity. Their in vitro tested inhibitory activities fall in the submicromolar range (Fig. 5.3), thus being high enough to be used in medical applications [36].

Binding mode of the above iron(II) mono- and bis-clathrochelates to the T7 RNAP system has been in silico studied [33–36] using molecular docking of their mono- and bis-cage molecules with known X-ray structures to the X-ray structure of this protein (Fig. 5.4). In all these cases,

Scheme 5.20 General scheme of preparation of α-dioxime-PNA conjugate [32]

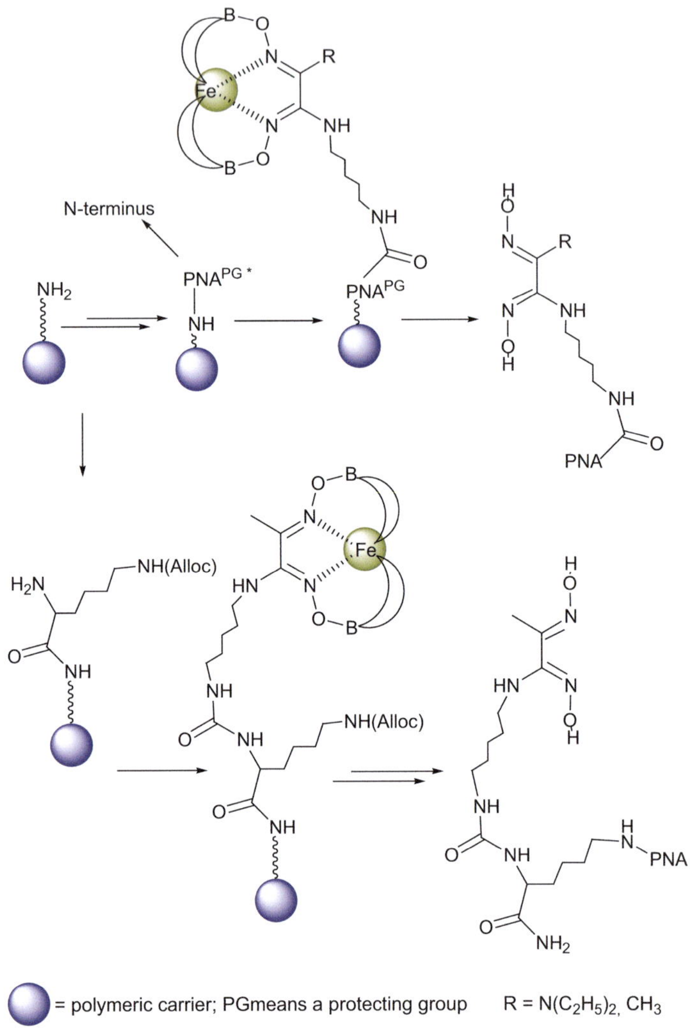

the inhibitor molecule was found to be located in the same region of a macromolecular complex, in the transcriptional "bubble," and is involved in the intermolecular contacts with protein residues as well as with DNA and RNA. The surfaces of DNA and RNA form the pocket for the inhibitor's binding, while two structural elements of the above protein interact with phenyl substituents of a clathrochelate or a bis-clathrochelate ligand that resulted in completing the binding pocket and trapping the inhibitor cage or bis-cage molecule. The bound bis-clathrochelate molecule (Fig. 5.5) was calculated [36] to be in quite different conformation than that in its X-rayed single crystal: the torsion angle ψ between the macrobicyclic fragments is 55.51° versus 160.85°. The inhibition of the above transcription is explained [33–36] by the blocking of further strand separation *via* the binding of cage or bis-cage inhibitor in the "transcription fork" to give the quaternary complex (Fig. 5.6) that can interfere with translocation step of this reaction; such quite unusual inhibition mode has been experimentally improved in [33, 34] by preincubation of the clathrochelate inhibitor with DNA, with T7

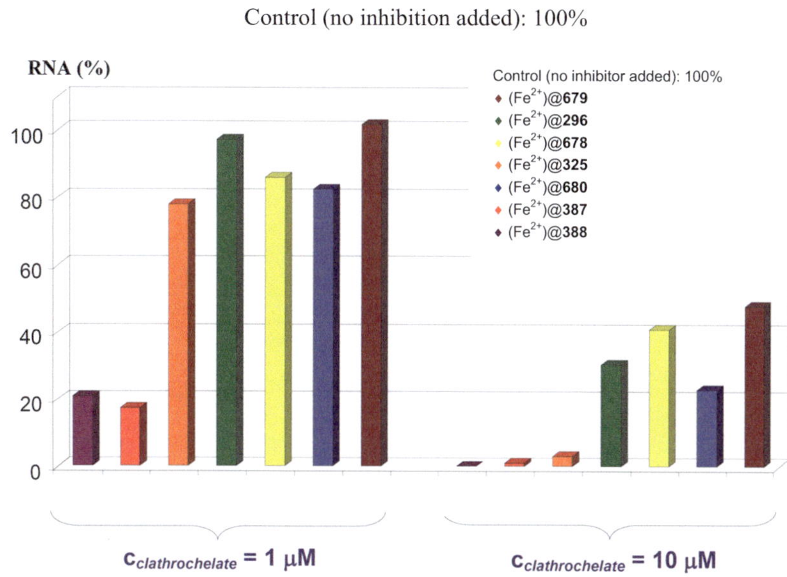

Fig. 5.2 Structure- and concentration-dependent inhibitor activities of several iron(II) monoclathrochelates [33]

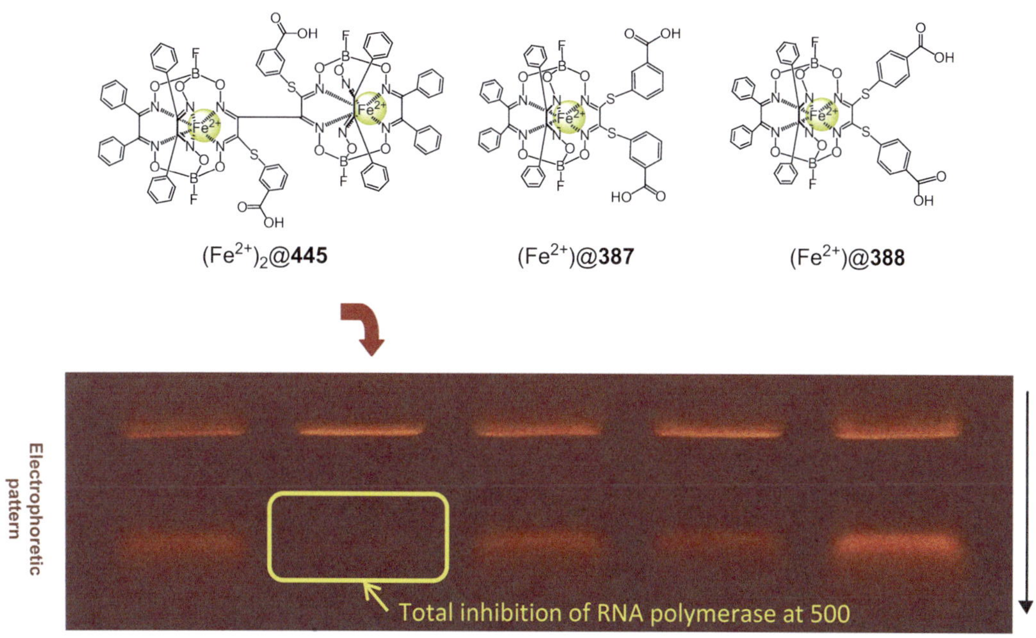

Fig. 5.3 The leader compounds: submicromolar mono- and bis-clathrochelate transcription inhibitors of T7 RNAP

RNAP, and with their mixture; the obtained experimental data were found to be in good agreement with the molecular docking results suggesting the DNA–RNAP complex (or yet DNA–RNA–RNAP binary supramolecular ternary macromolecular assembly) as the target for these inhibitors. The proposed mode of inhibition with the binding pocket formed by three interconnected macromolecules of the transcription complex has been attributed [33, 34, 36] to

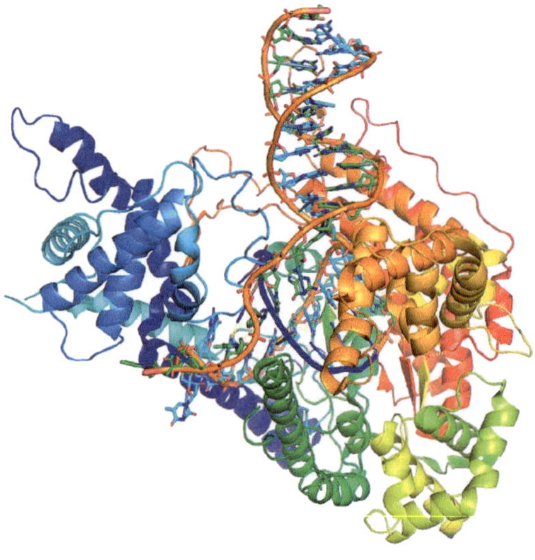

Fig. 5.4 The iron(II) clathrochelate (Fe^{2+})@**388** as a transcription inhibitor of T7 RNAP: molecular docking calculation of [33]

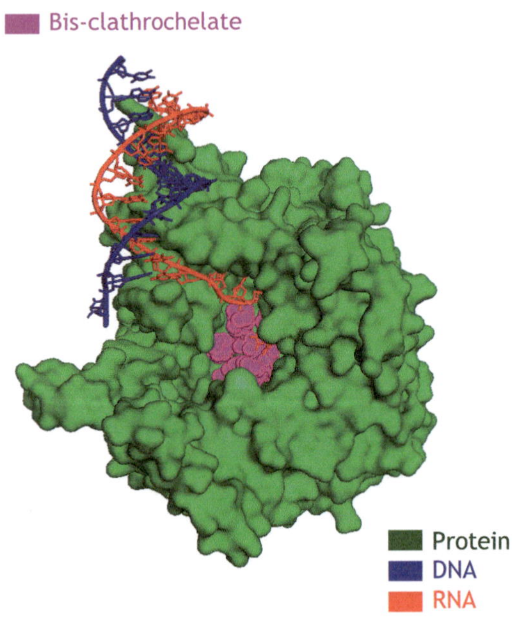

Fig. 5.5 The iron(II) bis-clathrochelate $(Fe^{2+})_2$@**445** as a transcription inhibitor of T7 RNAP: molecular docking calculation of [36]

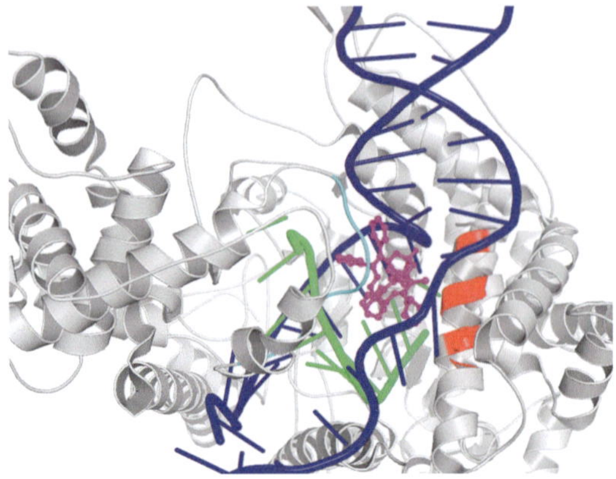

Fig. 5.6 Inhibition mode as predicted by molecular docking: the best-scoring solution for the clathrochelate (Fe^{2+})@**342** (DNA is shown in *blue*, RNA in *green*, helix in *red*, loop in *cyan*, inhibitor in *magenta*)

the unique structural properties of the mono- and bis-clathrochelate molecules: they are large enough to ensure the significant binding owing to van der Waals interactions mostly; this binding mode seems to be less probable for the small and planar guest molecules.

Neurodegenerative disorders, such as Alzheimer's, Parkinson's, and Creutzfeldt–Jakob diseases, type II diabetes, and amyloidosis, are reported in the literature to be caused by deposition of insoluble protein β-pleated formations (i.e., the aggregates or the amyloid fibrils) both in the cells and in the extracellular space of various organs or tissues. Fluorescent inhibitory assay, AFM, and flow cytometry methods have been used [37] to study the antifibrillogenic activity of the iron(II) mono- and bis-clathrochelates (Fe^{2+})@**387**, (Fe^{2+})@**388**, and $(Fe^{2+})_2$@**445** (Scheme 5.6). The presence of these cage intracomplexes affected the kinetics of insulin fibrillization reaction, reduced the amount of formed fibrils (up to 70%), and caused the decrease of their diameter from 5–12 nm (for free insulin fibrils) to 3–8 nm (for those formed in the presence of the above clathrochelates). The latter complexes also prevent the lateral aggregation of mature fibrils and the formation of superfibrillar clusters as well. So, these mono- and bis-clathrochelates have been proposed [37] as prospective and efficient agents able to suppress the protein aggregation processes; highest inhibitory activity was found for the monoclathrochelate (Fe^{2+})@**387** with its efficient inhibitory concentration of approximately 16 μM.

The in silico design of HIV protease inhibitors based on transition metal clathrochelates has been performed [38]. Their clathrochelate frameworks as molecular platforms have been fixed as a "rigid body" for all of the structures generated. Each of these macrobicyclic molecules tested in silico has been generated by placement of the ribbed (40 variants) and apical (20 variants) substituents on these molecular platforms, and then the geometry of the functionalized cage molecule has been optimized; the molecules with best estimated binding constants have been additionally examined [38]. It was found that there is more than one way to bind to the protease active site, and, unexpectedly, the cage complexes with best estimated binding energy (–13 kcal mol^{-1}), corresponding to picomolar inhibition constant, have demonstrated unusual binding mode. That for the leader iron(II) cage complex (Fe^{2+})@**683**, docked to the HIV protease active site, is shown in Fig. 5.7. Such clathrochelate inhibitor is deviated from the center of this active site and shifted to its periphery, thus giving the supramolecular complex with HIV protease (mainly *via* its van der Waals interactions with hydrophobic surface of such site rather than its electrostatic interactions with the catalytic aspartate dyad) [38].

5.1.4 Radiodiagnostic and Radiotherapeutic Cage Complexes with (Pseudo) Encapsulated Metal Radionuclides

5.1.4.1 64-Copper and 68-Gallium Compounds

Among metal ions that can be encapsulated by a macrobicyclic ligand to produce a clathrochelate, those that provide radioisotopes useful for this purpose include copper(II) and gallium(III) cations. In particular, nuclear medicine uses radiation to provide diagnostic information on how specific organs function or to treat them. A recent development of nuclear medicine, PET [6, 39, 40], is a more precise functional imaging technique used mostly for identifying the medical conditions (including cancers, heart diseases, and brain disorders) or following the progress of their treatment. Isotopes of copper ^{60}Cu, ^{61}Cu, ^{62}Cu, and ^{64}Cu are positron emitters with half-life ranging from relatively short (9.74 min for ^{62}Cu) to relatively long (12.7 h for ^{64}Cu) [6, 41]. The long half-life of ^{64}Cu allows using this isotope for PET imaging as well as for radiotherapy; selectivity and target delivery of the ^{64}Cu isotope to a biological organ are very important for its radiopharmaceutical applications. Biodistribution of $^{64}Cu^{2+}$ complexes is described [6] to be governed by their stability, size, shape, charge, lipophilicity,

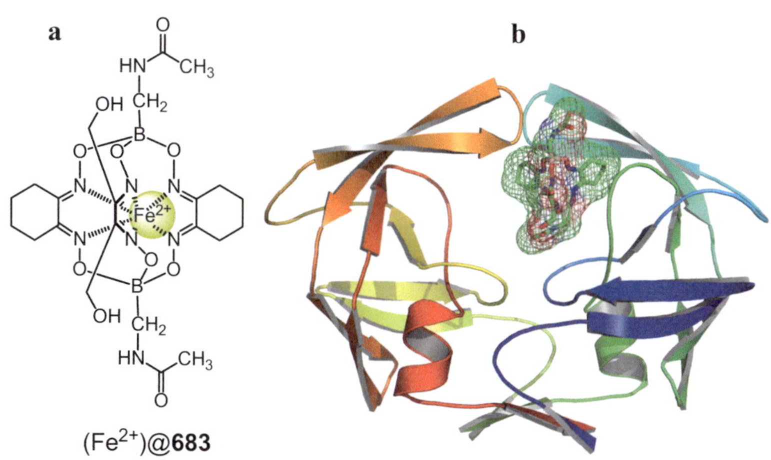

Fig. 5.7 The in silico designed topological clathrochelate inhibitor (Fe^{2+})@**683** of HIV protease (**a**) and its binding to the active site of this protein (**b**) [38]

and redox properties, and further selectivity can be achieved using vector substituents that bind to specific molecular targets in vivo. Among copper(II) clathrochelates, the complexes of polyamine caging ligands functionalized with biologically relevant groups have been recognized [42–44] as promising candidates for future PET imaging agents.

The functionalization of the encapsulating ligand **121** by pendant cyclic peptide residue (Scheme 5.21) has been used [45] for target delivery of its $^{64}Cu^{2+}$ cage complex to a given biological system.

$^{64}Cu^{2+}$ complex of the bifunctionalized polyamine ligand **124** (Scheme 5.22) is reported [46] to exhibit an uptake by tumor cells exceeding that for its monofunctionalized analog **121**, while $^{68}Ga^{3+}$ cage complex of **124** has been proposed as promising non-invasive PET imaging agent.

The apically mono- and difunctionalized polyamine macrobicyclic ligands **125–128** with various peptide substituent(s) at their capping carbon atoms have been used [47] for the preparation of $^{64}Cu^{2+}$ clathrochelates shown in Scheme 5.23, which are stable for a long time under the conditions of in vivo experiments.

The clathrochelate-based bioconjugate ($^{64}Cu^{2+}$)@**684** (Scheme 5.24 [48]), obtained by 1,3-dipolar cycloaddition of the dibenzocyclooctyne-functionalized macrobicyclic 64-copper(II) complex to the corresponding azide-terminated cyclic peptide as a ligand synthon, has been tested [48] using both the in vitro and in vivo experimental approaches.

According to in vivo biodistribution and PET imaging data of [49], the ^{64}Cu-labeled clathrochelate-based radioimmunoconjugates ($^{64}Cu^{2+}$)@**685** and ($^{64}Cu^{2+}$)@**686** with a

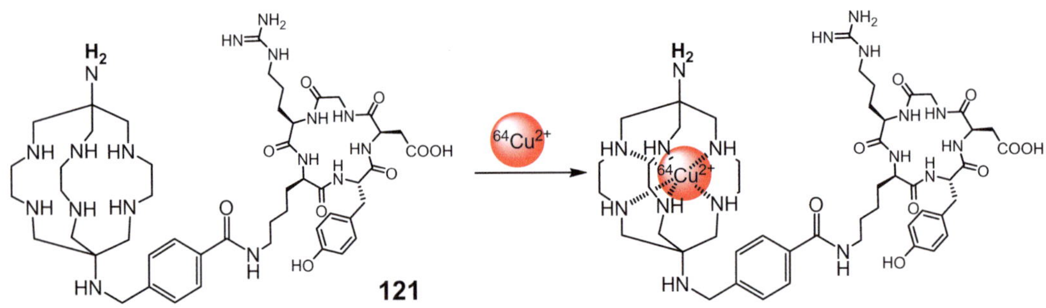

Scheme 5.21 Encapsulation of $^{64}Cu^{2+}$ ion by the caging ligand **121** [45]

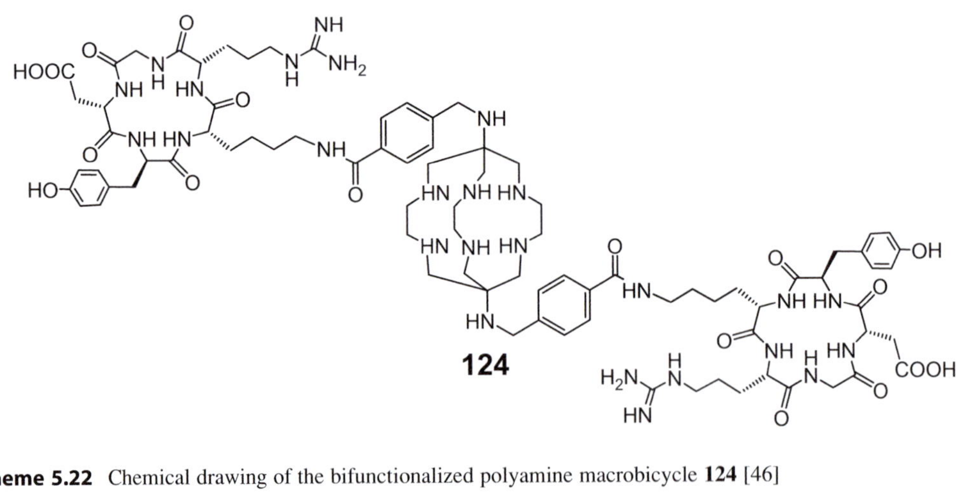

Scheme 5.22 Chemical drawing of the bifunctionalized polyamine macrobicycle **124** [46]

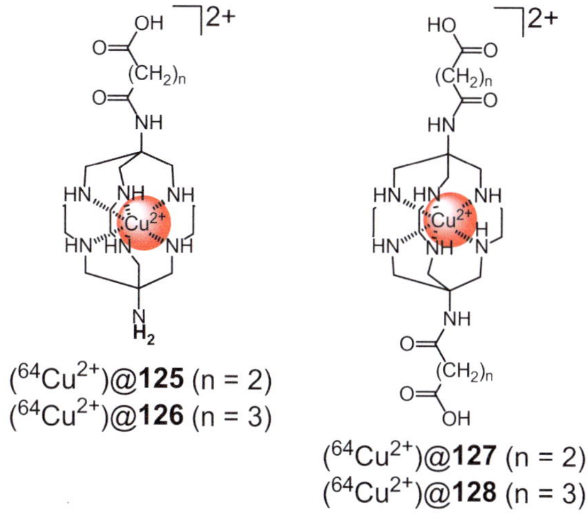

Scheme 5.23 Chemical drawings of the peptide-terminated 64-copper(II) clathrochelates

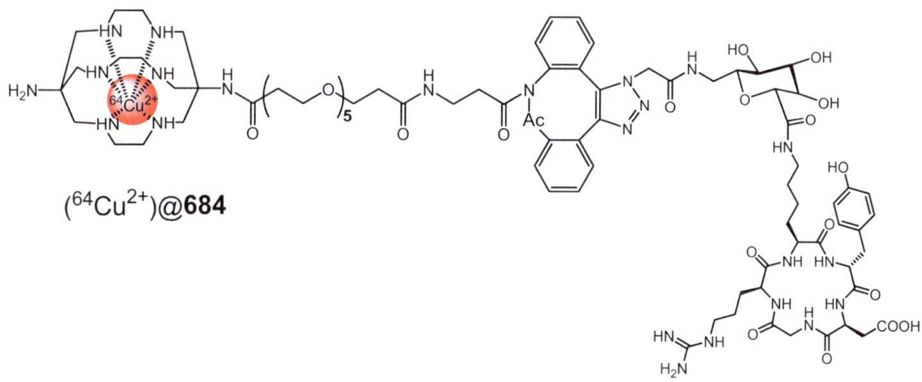

Scheme 5.24 Chemical drawing of the bioconjugate ($^{64}Cu^{2+}$)@**684** [48]

HER2/neu-targeting antibody (Scheme 5.25 [49]), undergoing their high and selective uptake in a HER2-positive cancer cell line, are highly stable under physiological conditions.

Cage complexes of the homo- and heteromultifunctionalized polyamine ligands **687–690** (Scheme 5.26 [50]) with an encapsulated $^{64}Cu^{2+}$ ion also have a good in vitro and in vivo stability, thus making these multifunctional chelators a versatile molecular platform for multivalent/multimodality probes for both the imaging and therapy [50].

^{64}Cu-labeled complex of the dipeptide encapsulating ligand **691** (Scheme 5.27 [51]), obtained by dipolar cycloaddition of its diazide-terminated macrobicyclic precursor to the corresponding alkyne-containing peptide ligand synthon, has been used [51] for microPET imaging of the receptor CD13 (a well-known tumor vasculature biomarker) expressed in vitro and in vivo in living mice; this cage complex showed a good binding affinity and CD13 specificity with given tumor cells and an excellent tumor uptake as well.

The $^{64}Cu^{2+}$ cage complexes of the exendin-4-functionalized macrobicyclic chelators **692** and **693** (Scheme 5.28) [52] with high kidney radioactivity levels are reported [52] to demonstrate their persistent and specific uptake in a given

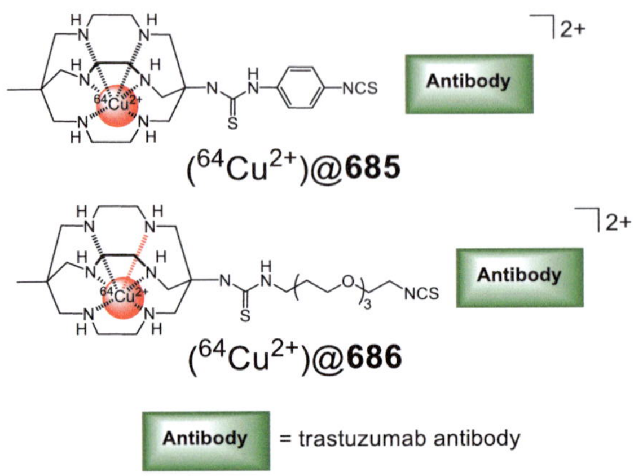

Scheme 5.25 ^{64}Cu-labeled clathrochelate-based radioimmunoconjugates [49]

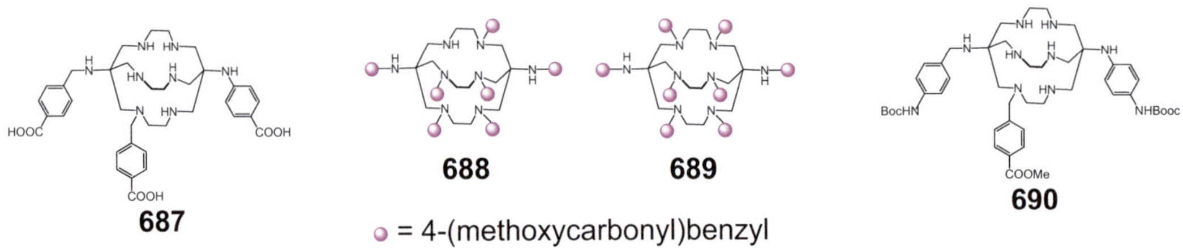

Scheme 5.26 Chemical drawings of a series of the homo- and heterofunctionalized polyamine caging ligands

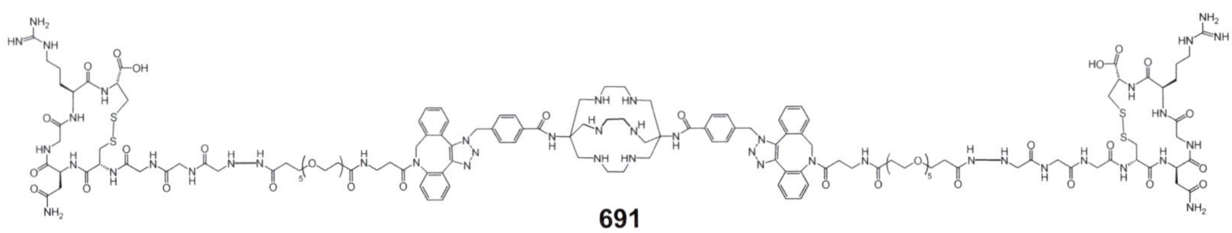

Scheme 5.27 Chemical drawing of the dipeptide encapsulating ligand **691** [51]

insulinoma model; an increased tumor uptake in vivo has been also observed for a metal-free macrobicyclic compound **693**. The above ligands and their $^{64}Cu^{2+}$ cage complexes have been proposed [52] as promising agents for PET imaging of the glucagon-like peptide receptor expression. This receptor has an elevated expression profile in pancreatic islets, insulinoma, and the cardiovascular system as well, and its accurate visualization and quantification of β-cell mass are critical for improved understanding, diagnosis, and treatment of type 1 diabetes and insulinoma [52].

Functionalization of ^{64}Cu-labeled clathrochelate ($^{64}Cu^{2+}$)@**694** (Scheme 5.29 [53]) with both the near-IR fluorescent dye fragment and the peptide's exendin-4 group has been used [53] for a bimodal PET–fluorescence imaging of the glucagon-like peptide 1 expression in pancreas and in pancreatic islet cell tumors. This complex showed a good performance both in vivo and ex vivo, visualizing small xenografts with PET and pancreatic β-cell mass by phosphor autoradiography. Its fluorescence has been also used [53] for detection of the individual pancreatic islets, thus confirming specific binding to a

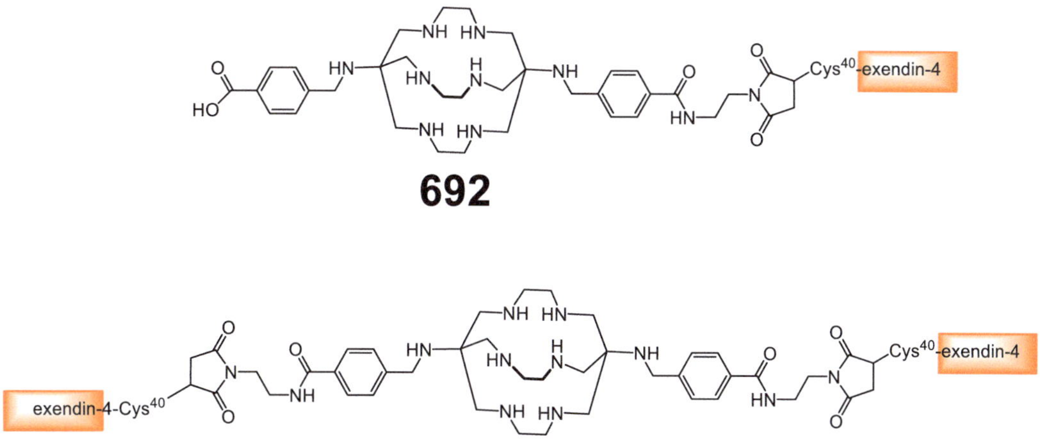

692

693

Scheme 5.28 Chemical drawings of the exendin-4-functionalized ligands **692** and **693**

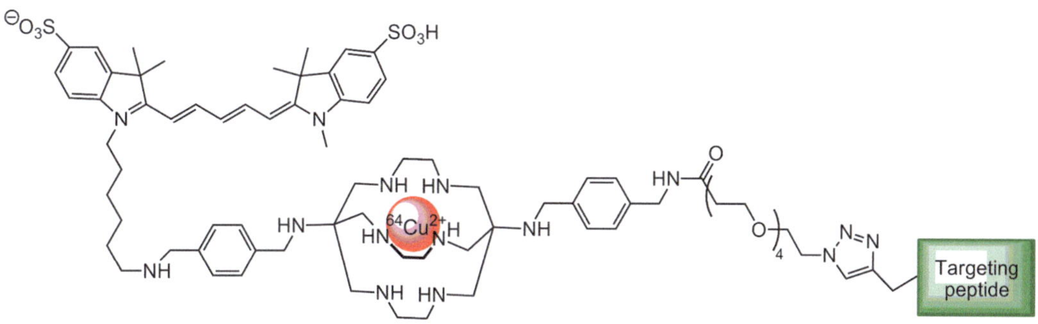

($^{64}Cu^{2+}$)@**694**

Scheme 5.29 Chemical drawing of a clathrochelate compound for a bimodal PET–fluorescence imaging

target peptide with the sensitivity surpassing that of a radioactive label. These dual clathrochelate-based imaging probes have been proposed [53] for the diagnosis of the primary growths and metastases as well as for the detection of the tumor margins, infiltrative growth, and residual tumor cells.

The heterofunctionalized cage complex ($^{64}Cu^{2+}$)@**695** shown in Scheme 5.30 has been designed [54] as promising candidate for further preparation of various tumor-targeted dual-modality imaging probes.

^{64}Cu-labeled radiopharmaceuticals, prepared starting from the difunctionalized caging ligands **696** and **697** and their peptide-containing derivatives **698** and **699** (Scheme 5.31), are reported [55] to exhibit a high tumor uptake and the tumor-to-normal tissue ratios.

$^{64}Cu^{2+}$ cage complex of the macrobicyclic ligand **700** (Scheme 5.32) is a rather stable when incubated with *L*-cysteine, while low levels of its decomplexation have been observed [56] with *L*-histidine as a concurrent *N*-donor ligand.

The in vitro and in vivo experiments of [57] showed that $^{64}Cu^{2+}$ complex of the apically functionalized encapsulating ligand **701** with functionalizing tumor-targeting peptide Tyr3-octreotate residue (Scheme 5.33) has a high

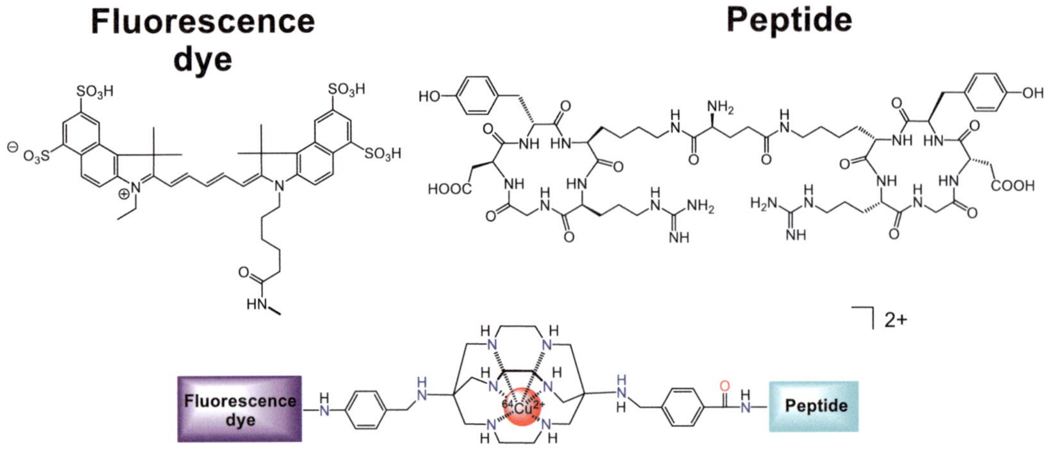

($^{64}Cu^{2+}$)@**695**

Scheme 5.30 Chemical drawing of a clathrochelate compound for tumor-targeted dual-modality probes

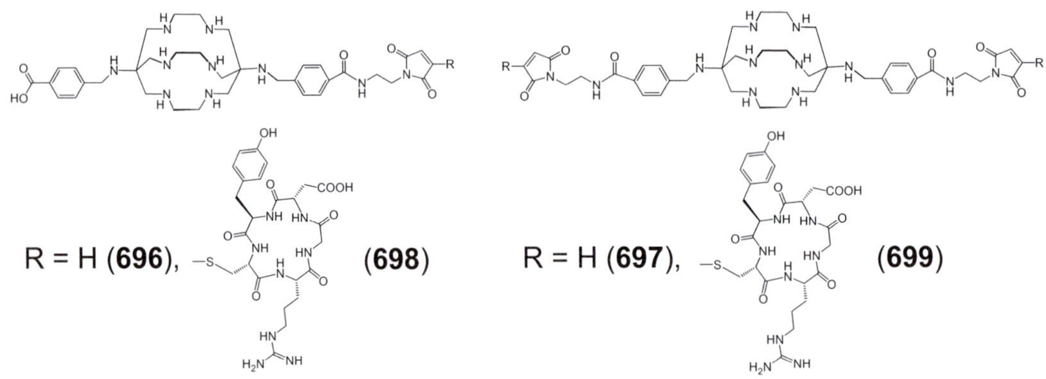

R = H (**696**), (**698**) R = H (**697**), (**699**)

Scheme 5.31 Chemical drawings of a series of the polyamine macrobicyclic ligands designed for a high tumor uptake [55]

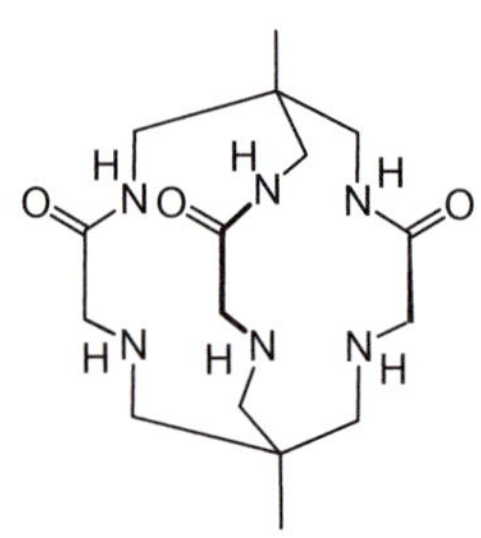

700

Scheme 5.32 Chemical drawing of a mixed triamine triamide macrobicyclic ligand **700**

selectivity for tumor cells expressing somatostatin receptor 2 (sstr2) and an excellent uptake in sstr2-positive tumors. $^{64}Cu^{2+}$- and $^{67}Cu^{2+}$-labeled cage complexes of this ligand have been proposed [57] for the combined imaging and therapeutic treatment of neuroendocrine tumors.

The conjugate of a single-chain antibody and the monofunctionalized $^{64}Cu^{2+}$ clathrochelate ($^{64}Cu^{2+}$)@**702** (Scheme 5.34) has been used [58] for diagnostic PET imaging of activated platelets. This conjugate accumulates in an injured vessel with its high and specific uptake,

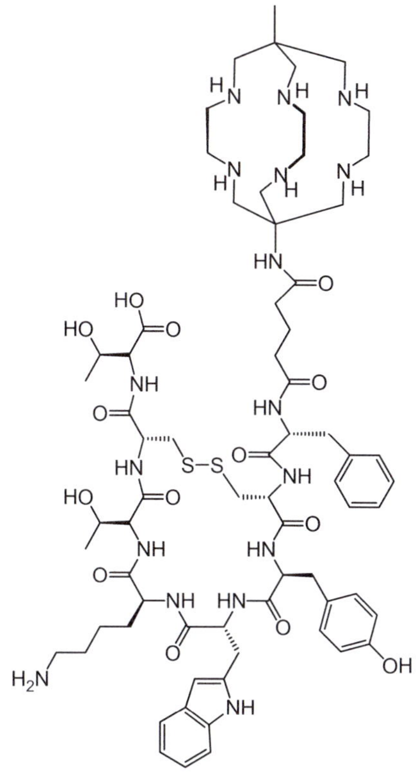

701

Scheme 5.33 Chemical drawing of the peptide-terminated macrobicyclic ligand **701**

thus paving the way toward a highly sensitive in vivo detection of activated platelets and early diagnosis of acute thrombotic events.

Similarly, a diagnostic potential for carotid artery thrombosis using PET method has been demonstrated [59] using the clathrochelate-based conjugate ($^{64}Cu^{2+}$)@**703** (Scheme 5.35 [59]) obtained *via* enzyme-mediated site-specific bioconjugation of the polyamine macrobicyclic chelator **704** with a single-chain antibody *91*. A fragment of this antibody is described to target the ligand-induced binding sites on a glycoprotein receptor GPIIb/IIIa found on the activated platelets. This conjugate, radiolabeled with the positron-emitting ^{64}Cu ions, selectively binds these platelets in an in vivo model [59].

A series of the conjugates formed by $^{64}Cu^{2+}$ complexes of the macrobicyclic chelators shown in Scheme 5.36 has been probed [60] for their radiolabeling efficiency, in vitro stability, and biodistribution. Among those, $^{64}Cu^{2+}$-labeled clathrochelate ($^{64}Cu^{2+}$)@**705** conjugated with anti-CD20 antibody was found to be extremely stable in vivo and to achieve high specific activity under very diluted conditions. The peptide-functionalized cage complex ($^{64}Cu^{2+}$)@**706** has been also proposed [61] as an efficient probe for non-invasive PET imaging of tumor-associated

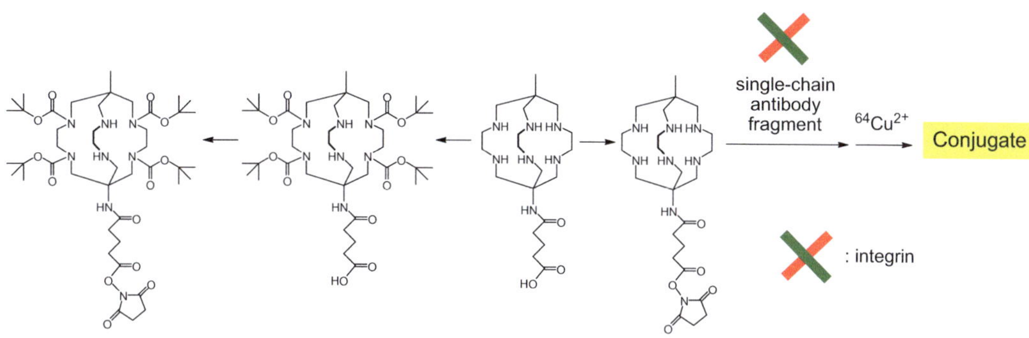

702

Scheme 5.34 Formation of the clathrochelate ($^{64}Cu^{2+}$)@**702**—single-chain antibody conjugate [58]

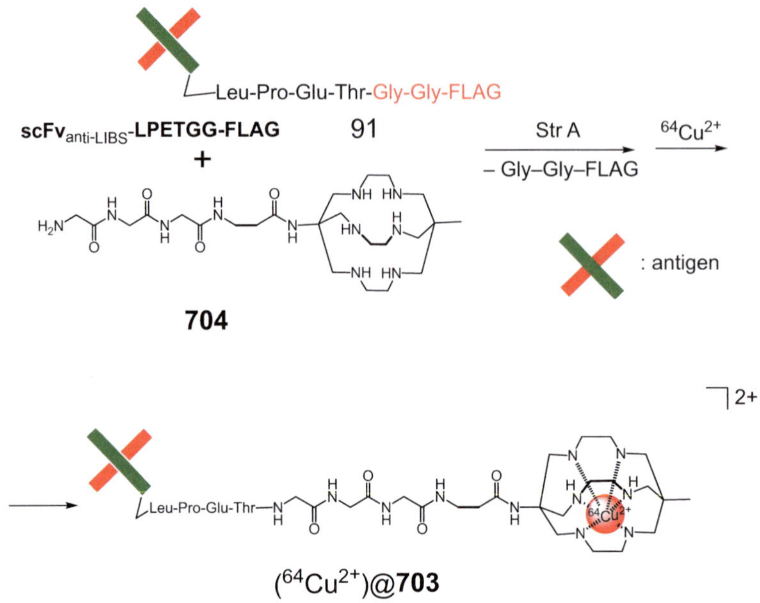

Scheme 5.35 Formation of the clathrochelate ($^{64}Cu^{2+}$)@**703**—single-chain antibody conjugate [59]

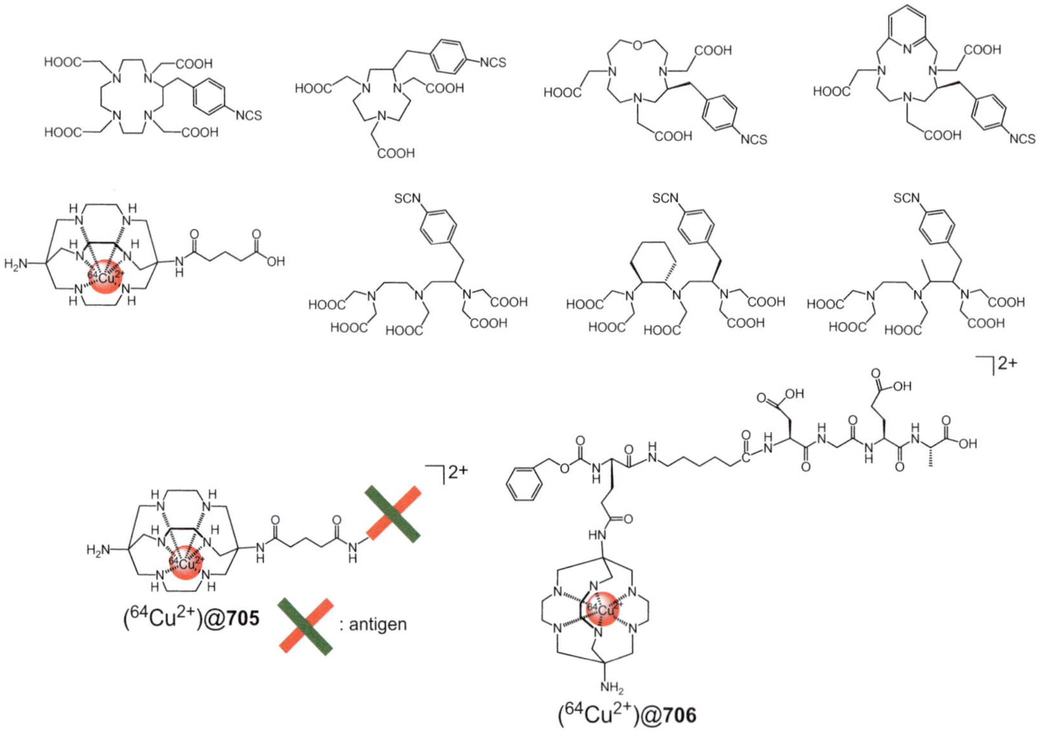

Scheme 5.36 Formation of the clathrochelate ($^{64}Cu^{2+}$)@**706**—anti-CD20 conjugate [60]

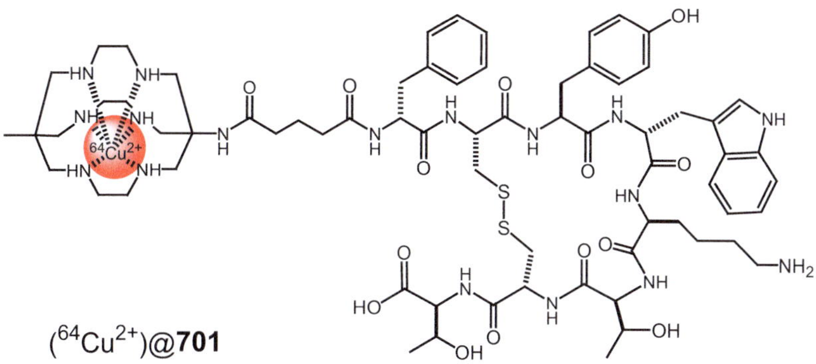

($^{64}Cu^{2+}$)@**701**

Scheme 5.37 Chemical drawing of the clathrochelate-based conjugate ($^{64}Cu^{2+}$)@**701**

$\alpha_2\beta_1$ expression, thus being an useful tool for identifying of prostate cancer.

The heterodifunctionalized $^{64}Cu^{2+}$ cage complexes have been used [62] for the design of long-circulating nanotherapeutics. Nanoparticles which are formed by polypeptides, modified by these clathrochelates *via* site-specific conjugation, retain their activity in vivo over several days, thus allowing a microPET imaging.

The macrobicyclic ligand **707** with functionalizing terminal carboxylic group is reported [63] to undergo coupling to *N*-terminus of Tyr^3-octreotate immobilized on the solid phase, thus giving the clathrochelate–peptide conjugate **701** shown in Scheme 5.37. This conjugate has been then labeled with $^{64}Cu^{2+}$ ion, thus giving the corresponding cage complex with promising pharmacokinetic properties for in vivo imaging of SSTr2-positive tumors [63].

5.1.4.2 Technetium and Rhenium Compounds and Other Encapsulated Radioemmiter Metal Ions

^{186}Re, ^{188}Re, and ^{99m}Tc radioisotopes are widely used for radiodiagnostics and radiotherapy in oncology, nuclear medicine, and cardiology [64]. Two main forms of these radiolabeles and radiotherapeutics, and their stable isotopes as models, are reported to date giving the cage and semiclathrochelate complexes. Their highly oxidized forms, perrhenate and pertechnetate oxoanions, typically form the complexes of this type with polyamine macrobicyclic ligands and their protonated forms as well [65], while the reduced forms, rhenium and technetium(III) cations, are known to give the corresponding tris-dioximate and dioximato-iminooximate semiclathrochelates. This class of the rhenium and technetium radiopharmaceuticals and their models is based on a quasiencapsulation of these metal(III) ions (typically generated in situ by reduction of the parent oxoanions) by boron- and tin-capped tris-dioximate pseudoclathrochelate ligands (Scheme 5.38a, b) [66–69]. Such template cross-linking with Lewis-acidic boron- and tin (IV)-capping agents has been complicated by side reduction reactions of these ligands giving their iminooxime derivatives (Scheme 5.38c) [70]; the pseudoclathrochelates with apical functionalizing nitroimidazole and isothiocyanate substituents also shown in this Scheme have been used [71–73] for technetium labeling of monoclonal antibodies.

More recent synthetic approach for the target delivery of such pseudocage complexes used template condensation by Scheme 5.39 of an appropriate functionalized phenylboronic acid with the corresponding α-dioxime on a rhenium(III) ion as a matrix, thus giving the semiclathrochelate rhenium(III) tris-dioximate with terminal reactive group as a metallocomplex precursor (building block) [74, 75]. Further functionalization of this precursor with a given vector substituent, containing the complimentary reactive fragment, allowed obtaining a series of the pseudomacrobicyclic rhenium(III) compounds for their target delivery to a given biological system or organ. So,

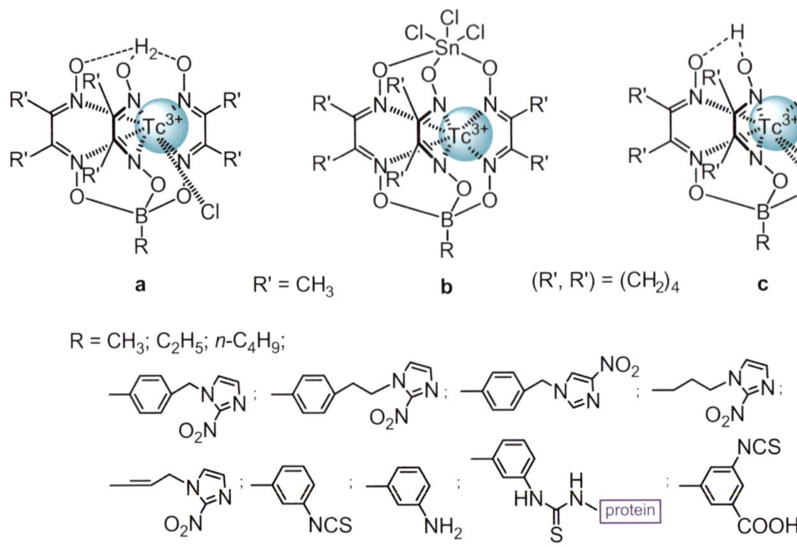

Scheme 5.38 Chemical drawings of a series of the boron- and tin-capped technetium(III) pseudoclathrochelates

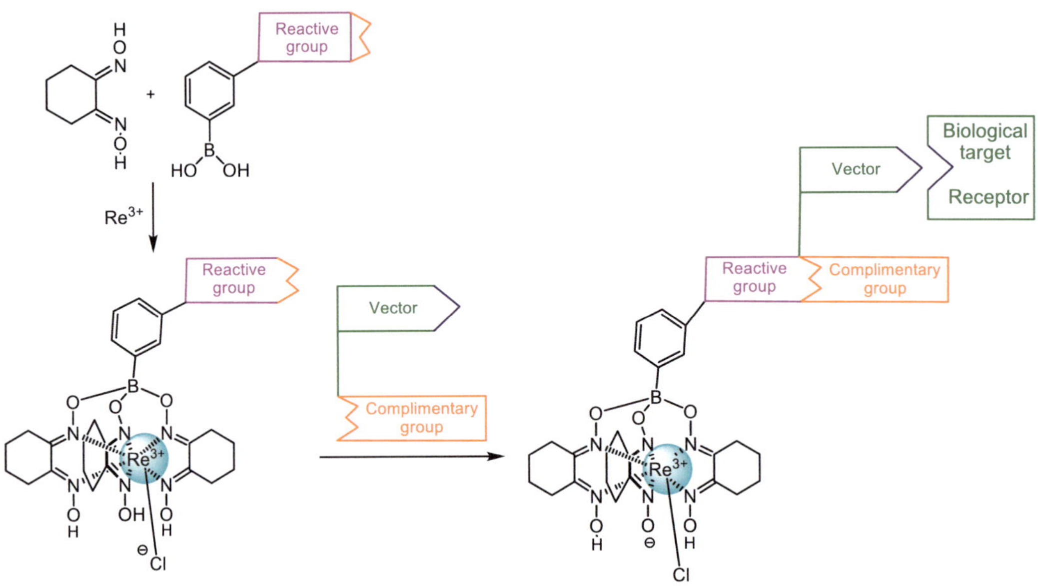

Scheme 5.39 Apical functionalization of a rhenium(III) pseudoclathrochelate with a vector substituent

such "molecular nanoconstructor" can be used for molecular industry of such conjugated radiopharmaceuticals with various vector substituents, thus designed for radiodiagnostics and radiotherapy of a wide range of diseases [74, 75].

The S_6-cage compounds **116**, **117**, and **119** (see Sect. 2.2.1) have been proposed [76] as potent encapsulating ligands for $^{111}Ag^+$ cation allowing us to perform radioimmunotherapy with this radioemitter.

5.1.5 Boron-Enriched Clathrochelate-Based Targets for ^{10}B-NCT Treatment

Higher polyhedral boron hydrides $B_nH_n^{2-}$ ($n = 6$–12) and the corresponding functionalized boron clusters are available, which are chemically robust and low-toxic compounds. Hence, the derivatives of the decahydro-*closo*-decaborate ($B_{10}H_{10}^{2-}$) and dodecahydro-*closo*-dodecaborate ($B_{12}H_{12}^{2-}$) anions seem to be most prospective precursors of the boron-enriched radiopharmaceuticals: their derivatives have been intensively studied as prospective targets for ^{10}B-NCT of tumor. This approach is based on the nuclear reaction of the stable isotope ^{10}B with thermal neutrons, and the appropriate properties of such boron-enriched targets are necessary for a successful ^{10}B-NCT treatment; a key problem is their target delivery to the tumor cells that allows to decrease a dose requiring for effective therapeutic treatment, thus preventing an intoxication of organism.

Ribbed functionalization of the iron(II) tris-dioximate clathrochelates as molecular scaffolds has been performed [77, 78] to give the polyfunctional and polytopic hybrid *closo*-boratoclathrochelates (Fe^{2+})@**320** and (Fe^{2+})@**438** (Scheme 5.40): the apical substituents of their cage analogs can play a role of vectors, while the ribbed substituents are enriched by boron clusters. The analogous semi- and clathrochelate iron(II) oximehydrazonates *28*, (Fe^{2+})@**496**, and (Fe^{2+})@**497** with apical functionalizing carboranyl substituent are described [79]. C-carboranylboron-capped tris-dioximate iron(II) clathrochelates (Fe^{2+})@**245** and (Fe^{2+})@**246** as well as the iron(II) carboranoclathrochelate (Fe^{2+})@**483** with an inherent ribbed carboranyl substituent have been also proposed [80, 81] as potent boron-enriched targets for ^{10}B-NCT treatment.

5.1.6 Clathrochelate-Based Paramagnetic Probes for NMR and MRI Applications

NMR and MRI have a well-deserved reputation as the most powerful techniques for studying biological systems. NMR spectroscopy is one of the main sources of information about their structural features; even being limited by the size of a biomacromolecule under study, it nevertheless allows determining its native conformation in solution. One of possible approaches to increase the accuracy of this structural information is the use of paramagnetic probes. However, the only report on the use of clathrochelates as relaxation agents dates back to 1992 [82]. For polyamine manganese(II), copper(II), and chromium(III) clathrochelates, the derivatives of the diaminomacrobicyclic encapsulating ligand **99**, a modest relaxivity in simple aqueous solutions was found, which dramatically increased upon binding of a cage complex to a protein.

Although regular *3d*-transition metal complexes have residual dipolar couplings [83], they are rarely used as shift agents due to their chemical instability and relatively low magnetic anisotropy resulting in smaller paramagnetic pseudocontact shifts and lower alignment ability. Clathrochelates eliminate most of the limitations caused by chemical stability or unwanted coordination; however small (for low-spin cobalt(II) complexes [84]) or nonexistent (for diamagnetic iron(II) clathrochelates) magnetic anisotropy precluded their use as shifting agents until 2010. Recently [85], the - high-spin cobalt(II) tris-dioximate clathrochelates with TP geometry were found to have extremely large paramagnetic shifts in NMR spectra and a very large anisotropy of the magnetic susceptibility (see Chap. 4). Their tris-pyrazoloximate cobalt(II) pseudoclathrochelate analog (Co^{2+})@**509** showed the magnetic anisotropy that was even larger [86]. Its estimated value not only surpassed those for most of the lanthanide-based paramagnetic tags but also appeared to be enough to result in single-ion magnet behavior [87] (see Sect. 5.3). This suggests a huge potential of the transition metal (pseudo)clathrochelates as new paramagnetic labels and contrast agents in biomolecular NMR and MRI [3]. These paramagnetic clathrochelates can be covalently bounded to a protein using either the apical substituent (s) with a vector terminal group for cysteine labeling or direct nucleophilic substitution of

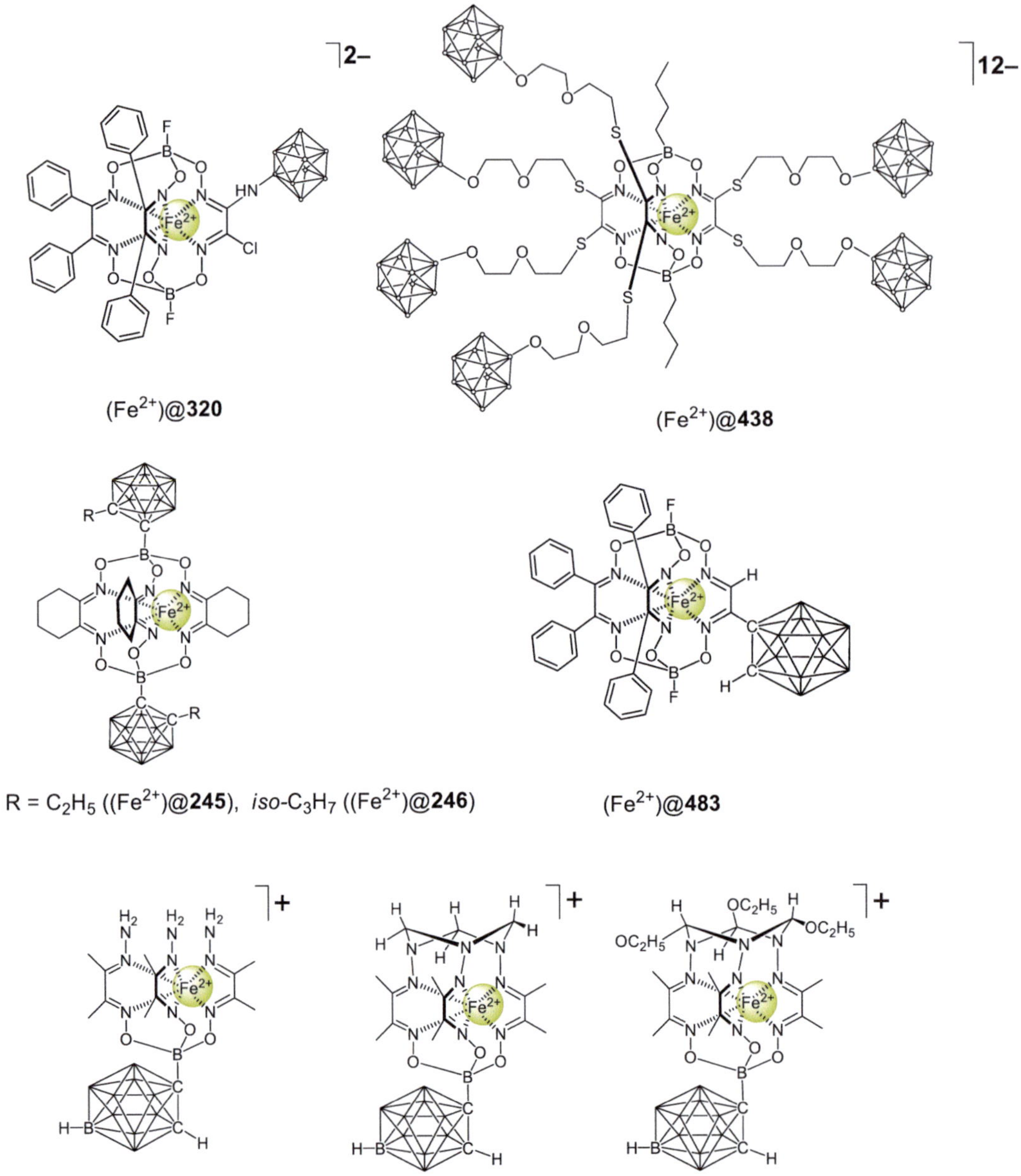

(Fe^{2+})@**320**

(Fe^{2+})@**438**

R = C_2H_5 ((Fe^{2+})@**245**), *iso*-C_3H_7 ((Fe^{2+})@**246**)

(Fe^{2+})@**483**

28

(Fe^{2+})@**496**

(Fe^{2+})@**497**

Scheme 5.40 Chemical drawings of a series of the iron(II) *closo*-borato- and carboranoclathrochelates

the ribbed chlorine atom(s) with generated in situ cysteine thiolate derivative, thus allowing them for single- or two-point attachments, depending on the nature of a protein molecule [3].

5.2 Clathrochelate-Based Catalysts and Precatalysts

For the first time, the boron-capped cobalt tris-dioximate clathrochelates shown in Scheme 5.41

have been proposed [88] as potent efficient electrocatalysts for hydrogen production from acidic media.

As follows from the electrochemical data of [89], the fluoroboron-capped tris-α-benzildioximate cobalt clathrochelate (Co^{2+})@**678** serves as a molecular precursor for electrodeposition of the cobalt-containing electrocatalytically active nanoparticles on the electrode surface. These efficiently catalyze a hydrogen evolution from aqueous solutions at pH 7 at a modest cathodic potential [89].

The boron-capped cobalt tris-dioximates (Co^{3+})@**668**, (Co^{3+})@**678**, and (Co^{2+})@**708** shown in Scheme 5.42 have been used [90] as precatalysts for catalytic HER from acidic aqueous solutions. The performed electrochemical study of these cage complexes revealed the formation of cobalt-based nanoparticles on a surface of the working electrode. The potential for their electrochemical deposition is reported [90] to be a function of the chemical nature of the substituents at a cage framework.

The cobalt(II) tris-dioximate hexachloroclathrochelates (Co^{2+})@**293** and (Co^{2+})@**294** are reported [91] to give the electrocatalytic systems, producing molecular hydrogen from H^+ ions at very low values of overpotentials of $2H^+/H_2$ process. The electrocatalytic properties of the cobalt clathrochelate-based electrocatalysts (Co^{2+})@**335** and (Co^{2+})@**378** of HER from acidic media have been studied [92]. The efficiency of such HER is reported to be increased by immobilization of these cage complexes with a terminal thiol group on a surface of working gold electrode. The macrobicyclic 2-thiopheneboron-capped iron and cobalt(II) - tris-dioximates (Fe^{2+})@**211** and (Co^{2+})@**211** are described [93] to be the components of highly active electrocatalytic systems for hydrogen production from H^+ ions. The boron-capped iron and cobalt(II) hexabromoclathrochelates (Fe^{2+})@**287** and (Co^{2+})@**287** have also been proposed as electrocatalytically active components for various hydrogen producing systems. The

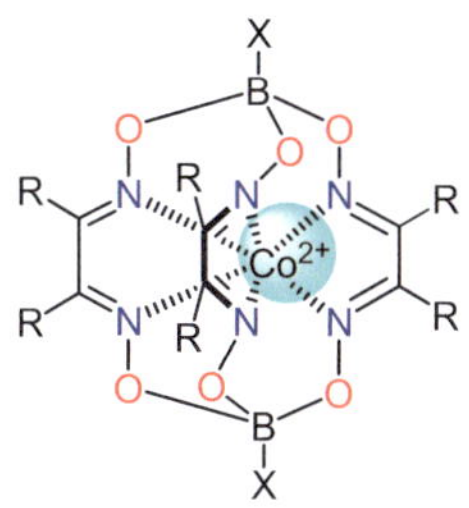

X = F   R = C_6H_5
X = F   R = CH_3
X = $C_6H_2F_3$ R = CH_3
X = C_6H_5 R = C_6H_5

Scheme 5.41 Chemical drawing of a series of the cobalt(II) clathrochelates

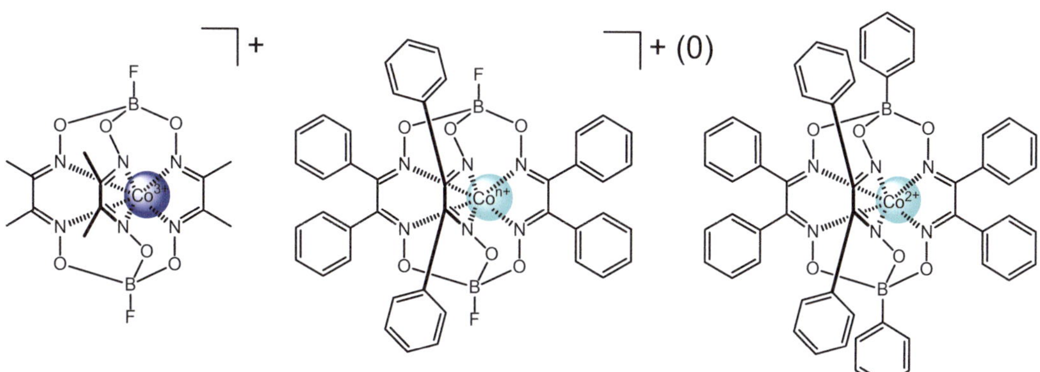

(Co^{3+})@**668**   (Co^{n+})@**678** n = 2, 3   (Co^{2+})@**708**

Scheme 5.42 Chemical drawings of the cobalt(II) and cobalt(III) cage complexes iron(II) α-oximehydrazonate bis-clathrochelates $(Fe^{2+})_2$@**507** and $(Fe^{2+})_2$@**508** have been tested [94] as potent electrocatalytically active compounds for hydrogen production from H^+ ions.

A series of the polyamine cobalt(III) cage complexes with tethered mercaptoethylamide-terminated oligopeptide apical substituents, allowing their immobilization of a surface of the working gold electrode (Scheme 5.43), have been prepared [95]. These clathrochelate-based systems are described to be potent heterogeneous electrocatalysts of HER and those for oxygen reduction to hydrogen peroxide.

The photodissociation of water using methanol as a sacrificial agent and rhodium-doped strontium titanate as a photomaterial as well as the surface modification of this material with the cobalt clathrochelates has been studied [96]. Such modification caused the charge of the band structure of perovskite and extended its light absorption range toward the visible range. Co-catalytic effect of metallic platinum nanoparticles and the immobilized cobalt clathrochelates in methanol oxidation has also been tested [96]. The presence of such cage complexes improves the kinetics of this reaction, but a loss of its performance with time has been observed. This result is explained [96] either by inhibition of methanol oxidation or by surface contamination with the products of photodegradation of these cobalt clathrochelates.

The use of the cobalt(III) clathrochelates $((Co^{3+})$@**648**$)_2(SO_4)_3$ and (Co^{3+})@**648**(Cl_3) as active and robust homogeneous oxidative catalysts is described [97]. These cage complexes catalyze an oxidation of styrene by H_2O_2 both in acetonitrile solution and in ionic liquid medium. Their superior catalytic activity in the first hours of the above reaction has been observed in acetonitrile medium; this result is explained [97] by solubility of its components in acetonitrile. In contrast, an induction period has been observed in an ionic liquid due to the immiscibility of H_2O_2 in this medium containing the catalyst and the substrate. A complete conversion of styrene proceeded in the presence of these cobalt clathrochelates after 24 h, and near complete conversion was also achieved using acetonitrile as a solvent. The selectivity of this oxidation in the above acetonitrile solution is described [97] to be higher than that in ionic liquid medium: in the former case, only benzaldehyde (80%) and a small amount of acetophenone (12%) have been formed in the presence of $((Co^{3+})$@**648**$)_2(SO_4)_3$. On the other hand, the use of ionic liquid as a solvent allowed recycling for consecutive cycles without substantial loss of activity [97].

1:1 cobalt(II) cage complex of the heteroditopic N_5O_3-macrobicyclic ligand **77** is reported [98] to be an efficient oxygenation catalyst toward a number of olefin and benzyl compounds as substrates, working at room temperature and using O_2 at atmospheric pressure or 2-methylpropanal as an oxidant. In this macrobicyclic molecule, the upper vacant part of its cavity can be occupied by a guest molecule, thus suggesting the above catalytic activity of (Co^{2+})@**77**. The olefins (including electrophilic)

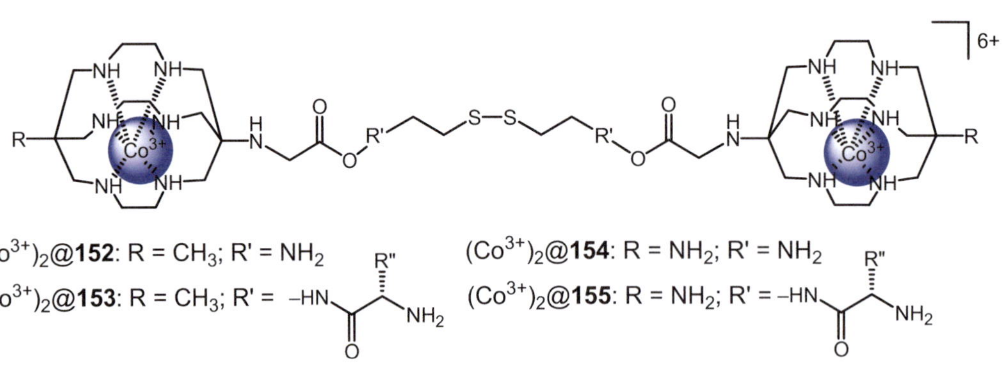

Scheme 5.43 Chemical drawing of a series of the cobalt(III) bis-clathrochelates

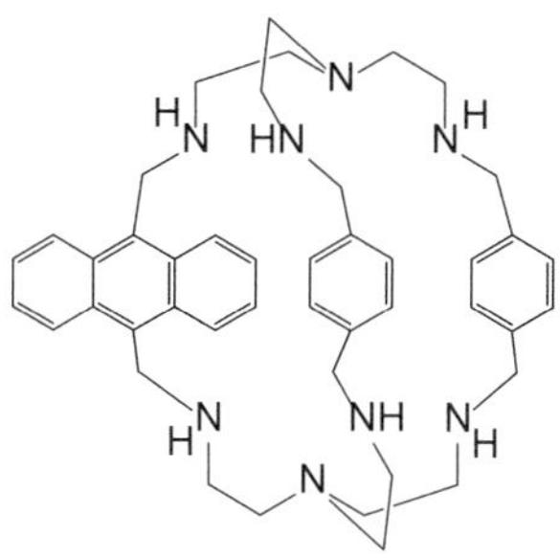

560

Scheme 5.44 Chemical drawing of the anthracene-containing ligand **560**

were found [98] to undergo their oxidation under such mild reaction conditions, thus forming epoxides, while the benzyl substrates have been converted in high yields to the corresponding ketones.

Photoinduced catalytic oxygen production in aqueous acetonitrile medium with 1:1 copper(II) cage complex of the fluorescent-active anthracene-containing polyamine macrobicyclic ligand **560** (Scheme 5.44) as a photocatalyst has been performed [99]. The proposed pathway of this homogeneous process is as follows:

$$\mathrm{An} \xrightarrow{h\nu^*} \mathrm{An}$$
$$Cu^{2+} + {}^*\mathrm{An} \rightarrow Cu^{+} + \mathrm{An}^{+}$$
$$\mathrm{An}^{+} + 1/2\mathrm{H_2O} \rightarrow \mathrm{An} + 1/4\mathrm{O_2} + \mathrm{H}^{+}$$
$$\mathrm{An} = \text{anthracene fragment of } \mathbf{560}$$

and includes several stages. An electron has been firstly exited from the ground state of the anthracene fragment of **560** to its exited state *An under the irradiation with light. The electron in the excited-state *An moiety has been then transferred to an encapsulated copper(II) ion, thus giving the copper(I) complex (Cu^+)@**560** and an An^+ cation as well. Finally, the latter An^+ cation abstracts one electron from water to form An, thus causing water oxidation to O_2 [99]. As a result, an overall photoinduced catalytic reaction is as follows:

$$\mathrm{Cu^{2+}} + 1/2\mathrm{H_2O} \xrightarrow{h\nu} \mathrm{Cu^{+}} + 1/4\mathrm{O_2} + \mathrm{H^{+}}$$

So, the photoinduced $Co^{2+/+}$ reduction by a fluorescent-active macrobicyclic ligand **560** with anthracene fragment has been observed [99]; on the next stage, the water molecules play a role of the sacrificial electron donor, undergoing an oxidation to O_2. Such photoinduced reduction of a copper(II) ion by water to produce a copper(I) cation and O_2 is described to be quite unique. Moreover, the caging ligand **560** with fluorescent-active anthracene fragment has been proposed [99] as potent water-splitting catalyst for hydrogen production.

Benzoyl peroxide-, lauryl peroxide-, and AIBN-initiated free-radical polymerization of MMA in bulk polymer and in its solution has been studied [100] in the presence of the ferrocenylboron-capped iron(II) tris-nioximate (Fe^{2+})@**199** (Scheme 5.45). This cage complex was found to form efficient initiating systems with the above peroxides, whereas in the case of AIBN-containing system, it did not affect the kinetic parameters of this polymerization process and the properties of its polymeric product as well. Such a mixed clathrochelate–peroxide initiator is reported to accelerate the polymerization of MMA within a wide temperature range and to decrease the average molecular mass of the polymer formed.

The ferrocenylboron-capped iron(II) oximehydrazonate semi- and clathrochelates *25* and (Fe^{2+})@**495** as well as their tris-dioximate analogs (Fe^{2+})@**199** and (Fe^{2+})@**200** shown in Scheme 5.45 have been studied [101] as the components of initiating systems for radical polymerization of MMA with benzoyl peroxide as a radical source. High efficiency and stability of these semi- and clathrochelate-based initiating systems, causing an increase in the rate of polymerization and a decrease in its apparent activation energy as well as in decomposition of benzoyl peroxide, are explained [101] by stabilization of the formed radical species *via* delocalization of their unpaired electron both on a highly conjugated quasiaromatic clathrochelate framework and on the ferrocenyl substituent(s) at this framework as well. Other radical initiators are also able to form the supramolecular assemblies with these semi- and clathrochelate complexes, thus decreasing energy of their O–O bonds; such CT assemblies have been spectroscopically

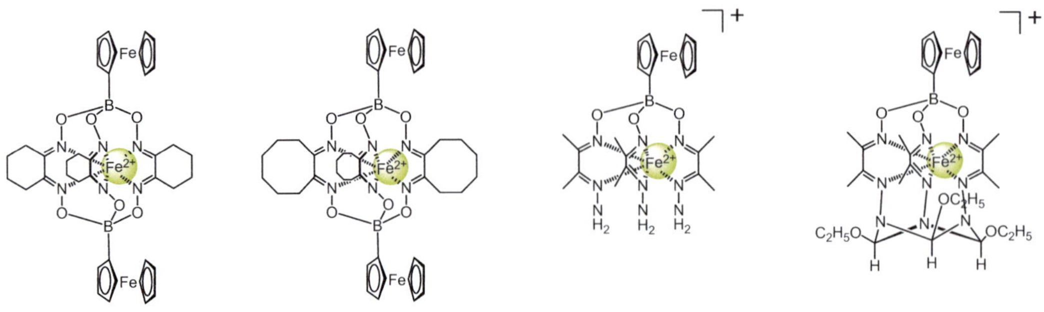

(Fe^{2+})@**199** (Fe^{2+})@**200** *25* (Fe^{2+})@**495**

Scheme 5.45 Chemical drawings of a series of the ferrocenylboron-capped iron(II) (semi)clathrochelates

detected [101]. The use of these complexes also allowed to control main characteristics of the polymeric products (i.e., their average molecular weights, polydispersity, microstructure, heat resistance, etc.). These semi- and clathrochelate complexes cause not only activation and stabilization of this radical process on its initiation stage, but also affect the chain propagation and termination stages, allowing performing it with high rate at relatively low, close to room, temperatures without a substantial decomposition of the radical initiation to give the polymeric products with improved molecular-weight characteristics [101].

The occurrence of radical polymerization of MMA, initiated by various radical initiators in the presence of the ferrocenylboron-capped iron(II) clathrochelates, has been studied [102]. The iron(II) cage complexes shown in Scheme 5.46 have been also tested [103, 104] as the components of the initiating systems for complex-radical polymerization of MMA. These clathrochelates were found to form efficient initiating systems with benzoyl peroxide as a radical initiator, thus allowing to perform MMA polymerization at a high rate with decreased decomposition of this initiator and improved molecular-mass characteristics of the polymeric products. When AIBN has been used as an initiator of MMA polymerization, the same clathrochelates affected mainly the chain propagation stage of this radical process [103, 104].

A CuBr-catalyzed polymerization of styrene in the presence of the N_3S_3-macrobicyclic ligand **101** (Scheme 5.47), effectively encapsulating a copper(I) ion, has been studied [105]. Such caging of this redox-active metal ion activates the radical intermediates through an outer-sphere ET process.

MMA polymerization in the presence of the same copper(I) clathrochelate has been studied [106]. This cage complex initiates the above polymerization process at room temperature in the presence of RAFT agent, allowing to control the molecular weight distribution. It is described [106] to show ideal "living" radical behavior, in which the molecular weight polydispersity for a range of different molecular weight targets is less than 1.1.

Various types of the boron-, aluminum-, and titanium-capped transition metal tris-dioximates and their analogs have been patented [107] as olefin polymerization catalyst components allowing the fine-tuning of catalyst's reactivity and improving of the characteristics of the polyolefin products as well.

The palladium(II) complexes PdCl·**353** and PdCl·**351**, the derivatives of the *N,S*-donor clathrochelate ligands, have been studied [108] as catalysts of the model Suzuki cross-coupling reaction of iodobenzene with phenylboronic acid by Scheme 5.48 in DMF media using K_3PO_4 as inorganic base. In both these cases, very high conversion of iodobenzene to

Scheme 5.46 Chemical drawings of a series of the olefinboron-capped iron(II) clathrochelates

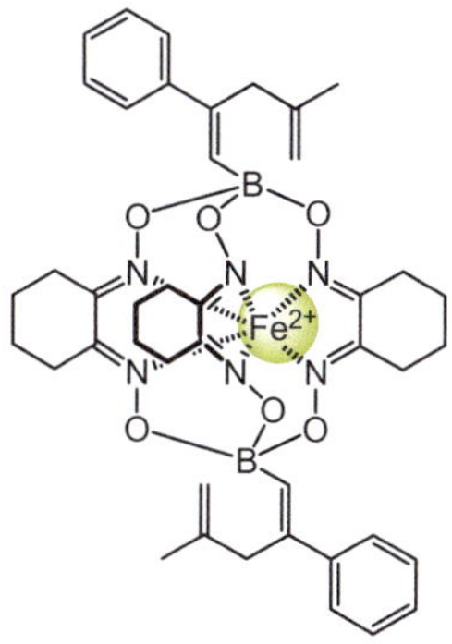

(Fe^{2+})@**194**

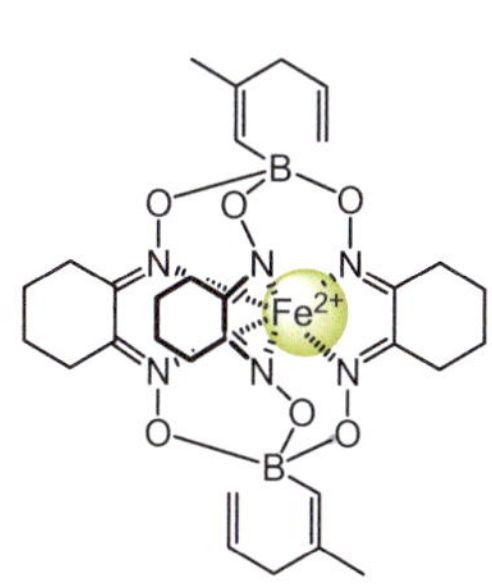

(Fe^{2+})@**195**

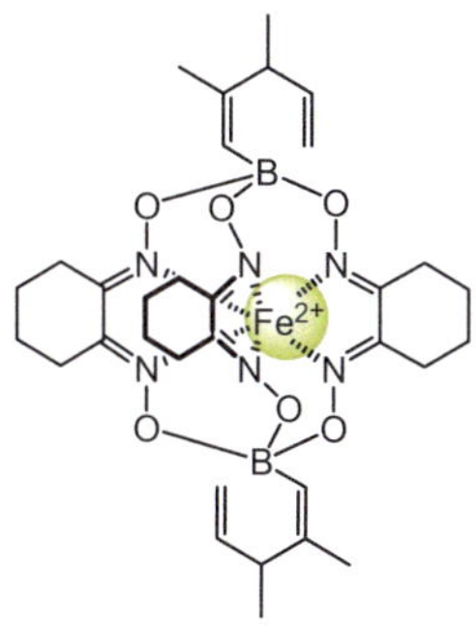

(Fe^{2+})@**197**

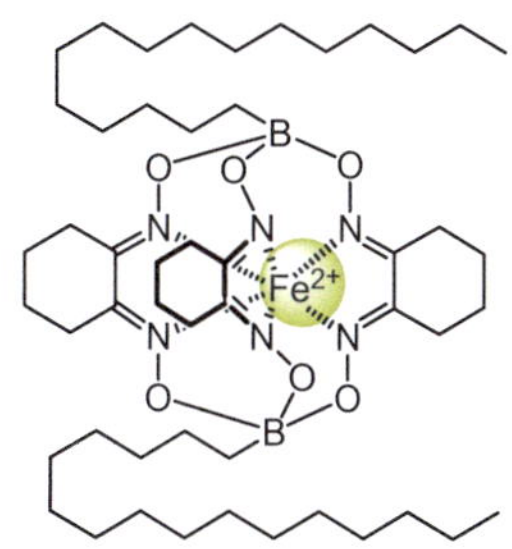

(Fe^{2+})@**237**

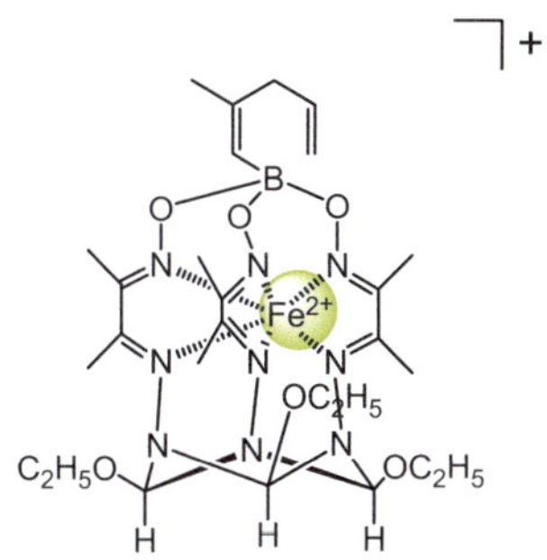

(Fe^{2+})@**500**

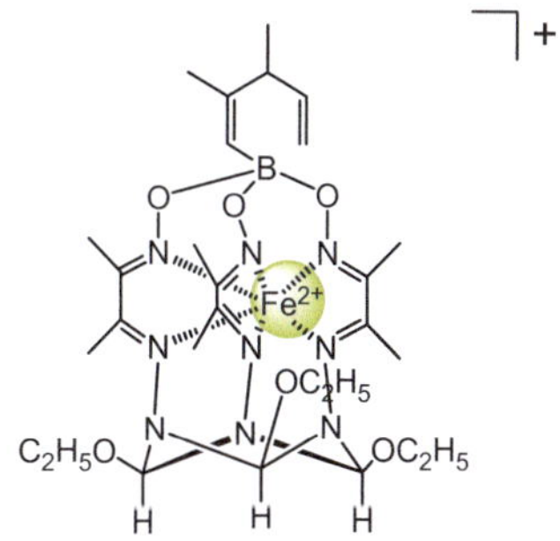

(Fe^{2+})@**501**

biphenyl has been observed. To further evaluate their catalytic activity, a less reactive bromobenzene has been used as a substrate. The complex PdCl·**353** with ethylene spacer fragment, forming a six-membered PdN_2C_3 chelate cycle, was found [108] to cause a better conversion in this system as compared with its methylene-containing analog PdCl·**351**, having the five-membered PdN_2C_2 cyclic moiety. Both these Pd,Fe-binuclear 1:1 complexes of the above clathrochelate ligands have been proposed [108] as prospective efficient and structure-dependent catalysts of various cross-coupling reactions.

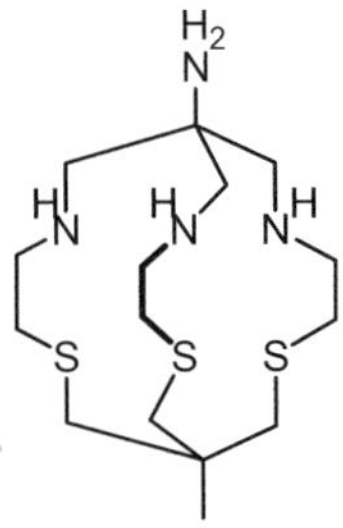

101

Scheme 5.47 Chemical drawing of the N_3S_3-macrobicyclic ligand **101**

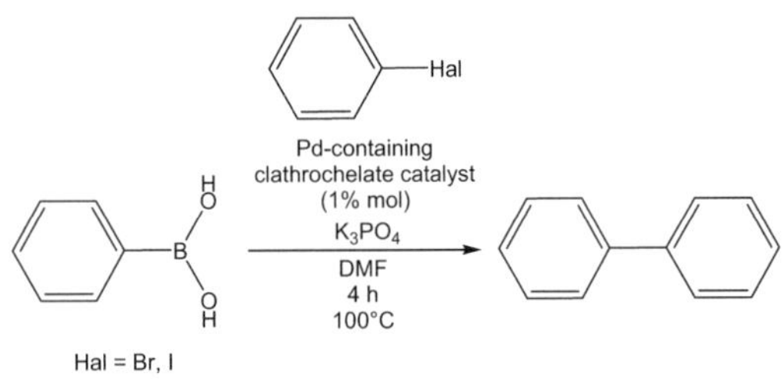

Scheme 5.48 Model Suzuki reaction

5.3 Molecular and Supramolecular (Photo)Electronic and Mechanic Devices, Single-Molecule Magnets, Mediators of Electron Transfer, Sensors and Probes, Macropolycyclic Building Blocks, and Clathrochelate-Modified and Clathrochelate-Based Materials

The polyamine cobalt cage complexes have been widely used as artificial inert outer-sphere electron donors and mediators for various biologically relevant redox processes, such as fermentative production of propionic acid [109], biocatalytic reduction of NAD to NADH [110], redox reactions of the single iron protein rubredoxin [111], electromicrobial reduction of chloropyruvate [112], ω-hydroxylation of *para*-nitrophenoxydodecanoic acid with zinc dust – (Co^{3+})@**648** system using cytochrome P450 enzyme as a catalyst [113] and for improving these monooxygenases using the above system as an alternative and cost-effective cofactor [114] as well as for protein redox potentiometry [115], in construction of microsomal nanobiosensor for the determination of 2,4-dichlorophenol as an endocrine disruptor compound [116], for electrochemically driven biocatalysis with cytochrome monooxygenase P450cin in the presence of protein cindoxin [117], bioelectrochemical conversion of *para*-xylene to 2,5-dimethylphenol [118], and the design of a microtiter plate for a fast development of electroenzymatic processes with cytochrome P450s [119].

Electrocatalytically driven ω-hydroxylation of fatty acids using cytochrome P450 4A1 with electrochemically generated cobalt(II) clathrochelate (Co^{2+})@**648** as an electron mediator, replacing of NADPH, has been performed [120]; the rates of formation of 12-hydroxydodecanoic acid by electrochemical method are comparable to those obtained using NADPH as an electron donor. The results of [120] suggest the prospective practical applications of such electrocatalytically dependent bioreactors containing different P450s as catalysts and the above cobalt cage complex as a mediator for the large-scale synthesis of stereo- and regioselectively hydroxylated chemical products. The electrochemically reduced mediator, (Co^{2+})@**648**, is reported [121] requiring the presence of a flavoprotein for the rapid transfer of electrons to such enzyme. The use of P450 monooxygenases for preparative chemical syntheses in cell-free systems is hampered by their requirement for cofactors, such as NAD(P)H [122]. In order to bioengineer P450 BM3 for cost-effective process conditions in vitro, a validated medium-throughput screening system based on cheap zinc dust as an electron source and (Co^{3+})@**648** as a mediator is described [122] as an alternative cofactor system. The obtained total turnover numbers for this system, however, were found [122] to be one to two orders of magnitude lower that those requiring for industrial applications.

Direct ET from the modified glassy carbon electrode with covalently immobilized cobalt diaminoclathrochelate (Co^{3+})@**99** to a dissolved viologen-accepting pyridine nucleotide oxidoreductase and dihydrolipoamide dehydrogenase has been tested [123] using the formation of NADH *via* electrochemically induced reaction, proceeding in a potentiostatically controlled electrochemical cell:

$$2(Co^{2+})@\mathbf{99} + NAD^{+} + \mathrm{H}^{+} \rightleftharpoons \mathrm{NADH} + 2(Co^{3+})@\mathbf{99}$$

NADH formed has been used to reduce pyruvate as a substrate to *L*-lactate; this reaction has been catalyzed by *L*-lactate dehydrogenase [123].

Reactivity of the terminal hydroxyl groups of the apically functionalized polyamine clathrochelates (Co^{3+})@**103** and (Co^{3+})@**104** is described [124] to allow their surface immobilization and the protein grafting with these cage complexes as well.

The cobalt(III) clathrochelate trication (Co^{3+})@**648** has been used [125] as a mediator of ET in photodecarboxylation reaction of citrate anion through the formation of 1:1, 1:2, and 1:3 ionic pairs. Irradiation at maxima of the characteristic CTB of this cage complex at different pH caused the decarboxylation of the above anion, thus giving CO_2, and the formation of the cobalt(II) macrobicyclic complex (Co^{2+})@**648** as well. In the presence of colloidal platinum, the evolution of both CO_2 and H_2 has been observed [125].

The potential of the clathrochelate redox couple $(Co^{2+/3+})$@**648** has been estimated [126] using the temperature plots of its CVs in aqueous and various nonaqueous solutions. The transfer free energy of each redox couple from water to nonaqueous solvents has been calculated from the solvent dependence of this potential at 25 °C, while the values of the reaction entropy of the above clathrochelate-based redox couple in each solvent have been obtained from the temperature dependence of its potential using a non-isothermal cell arrangement. This redox potential is reported [127] to be affected by immobilization of the above cobalt cage complex on silica support and chemically modified either with zirconium(IV) oxide or with zirconium(IV) phosphate. The electronic density on its encapsulated cobalt ion was found to be affected by both the contact ion-pairing and hydrogen bonding between an anion of the supporting electrolyte and this macrobicyclic cation. As a result, the shifts in its redox potentials are described [127] to affect both the nature and the concentration of this anion.

Heterogeneous kinetics of interfacial ET at single-layer graphene electrodes in an aqueous solution of the cobalt clathrochelate $(Co^{2+/3+})$@**648** as a redox mediator exhibiting a quasireversible ET behavior has been studied [128]; SECM method has been used to probe the apparent standard ET rate constant for this process. The absence of negative-like feedback on the SECM curves at all graphene overpotentials and the positive-like approach curves at open circuit allowed the authors of [128] to deduce a fast kinetics of the above ET process. The same clathrochelate redox mediator, undergoing a one-electron ET process and, thus, being a convenient redox mediator, is described [129] to be a suitable model system for testing of the spectroelectrochemical cell, designed [128] for in situ XANES measurements.

The rate constants of *ortho*- and *para*-positronium spin conversion reactions caused by high-spin cobalt(II) polyamine clathrochelates (Co^{2+})@**98**, (Co^{2+})@**99**, and (Co^{2+})@**648** as well as their temperature plots have been measured [130].

The surface-enhanced Raman spectra in the range of the vibrations of a clathrochelate framework have been measured [131] for the $(Co^{2+/3+})$@**648** redox pair adsorbed on silver and copper electrodes. The nature of a supporting electrolyte affected the adsorption of the clathrochelate trication (Co^{3+})@**648** on a silver electrode, thus suggesting that the above process proceeds *via* its ion-pairing with the anions, specifically adsorbed on the surface of this electrode. The Raman intensity of the symmetrical stretching vibration of the methylene groups of **648** has been used [131] to estimate the pK_a value for its cobalt(III) cage complex.

Correlations between the outer-sphere self-exchange rates and the corresponding ET reaction entropies have been evaluated [132] using the same encapsulated redox pair as an appropriate model of purely outer-sphere redox process. Such character of ET in the case of the $(Co^{2+/3+})$@**648** redox couple has been used [133] for an outer-sphere reduction of the hydrogen-bonded cobalt(III) bis-dimethylglyoximate and its BF_2-cross-linked macrocyclic derivative with the cobalt(II) cage complex (Co^{2+})@**648** as a reductant. The obtained reaction rates allowed the authors of [133] to estimate the cobaloxime(III)/(II) electron self-exchange rates using the Marcus cross relationship.

Applicability of a cross relation in volumes of activation for ET reactions, such as the reduction of aqueous trication $[Fe(H_2O)_6]^{3+}$ by (Co^{2+})@**648** as a reductant, has been tested [134]. The obtained data suggested the usefulness of the volume cross relation as a mechanistic criterion, at least for the reactions with moderate ΔE.

Nanotubes that selectively conduct the ions across the membranes, thus giving the ionic devices with transport characteristics similar to those of biological ionic channels and semiconductor electronic devices, are described [135]. The transport properties of uniformly charged conical nanopores and the cation versus anion selectivity have been changed using various polyvalent cations, such as the cobalt clathrochelate trication (Co^{3+})@**648**. It was found [135] to produce localized charge inversion allowing to reversibly change the effective charge profile of the pore surfaces from negative to positive. The same cobalt(III) cage complex has been used in [136] for optical inspection and electrochemical characterization of Hg/Pt hemispherical ultramicroelectrodes.

Photocurrent enhancement by multilayered porphyrin sensitizers in a photoelectrochemical cell in the presence of various ET mediators has been performed [137]. The photoelectrochemical responses of multilayer films of zinc(II) 5,10,15,20-tetra(4-ethynylphenyl)porphyrin assembled on planar ITO substrates operating as a *p*-type dye-sensitized solar cells using different types of the redox mediators, such as the clathrochelate $(Co^{2+/3+})$@**648** redox couple, are reported [137]. The use of such clathrochelate redox mediator was found to generate relatively high open circuit potential values but at the expense of relatively low photocurrents; the clathrochelate-containing cells reached a maximal power output at three layers of the above sensitizer.

UV-vis measurements of (Co^{3+})@**648**–I^- supramolecular system in its aqueous solution in the range of the ion-pair CTB have been performed [138] to deduce the effect of such self-assembly on the photochemical behavior of the above system. The obtained reliable values of its association constant and its molar absorptivity have been used [138] to evaluate the correct quantum yields for the corresponding photochemical redox reaction.

A "molecular conduit" effect of the adsorbed cobalt(III) clathrochelate (Co^{3+})@**648** is reported [139] to substantially enhance the photocurrents at WO_3–ITO nanoelectrodes: this cage complex was found to speed up the interparticle electron conduction, thus allowing hopping of electrons from particle to particle without "draining." The same cage complex has been tested [112] as an electron mediator for electromicrobial reduction of chloropyruvate to chlorolactate; its mediator properties have been compared with those of the viologens. The following advantages of this clathrochelate mediator are described [112]: (1) its reactivity against of O_2 is seven orders of magnitude lower than that of the viologens, (2) it is also less poisonous, and (3) its redox potential can be tuned using both the pH value and the type of buffer, whereas those of the viologens have not been affected by these factors. On the other hand, the hydroxycarboxylate viologen oxidoreductase activity of (Co^{3+})@**648** was found [112] to be lower in most of the common buffers than that of 1,1′-carbamoylmethyl viologen in phosphate buffer.

The reduction of horse heart myoglobin, horseradish peroxidase, and horse heart cytochrome by cobalt(II) clathrochelate (Co^{2+})@**648** has been studied [140]. Similar patterns of kinetic behavior have been observed in all these cases [i.e., the direct reduction of cyanide and imidazole adducts of the iron(III) proteins and the indirect (*via* dissociation) reduction of the fluoride adduct]. In the case of a reduction of horseradish ferriperoxidase by this cobalt(II) cage complex, its three steps have been observed and the spectral properties of the intermediate(s) as well as their kinetic behavior are described [140]. The spectral properties of the final product, ferroperoxidase, and its behavior on an oxygenation have also been studied. Reductions of cytochrome $c^{(III)}$ and that of Hipip by (Co^{2+})@**648** are reported [140] to be the uniphasic reactions, and the corresponding second-order rate constants have been also determined.

The reduction of manganese(III) porphyrins shown in Scheme 5.49 by the same clathrochelate reductant has been studied [141]. The reduction of (Mn^{3+})(*92*)$(H_2O)_2$ with (Co^{2+})@**648** is strongly affected by ionic strength of the medium as well as catalyzed by chloride ion; the kinetic data for this reaction suggested an attack of the above macrobicyclic reductant at the periphery of the porphyrin ring of (Mn^{3+})(*92*)$(H_2O)_2$. With (Mn^{3+})(*93*)$(H_2O)_2$ as a substrate, such outer-sphere reduction is reported [141] to be biphasic, thus suggesting two parallel reaction pathways: (1) an attack of (Co^{2+})@**648** at the periphery of its macrocyclic ring and (2) that of its axially coordinated water molecules. The cuboidal mixed-valence cluster $[Mo_4S_4(H_2O)_{12}]^{5+}$, each of the four equivalent molybdenum ions of which was assumed a mean oxidation state of 3.25, has been reduced [142] to its isostructural molybdenum(III)-containing tetramer $[Mo_4S_4(H_2O)_{12}]^{4+}$ using an equimolar quantity of the above cobalt(II) cage complex as an outer-sphere reductant; the corresponding redox reaction rate is independent on the activity of H^+ ions in the range of concentrations of mineral acids from 0.5 to 2.0 M.

Δ-vs.-Λ stereoselectivity of an outer-sphere oxidation of the cobalt(II) clathrochelate (Co^{2+})@**648** by $[Co(edta)]^-$ has been examined [143] under various reaction conditions. The observed significant stereoselectivity of approximately 18% correlates with ion-pairing stereoselectivity for the above cobalt macrobicyclic dication, thus suggesting an important role for its chemical structure. The observed *Δ-vs.-Λ* stereoselectivity of this redox reaction is explained [143] in terms of a balance between *ΔΔ* and *ΔΛ* interactions along the molecular C_3 and C_2 axes of the above clathrochelate reductant, respectively, which are affected by hydrogen bonding and steric effects.

Critical micelle concentration of the methoxy polyethylene glycol-based macromonomer and the corresponding partition coefficient have

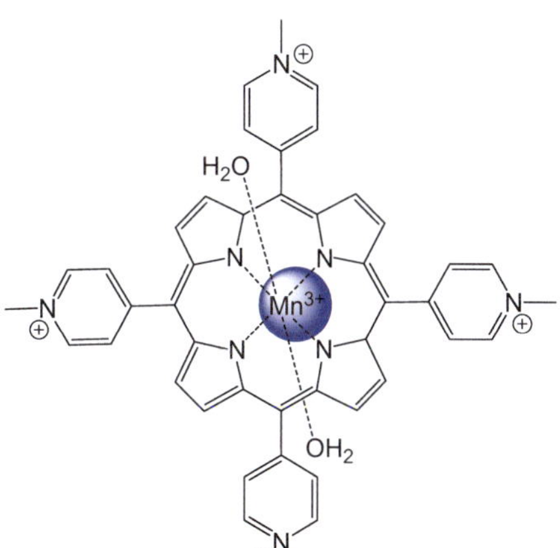

(Mn^{3+})(*92*)

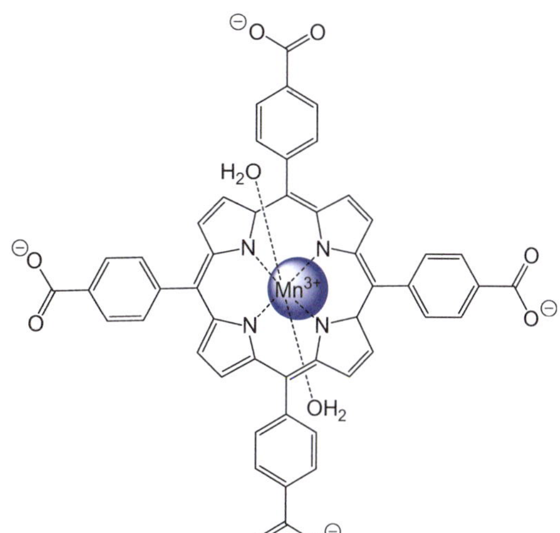

(Mn^{3+})(*93*)

Scheme 5.49 Chemical drawings of manganese(III) porphyrins, the derivatives of *92* and *93*

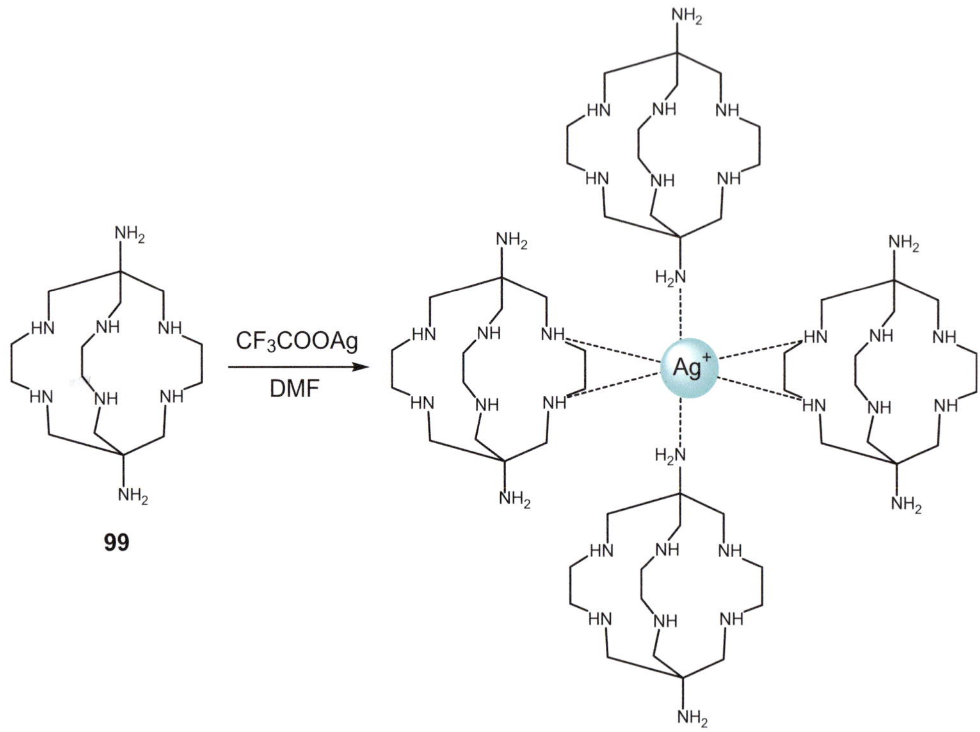

Scheme 5.50 Silver nanoparticles modified with polyamine macrobicycle **99**

been determined [144] by CV method using the cobalt(III) clathrochelate (Co^{3+})@**648** as an electrochemical probe.

Polyamine macrobicycle-modified silver nanoparticles have been prepared [145] by Scheme 5.50 using a reduction of silver trifluoroacetate with the diaminoclathrochelate **99** as a reductant. The resulting nanoparticles displayed the characteristic peak of **99** in their FT-IR spectra, thus evidencing its immobilization on a surface of these nanomaterials.

Cage metal complexes bearing ribbed spacer thiol-terminated substituent(s) are suitable for their immobilization on a surface of the gold electrode; the examples of such immobilization are shown in Schemes 5.51 and 5.52. The immobilization by Scheme 5.53 is reported [146] to be prospective for the design of molecular electronic devices, electrochemical sensors, and new types of clathrochelate-modified reference electrodes using one-electron encapsulated metal-centered redox process and supramolecular architecture of the clathrochelate molecules in SAMs on an electrode surface; in particular, the gold-immobilized monofunctionalized iron(II) cage complex (Fe^{2+})@(**378**–H^+) has been used [84] as a redox mediator for efficient amperometric determination of trace amounts of hydrogen peroxide (including those under physiological conditions). The calibration curve for hydrogen peroxide concentrations was found [146] to be linear in the range 2.0×10^{-6} -1.6×10^{-5} mol cm^{-3} with a limit of its detection equal to 1.0×10^{-6} mol cm^{-3}. So, this clathrochelate SAM is described [146] to provide precise and sensitive amperometric determination of hydrogen peroxide in diluted aqueous solutions and in physiological media as well.

Single molecular switch based on the thiol-terminated iron(II) clathrochelate (Fe^{2+})@**378** is described [147]. A mixed SAM of DDT and this cage complex shown in Scheme 5.54 has been assembled on a gold film; the electrochemical properties and electronic structure of the clathrochelate molecules inserted into DDT monolayer have been studied using voltammetric,

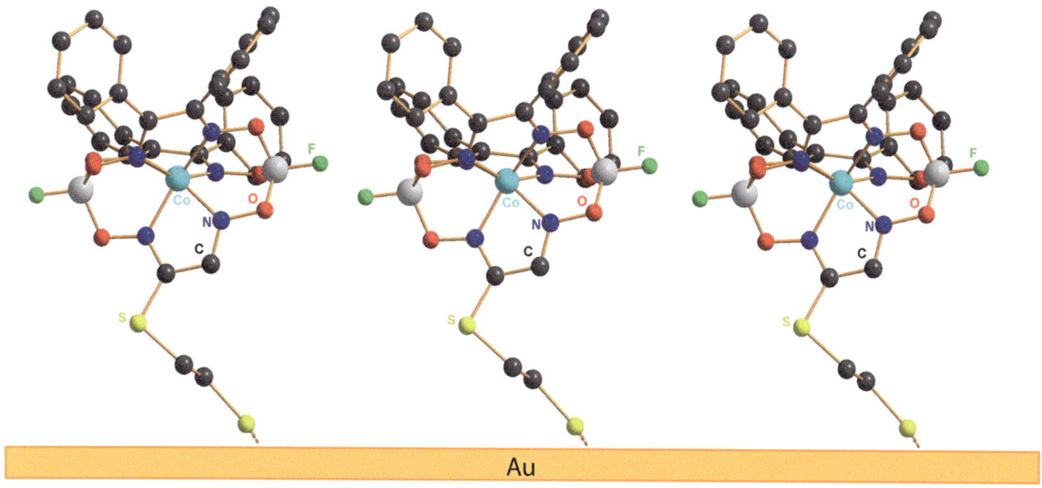

Scheme 5.51 Gold surface modified with thiol-terminated cobalt(II) clathrochelate

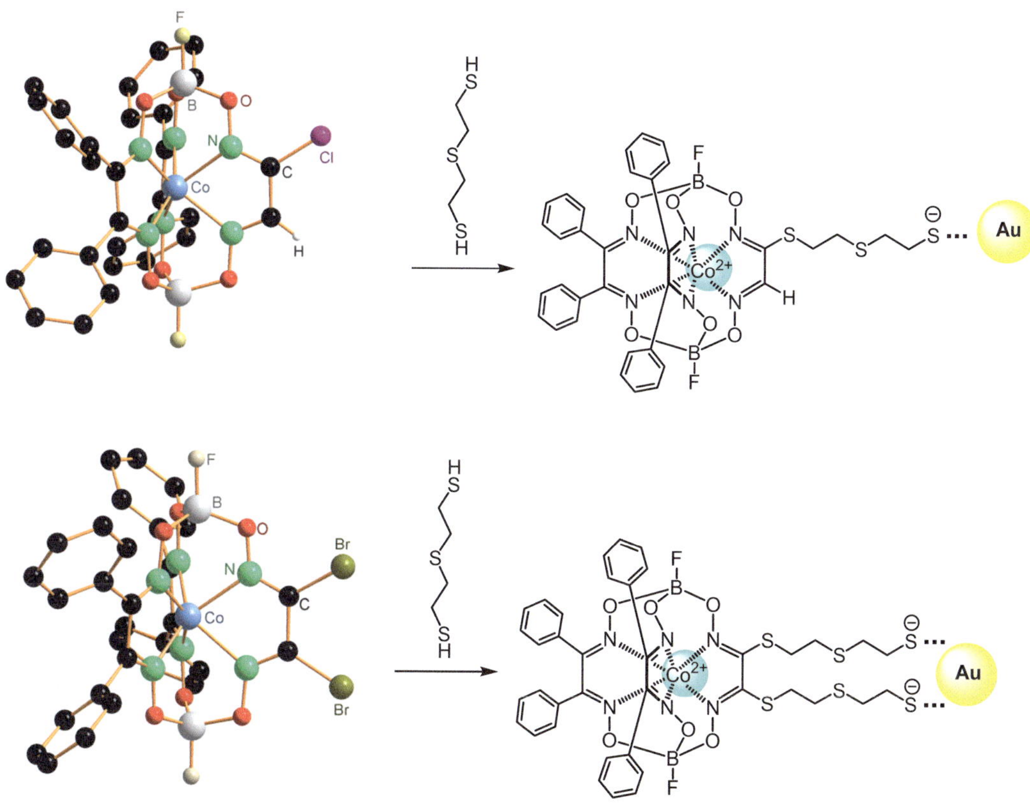

Scheme 5.52 Immobilization of the thiol-terminated cobalt(II) clathrochelates on a gold surface

EIS, STM, and cross-wire tunneling techniques. The voltage-triggered switching behavior of such cage molecules can be illustrated by Figs. 5.8 and 5.9. Deposition of polyaniline on these redox sites on a gold surface using electrochemical polymerization of aniline has been performed [147] in order to confirm that this mixed monolayer can act as a nano-patterned semiconducting

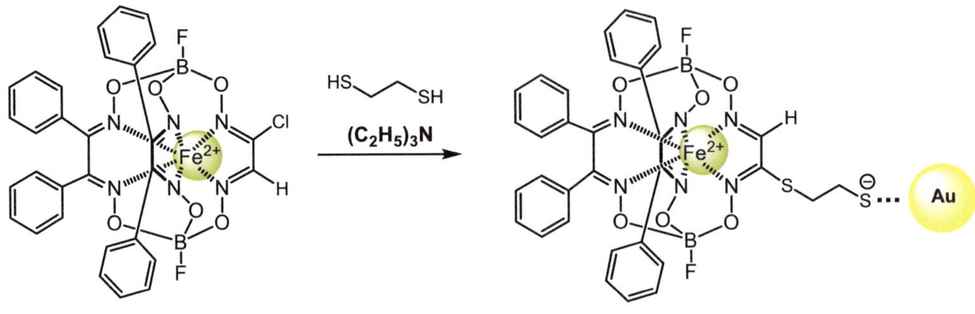

Scheme 5.53 Immobilization of the thiol-terminated iron(II) clathrochelate on a gold surface

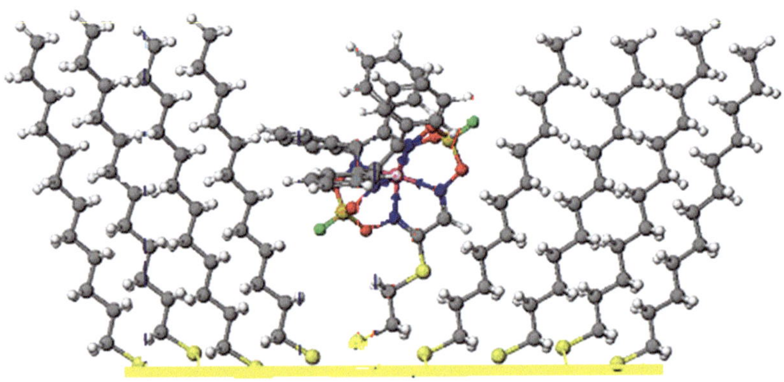

Scheme 5.54 Formation of a mixed dodecanethiol—iron(II) clathrochelate (Fe^{2+})@**378** monolayer [147]

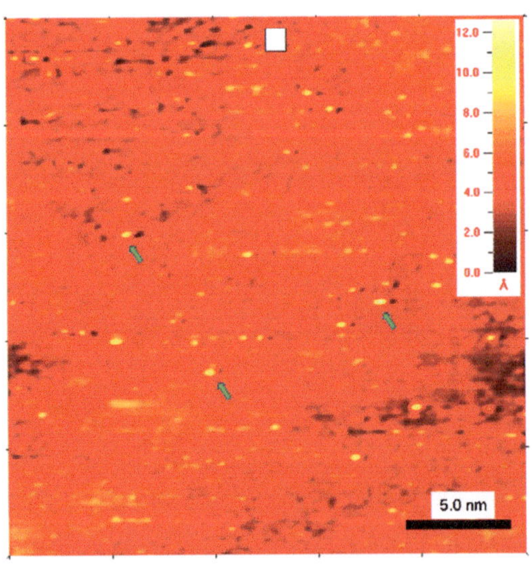

Fig. 5.8 STM height image of the molecules (Fe^{2+})@**378** (bright points) in high conductance state [147]

electrode surface. So, these mixed clathrochelate-containing SAMs have been proposed [147] as suitable transducer materials for nano-biosensors and molecular electronics.

A series of the clathrochelate metal(II) tris-nioximates (M^{2+})@**206** with redox-active apical dithiophene substituents is reported [148] to undergo electropolymerization by Scheme 5.55, thus giving the corresponding metal-containing conducting polymers. These materials were found to display metal-dependent electrochemical properties as evidenced [148] by different redox windows in the case of the corresponding cobalt-, iron-, and ruthenium-containing metallopolymers. Consecutive electropolymerization of two clathrochelate monomers, the derivatives of different metal ions, has been used [148] for preparation of the bilayer materials having different metallocenters in each of their layers.

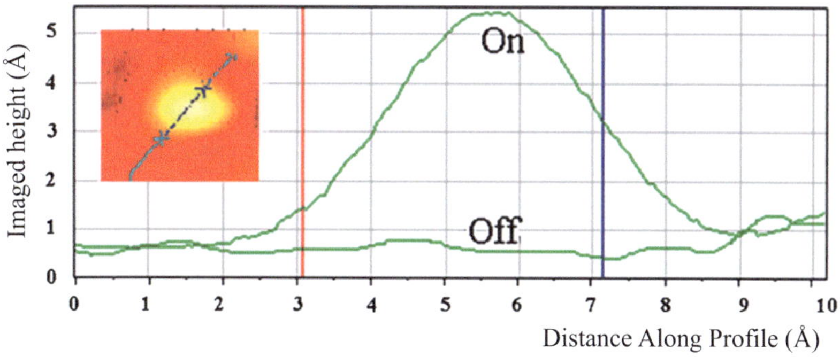

Fig. 5.9 Apparent heights of the molecule (Fe^{2+})@**378** at its "ON" and "OFF" states [147]

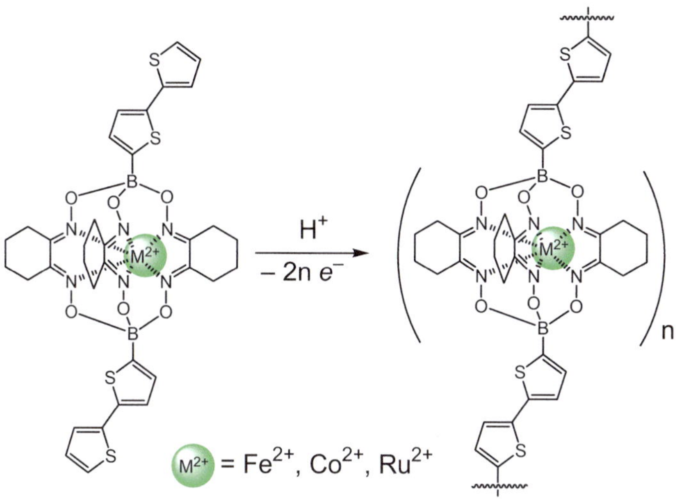

Scheme 5.55 Electropolymerization of the clathrochelates (M^{2+})@**206**

Charge injection and transport in the iron and cobalt clathrochelate-based conducting metallopolymers have been studied [149]. Electropolymerization of the tris-nioximate and tris-α-benzildioximate cage complexes (M^{2+})@**206** and (M^{2+})@**207** shown in Scheme 5.56 allowed to deposit such conducting polymers directly onto an electrode surface as uniform films. As follows from electrochemical, spectral, and SEM data of [149], proceeding of an injection—removal of electrons for them through the different pathways with the different rates has been deduced [149]. An interplay between the electroactive metallocenters and π-conjugated organic polymers in these materials allowed to obtain the redox properties of these electropolymers that cannot be realized by their individual components. The design of such conductive hybrid materials is based on the metallocenters and the organic components having the similar oxidation potentials that facilitate a charge hopping. In particular, a rapid injection of holes to the cobalt clathrochelate-based electropolymer has been observed [149]. Upon injection of electrons, however, its organic component was found to turn redox silent, which is a sign of a metal-localized electrochemical oxidation and reduction, giving the cobalt(III) and cobalt(I) cage complexes, respectively. This allowed the authors of [149] to elucidate the different pathways of charge transport within this polymer. In particular, a reversible one-electron

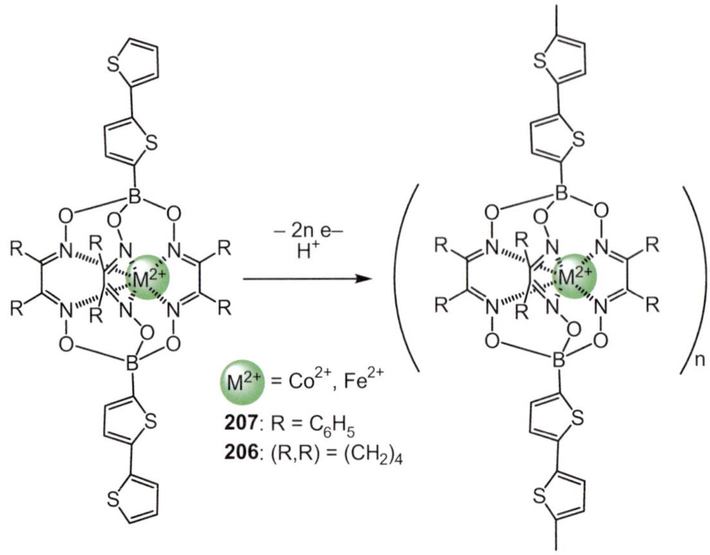

Scheme 5.56 Electropolymerization of the clathrochelates (M^{2+})@**206** and (M^{2+})@**207**

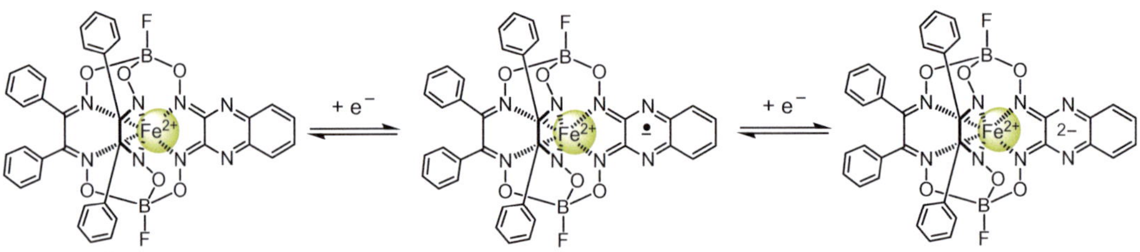

(Fe^{2+})@**492**

Scheme 5.57 Redox transformation of the clathrochelate (Fe^{2+})@**492**

cobalt-localized reduction has been used to probe the structural and functional integrity of the corresponding metallopolymer and its charge transport properties. For this purpose, a reduced cobalt(I) cage complex has been prepared (see Sect. 5.4) and characterized as a structural model for the repeating unit of the above *n*-doped metallopolymer. The structurally well-defined nature of this polymer and its capability to deliver electrons to the metallocenters allowed to propose the above clathrochelate-based materials as potent surface-immobilized electrocatalyts [149].

The iron(II) cage complex (Fe^{2+})@**492** with annelated quinoxaline fragment is described [150] to undergo the chemical and electrochemical redox transformations shown in Scheme 5.57 both in solution and in thin film; it is reported to be prospective macrobicyclic precursor of the clathrochelate-based electrochromic materials.

Fluorescence signaling system on the simultaneous presence of transition and alkali metal ions, based on the polytopic ligand **61**, has been evaluated [151] as potent *AND* logic gate, the molecular action of which can be induced by one of the ionic inputs, such as a transition metal ion.

Exocyclic coordination and *in*-to-*out* translocation (including that of anion driven) of the encapsulated metal ions in their cage complexes with laterally non-symmetric N_5O_3-macrobicyclic ligands **50**, **51**, and **58** have been used [152] for the design of (photo)electronic and chemical logic devices, for modulation of quadratic optical nonlinearity and fluorescence signaling by

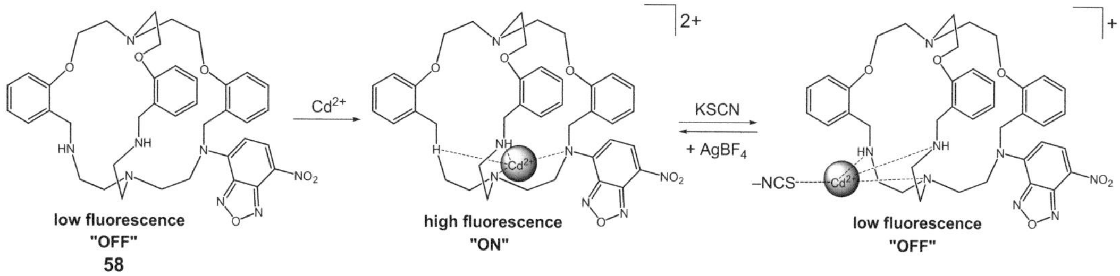

Scheme 5.58 Cadmium(II)-switched fluorescence of the macrobicyclic ligand **58**

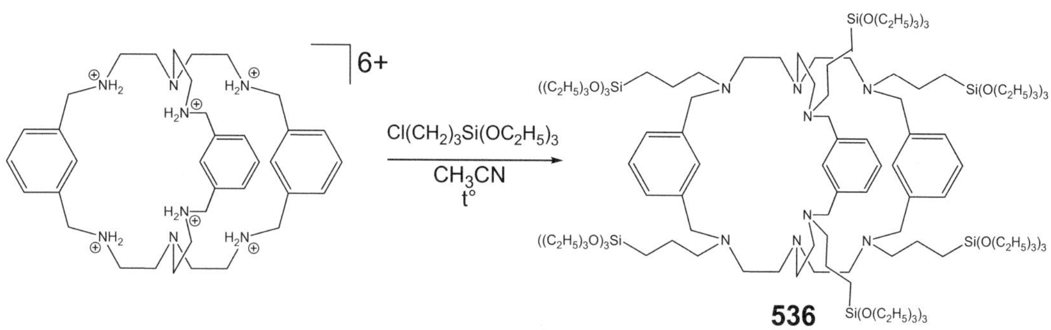

Scheme 5.59 Synthesis of the silicated macrobicycle **536**

Scheme 5.58, as well as for the controlled mechanical movements of the side groups resulted in signaling by both the monomer enhancement and the exciplex emission. Depending on nature of the substituents at a cage framework, these clathrochelate-based systems allowed ON/OFF fluorescence modulation of optical quadratic nonlinearity. The above translocation can be engineered by variation of the donor properties of the counteranions, thus offering possibilities of both the cationic and the anionic sensing, while binding of a metal ion from outside of its cavity can also trigger the controlled and reversible movements of one of the pendant substituents at their cage frameworks [152]. Redox-driven translocation of an encapsulated copper ion within the cavities of the N_5S_3-macrobicyclic ligands **62** and **63** has been used [153] to induce a reversible ON–OFF fluorescence in the fluorescence signaling systems.

The laterally non-symmetric macrobicycles **60** and **71** have been functionalized [154] with electron-withdrawing fluorophoric 7-nitrobenz-2-oxa-1,3-diazole group(s), thus giving the corresponding mono-, bis-, and tris-substituted encapsulating ligands. They exhibited a high fluorescence enhancement in the presence of heavy metal ions, such as a mercury(II) cation. In the case of the caging ligand **58**, this ion caused the largest enhancement of fluorescence together with its highest binding selectivity. The observed high signal output resulted from a very effective PET process from the donor nitrogen atoms of **58** to the fluorophore substituent(s) at its framework with very low metal-fluorophoric interactions. So, such macrobicyclic ligands with electron-withdrawing fluorophores are described [154] to cause a higher enhancement of fluorescence after encapsulation of Hg^{2+} ion as compared with their non-functionalized macrobicyclic precursors.

The binding properties and optical sensing capability toward the copper(I) ions of the sol–gel materials with covalently bonded silicated caging ligand **536** (Scheme 5.59) have been studied [155]. This immobilized ligand was found to co-encapsulate two copper(II) ions, thus forming its homodinuclear cage complex. A large number of the encapsulation–extraction–re-encapsulation cycles involving the formation of such immobilized copper(I) clathrochelate and their decomposition with an excess of acid have been performed [155], thus suggesting a potent practical application of these functionalized sol–gel materials.

1:1 europium(III) cage complexes of the tris-bipyridine macrobicyclic ligands **709** and **624** have been incorporated [156] into the thin films of sol–gel zirconia oxide and those of the organically modified sol–gel hybrid zirconia–silica–urethane material. The fluorescence outputs of these clathrochelates (Eu^{3+})@**709** and (Eu^{3+})@**624** were found to be increased by the factors of 64 and of 2044, respectively, as compared with europium(II) oxide. This result is explained [156] by CT in these hybrid materials.

Photophysical properties of the heterobinuclear 1:1:1 lanthanide(III), zinc(II) cage complexes of the encapsulating ligand **554** have been studied [157]. A ligand-to-europium(III) ET was found to be efficient at low temperature, but a back transfer was implied in its deactivation process, especially at room temperature, because a triplet state of this caging ligand lies at a very low energy. Although an overall sensitization of luminescence of an encapsulated metal ion remained small, the corresponding mononuclear europium(III) clathrochelate displayed a much weaker emission because the oxygen atoms of the phenolic groups of its caging ligand are protonated [157]. The encapsulated terbium(III) ion is weakly sensitized through the excited singlet state of this ligand. On the other hand, the low energy of its triplet state allowed an appreciable conversion of the visible light absorbed into IR light emitted by the caged neodymium and ytterbium(III) ions; this is prospective for the design of the clathrochelate-based light-converting devices [157].

The mono- and dinuclear phenolic lanthanide(III) cage complexes have been proposed [158] for the design of light-converting devices, luminescent stains, and the contrast agents as well, while the predeuterated lanthanide(III) tris-bipyridinate macrobicyclic complexes shown in

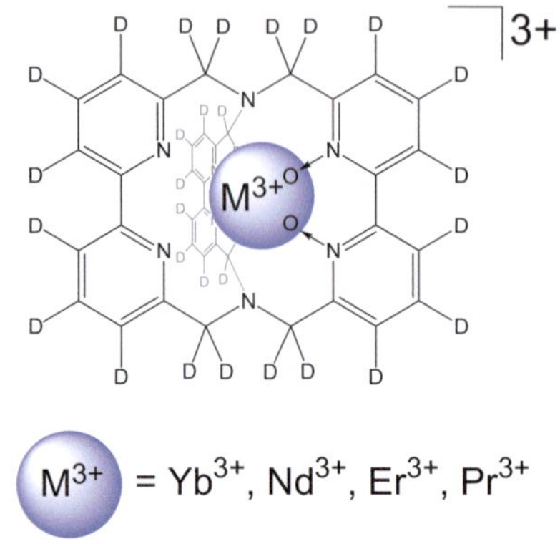

Scheme 5.60 Chemical drawing of a series of the tris-bipyridinate lanthanide(III) clathrochelates

Scheme 5.60 are reported [159] to be extraordinarily bright near-IR molecular luminophores. The highly luminescent, NIR-emitting 1:1 ytterbium, neodymium, erbium, and praseodymium(III) cage complexes of the perdeuterated macrobicyclic tris-bipyridinate **622** are also promising luminophores [160]. In the case of these lanthanoid ions, their encapsulation by **622** led to substantial minimization of nonradiative deactivation processes and to very efficient sensitization of lanthanoid luminescence, giving it very long lifetimes in deuterated CD_3CN and CD_3OD solvents and even in non-deuterated acetonitrile solutions. These cage complexes were found [160] to have a high thermodynamic stability in solution even under vigorous conditions, thus making them quite unique among the NIR luminophores.

The clathrochelate (Fe^{2+})@**321** with ribbed dopamine substituents has been proposed [161] as efficient receptor for boric acid, giving the complexes of various stoichiometry shown in Scheme 5.61.

Cation-receptor properties of the iron(II) di- and hexachlorine-containing mono- and bis-clathrochelates and their oximehydrazonate, dimethyl-, and diamine-containing macrobicyclic analogs (Scheme 5.62) and the cation-receptor properties of such cage compounds toward alkali metal cations (as individual ions or the mixtures of these cations) as well as their concurrent complexation in the presence of 18-crown-6 and ionization pathways have been studied [162]

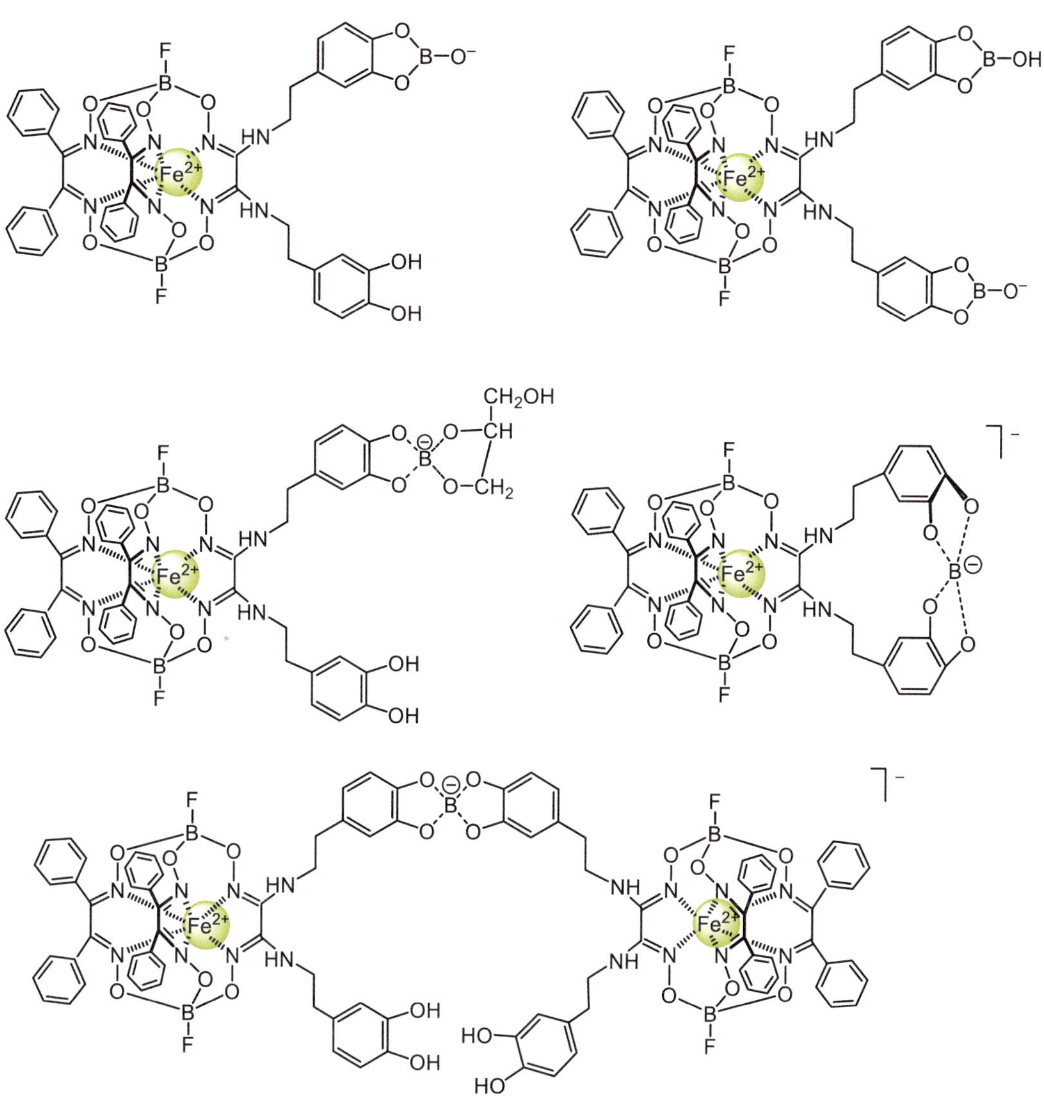

Scheme 5.61 Chemical drawings of the dopamine-containing iron(II) clathrochelate ligands and their complexes

using ESI-MS method. ESI-MS ionization of such iron(II) mono- and bis-clathrochelates proceeded in different pathways. If that of the ionic-associate macrobicyclic iron(II) oximehydrazonate (Fe^{2+})@**710** proceeded through its heterolytic dissociation, the ionization pathways for the cage and bis-cage iron(II) dioximate intracomplexes are substantially affected by the nature of ribbed substituents at their cage frameworks. The methyl- and amine-containing macrobicyclic complexes (Fe^{2+})@**296** and (Fe^{2+})@**323** easily form the corresponding anion-radical species by their one-electron oxidation, while in the case of the iron(II) dihalogenoclathrochelates (Fe^{2+})@**302**, (Fe^{2+})@**309**, and (Fe^{2+})@**310** most intensive peaks belong to their ionic associates with alkali metal cations. ESI-MS for the iron(II) hexachloroclathrochelate (Fe^{2+})@**293** did not contain the detectable peaks of the cationic species resulted either from its oxidation to the corresponding cation-radical or from its ionic association with alkali metal cations. In the negative range of this spectrum, the intensive peak of the anion-radical product of

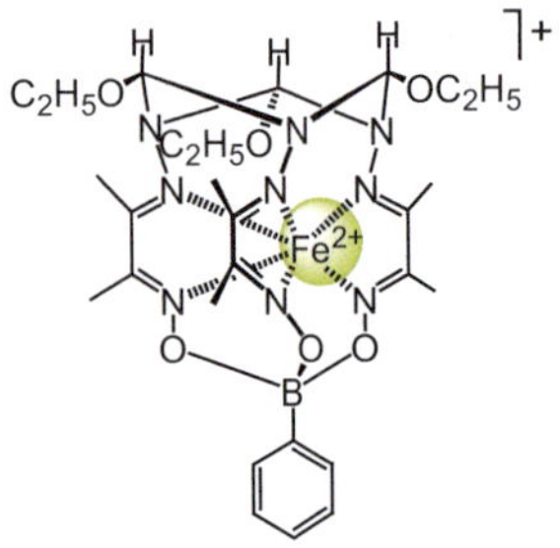

(Fe^{2+})@**710**

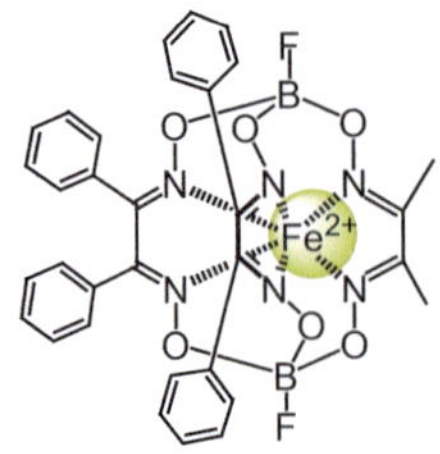

(Fe^{2+})@**296**

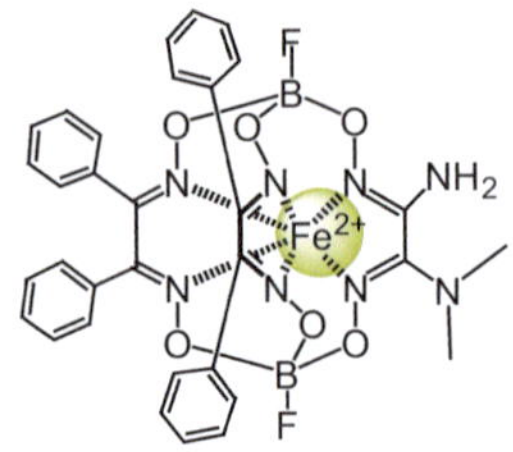

(Fe^{2+})@**323**

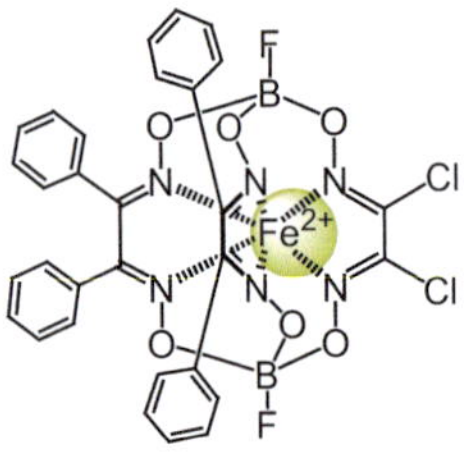

(Fe^{2+})@**302**

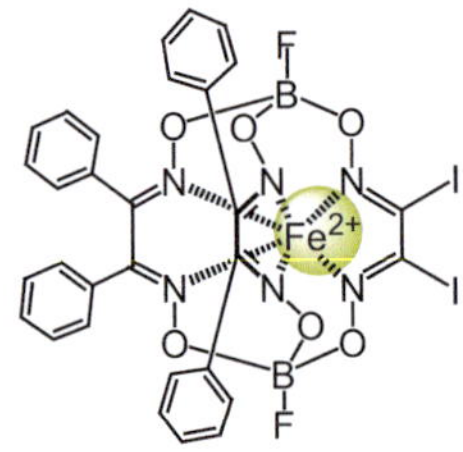

(Fe^{2+})@**310**

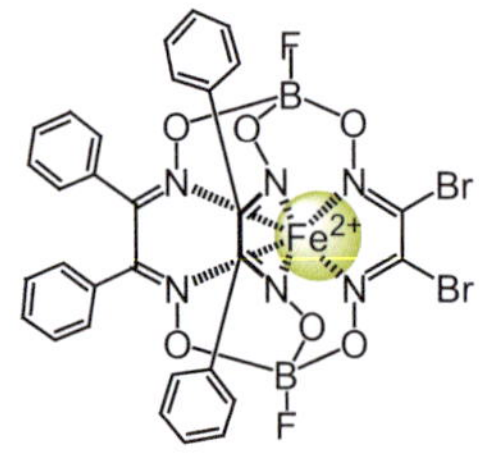

(Fe^{2+})@**309**

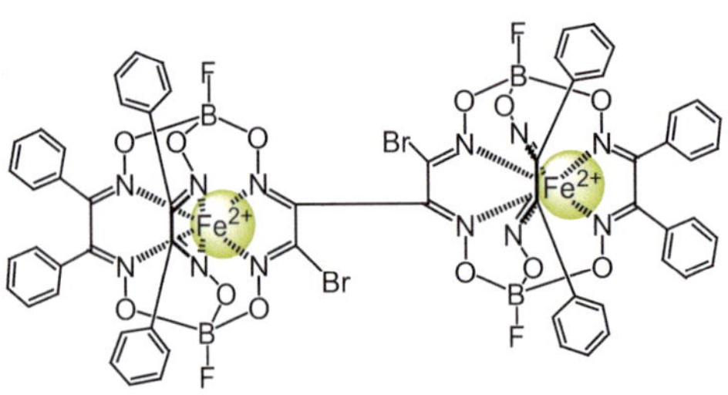

$(Fe^{2+})_2$@**441**

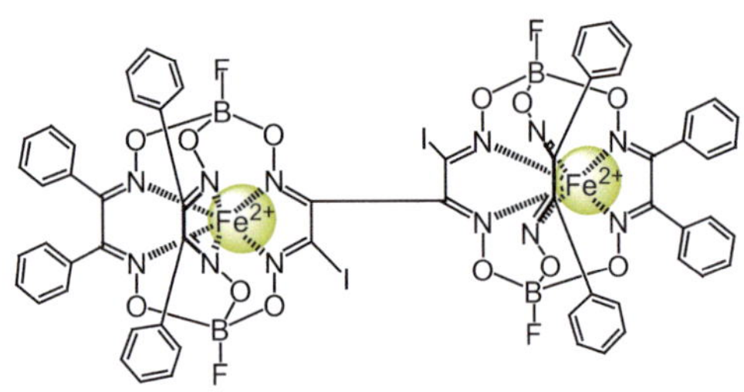

$(Fe^{2+})_2$@**484**

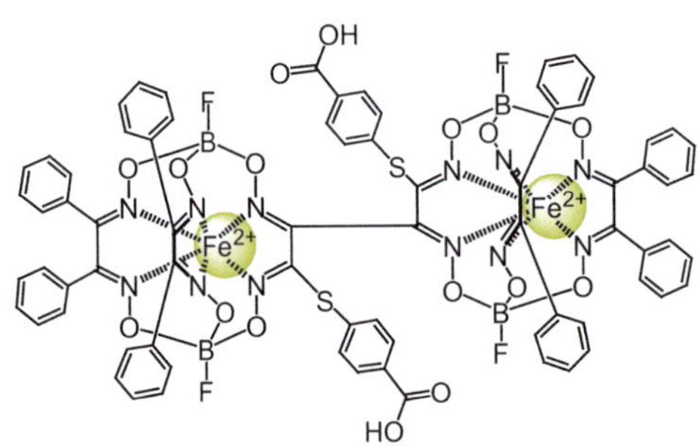

$(Fe^{2+})_2$@**444**

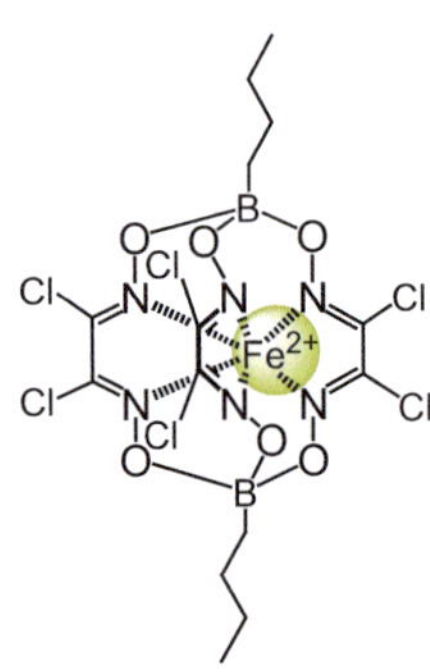

(Fe^{2+})@**293**

Scheme 5.62 Chemical drawings of a series of the iron(II) mono- and bis-clathrochelates

its one-electron reduction has been observed [162]. As ESI mass spectrometry allows maintaining precise concentrations of the complex under study and that of "*cationization agents*" (i.e., the compounds that promote formation of cationic species of the analyte and can be used in the case of low-ionizable complexes), alkali metal tetraphenylborates have been used

[162] as agents to study the cation–receptor properties of the iron(II) mono- and bis-clathrochelates with ribbed substituents of different nature toward these metal ions. The above iron(II) monoclathrochelates were found to form both 1:1 and 2:1 ionic associates with alkali metal cations, being more and less intensive, respectively, in their mass spectra, whereas the spectra of the bis-cage complexes $(Fe^{2+})_2$@**441**, $(Fe^{2+})_2$@**444**, and $(Fe^{2+})_2$@**484** contained the intensive peaks of only their 1:1 ionic associates with alkali metal cations, showing a high affinity toward Cs^+ ion [162].

The template synthesis and characterization of the gallophosphates, $GaPO_4$-based molecular sieves, using the apically permethylated cobalt(III) diammoniumclathrochelate (Co^{3+})@**675** (Scheme 5.63) as a matrix, have been performed [163].

Two types of the clathrochelate-functionalized zincophosphates have been prepared [164] using hydrothermal synthetic procedures in the presence of the chiral cobalt(III) diammoniumclathrochelate pentacation (Co^{3+})@(**99**·$2H^+$) as a structure-directing agent. The inorganic part of the first of them consists of pearl-like chains of alternating $[ZnO_4]$ and $[PO_4]$ tetrahedra, which are connected to this incorporated macrobicyclic cation through the numerous of hydrogen bonds. The second type of these zincophosphates has been prepared under similar reaction conditions but at higher temperature. Their crystal structures are closely related, although the systematic disorder in the latter case and its lower symmetry resulted in a unit cell twice as large as a unit cell of the former zincophosphate. In both these clathrochelate-functionalized compounds, only one enantiomer of the above cobalt(III) cage complex has been detected [164], despite the fact that its racemic mixture has been used for their synthesis.

Surface modification of 3D cubic mesoporous silica has been performed [165] using the apically functionalized polyamine cobalt(III) diaminoclathrochelates (Co^{3+})@**99**, (Co^{3+})@**711**, and (Co^{3+})@**712**, which have been covalently grafted onto a silica surface through the silication with sylanol group by Scheme 5.64.

The metallosurfactant (Co^{3+})@**151** is reported [166] to form the wormlike micelles with unusual cylindrical structure in aqueous solutions; these micelles coexist with or easily interconvert to the vesicle structures.

An aqueous-phase method for preparation of the stable gold nanoparticles, based on the use of a metal-free macrobicyclic compound **713** both as a reductant and a stabilizer, has been developed [167]. Reduction of $HAuCl_4$ with **713** at room temperature easily gave these nanoparticles within a short time. The macrobicyclic compound **536** has been used [155] for the preparation of the monolithic samples of sol–gel materials in which this amine covalent capsule is covalently bonded to the silica network.

The in-plane packing of pillars in smectites as a function of relative pillar charge (pillaring ion charge vs. clay surface charge ratio) has been studied [168] using sodium montmorillonite and a highly charged synthetic fluorhectorite, pillared with the iridium(III) and mercury(II) diaminoclathrochelates (Ir^{3+})@**99** and (Hg^{2+})@**99** (in the latter case, also with its diprotonated tetracationic form (Hg^{2+})@**99**·$2H^+$) to give well-oriented samples. Pillared clays with controlled interpillar distances have been prepared [169] by systematically varying the pillar and clay charges in clathrochelate–clay systems. The mercury(II) diaminoclathrochelate (Hg^{2+})@**99** (Scheme 5.65) has been proposed as a principal pillaring agent to optimize the in-plane X-ray diffraction. The study of nitrogen and hydrogen adsorptions suggested that in (Hg^{2+})@**99**–montmorillonite I system the interpillar distances (15 Å) are big enough that the size of the adsorbate did not substantially affect the specific surface. Its surface area, measured

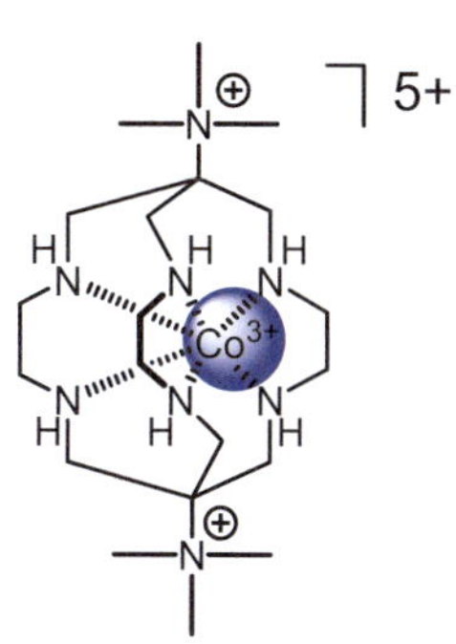

(Co^{3+})@**675**

Scheme 5.63 Chemical drawing of the clathrochelate (Co^{3+})@**675**

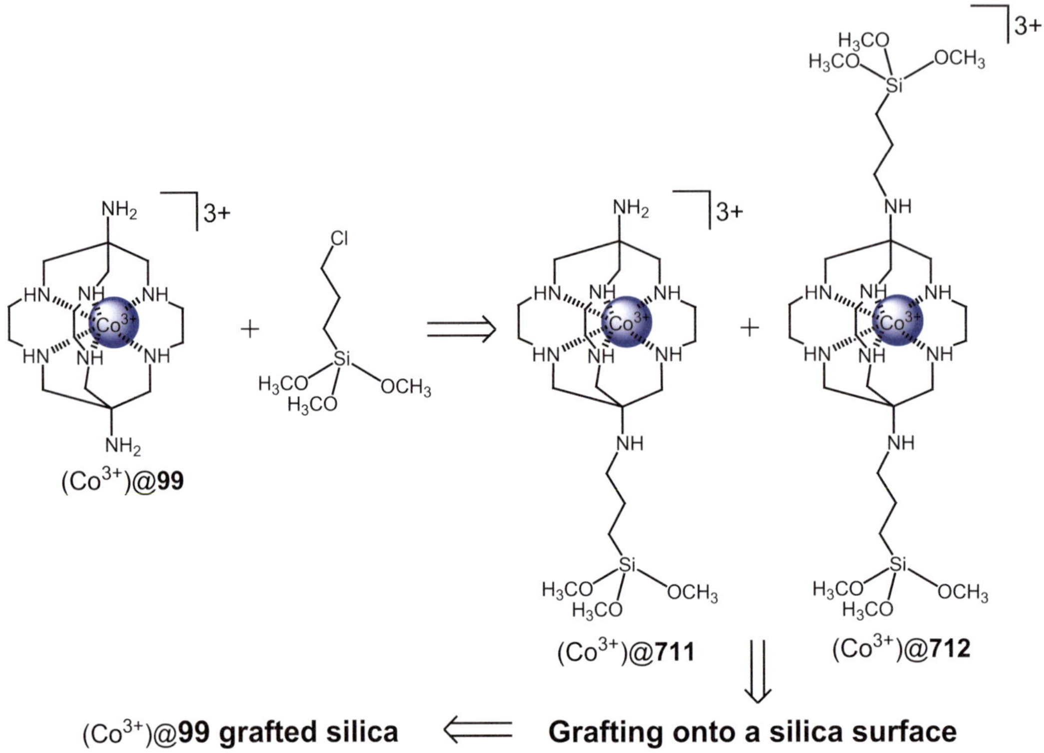

Scheme 5.64 Silication of the clathrochelate (Co^{3+})@**99** and grafting of the macrobicyclic products on SiO_2 surface [165]

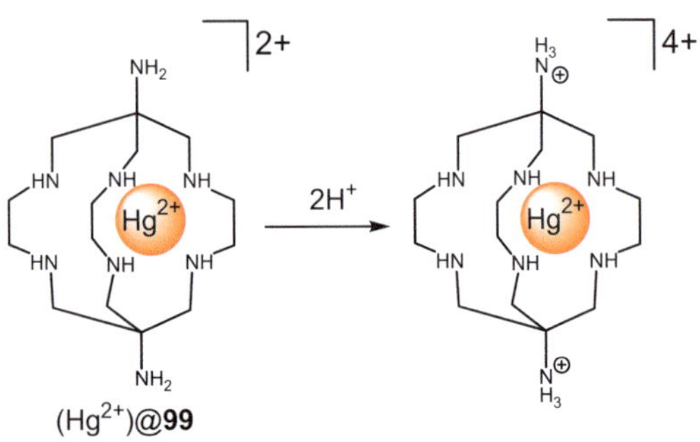

Scheme 5.65 Protonation of the complex (Hg^{2+})@**99**

[169] by hydrogen adsorption, was found to be almost equal to that measured with nitrogen. The above hydrogen adsorption is affected by the charge of the terminal amino groups of this pillaring cage complex [169]. The cobalt(III) clathrochelate (Co^{3+})@**648** has been studied [170] as a potent pillaring agent for Na^+-magadiite, a synthetic layered sodium silicate. Reaction of Na^+-magadiite with aqueous solution of this cage complex at 25 °C resulted in its binding to the external crystalline surface of this layered silicate, while a clathrochelate-containing intercalate, resulting from its penetration between the layers of magadiite and exhibiting a 17.6 Å basal spacing, has been prepared [170] at 100 °C. On the other hand, the above magadiite structure has been retained even after a prolonged reaction with (Co^{3+})@**648** at 100 °C. So, this clathrochelate trication is reported [170] to be not suitable for the pillaring of magadiite.

The 1,4-pentadienylboron-capped iron(II) clathrochelates (Fe^{2+})@**193–197** have been proposed [171] as prospective macrobicyclic monomers for the preparation of the corresponding functionalized metal-containing linear and cross-linked carbochain polymers.

The macrobicyclic ligands with terminal donor groups have been widely used as building blocks for the design of clathrochelate-based metallamacrocycles, MOFs, and coordination cages (Fig. 5.10; see Sect. 2.4). The use of

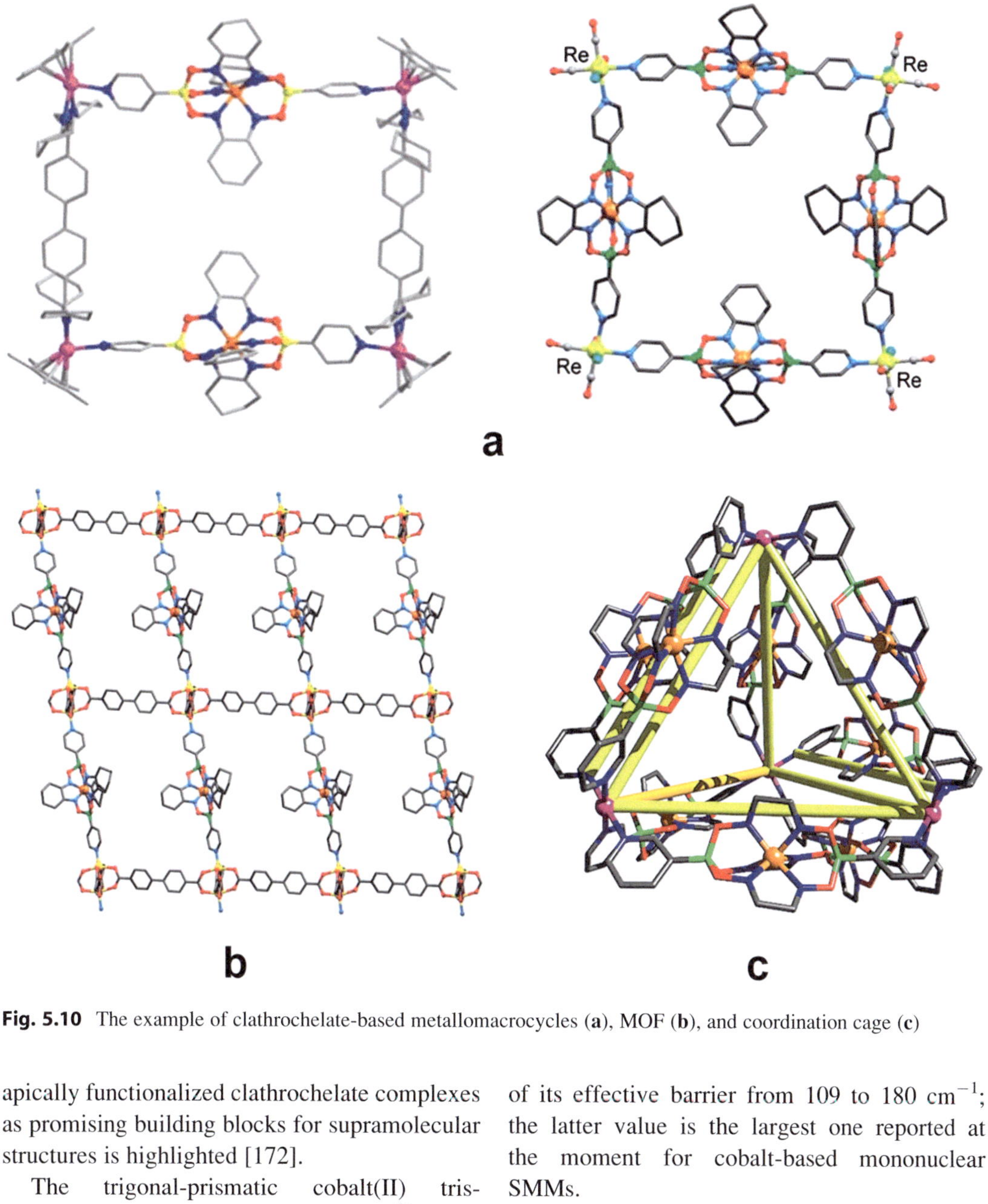

Fig. 5.10 The example of clathrochelate-based metallomacrocycles (**a**), MOF (**b**), and coordination cage (**c**)

apically functionalized clathrochelate complexes as promising building blocks for supramolecular structures is highlighted [172].

The trigonal-prismatic cobalt(II) tris-pyrazoloximate pseudoclathrochelate (Co^{2+})@**509** is reported [87] to show a SMM behavior with very high Orbach relaxation barrier of 152 cm^{-1} (Fig. 5.11).

Two polymorphs (α and β) of the hexadecylboron-capped cobalt(II) pseudoclathrochelate (Co^{2+})@**510** shown in Scheme 5.66 are described [173] to have substantially different SMM behavior. This resulted in an increase of its effective barrier from 109 to 180 cm^{-1}; the latter value is the largest one reported at the moment for cobalt-based mononuclear SMMs.

The clathrochelate-based molecular nanomagnet dimers *63–65* (see Sect. 2.4), which have been self-assembled [174] using spin quibit Cr_7Ni-building blocks, showed the long coherence time and allowed to control over the inter-qubit quantum interactions (see Chap. 4). Those can be used to mediate two-quibit quantum gates using the clathrochelate linkers of an appropriate size [175].

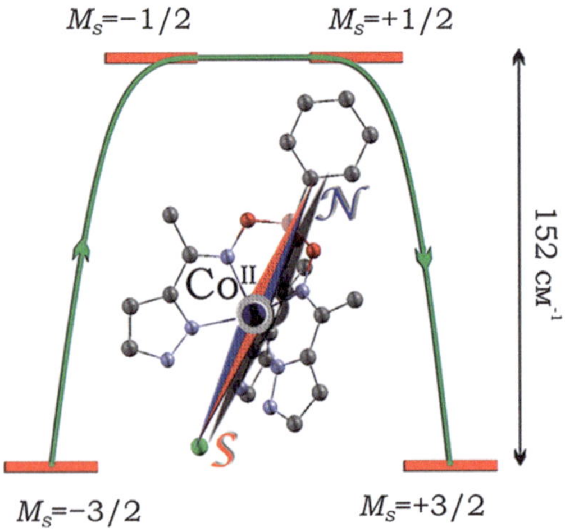

Fig. 5.11 The phenylboron-capped cobalt(II) pseudoclathrochelate (Co^{2+})@**509** as a mononuclear SMM [87]

5.4 Stabilization of Unusual Oxidation States of an Encapsulated Metal Ion

For the first time, the use of encapsulation of a metal ion by appropriate caging ligand for stabilization of its given oxidation state has been proposed [176]. "Oxidation state" is one of most fundamental concepts in chemistry, and the stabilization of previously unknown or unusual oxidation states of metal ions allows one to obtain the results that are important both for theoretical chemistry and for various practical applications. The metal clathrochelates are described [176] to be undoubtedly promise for the stabilization of unusual oxidation states of their caged ions, because their encapsulated and coordinatively saturated metal ions are completely isolated from external factors. The size of a three-dimensional ligand's cavity may be controlled by changing the nature of capping and chelate fragments of an encapsulating ligand, while its appropriate donor atoms (groups) may be used for the stabilization of a given electronic configuration of this caged ion. Varying the substituents at a quasiaromatic clathrochelate framework is proposed [176] for the stabilization of unusual oxidation states of an encapsulated metal ion: if the electronic effects of the apical group(s) in its capping fragment(s) are negligible, the chelate-ribbed electron-donor (electron-withdrawing) substituents in its ribbed azomethine fragments, forming a quasiaromatic cage framework, can do this. So, the stabilization of lowest or highest (including unusual and previously unknown) oxidation states of an encapsulating metal ion can be performed by introducing electron-withdrawing (in particular, halogenide, alkoxyl, or aryloxyl) substituents or electron-donor (in particular, amine) groups in the chelate fragments of a quasiaromatic macrobicyclic ligand [176]. In particular, the *n*-butylboron-capped cobalt(II) hexachloroclathrochelate (Co^{2+})@**293** was found to easily undergo its chemical reduction with metallic silver in the presence of tetraalkylammonium or tetra(dimethylamido)phosphonium halogenides by Scheme 5.67, thus giving the clathrochelate monoanion (Co^{+})@**293** in a high yield. The fine-crystalline samples of tetraalkylammonium and tetra(dimethylamido) phosphonium salts of this clathrochelate anion with an encapsulated cobalt(I) ion were found [176] to be air stable for several months.

Reduction of the dithiopheneboron-capped cobalt(II) tris-α-benzildioximate (Co^{2+})@**207** with cobaltocene as a reductant by Scheme 5.68 under PTC conditions with $((n\text{-}C_4H_9)_4N)Br$ as a phase transfer agent has been performed [149] to give the corresponding cobalt(I)-encapsulating clathrochelate monoanion that was isolated and X-ray structurally characterized in the form of its ionic associate with *n*-tetrabutylammonium cation.

Reduction of the cobalt(II) tris-perfluoro-α-benzildioximate (Co^{2+})@**291** by Scheme 5.69 using an excess of metallic silver in the presence of $((CH_3)_4N)Cl$ has been used [177] for the preparation of the corresponding cobalt(I)-encapsulating clathrochelate anion $[(Co^{+})@\mathbf{291}]^{-}$.

The cobalt(II) clathrochelate–zinc porphyrin assemblies are reported [178] to photogenerate

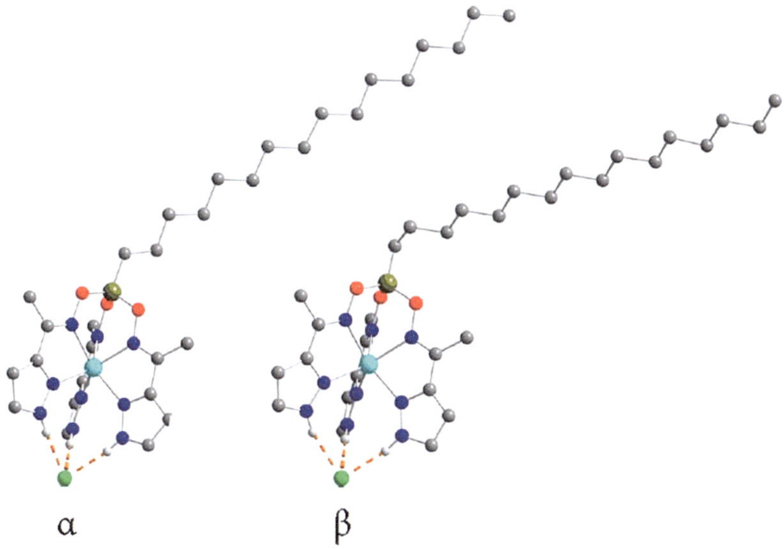

Scheme 5.66 Two polymorphs of the cobalt(II) pseudoclathrochelate (Co^{2+})@**510**

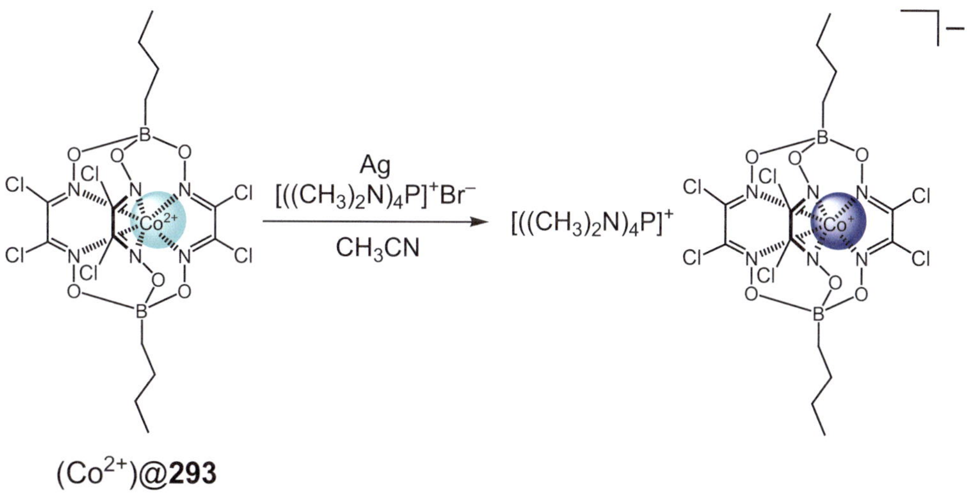

Scheme 5.67 Chemical reduction of the hexachloroclathrochelate (Co^{2+})@**293**

the corresponding cobalt(I)-containing species, which are stable during several milliseconds in their aqueous solutions. The formation of these long-lived cobalt(I) species has been confirmed [178] using laser flash photolysis experiments in the presence of ascorbate anion as an electron donor.

The macrobicyclic ligands **644–647** and **714** shown in Scheme 5.70 have been designed [179] for stabilization of unusual 2+ oxidation state of an europium ion by its encapsulation within the cavity of the corresponding caging ligand. The presence of *ortho*-phenylene-ribbed fragment(s) in such macrobicyclic ligand is described [179]

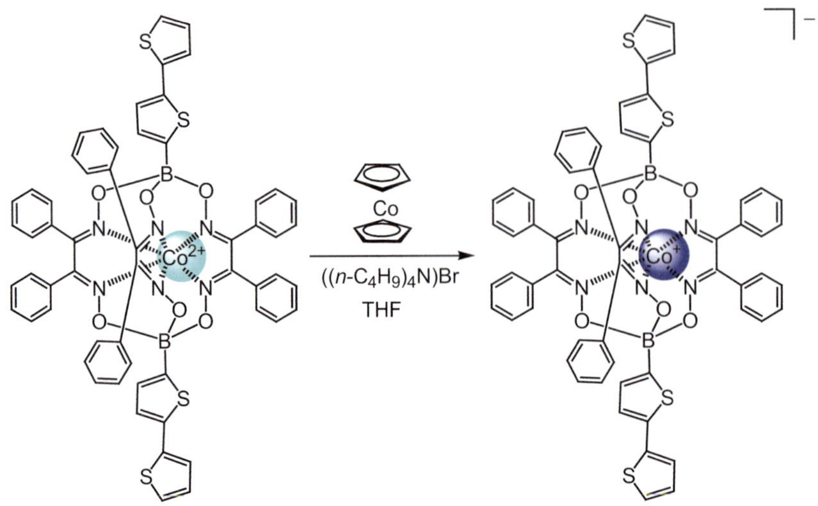

(Co^{2+})@**287**

Scheme 5.68 Chemical reduction of the tris-α-benzildioximate clathrochelate (Co^{2+})@**287**

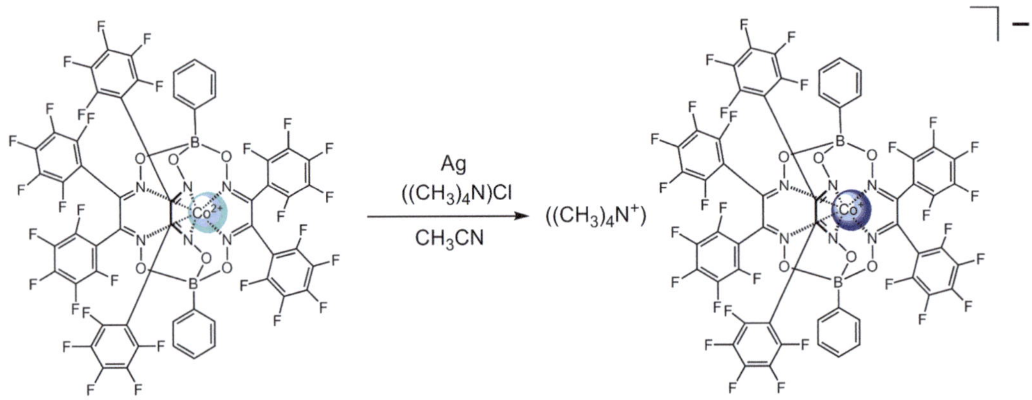

(Co^{2+})@**291**

Scheme 5.69 Chemical reduction of the tris-perfluoro-α-benzildioximate clathrochelate (Co^{2+})@**291**

to stabilize the corresponding europium(II) clathrochelates.

One-pot template reaction of iron(III) ions with oxalodihydrazide ligand synthon *94* and formaldehyde by Scheme 5.71 in the presence of air oxygen in alkali aqueous solution is reported [180] to unexpectedly give a clathrochelate iron(IV)-encapsulating dianion, the derivative of strongly electron-donating macrobicyclic hexahydrazide ligand **715**, stabilizing such unusually high oxidation state of an encapsulated metal cation.

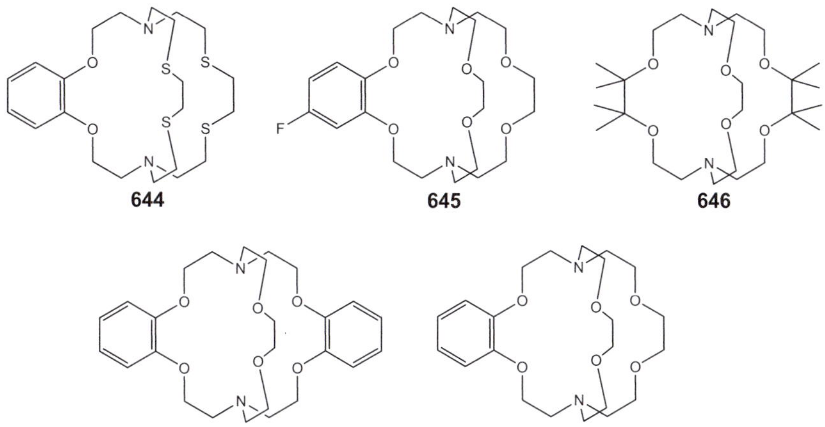

Scheme 5.70 Chemical drawings of a series of the O_6-donor macrobicyclic ligands

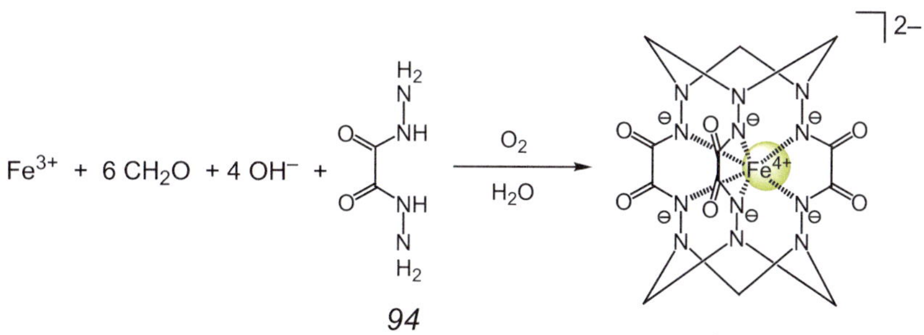

Scheme 5.71 Preparation of the tris-oxalodihydrazide iron(IV) clathrochelate (Fe^{4+})@**715**

References

1. Voloshin YZ, Varzatskii OA, Bubnov YN (2007) Cage complexes of transition metals in biochemistry and medicine. Russ Chem Bull Int Ed 56:577–605
2. Varzatskii O, Shul'ga S, Novikov V, Belov A, Makarenko I, Dubey I, Krivorotenko D, Negrutska V, Zhizhin K, Kuznetsov N, Bubnov Y (2010) Clathrochelate complexes as promising molecular scaffolds for new biomimetic systems and (radio)pharmaceuticals. In: Kessissoglou D, Salifoglou T (eds) Proceedings of 10th European biological inorganic chemistry conference, EUROBIC 10, Thessaloniki, 2010
3. Voloshin YZ, Novikov VV, Nelyubina YV (2015) Recent advances in biological applications of cage metal complexes. RSC Adv 5:72621–72637
4. Wagner CE, Mohler ML, Kang GS, Miller DD, Geisert EE, Chang YA, Fleischer EB, Shea KJ (2003) Synthesis of 1-boraadamantaneamine derivatives with selective astrocyte vs C6 Glioma antiproliferative activity. A novel class of anti-hepatitis C agents with potential to bind CD81. J Med Chem 46:2823–2833
5. Papanastasiou I, Tsotinis A, Kolocouris N, Prathalingam SR, Kelly JM (2008) Design, synthesis, and trypanocidal activity of new aminoadamantane derivatives. J Med Chem 51:1496–1500
6. Donnelly PS (2011) The role of coordination chemistry in the development of copper and rhenium radiopharmaceuticals. Dalton Trans 40:999–1010
7. Millet P, Mbemba N, Grigoriev SA, Fateev VN, Aukauloo A, Etiévant C (2011) Electrochemical performances of PEM water electrolysis cells and perspectives. Int J Hydrog Energy 36:4134–4142
8. Nguyen MTD, Ranjbari A, Catala L, Brisset F, Millet P, Aukauloo A (2012) Implementing molecular catalysts for hydrogen production in proton exchange membrane water electrolysers. Coord Chem Rev 256:2435–2444
9. Islamova RM (2016) Iron compounds in controlled radical polymerization: ferrocenes, (clathro)

chelates, and porphyrins. Russ J Gen Chem 86:125–143
10. Fletcber TM, Walmsley JA (1997) Stereospecific interactions of chiral cobalt(III)-amine complexes with guanosine 5′-monophosphate. J Inorg Biochem 68:239–249
11. Peter SK, Lindoy LF, Walker GW, Everett GW (1993) Supramolecular transport of metal amine complexes through liquid membranes by the ionophore lasalocid A. Pure Appl Chem 65:521–526
12. Losytskyy MY, Kovalska VB, Varzatskii OA, Sergeev AM, Yarmoluk SM, Voloshin YZ (2013) Interaction of the iron(II) cage complexes with proteins: protein fluorescence quenching study. J Fluoresc 23:889–895
13. Deng H, Bloomfield VA (1999) Structural effects of cobalt-amine compounds on DNA condensation. Biophys J 77:1556–1561
14. Bardajee GR, Hooshyar Z, Shafagh P, Ghiasvand S, Kakavand N (2014) Combined spectroscopic and molecular docking techniques to study interaction of Zn (II) DiAmsar with serum albumins. J Lumin 156:55–62
15. Hooshyar Z, Bardajee GR, Shafagh P, Ghiasvand S, Khanjari M, Dianatnejad N (2014) In vitro study on the interaction of Mn(II)-DiAmsar with human serum albumin (HSA) and bovine serum albumin (BSA) by spectroscopic and molecular docking methods. J Iran Chem Soc 12:715–725
16. Hooshyar Z, Rezanejade BG, Kakavand N, Khanjari M, Dianatnejad N (2015) Investigations on the interactions of DiAmsar with serum albumins: Insights from spectroscopic and molecular docking techniques. Luminescence 30:538–548
17. Voloshin YZ, Belov AS, Lebedev AY, Varzatskii OA, Antipin MY, Starikova ZA, Kron TE (2004) Synthesis and structure of the first ribbed-functionalized quinoxaline clathrochelate: design of cage complexes for efficient intercalation into DNA structure. Russ Chem Bull Int Ed 53:1218–1222
18. Voloshin YZ, Varzatskii OA, Belov AS, Starikova ZA, Suponitsky KY, Novikov VV, Bubnov YN (2008) Interaction of dichloride iron (II) clathrochelate with dimercaptomaleodinitrile: synthesis of the precursor of monoribbed-functionalized phthalocyaninoclathrochelates and the unexpected formation of a new thiophene-containing heterocyclic system in the ribbed chelate fragment of the clathrochelate framework. Inorg Chem 47:2155–2161
19. Voloshin YZ, Varzatskii OA, Vorontsov II, Antipin MY, Lebedev AY, Belov AS, Palchik AV (2003) Template synthesis and structures of apically functionalized iron(II) clathrochelates. Russ Chem Bull Int Ed 52:1552–1561
20. Hammershøi A, Nord G, Skov LK, Rowatt E, Williams RJP (1993) The binding of a complex cobalt ruthenium polyamine by deoxyribonucleic acid and a lipopolysaccharide: a model for a novel class of drugs. J Inorg Biochem 49:295–304
21. Zelinskii GE, Belov AS, Lebed EG, Vologzhanina AV, Novikov VV, Voloshin YZ (2016) Synthesis, structure and reactivity of iron(II) clathrochelates with terminal formyl (acetal) groups and their further functionalization for target delivery to biological systems. Inorg Chim Acta 440:154–164
22. Yang LN, Xu F, Sun LX, Zhao ZB (2008) A microcalorimetric study of the toxicity of two cobalt compounds on *Escherichia coli* DH5α growth. J Therm Anal Calorim 93:583–588
23. Walker GW, Geue RJ, Sargeson AM, Behm CA (2003) Surface-active cobalt cage complexes: synthesis, surface chemistry, biological activity, and redox properties. Dalton Trans 15:2992–3001
24. Sánchez-Moreno M, Marín C, Navarro P, Lamarque L, García-España E, Miranda C, Huertas O, Olmo F, Gómez-Contreras F, Pitarch J, Arrebola F (2012) In vitro and in vivo trypanosomicidal activity of pyrazole-containing macrocyclic and macrobicyclic polyamines: Their action on acute and chronic phases of Chagas disease. J Med Chem 55:4231–4243
25. Navarro P, Sánchez-Moreno M, Marín C, García-España E, Ramírez-Macías I, Olmo F, Rosales MJ, Gómez-Contreras F, Yunta MJR, Gutierrez-Sánchez R (2014) In vitro leishmanicidal activity of pyrazole-containing polyamine macrocycles which inhibit the Fe-SOD enzyme of *Leishmania infantum* and *Leishmania braziliensis* species. Parasitology 141:1031–1043
26. Anaud-Neu F, Fuangswasdi S, Maubert B, Nelson J, McKee V (2000) Binding properties of octaamino-cryptands. Inorg Chem 39:573–579
27. Funahashi Y, Shirota T, Kinoshita W, Toyama T, Nagata K, Nishikawa T, Fukui K, Inomata T, Ozawa T, Masuda H (2012) Synthesis of biomimetic models of multinuclear copper sites for O_2-activation. Paper presented at the 40 International conference on coordination chemistry, Valencia Conference Centre, Valencia, 9–13 Sept 2012
28. Murase M, Inomata T, Ozawa T, Masaoka S, Funahashi Y, Masuda H (2012) Structure and reactivity of a cage-type Zn(II) complex as a functional model of alcohol dehydrogenase. Paper presented at the 40 International conference on coordination chemistry, Valencia Conference Centre, Valencia, 9–13 Sept 2012
29. Xu J, Corneillie TM, Moore EG, Law GL, Butlin NG, Raymond KN (2011) Octadentate cages of Tb (III) 2-hydroxyisophthalamides: a new standard for luminescent lanthanide labels. J Am Chem Soc 133:19900–19910
30. Bazin H, Benech JM, Bleriot A, Bourrier E, Mathis G (2005) Luminescent lanthanides cryptates: from design to applications. Paper presented at the XXX

international symposium on macrocyclic and supramolecular chemistry, Dresden, 17–21 July 2005
31. Wahsner J, Seitz M (2012) Novel isotope-coded and metal-containing mass tags for quantitation approaches. Paper presented at the 40 International conference on coordination chemistry, Valencia Conference Centre, Valencia, 9–13 Sept 2012
32. Mokhir A, Krämer R, Voloshin YZ, Varzatskii OA (2004) Synthesis and DNA binding properties of dioxime–peptide nucleic acids. Bioorg Med Chem Lett 14:2927–2930
33. Novikov VV, Varzatskii OA, Negrutska VV, Bubnov YN, Palchykovska LG, Dubey IY, Voloshin YZ (2013) Size matters, so does shape: inhibition of transcription by cage iron(II) compounds. J Inorg Biochem 124:42–45
34. Voloshin YZ, Novikov VV, Varzatskii OA, Negrutska VV, Palchykovska LG, Dubey IY, Bubnov YN (2013) RU Patent 2499832, 27 Nov 2013
35. Belov AS, Vologzhanina AV, Novikov VV, Negrutska VV, Dubey IY, Mikhailova ZA, Lebed EG, Voloshin YZ (2014) Synthesis of the first morpholine-containing iron (II) clathrochelates: a new class of efficient functionalized transcription inhibitors. Inorg Chim Acta 421:300–306
36. Varzatskii OA, Novikov VV, Shulga SV, Belov AS, Vologzhanina AV, Negrutska VV, Dubey IY, Bubnov YN, Voloshin YZ (2014) Copper-promoted reductive homocoupling of quasi-aromatic iron(II) clathrochelates: boosting the inhibitory activity in a transcription assay. Chem Commun 50:3166–3168
37. Kovalska VB, Losytskyy MY, Varzatskii OA, Cherepanov VV, Voloshin YZ, Mokhir AA, Yarmoluk SM, Volkov SV (2014) Study of anti-fibrillogenic activity of iron(II) clathrochelates. Bioorg Med Chem 22:1883–1888
38. Voloshin YZ, Varzatskii OA, Novikov VV, Bubnov YN (2008) Biochemical and medicinal application of cage transition metal complexes: the design of new HIV protease inhibitors. In: Kozlowski H (ed) Proceedings of 9th European biological inorganic chemistry conference, EUROBIC 9, Wroclaw, 2008
39. Bailey DL, Townsend DW, Valk PE, Maisey MN (2005) Positron emission tomography: basic sciences. Springer, Secaucus
40. Amendola V, Bonizzoni M, Esteban-Gómez D, Fabbrizzi L, Licchelli M, Sancenón F, Taglietti A (2006) Some guidelines for the design of anion receptors. Coord Chem Rev 250:1451–1470
41. Blower PJ, Lewis JS, Zweit J (1996) Copper radionuclides and radiopharmaceuticals in nuclear medicine. Nucl Med Biol 23:957–980
42. Smith SV (2004) Molecular imaging with copper-64. J Inorg Biochem 98:1874–1901
43. Di Bartolo N, Sargeson AM, Smith SV (2006) New ^{64}Cu PET imaging agents for personalised medicine and drug development using the hexa-aza cage, SarAr. Org Biomol Chem 4:3350–3357
44. Voss SD, Smith SV, Di Bartolo N, McIntosh LJ, Cyr EM, Bonab AA, Carter EA, Fischman AJ, Treves ST, Gillies SD, Sargeson AM, Huston JS, Packard AB (2007) Positron emission tomography (PET) imaging of neuroblastoma and melanoma with 64Cu-SarAr immunoconjugates. Proc Natl Acad Sci USA 104:17489–17493
45. Cai H, Fissekis J, Conti PS (2009) Synthesis of a novel bifunctional chelator AmBaSar based on sarcophagine for peptideconjugation and ^{64}Cu radiolabelling. Dalton Trans 27:5395–5400
46. Ma MT, Neels OC, Denoyer D, Roselt P, Karas JA, Scanlon DB, White JM, Hicks RJ, Donnelly PS (2011) Gallium-68 complex of a macrobicyclic cage amine chelator tethered to two integrin-targeting peptides for diagnostic tumor imaging. Bioconjug Chem 22:2093–210337
47. Ma MT, Cooper MS, Paul RL, Shaw KP, Karas JA, Scanlon DB, White JM, Blower PJ, Donnelly PS (2011) Macrobicyclic cage amine ligands for copper radiopharmaceuticals: a single bivalent cage amine containing two Lys^{3}-bombesin targeting peptides. Inorg Chem 50:6701–6710
48. Chen K, Wang X, Lin WY, Shen CKF, Yap LP, Hughes LD, Conti PS (2012) Strain-promoted catalyst-free click chemistry for rapid construction of ^{64}Cu-labeled PET imaging probes. ACS Med Chem Lett 3:1019–1023
49. Paterson BM, Buncic G, McInnes LE, Roselt P, Cullinane C, Binns DS, Jeffery CM, Price RI, Hicks RJ, Donnelly PS (2015) Bifunctional ^{64}Cu-labelled macrobicyclic cage amine isothiocyanates for immuno-positron emission tomography. Dalton Trans 44:4901–4909
50. Liu S, Li Z, Conti PS (2014) Development of multifunctional chelators based on sarcophagine cages. Molecules 19:4246–4255
51. Li G, Wang X, Zong S, Wang J, Conti PS, Chen K (2014) MicroPET imaging of CD13 expression using a ^{64}Cu-labeled dimeric NGR peptide based on sarcophagine cage. Mol Pharm 11:3938–3946
52. Wu Z, Liu S, Nair I, Omori K, Scott S, Todorov I, Shively JE, Conti PS, Li Z, Kandeel F (2014) ^{64}Cu labeled sarcophagine exendin-4 for microPET imaging of glucagon like peptide-1 receptor expression. Theranostics 4:770–777
53. Brand C, Abdel-Atti D, Zhang Y, Carlin S, Clardy SM, Keliher EJ, Weber WA, Lewis JS, Reiner T (2014) In vivo imaging of GLP-1R with a targeted bimodal PET/fluorescence imaging agent. Bioconjug Chem 25:1323–1330
54. Liu S, Li D, Huang CW, Yap LP, Park R, Shan H, Li Z, Conti PS (2012) Efficient construction of PET/fluorescence probe based on sarcophagine cage: an opportunity to integrate diagnosis with treatment. Mol Imaging Biol 14:718–724

55. Liu S, Li D, Huang CW, Yap LP, Park R, Shan H, Li Z, Conti PS (2012) The efficient synthesis and biological evaluation of novel bi-functionalized sarcophagine for ^{64}Cu radiopharmaceuticals. Theranostics 2:589–596
56. Tan KV, Pellegrini PA, Skelton BW, Hogan CF, Greguric I, Barnard PJ (2014) Triamidetriamine bearing macrobicyclic and macrotricyclic ligands: Potential applications in the development of copper-64 radiopharmaceuticals. Inorg Chem 53:468–477
57. Paterson BM, Roselt P, Denoyer D, Cullinane C, Binns D, Noonan W, Jeffery CM, Price RI, White JM, Hicks RJ, Donnelly PS (2014) PET imaging of tumours with a ^{64}Cu labeled macrobicyclic cage amine ligand tethered to Tyr^3-octreotate. Dalton Trans 43:1386–1396
58. Alt K, Paterson BM, Ardipradja K, Schieber C, Buncic G, Lim B, Poniger SS, Jakoby B, Wang X, O'Keefe GJ, Tochon-Danguy HJ, Scott AM, Ackermann U, Peter K, Donnelly PS, Hagemeyer CE (2014) Single-chain antibody conjugated to a cage amine chelator and labeled with positron-emitting copper-64 for diagnostic imaging of activated platelets. Mol Pharm 11:2855–2863
59. Paterson BM, Alt K, Jeffery CM, Price RI, Jagdale S, Rigby S, Williams CC, Peter K, Hagemeyer CE, Donnelly PS (2014) Enzyme-mediated site-specific bioconjugation of metal complexes to proteins: Sortase-mediated coupling of copper-64 to a single-chain antibody. Angew Chem Int Ed 53:6115–6119
60. Cooper MS, Ma MT, Sunassee K, Shaw KP, Williams JD, Paul RL, Donnelly PS, Blower PJ (2012) Comparison of ^{64}Cu-complexing bifunctional chelators for radioimmunoconjugation: Labeling efficiency, specific activity, and in vitro/in vivo stability. Bioconjug Chem 23:1029–1039
61. Huang CW, Li Z, Cai H, Chen K, Shahinian T, Conti PS (2011) Design, synthesis and validation of integrin α2β 1-targeted probe for microPET imaging of prostate cancer. Eur J Nucl Med Mol Imaging 38:1313–1322
62. Janib SM, Liu S, Park R, Pastuszka MK, Shi P, Moses AS, Orosco MM, Lin YA, Cui H, Conti PS, Li Z, MacKay JA (2013) Kinetic quantification of protein polymer nanoparticles using non-invasive imaging. Integr Biol 5:183–194
63. Paterson BM, Denoyer D, Cullinane C, Roselt P, White JM, Hicks RJ, Donnelly PS (2012) A macrobicyclic ligand for targeted cancer imaging. Paper presented at the 40 international conference on coordination chemistry, Valencia Conference Centre, Valencia, 9–13 Sept 2012
64. Amendola V, Alberti G, Bergamaschi G, Biesuz R, Boiocchi M, Ferrito S, Schmidtchen FP (2012) Cavity effect on perrhenate recognition by polyammonium cages. Eur J Inorg Chem 2012:3410–3417
65. Voloshin Y, Belaya I, Krämer R (2016) The encapsulation phenomenon: synthesis, reactivity and applications of caged ions and molecules. Springer International Publishing, Switzerland
66. Treher EN, Gougoutas J, Malley M, Nunn AD, Unger SE (1986) New technetium radiopharmaceuticals: boronic acid adducts of vicinal dioxime complexes. J Label Compd Radiopharm 23:1118–1120
67. Linder KE, Malley MF, Gougoutas JZ, Unger SE, Nunn AD (1990) Neutral, seven-coordinate dioxime complexes of technetium(III): synthesis and characterization. Inorg Chem 29:2428–2434
68. Linder KE, Malley MF, Gougoutas JZ, Unger SE, Nunn AD (1989) Chemistry of $Tc(oxime)_3(\mu\text{-}OH)SnCl_3$ and $TcCl(oxime)_3$: neutral intermediates in the synthesis of boron-caped 99-TcOXIME complexes. J Label Compd Radiopharm 26:54–56
69. Treher EN, Francesconi LC, Gougoutas JZ, Malley MF, Nunn AD (1989) Monocapped tris(dioxime) complexes of technetium(III): synthesis and structural characterization of $TcX(dioxime)_3B\text{-}R$ (X = Cl, Br; dioxime = dimethylglyoxime, cyclohexanedione dioxime; R = CH_3, C_4H_9). Inorg Chem 28:3411–3416
70. Linder KE, Nowotnik DP, Malley MF, Gougoutas JZ, Nunn AD (1991) An unexpected by-product obtained during the preparation of technetium(III) boronic acid adducts of dioximes. The single crystal structure of $TcCl(DMG)_2(BDI)BEt$ (DMG=dimethylglyoxime, BDI=butane-2, 3-dione imine-oxime). Inorg Chim Acta 190:249–255
71. Linder KE, Chan YW, Cyr JE, Nowotnik DP, Eckelman WC, Nunn AD (1993) Synthesis, characterization, and in vitro evaluation of nitroimidazole-BATO complexes: new technetium compounds designed for imaging hypoxic tissue. Bioconjug Chem 4:326–333
72. Linder KE, Wen MD, Nowotnik DP, Ramalingam K, Sharkey RM, Yost F, Narra RK, Nunn AD, Eckelman WC (1991) Technetium-labeling of monoclonal antibodies with functionalized BATOs: 2. $TcCl(DMG)_3CPITC$ (CPITC=carboxyphenylisothiocyanate) labeling of B72.3 and NP-4 whole antibodies and NP-4 F$(ab')_2$. Bioconjug Chem 2:407–414
73. Linder KE, Wen MD, Nowotnik DP, Malley MF, Gougoutas JZ, Nunn AD, Eckelman WC (1991) Technetium labeling of monoclonal antibodies with functionalized BATOs. 1. $TcCl(DMG)_3PITC$ [phenyl isothiocyanate]. Bioconjug Chem 2:160–170
74. Voloshin YZ, Kuznetsov EV, Varzatskii OA, Belaya IG, Tananaev IG, Vologzhanina AV, Belov AS, Novikov VV, Myasoedov BF (2013) Functionalized rhenium(III) semi-clathrochelates as prototypes of the 99m-technecium-containing radiopharmaceuticals. Paper presented at the XII International

symposium on inorganic biochemistry, Wroclaw, 28 Aug–1 Sept 2013

75. Belov AS, Varzatskii OA, Novikov VV, Lebed EG, Vologzhanina AV, Voloshin YZ (2014) Template synthesis and structure of rhenium (III) semiclathrochelates as prototypes of radiopharmaceutical (diagnostic) compounds with a target delivery to the biological systems. Paper presented at the XXVI international Chugaev conference on coordination chemistry, Kazan, 6–10 Oct 2014
76. Alberto R, Angst D, Ortner K, Abram U, Schubiger PA, Kaden TA (2007) Syntheses of a series of S_6 thioether cages and their coordination chemistry with Ag^+. New J Chem 31:409–417
77. Voloshin YZ, Varzatskii OA, Zhizhin KY, Kuznetsov NT, Bubnov YN (2006) Cage complexes as a molecular scaffold for polyfunctional and polytopic systems: synthesis of the first closo-borate iron(II) clathrochelate. Russ Chem Bull Int Ed 55:22–25
78. Kuznetsov NT, Belaya Makarenko IG, Dolganov AV, Zelinskii GE, Matveev EY, Zhizhin KY, Voloshin YZ, Bubnov YN (2011) First example of the ribbed-functionalized iron(ii) clathrochelate with six pendante *closo*-borate substituents. Russ Chem Bull Int Ed 60:2518–2521
79. Voloshin YZ, Erdyakov SY, Makarenko IG, Lebed' EG, Potapova TV, Svidlov SV, Starikova ZA, Pol'shin EV, Gurskii ME, Bubnov YN (2007) Macrocyclization of the semiclathrochelate *o*-carboranylboronate and *n*-butylboronate iron (II) oximehydrazonates: synthesis and structure of clathrochelate products and unexpected allosteric effect of the apical substituent. Russ Chem Bull Int Ed 56:1787–1794
80. Erdyakov SY, Voloshin YZ, Makarenko IG, Lebed EG, Potapova TV, Ignatenko AV, Vologzhanina AV, Gurskii ME, Bubnov YN (2009) New *o*-carboranyl-containing capping agents for *d*-metal tris-dioximates and first bis-C-carboranylboron-capped iron(II) clathrochelates: synthesis and X–ray structure. Inorg Chem Commun 12:135–139
81. Svidlov SV, Varzatskii OA, Potapova TV, Vologzhanina AV, Bukalov SS, Leites LA, Voloshin YZ, Bubnov YN (2014) C-carboranylation of a quasi-aromatic iron(II) cage complex and its organic aromatic analog by the metal-promoted (catalyzed) cross-coupling reactions. Inorg Chem Commun 43:142–145
82. Szczepaniak LS, Sargeson A, Creasei II, Geue RJ, Tweedle M, Bryant RG (1992) Nuclear magnetic spin-lattice relaxation of water protons caused by metal cage compounds. Bioconjug Chem 3:27–31
83. Bothner-By AA, Domaille PJ, Gayathri C (1981) Ultra-high field NMR spectroscopy: observation of proton-proton dipolar coupling in paramagnetic bis [tolyltris(pyrazolyl)borato]cobalt(II). J Am Chem Soc 103:5602–5603
84. Voloshin YZ, Belov AS, Varzatskii OA, Vologzhanina AV, Viswanathan S, Radecki J, Bubnov YN (2009) Synthesis, structure and electron-mediator properties of the mono- and difunctionalized macrobicyclic iron(II) tris-dioximates with thiol terminated ribbed spacer substituents. Inorg Chim Acta 362:2982–2988
85. Voloshin YZ, Varzatskii OA, Novikov VV, Strizhakova NG, Vorontsov II, Vologzhanina AV, Lyssenko KA, Romanenko GV, Fedin MV, Ovcharenko VI, Bubnov YN (2010) Tris-dioximate cobalt(I, II, and III) clathrochelates: stabilization of different oxidation and spin states of an encapsulated metal ion by ribbed functionalization. Eur J Inorg Chem 2010:5401–5415
86. Varzatskii OA, Penkova LV, Kats (Menkach) SV, Dolganov AV, Vologzhanina AV, Pavlov AA, Novikov VV, Bogomyakov AS, Nemykin VN, Voloshin YZ (2014) Chloride ion-aided self-assembly of the pseudo-clathrochelate metal tris-pyrazoloximates. Inorg Chem 53:3062–3071
87. Novikov VV, Pavlov AA, Nelyubina YV, Boulon ME, Varzatskii OA, Voloshin YZ, Winpenny REP (2015) A trigonal prismatic mononuclear cobalt (II) complex showing single-molecule magnet behavior. J Am Chem Soc 137:9792–9795
88. Pantani O, Naskar S, Guillot R, Millet P, Anxolabéhère-Mallart E, Aukauloo A (2008) Cobalt clathrochelate complexes as hydrogen-producing catalysts. Angew Chem Int Ed 47:9948–9950
89. Anxolabéhère-Mallart E, Costentin C, Fournier M, Nowak S, Robert M, Savéant JM (2012) Boron-capped tris(glyoximato) cobalt clathrochelate as a precursor for the electrodeposition of nanoparticles catalyzing H_2 evolution in water. J Am Chem Soc 134:6104–6107
90. El Ghachtouli S, Fournier M, Cherdo S, Guillot R, Charlot MF, Anxolabéhère-Mallart E, Robert M, Aukauloo A (2013) Monometallic cobalt–trisglyoximato complexes as precatalysts for catalytic H_2 evolution in water. J Phys Chem C 117:17073–17077
91. Voloshin YZ, Dolganov AV, Varzatskii OA, Bubnov YN (2011) Efficient electrocatalytic hydrogen production from the H^+ ions using the specially designed boron-capped cobalt clathrochelates. Chem Commun 47:7737–7739
92. Voloshin YZ, Belov AS, Vologzhanina AV, Aleksandrov GG, Dolganov AV, Novikov VV, Varzatskii OA, Bubnov YN (2012) Synthesis, structure, properties and immobilization on a gold surface of the monoribbed-functionalized tris-dioximate cobalt (II) clathrochelates and an electrocatalytic hydrogen production from H^+ ions. Dalton Trans 41:6078–6093
93. Dolganov AV, Belov AS, Novikov VV, Vologzhanina AV, Mokhir A, Bubnov YN, Voloshin YZ (2013) Iron *vs* Cobalt clathrochelate electrocatalysts of HER: first example on a cage iron complex. Dalton Trans 42:4373–4376

94. Dolganov AV, Belaya IG, Voloshin YZ (2014) Binuclear iron(II) cage complexes as electrocatalysts of HER in different hydrogen-producing systems. Electrochim Acta 125:302–306
95. Harrowfield JM, Koutsantonis GA, Kraatz HB, Nealon GL, Orlowski GA, Skelton BW, White AH (2007) Cages on surfaces: Thiol functionalisation of Co^{III} sarcophagine complexes. Eur J Inorg Chem 2007:263–278
96. Villagra A, Ranjbari A, Kudo A, Millet P (2015) Surface modification of rhodium-doped strontium titanate by adsorption of cobalt clathrochelates for water photo-dissociation. Paper presented at the 3rd Southern African solar energy conference, Skukuza Rest Camp, Kruger National Park, 11–13 May 2015
97. Ribeiro S, Cunha-Silva L, Balula SS, Gago S (2014) Cobalt(III) sepulchrate complexes: application as sustainable oxidative catalysts. New J Chem 38:2500–2507
98. Chand DK, Bharadwaj PK (1997) A cobalt (II) cryptate of a heteroditopic cryptand L as an efficient oxygenation catalyst of organic substrates using molecular oxygen and 2-methylpropanal. Inorg Chem 36:5658–5660
99. Hao HG, Zheng XD, Lu TB (2010) Photoinduced catalytic reaction by a fluorescent active cryptand containing an anthracene fragment. Angew Chem Int Ed 49:8148–8151
100. Islamova RM, Sadykova GR, Ionova IA, Voloshin YZ, Bubnov YN, Makarov IS, Monakov Y (2009) Free-radical polymerization of methyl methacrylatein the presence of a ferrocenyl-containing iron(II) clathrochelate complex and initiators of various natures. Polym Sci Ser B 51:219–225
101. Islamova RM, Sadykova GR, Monakov YB, Voloshin YZ, Makarov IS, Bubnov YN (2009) Preparation of poly(methyl methacrylate) in the presence of ferrocenyl-containing Fe(II) semi- and clathrochelates. Russ J Appl Chem 82:1467–1471
102. Monakov YB, Islamova RM, Sadykova GR, Voloshin YZ, Bubnov YN, Makarov IS (2009) Radical polymerization of methyl methacrylate in the presence of ferrocenyl-containing clathrochelates of Fe(II) and various initiators. In: Monakov YB (ed) Metal compounds for controlled radical polymerization of vinyl monomers and ion-coordinating polymerization of dienes. Transworld Research Network, pp 29–38
103. Monakov YB, Islamova RM, Sadykova GR, Voloshin YZ, Makarenko IG, Lebedev AY, Bubnov YN (2010) Clathrochelates as components of new initiating systems for complex–radical polymerization of methyl methacrylate. Dokl Chem 431:89–93
104. Monakov YB, Islamova RM, Sadykova GR, Voloshin YZ, Makarov IS, Bubnov YN (2008) RU Patent 2394045, 04 Aug 2008
105. Bell CA, Whittaker MR, Gahan LR, Monteiro MJ (2008) Outer-sphere electron transfer metal-catalyzed polymerization of styrene using a macrobicyclic ligand. J Polym Sci A Polym Chem 46:146–154
106. Bell CA, Sun Q, Zhang H, Smith SC, Bernhardt PV, Monteiro MJ (2010) Methyl acrylate polymerizations in the presence of a copper/N_3S_3 macrobicyclic cage in DMSO at 25°C. Polym Chem 1:207–212
107. Nagy S, Tsuie BM (2004) Clathrochelates as olefin polymerization catalyst components. US Patent 10/779,939, 17 Feb 2004
108. Artyushin OI, Odinets IL, Matveeva EV, Vologzhanina AV, Varzatskii OA, Lubimov SE, Voloshin YZ (2014) Clathrochelates meet phosphorous. 1. Thiophosphorylation of an iron (II) dichlathrochelate precursor, synthesis of *N,S*-donor clathrochelate ligands and their palladium (II) complexes as potent catalysts of Suzuki cross-coupling reaction. Dalton Trans 43:9677–9689
109. Schuppert B, Schink B, Troesch W (1992) Batch and continuous production of propionic acid from whey permeate by Propionibacterium acidi-propionici in a three-electrode amperometric culture system. Appl Microbiol Biotechnol 37:549–553
110. Gunther H, Simon H (1995) Artificial electron carriers for preparative biocatalytic redox reactions forming reversibly carbon hydrogen bonds with enzymes present in strict or facultative anaerobes. Biocatal Biotrans 12:1–26
111. Im SC, Sykes AG (1996) Kinetic studies on the redox reactions of *Clostridium pasteurianum* rubredoxin. J Chem Soc Dalton Trans 2219–2222
112. Andersson M, Holmberg H, Adlercreutz P (1997) Evaluation of electron mediators for the electromicrobial reduction of chloropyruvate by *Proteus vulgaris* cells. Enzym Microb Technol 20:150–156
113. Schwaneberg U, Appel D, Schmitt J, Schmid RD (2000) P450 in biotechnology: zinc driven ω-hydroxylation of *p*-nitrophenoxydodecanoic acid using P450 BM-3 F87A as a catalyst. J Biotechnol 84:249–257
114. Nazor J, Schwaneberg U (2006) Laboratory evolution of P450 BM-3 for mediated electron transfer. ChemBioChem 7:638–644
115. Bernhardt PV, Chen KI, Sharpe PC (2006) Transition metal complexes as mediator-titrants in protein redox potentiometry. J Biol Inorg Chem 11:930–936
116. Hendricks NR, Waryo TT, Arotiba O, Jahed N, Baker PGL, Iwuoha EI (2009) Microsomal cytochrome P450-3A4 (CYP3A4) nanobiosensor for the determination of 2,4-dichlorophenol-an endocrine disruptor compound. Electrochim Acta 54:1925–1931

117. Çekiç SZ, Holtmann D, Gueven G, Mangold KM, Schwaneberg U, Schrader J (2010) Mediated electron transfer with P450cin. Electrochem Commun 12:1547–1550
118. Tosstorff A, Dennig A, Ruff AJ, Schwaneberg U, Sieber V, Mangold KM, Schrader J, Holtmann D (2014) Mediated electron transfer with monooxygenases – insight in interactions between reduced mediators and the co- substrate oxygen. J Mol Catal B Enzym 108:51–58
119. Ley C, Schewe H, Strohle FW, Ruff AJ, Schwaneberg U, Schrader J, Holtmann D (2013) Coupling of electrochemical and optical measurements in a microtiter plate for the fast development of electro enzymatic processes with P450s. J Mol Catal B Enzym 92:71–78
120. Faulkner KM, Shet MS, Fisher CW, Estabrook RW (1995) Electrocatalytically driven ω-hydroxylation of fatty acids using cytochrome P450 4A1. Proc Natl Acad Sci USA 92:7705–7709
121. Estabrook RW, Shet MS, Fisher CW, Jenkins CM, Waterman MR (1996) The interaction of NADPH-P450 reductase with P450: an electrochemical study of the role of the flavin mononucleotide-binding domain. Arch Biochem Biophys 333:308–315
122. Nazor J, Dannenmann S, Adjei RO, Fordjour YB, Ghampson IT, Blanusa M, Roccatano D, Schwaneberg U (2008) Laboratory evolution of P450 BM3 for mediated electron transfer yielding an activity-improved and reductase- independent variant. Protein Eng 21:29–35
123. Paxinos AS, Günther H, Schmedding DJM, Simon H (1991) Direct electron transfer from modified glassy carbon electrodes carrying covalently immobilised mediators to a dissolved viologen accepting pyridine nucleotide oxidoreductase and dihydrolipoamide dehydrogenase. Bioelectrochem Bioenerg 25:425–436
124. Bernhardt PV, Harrowfield JM, Kim Y, Lee YH, Park YC, Sujandi S (2007) A functionalised macrobicycle complex available for surface immobilisation and protein grafting. Bull Kor Chem Soc 28:589–595
125. Sotomayor J, Costa JC, Mulazzani QG, Pina F (1989) Photodecarboxylation of citrate through ion pair photochemistry: the $Co(sepulchrate)^{3+}$·$citrate^{n-}$ (n = 1,2,3) system. J Photochem Photobiol A 49:195–202
126. Sahami S, Weaver MJ (1981) Entropic and enthalpic contributions to the solvent dependence of the thermodynamics of transition-metal redox couples. Part II Couples containing ammine and ethylenediamine ligands. J Electroanal Chem Interfacial Electrochem 122:171–181
127. Peixoto CRM, Kubota LT, Gushikem Y (1995) Matrix effect on electrochemical properties of $[Co(sepulchrate)]^{3+}$ immobilized on silica gel and modified silica gel surfaces. J Braz Chem Soc 6:83–87
128. Ritzert NL, Rodríguez-López J, Tan C, Abruña HD (2013) Kinetics of interfacial electron transfer at single-layer graphene electrodes in aqueous and nonaqueous solutions. Langmuir 29:1683–1694
129. Xuan GS, Jang S, Kwag G, Kim S (2005) Simple and convenient design of a spectroelectrochemical cell for in situ XANES measurements of adsorbed species in transmission mode. Bull Kor Chem Soc 26:671–674
130. Fantola-Lazzarini A, Lazzarini E (2002) On the positronium spin conversion reactions caused by some macrocyclic Co^{II} complexes. Radiat Phys Chem 64:403–409
131. Bertolino JR, Temperini MLA, Sala O (1989) Surface-enhanced Raman scattering of the redox pair $Co(sep)^{3+}/Co(sep)^{2+}$ on silver and copper electrodes. J Raman Spectrosc 20:725–728
132. Sutin N, Weaver MJ, Yee EL (1980) Correlations between outer-sphere self-exchange rates and reaction entropies for some simple redox couples. Inorg Chem 19:1096–1098
133. Wang K, Jordan RB (1996) Electron transfer kinetics of cobaloxime complexes. Can J Chem 74:658–665
134. Grace MR, Takagi H, Swaddle TW (1994) A cross relation in volumes of activation for electron-transfer reactions. Inorg Chem 33:1915–1920
135. He Y, Gillespie D, Boda D, Vlassiouk I, Eisenberg RS, Siwy ZS (2009) Tuning transport properties of nanofluidic devices with local charge inversion. J Am Chem Soc 131:5194–5202
136. Mauzeroll J, Hueske EA, Bard AJ (2003) Scanning electrochemical microscopy. 48. Hg/Pt hemispherical ultramicroelectrodes: fabrication and characterization. Anal Chem 75:3880–3889
137. Palomaki PKB, Civic MR, Dinolfo PH (2013) Photocurrent enhancement by multilayered porphyrin sensitizers in a photoelectrochemical cell. ACS Appl Mater Interfaces 5:7604–7612
138. Pina F, Sotomayor J, Moggi L (1990) Photochemistry of the $Co(Sep)^{3+}$–I^{-} system: evaluation of the quantum yield for the photochemical redox reaction of the ion pair. J Photochem Photobiol A 53:411–422
139. Ahmed S, Hassan IAI, Roy H, Marken F (2013) Photoelectrochemical transients for chlorine/hypochlorite formation at "Roll-On" nano-WO_3 film electrodes. J Phys Chem C 117:7005–7012
140. Balahura RJ, Wilkins RG (1983) Ligational effects on reduction of myoglobin and horseradish peroxidase by inorganic reagents. Biochim Biophys Acta 724:465–472
141. Balahura RJ, Kirby RA (1994) Kinetics and mechanism of the reduction of manganese porphyrins: reaction at the periphery of the porphyrin ring. Inorg Chem 33:1021–1025
142. Ghosh MC, Gould ES (1994) Electron transfer. 120. Reductions of the cuboidal mixed-valence molybdenum/sulfur cluster, $[Mo_4S_4(H_2O)_{12}]^{5+}$. Inorg Chim Acta 225:297–303
143. Osvath P, Lappin AG (1987) Stereoselectivity as a probe of reaction mechanism in the oxidation of [Co

$(en)_3]^{2+}$ and its derivatives by $[Co(edta)]^-$. Inorg Chem 26:195–202
144. Geetha B, Mandal AB (1997) Determination of the critical micelle concentration of the methoxy polyethylene glycol based macromonomer and partition coefficient of a new electrochemical probe using a cyclic voltammetric technique. Langmuir 13:2410–2413
145. Rezanejade Bardajee G, Hooshyar Z, Khanjari M (2013) Dye fluorescence quenching by newly synthesized silver nanoparticles. J Photochem Photobiol A 276:113–121
146. Belov AS, Varzatskii OA, Vologzhanina AV, Viswanathan S, Radecki J, Bubnov YN (2009) Synthesis, structure and electron-mediator properties of the mono- and difunctionalized macrobicyclic iron (II) tris-dioximates with thiol terminated ribbed spacer substituents. Inorg Chim Acta 362:2982–2988
147. Viswanathan S, Voloshin YZ, Radecka H, Radecki J (2009) Single molecular switch based on thiol tethered iron(II)clathrochelate on gold. Electrochim Acta 54:5431–5438
148. Liu W, Huang W, Pink M, Lee D (2010) Layer-by-layer synthesis of metal-containing conducting polymers: caged metal centers for interlayer charge transport. J Am Chem Soc 132:11844–11846
149. Liu W, Huang W, Chen CH, Pink M, Lee D (2012) Charge injection and transport in metal-containing conducting polymers: spectroelectrochemical mapping of redox activities. Chem Mater 24:3650–3658
150. Burdukov AB, Vershinin MA, Pervukhina NV, Boguslvaskii EG, Eltsov IV, Shundrin LA, Selector SL, Shokurov AV, Voloshin YZ (2014) Towards the clathrochelate-based electrochromic materials: the case study of the first iron(II) cage complex with an annelated quinoxaline fragment. Inorg Chem Commun 44:183–187
151. Bag B, Bharadwaj PK (2005) Fluorescence enhancement of a signaling system in the simultaneous presence of transition and alkali metal ions: a potential AND logic gate. Chem Commun:513–515
152. Bag B, Mukhopadhyay P, Bharadwaj PK (2006) Exocyclic coordination and translocation of metal ions in laterally non-symmetric aza cryptands. Curr Sci 91:1166–1175
153. Sadhu KK, Bharadwaj PK (2008) Translocation of copper within the cavity of cryptands: reversible fluorescence signaling. Chem Commun:4180–4182
154. Sadhu KK, Sen S, Bharadwaj PK (2011) Cryptand derived fluorescence signaling systems for sensing Hg(II) ion: a comparative study. Dalton Trans 40:726–734
155. Basallote MG, Blanco E, Blázquez M, Fernández-Trujillo MJ, Litrán R, Máñez MÁ, Ramírez del Solar M (2003) Exploring the properties and optical sensing capability of sol–gel materials containing a covalently bonded binucleating cryptand. Chem Mater 15:2025–2032
156. Reisfeld R, Saraidarov T, Gaft M, Pietraszkiewicz M (2007) Luminescence of cryptate-type Eu^{3+} complexes incorporated in inorganic and ormocer sol–gel matrices. Opt Mater 29:521–527
157. Rodríguez-Cortiñas R, Avecilla F, Platas-Iglesias C, Imbert D, Bünzli JCG, de Blas A, Rodríguez-Blas T (2002) Structural and photophysical properties of heterobimetallic 4f-Zn iminophenolate cryptates. Inorg Chem 41:5336–5349
158. Platas C, Avecilla F, de Blas A, Rodríguez-Blas T, Geraldes CFGC, Tóth E, Merbach AE, Bünzli JCG (2000) Mono- and bimetallic lanthanide(III) phenolic cryptates obtained by template reaction: solid state structure, photophysical properties and relaxivity. J Chem Soc Dalton Trans 4:611–618
159. Doffek C, Alzakhem N, Molon M, Seitz M (2012) Perdeuterated lanthanoid cryptates: extraordinarily bright near-IR luminophores. Paper presented at the 40 international conference on coordination chemistry, Valencia Conference Centre, Valencia, 9–13 Sept 2012
160. Doffek C, Alzakhem N, Molon M, Seitz M (2012) Rigid, perdeuterated lanthanoid cryptates: extraordinarily bright near-IR luminophores. Inorg Chem 51:4539–4545
161. Voloshin YZ, Varzatskii OA, Belov AS, Bubnov YN (2006) Cage complexes as a molecular "scaffold" for assembling of polyfunctional and multicentered systems: synthesis of ribbed-functionalized dopamine-containing iron(II) clathrochelate and its properties as a receptor for boric acid. Russ Chem Bull Int Ed 55:1119–1125
162. Varzatskii OA, Shul'ga SV, Lebed EG, Kozlowski H, Gumienna-Kontecka E, Szyrwiel Ł, Voloshin YZ (2014) ESI-MS study of ionization pathways and cation-receptor properties of the iron(II) mono- and bis-clathrochelates. Macroheterocycles 7:302–310
163. Balkus Jr KJ, Kim SJ, Sargeson AM (1997) Synthesis and characterization of GaPO4 molecular sieves using metal complexes as templates. Mater Res Soc Symp Proc 454:217–224
164. Klingelhöfer S, Wiebcke M, Behrens P (2007) Characterization of two zincophosphates (UH-6 and UH-10) synthesized using the cobalt(III) sarcophagine complex as structure-directing agent. Z Anorg Allg Chem 633:113–119
165. Han SC, Sujandi S, Park SE (2005) Fuctionalization of SBA-16 mesoporous materials with cobalt(III) cage amine complex. Bull Kor Chem Soc 26:1381–1384
166. Koutsantonis GA, Nealon GL, Buckley CE, Paskevicius M, Douce L, Harrowfield JM, McDowall AW (2007) Wormlike micelles from a cage amine metallosurfactant. Langmuir 23:11986–11990
167. Lee KY, Hwang J, Lee YW, Kim J, Han SW (2007) One-step synthesis of gold nanoparticles using azacryptand and their applications in SERS and catalysis. J Colloid Interface Sci 316:476–481

168. Tsvetkov F, White J (1988) Aggregation of organometallic complex pillars in synthetic fluorhectorite and montmorillonite. J Am Chem Soc 110:3183–3187
169. Tsvetkov F, White J (1989) Structural order and orientation of organometallic pillars on synthetic fluorhectorite and montmorillonite surfaces. Chem Mater 1:149–158
170. Dailey JS, Pinnavaia TJ (1992) Intercalative reaction of a cobalt(III) cage complex, $Co(sep)^{3+}$, with magadiite, a layered sodium silicate. J Incl Phenom Mol Recognit Chem 13:47–61
171. Voloshin YZ, Lebed EG, Erdyakov SY, Makarenko IG, Lyssenko KA, Starikova ZA, Gurskii ME, Bubnov YN (2006) Application of the allylboration reaction of terminal acetylenes with allyldihaloboranes for the preparation of capping agents for the synthesis of precursors of polymeric iron(II) clathrochelates. Russ Chem Bull Int Ed 55:1971–1981
172. Wise MD, Severin K (2015) Functionalised clathrochelate complexes – new building blocks for supramolecular structures. Chimia 69:191–195
173. Pavlov AA, Nelyubina YV, Kats SV, Penkova LV, Efimov NN, Dmitrienko AO, Vologzhanina AV, Belov AS, Voloshin YZ, Novikov VV (2016) Polymorphism in a cobalt-based single-ion magnet tuning its barrier to magnetization relaxation. J Phys Chem Lett 7:4111–4116
174. Ardavan A, Bowen AM, Fernandez A, Fielding AJ, Kaminski D, Moro F, Muryn CA, Wise MD, Ruggi A, McInnes EJL, Severin K, Timco GA, Timmel CR, Tuna F, Whitehead GFS, Winpenny REP (2015) Engineering coherent interactions in molecular nanomagnet dimers. npj Quant Inf. doi:10.1038/npjqi.2015.12
175. McInnes EJL, Timco GA, Whitehead GFS, Winpenny REP (2015) Heterometallic rings: their physics and use as supramolecular building blocks. Angew Chem Int Ed 54:14244–14269
176. Voloshin YZ, Varzatskii OA, Vorontsov II, Antipin MY (2005) Tuning a metal's oxidation state: the potential of clathrochelate systems. Angew Chem Int Ed 44:3400–3402
177. Zelinskii G, Novikov V, Vologzhanina A, Voloshin Y, Bubnov Y, Mokhir A (2015) Template synthesis of iron and cobalt(II) (per)fluoroclathrochelates with superhydrophobic exterior. In: International symposia on metal complexes, Wrocław, June 24–28 2015
178. El Ghachtoulia S, Guillota R, Leibl W, Aukauloo A (2014) Intramolecular photoinduced electron transfer in cobalt(II) chlathrochelate-zinc porphyrin assemblies leading to long-lived Co(I) species. J Porphyrins Phthalocyanines 18:1125–1130
179. Gamage NDH, Mei Y, Garcia J, Allen MJ (2010) Oxidatively stable, aqueous europium(II) complexes through steric and electronic manipulation of cryptand coordination chemistry. Angew Chem Int Ed 49:8923–8925
180. Tomyn S, Shylin SI, Bykov D, Ksenofontov V, Gumienna-Kontecka E, Bon V, Fritsky IO (2017) Indefinitely stable iron(IV) cage complexes formed in water by air oxidation. Nat Commun. doi:10.1038/ncomms14099

Appendix A

List of Covalent and Coordination Capsules

1.

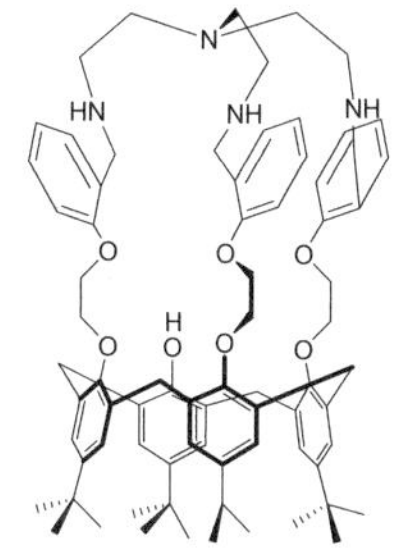

2.

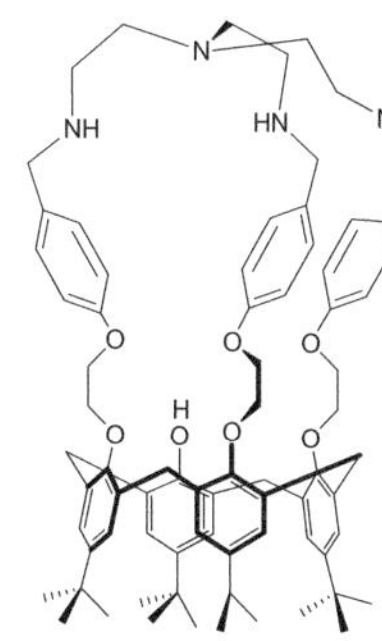

3.

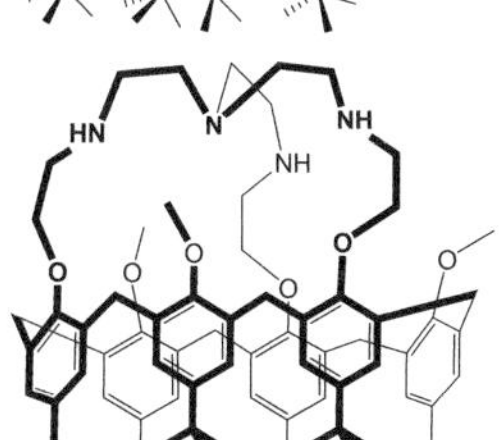

4.

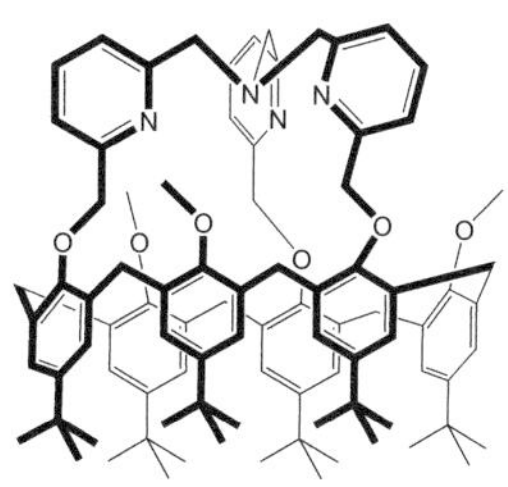

5.

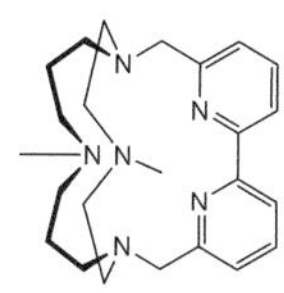

6.

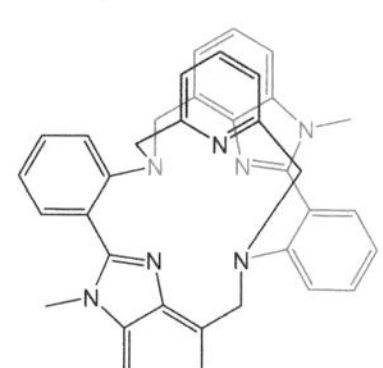

7.

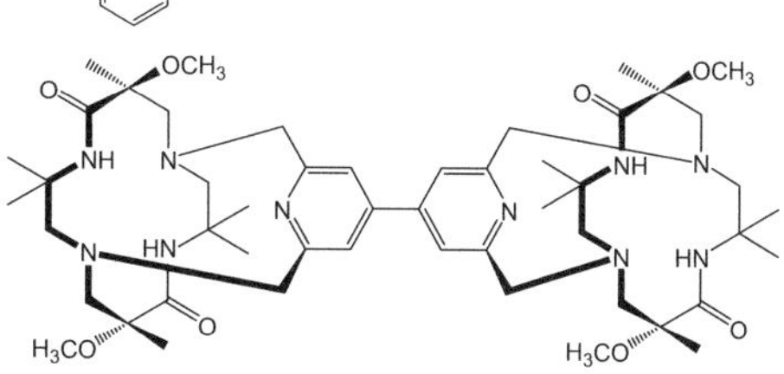

Y. Voloshin et al., *Cage Metal Complexes*, DOI 10.1007/978-3-319-56420-3

8.

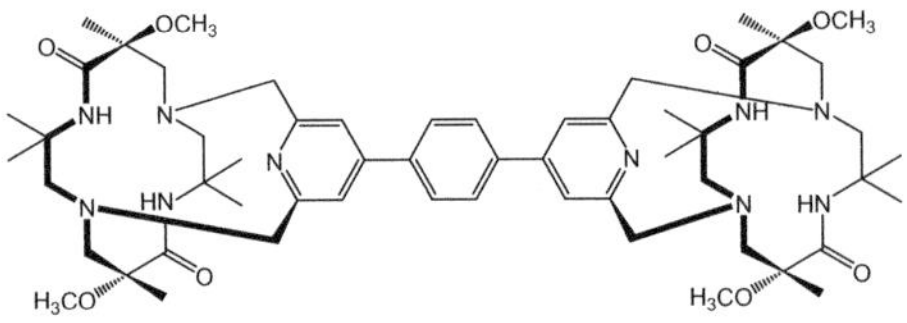

9.

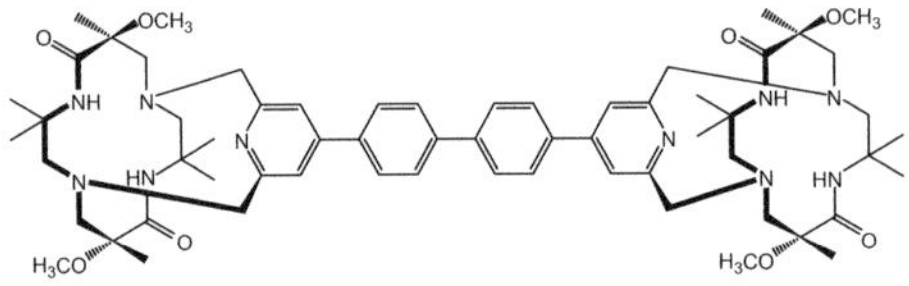

10.

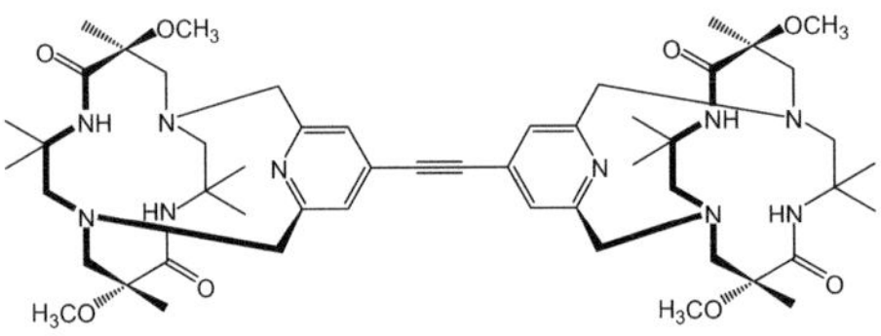

11.

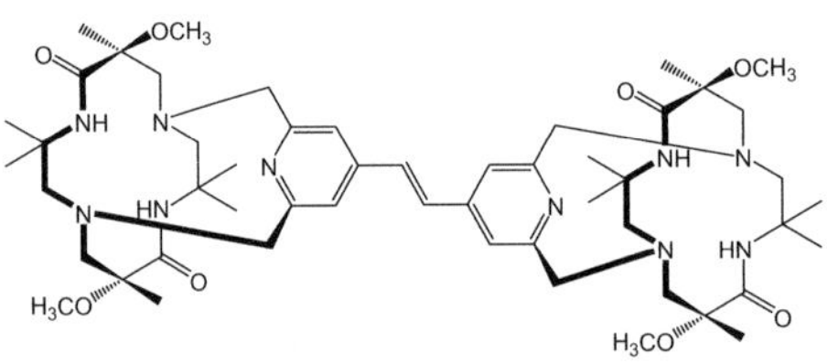

12.

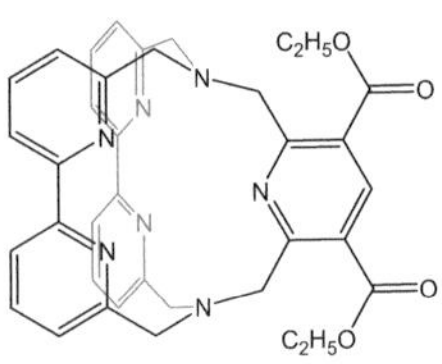

13.

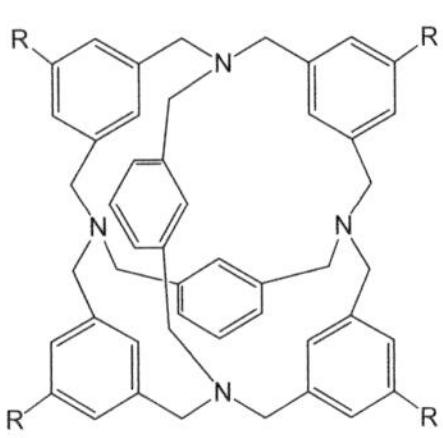

R = OCH_3

14.

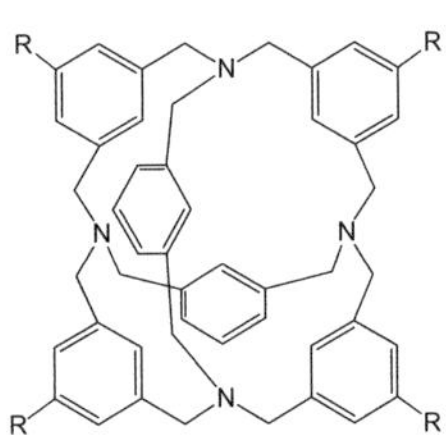

R = *tert*-C_4H_9

15.

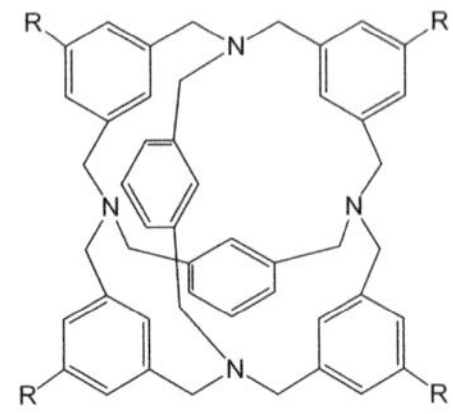

R = NO_2

16.

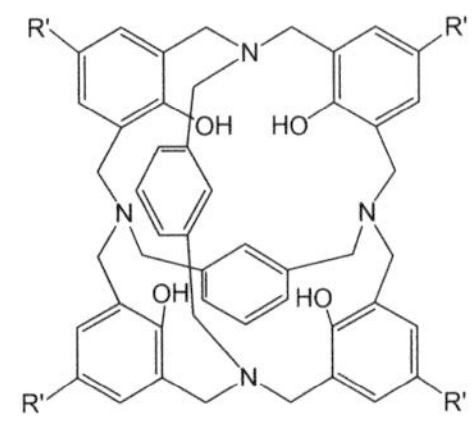

R' = Cl

17.

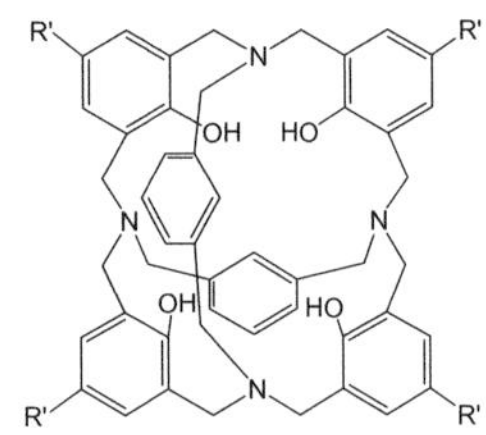

R' = Br

18.

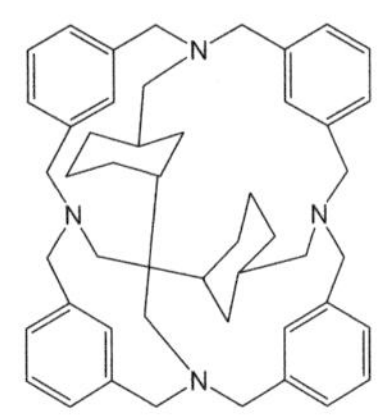

19.

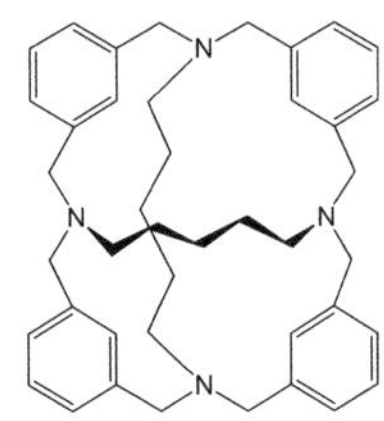

20.

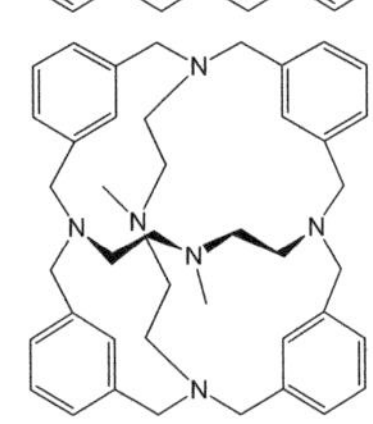

21.

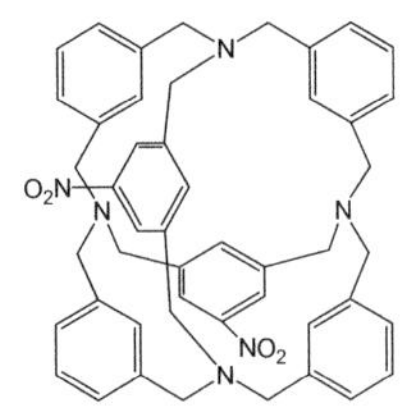

22.

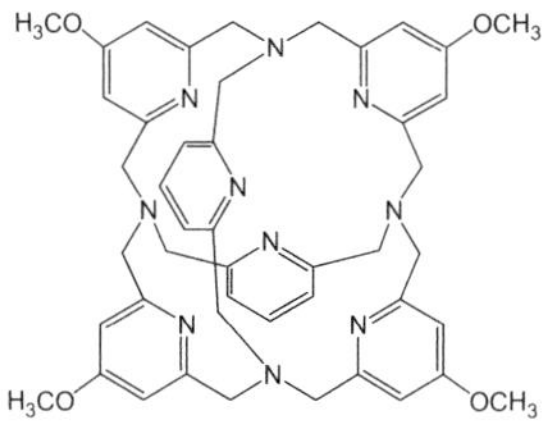

23.

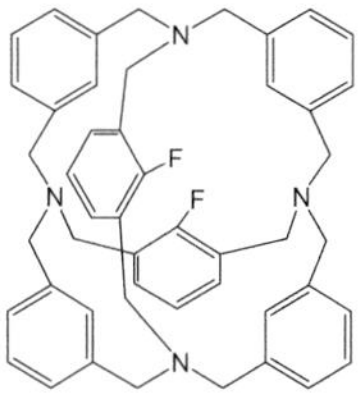

24.

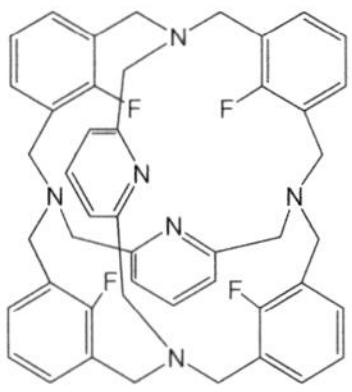

25.

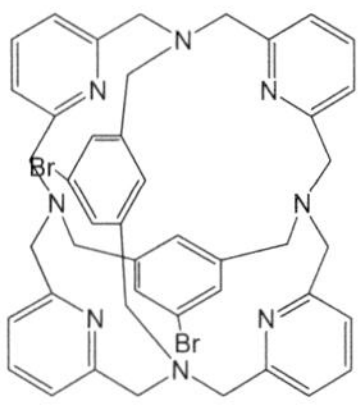

26.

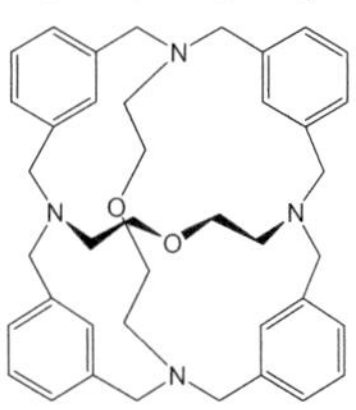

27.

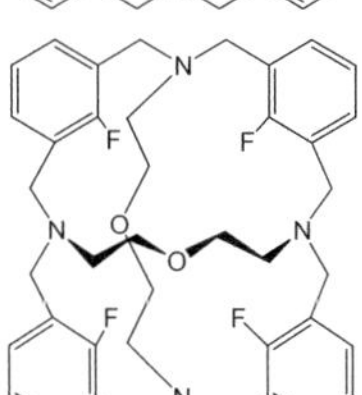

28.

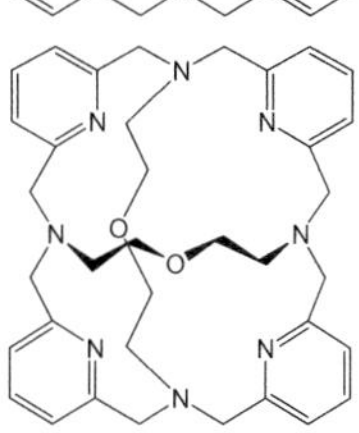

29.

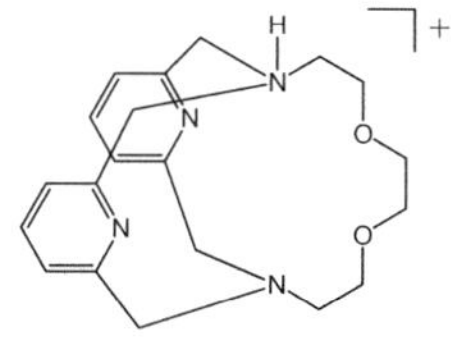

30.

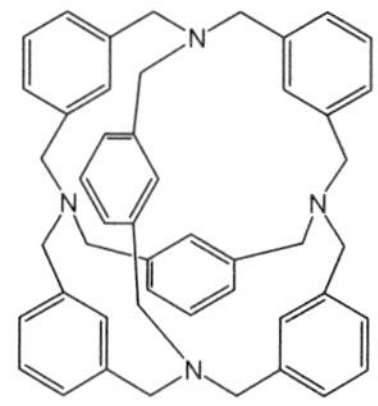

31.

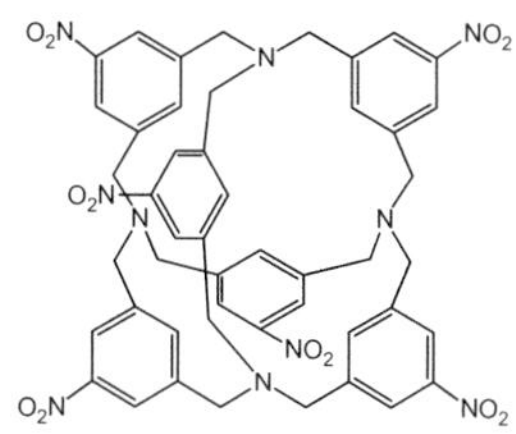

32.

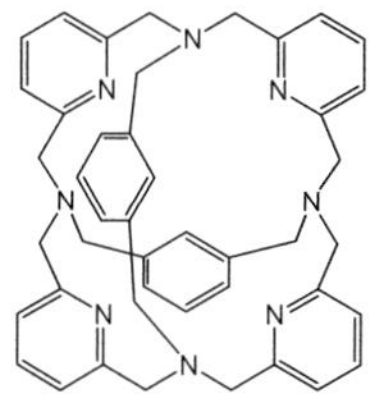

33.

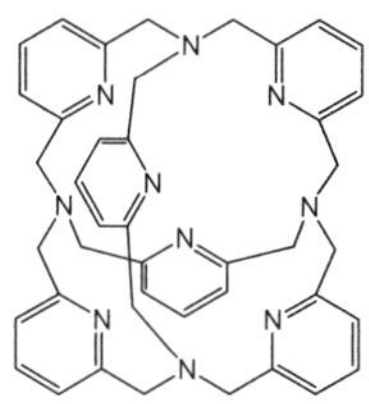

34.

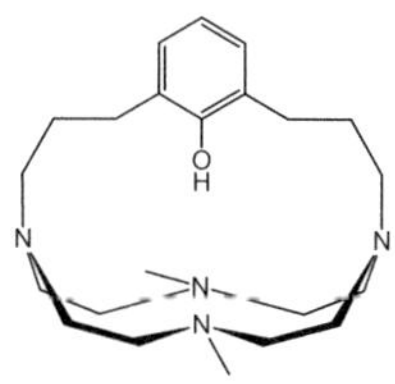

35.

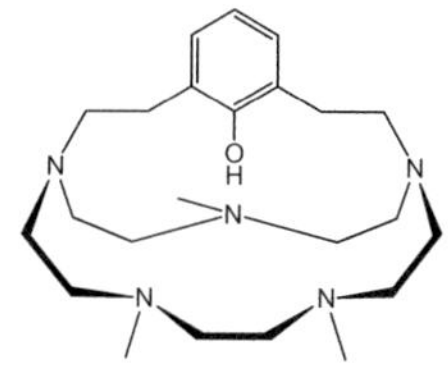

36.

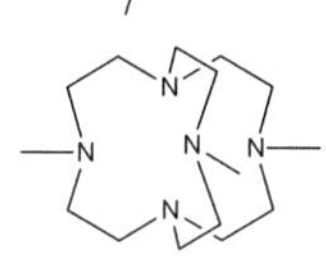

37.

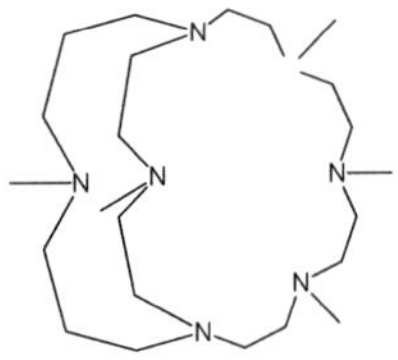

38.

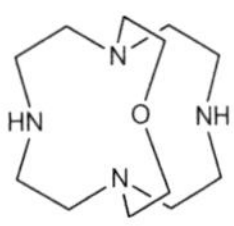

39.

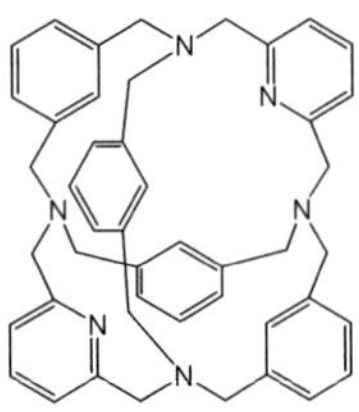

40.

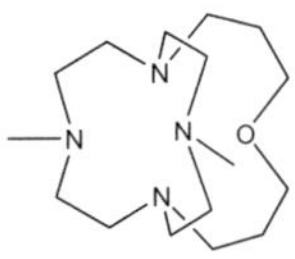

41.

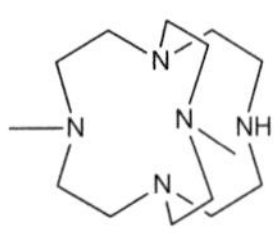

42.

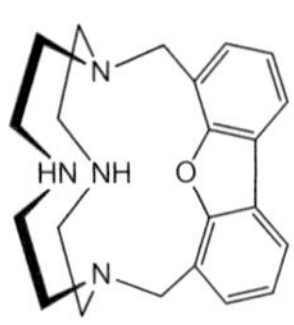

43.

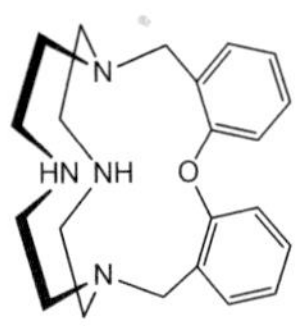

44.

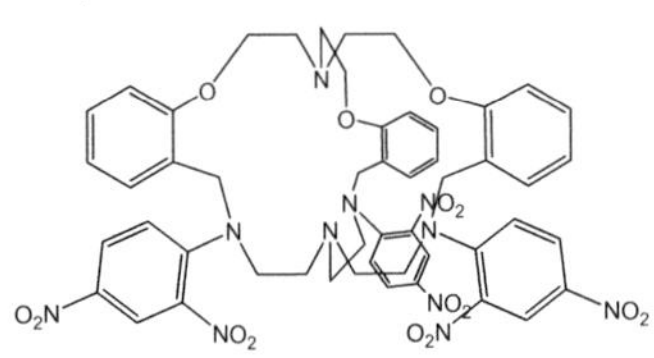

45.

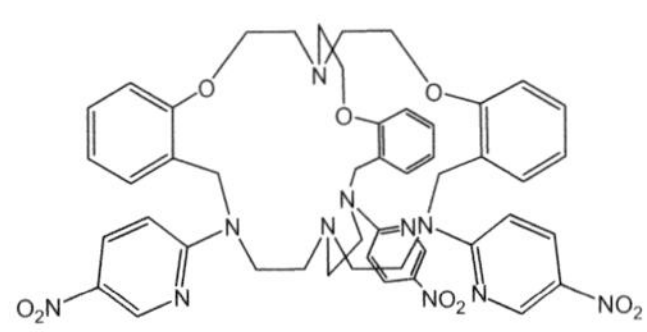

46.

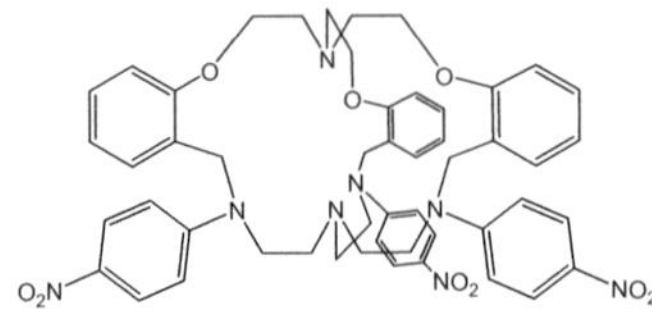

47.

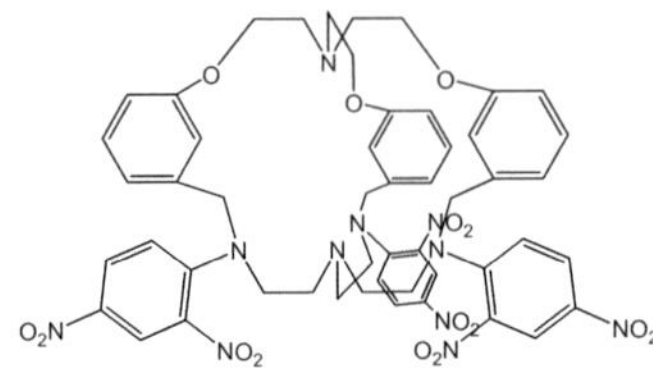

48.

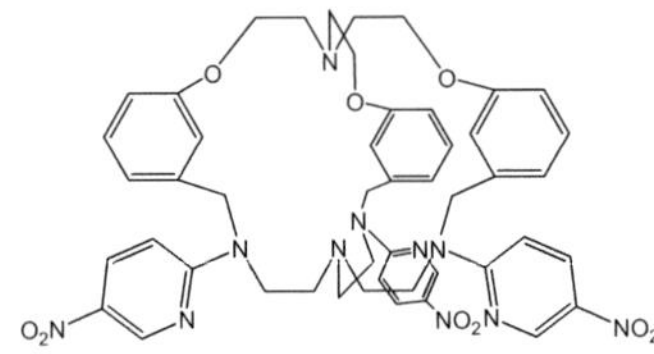

49.

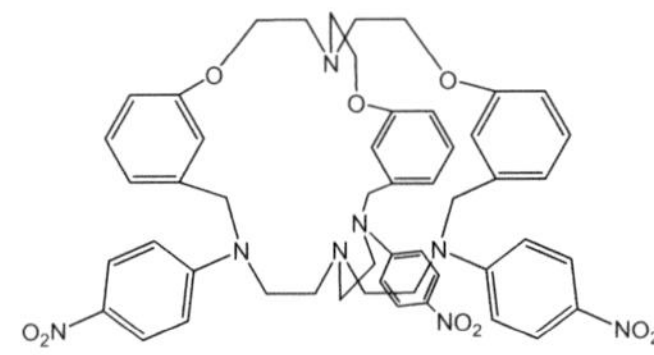

50.

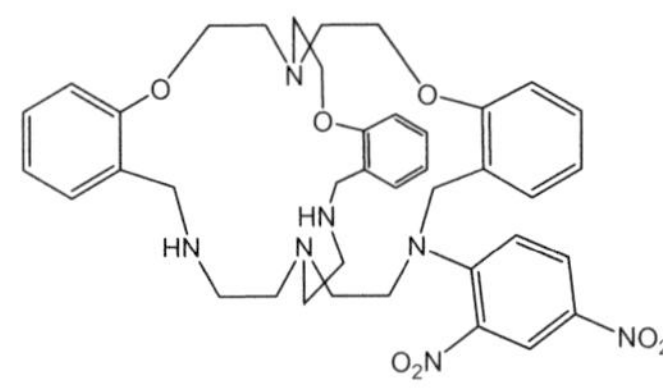

51.

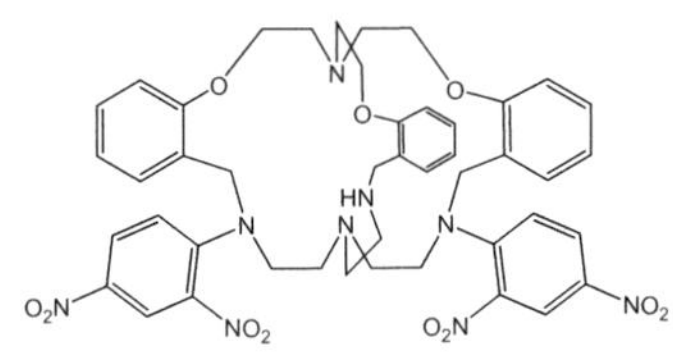

52.

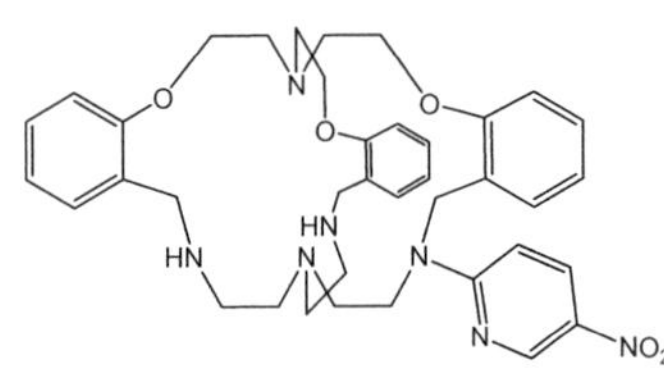

53.

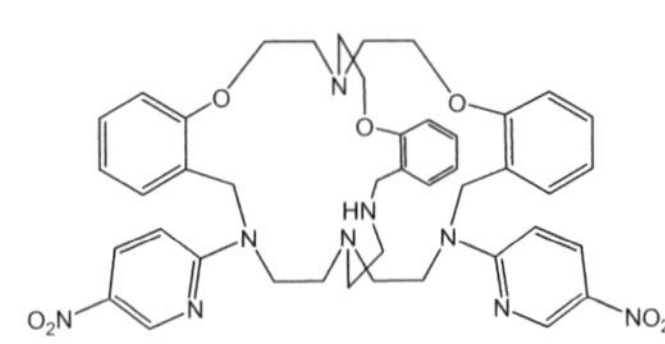

54. 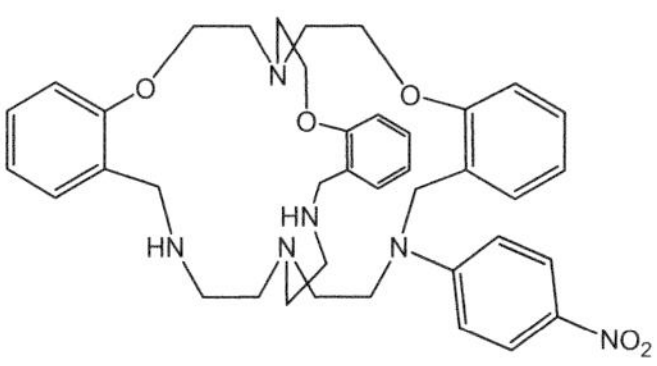

55. 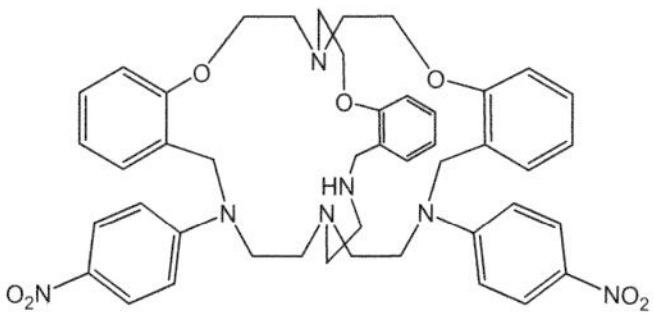

56. 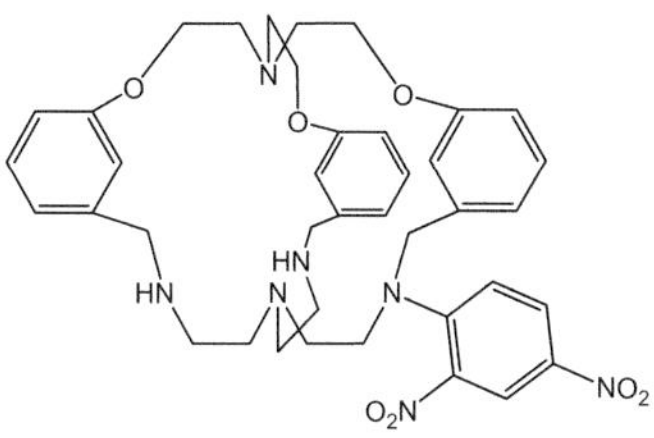

57. 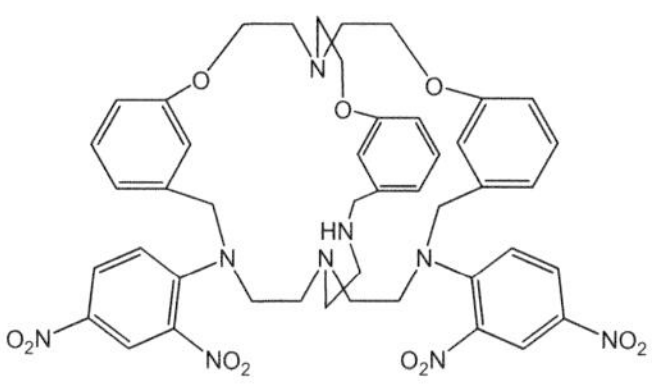

58. 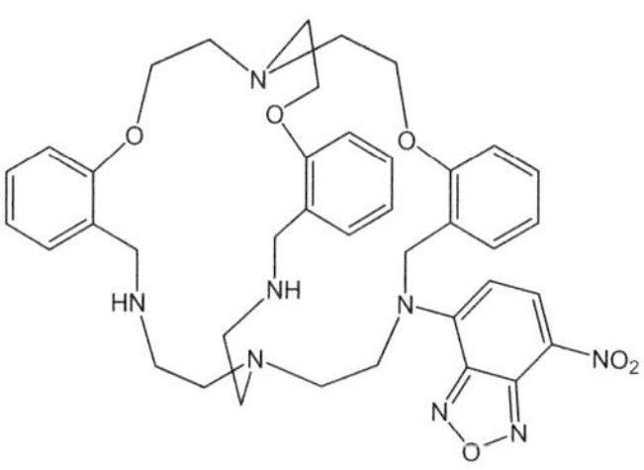

59. 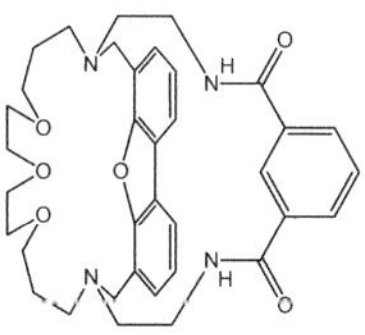

60. 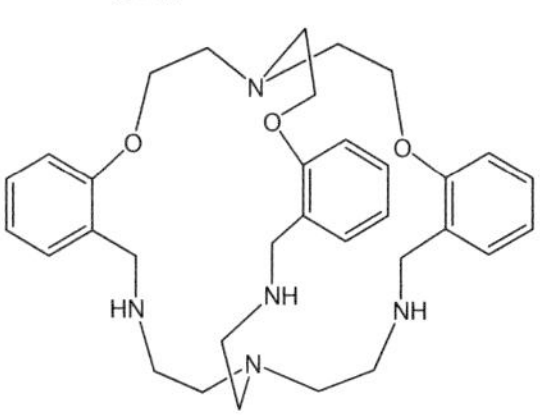

61. 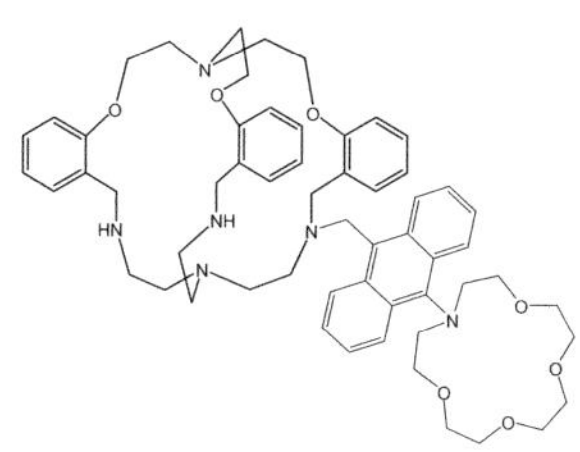

62. 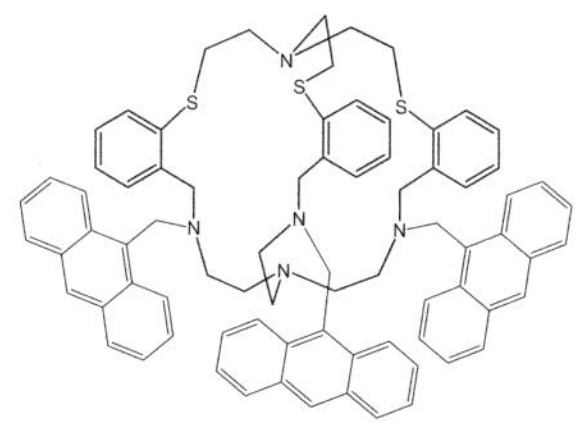

63. 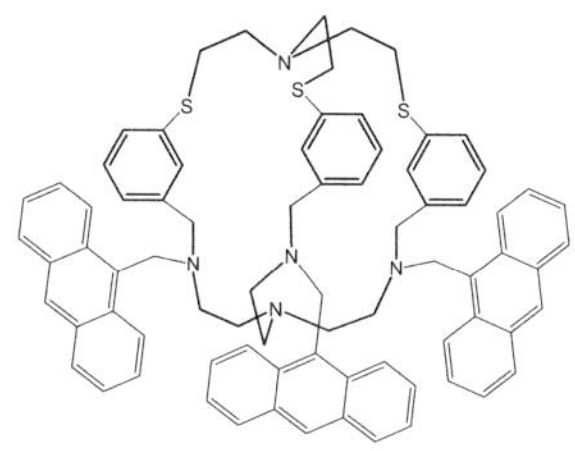

64. 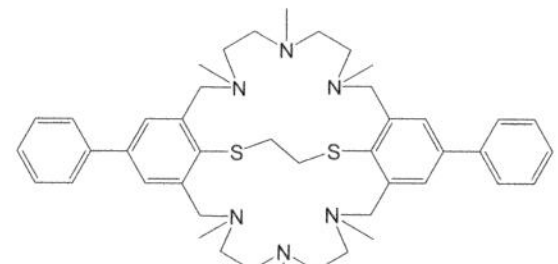

65. 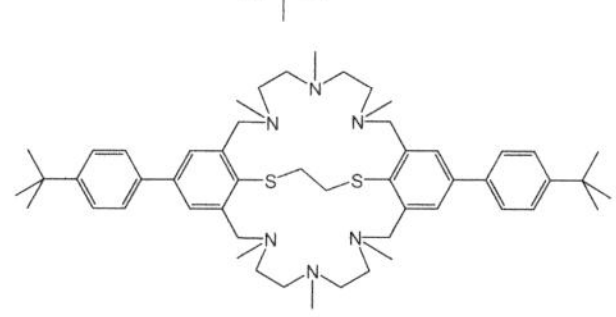

66. 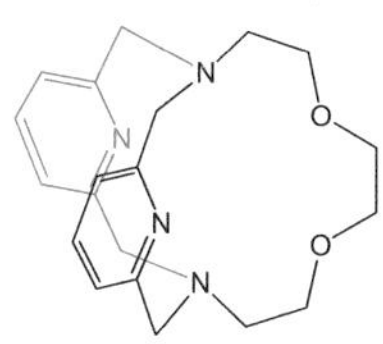

67. 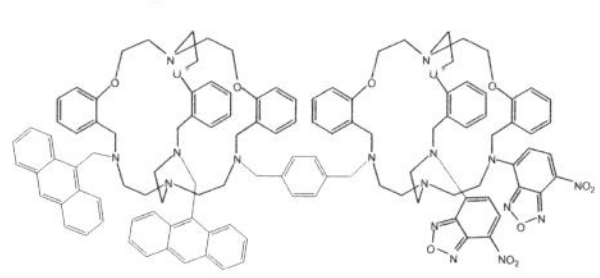

68. 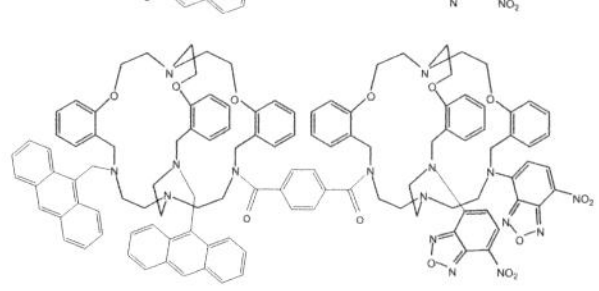

69.
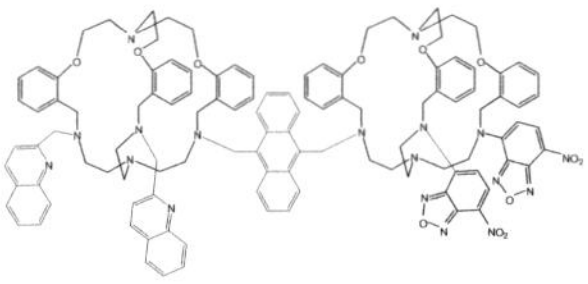

70.
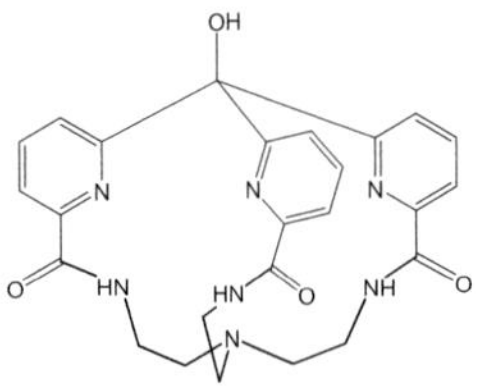

71.
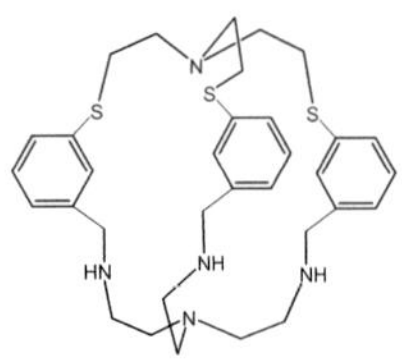

72.
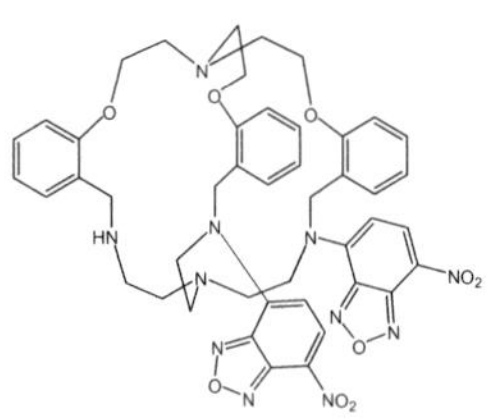

73.
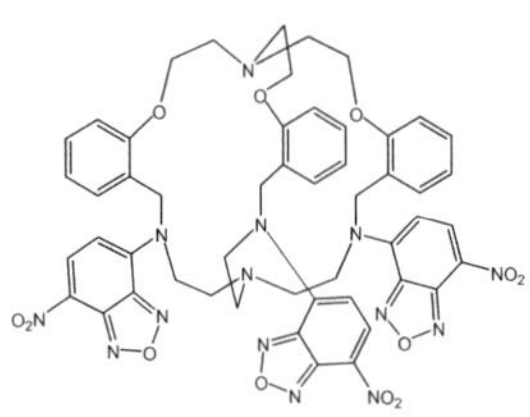

74.
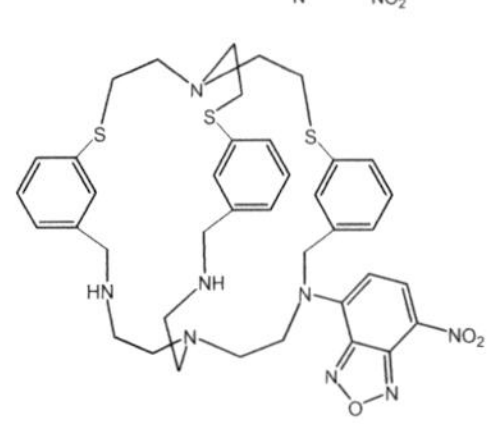

75.
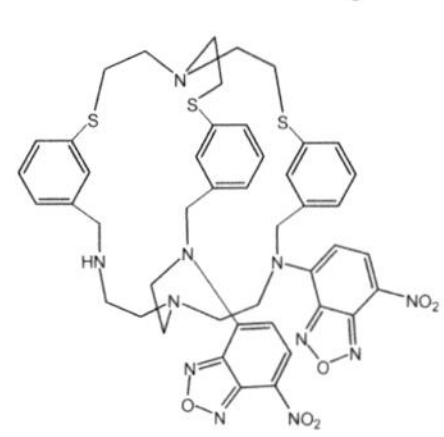

76.
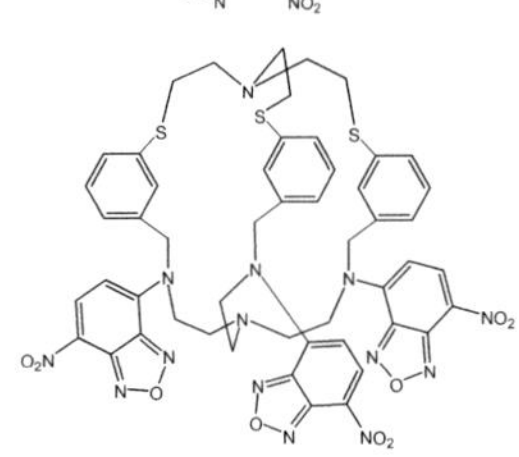

77.
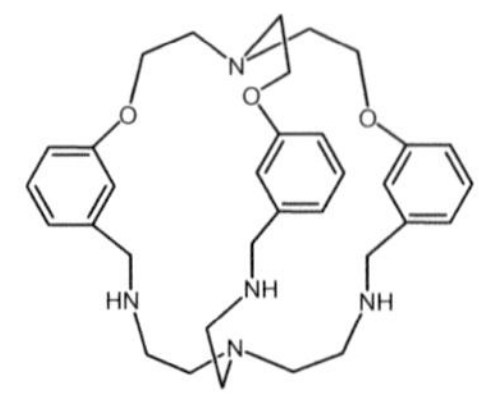

78.
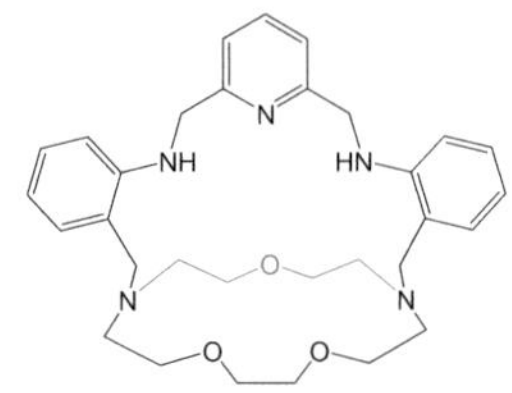

79.
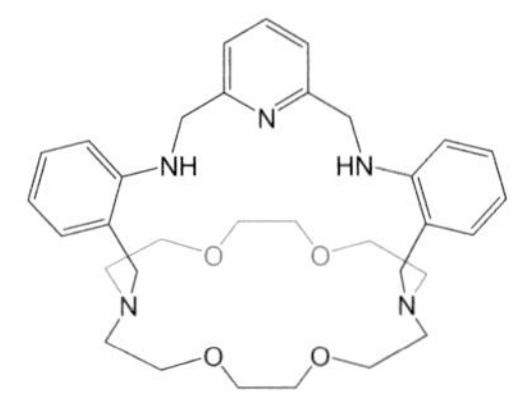

80.
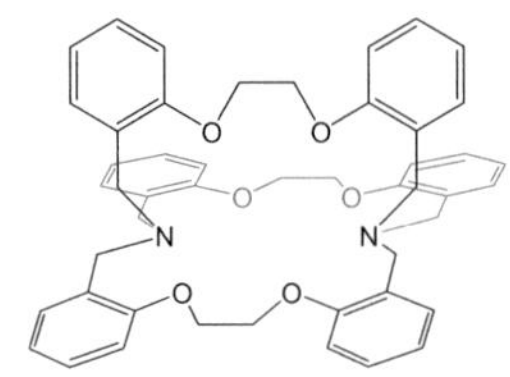

81.
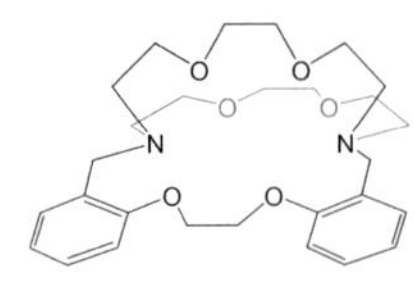

82.
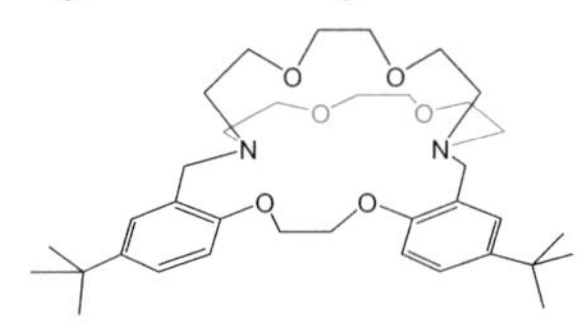

83.
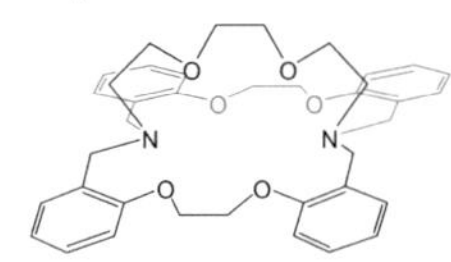

84.
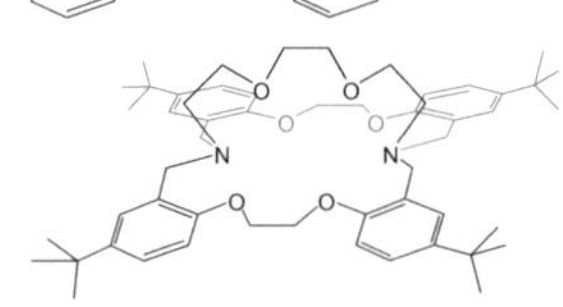

85.
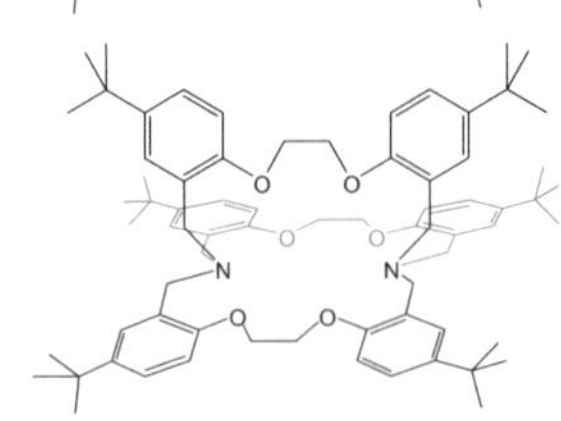

86.
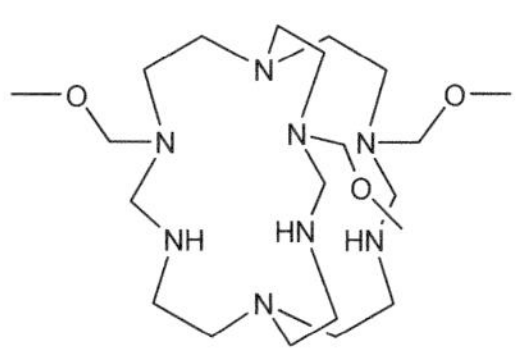

87.
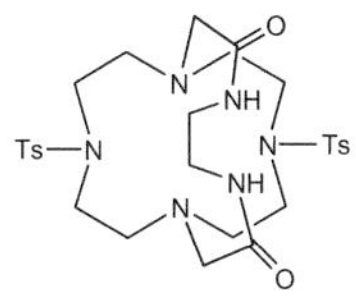

88.
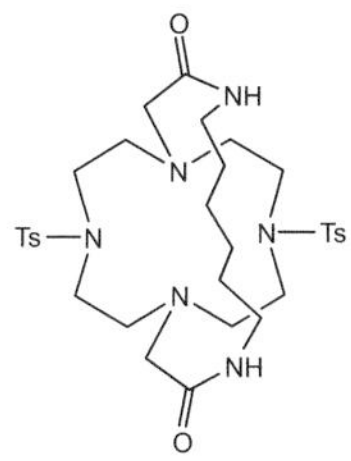

89.
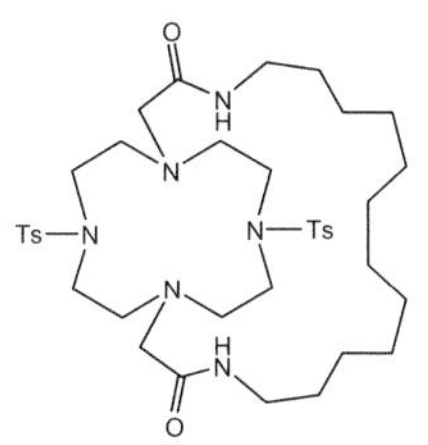

90.
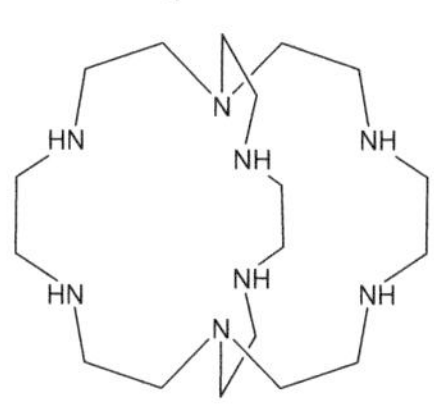

91.
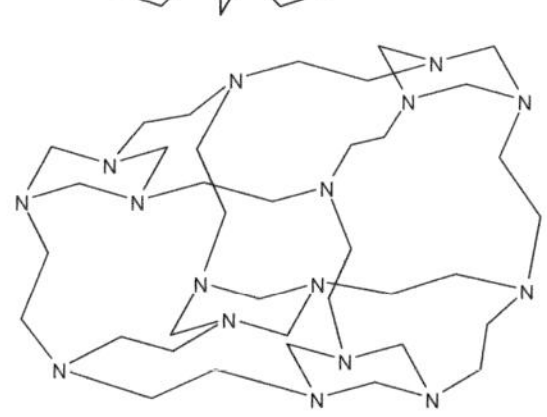

92.
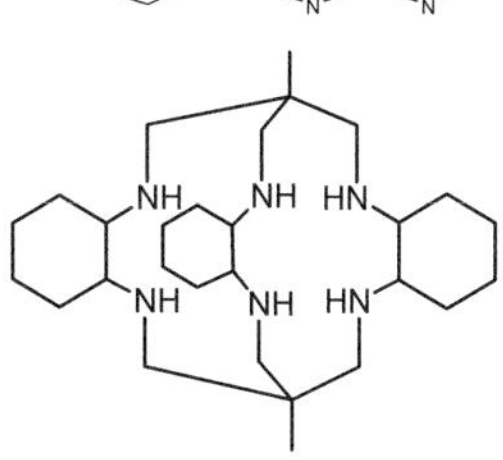

93.
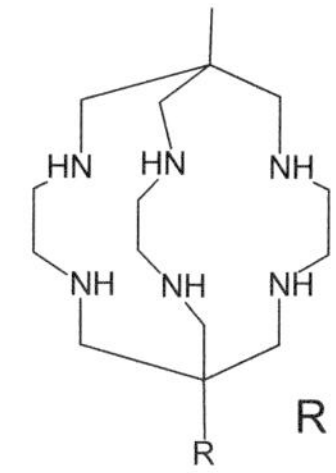
R = n-C_8H_{17}

94.
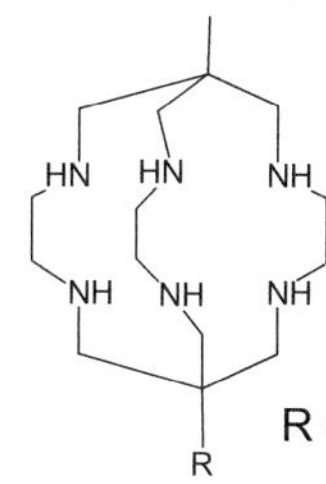
R = n-$C_{12}H_{25}$

95.
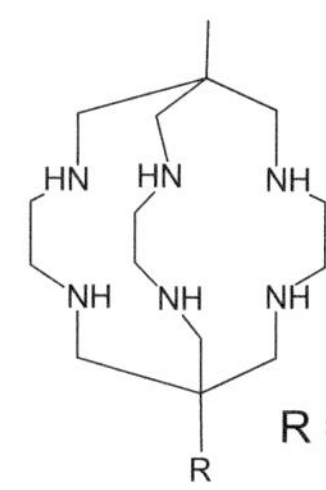
R = n-$C_{16}H_{33}$

96.
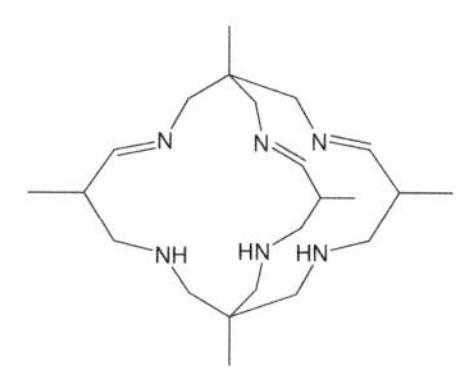

97.
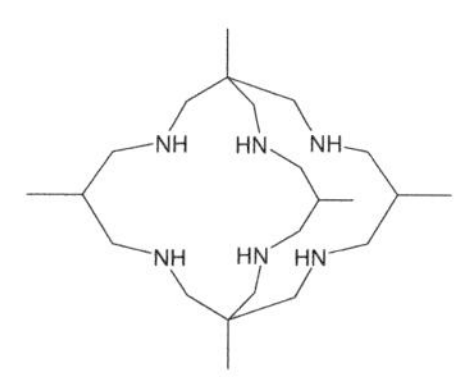

98.
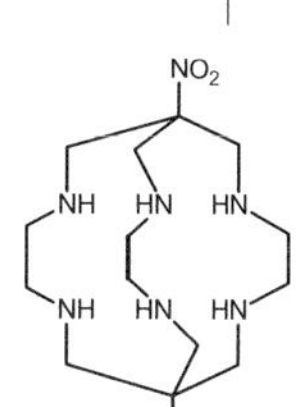

99.
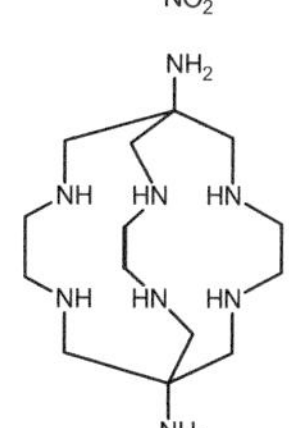

100.

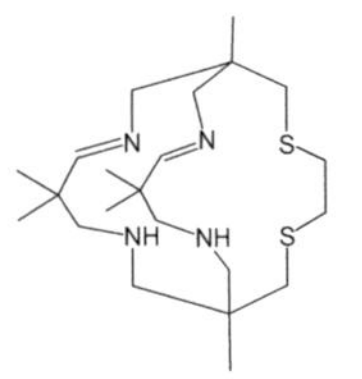

101.

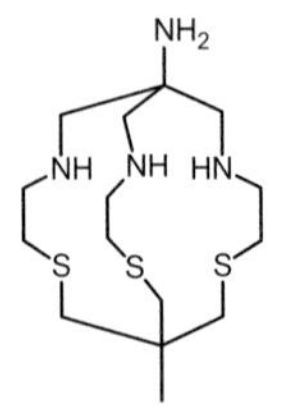

102.

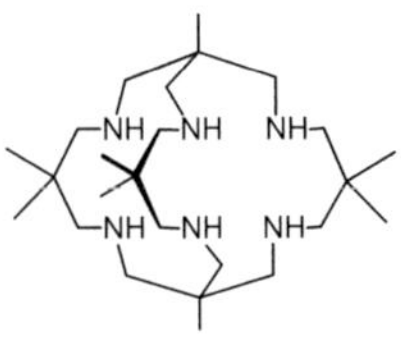

103.

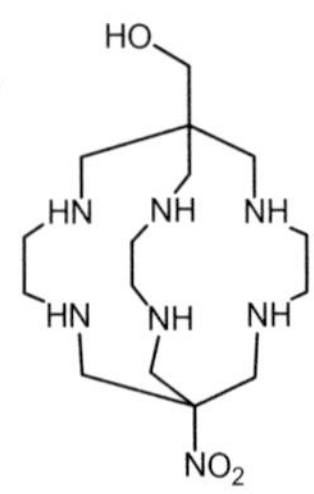

104.

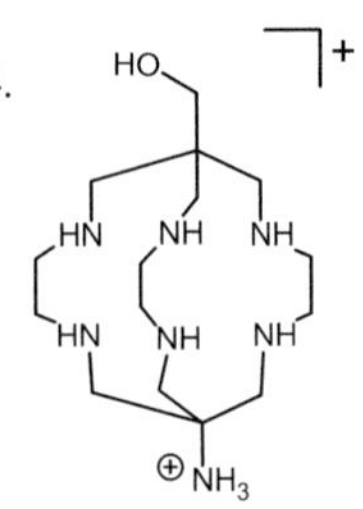

105.

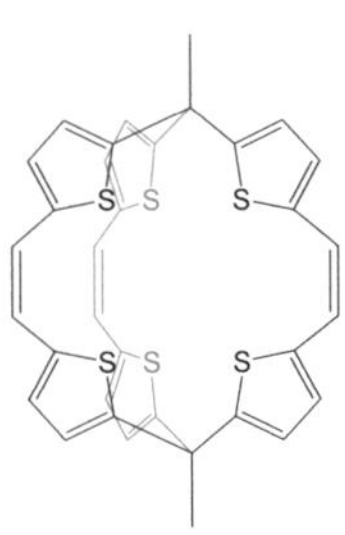

106.

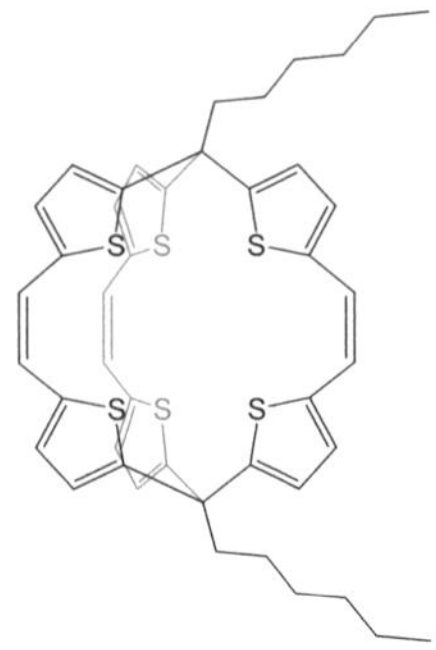

107.

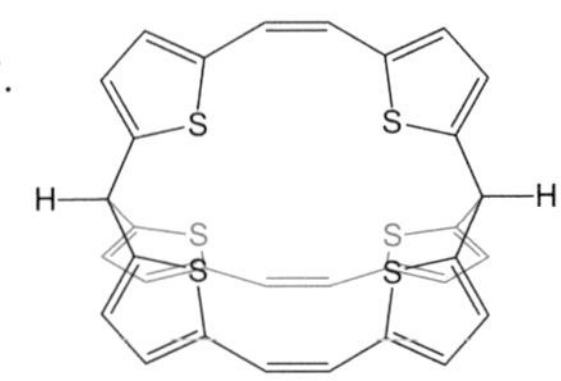

108.

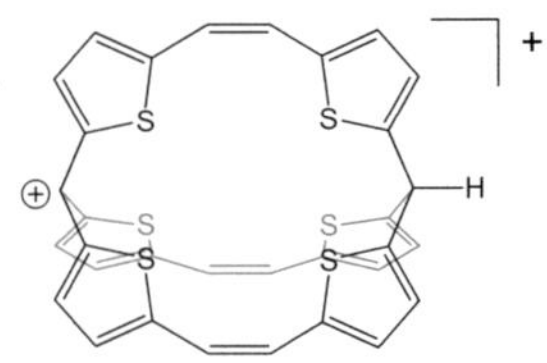

109.

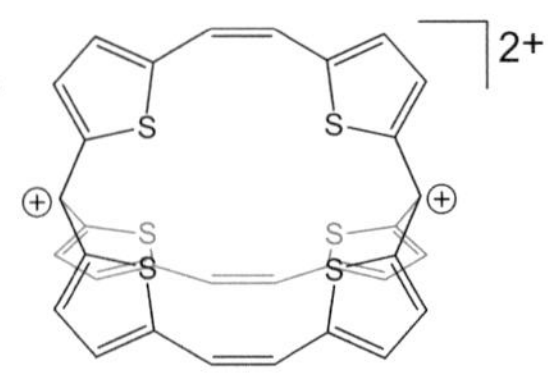

110.

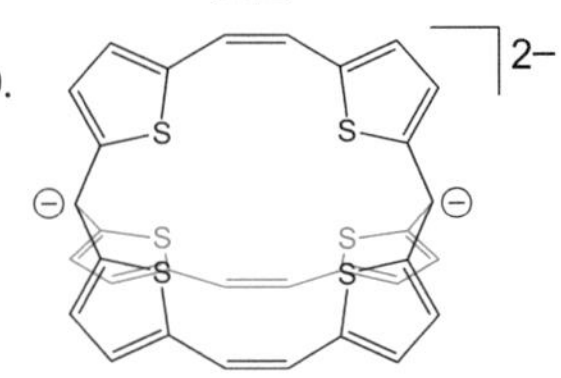

111.

112.

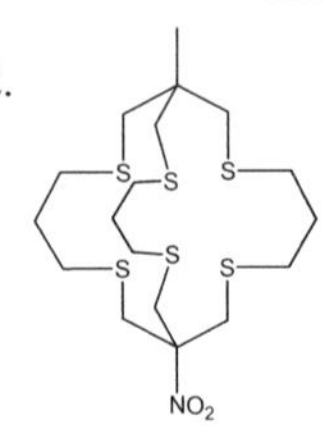

113.

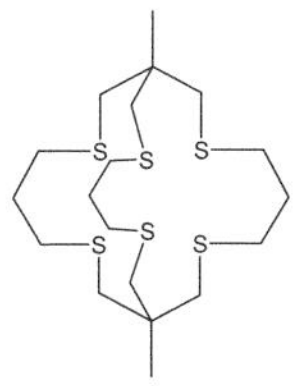

114.

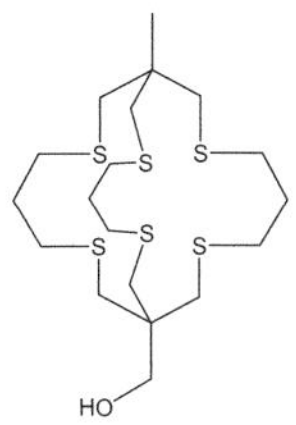

115.

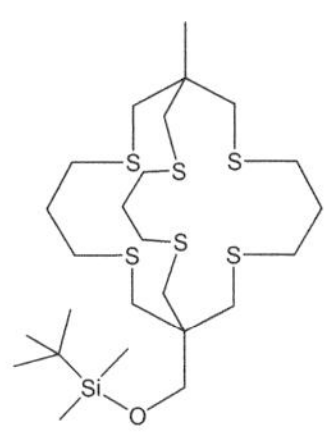

116.

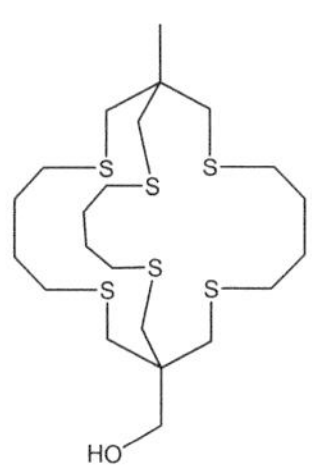

117.

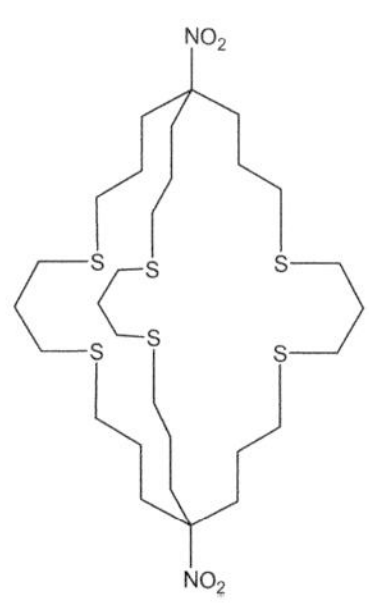

118.

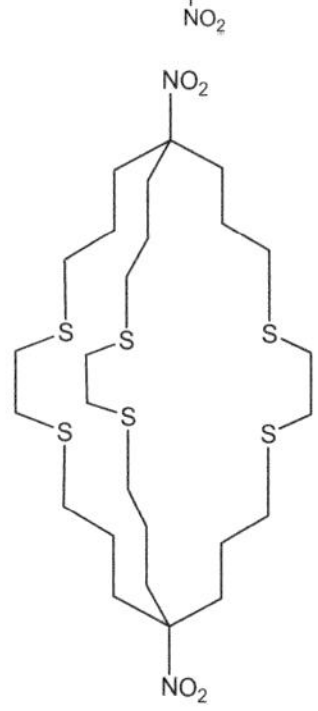

119.

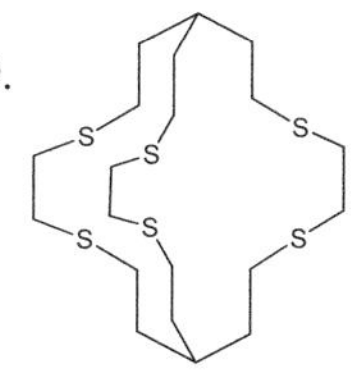

120.

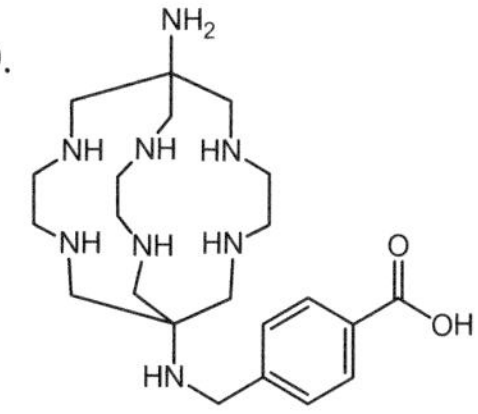

121.

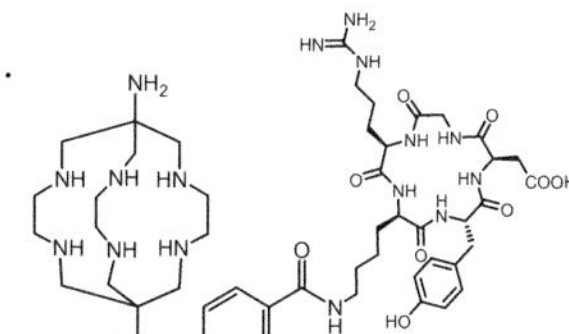

122.

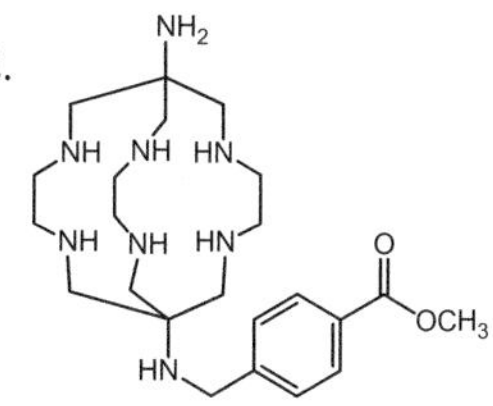

123.

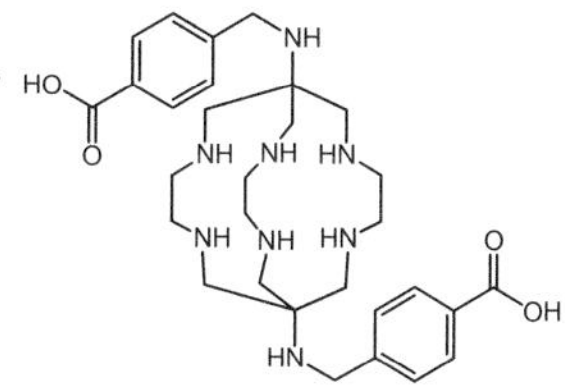

124.

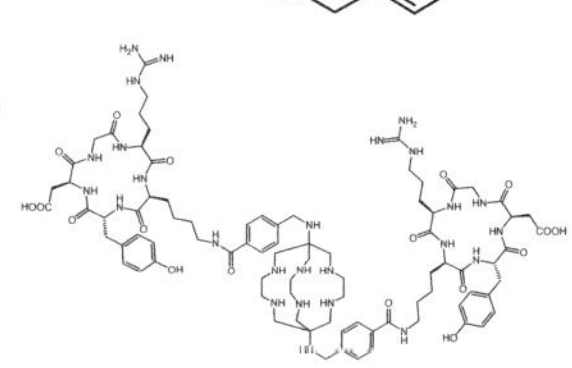

125.

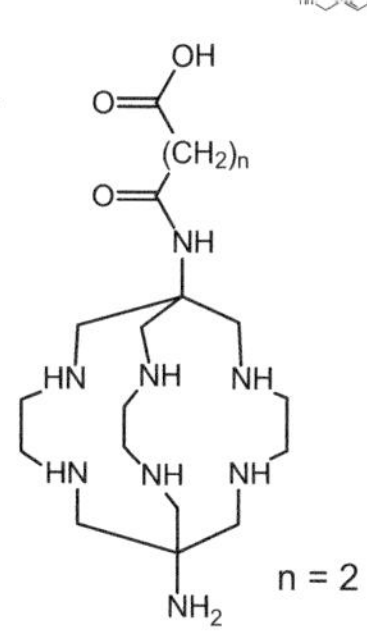

126.

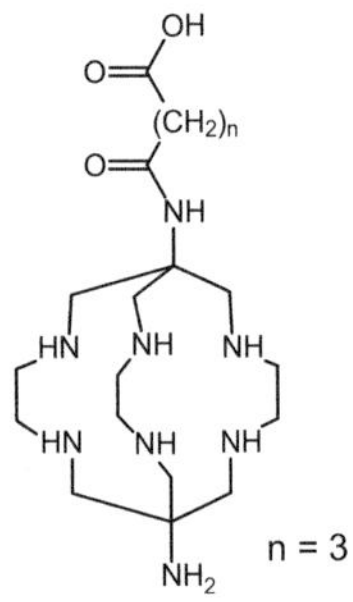

127.

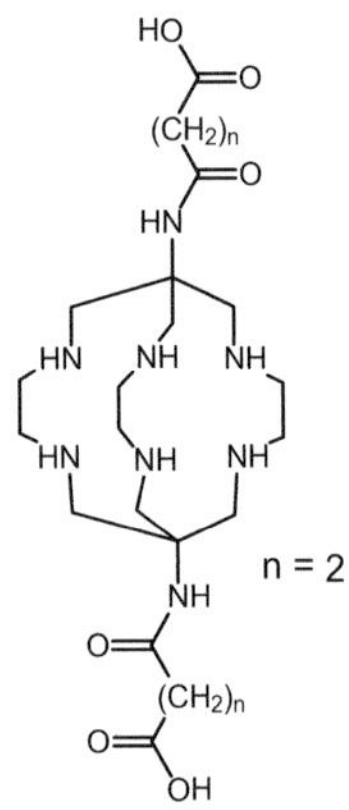

128.

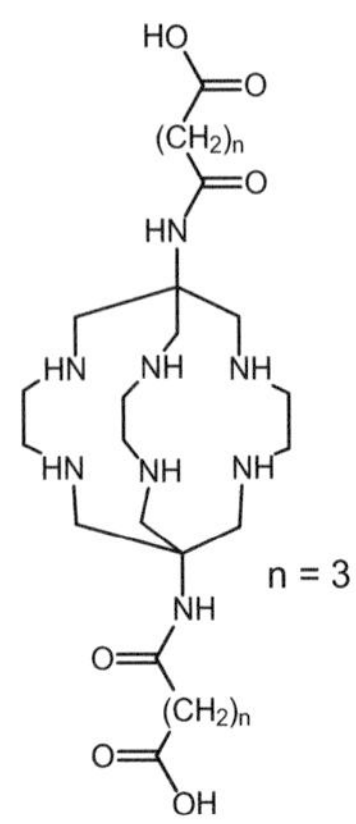

129.

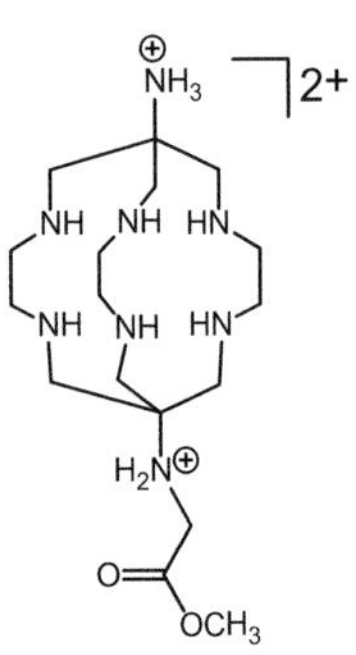

130.

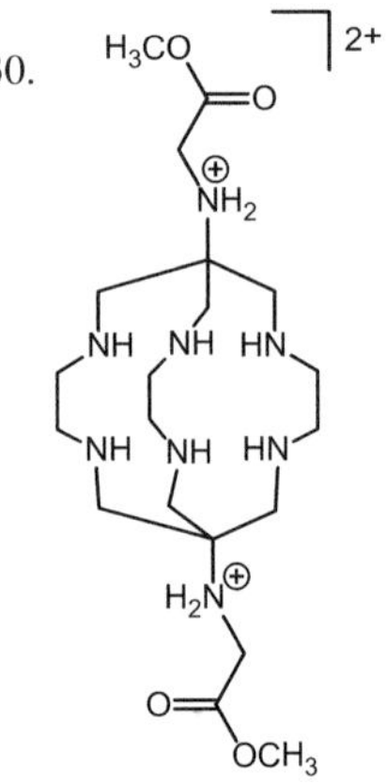

131.

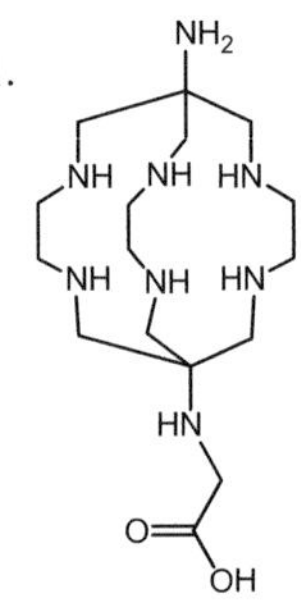

132.

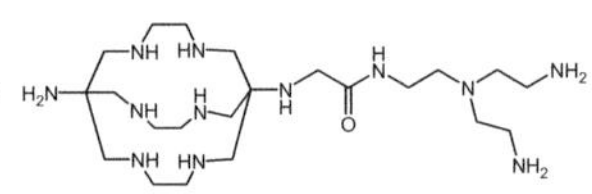

133.

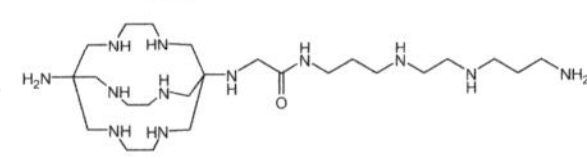

134.

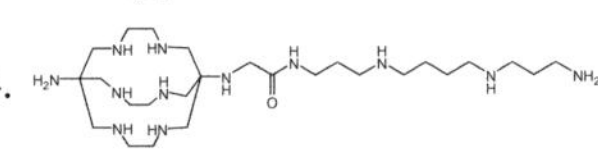

135.

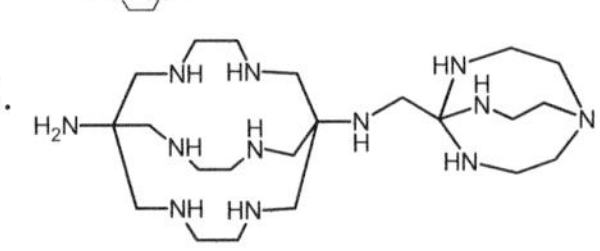

136.

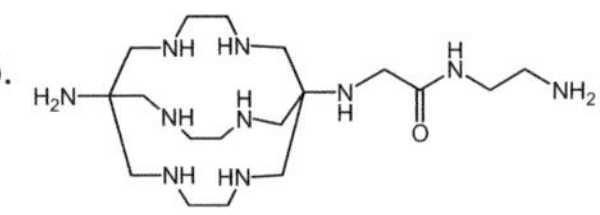

137.

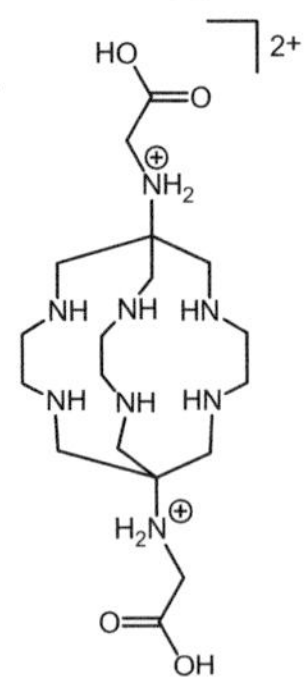

138.
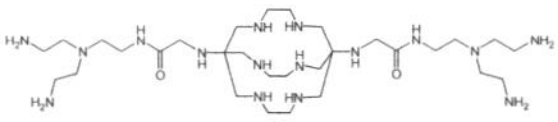

139.

140.
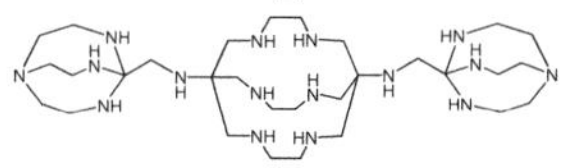

141.
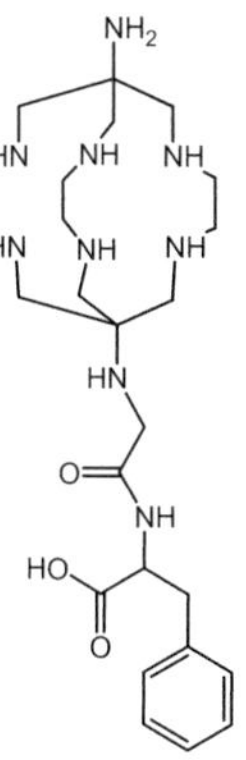

142.
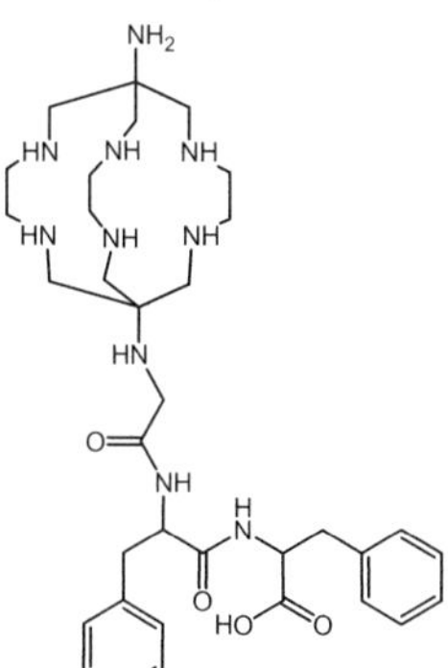

143.
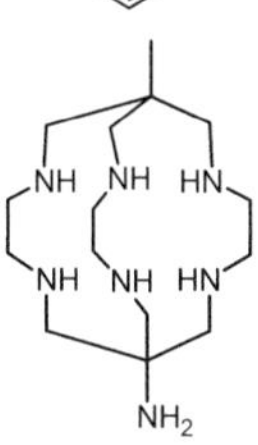

144.
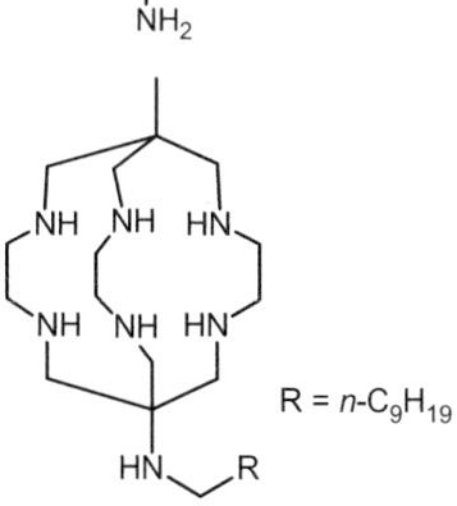

145.
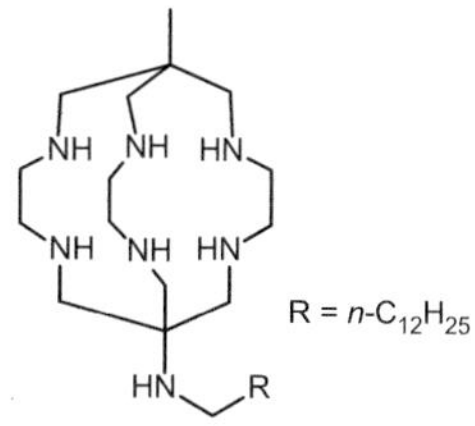

146.
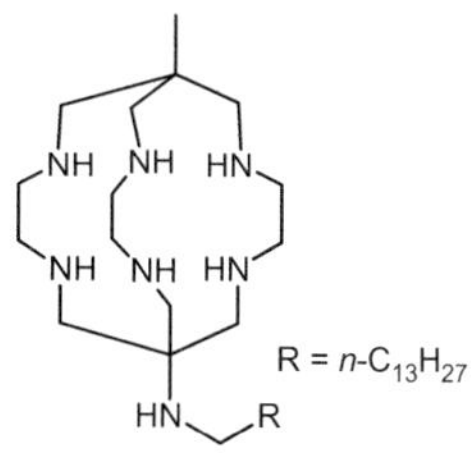

147.
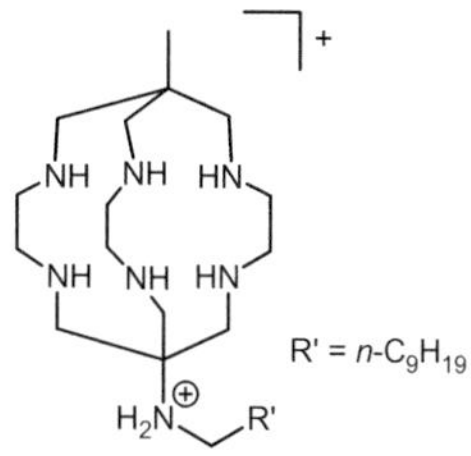

148.
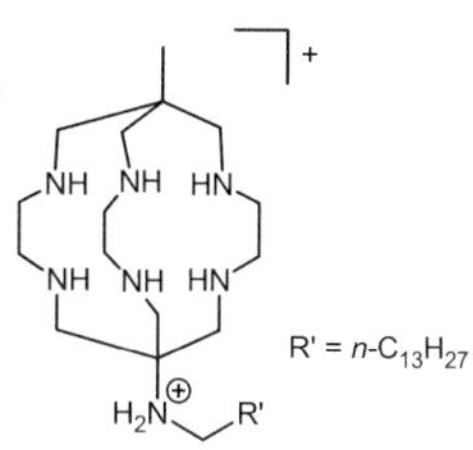

149.
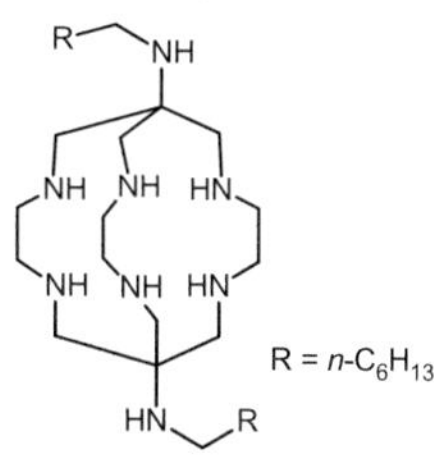

150.
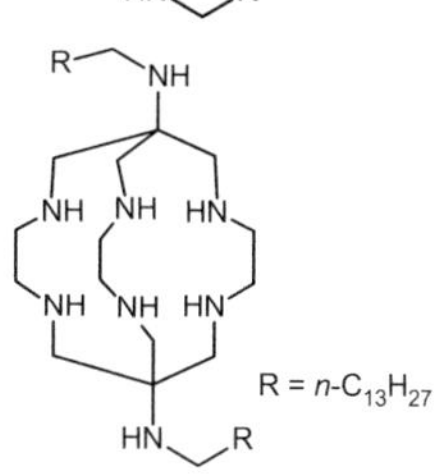

151.
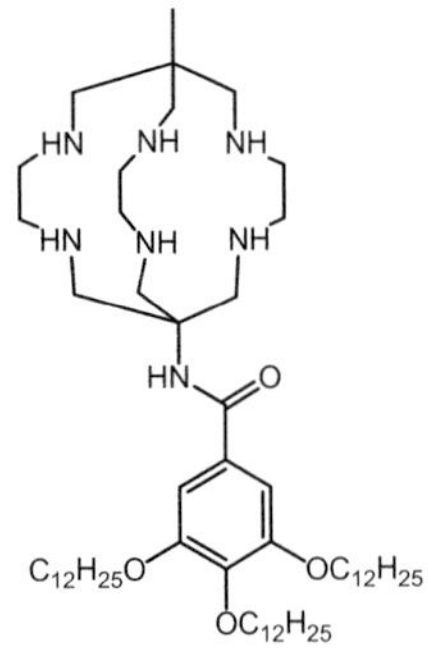

152.
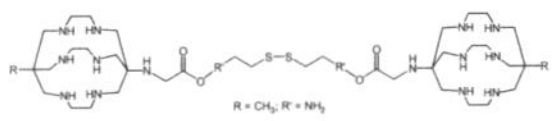

153.
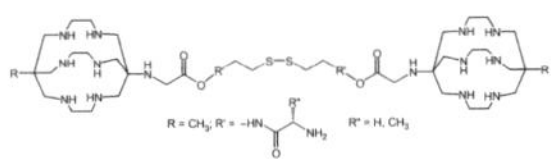

154.
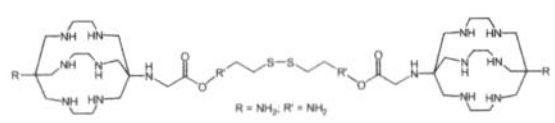

155.
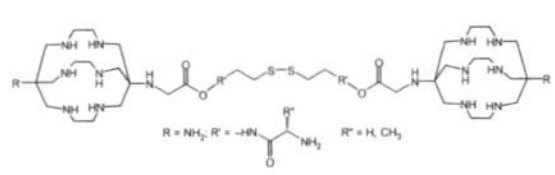

156.
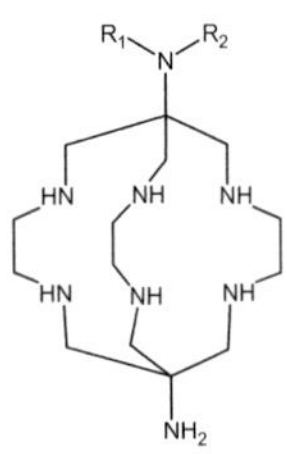

157.
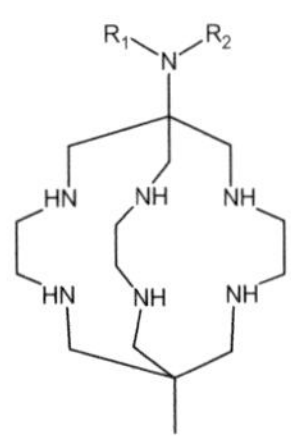

158.
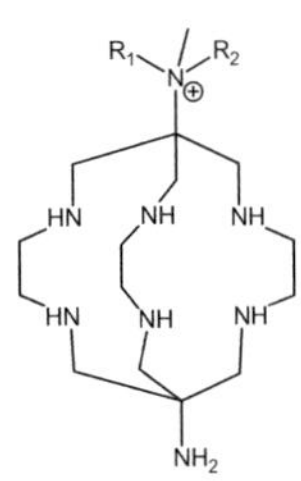

159.
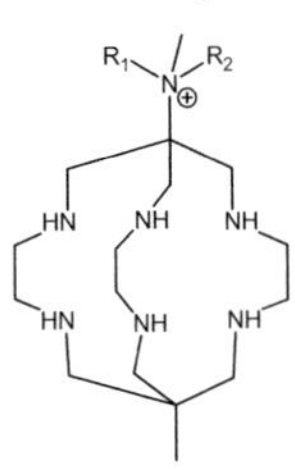

160.
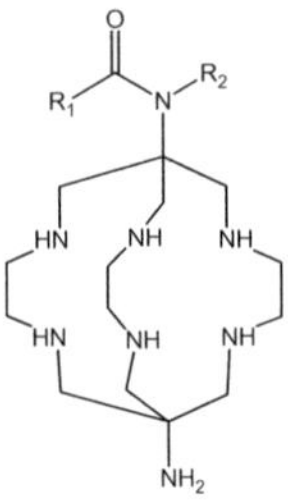

161.
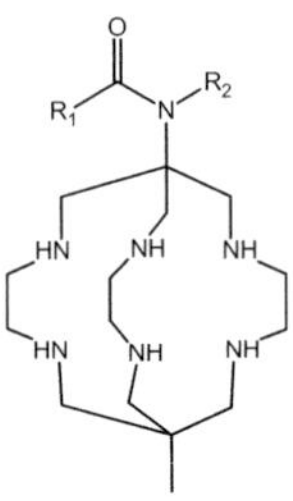

162.
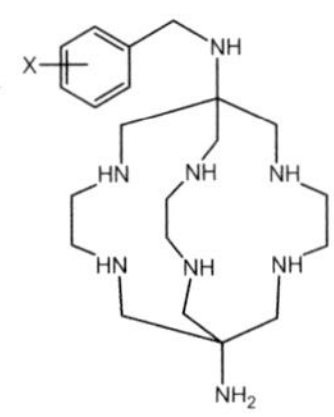

163.
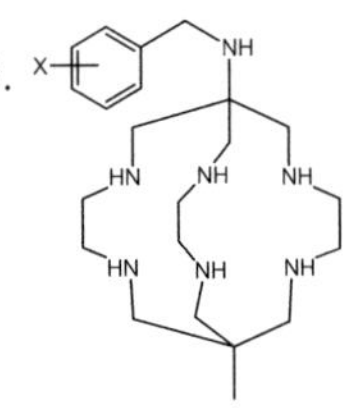

164.
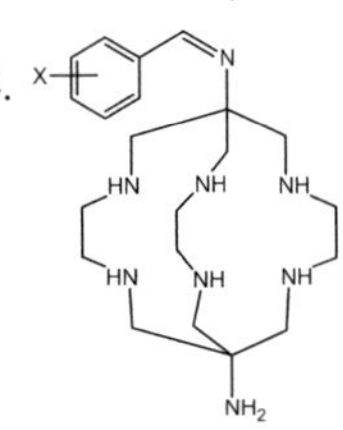

165.
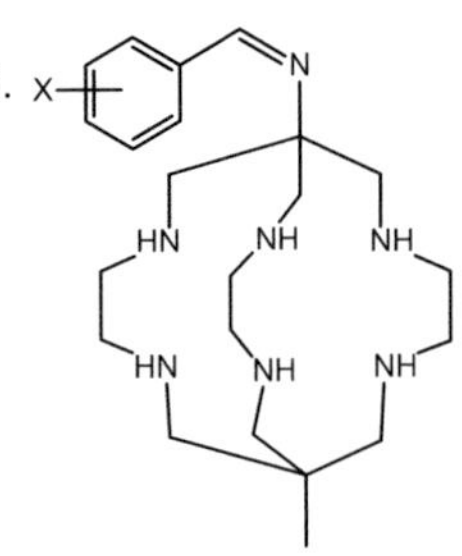

166.

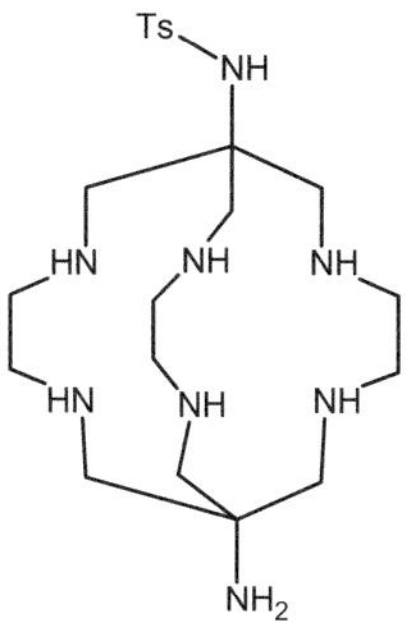

167.

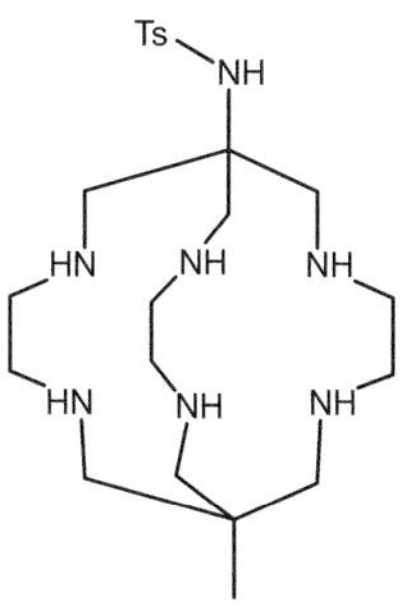

168.

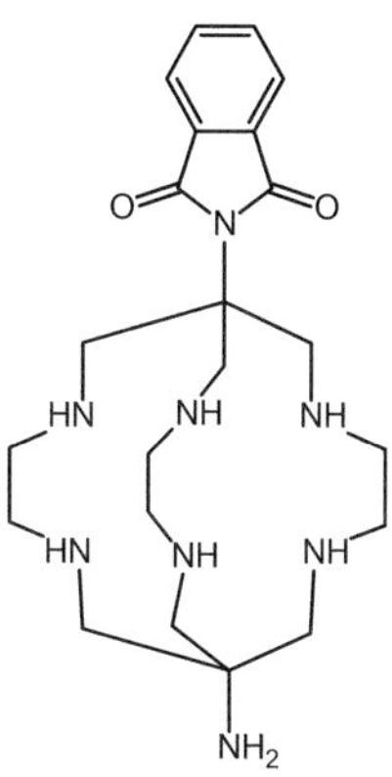

169.

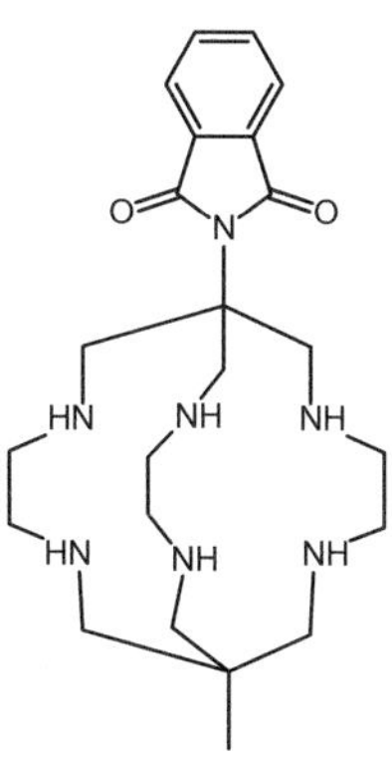

170.

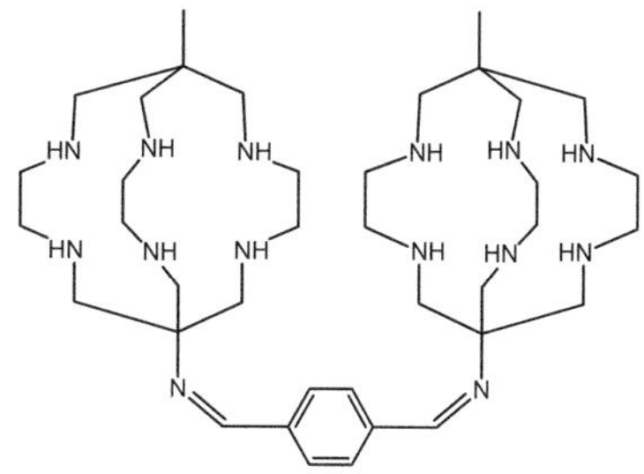

171.

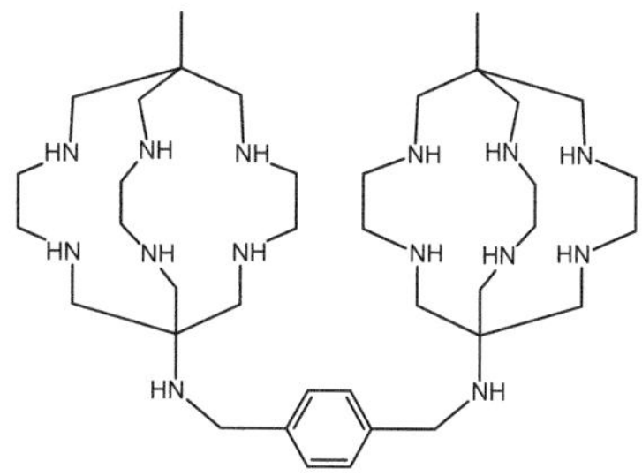

172.

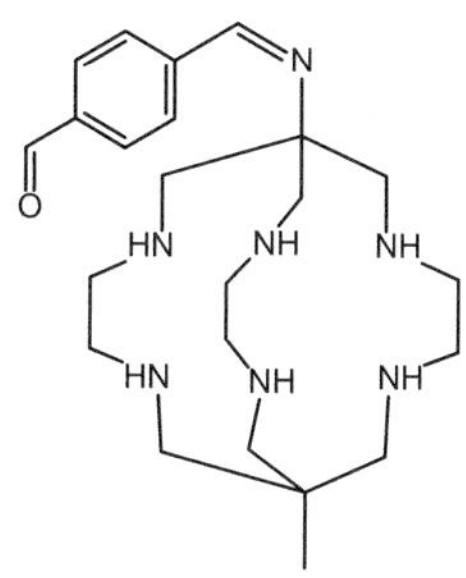

173.

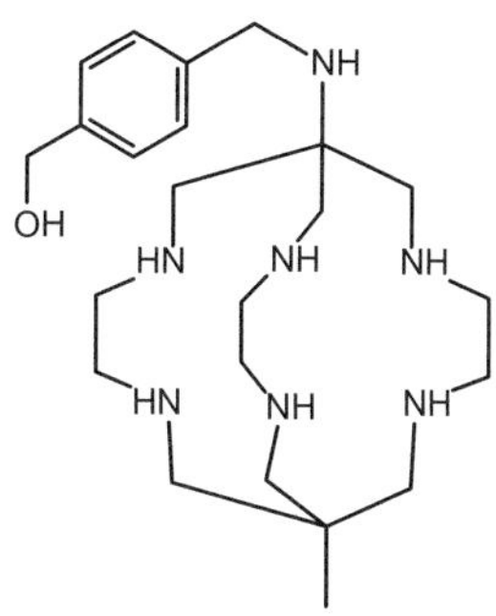

174.

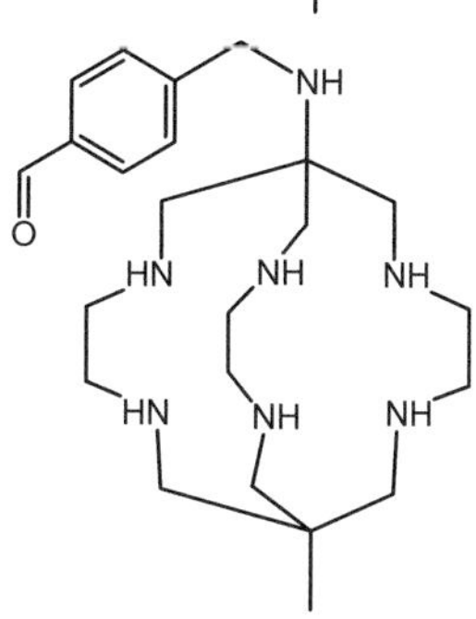

175.

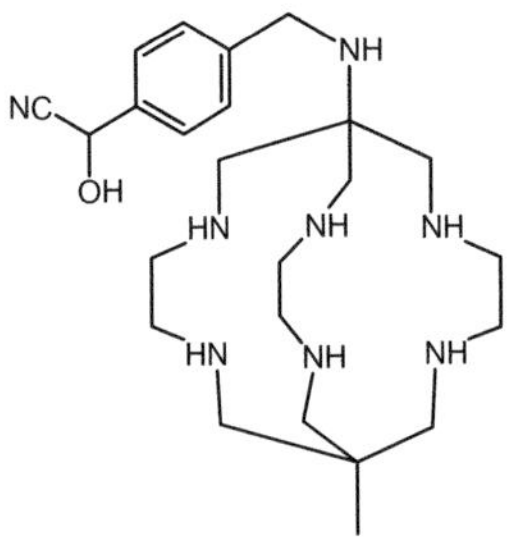

176.

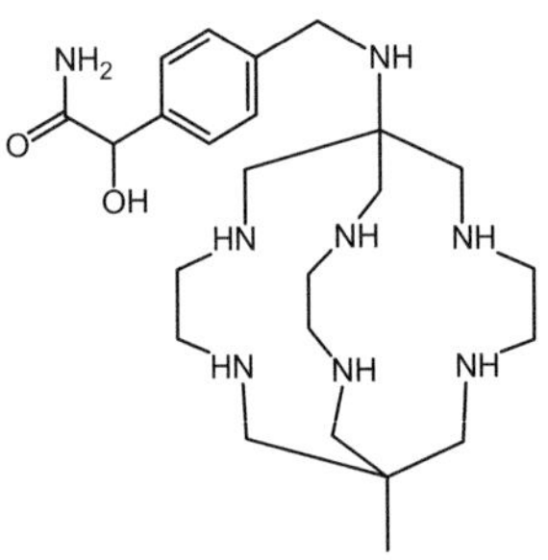

177.

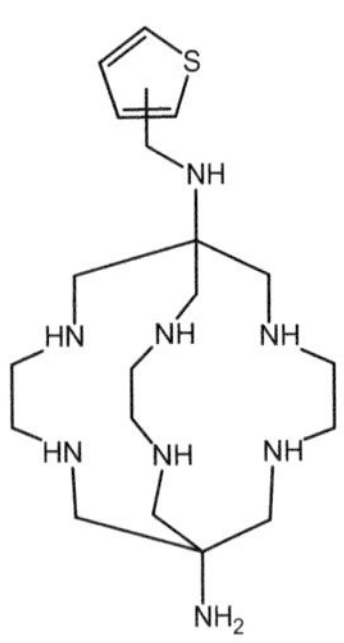

178.

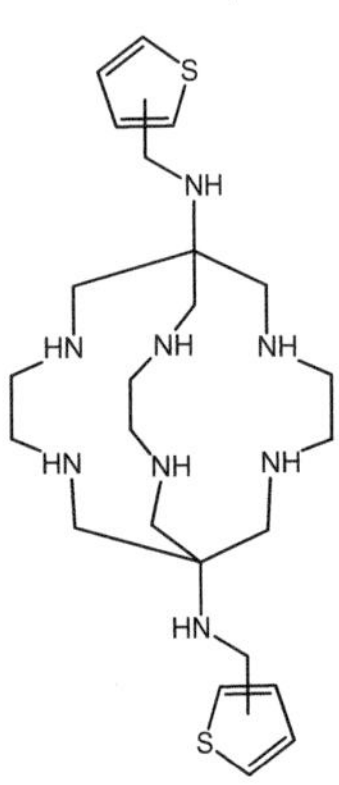

179.

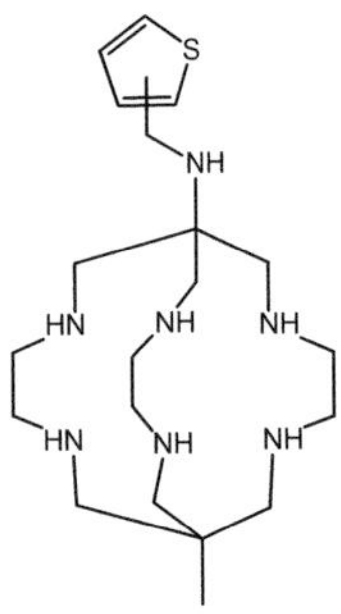

180.

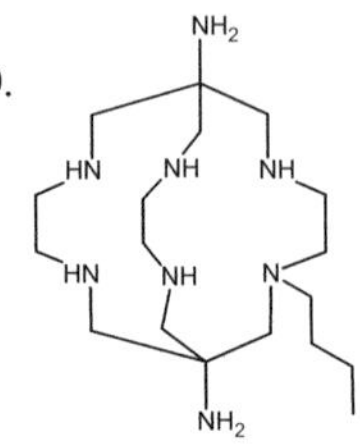

181.

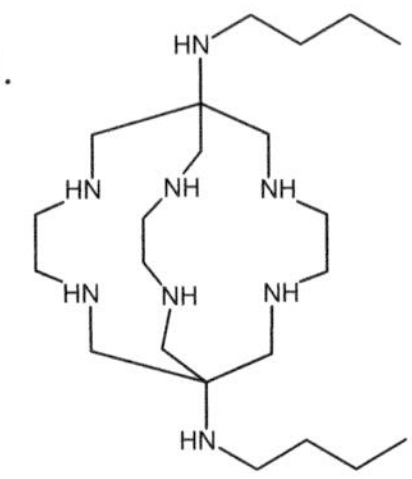

182.

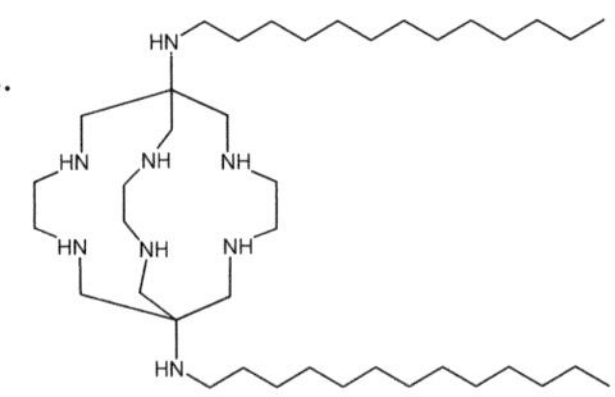

183.

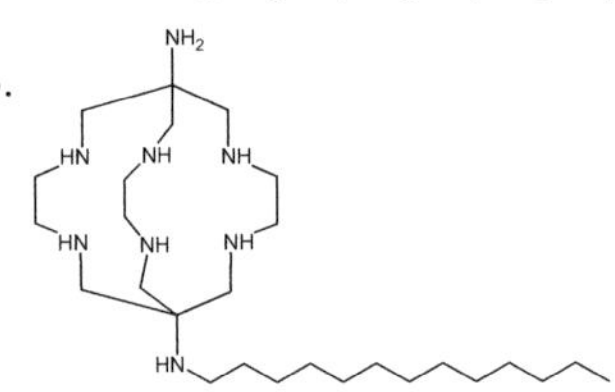

184.

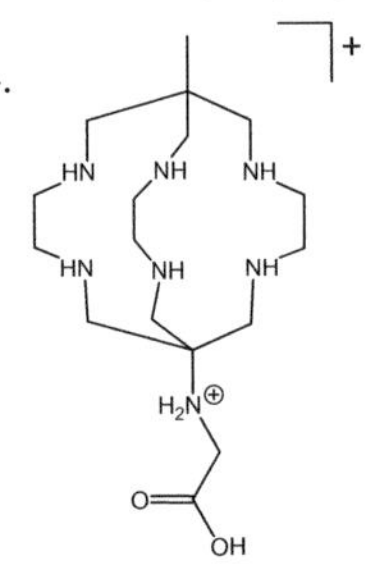

185.

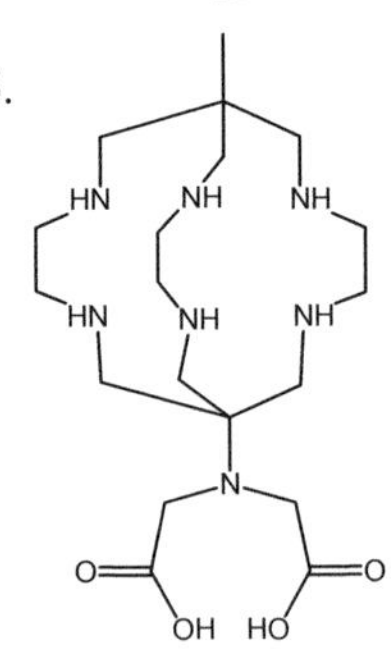

186.

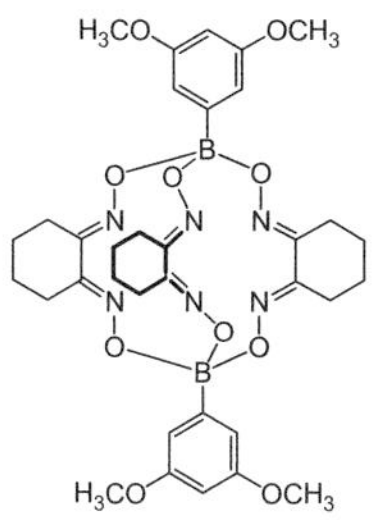

187.

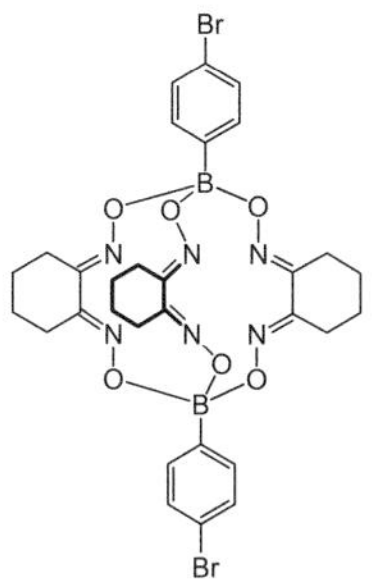

188.

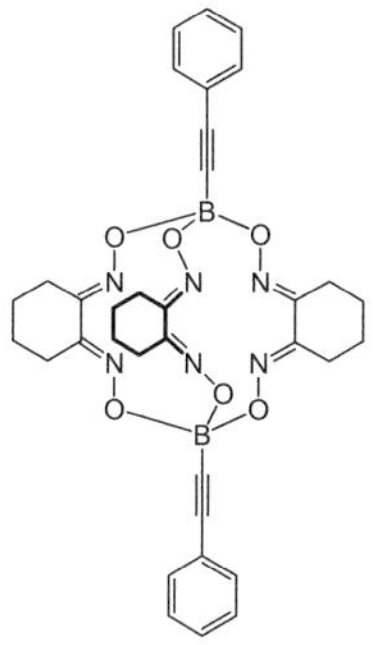

189.

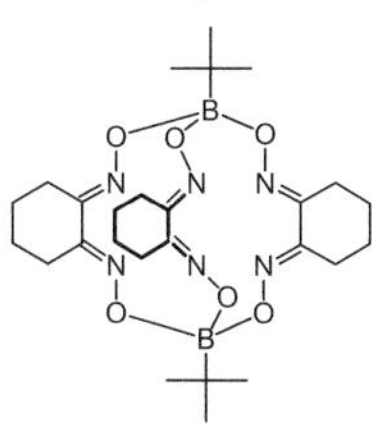

190.

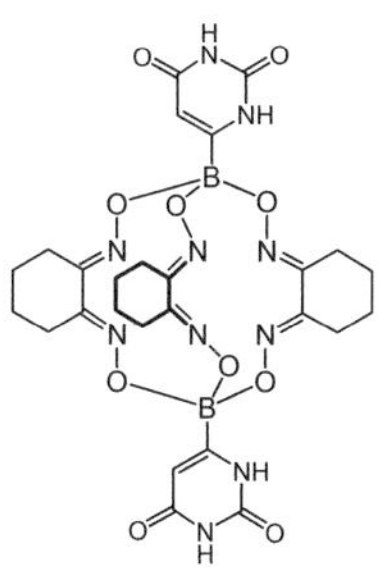

191.

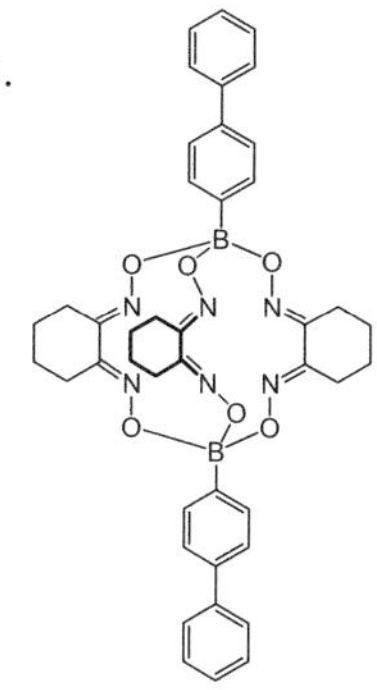

192.

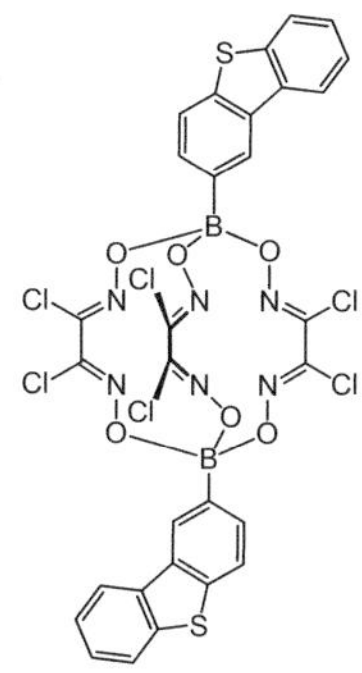

193.

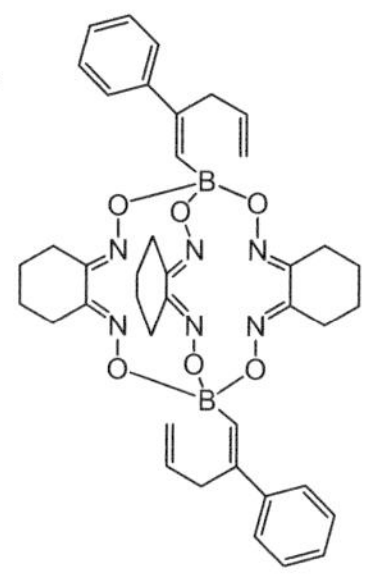

194.

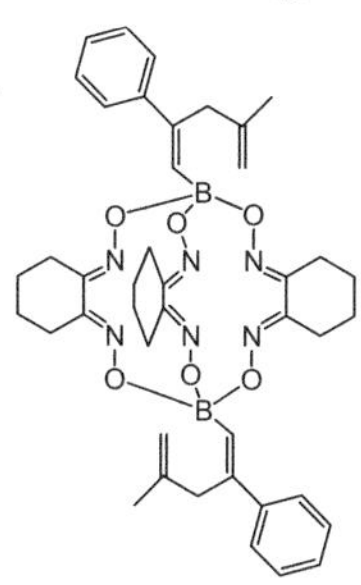

195.

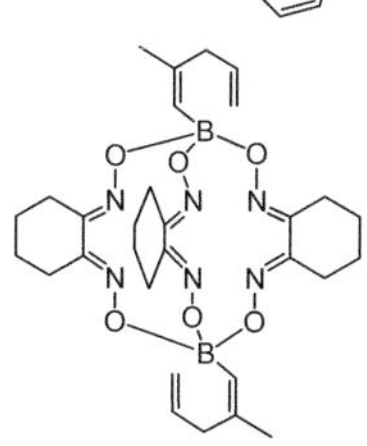

196.

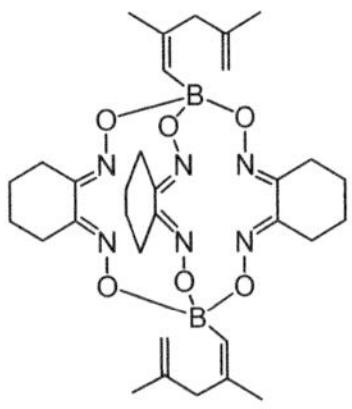

197.

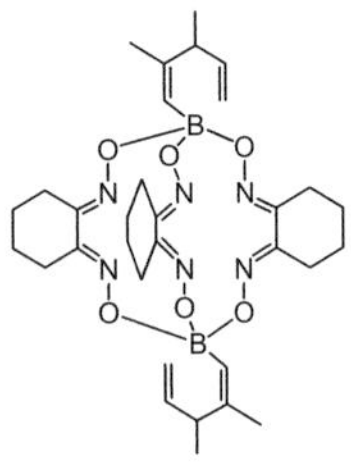

198.

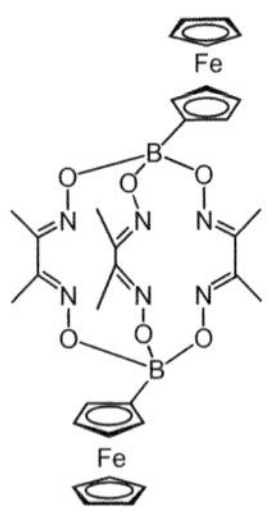

199.

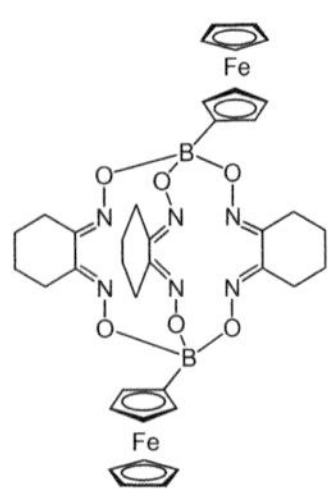

200.

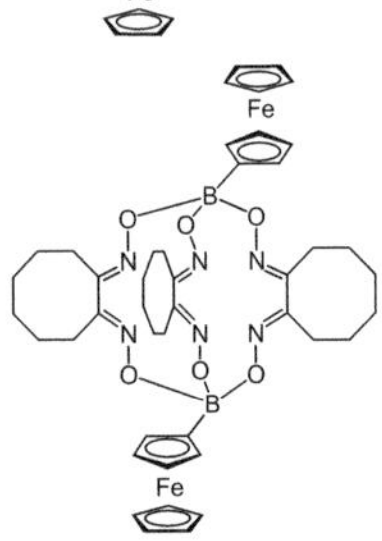

201.

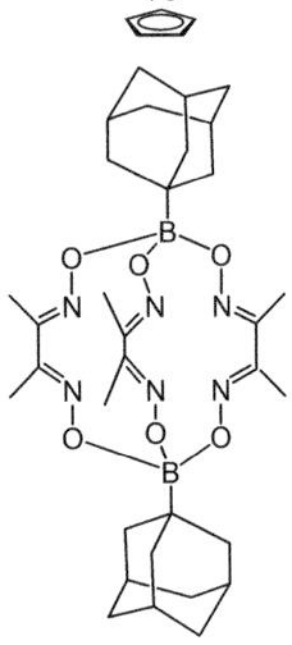

202.

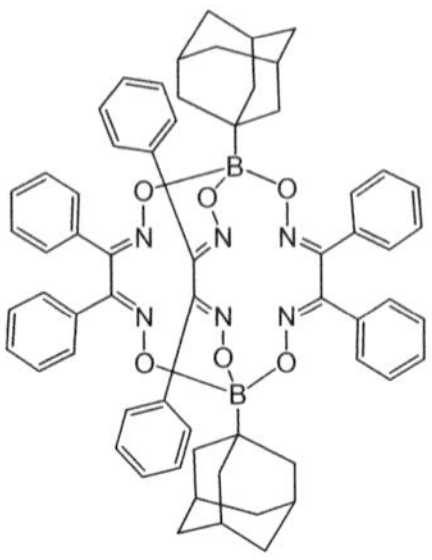

203.

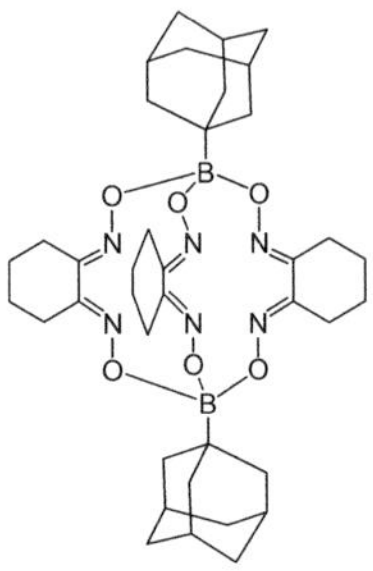

204.

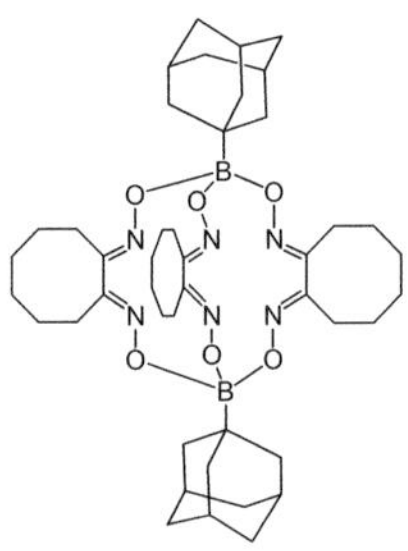

205.

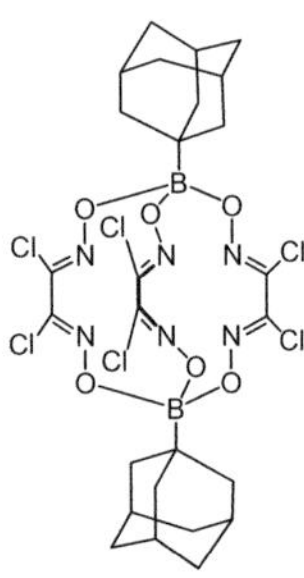

206.

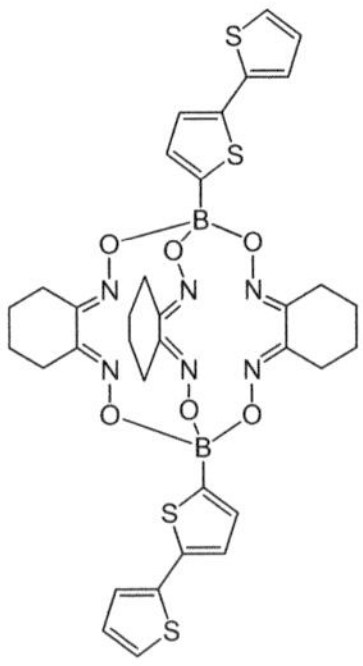

207.

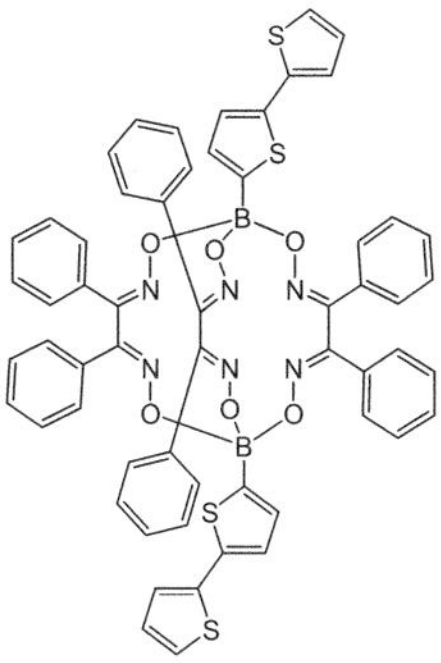

208.

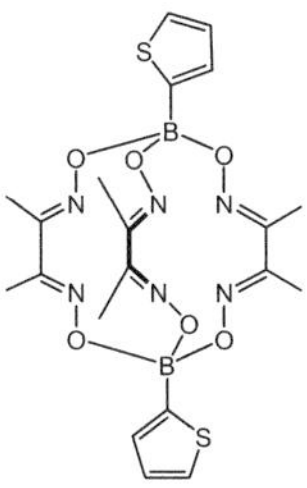

209.

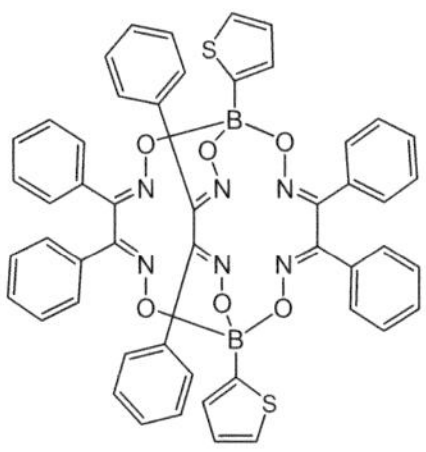

210.

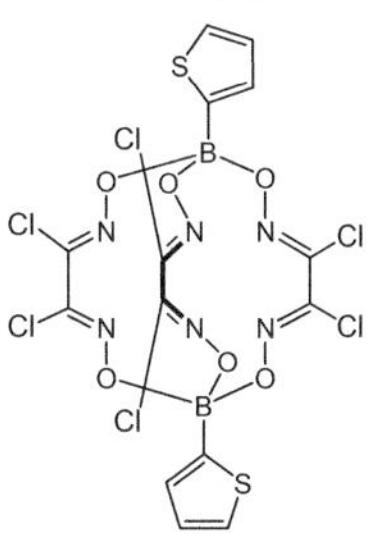

211.

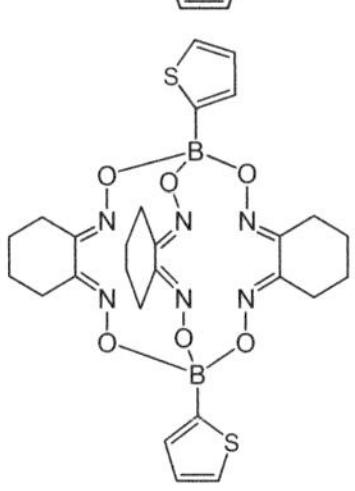

212.

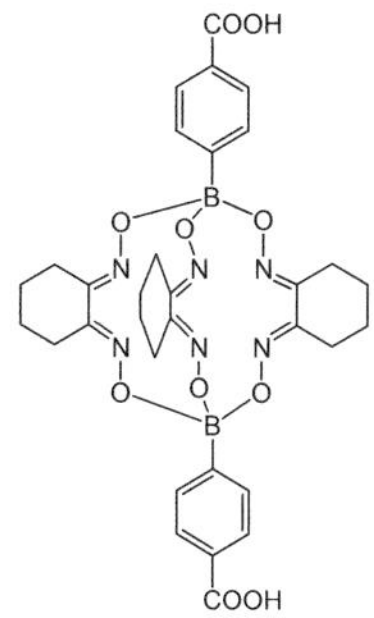

213.

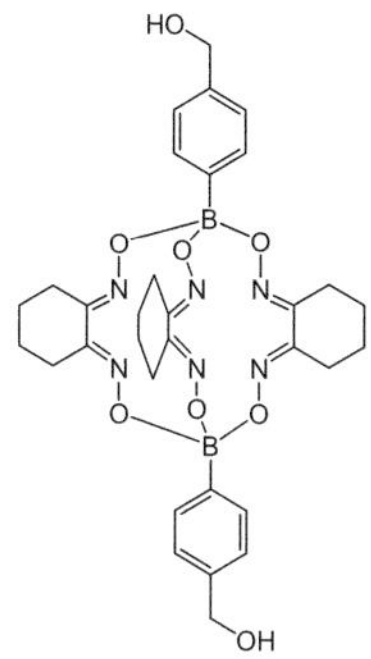

214.

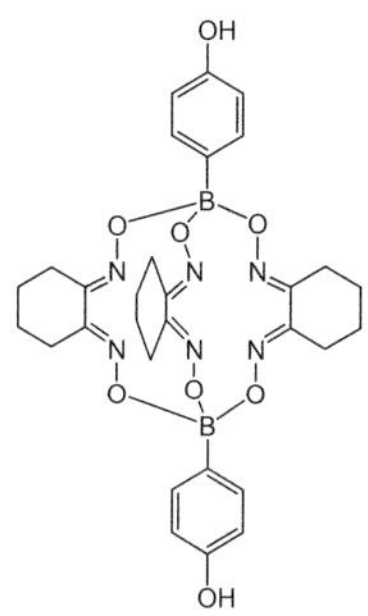

215.

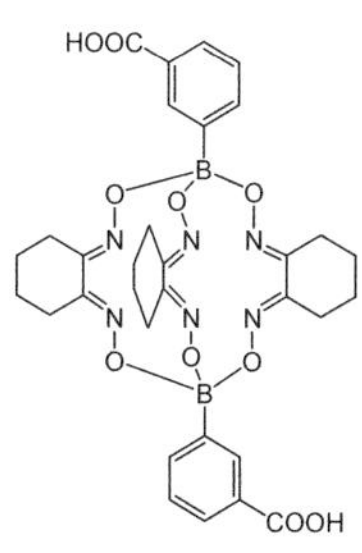

216.

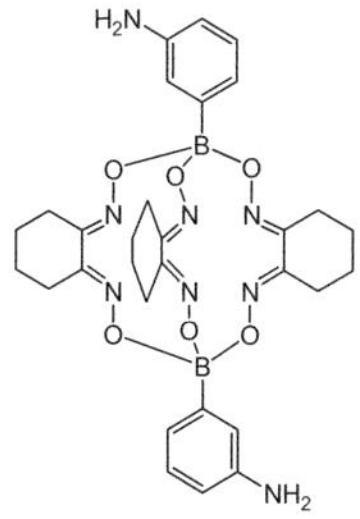

217.

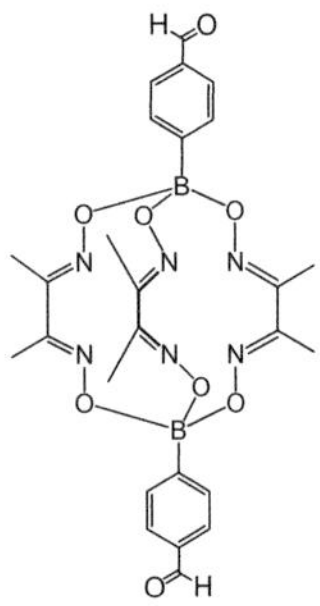

218.

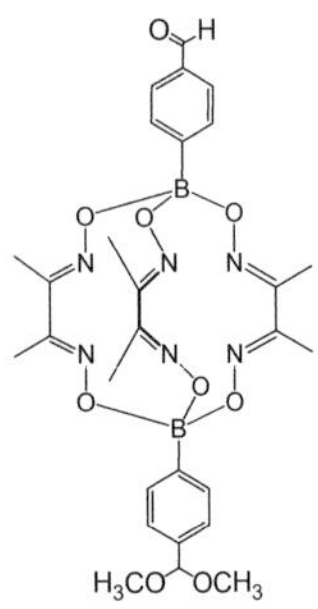

219.

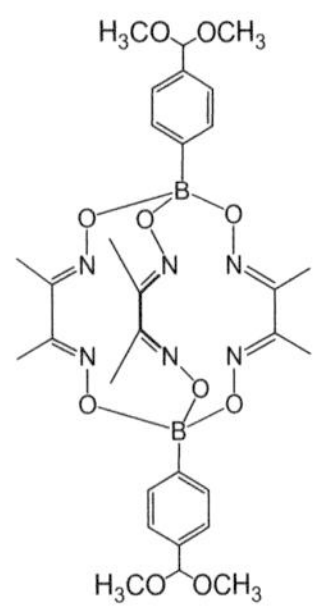

220.

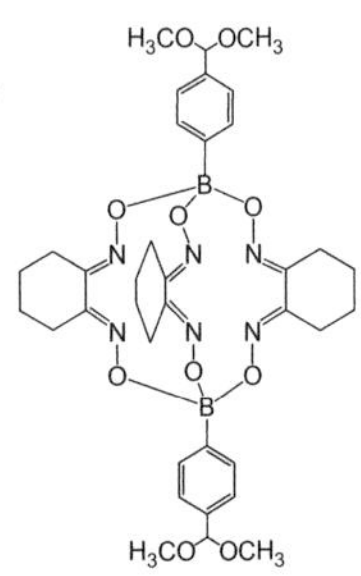

221.

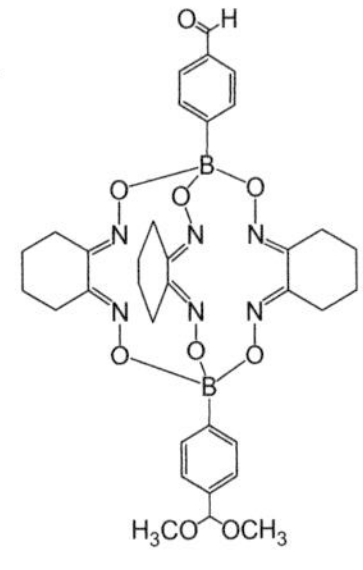

222.

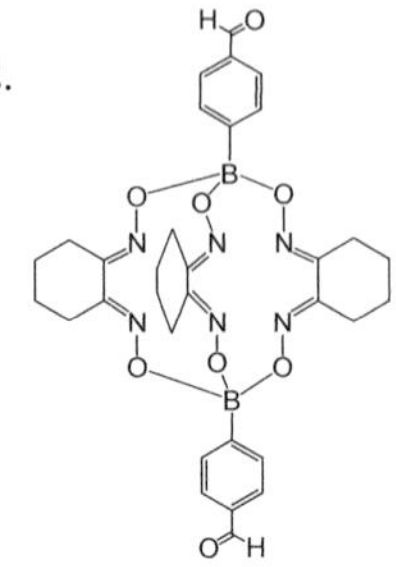

223.

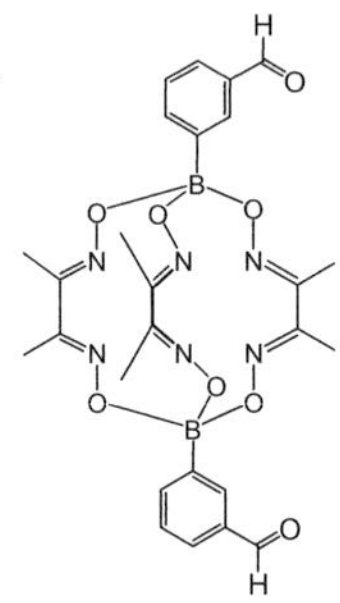

224.

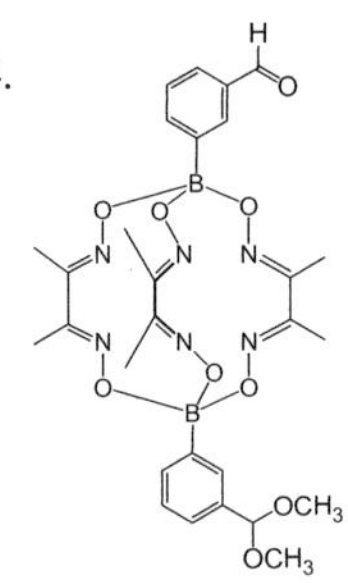

225.

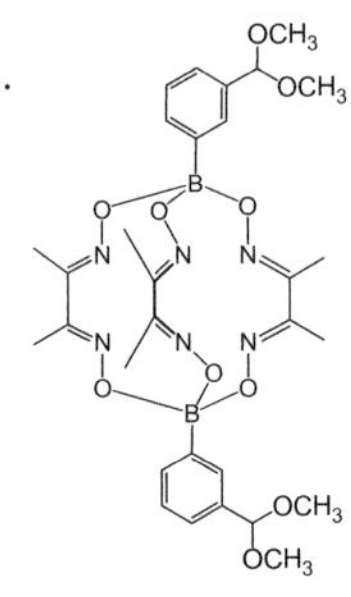

226.

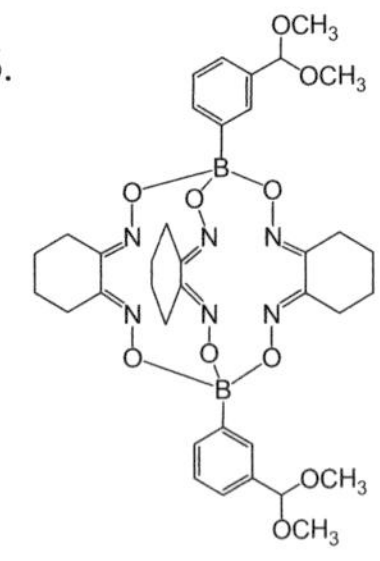

227.

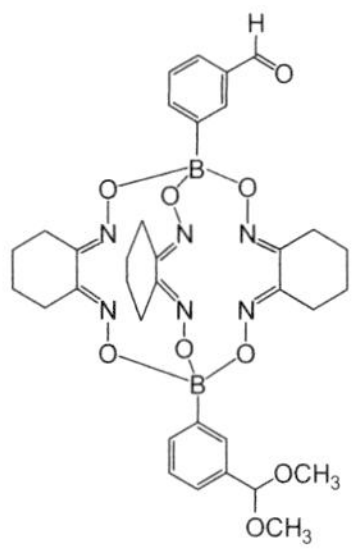

228.

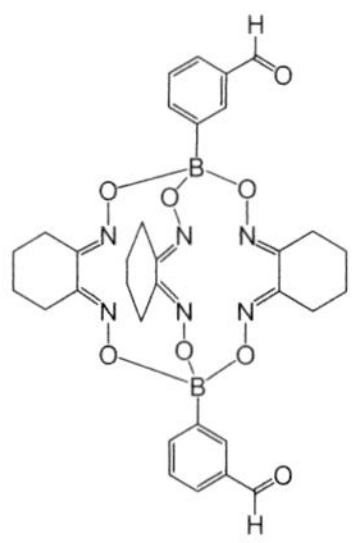

229.

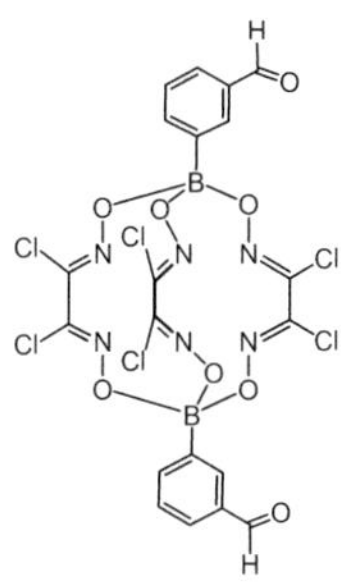

230.

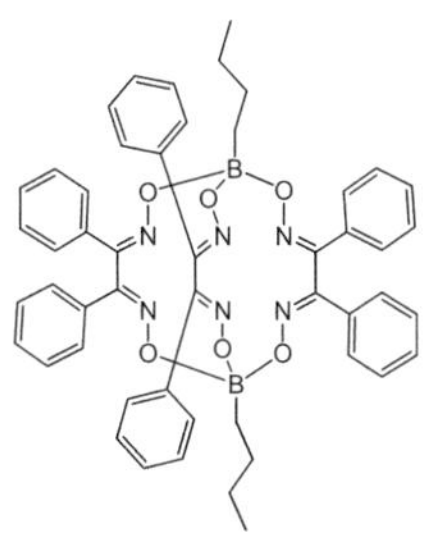

231.

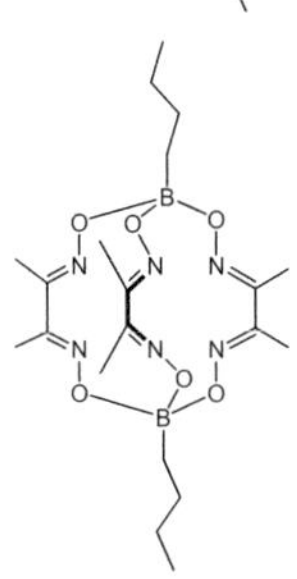

232.

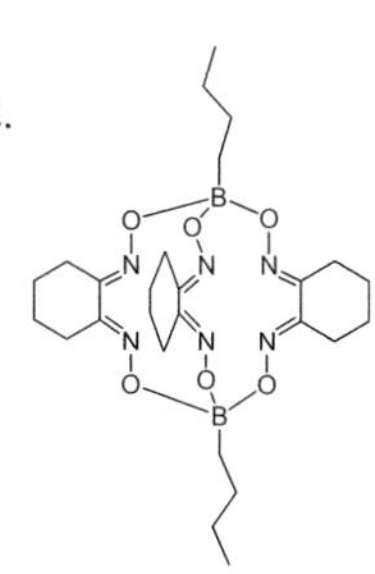

233.

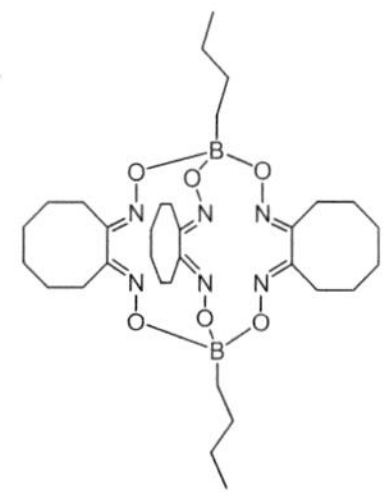

234.

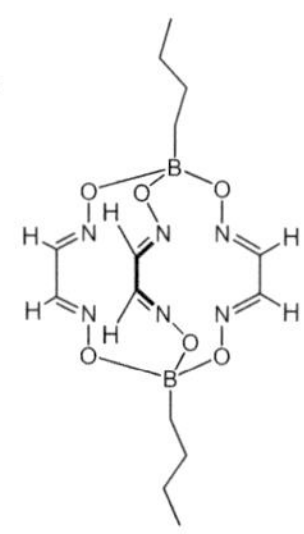

235.

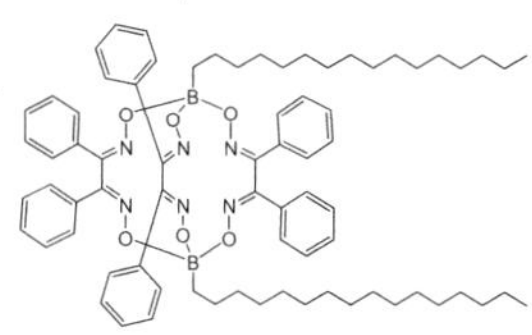

236.

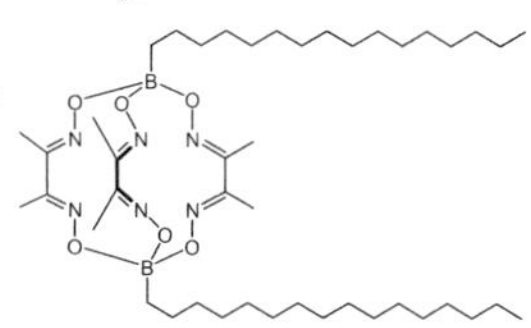

237.

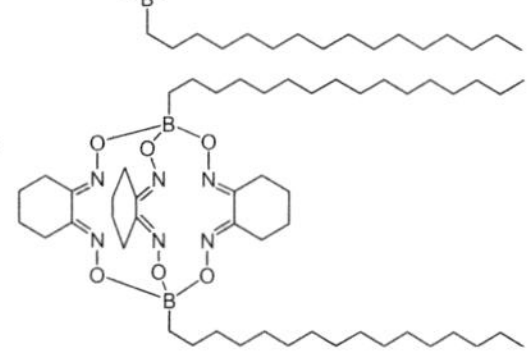

238.

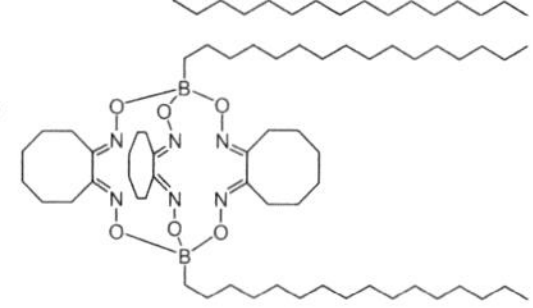

239.

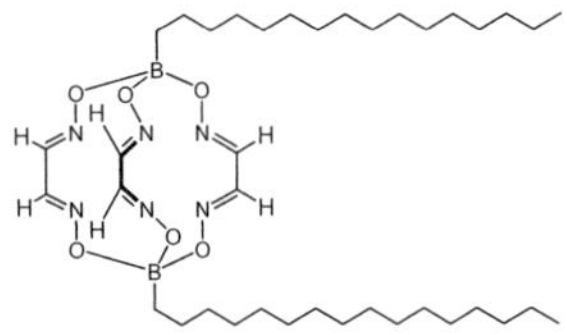

240.

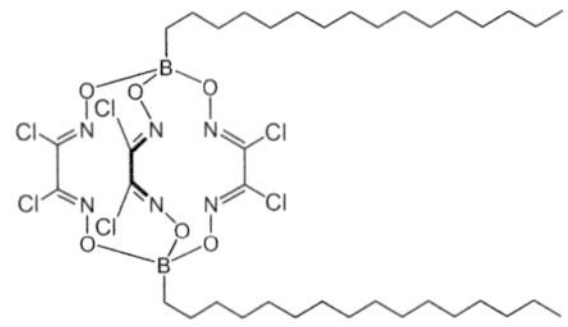

241.

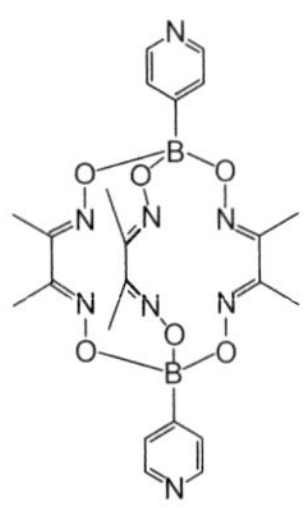

242.

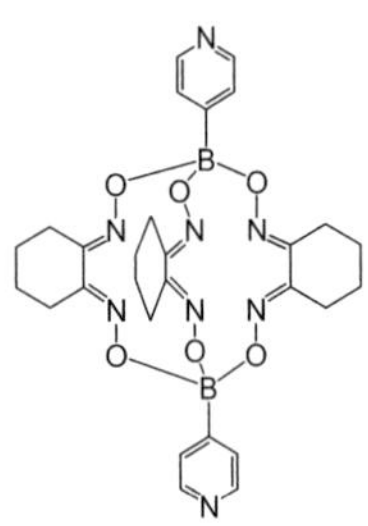

243.

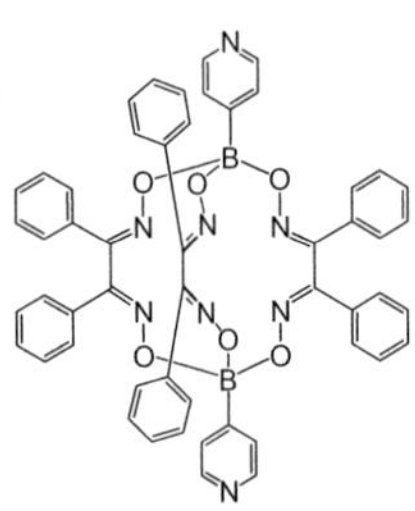

244.

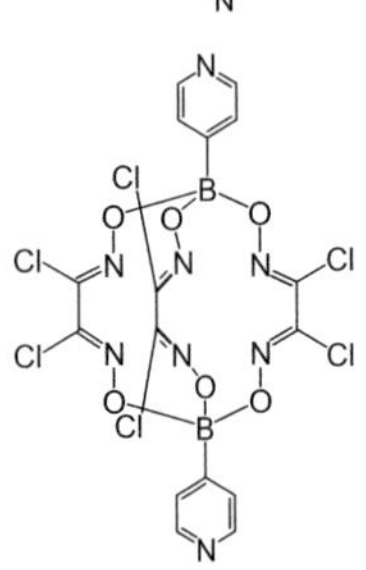

245.

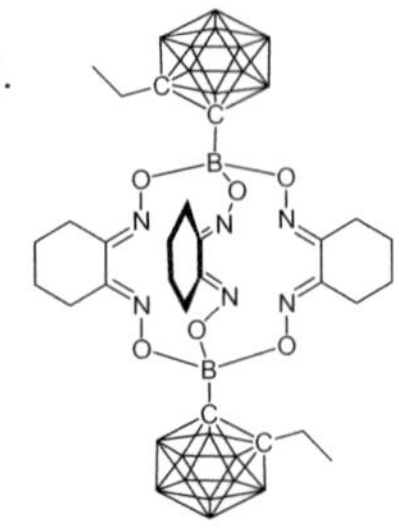

246.

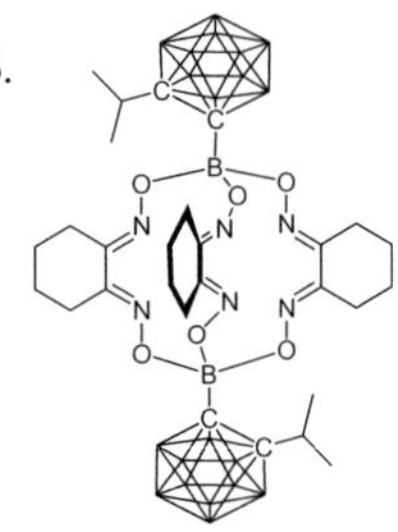

247.

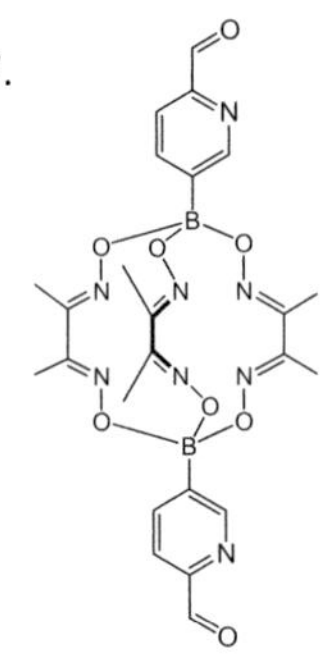

248.

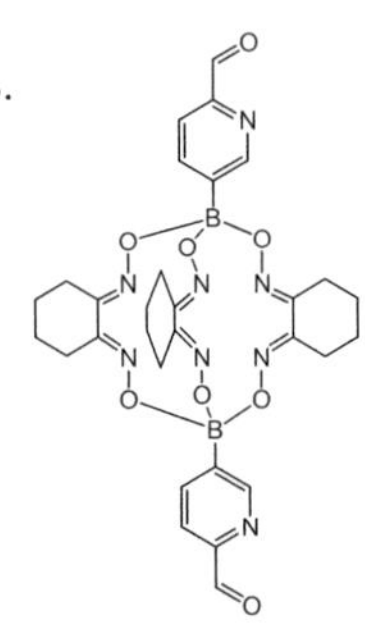

249.

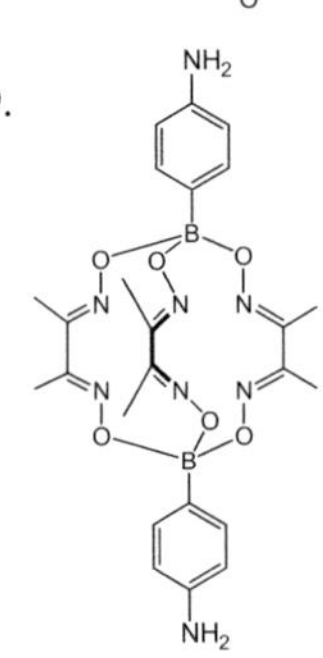

250.

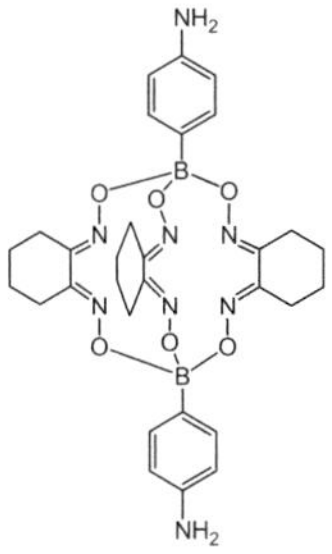

251.

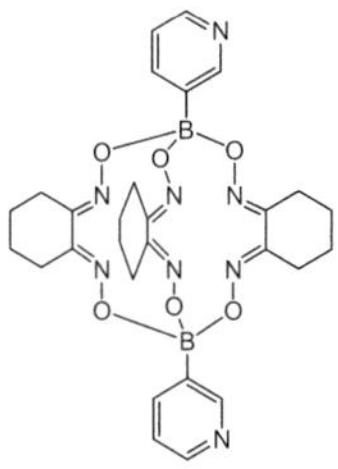

252.

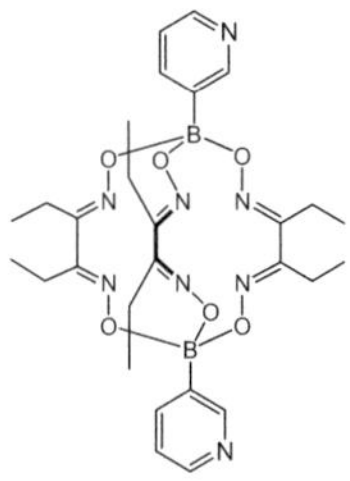

253.

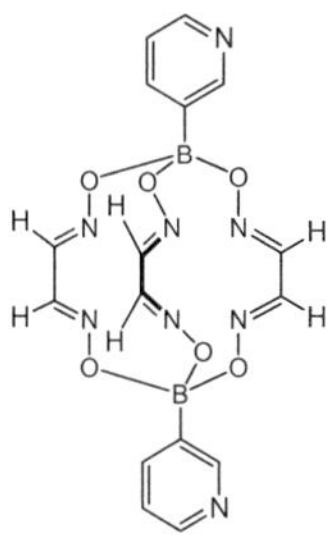

254.

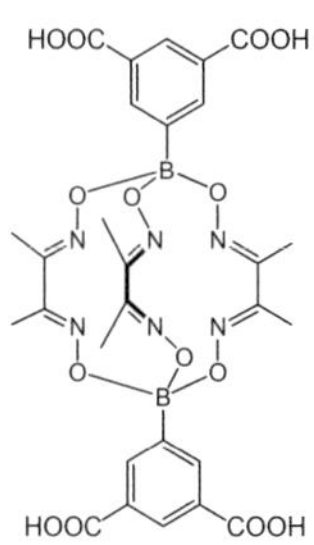

255.

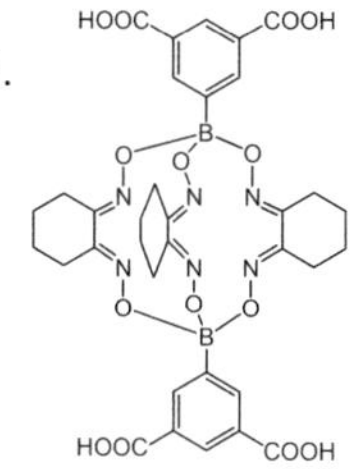

256.

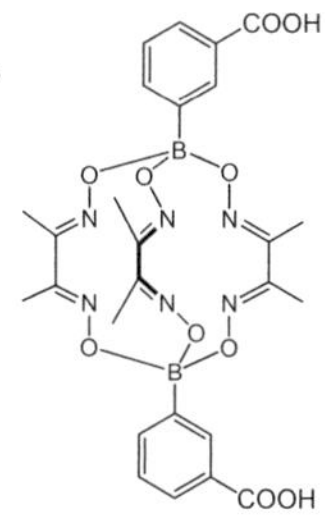

257.

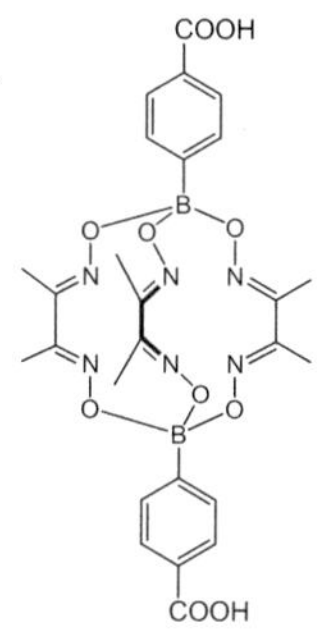

258.

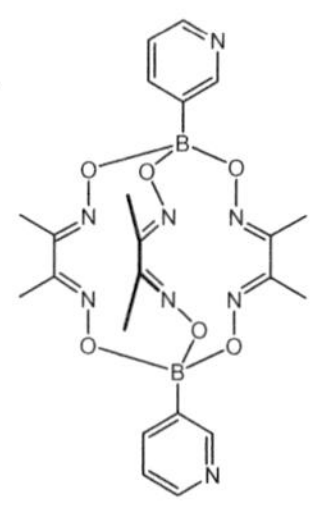

259.

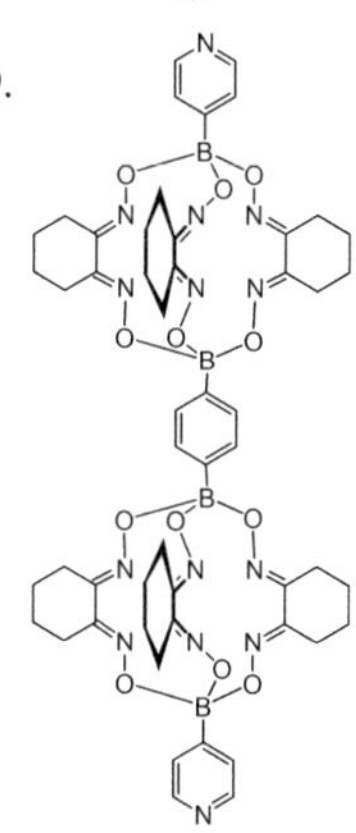

260.

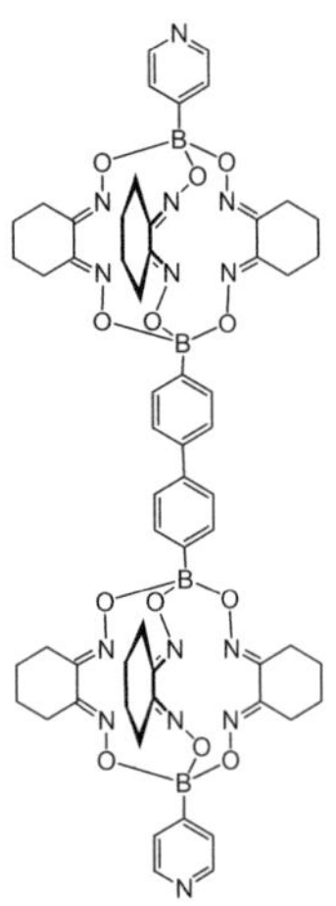

261.

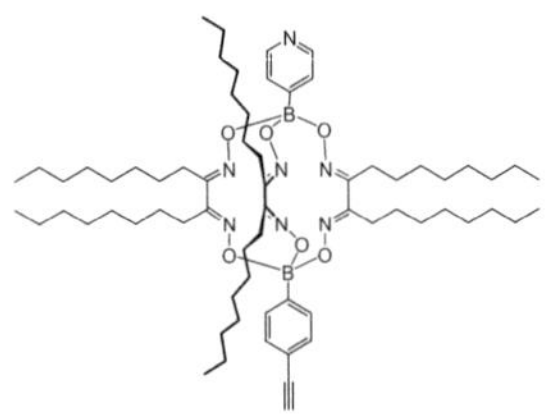

262.

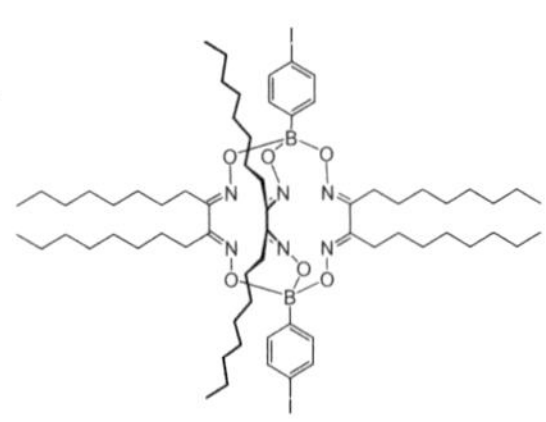

263.

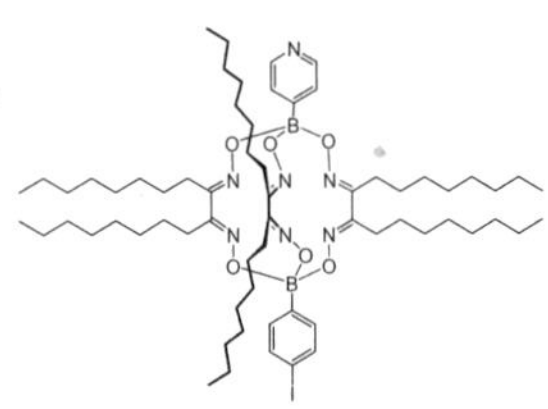

264.

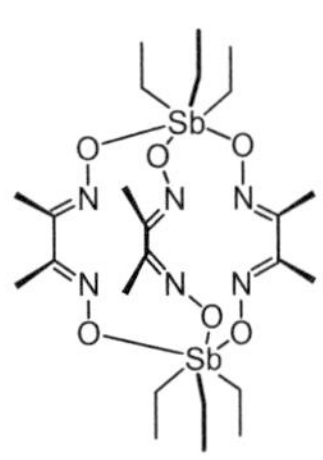

265.

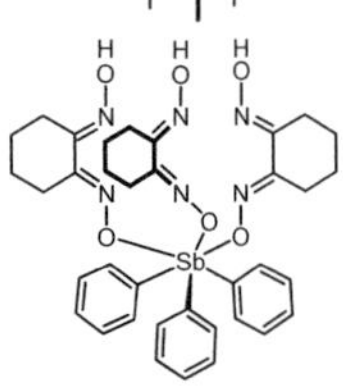

266.

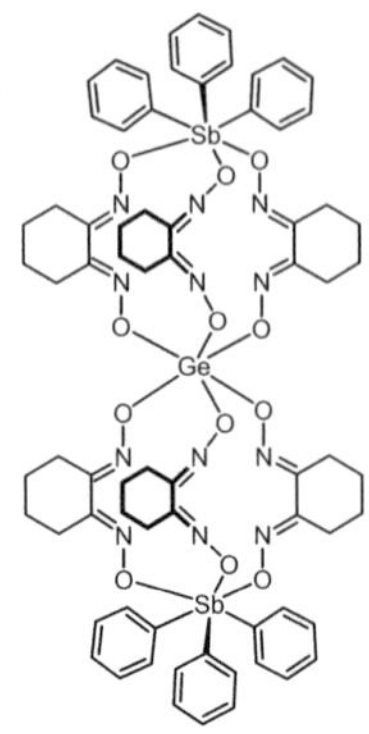

267.

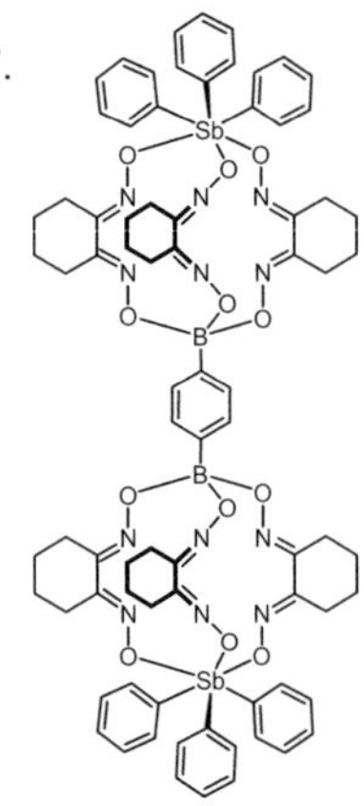

268.

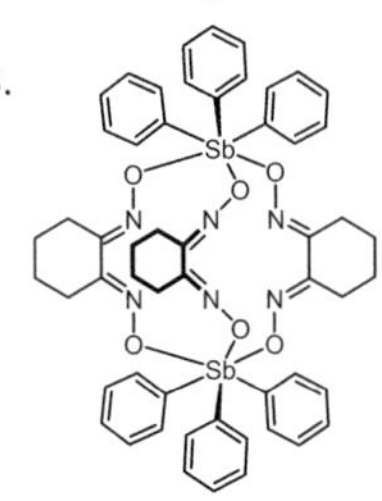

269.

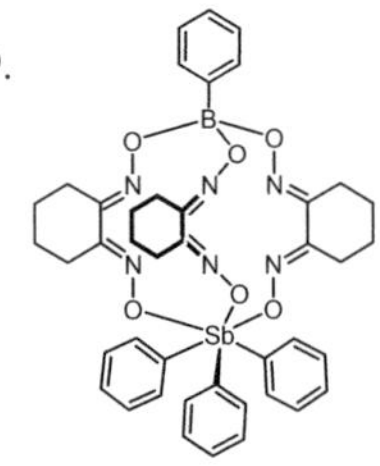

270.

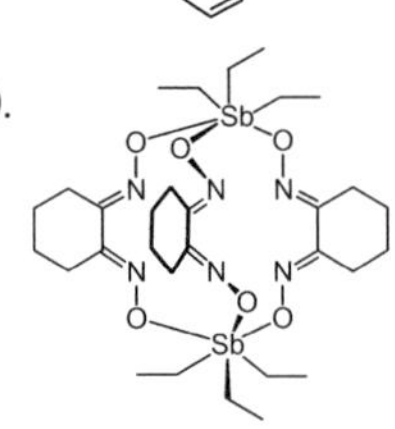

271.

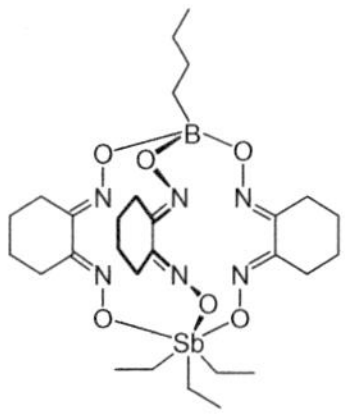

272.

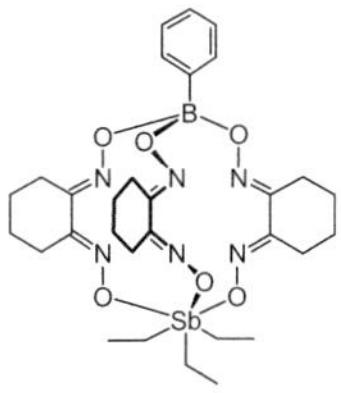

273.

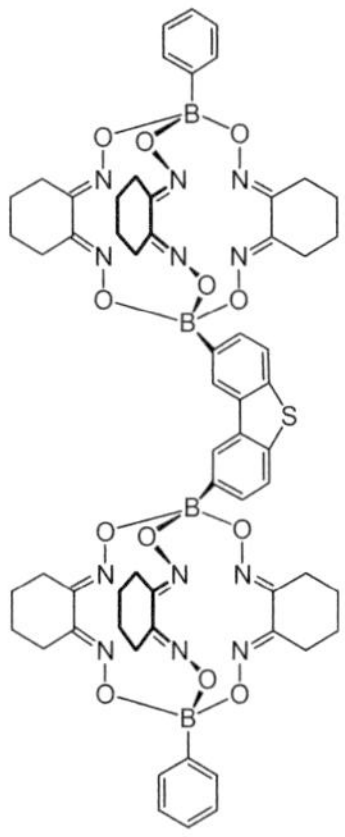

274.

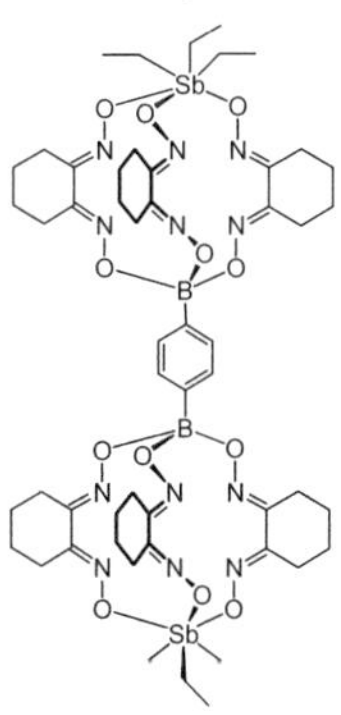

275.

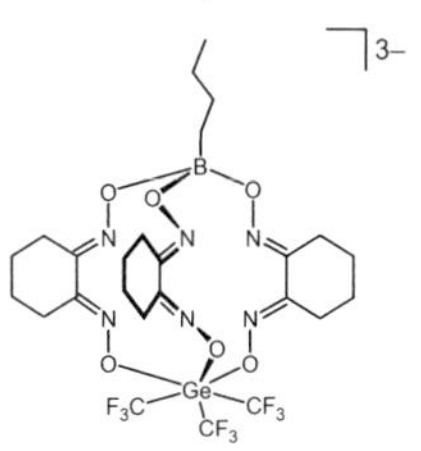

276.

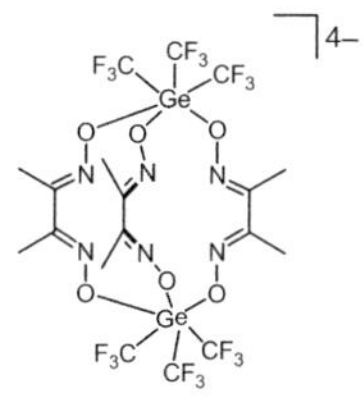

277.

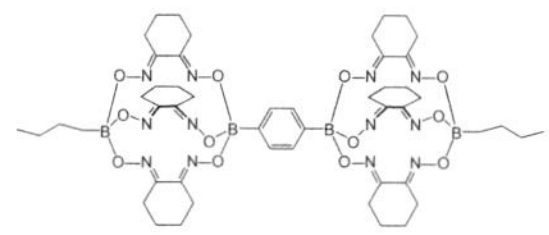

278.

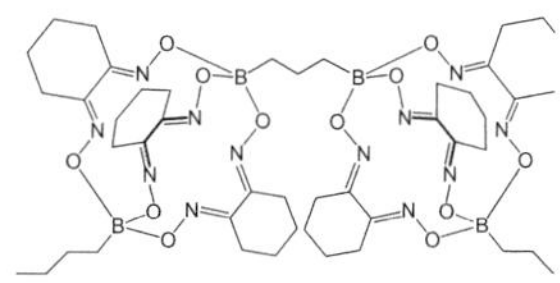

279.

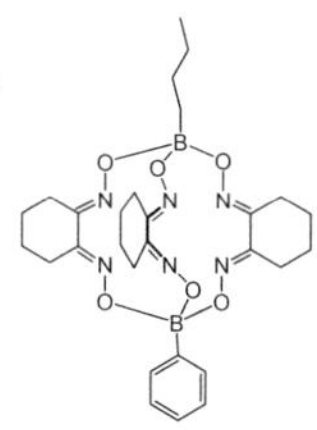

280.

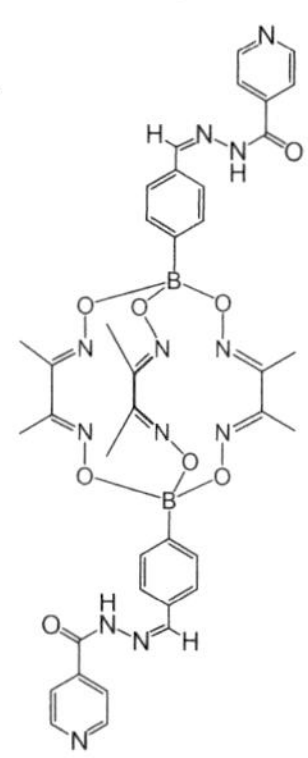

281.

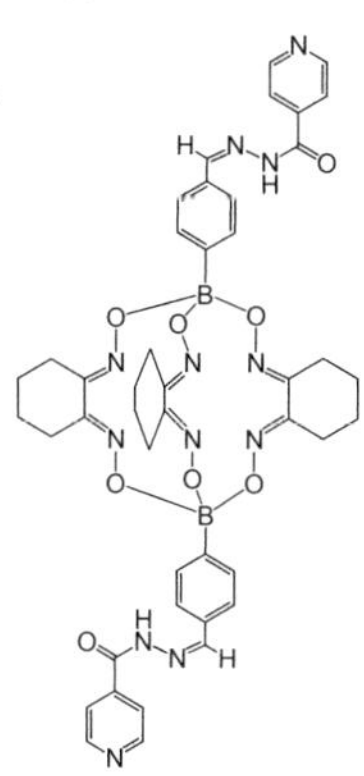

282.

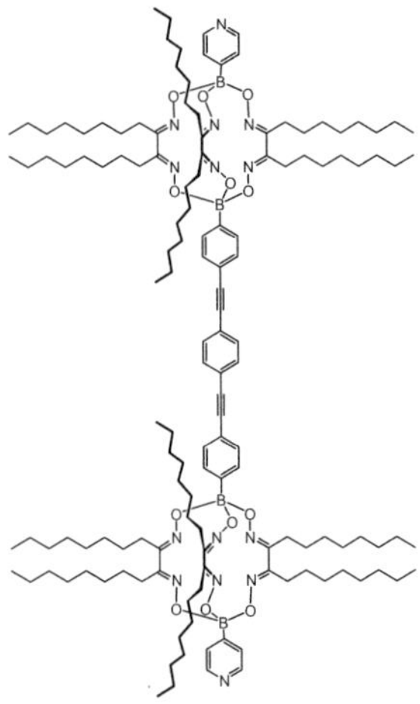

283.

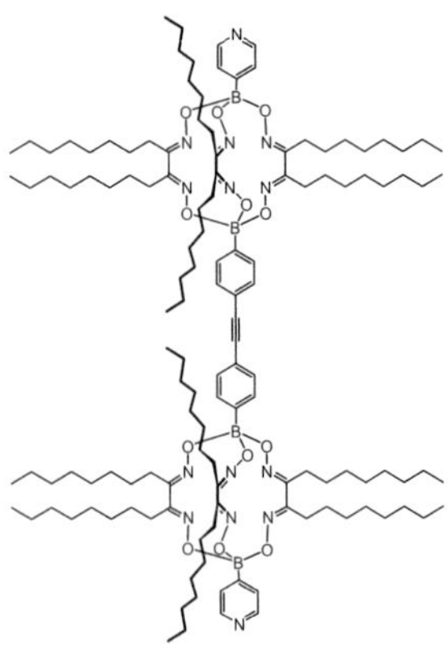

284.

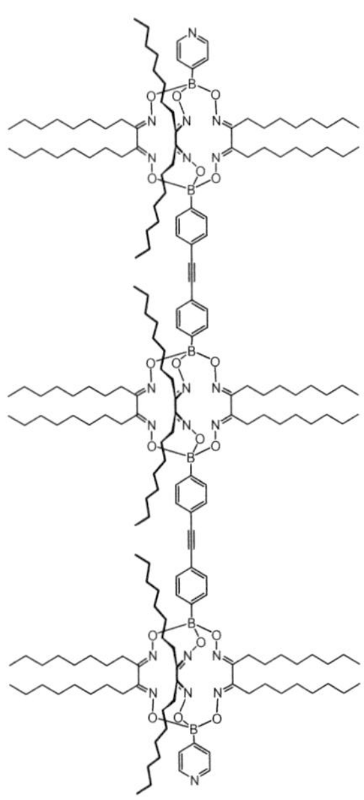

285.

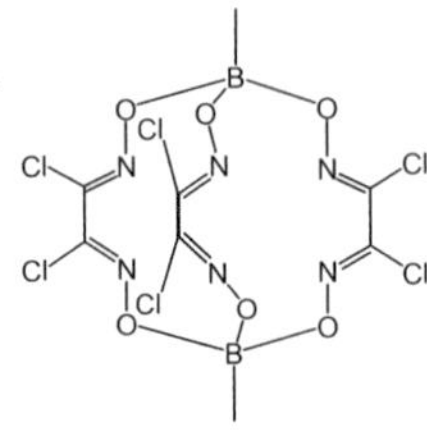

286.

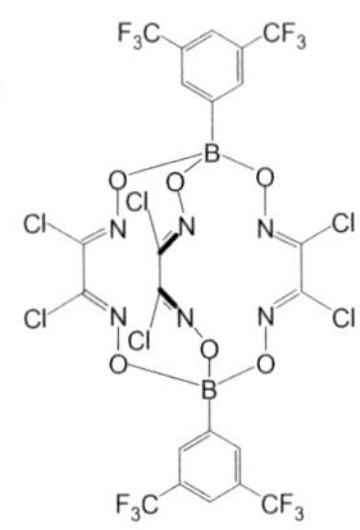

287.

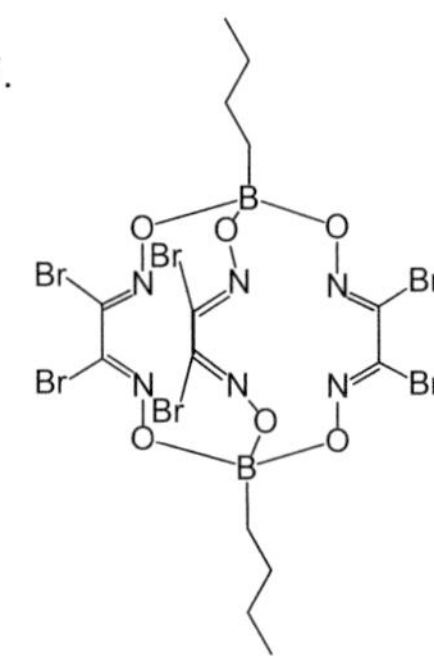

288.

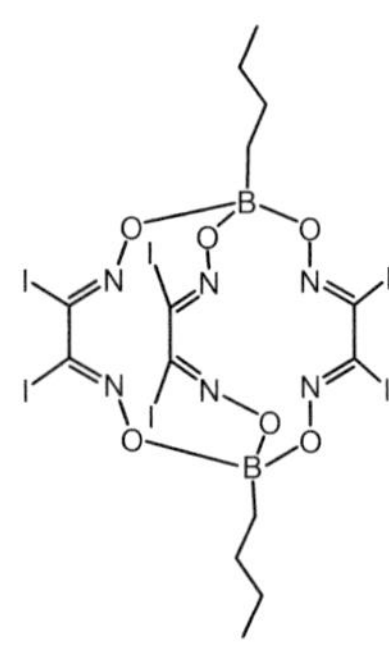

289.

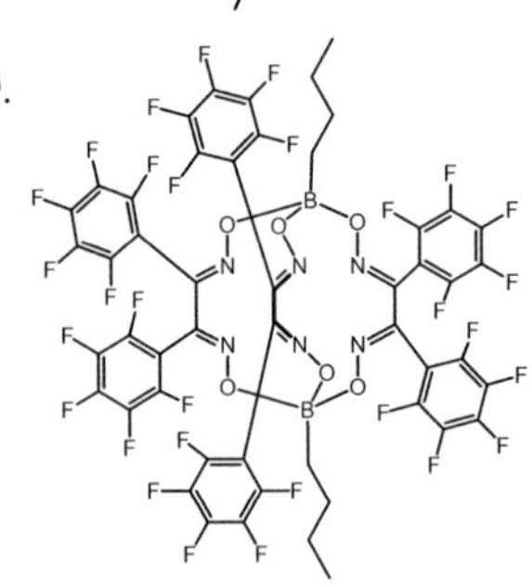

290.

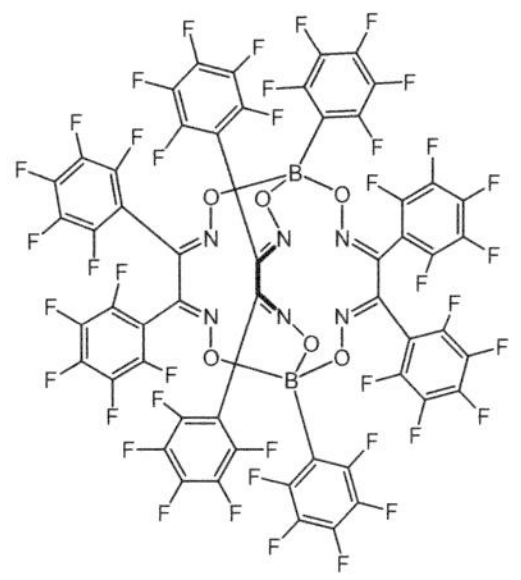

291.

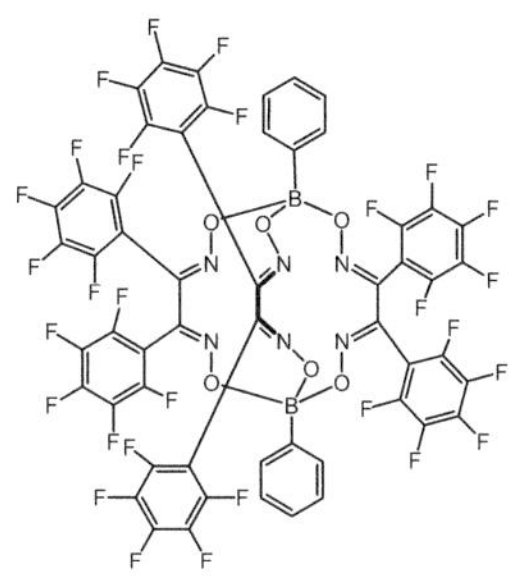

292.

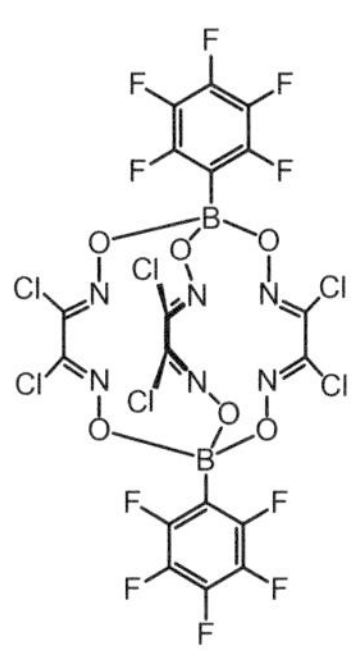

293.

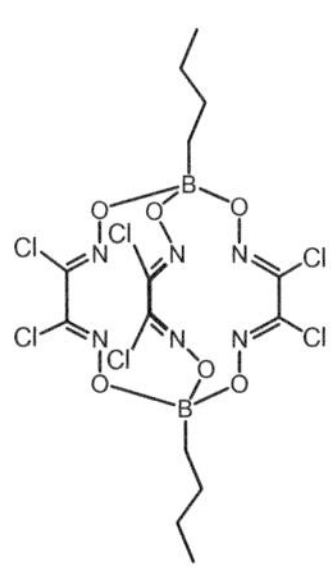

294.

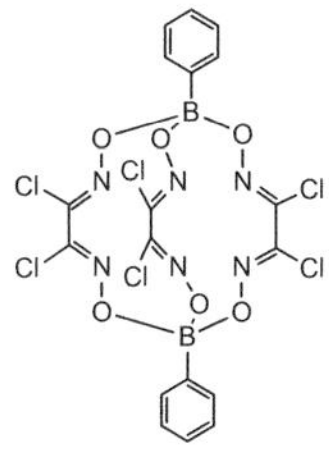

295.

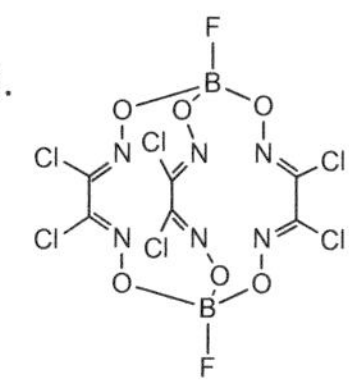

296.

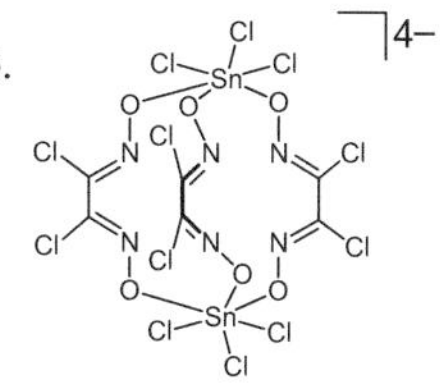

297.

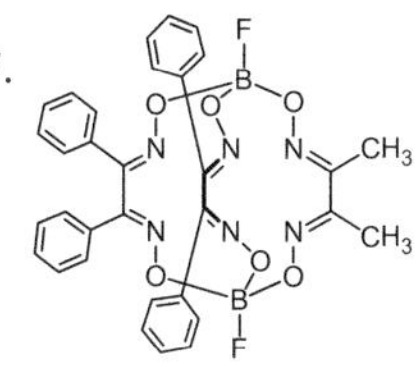

298.

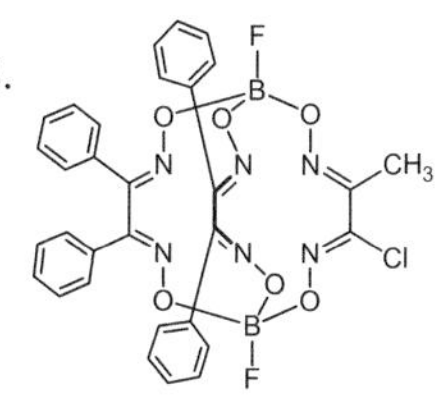

299.

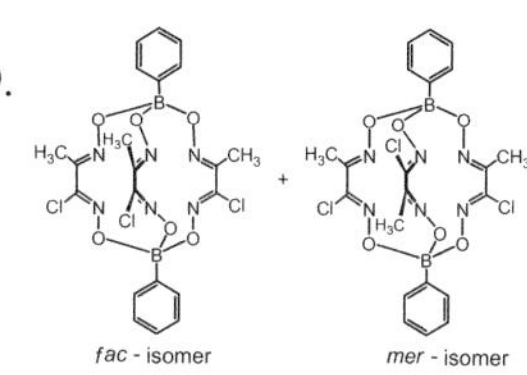

300.

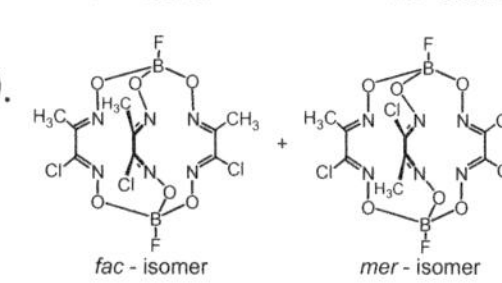

301.

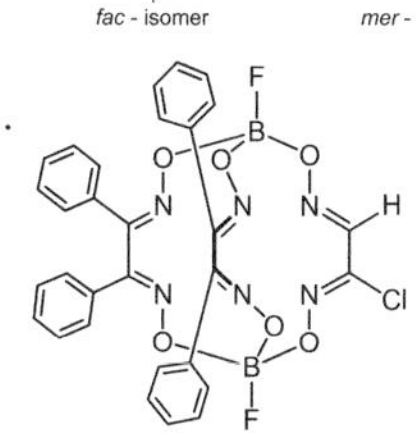

302.

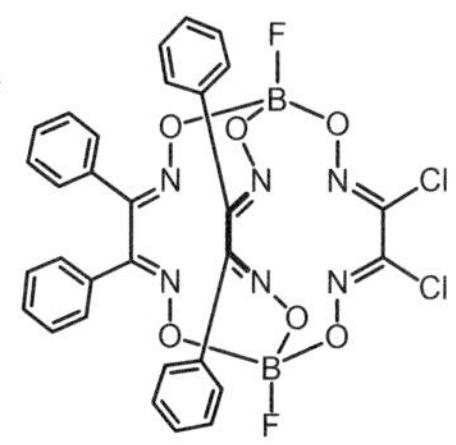

303.

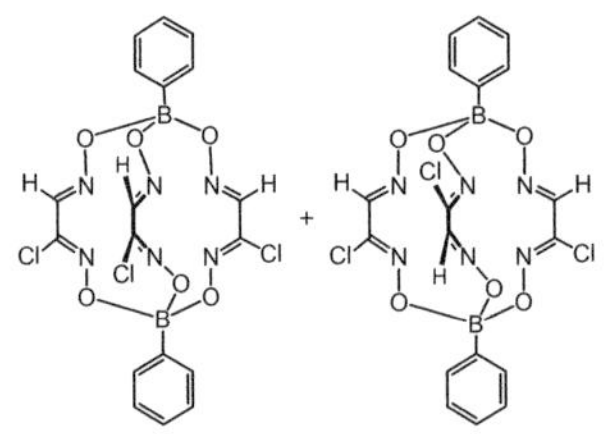

fac - isomer *mer* - isomer

304.

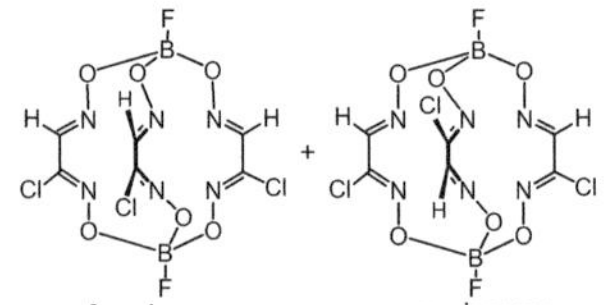

fac - isomer *mer* - isomer

305.

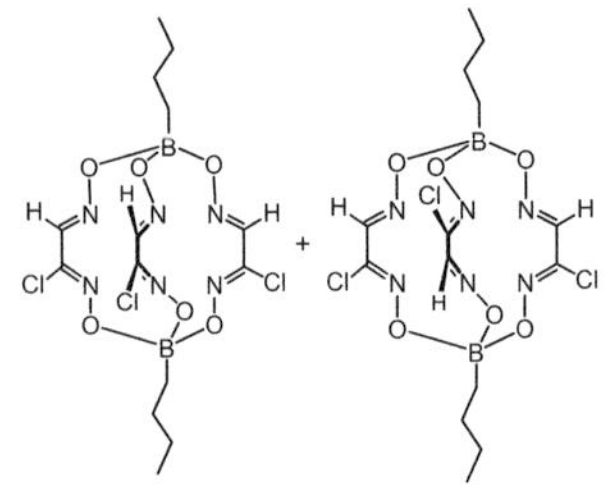

fac - isomer *mer* - isomer

306.

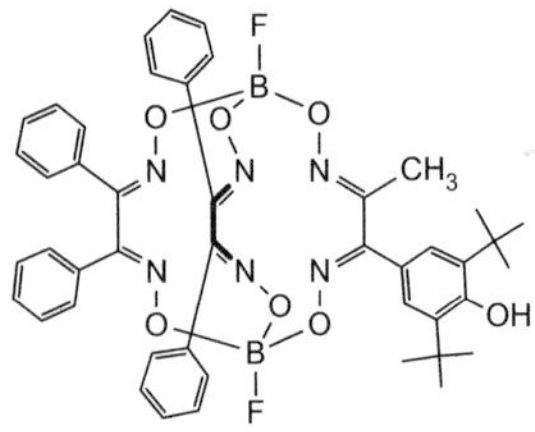

307.

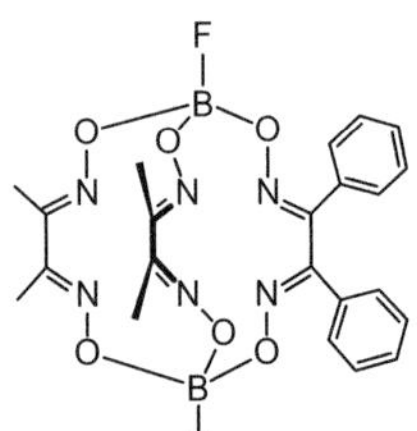

308.

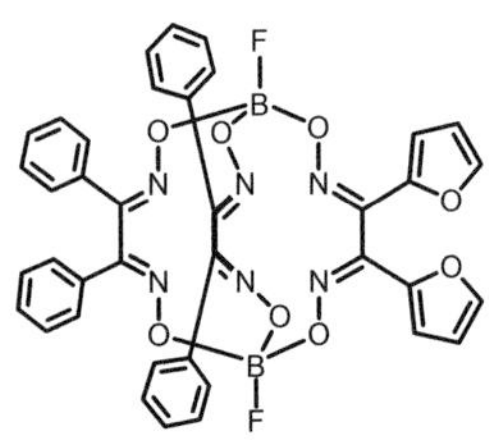

309.

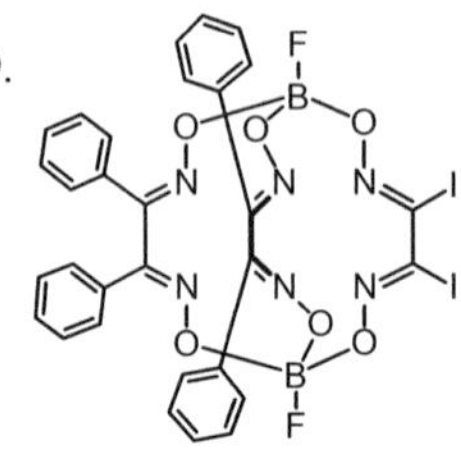

310.

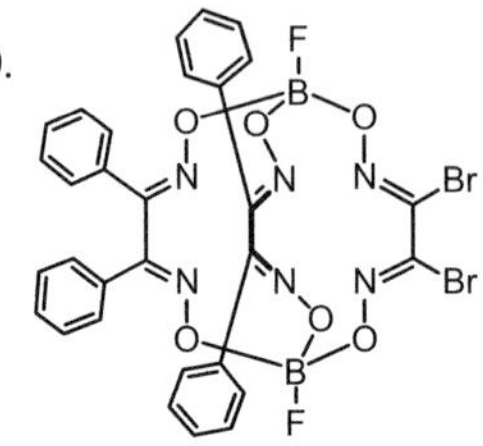

311.

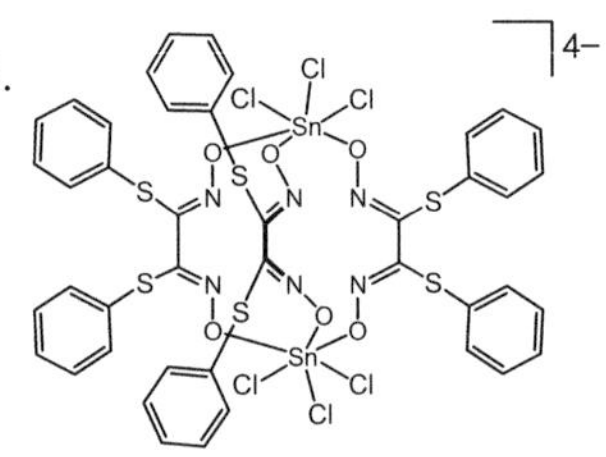

312.

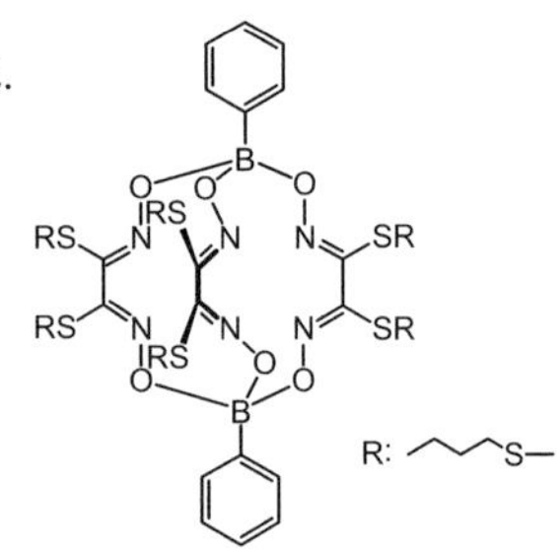

313.

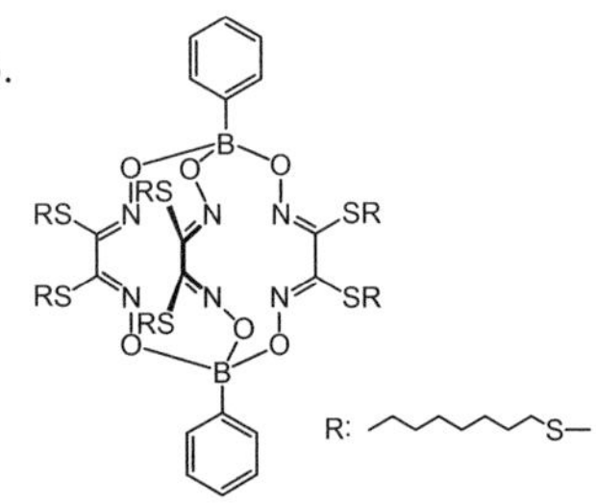

314.

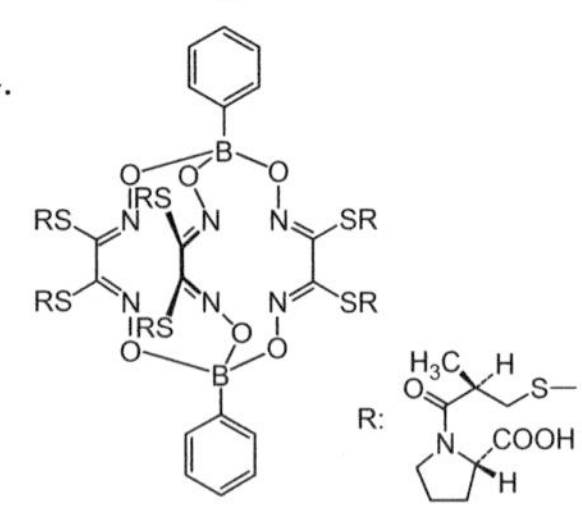

315.

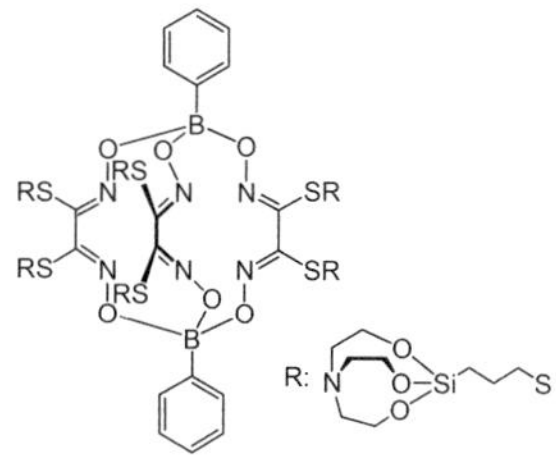

316.

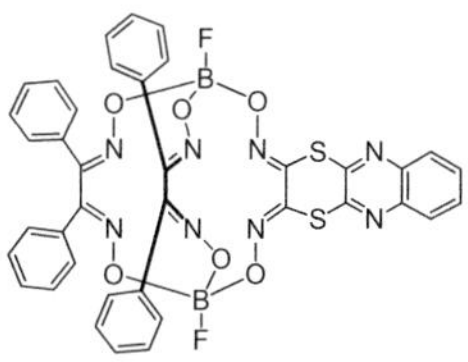

317.

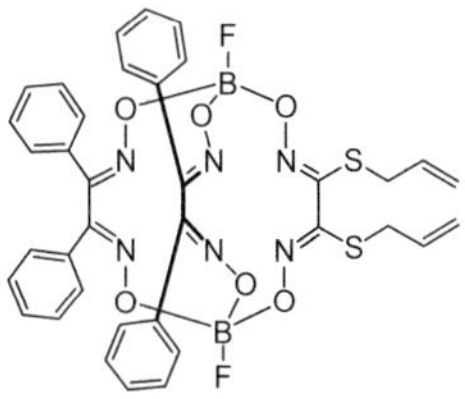

318.

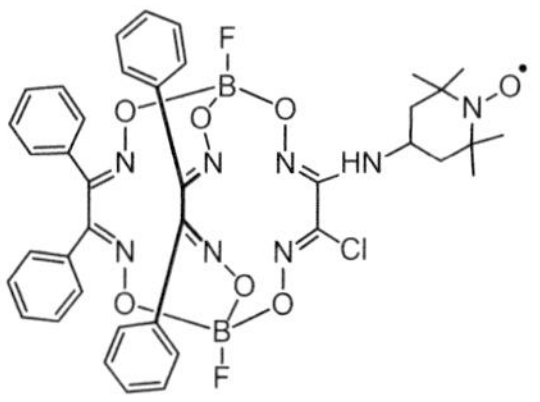

319.

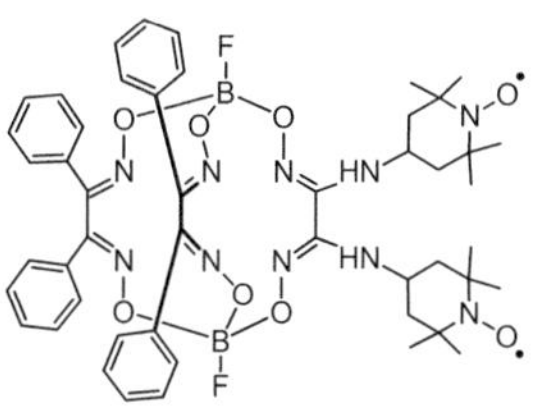

320.

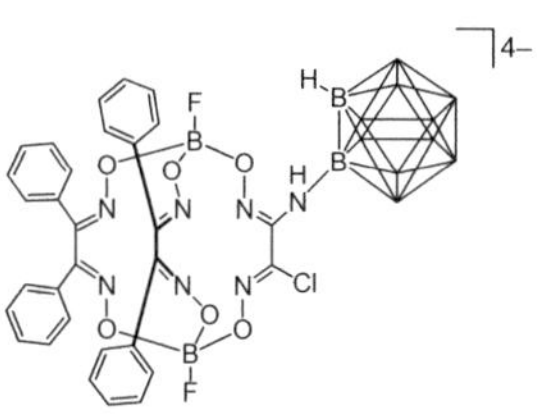

321.

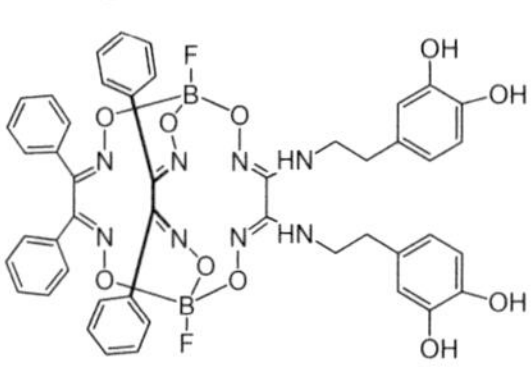

322.

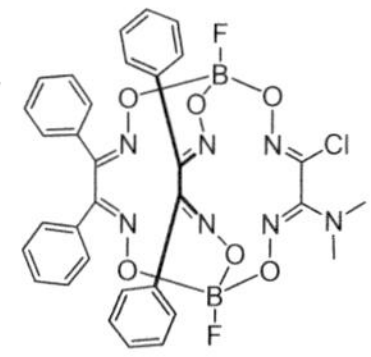

323.

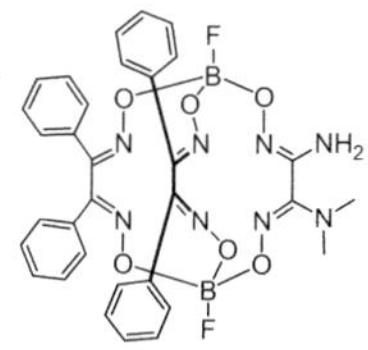

324.

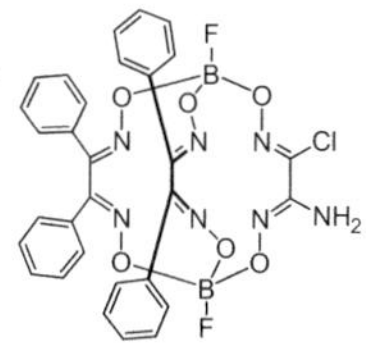

325.

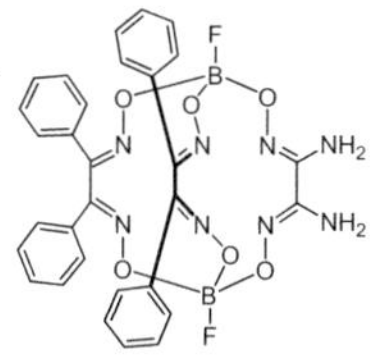

326.

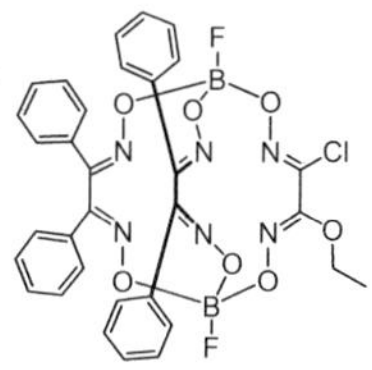

327.

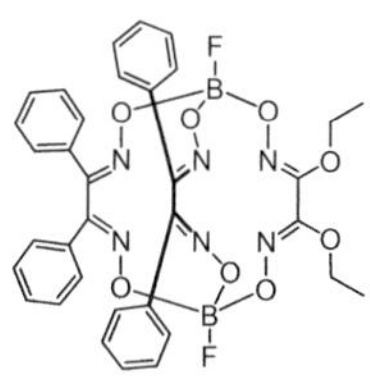

328.

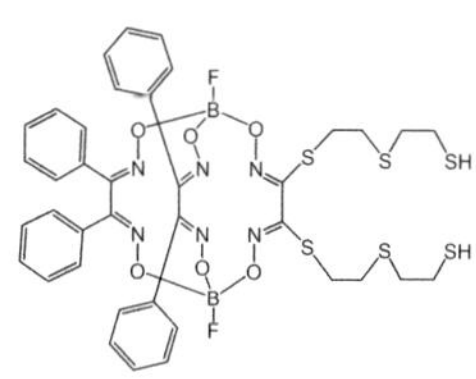

329.

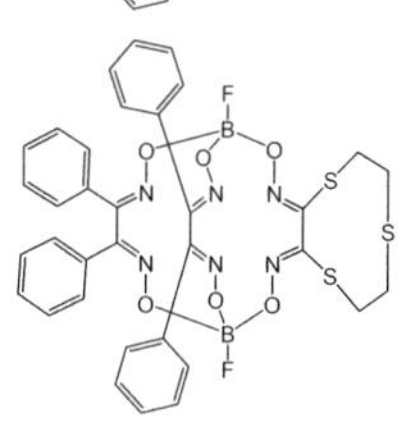

330.

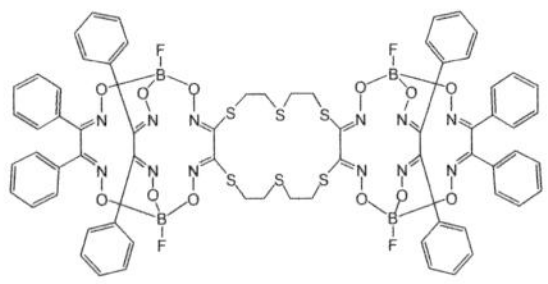

331.

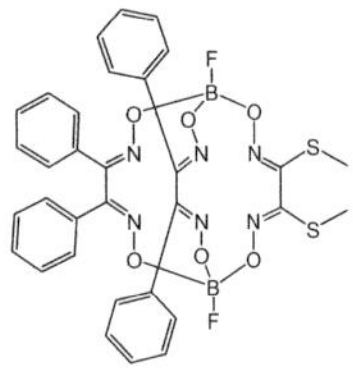

332.

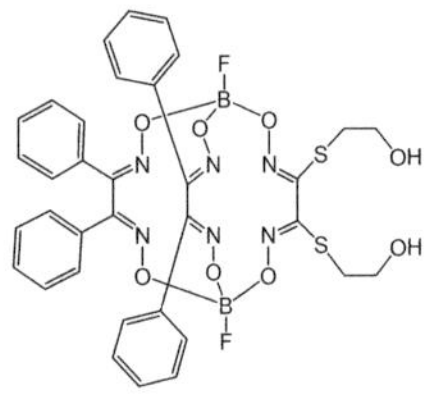

333.

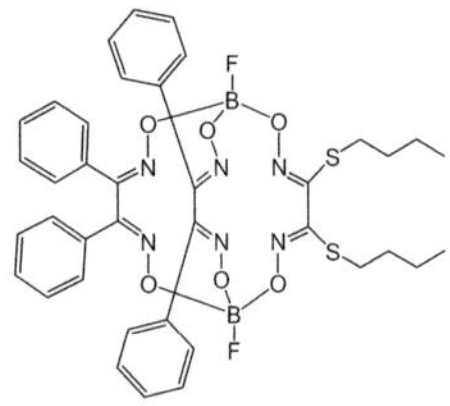

334.

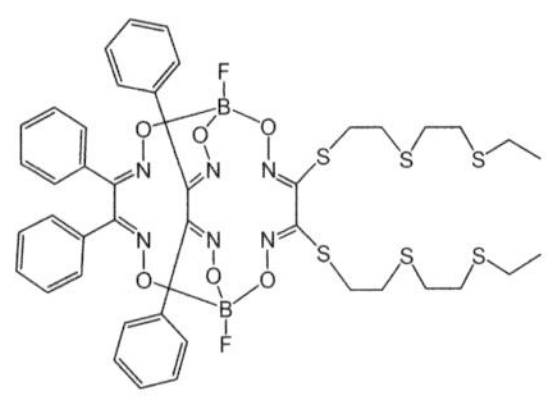

335.

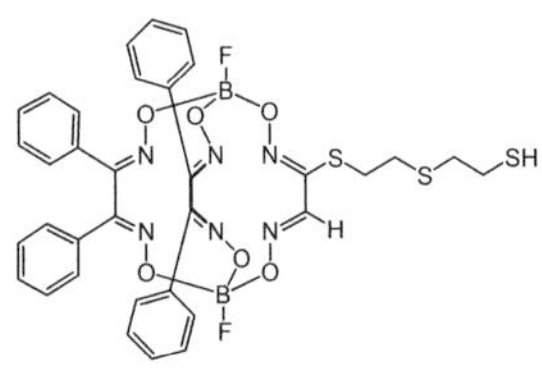

336.

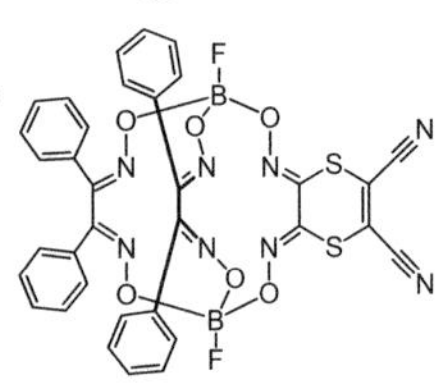

337.

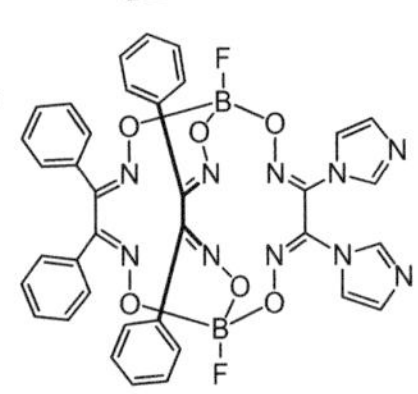

338.

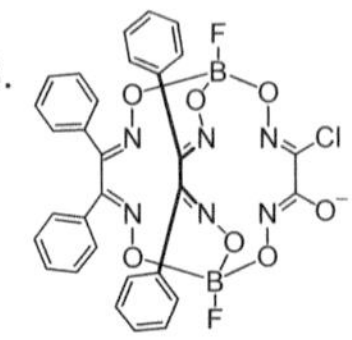

339.

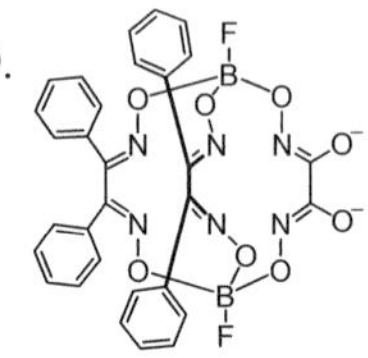

340.

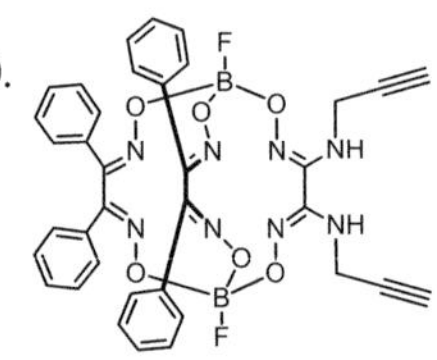

341.

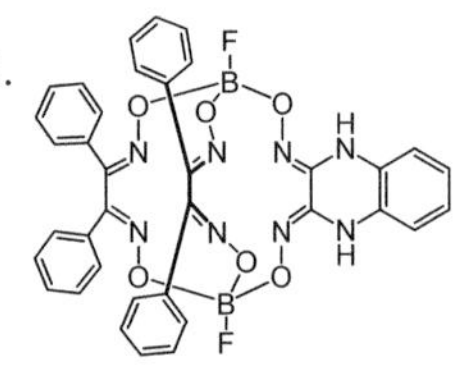

342.

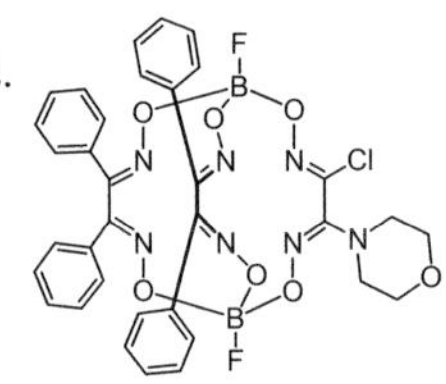

343.

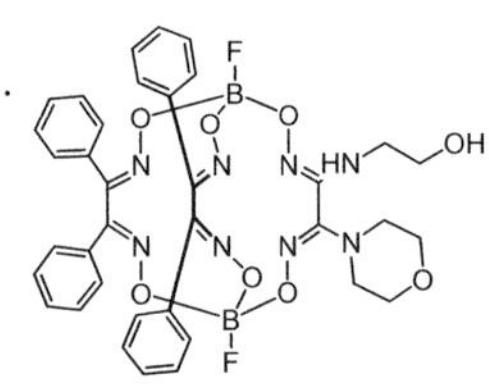

344.

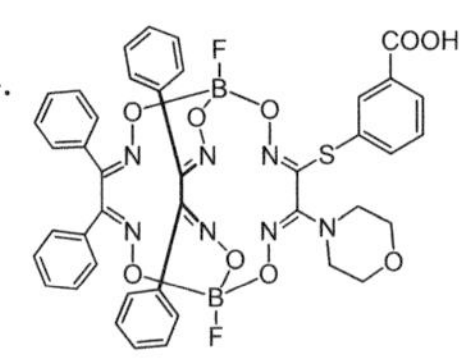

345.

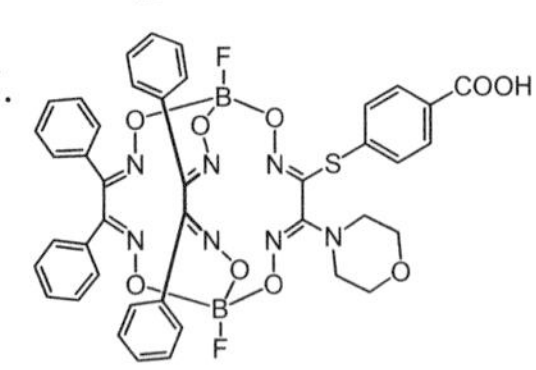

346. 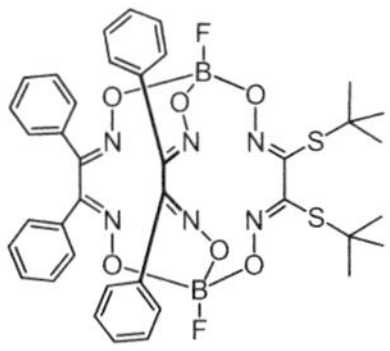

347. 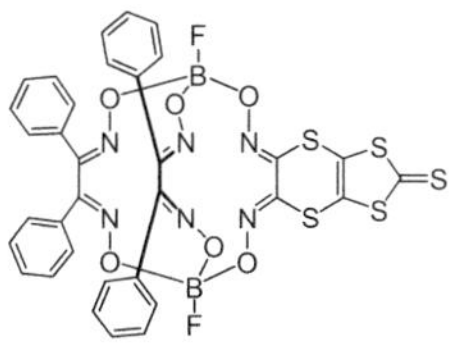

348. 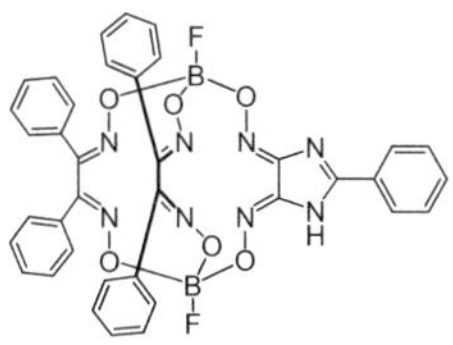

349. 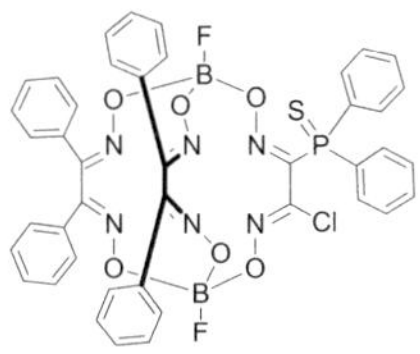

350. 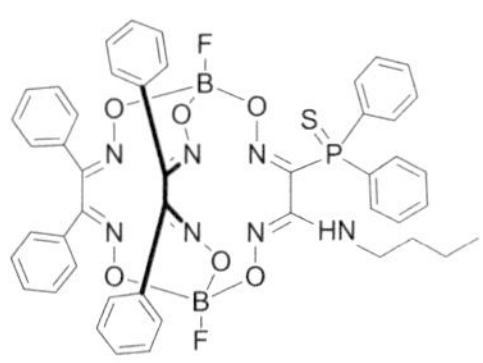

351. 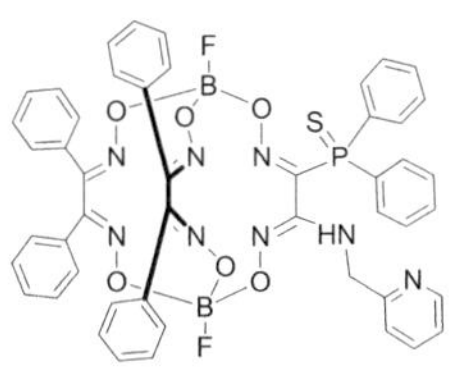

352. 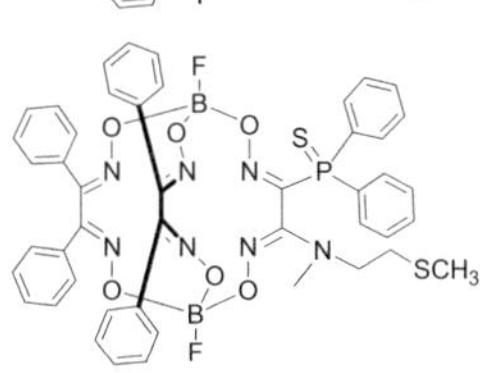

353. 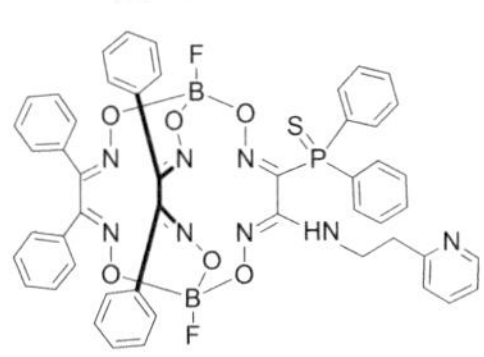

354. 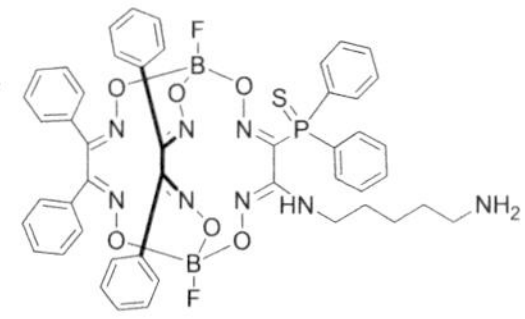

355. 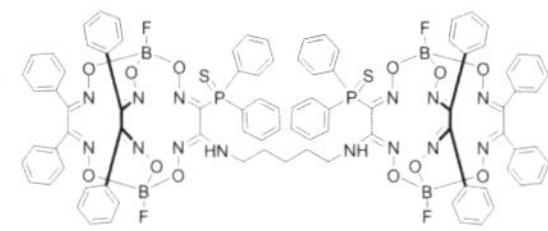

356. 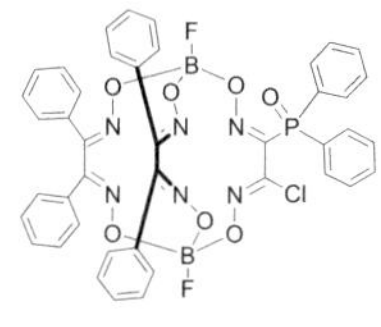

357. 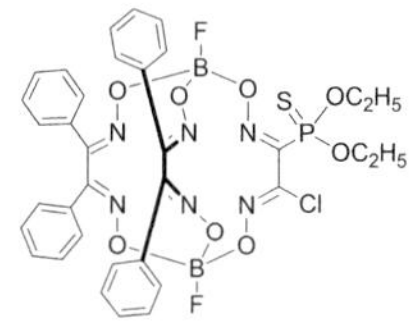

358. 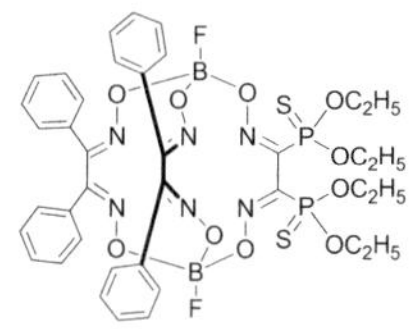

359. 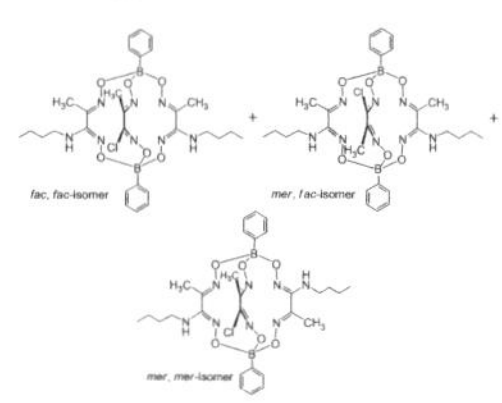

360. 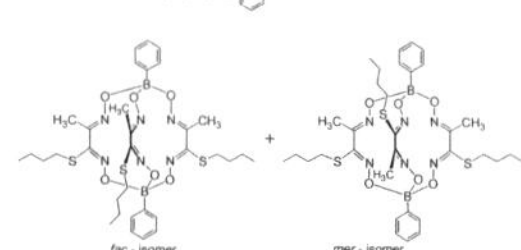

361. 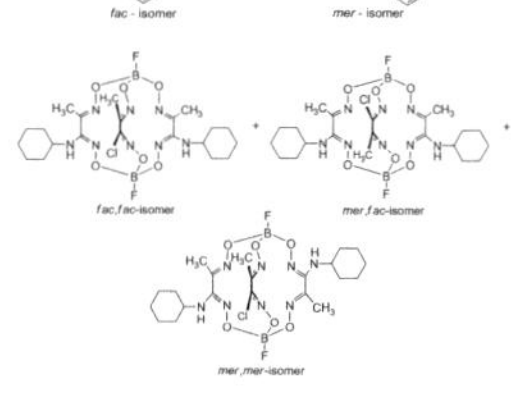

362. 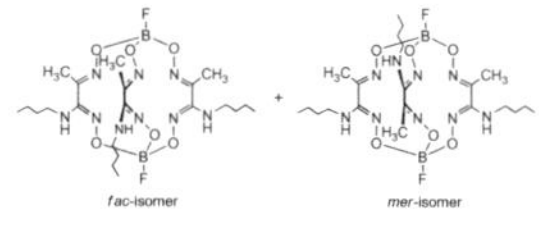

363. 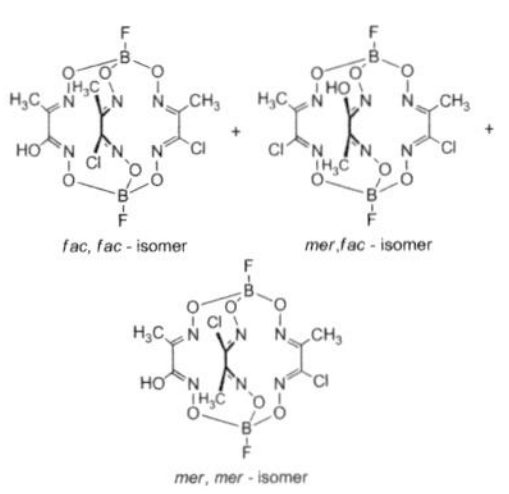

364. 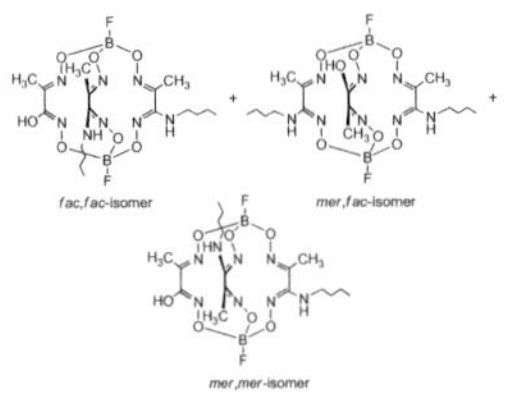

365. 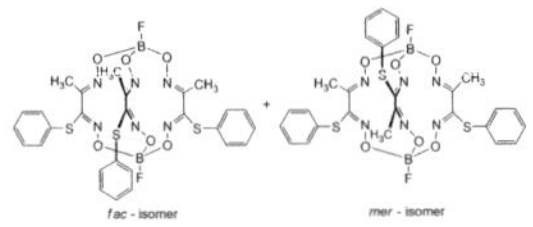

366. 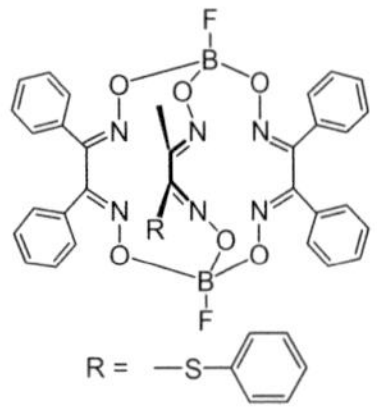

367. 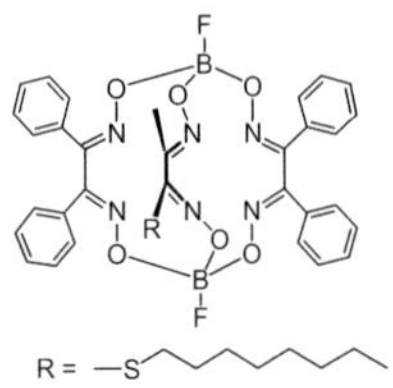

368. 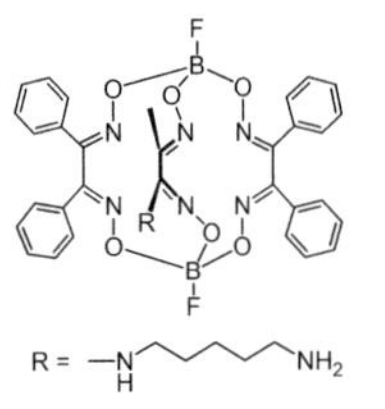

369. 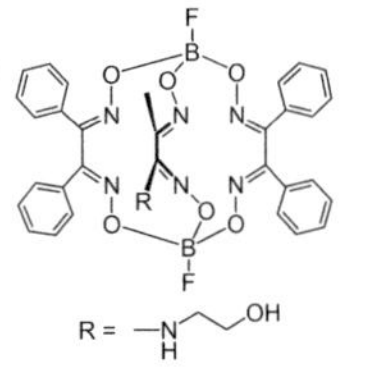

370. 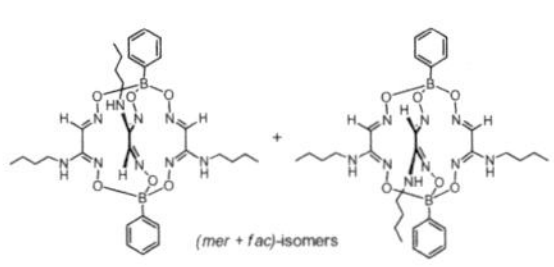

371. 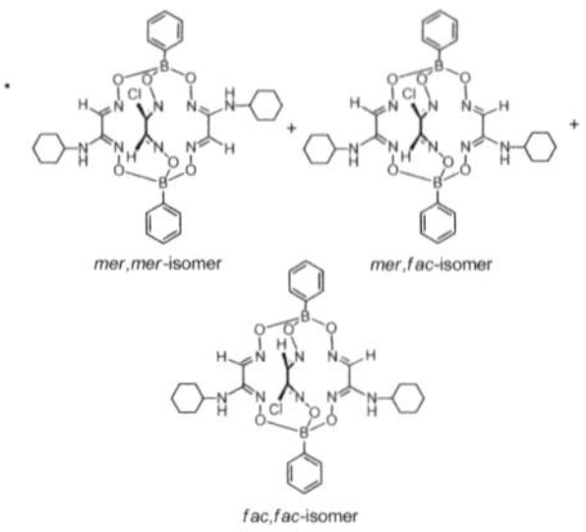

372. 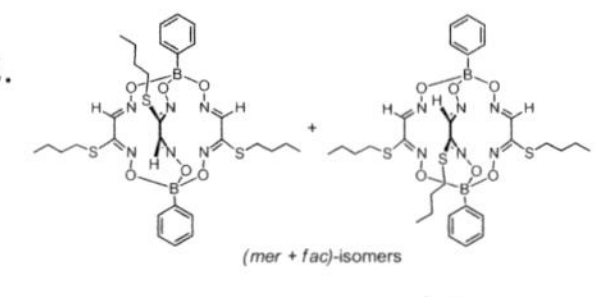

373. 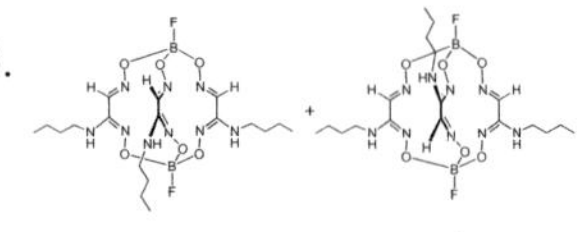

374. 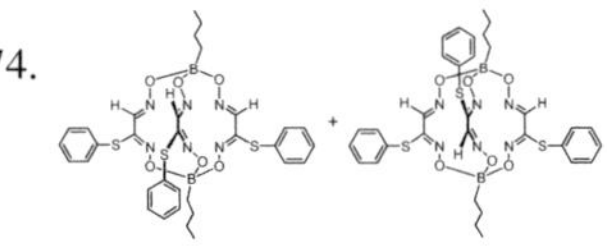

375. 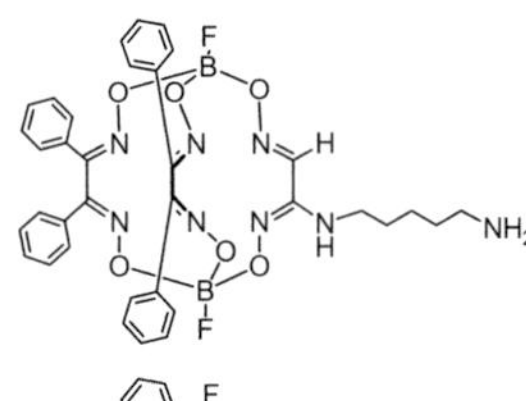

376. 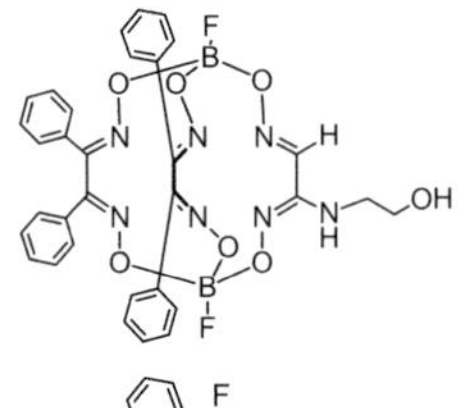

377. 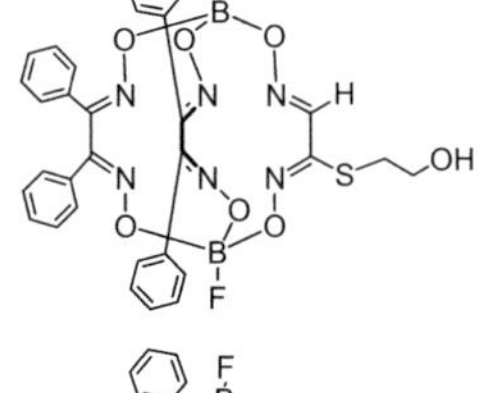

378. 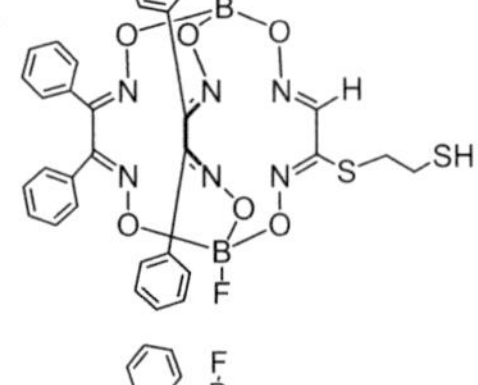

379. 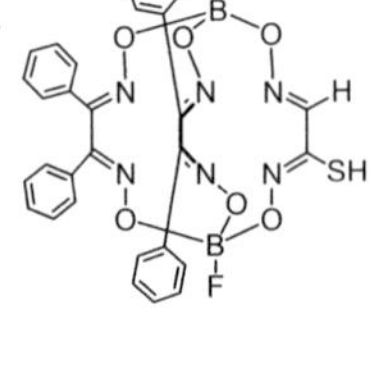

380. 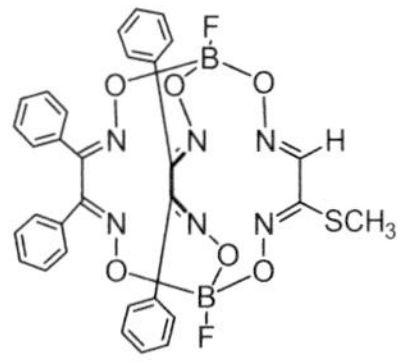

381. 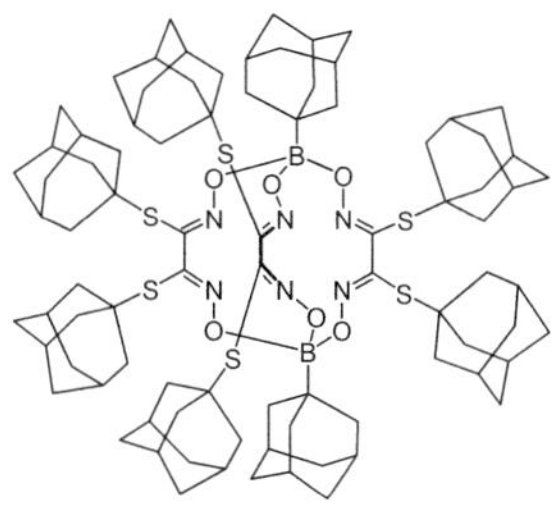

382. 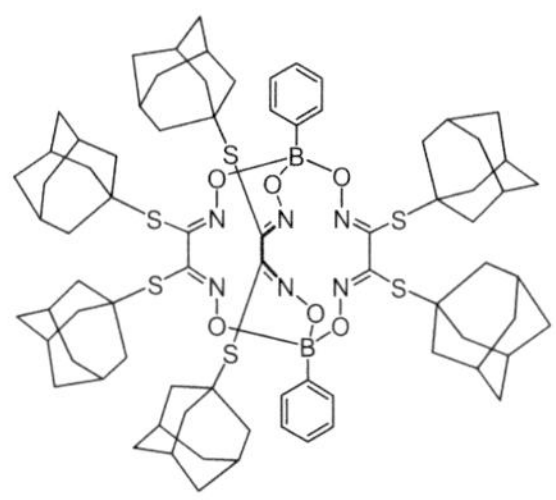

383. 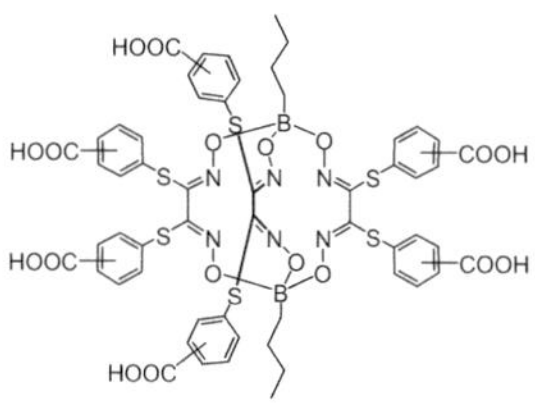

ortho-substituted

384. 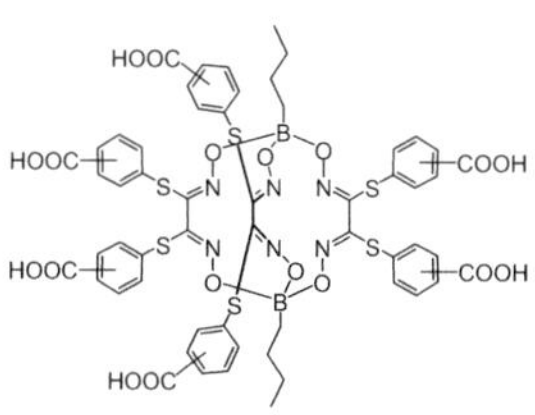

meta-substituted

385. 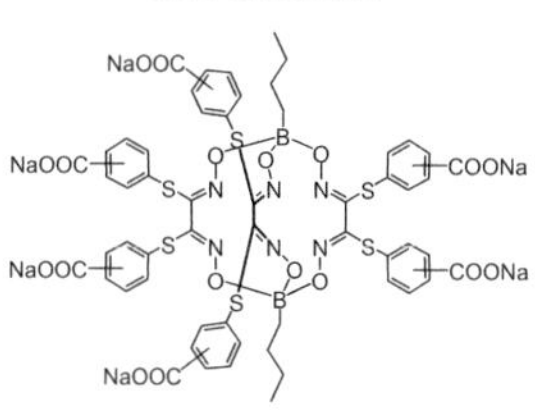

para-substituted

386. 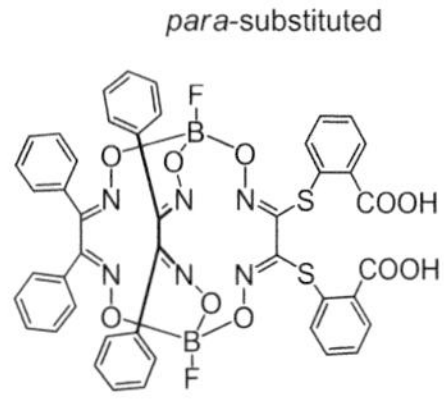

387. 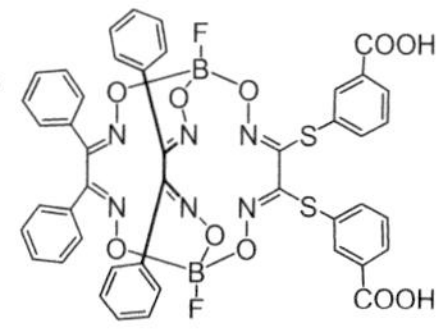

388. 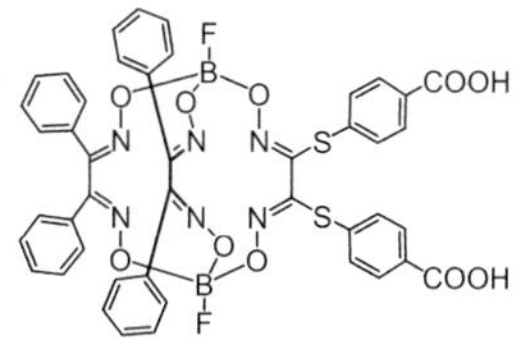

389. 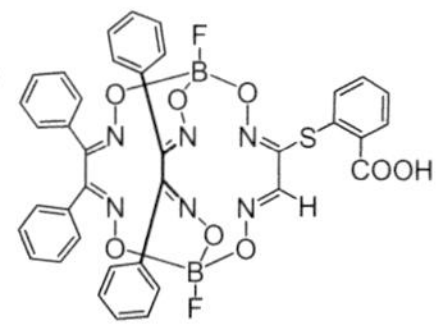

390. 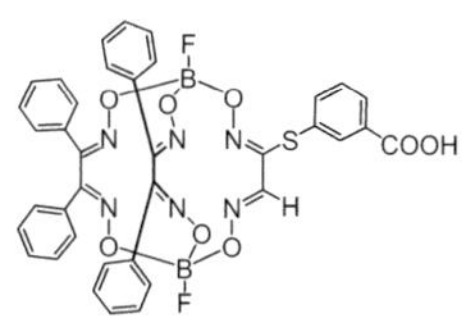

391. 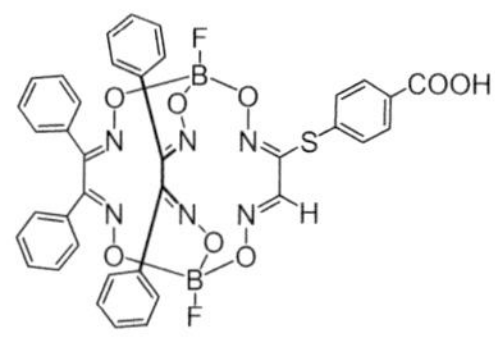

392. 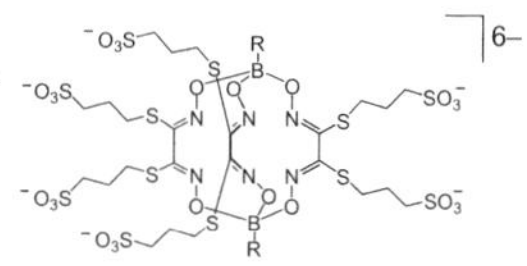

R= n-C_4H_9

393. 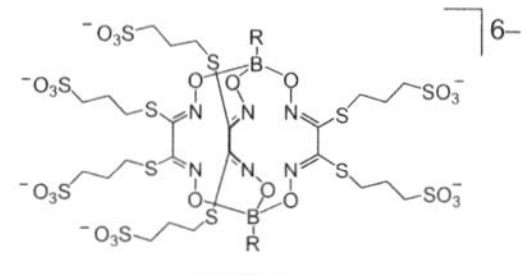

R= C_6H_5

394. 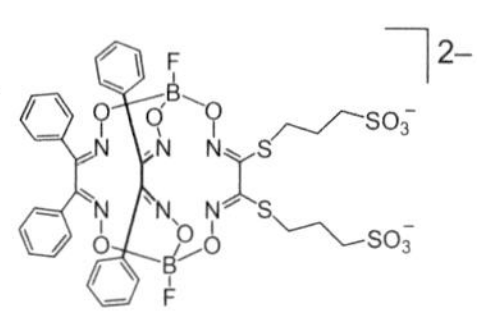

395. 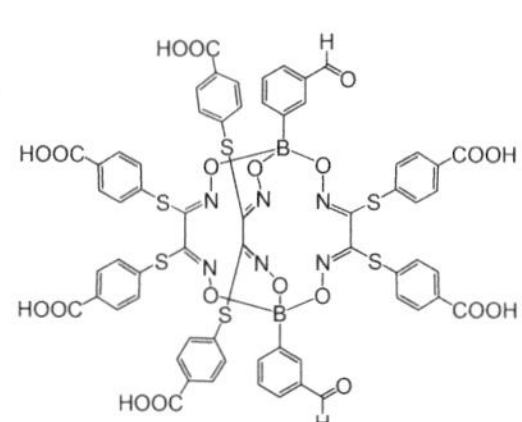

396.
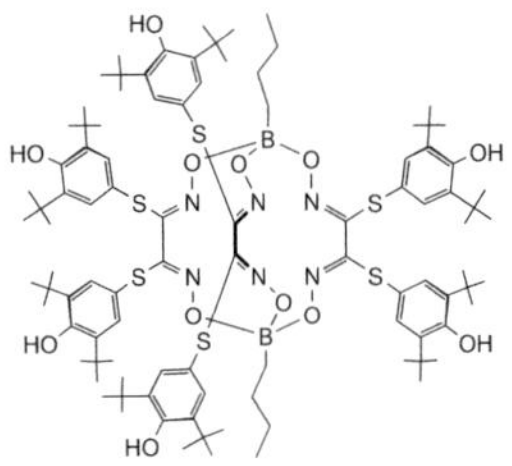

397.
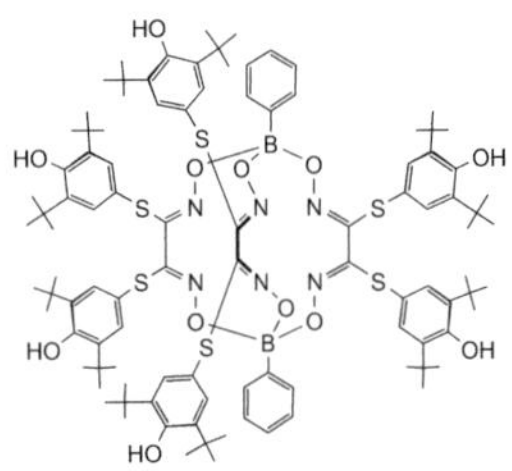

398.
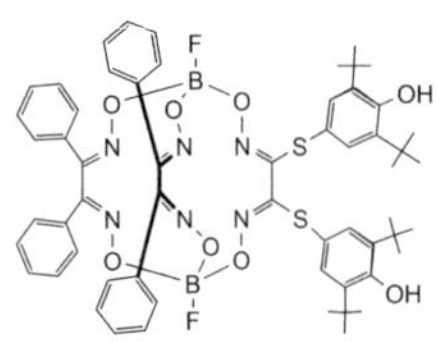

399.
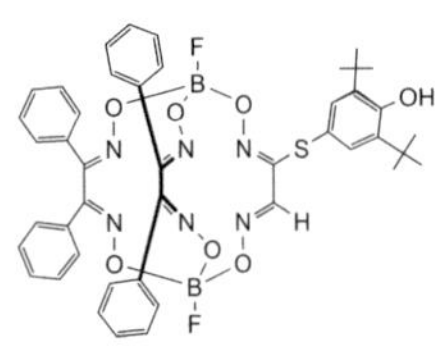

400.
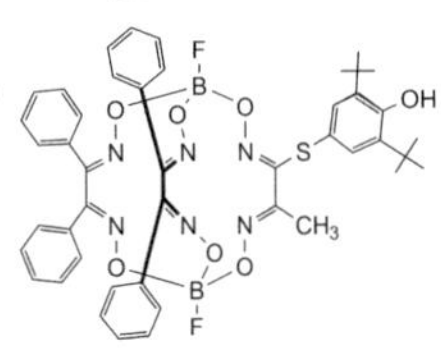

401.
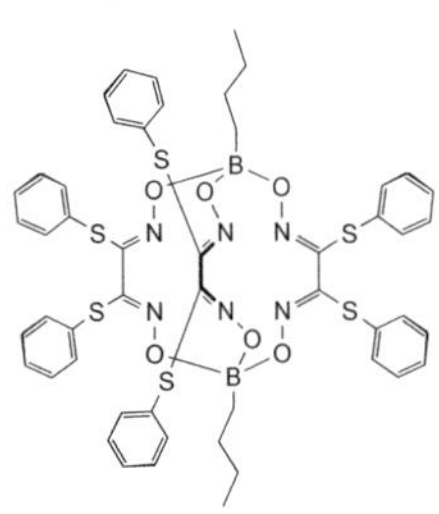

402.
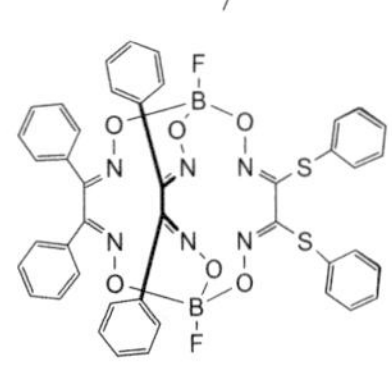

403.
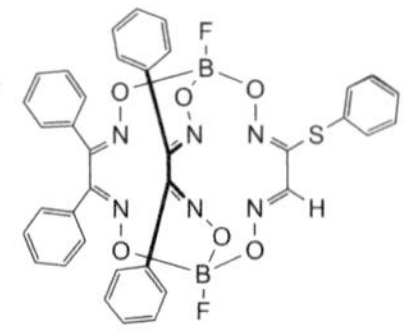

404.
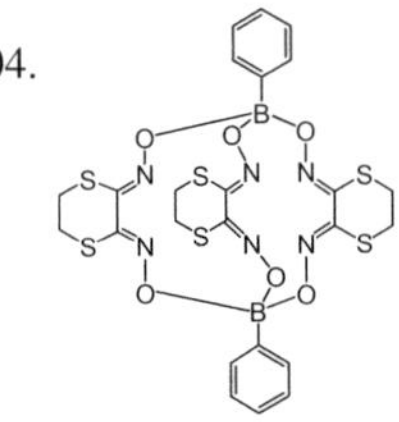

405.
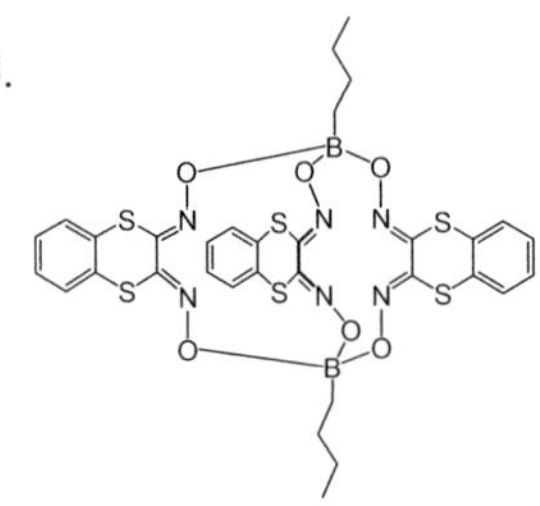

406.
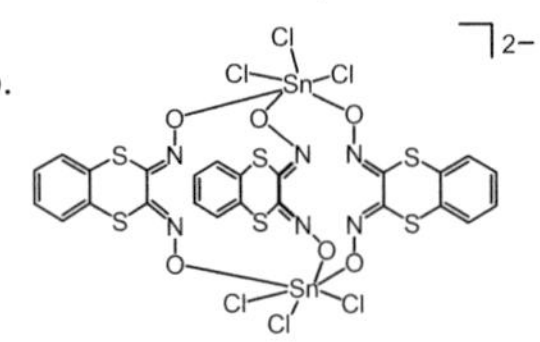

407.
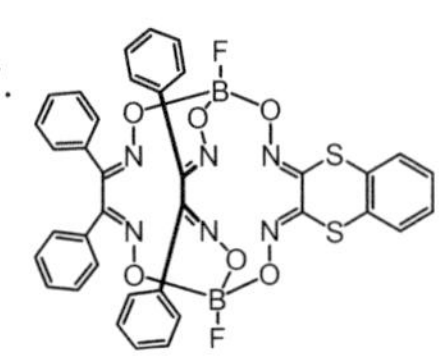

408.
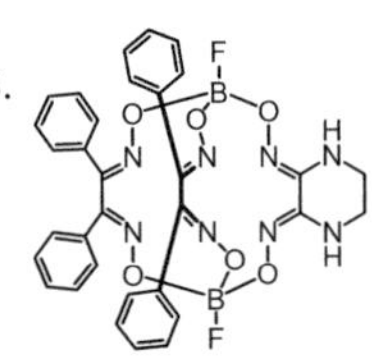

409.
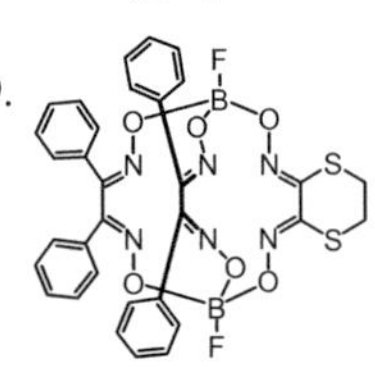

410.
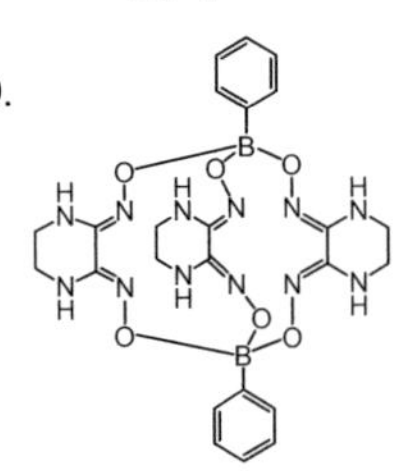

411.

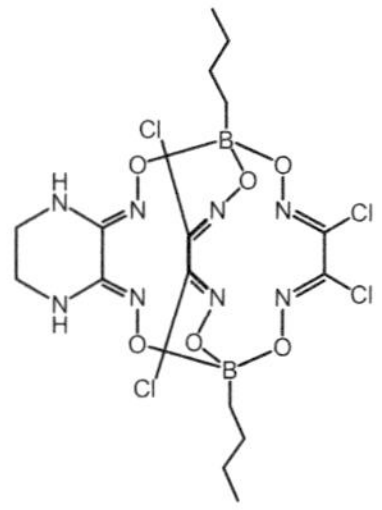

412.

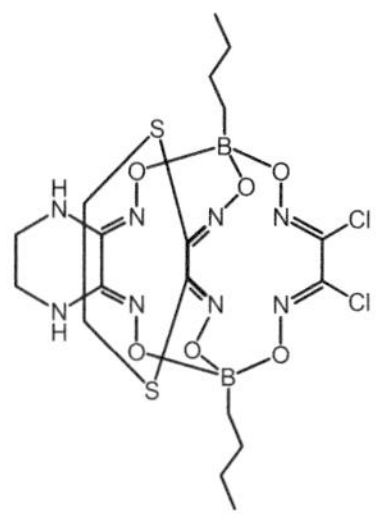

413.

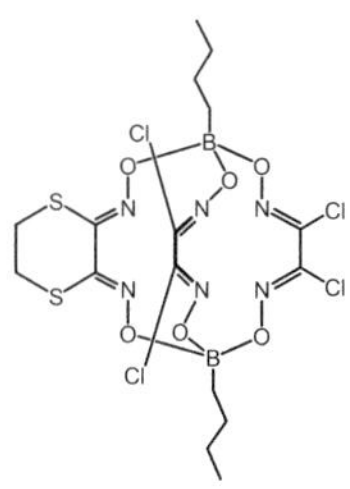

414.

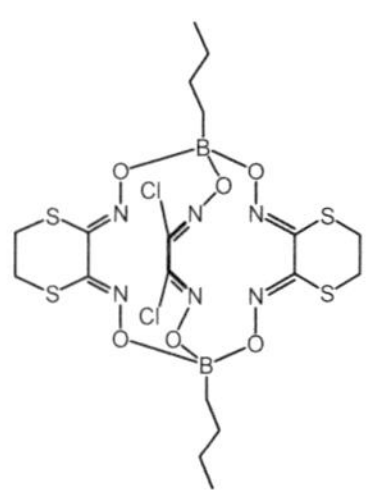

415.

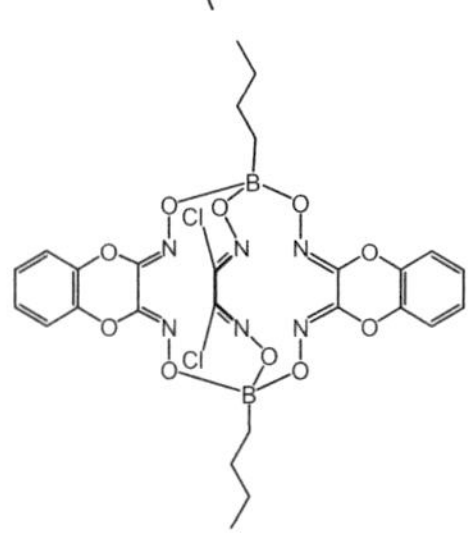

416.

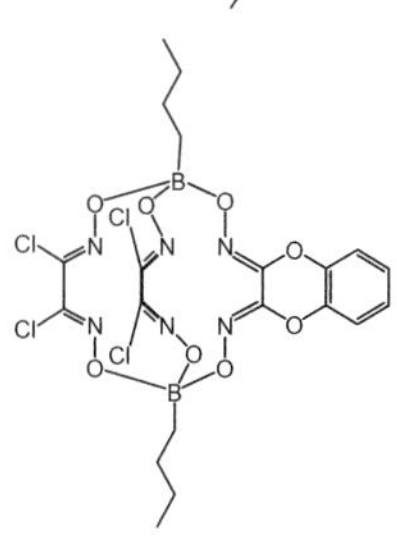

417.

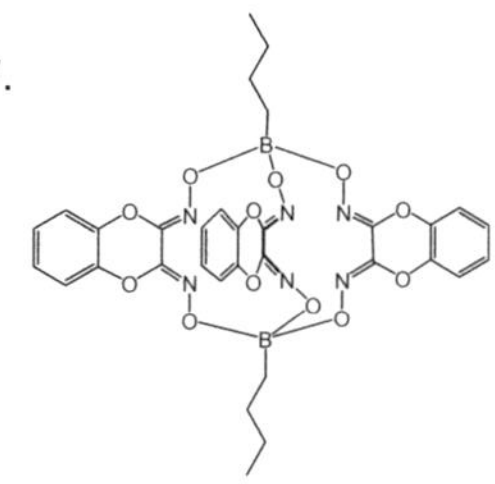

418.

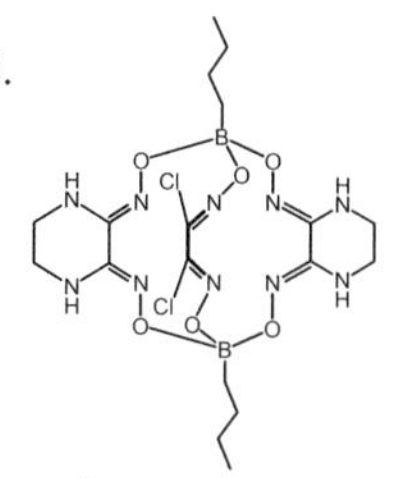

419. $[Co^{3+}((C_2H_5)_2N)_3(AlkNH)_3(Gm)_3(BC_6H_5)_2]^+$
mixture of isomers

420. $[Co^{3+}(\textit{tert}\text{-}C_4H_9NH)_4(n\text{-}C_4H_9NH)_2(Gm)_3(BC_6H_5)_2]^{3+}$
mixture of isomers

421.

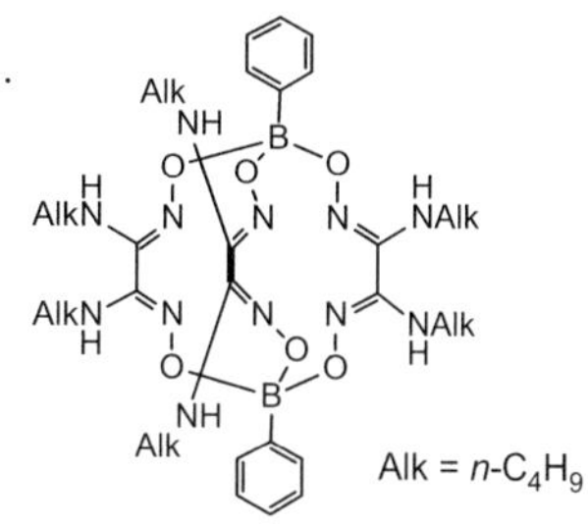

422.

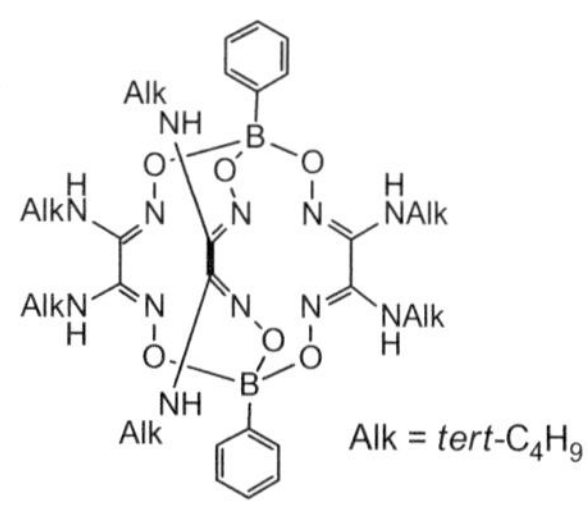

423.

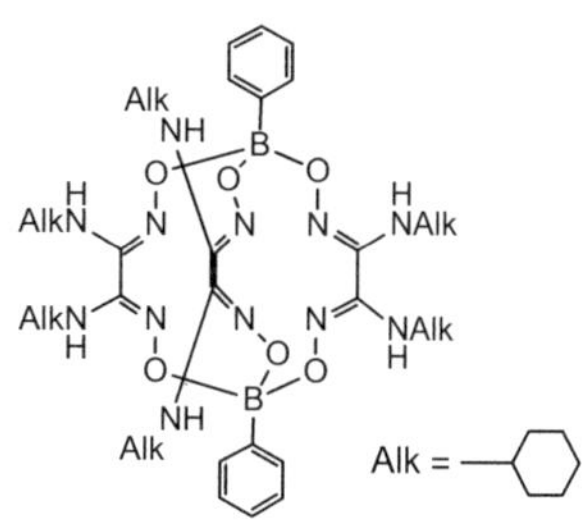

424.

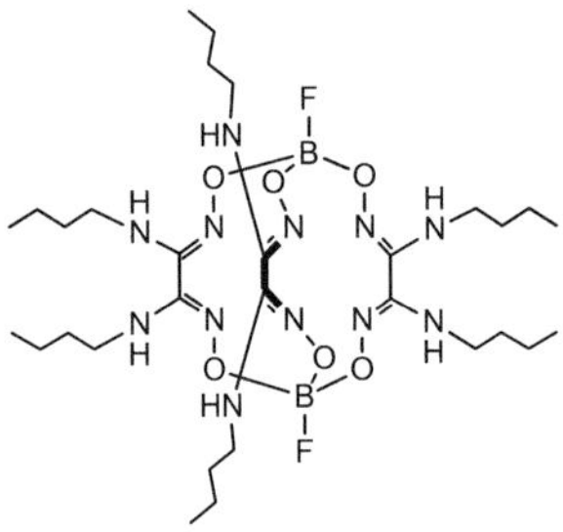

425.

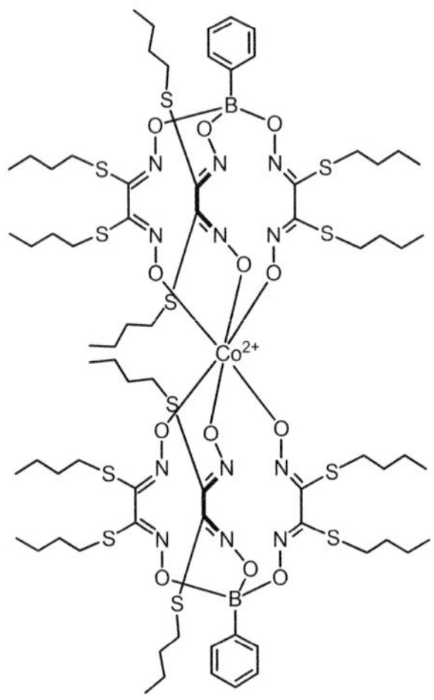

426.

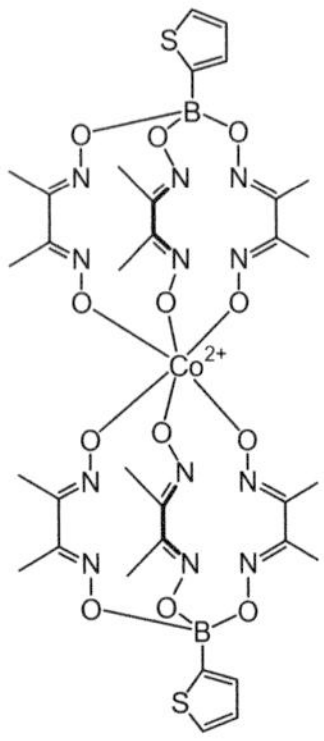

427.

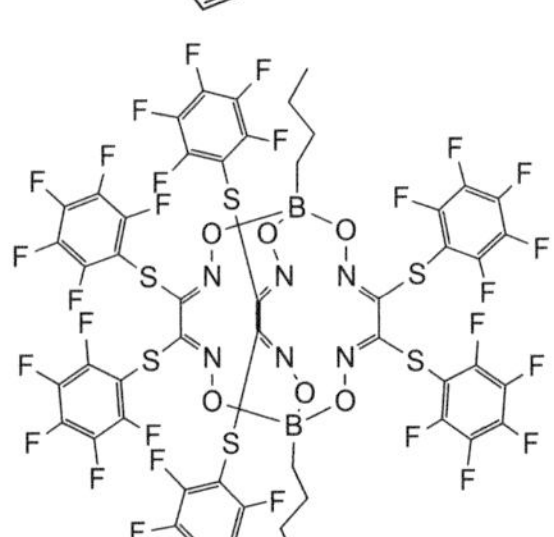

428.

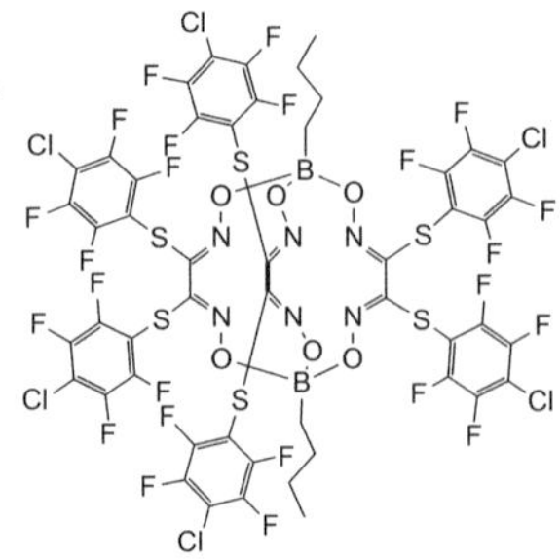

429.

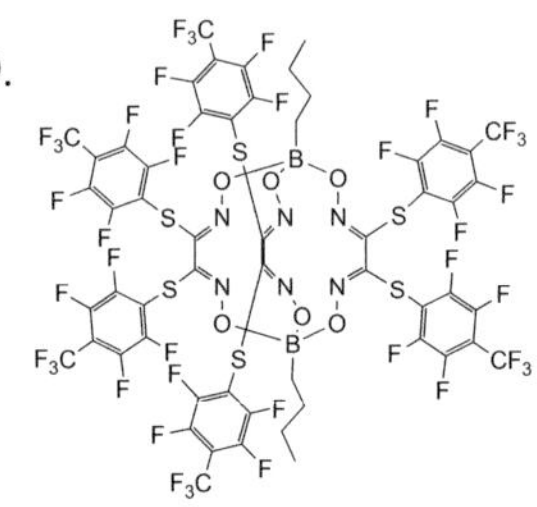

430.

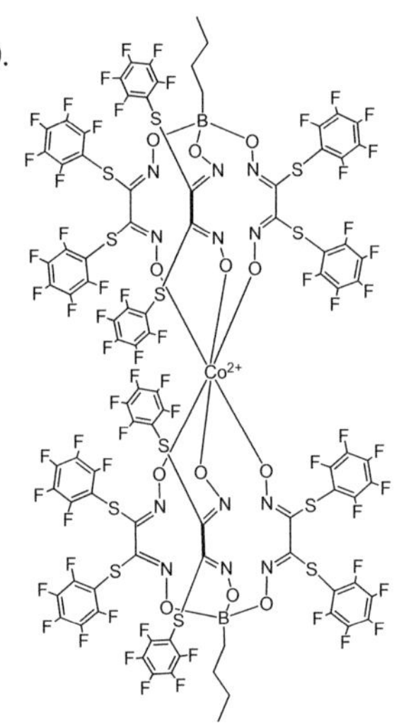

431.

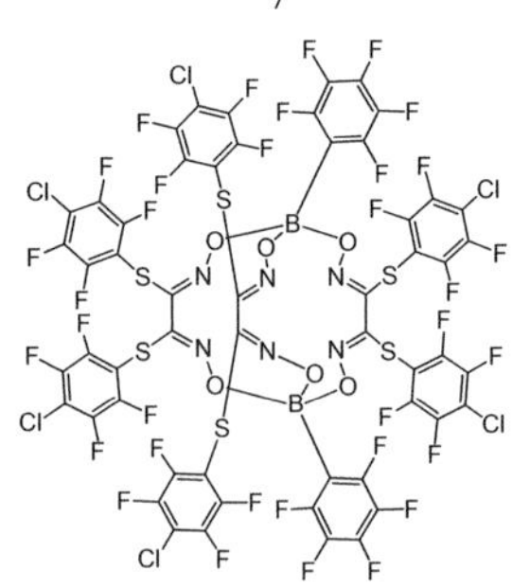

432.

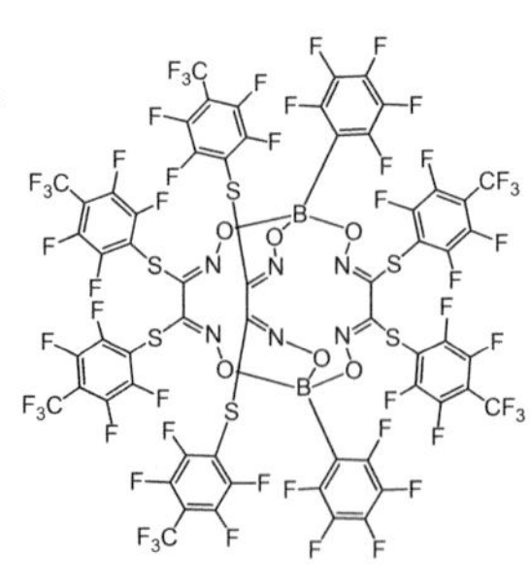

433.

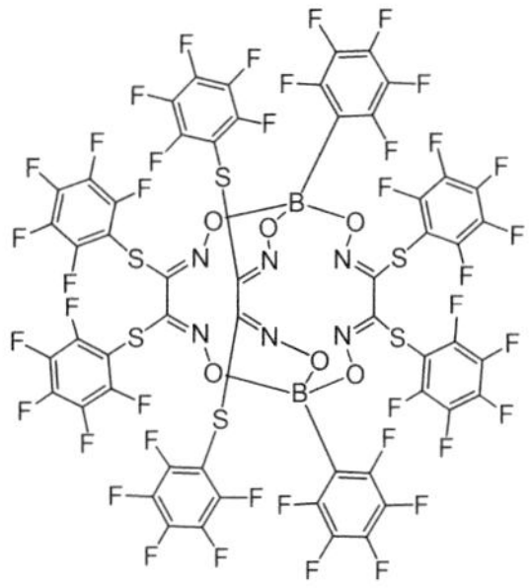

434.

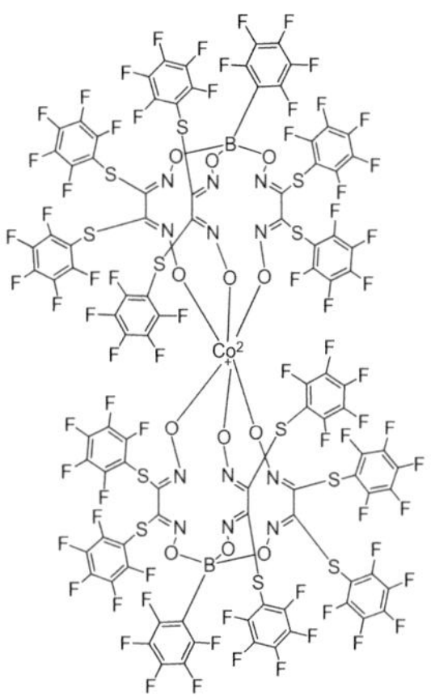

435.

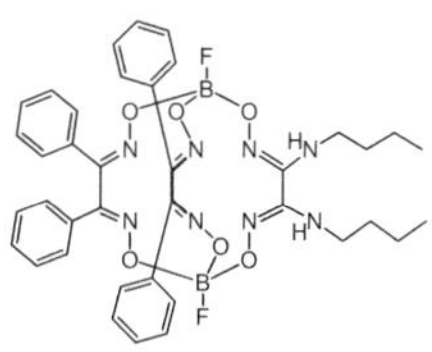

436.

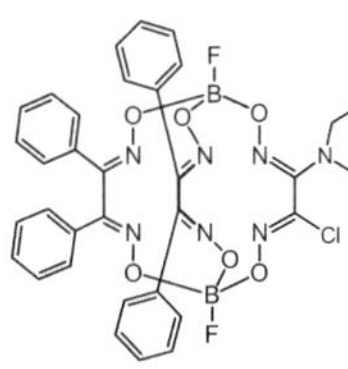

437.

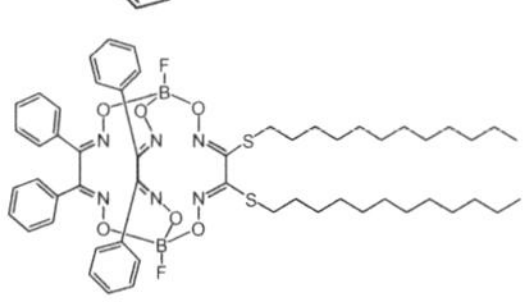

438.

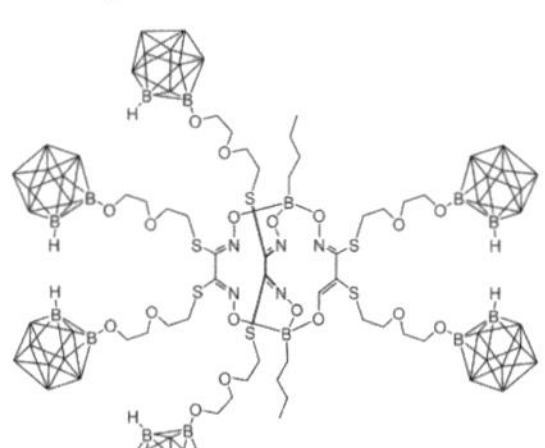

439.

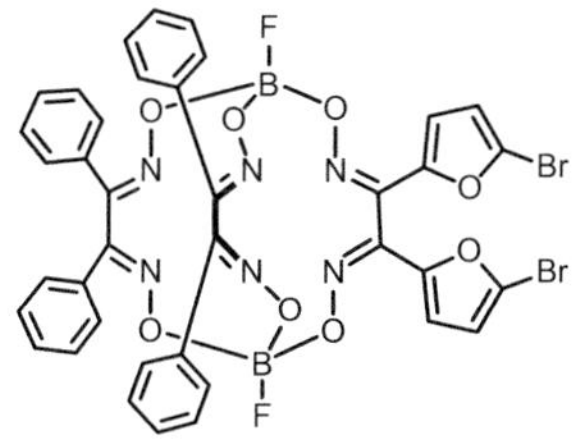

440.

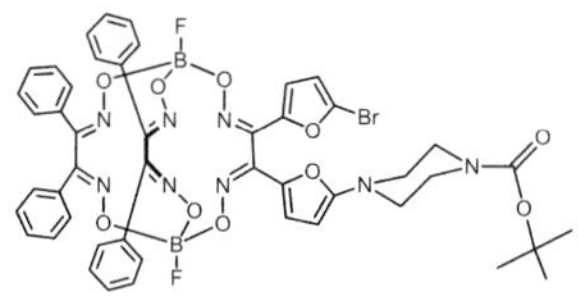

441.

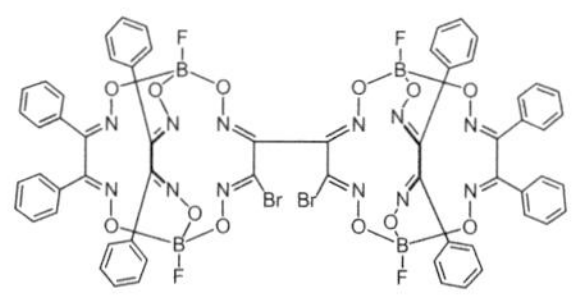

442.

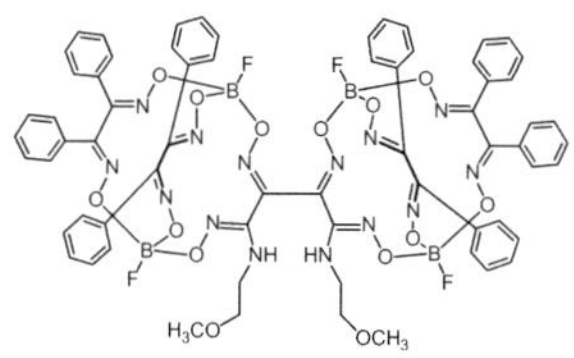

443.

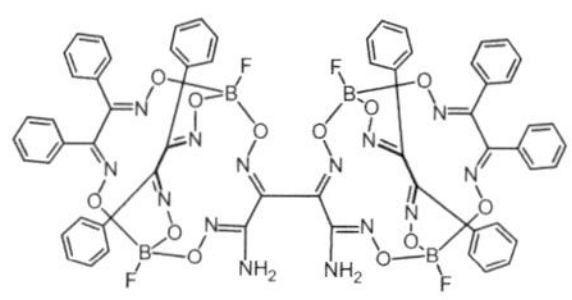

444.

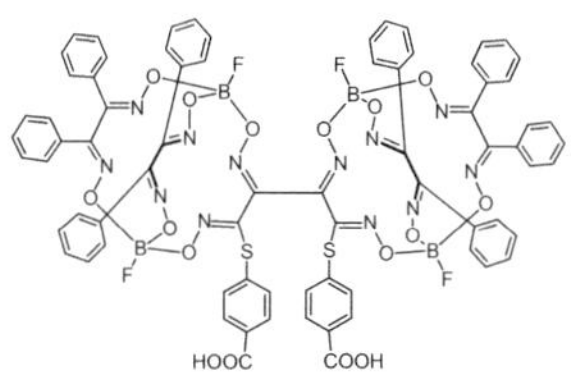

445.

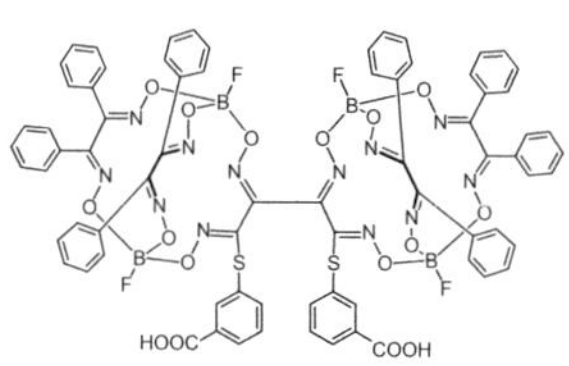

446.

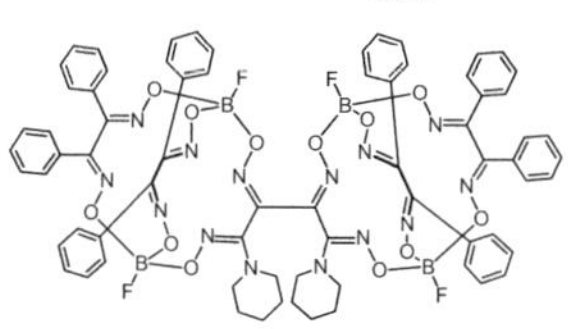

447.

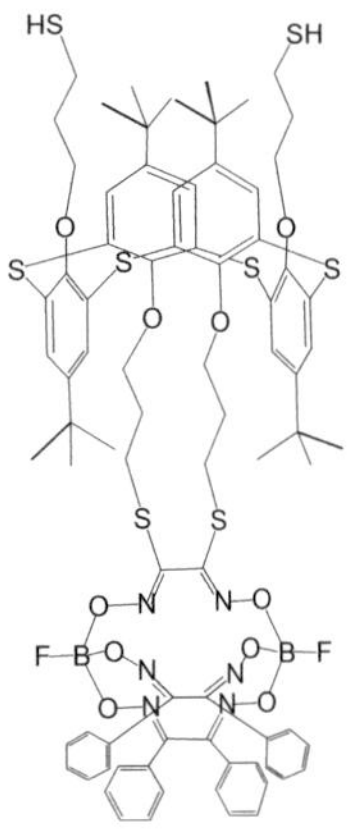

448.

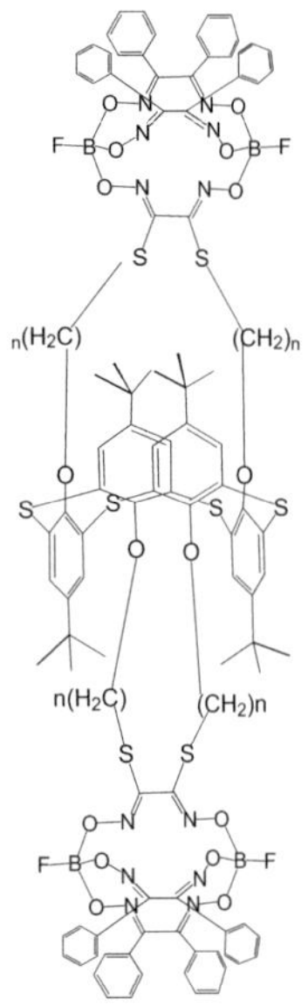

n=3

449.

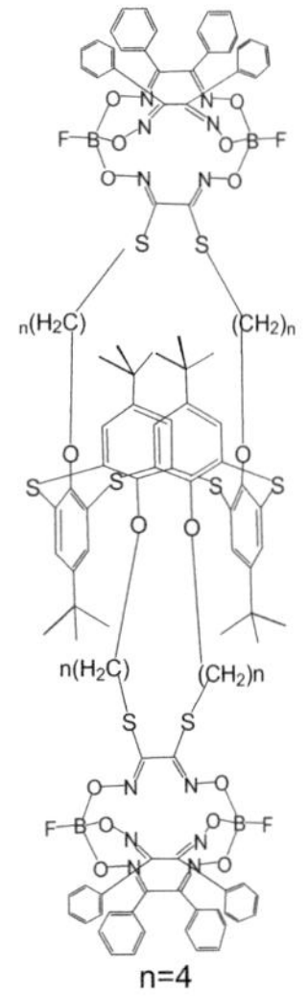

n=4

450.

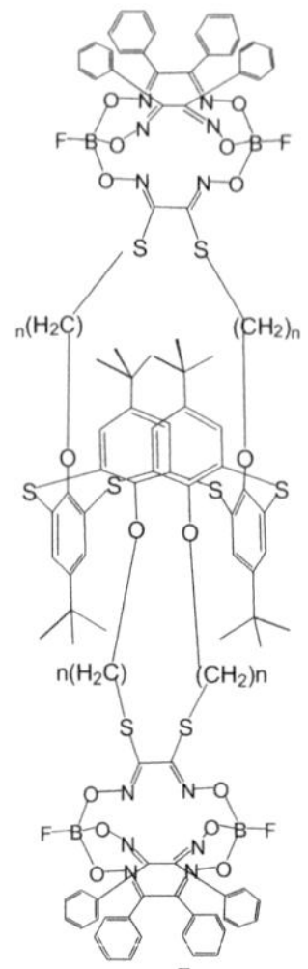

n=5

451.

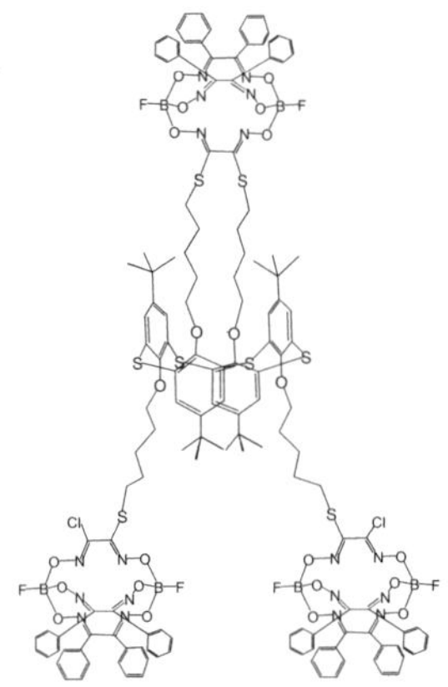

452.

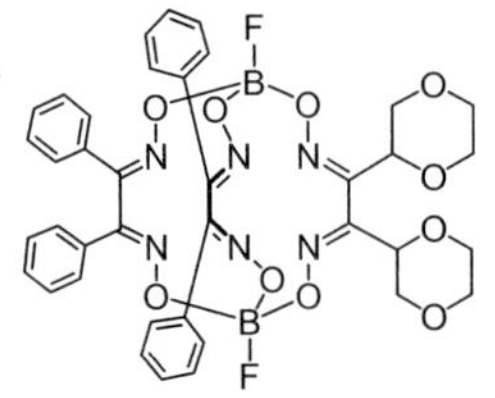

453.

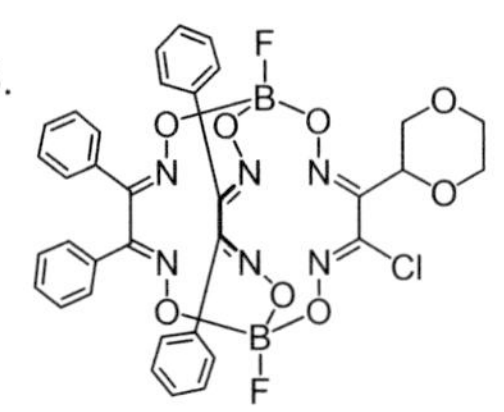

454.

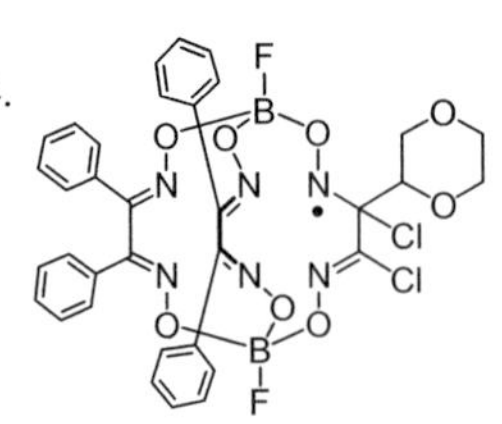

455.
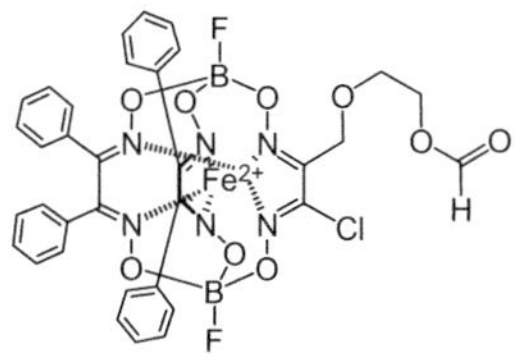

456.
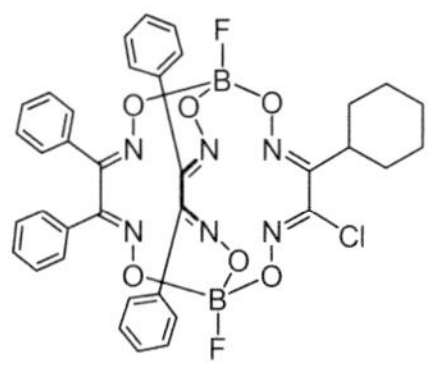

457.
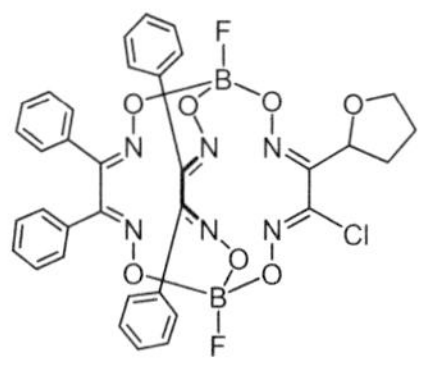

458.
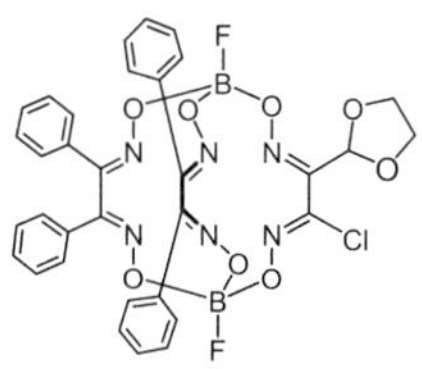

459.
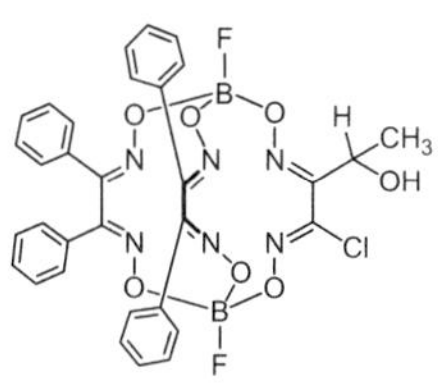

460.
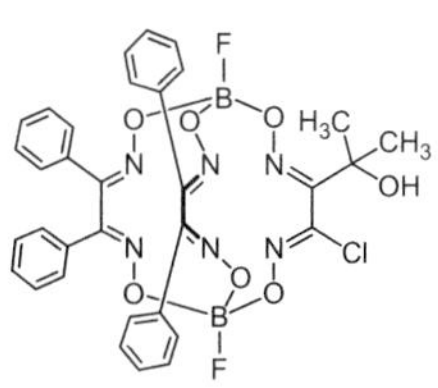

461.
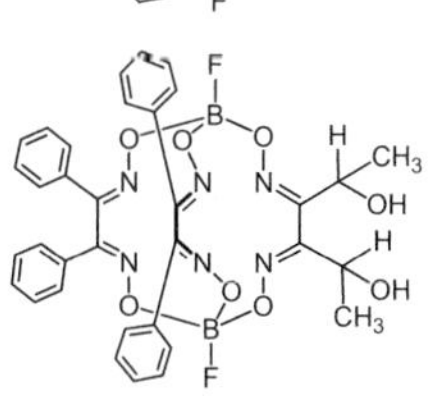

462.
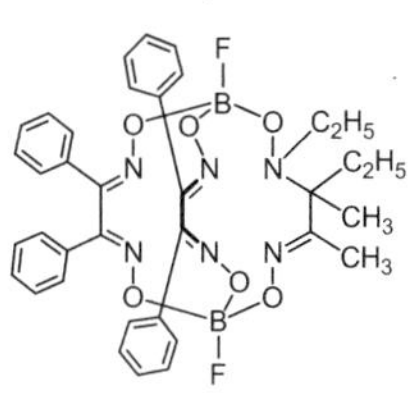

463.
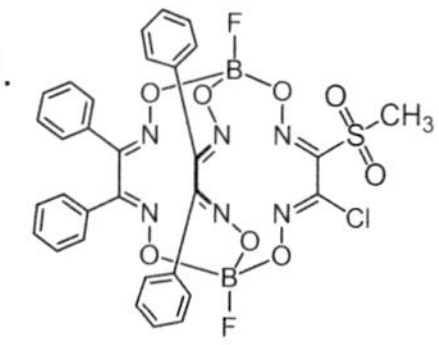

464.
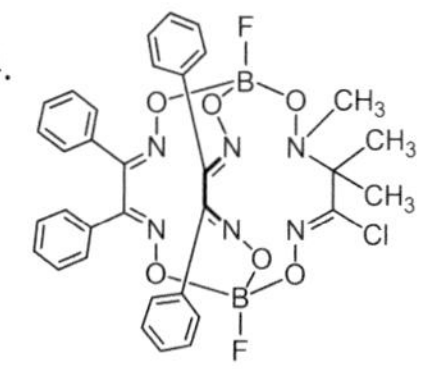

465.
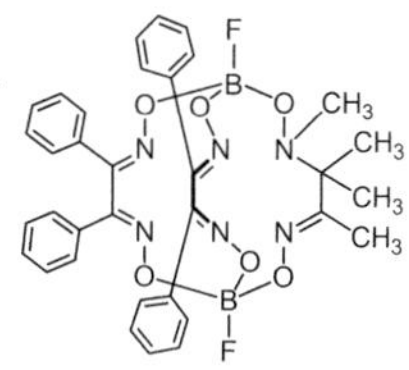

466.
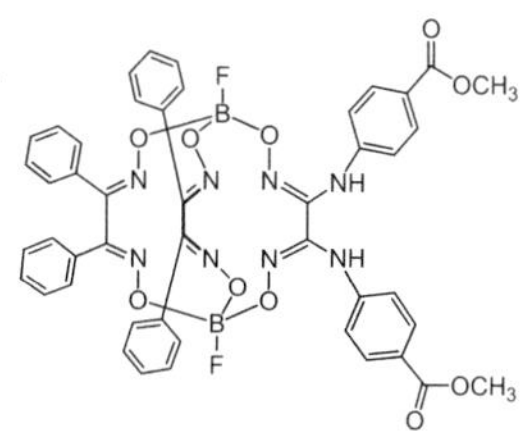

467.
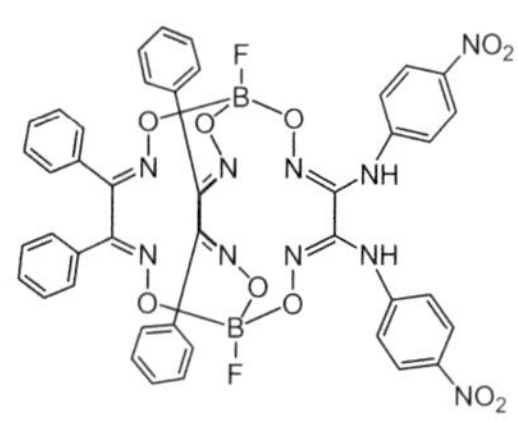

468.
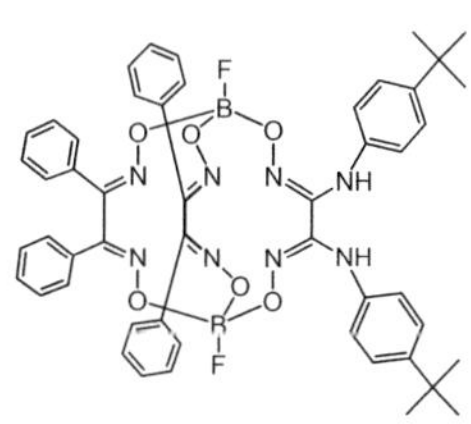

469.
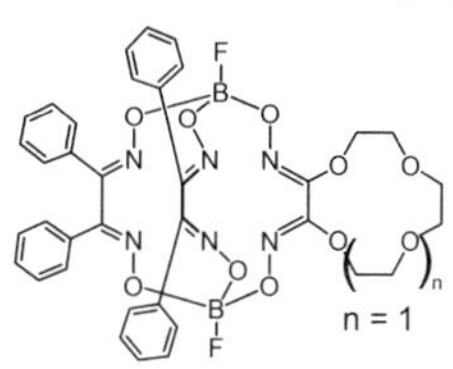

470.
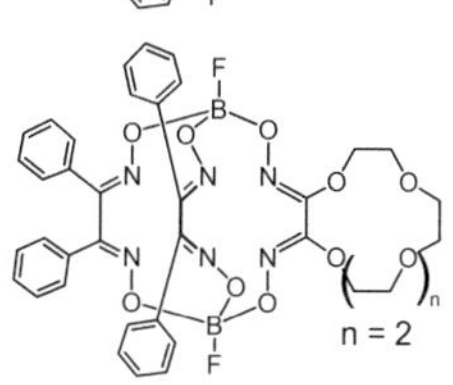

471.

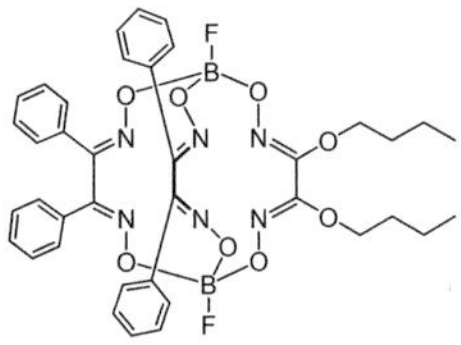

472.

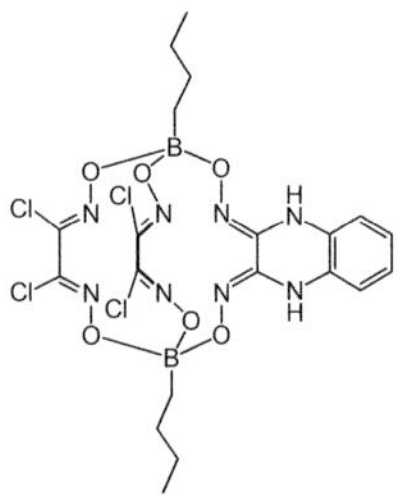

473.

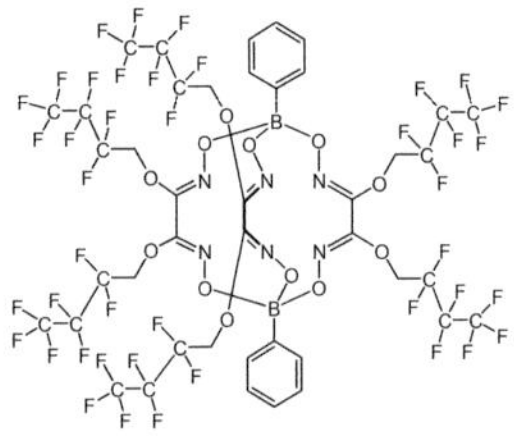

474.

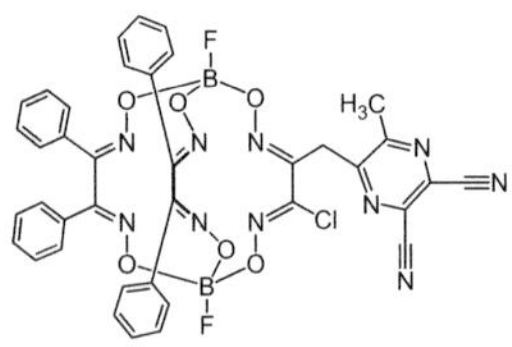

475.

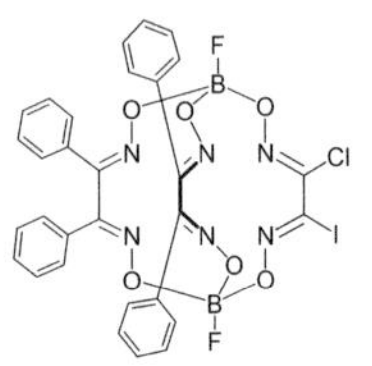

476.

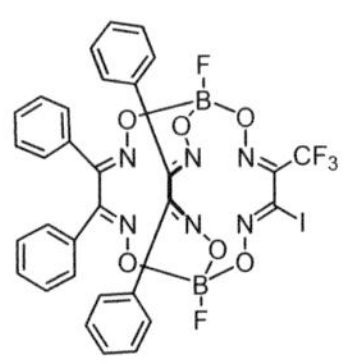

477.

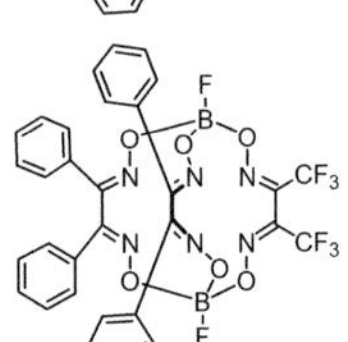

478.

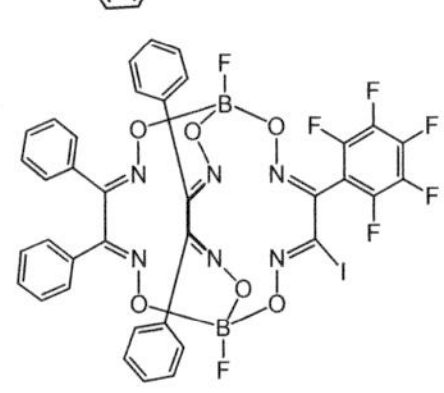

479.

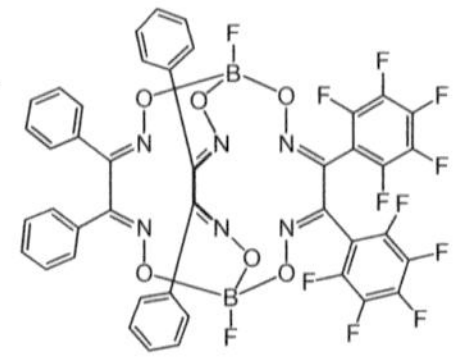

480.

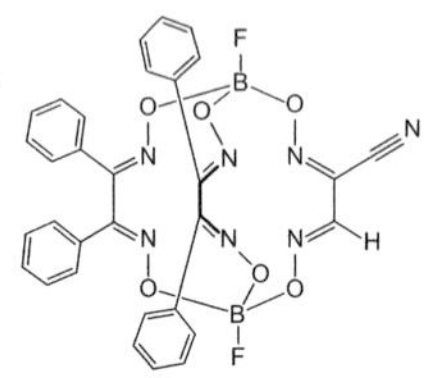

481.

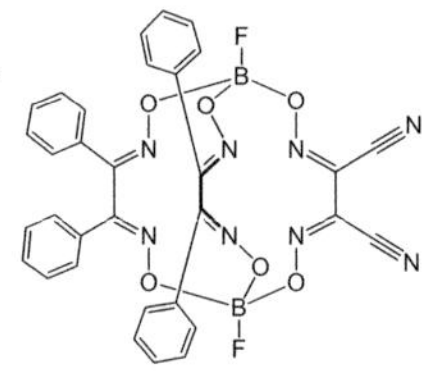

482.

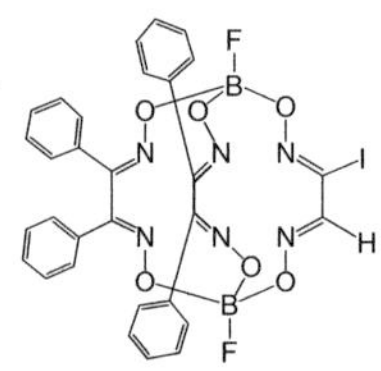

483.

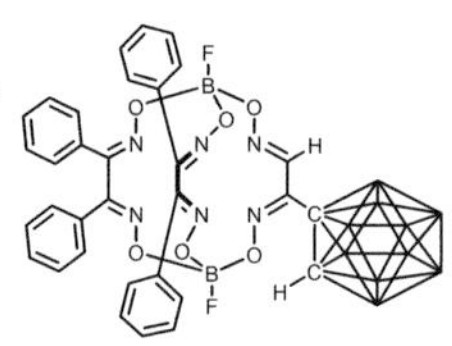

484.

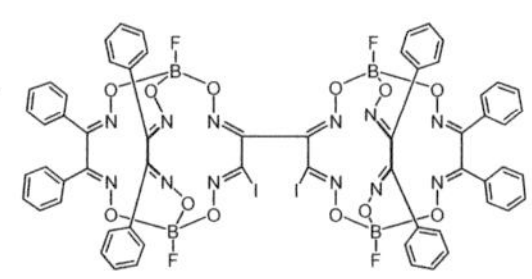

485.

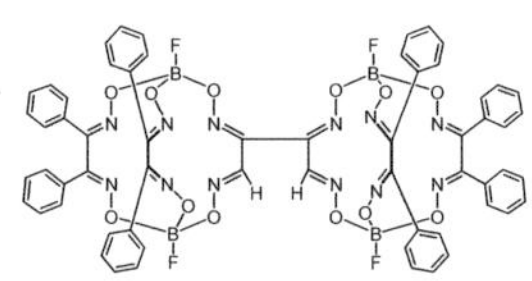

486.

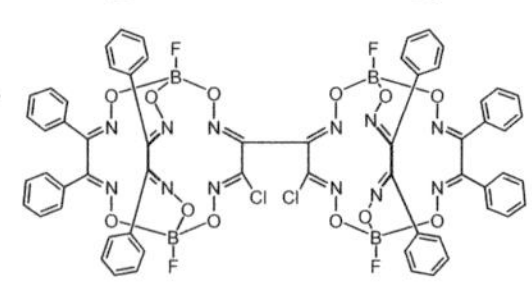

487.

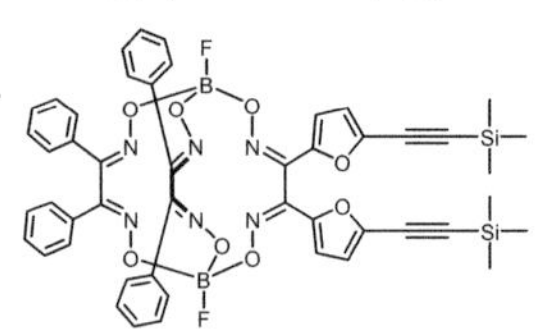

488.

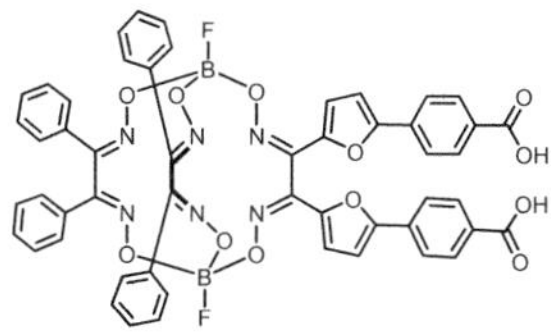

489.

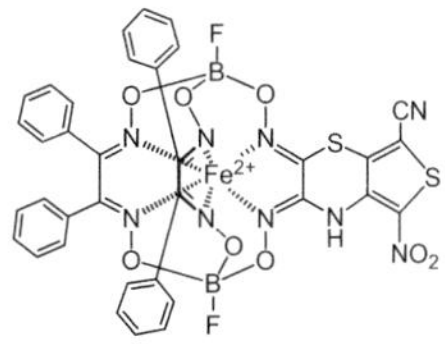

490.

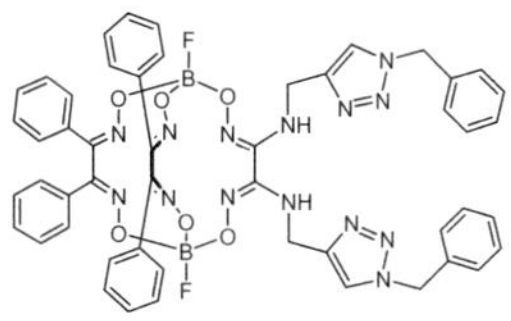

491.

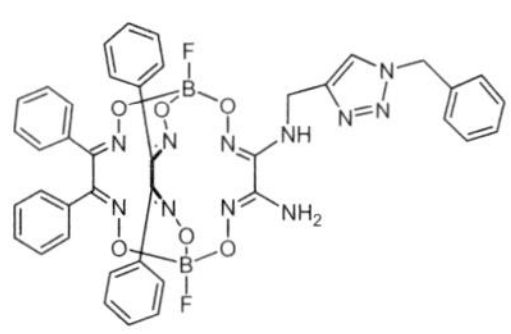

492.

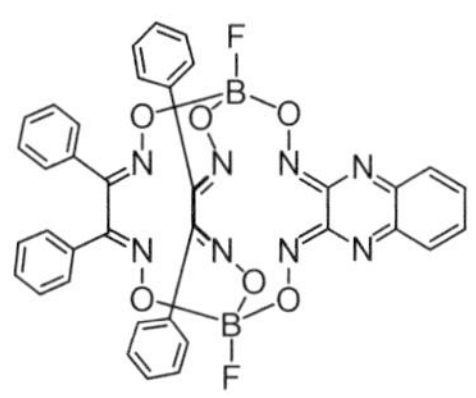

493.

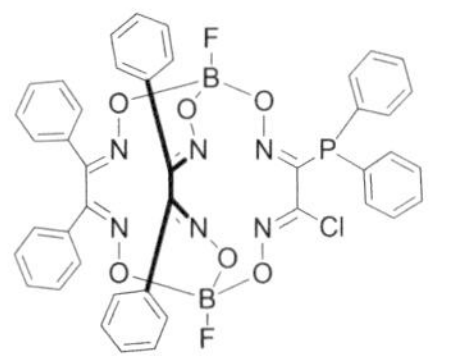

494.

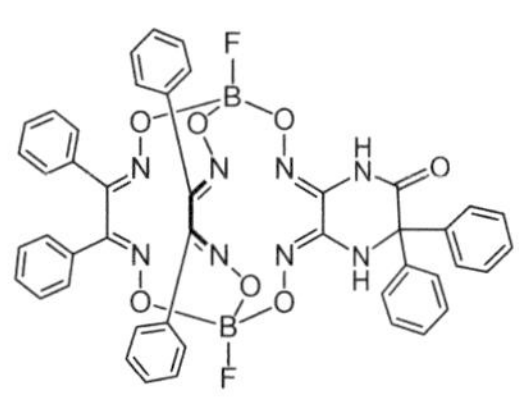

495.

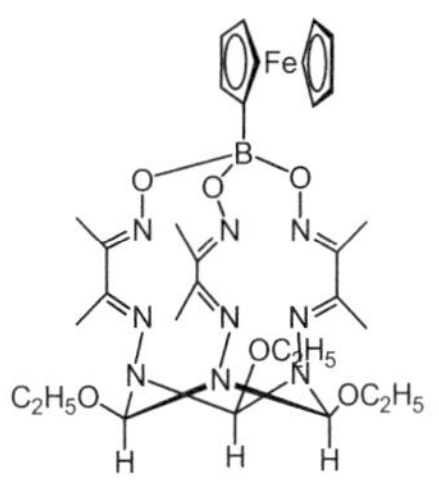

496.

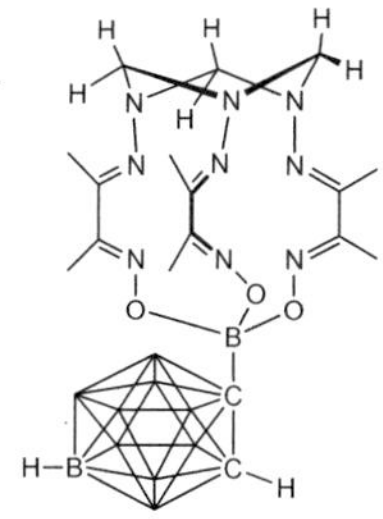

497.

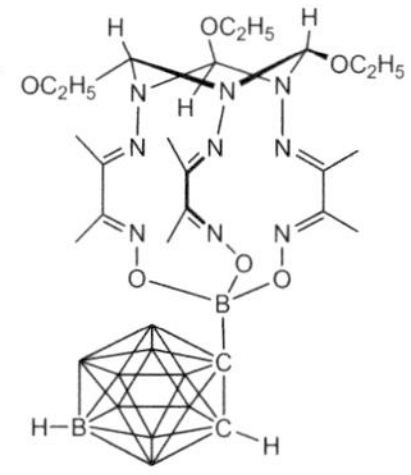

498.

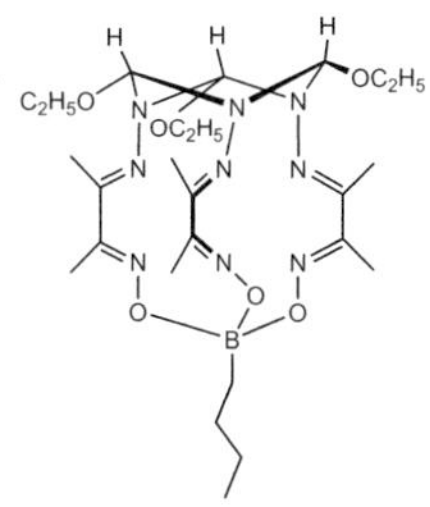

499.

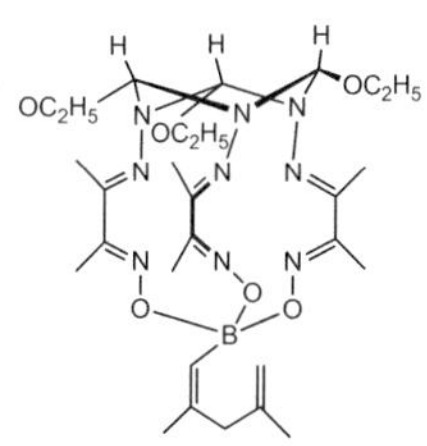

500.

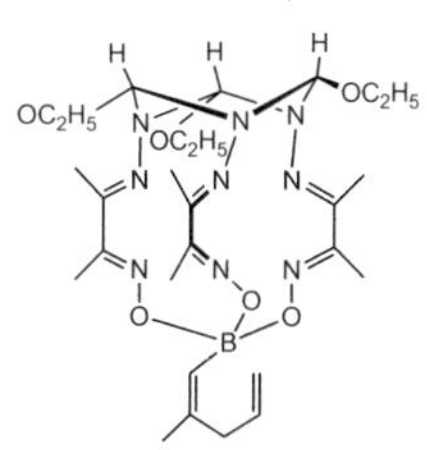

501.

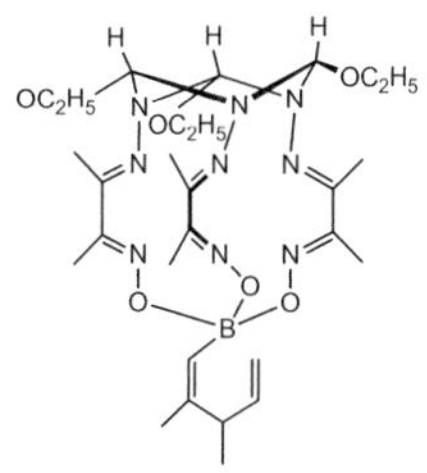

502. 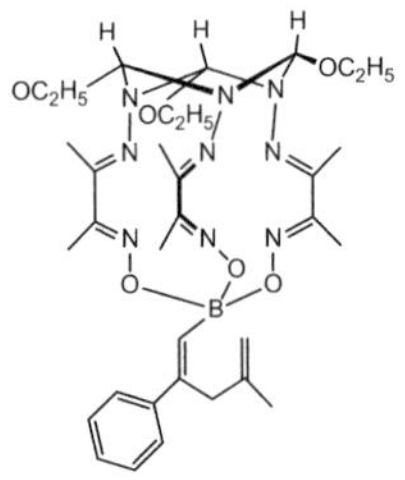

503. 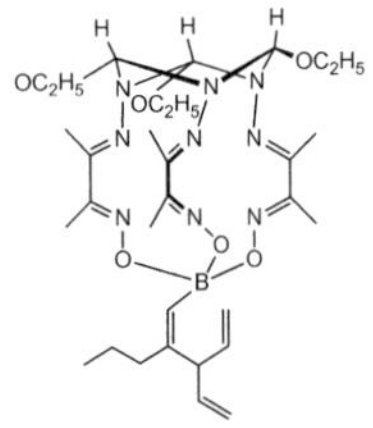

504. 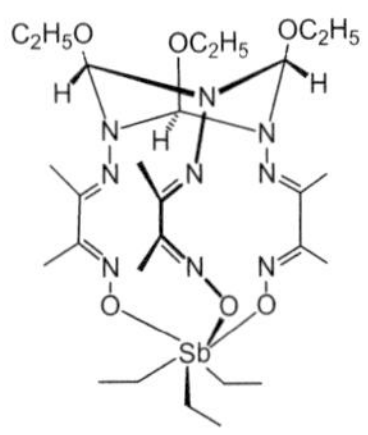

505. 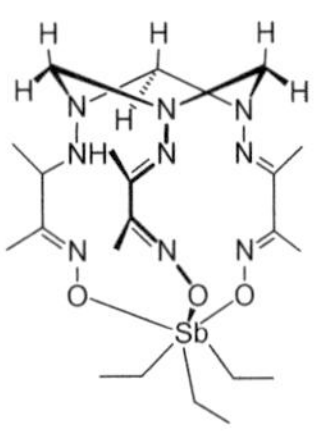

506. 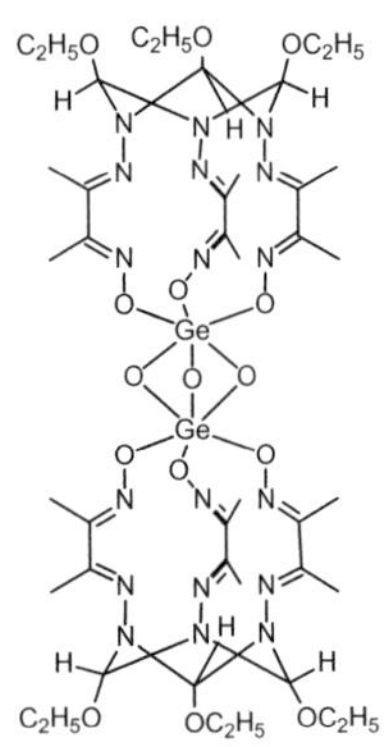

507. 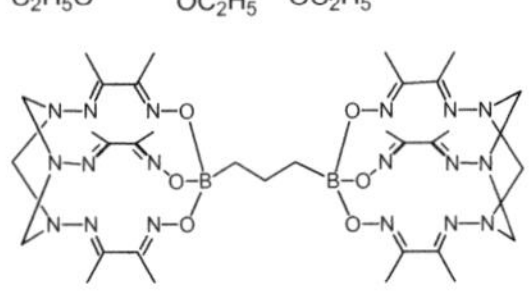

508. 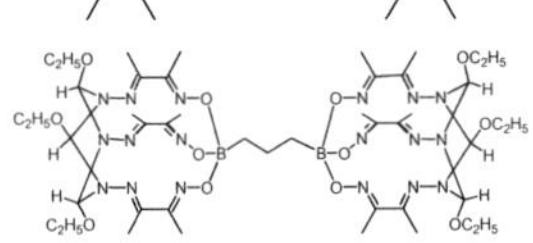

509. 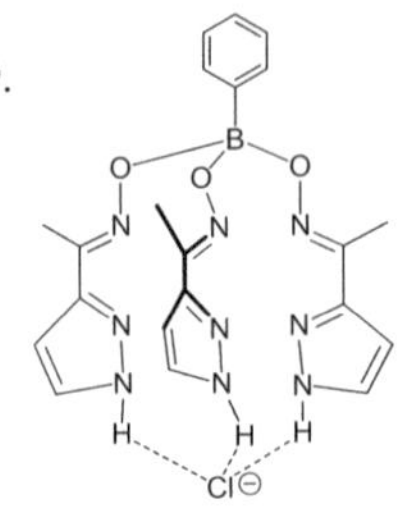

510. 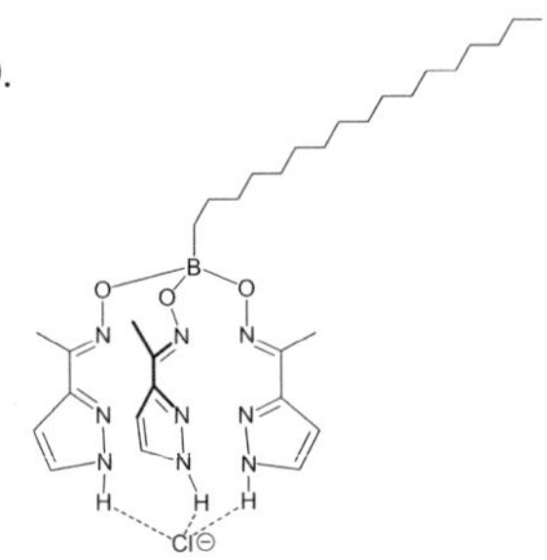

511. 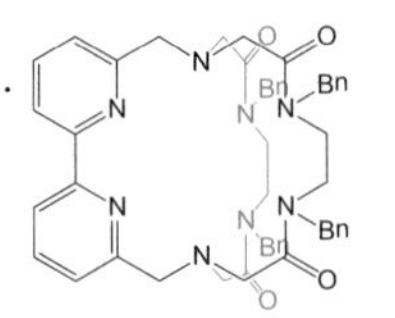

512. 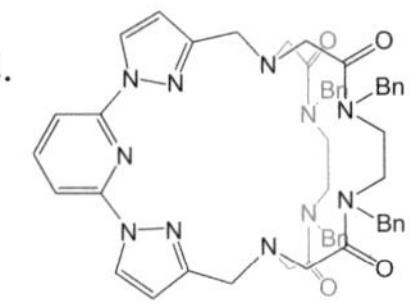

513. 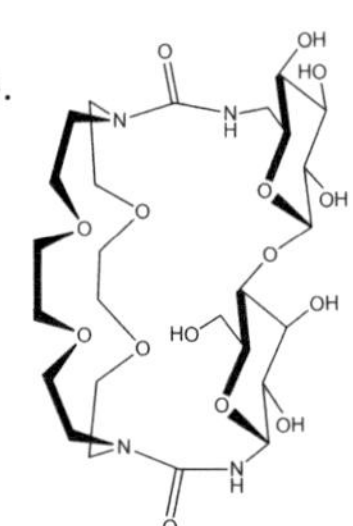

514. 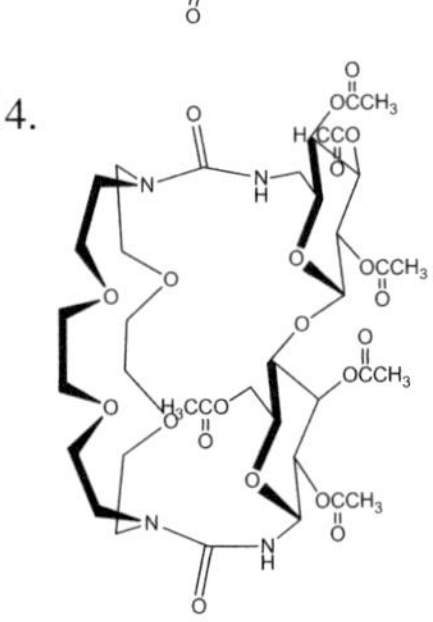

515.
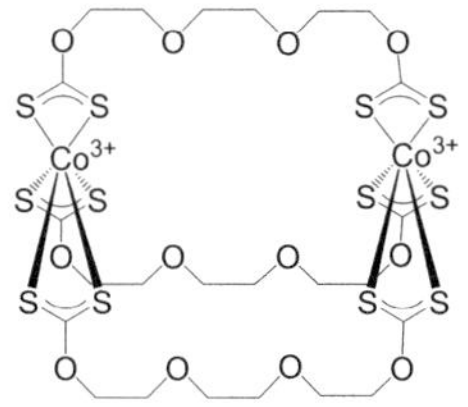

516.
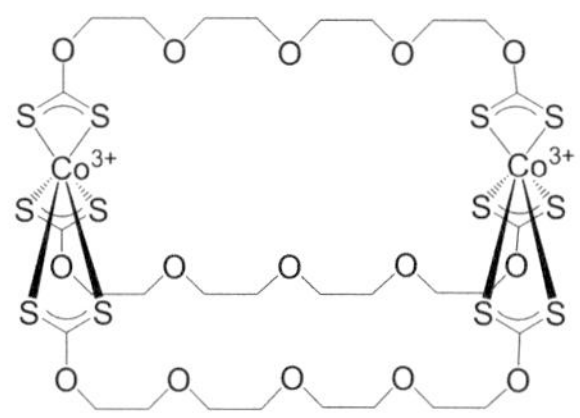

517.
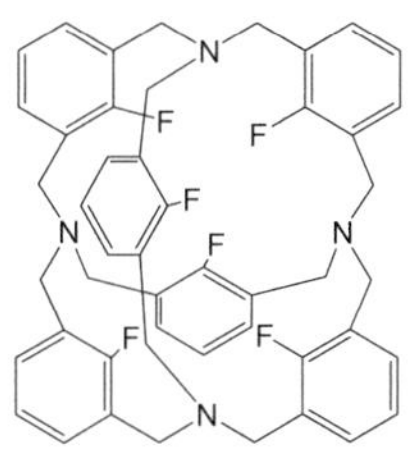

518.
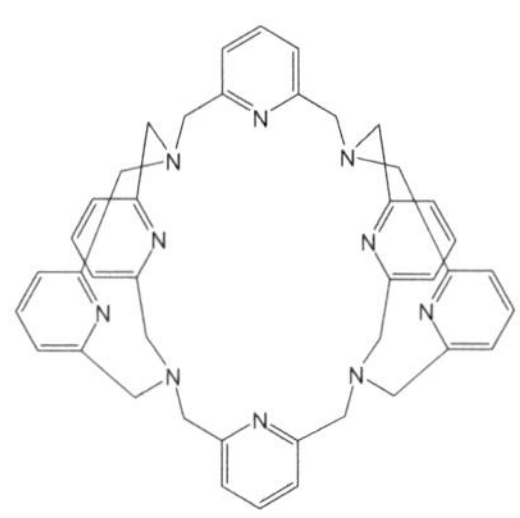

519.
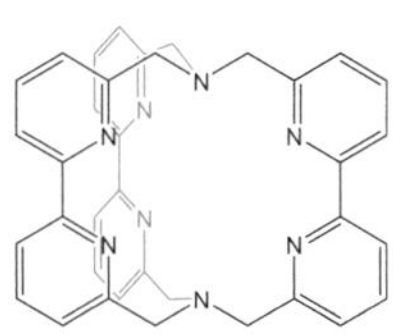

520.
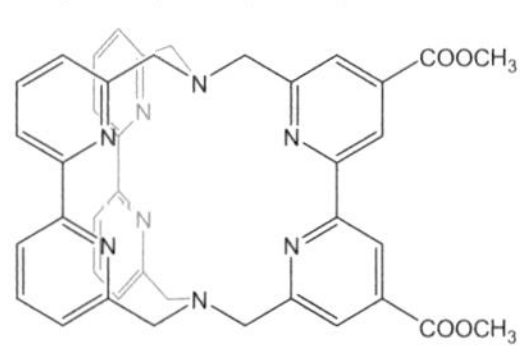

521.
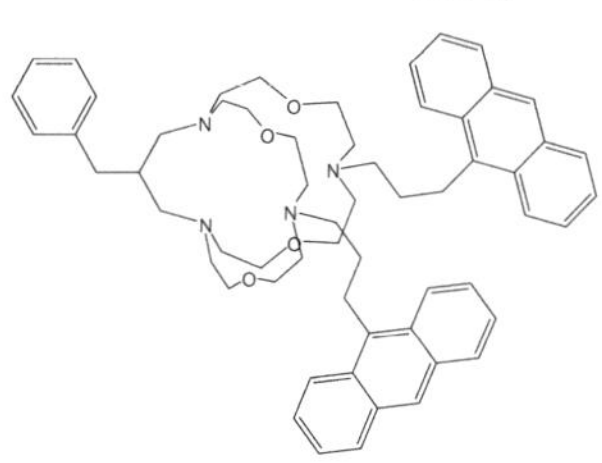

522.
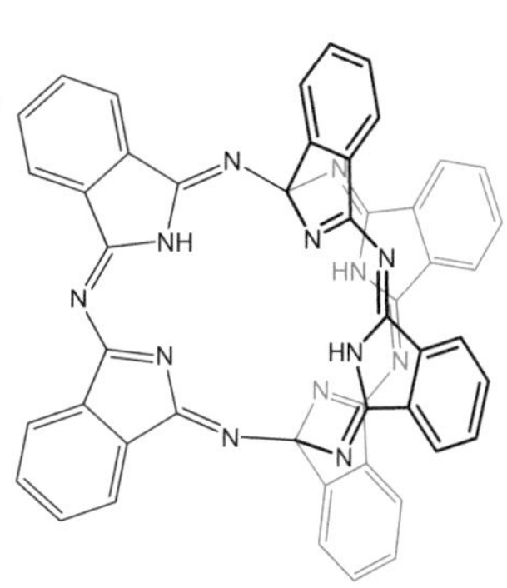

523.
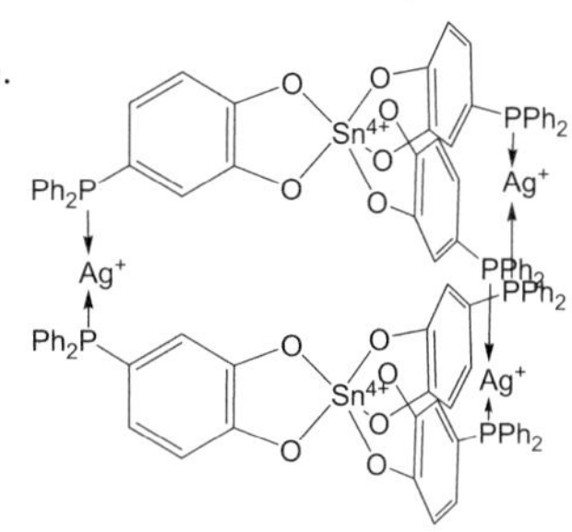

524.
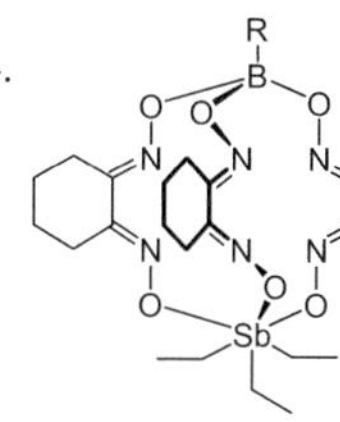

R=n-$C_{16}H_{33}$

525.
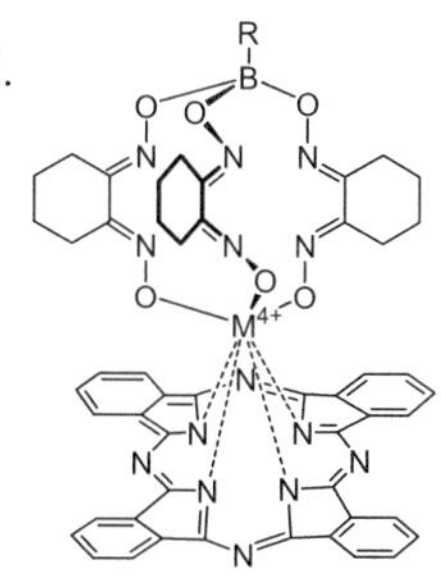

R=n-$C_{16}H_{33}$, M^{4+}=Zr^{4+}

526.
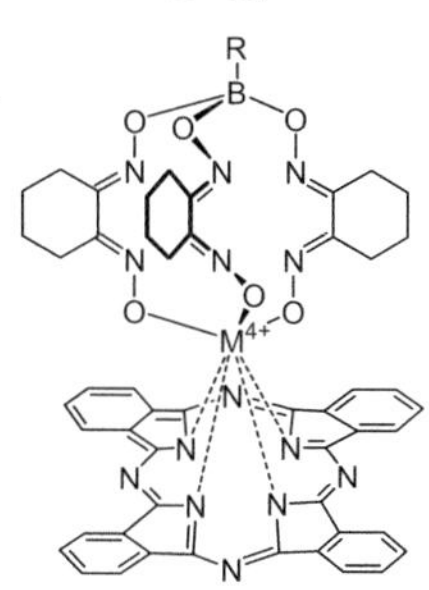

R=n-C_4H_9, M^{4+}=Zr^{4+}

527.

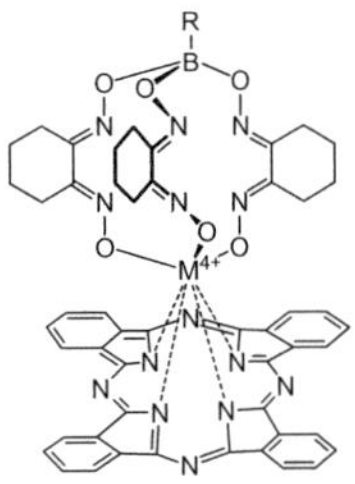

R=n-C_4H_9, M^{4+}=Hf^{4+}

528.

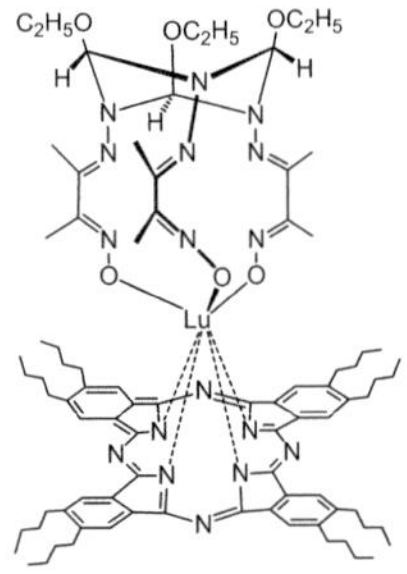

529.

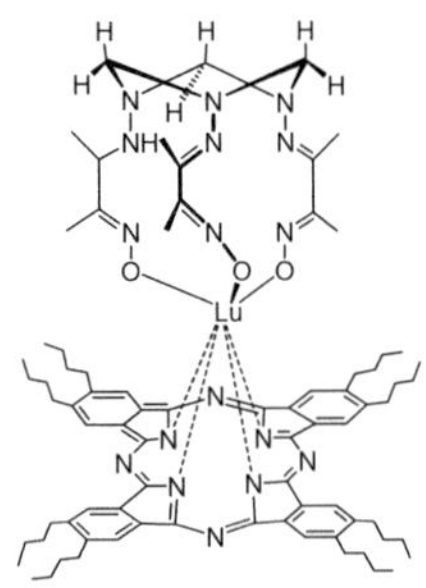

530.

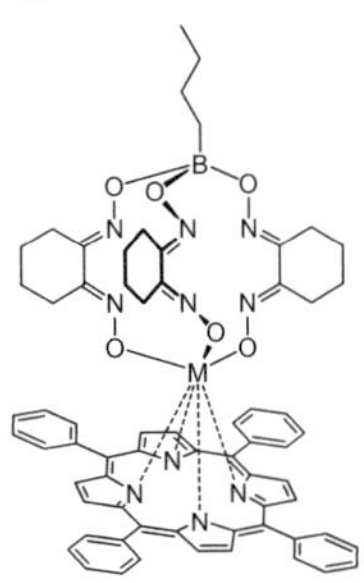

M = Zr

531.

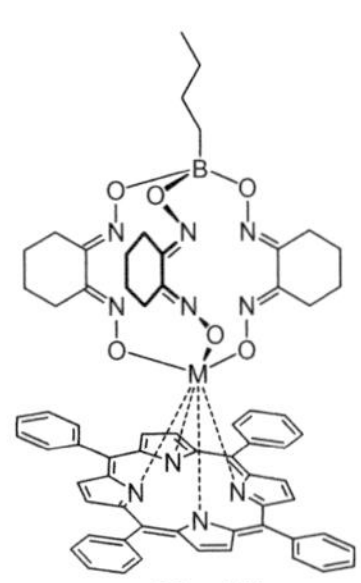

M = Hf

532.

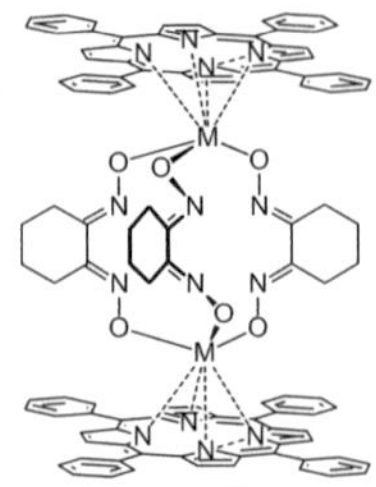

M = Zr

533.

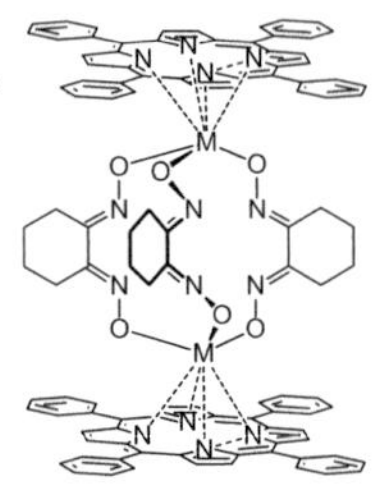

M = Hf

534.

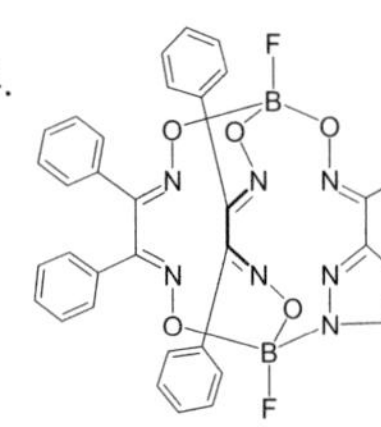

535.

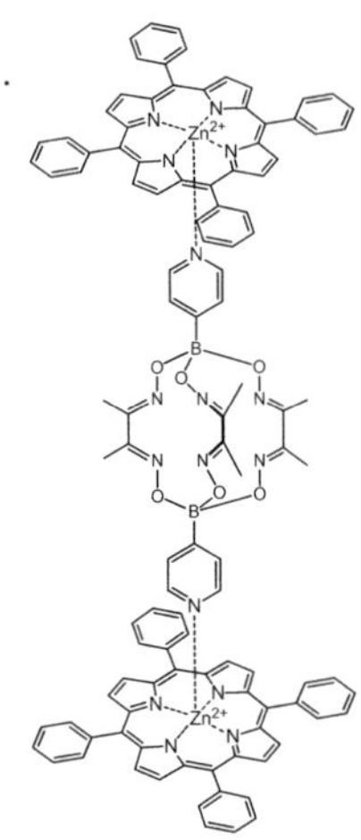

536.

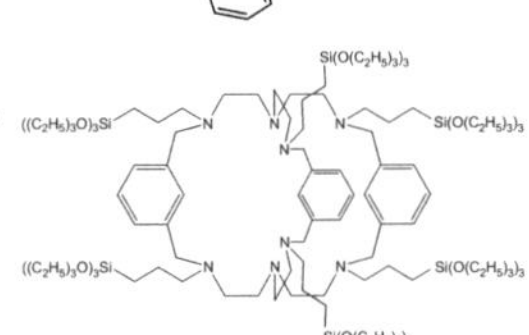

537. 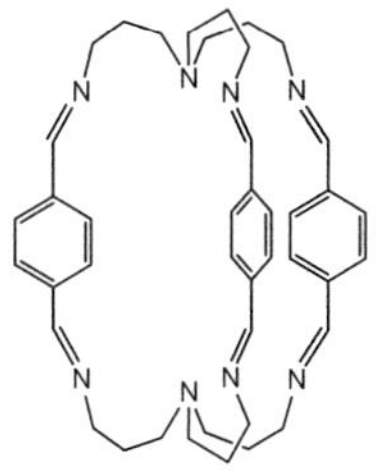

538. 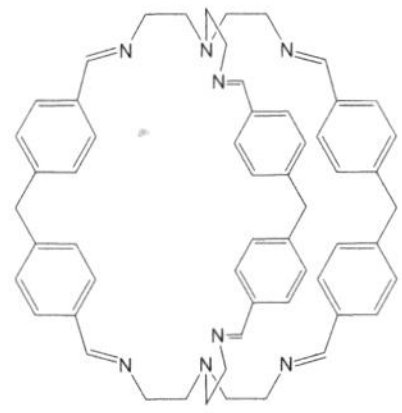

539. 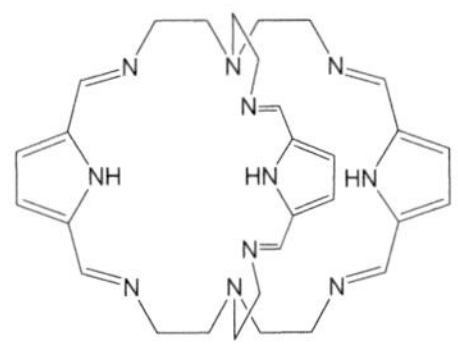

540. 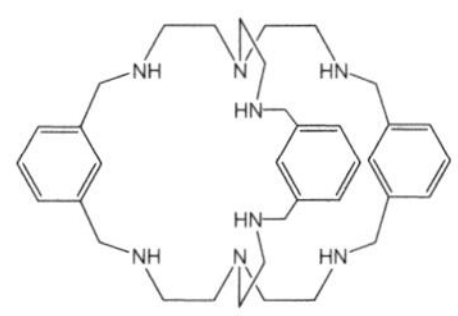

541. 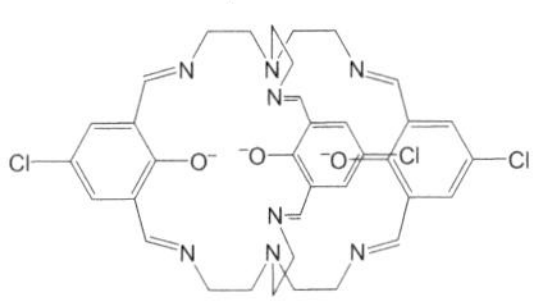

542. 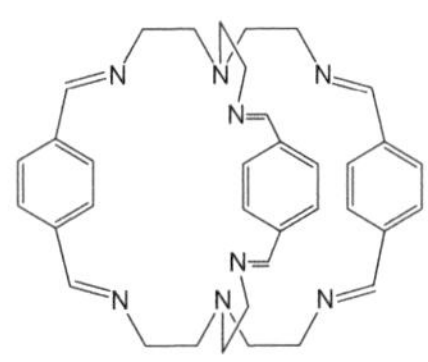

543. 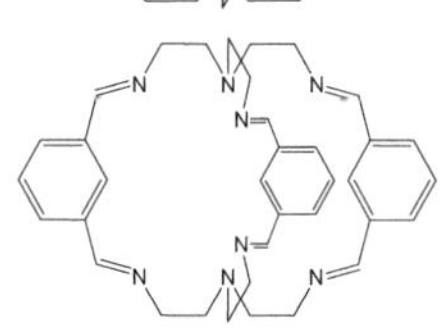

544. 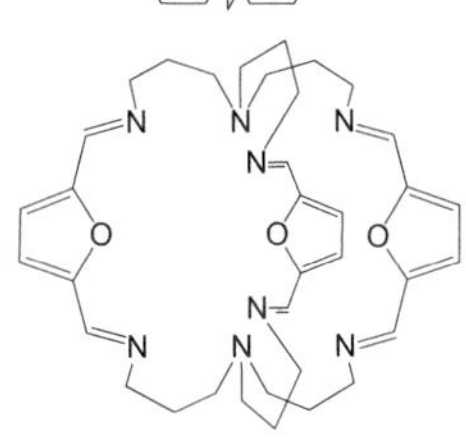

545. 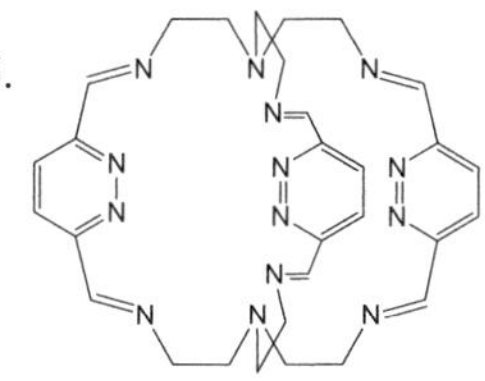

546. 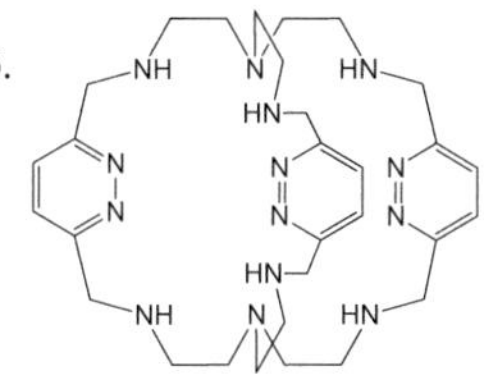

547. 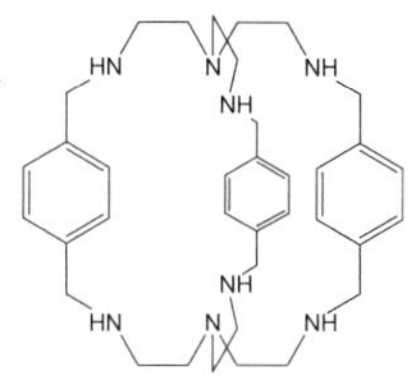

548. 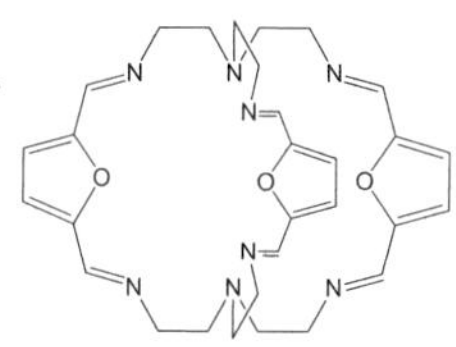

549. 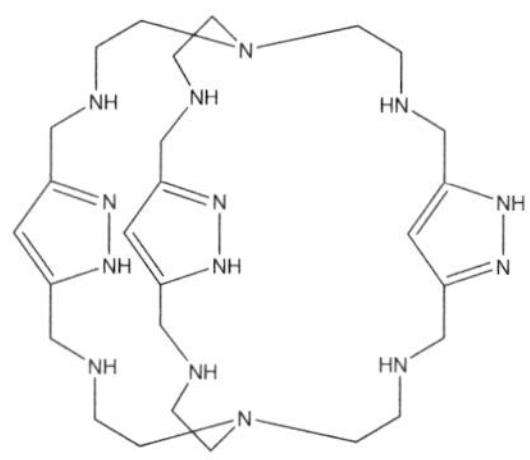

550. 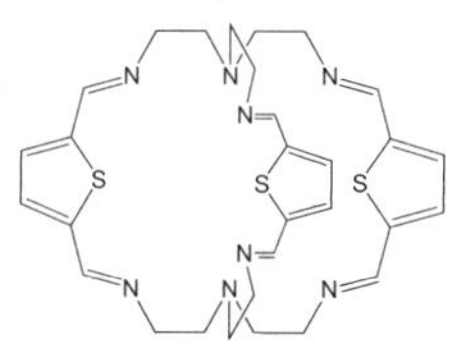

551. 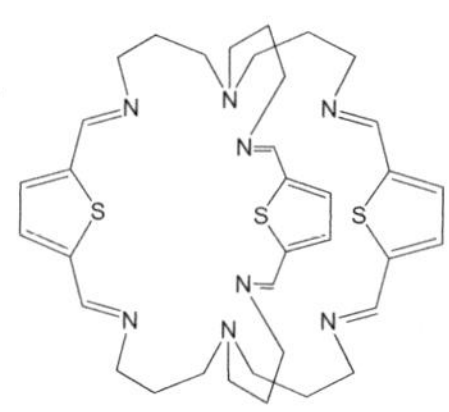

552 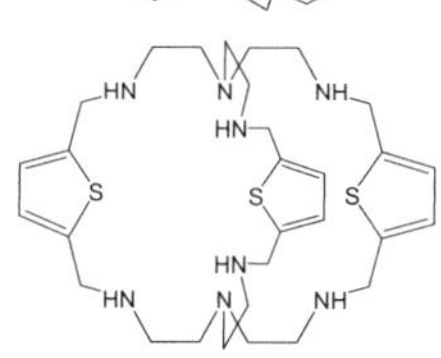

553.
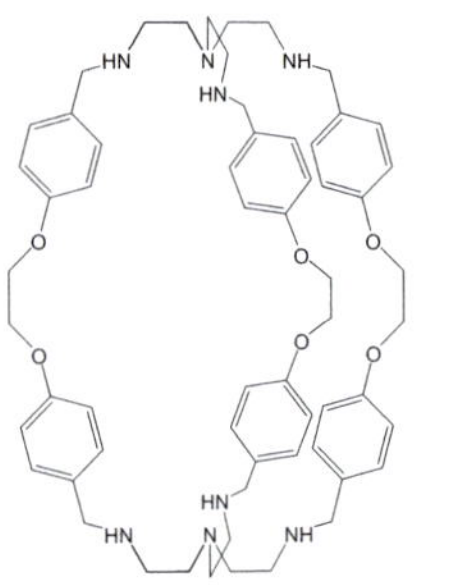

554.
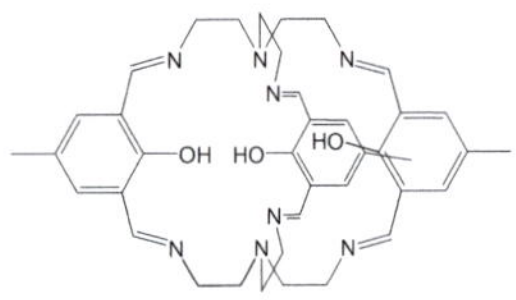

555.
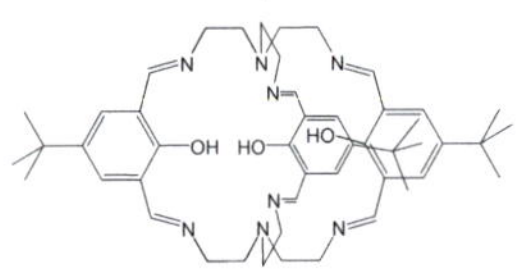

556.
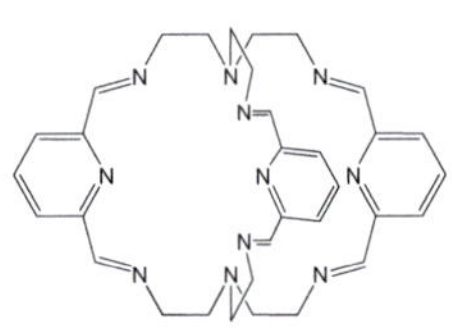

557.
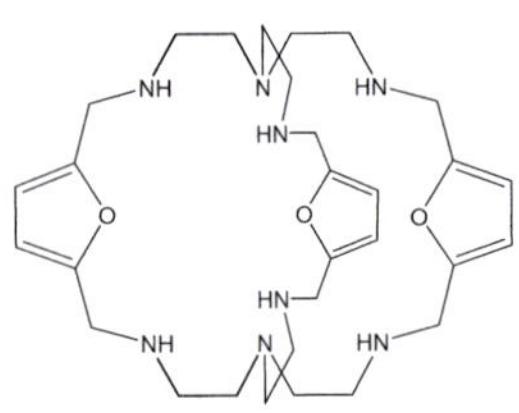

558.
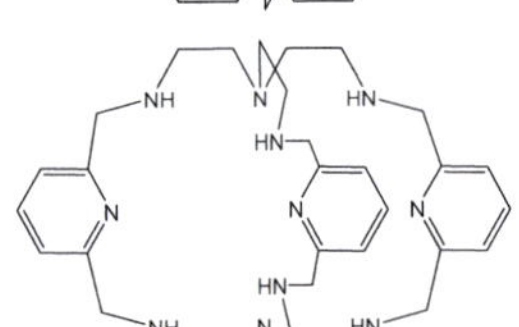

559.
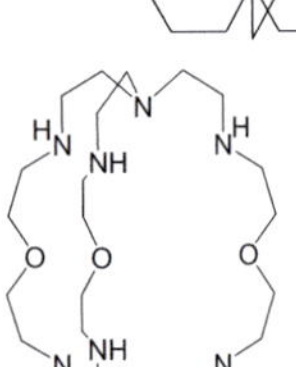

560.
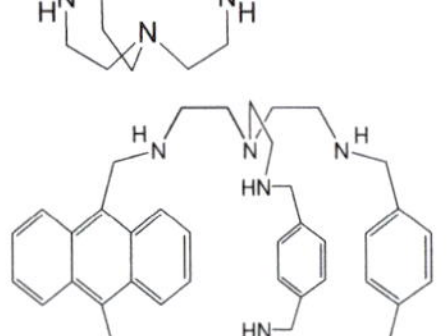

561.
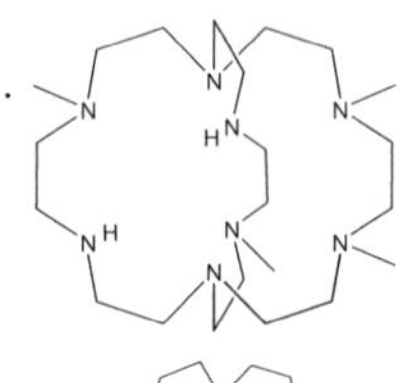

562.
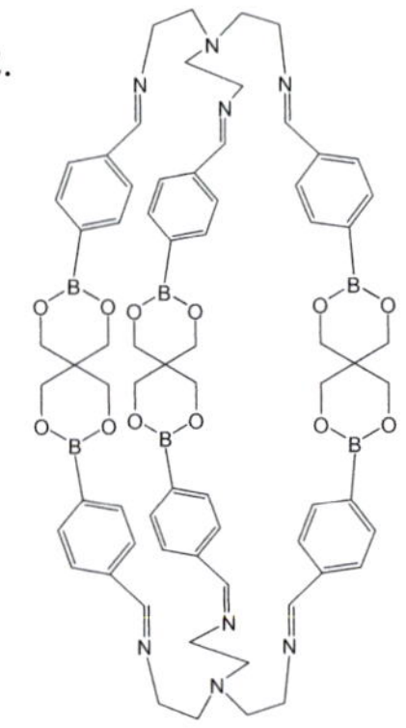

563.
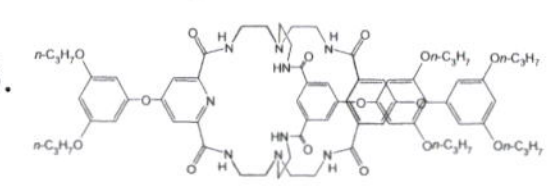

564.
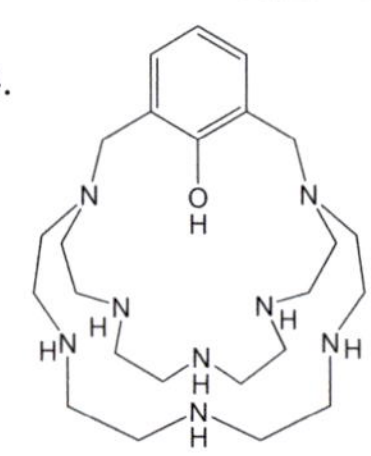

565.
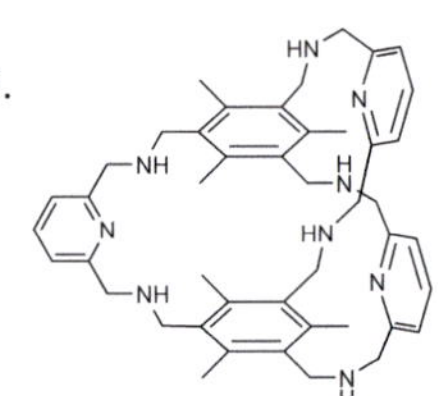

566.
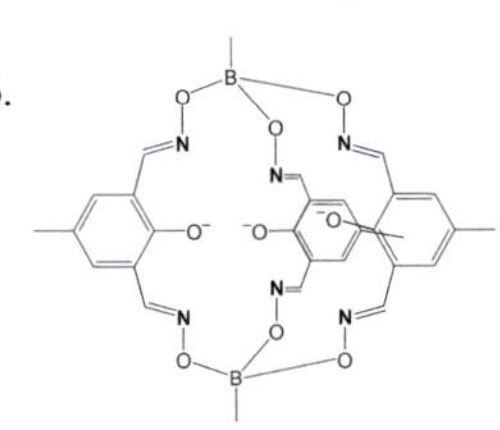

567.
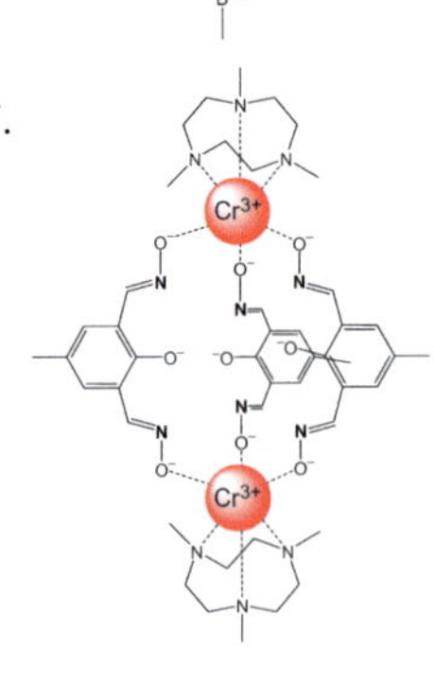

568.

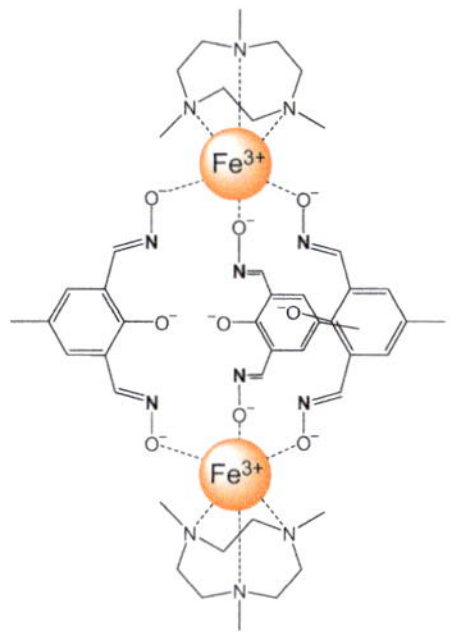

569.

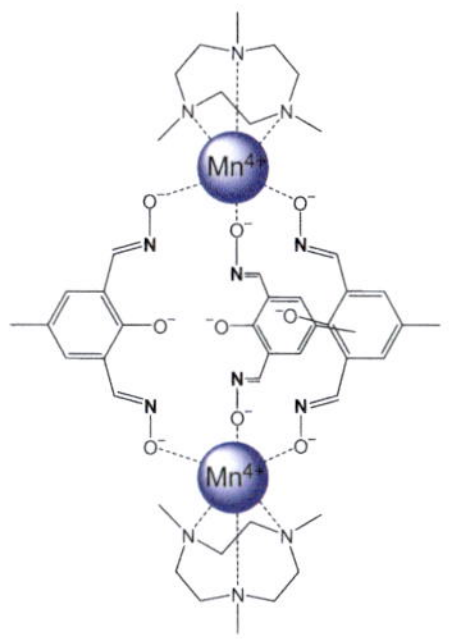

570.

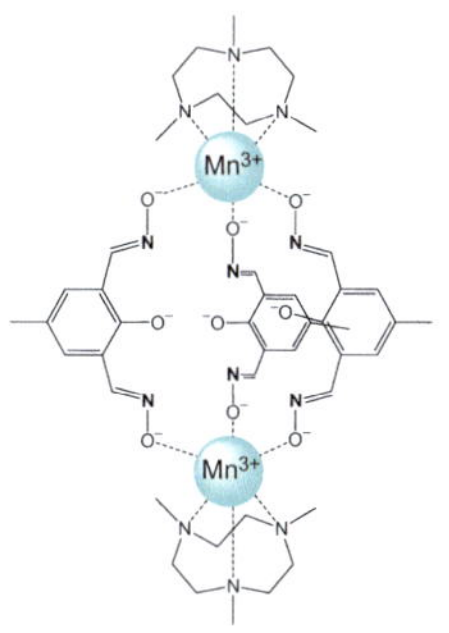

571.

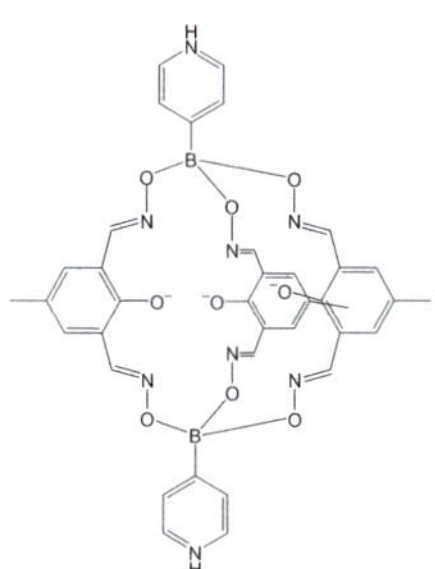

572.

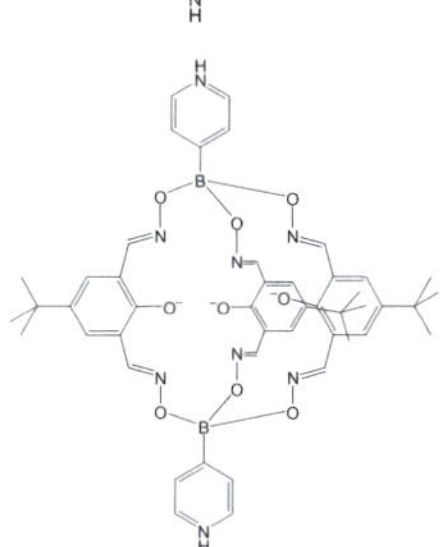

573.

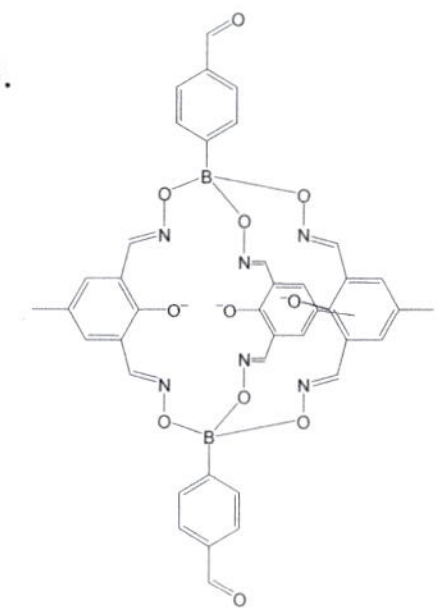

574.

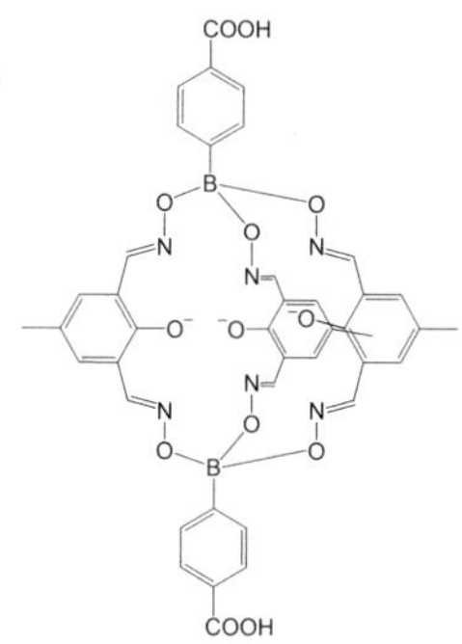

575.

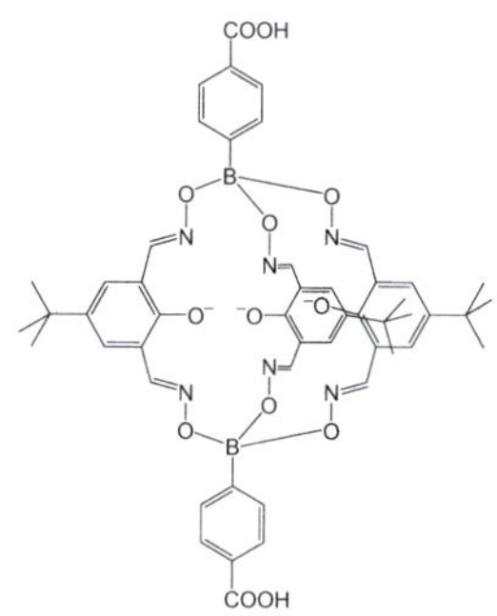

576.

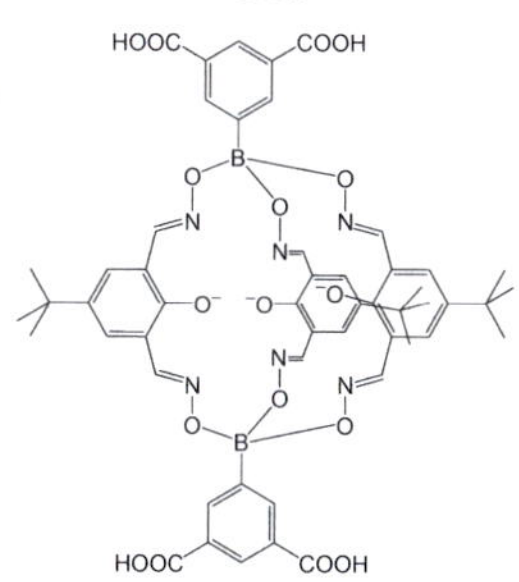

577.

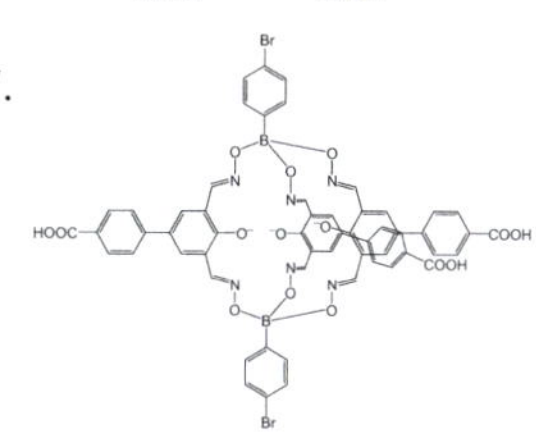

578.

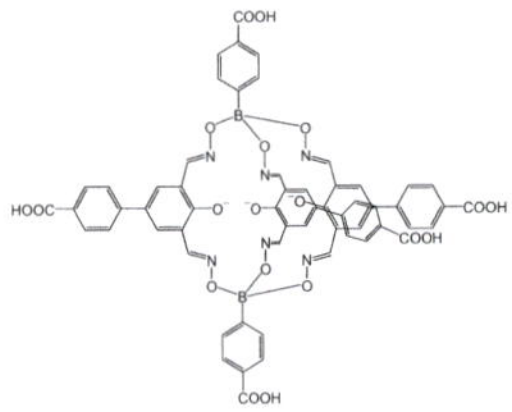

579.

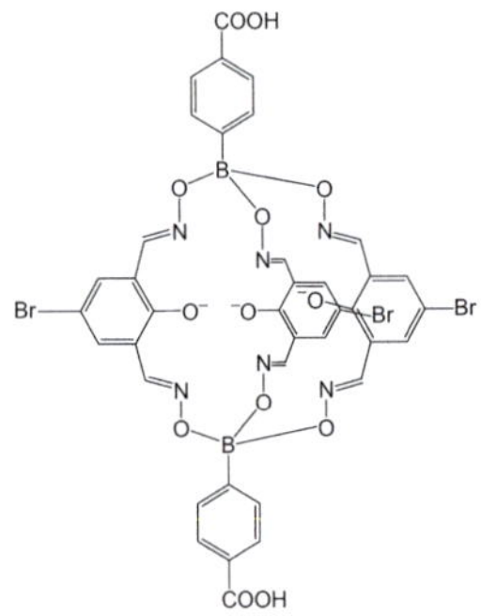

580.

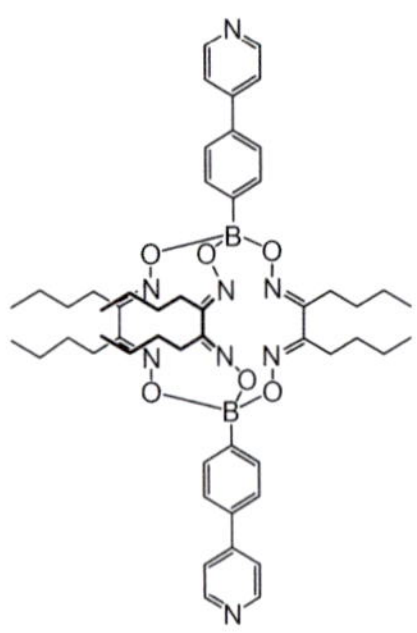

581.

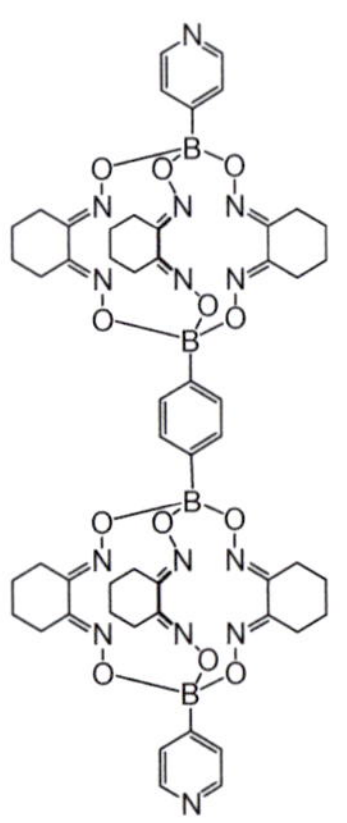

582.

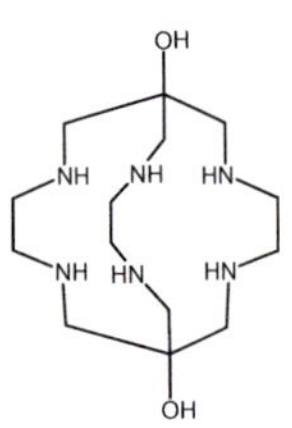

583.

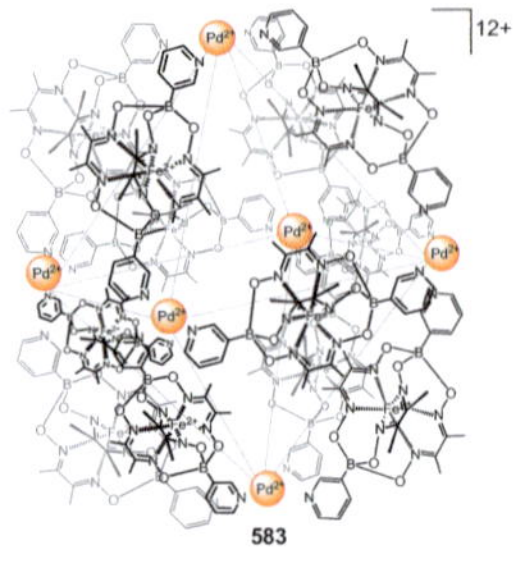

584.

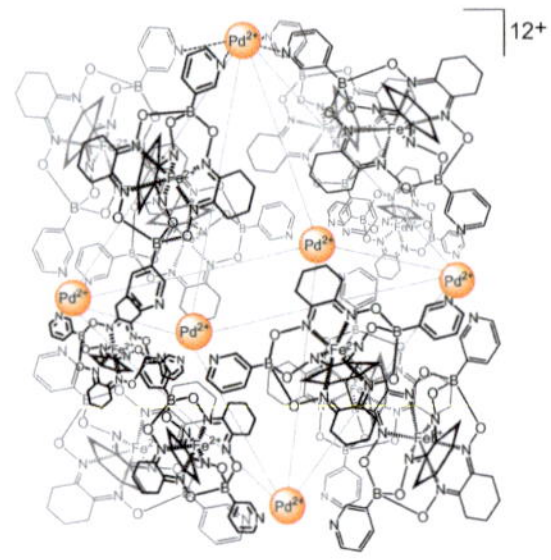

585.

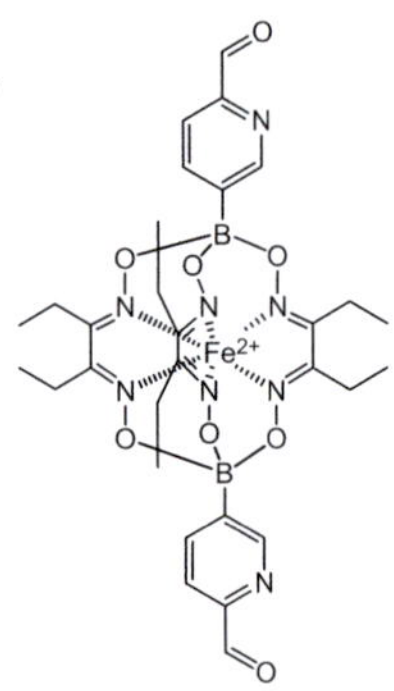

586.

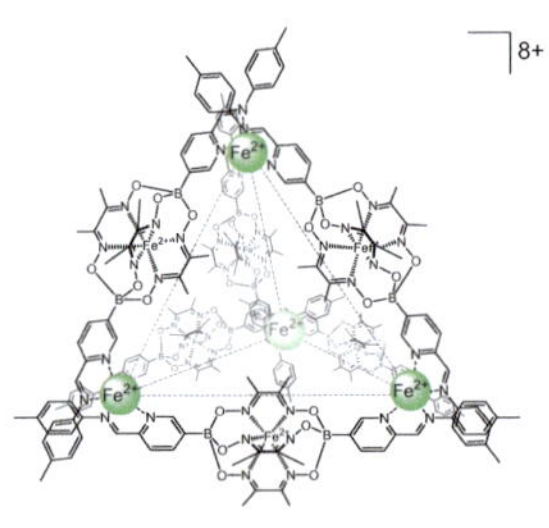

587.

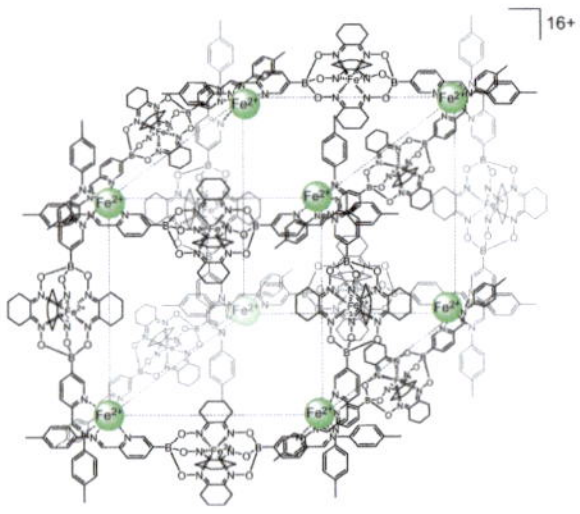

588.
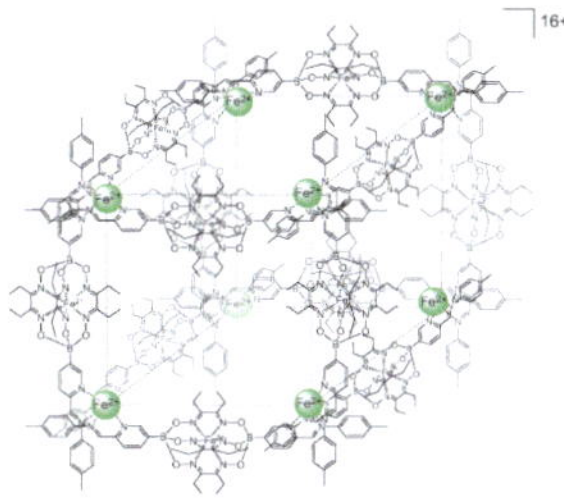

589.
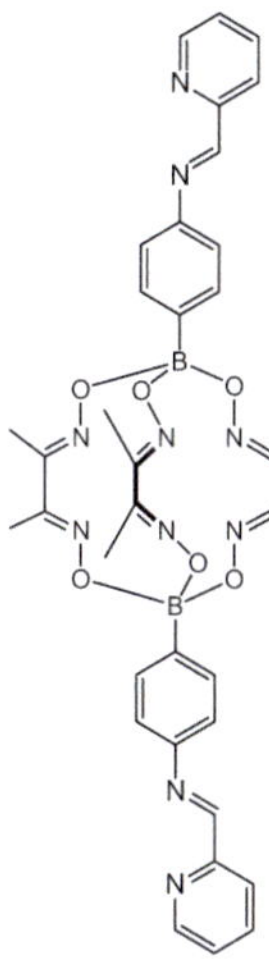

590.
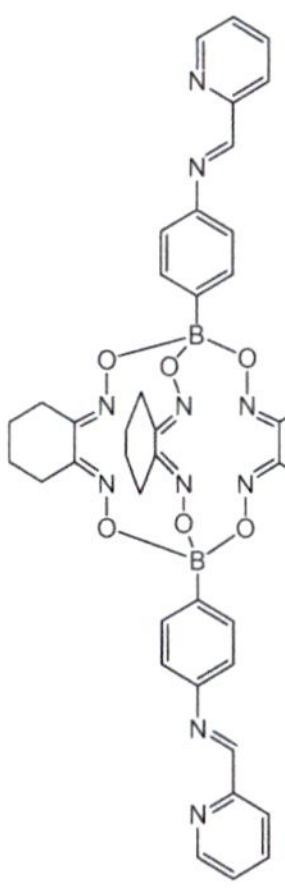

591.
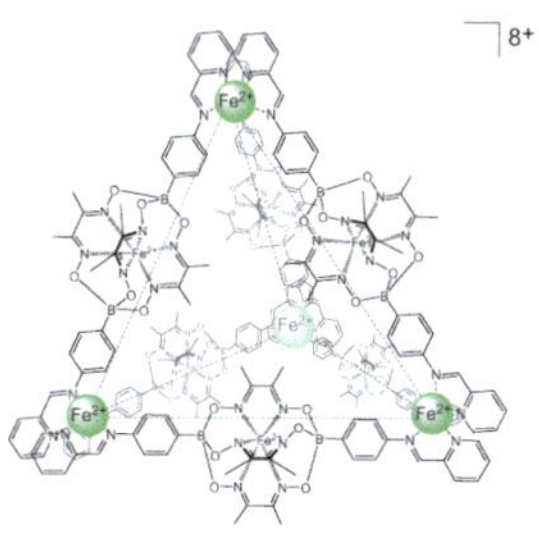

592.
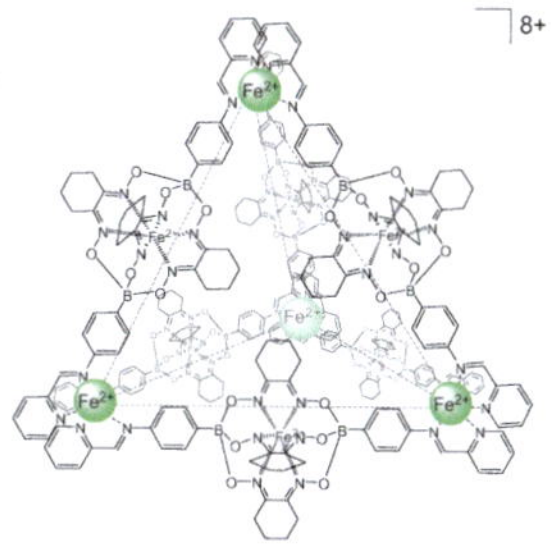

593.
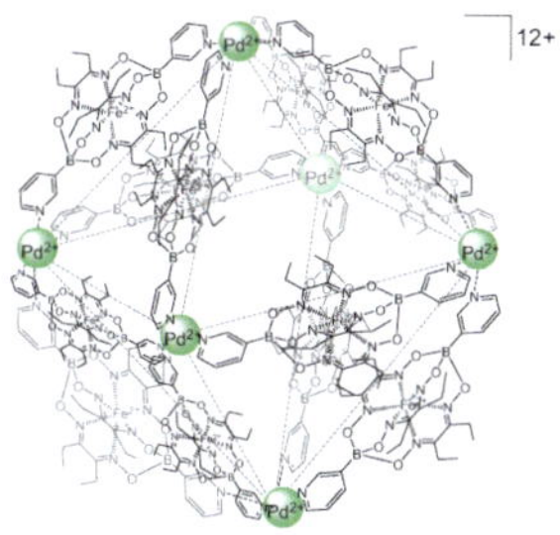

594.
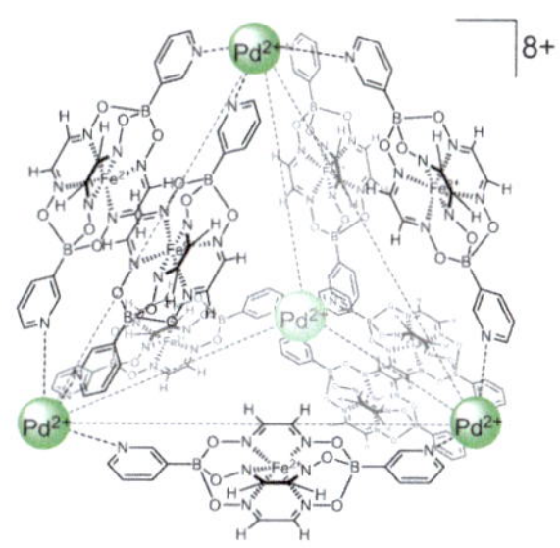

595.
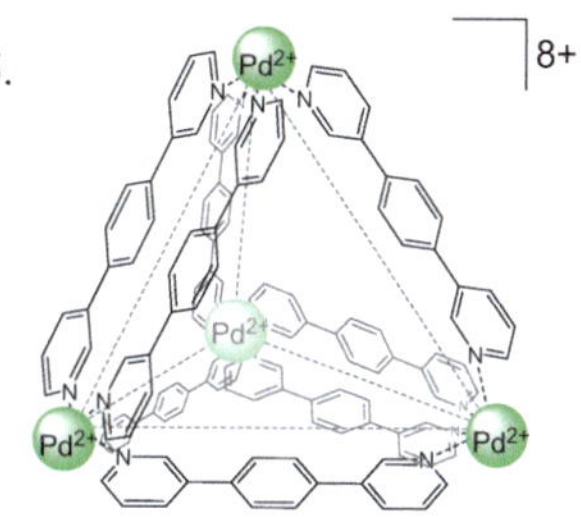

596.
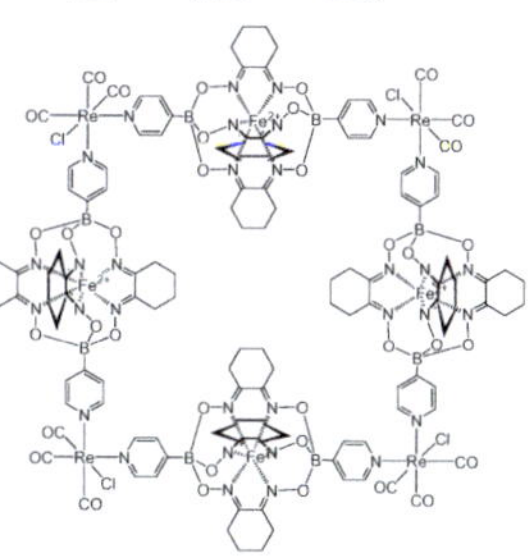

597.

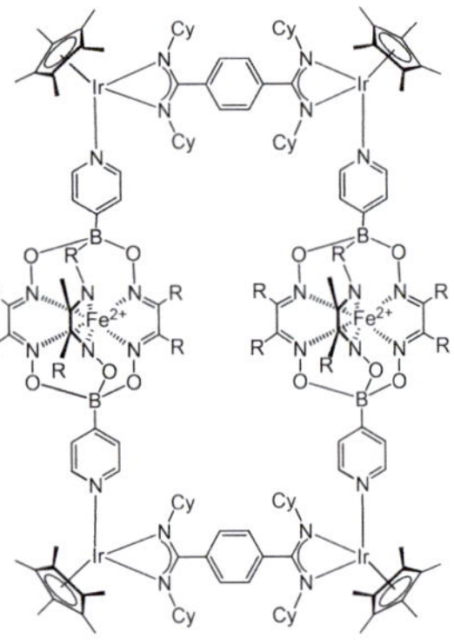

$R = CH_3$

598.

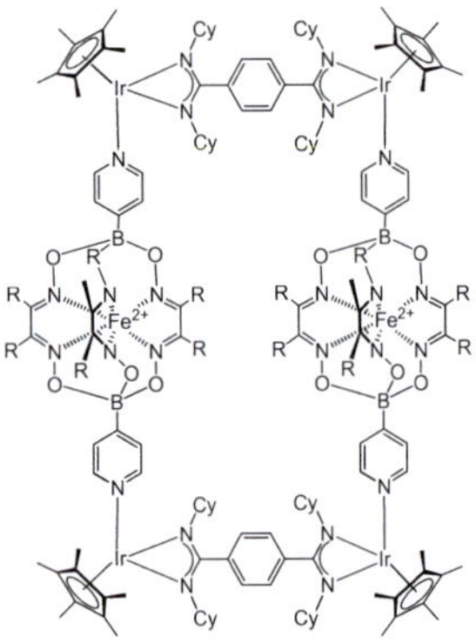

$(R,R) = (CH_2)_4$

599.

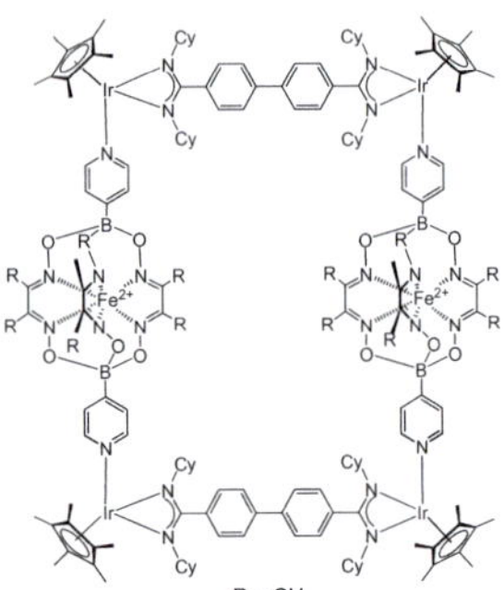

$R = CH_3$

600.

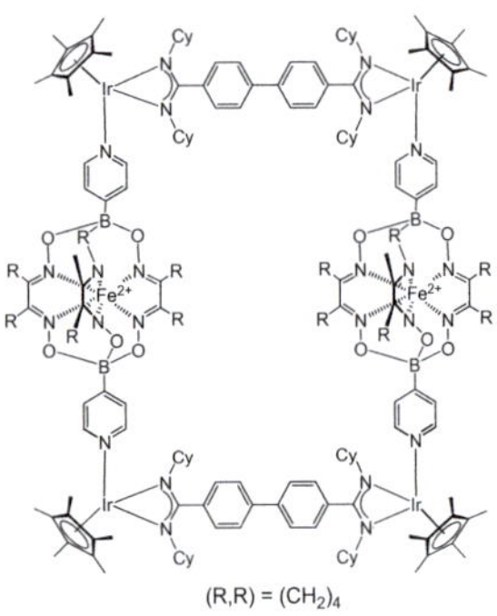

$(R,R) = (CH_2)_4$

601.

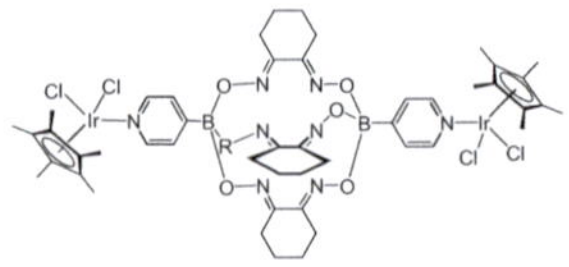

602.

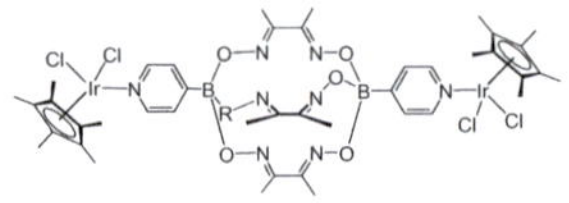

603.

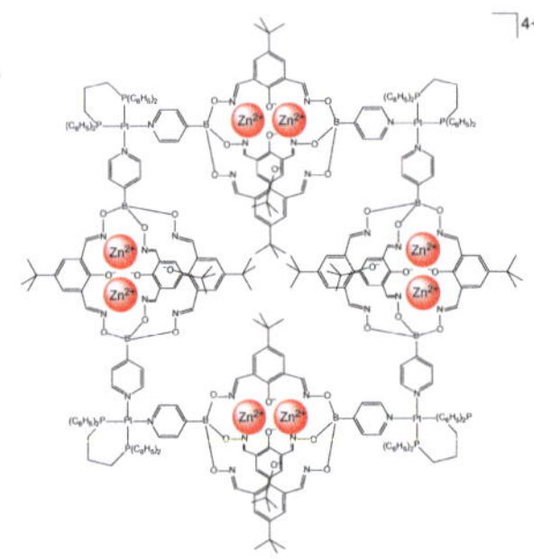

604.

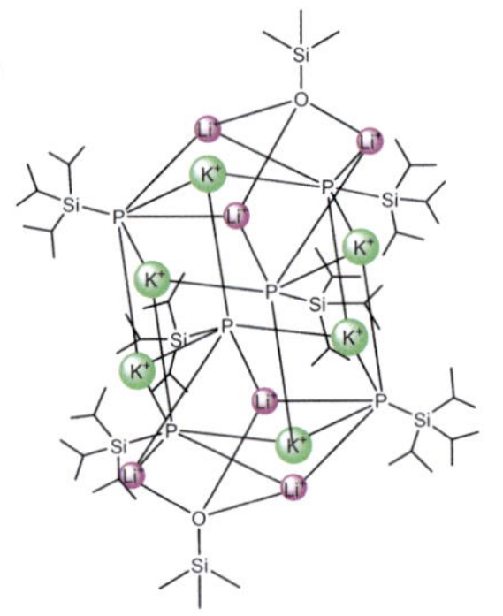

605.

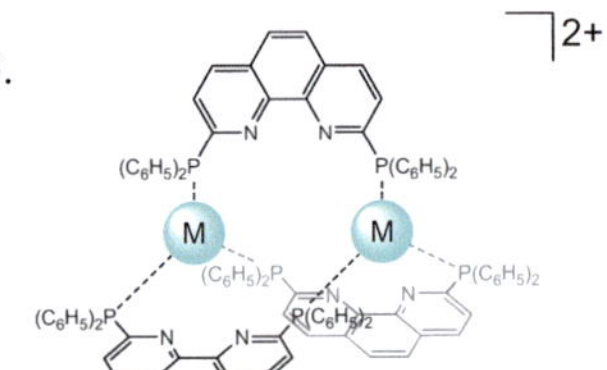

M = Pt

606.

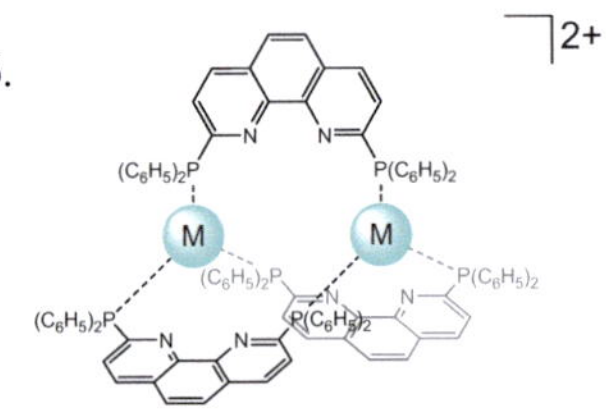

M = Pd

607.

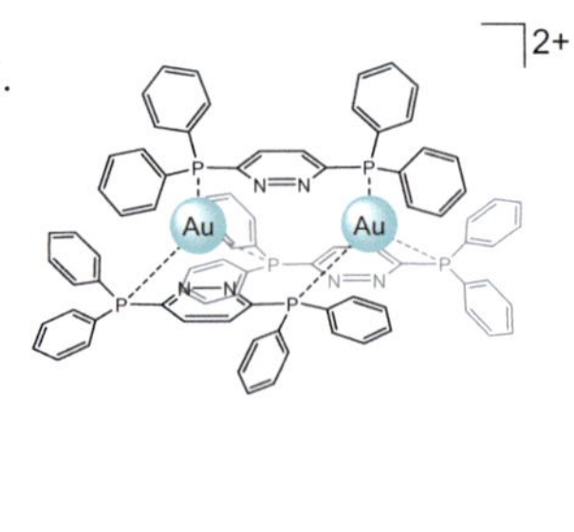

608.

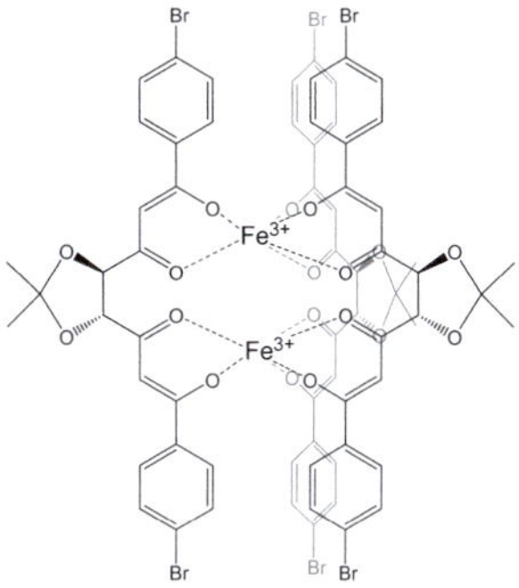

609.

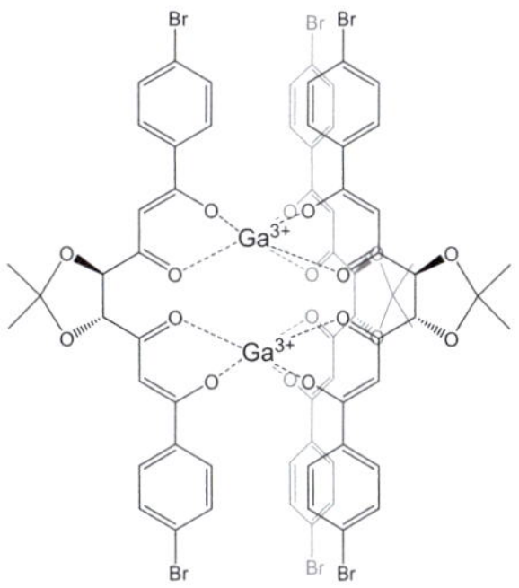

610.

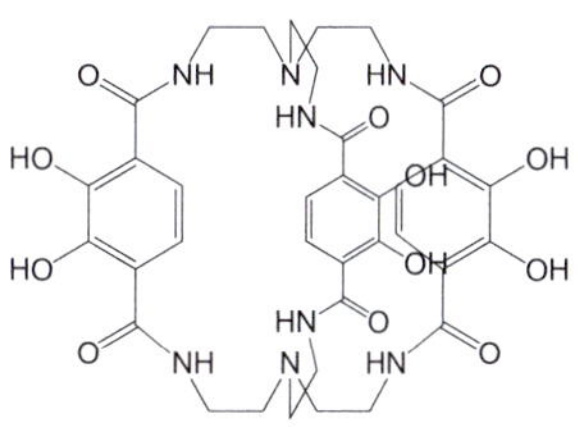

611.

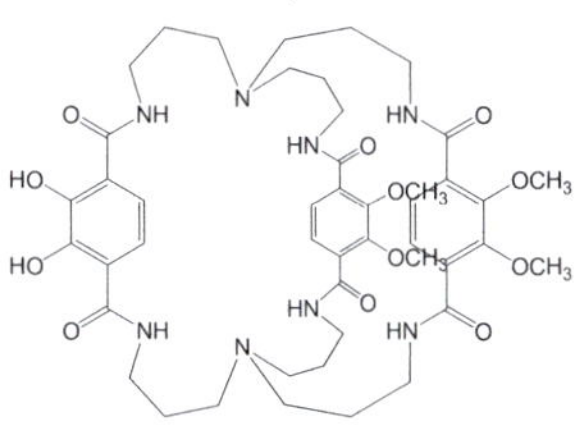

612.

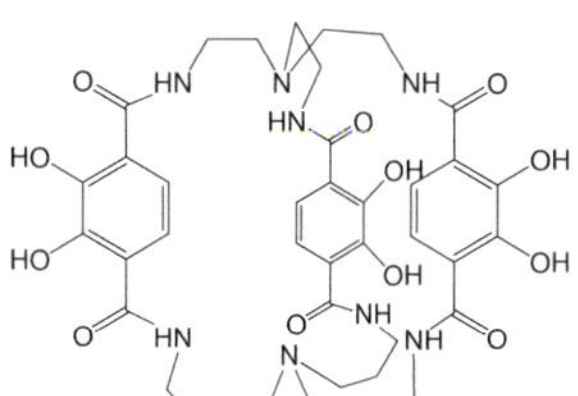

613.

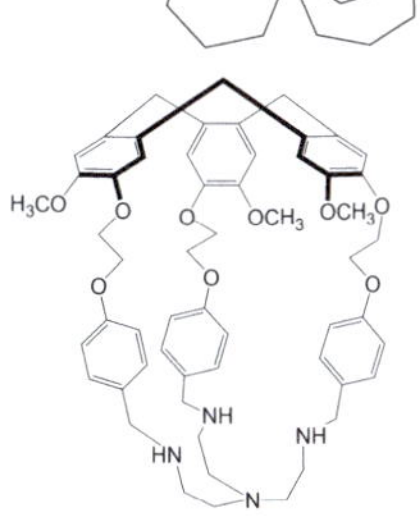

614.

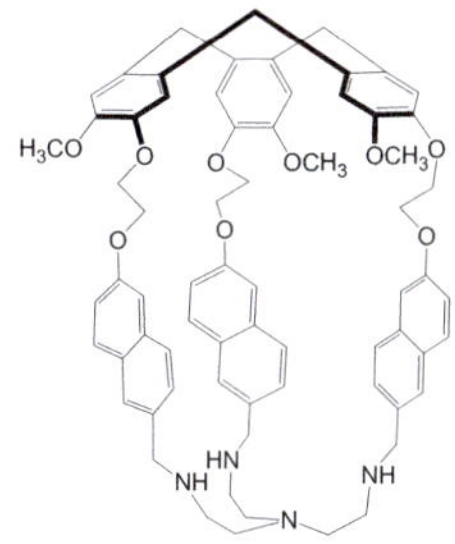

615.

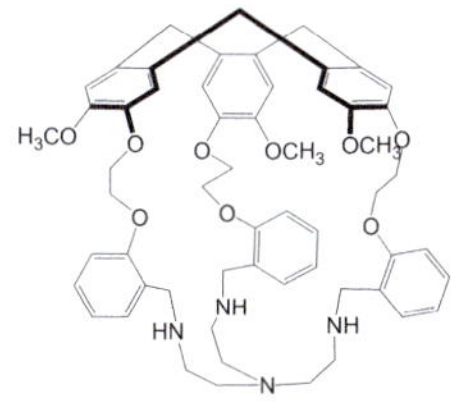

616.

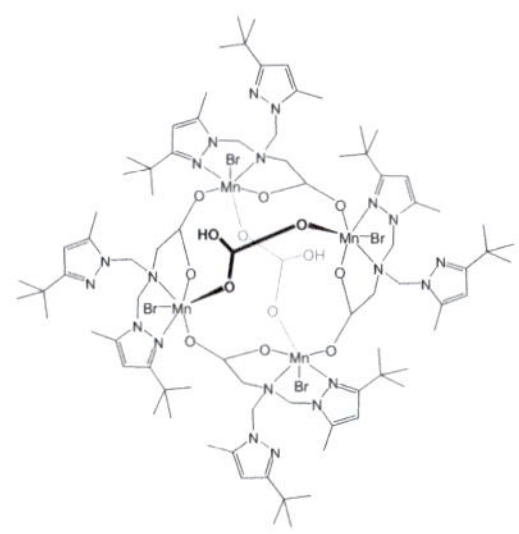

617.

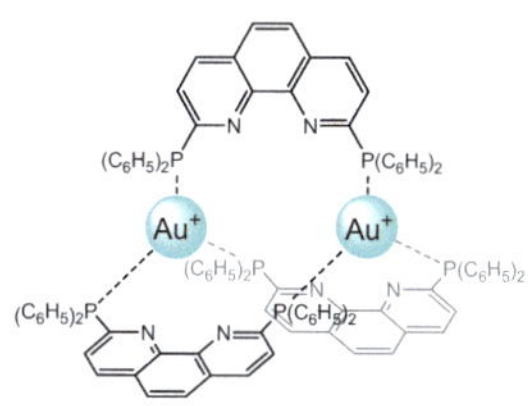

618.

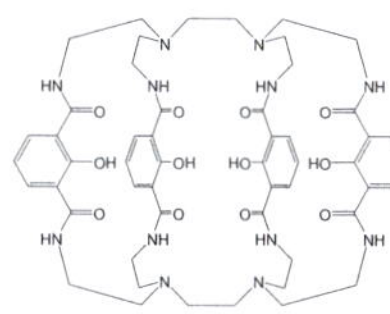

619.

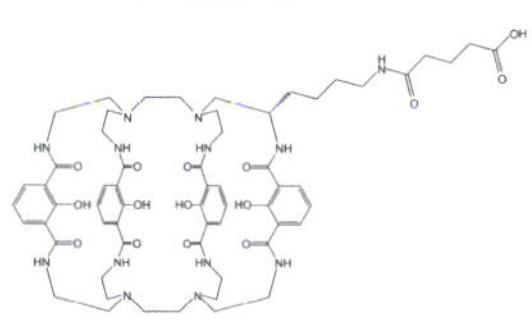

620.

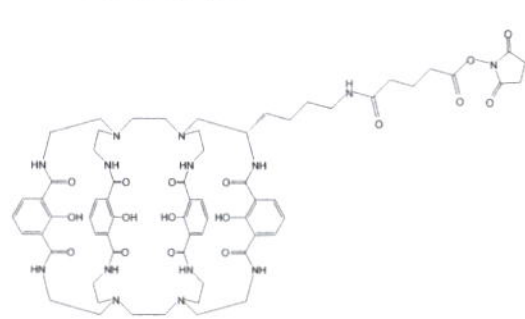

621.

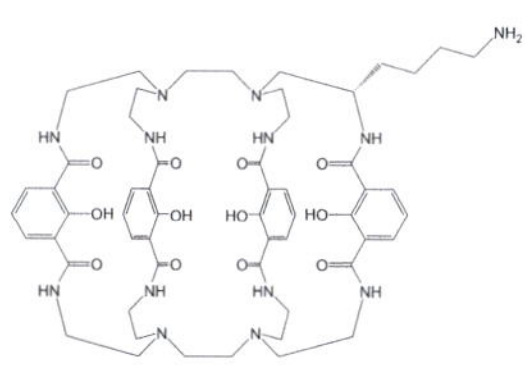

622.
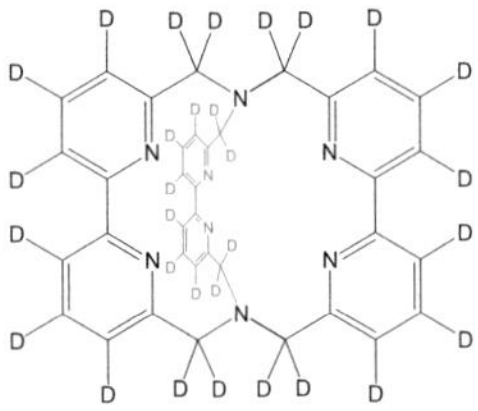

623.
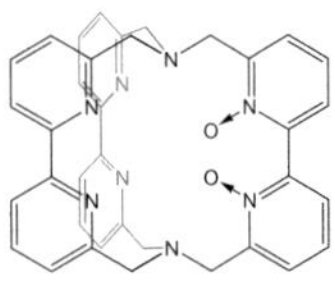

624.
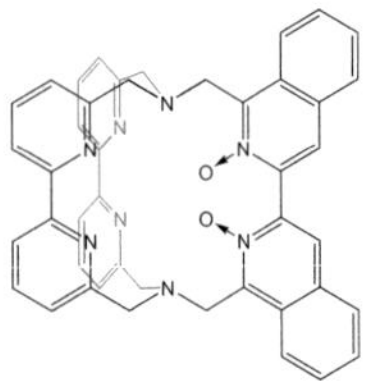

625.
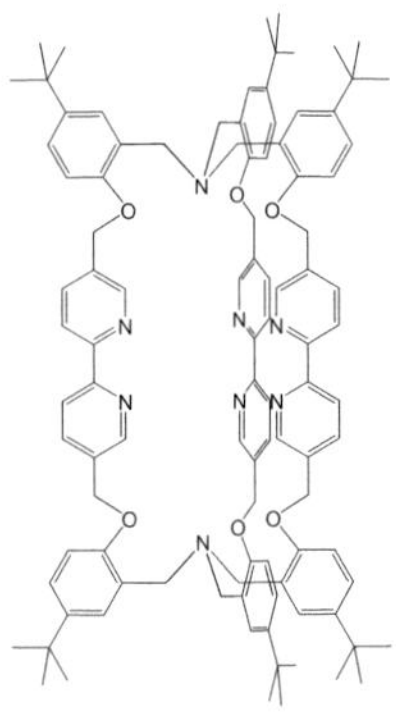

626.
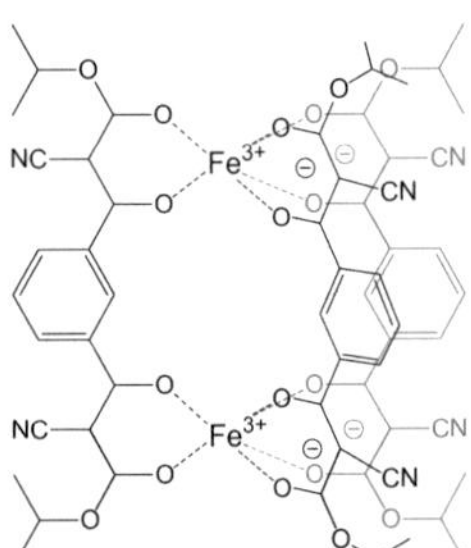

627.
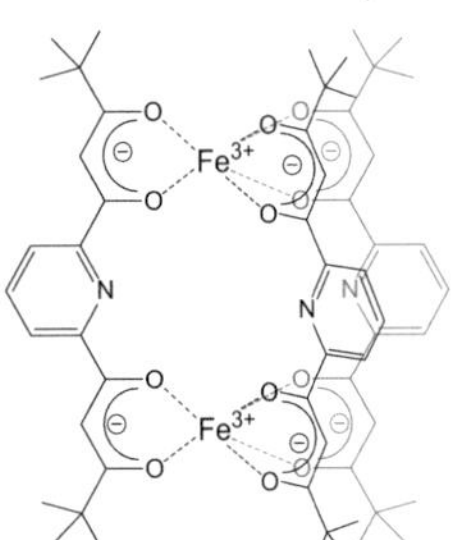

628.
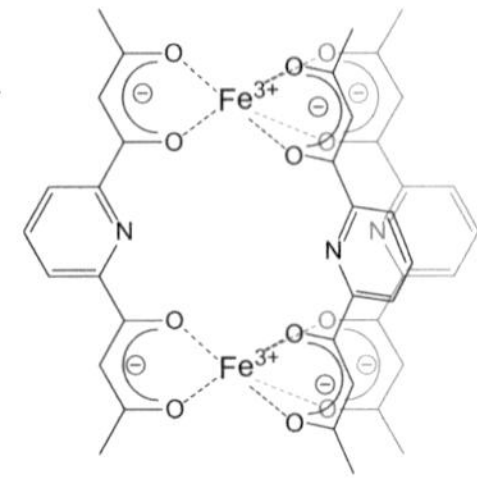

629.
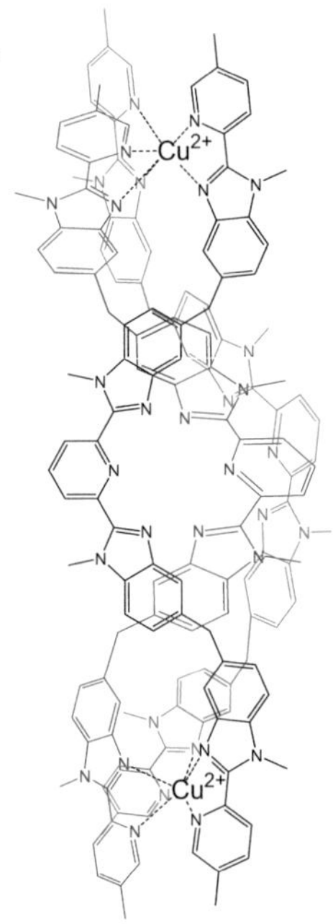

630.
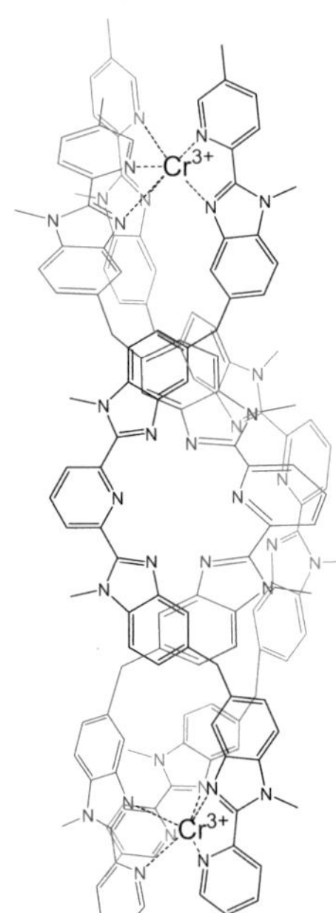

631. 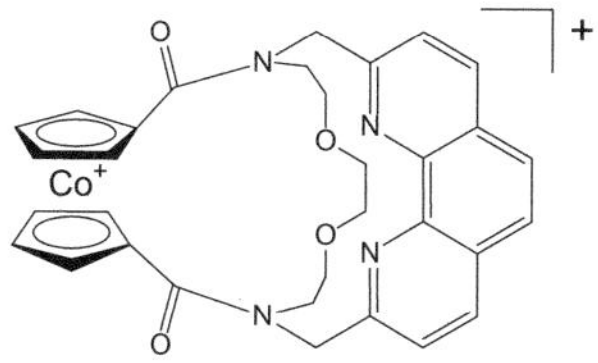

632. 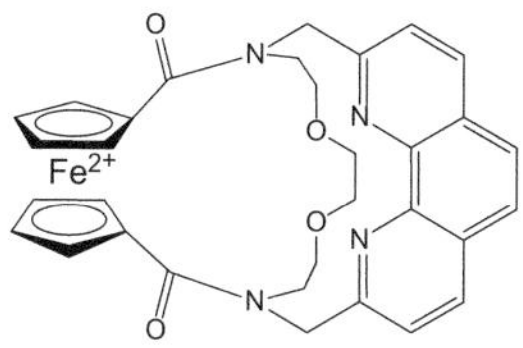

633. 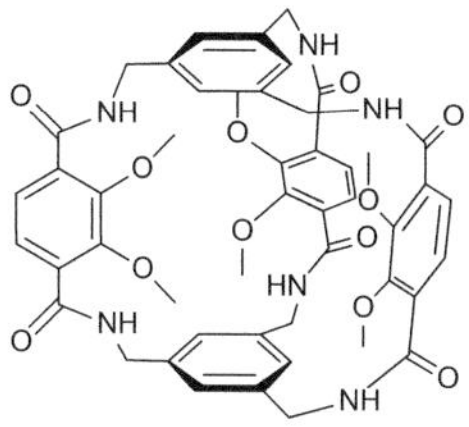

634. 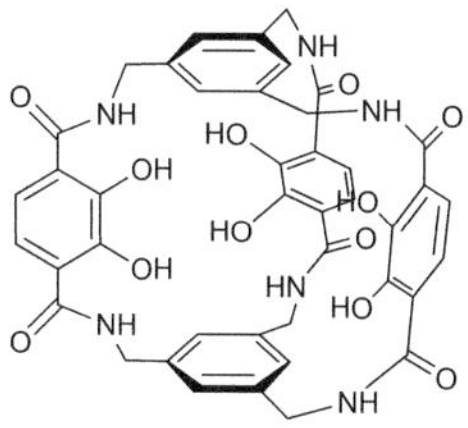

635. 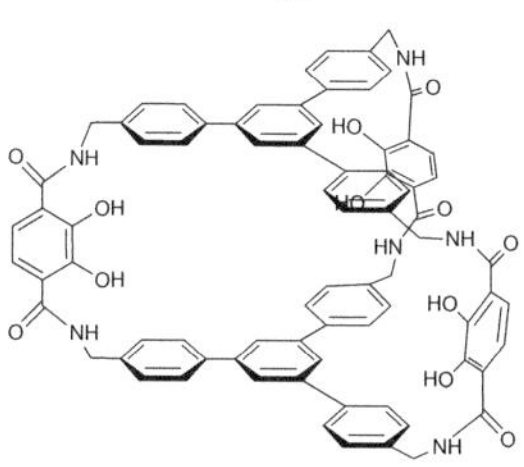

636. 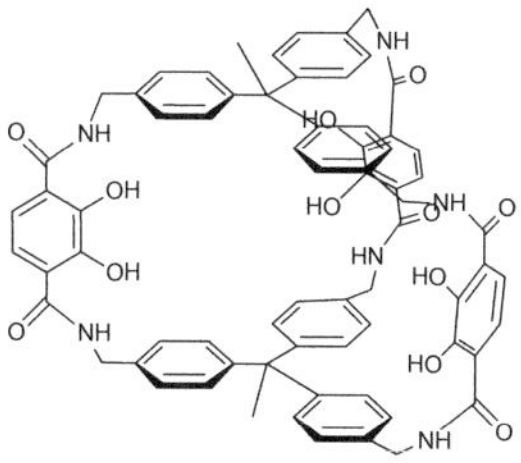

637. 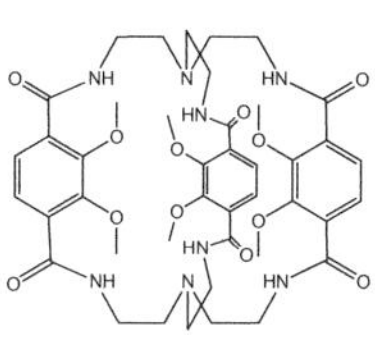

638. 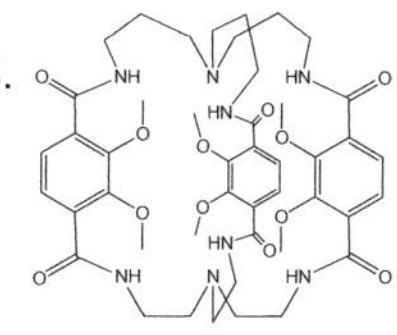

639. 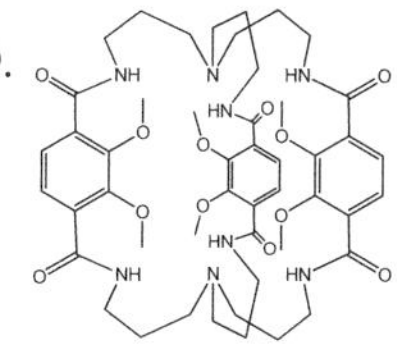

640. 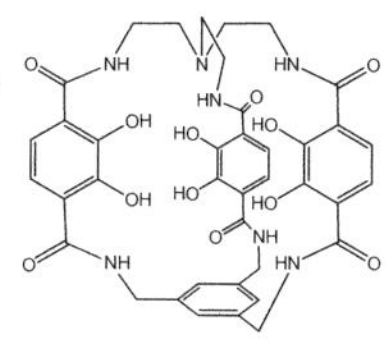

641. 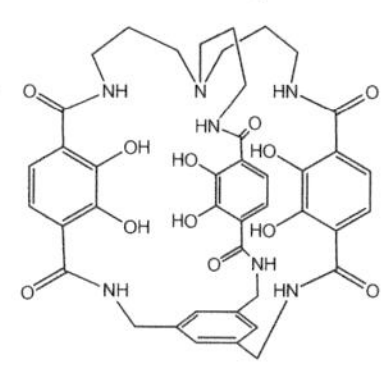

642. 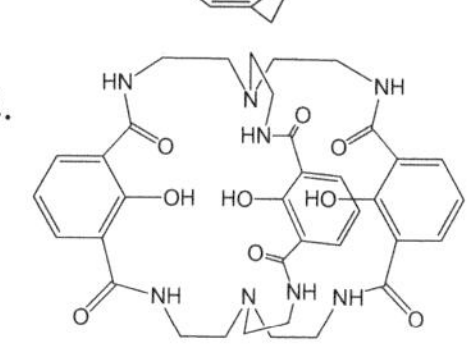

643. 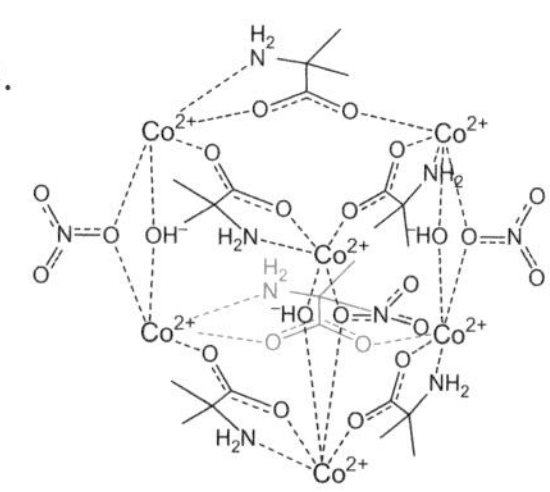

644. 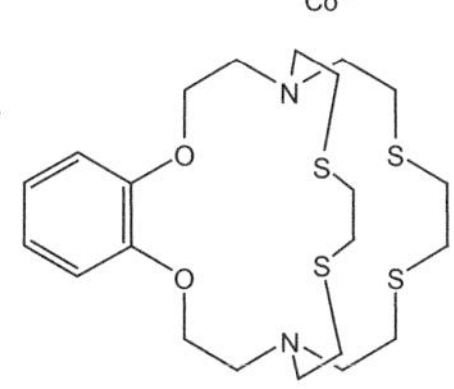

645. 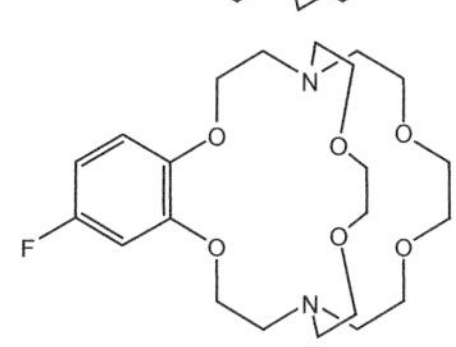

646.

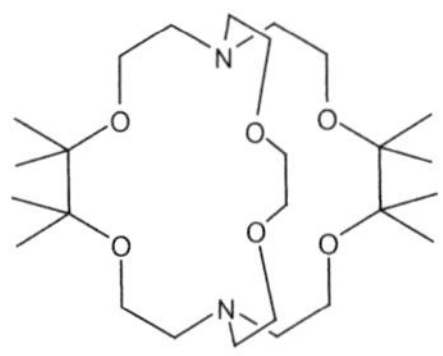

647.

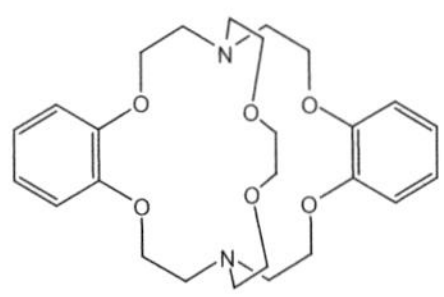

648.

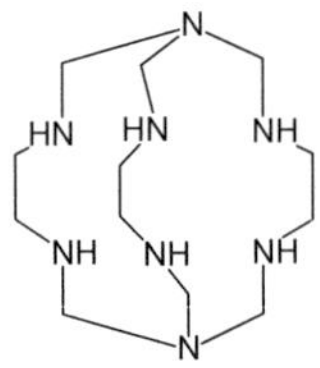

649.

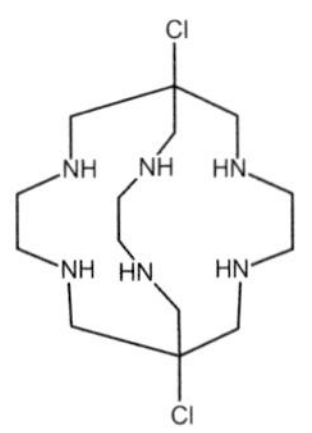

650.

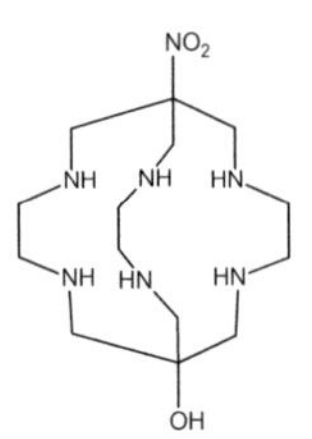

651.

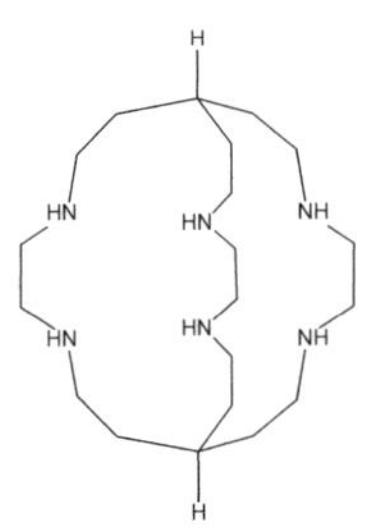

652.

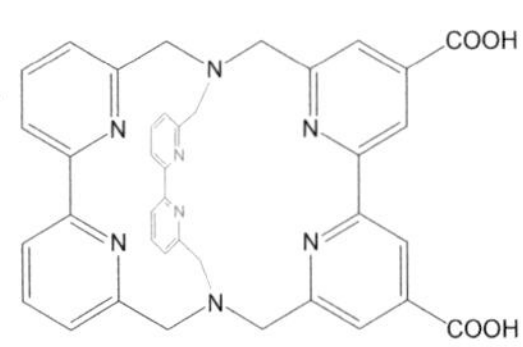

653.

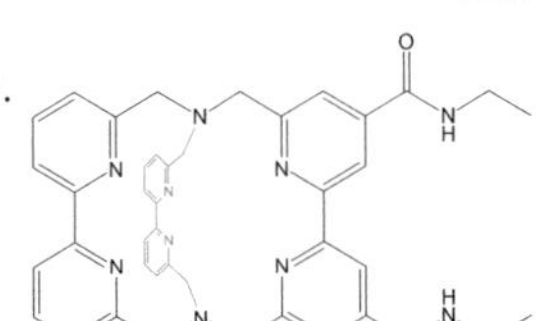

654.

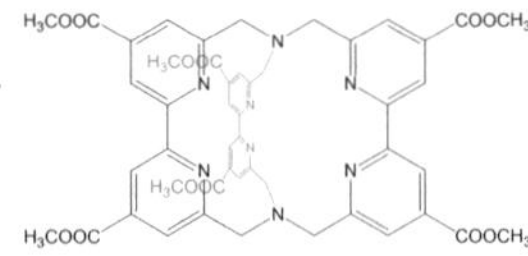

655.

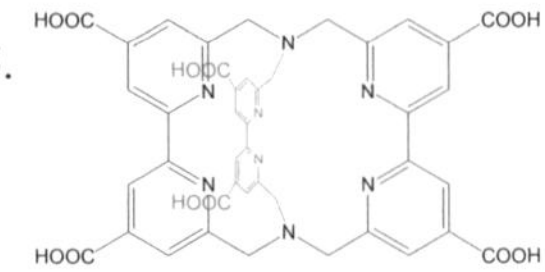

656.

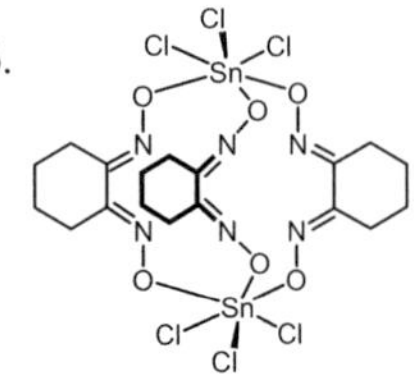

657.

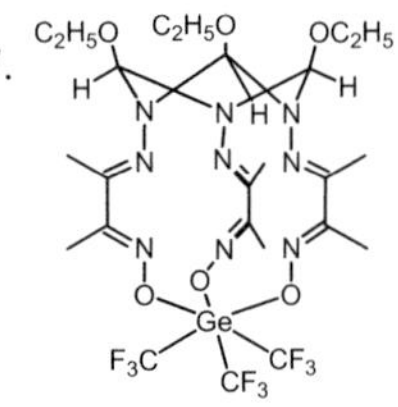

658.

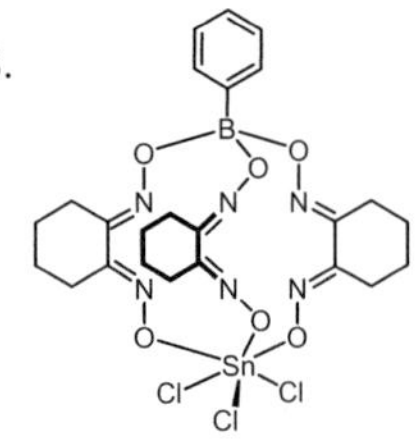

659.

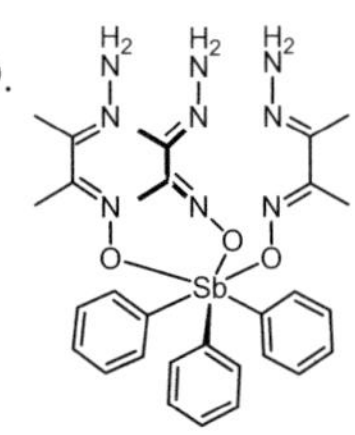

660.

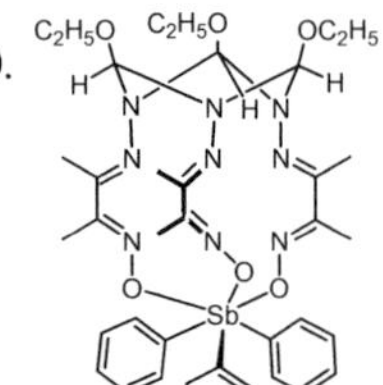

661.

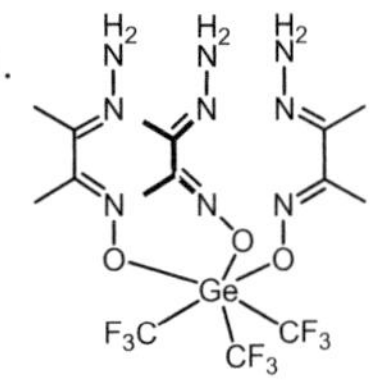

662.
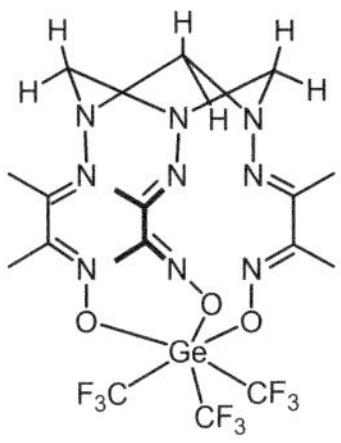

663.
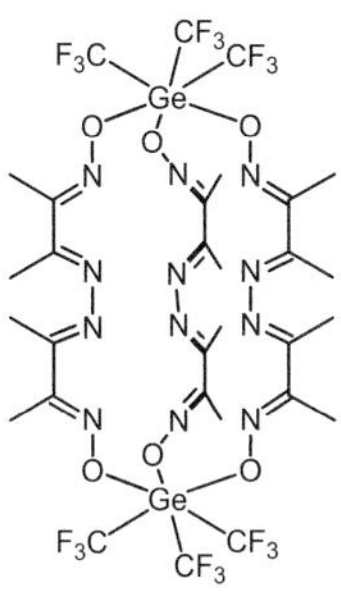

664.
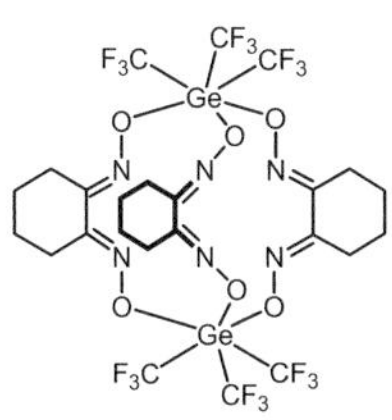

665.
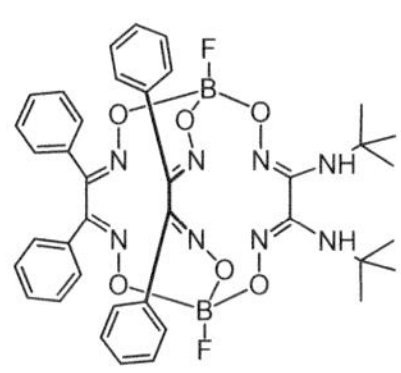

666.
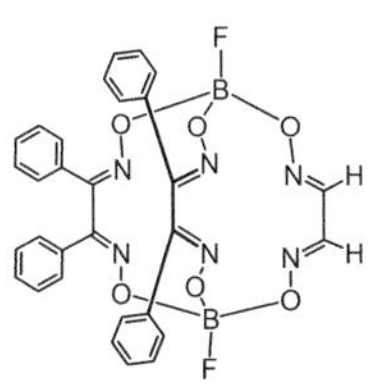

667.
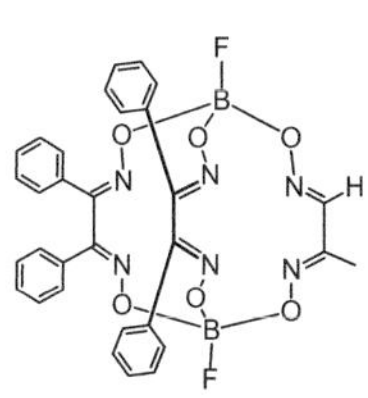

668.
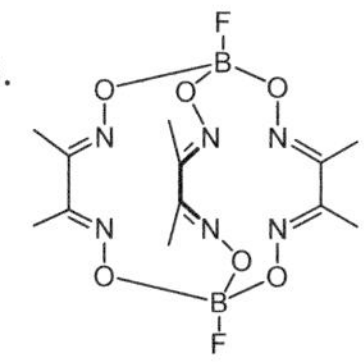

669.
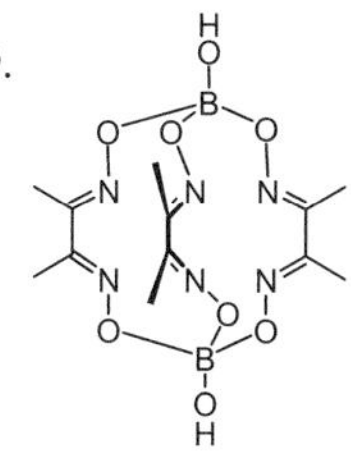

670.
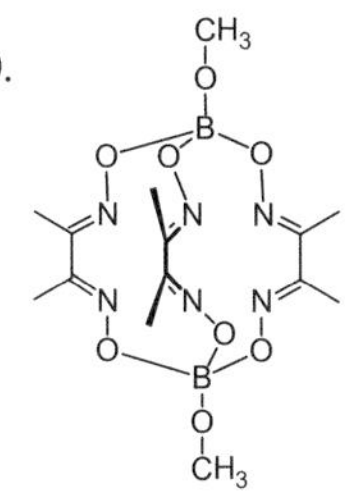

671.
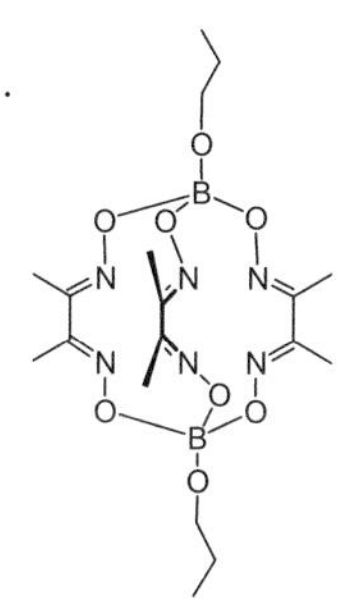

672.
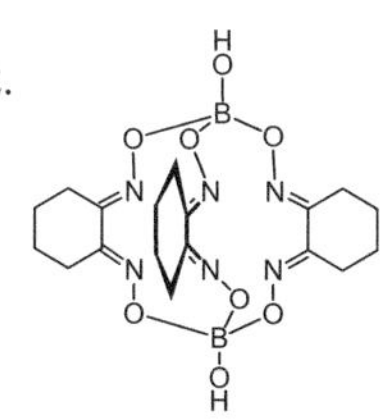

673.
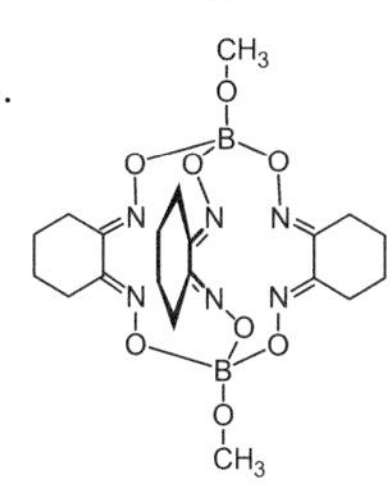

674. 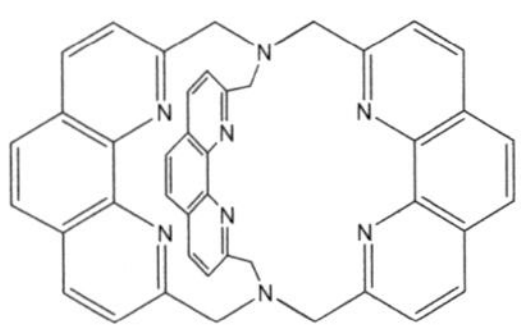

675. 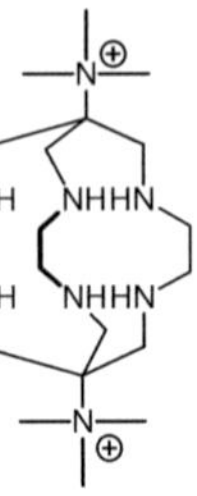

Appendix B

List of Synthons, Reagents, Non-macrocyclic Compounds, Polymers et al.

1. 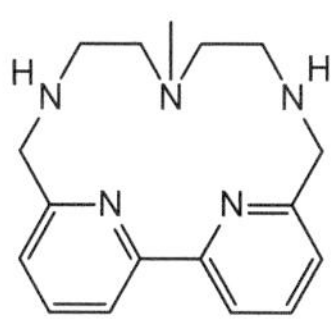

2. 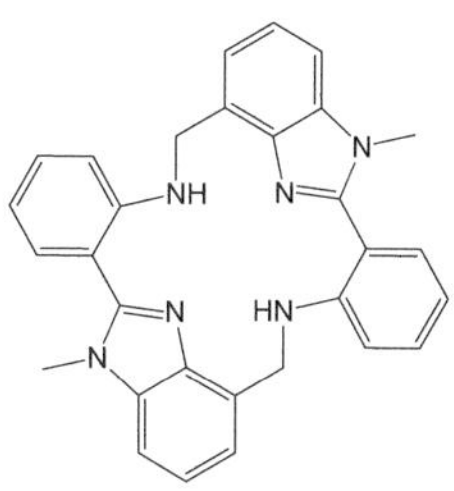

3. 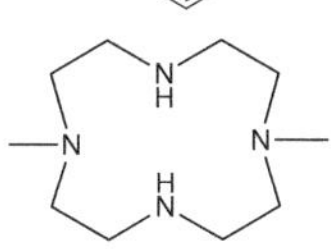

4. 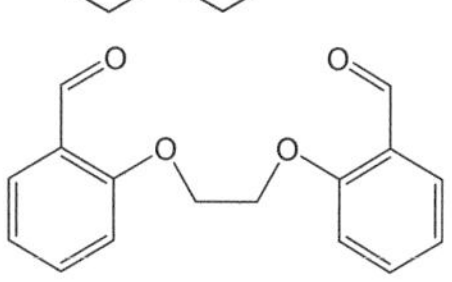

5. 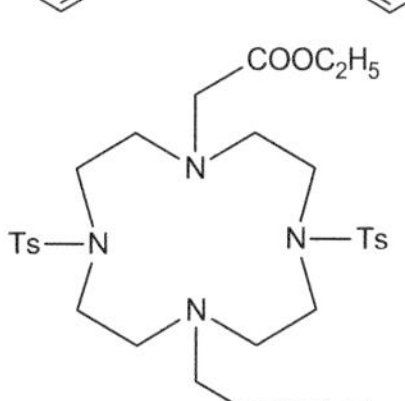

6. 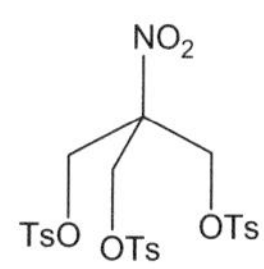

7. 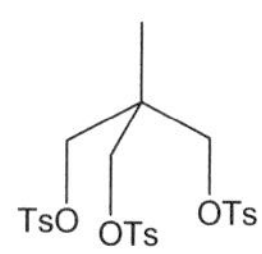

8. 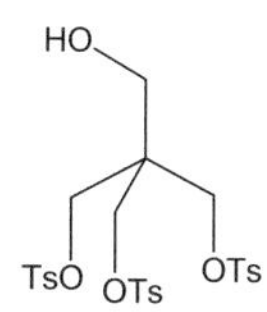

9. 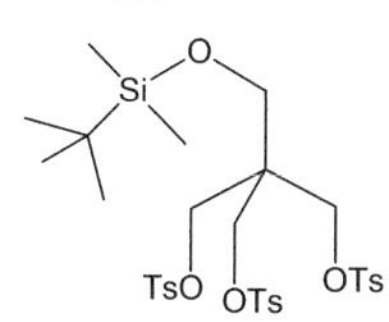

10. 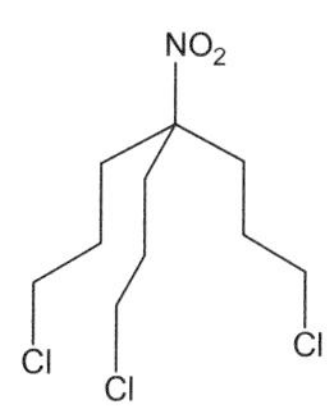

11. 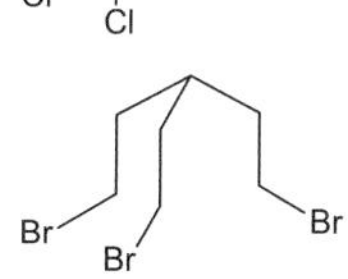

Y. Voloshin et al., *Cage Metal Complexes*, DOI 10.1007/978-3-319-56420-3

12. 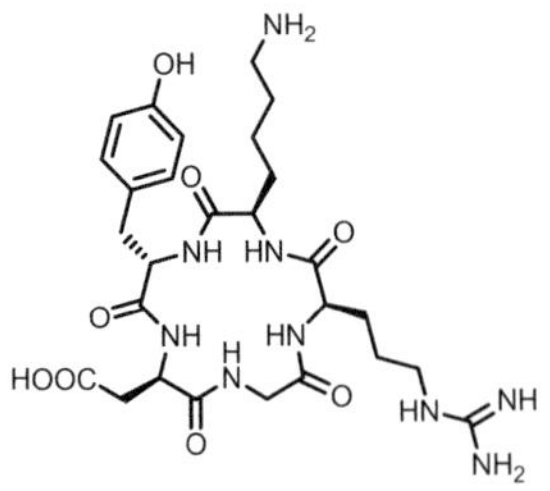

13. 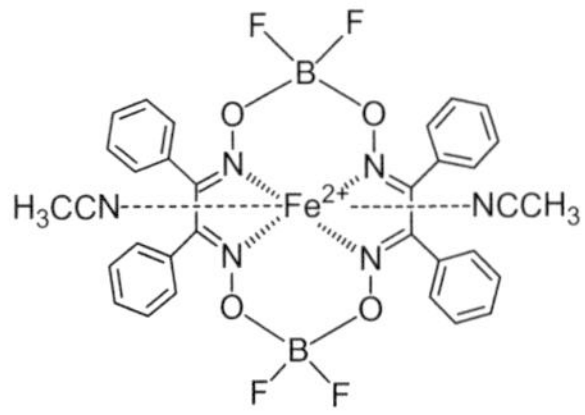

14. 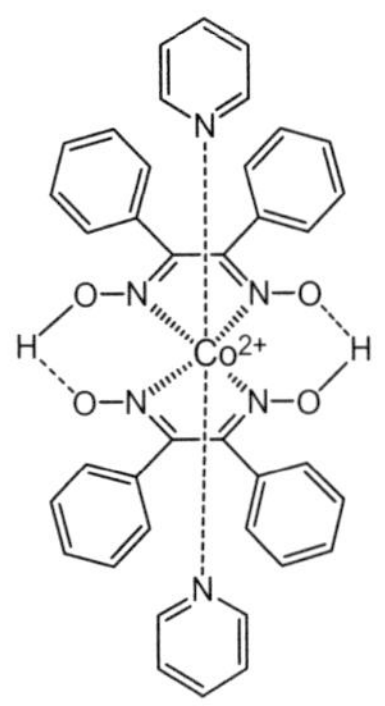

15. 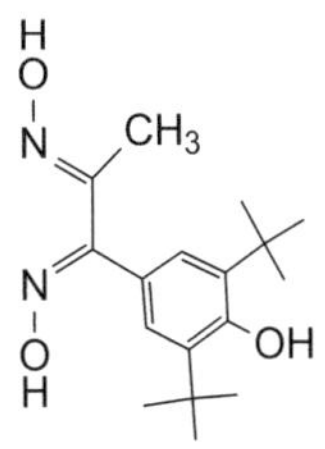

16. 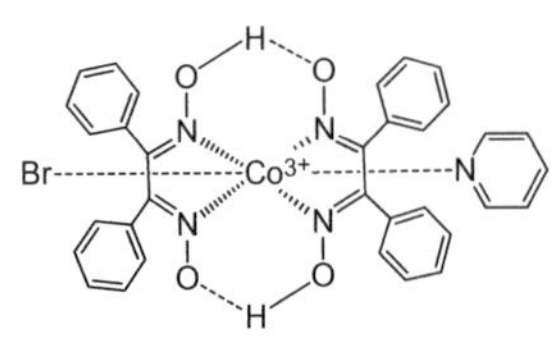

17. 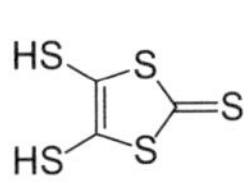

18. 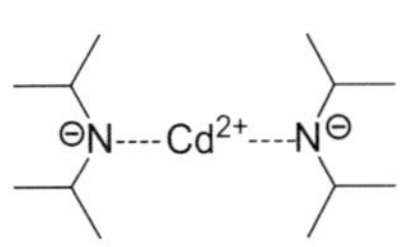

19. 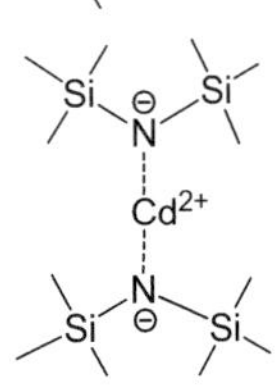

20. $Pd(P(C_6H_5)_3)_2(CH_3COO)_2$

21. 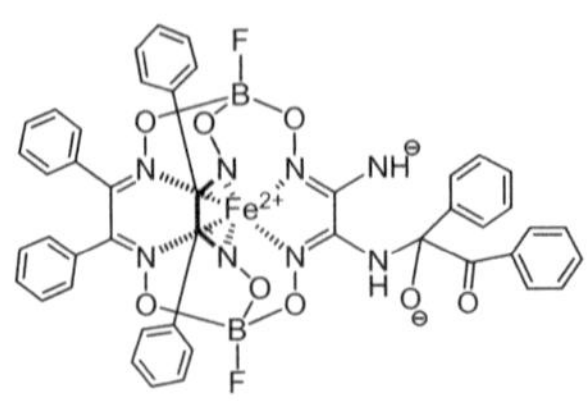

22. 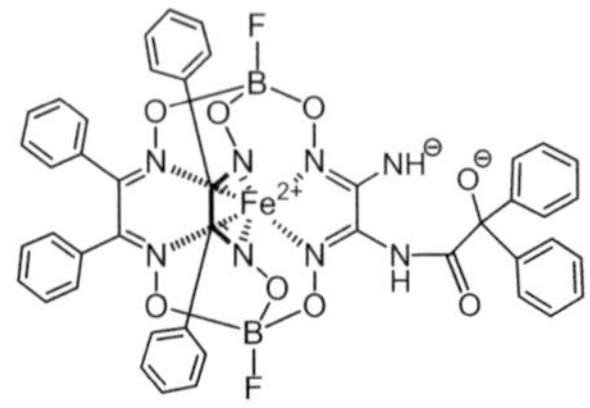

23. 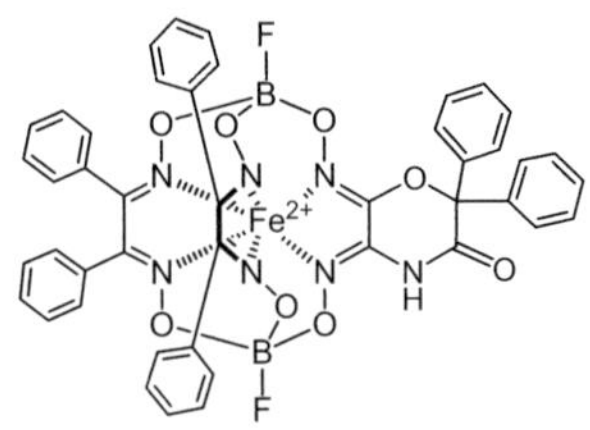

24. 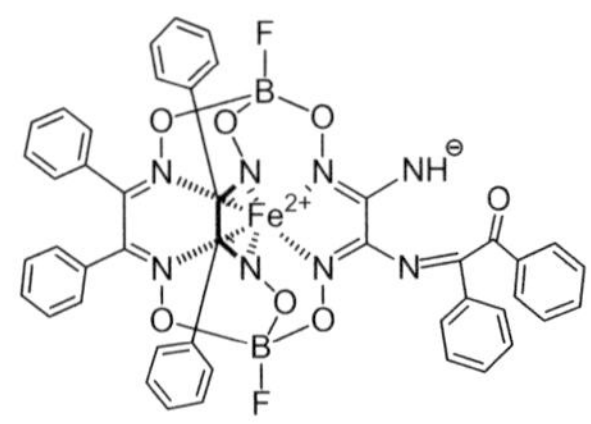

25. 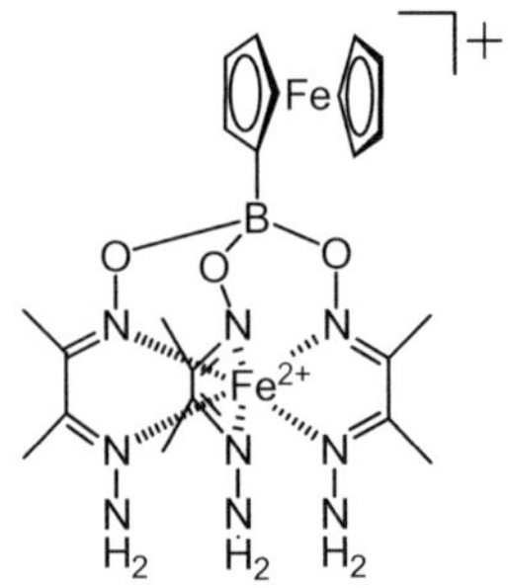

26. 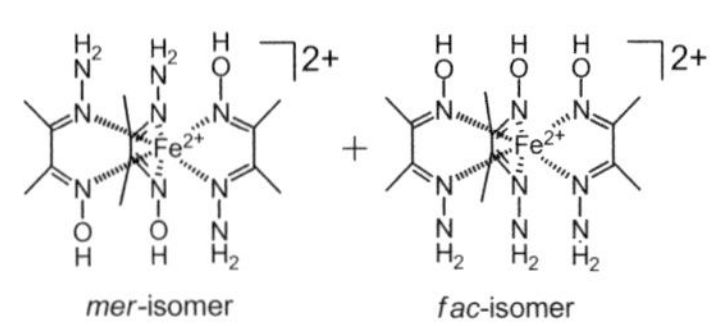

27. 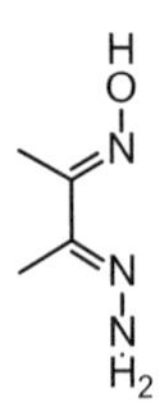

28.

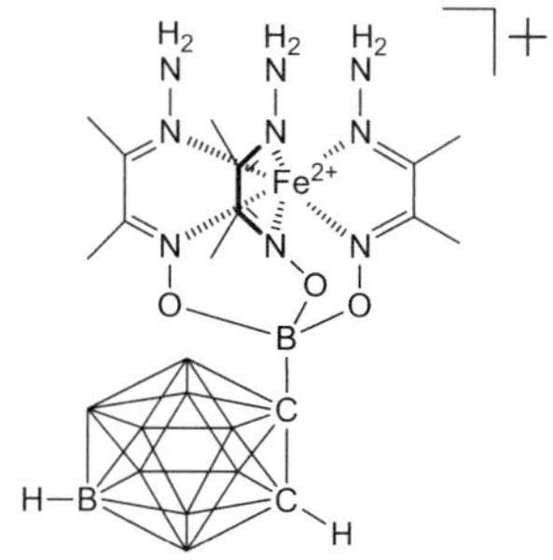

29.

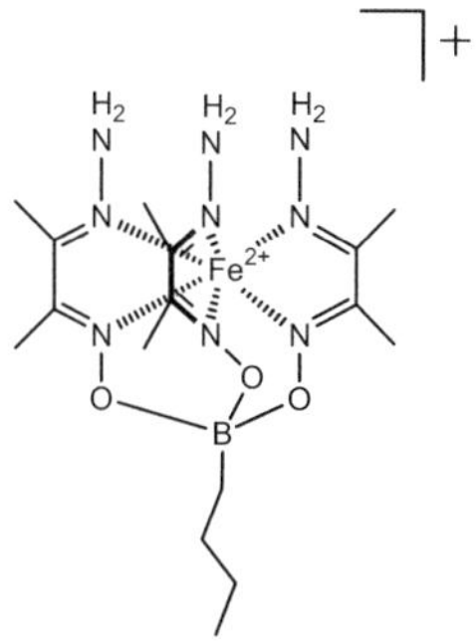

30.

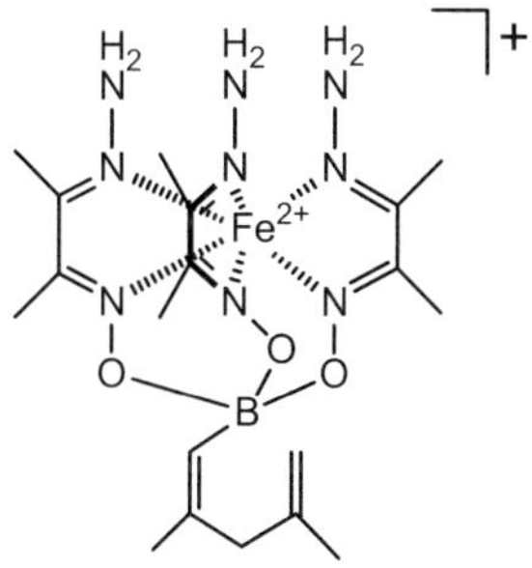

31.

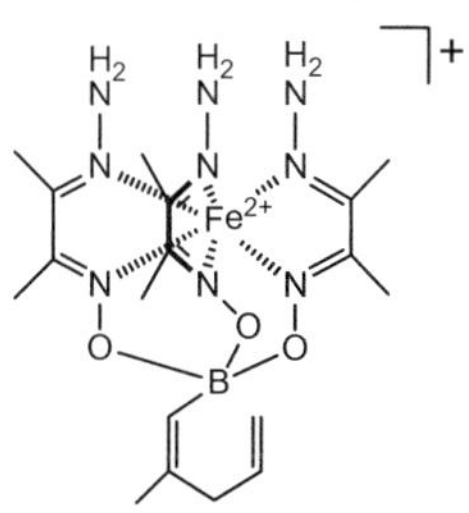

32.

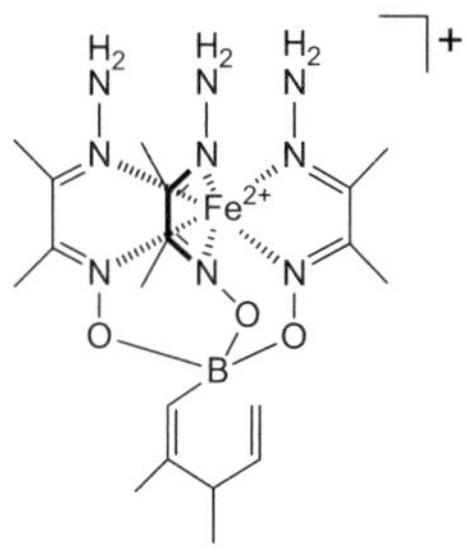

33.

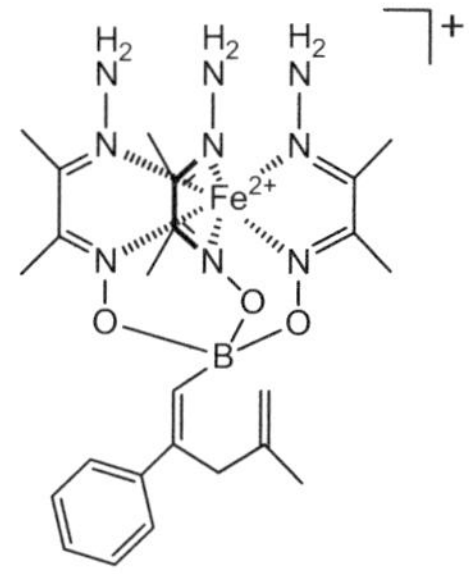

34.

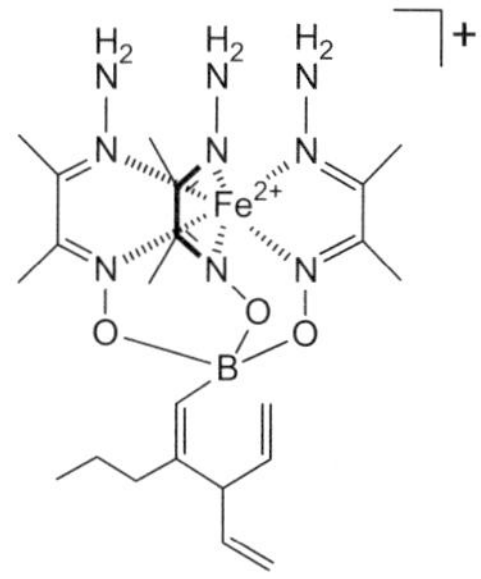

35.

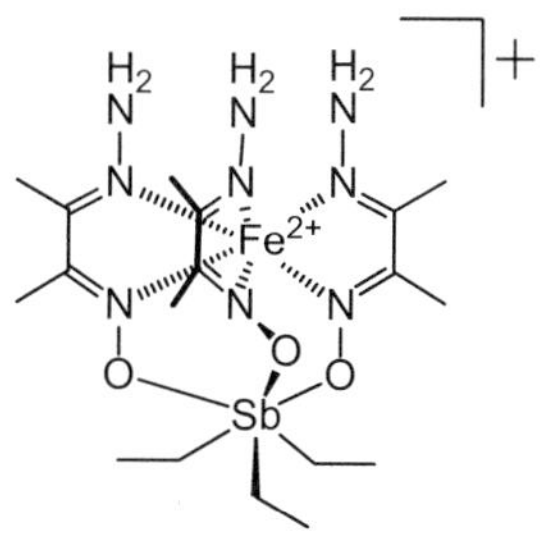

36.

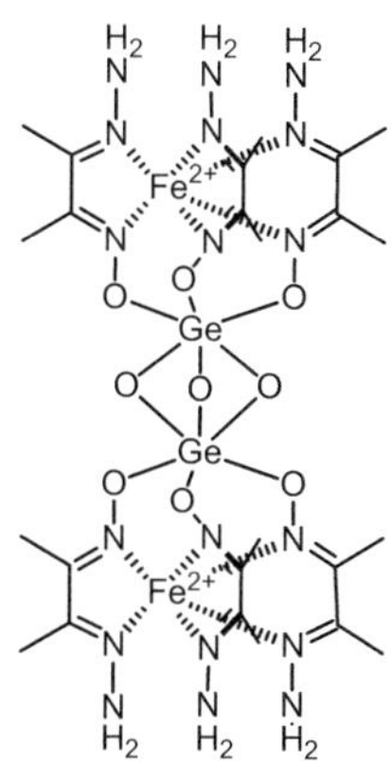

37.

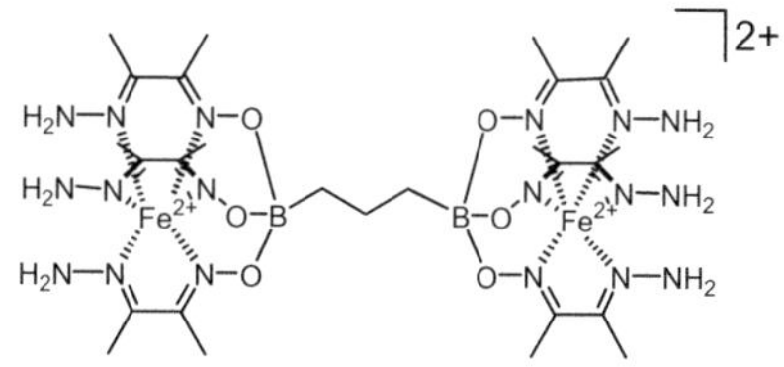

38. 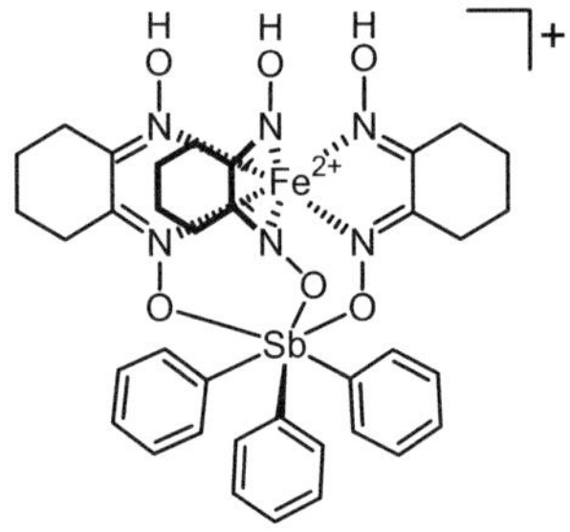

39. 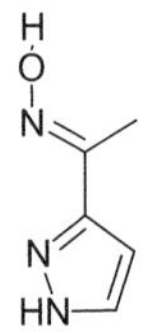

40. 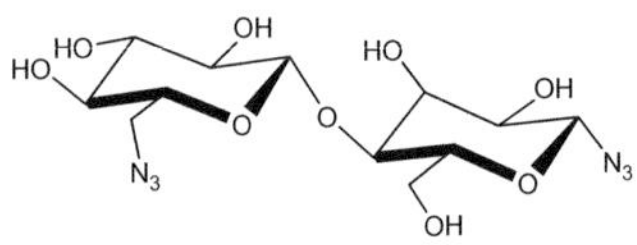

41. 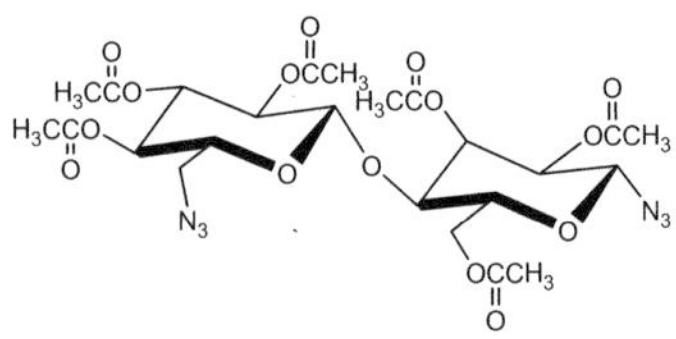

42. 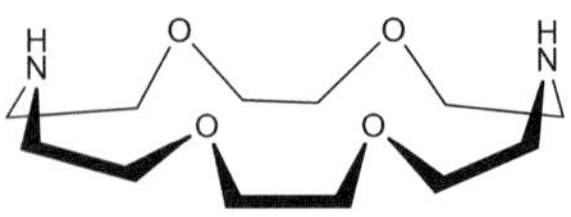

43. 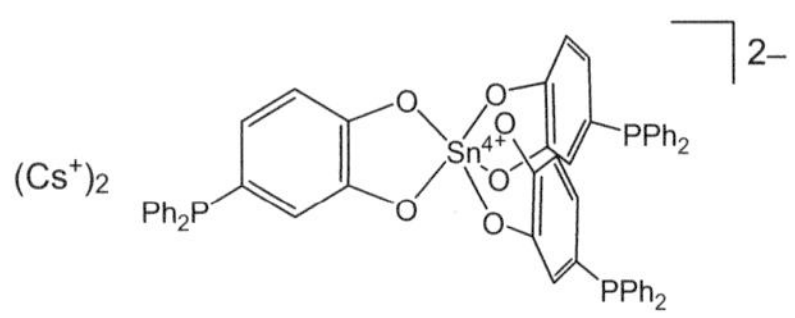

44. 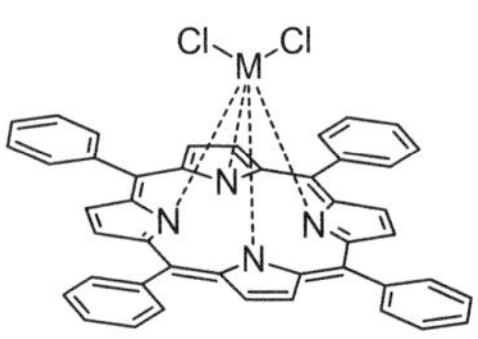

M = Zr

45. 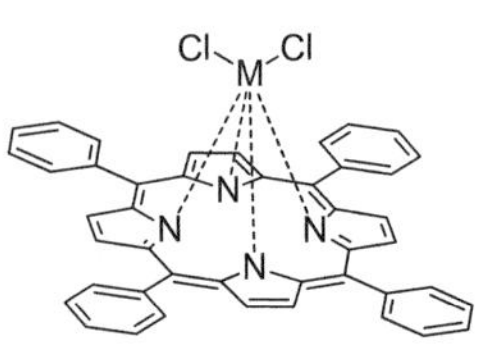

M = Hf

46. 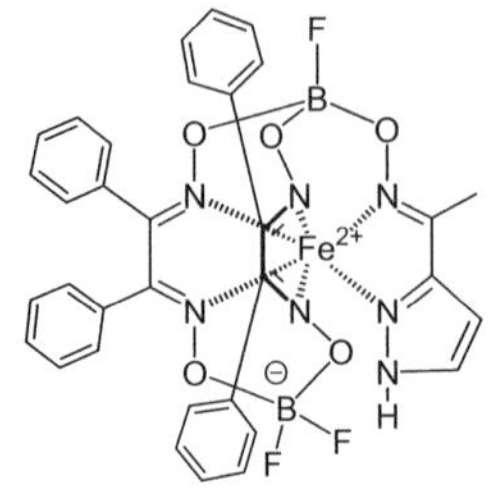

47. 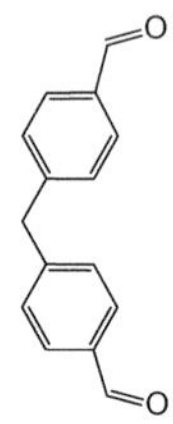

48. 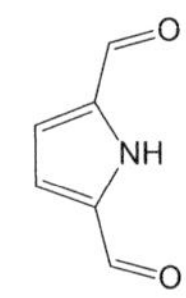

49. 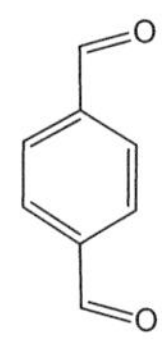

50. 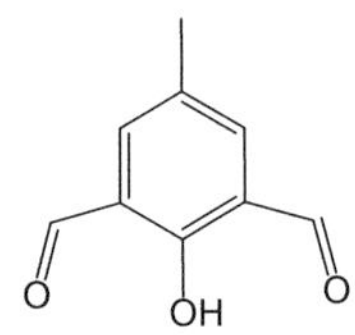

51. 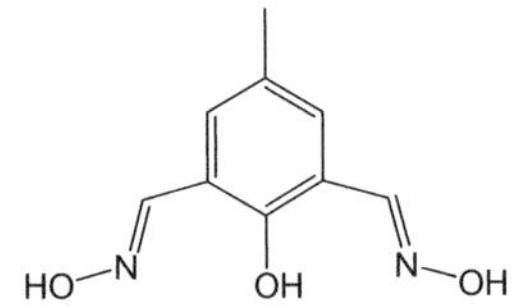

52. 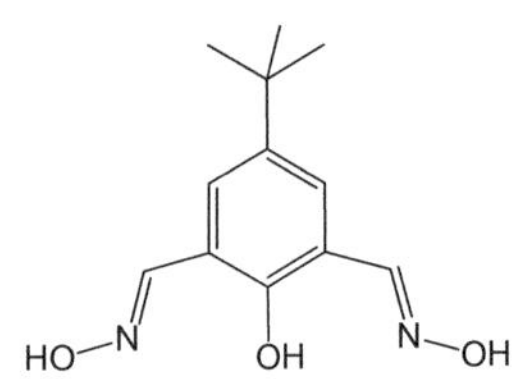

53. 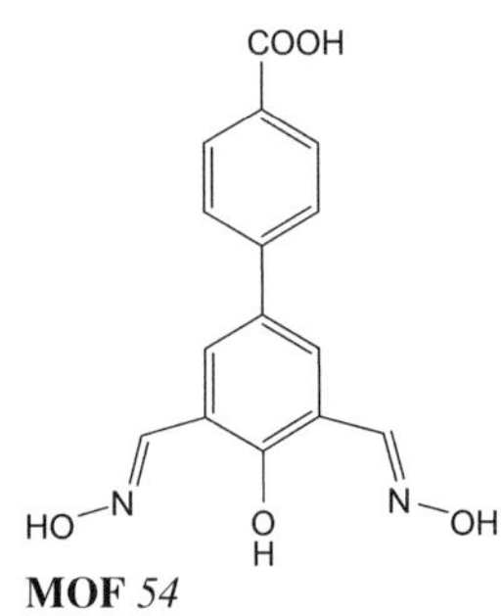

54. **MOF** *54*

55. **MOF** *55*
56. **MOF** *56*
57. **MOF** *57*
58. **MOF** *58*
59. **MOF** *59*
60. **MOF** *60*
61. **MOF** *61*
62. **MOF** *62*

63.
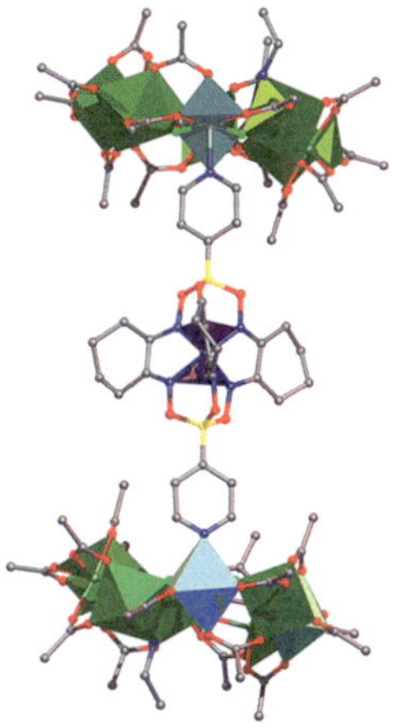

64.
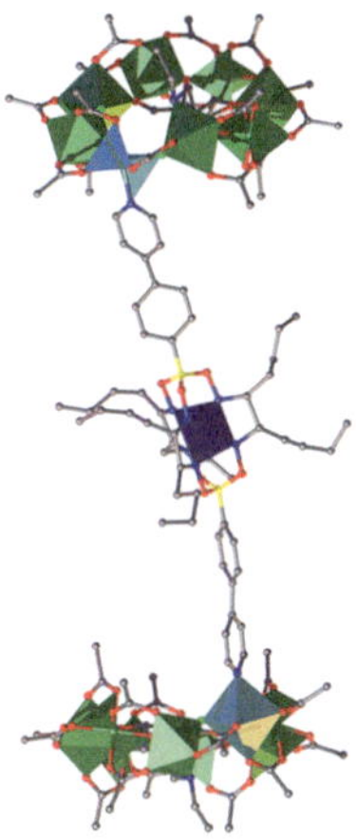

65.
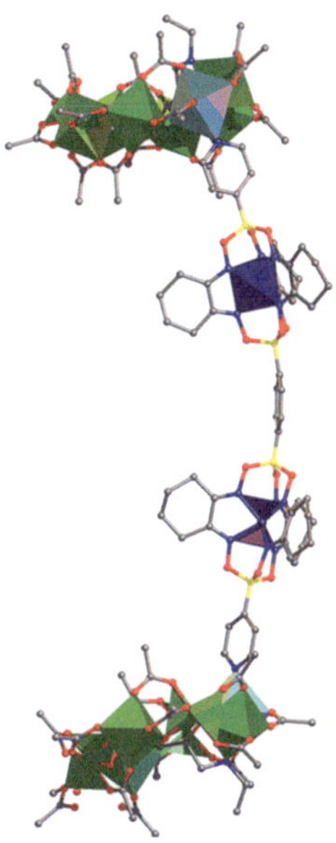

66.
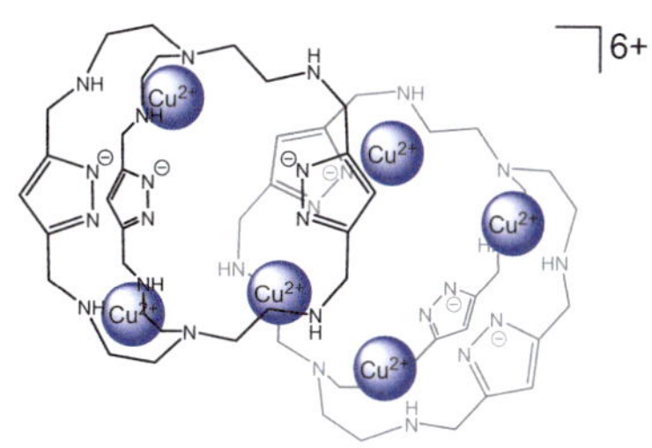

67.
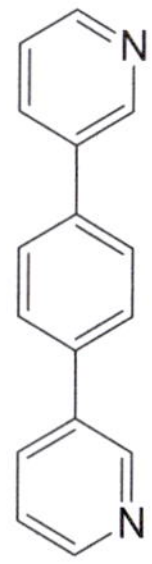

68.
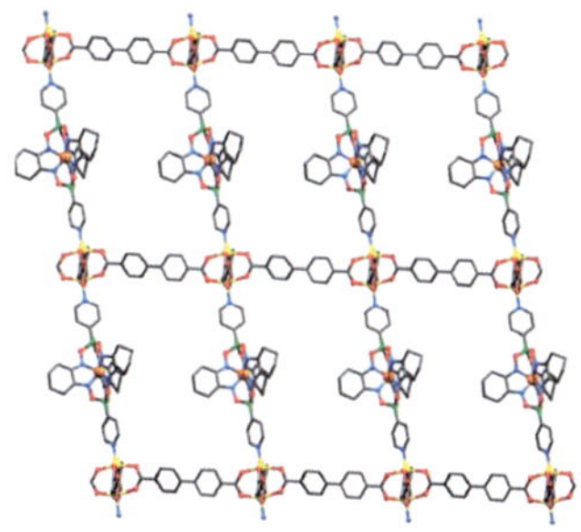

69.
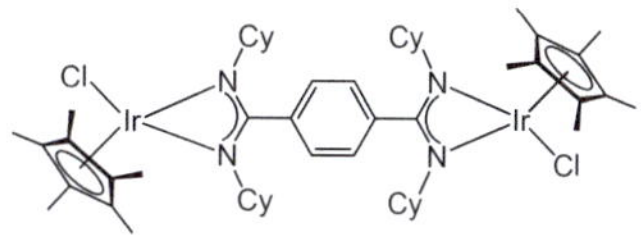

70.
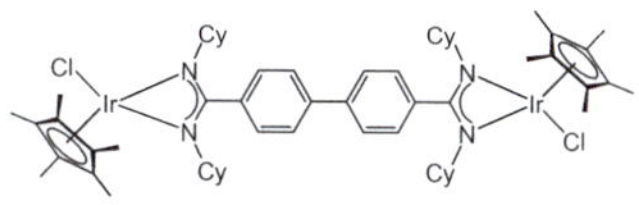

71.
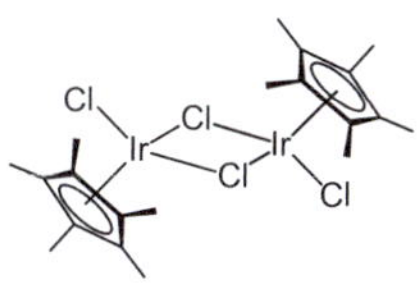

72.
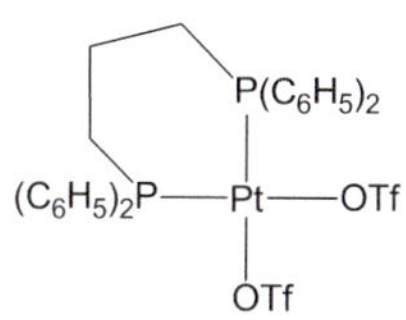

73.

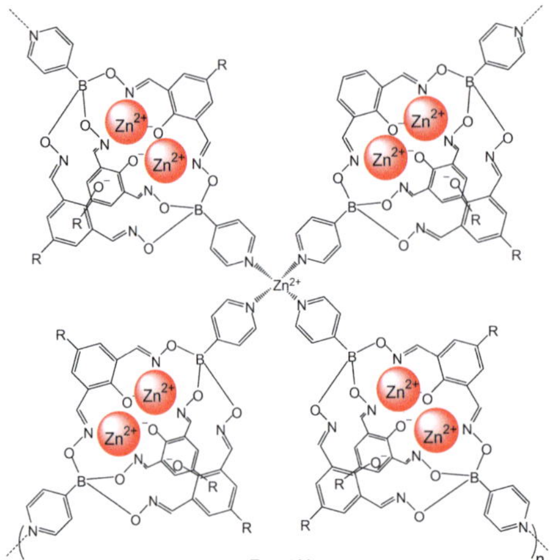

R = CH_3

74.

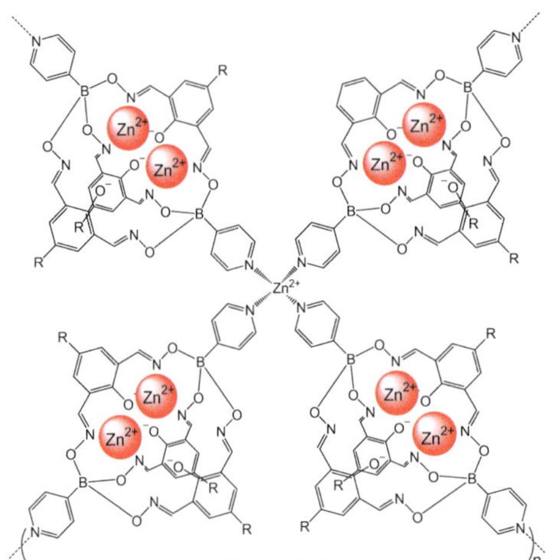

R = *tert*-C_4H_9

75.

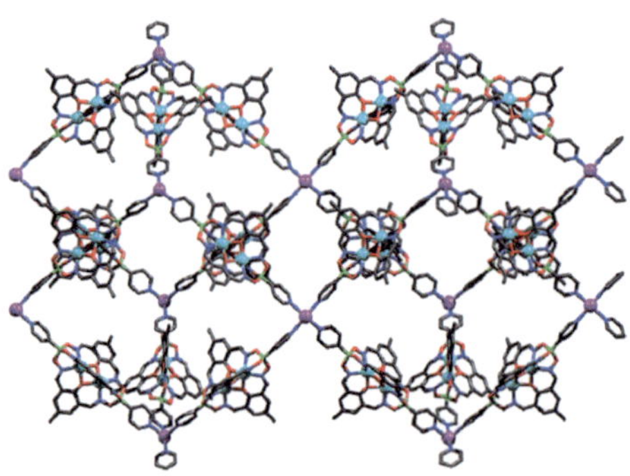

R = CH_3

76.

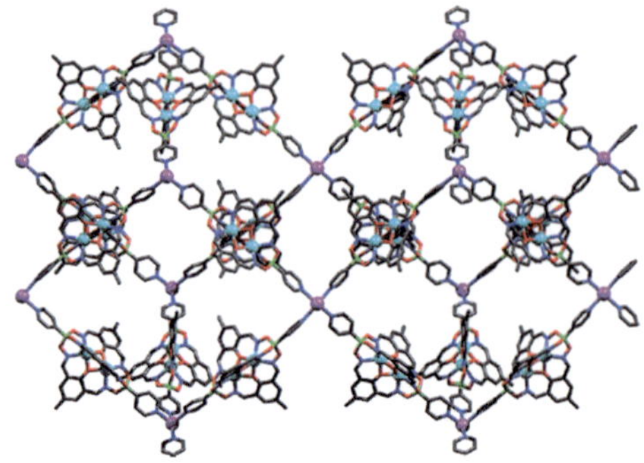

R = *tert*-C_4H_9

77.

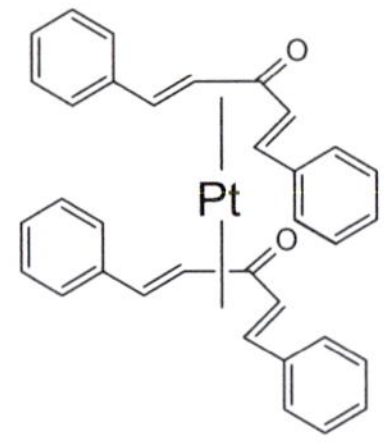

78.

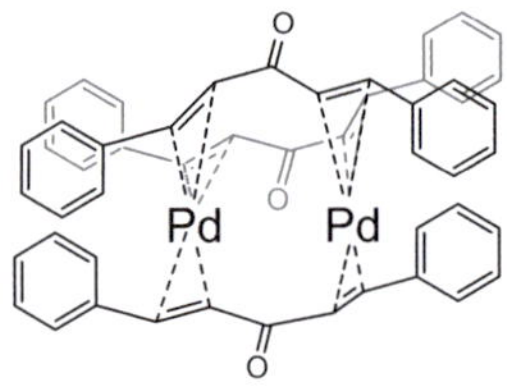

79.

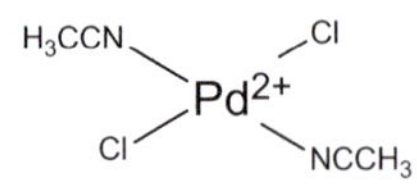

80.

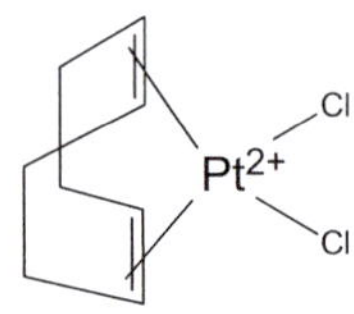

81.

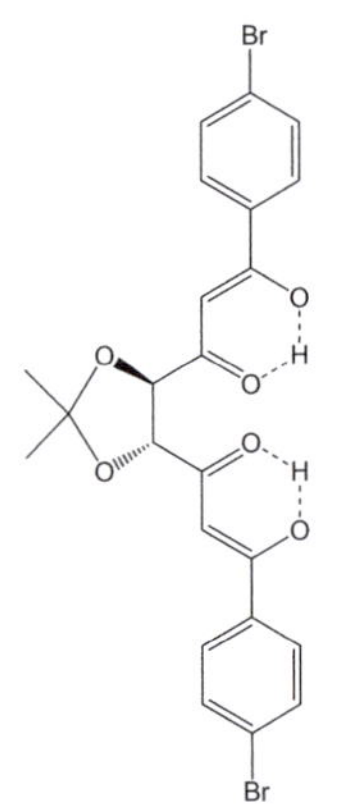

82.

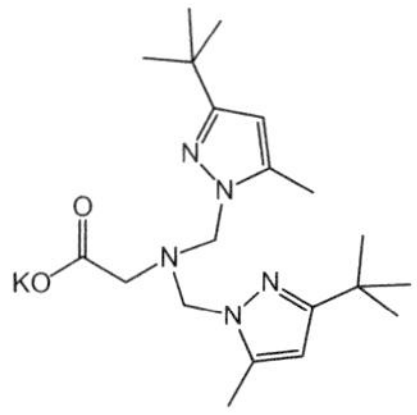

83.

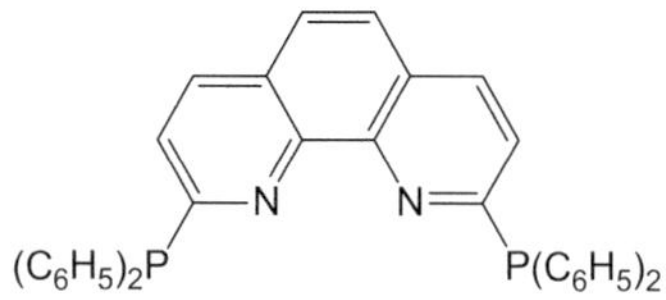

84.

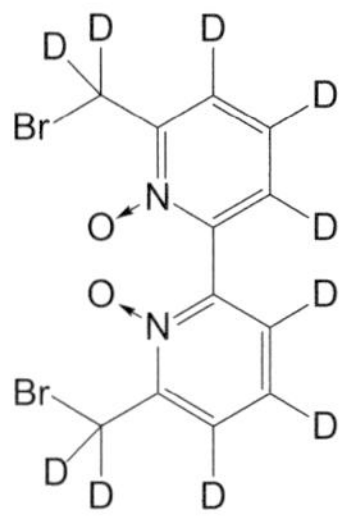

85.

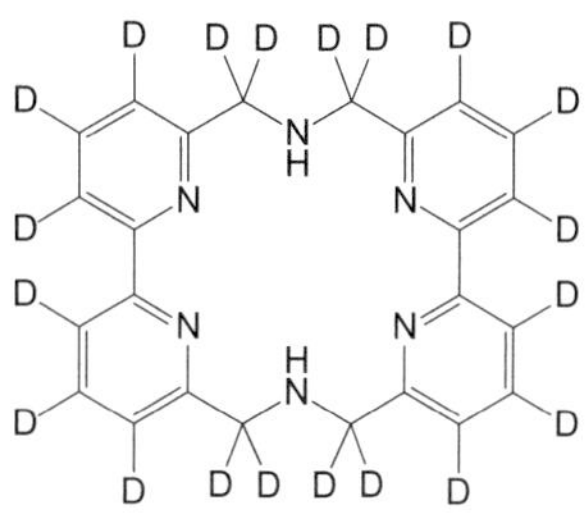

86.

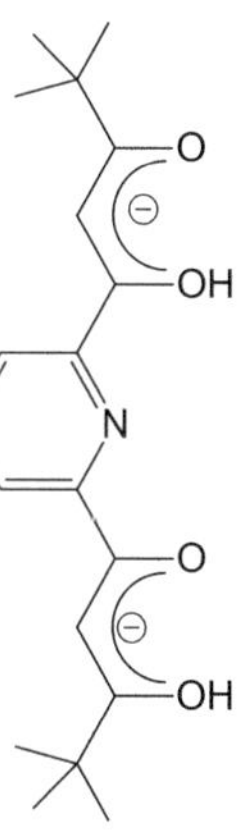

87.

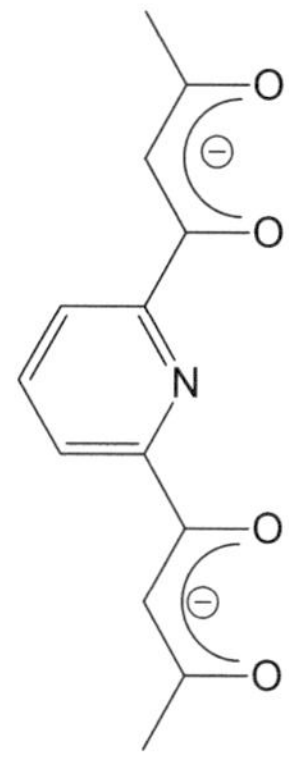

88.

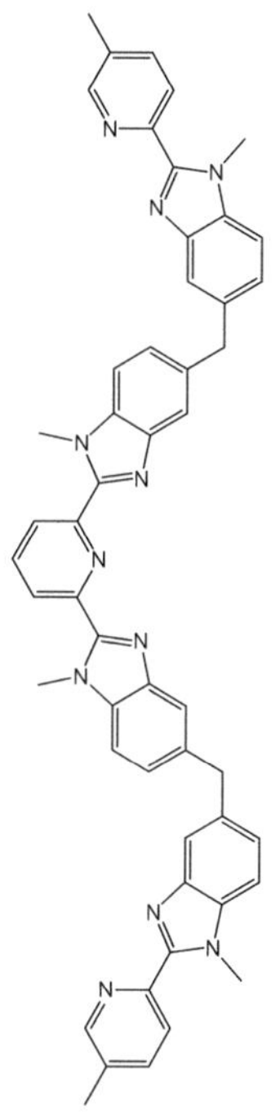

89.

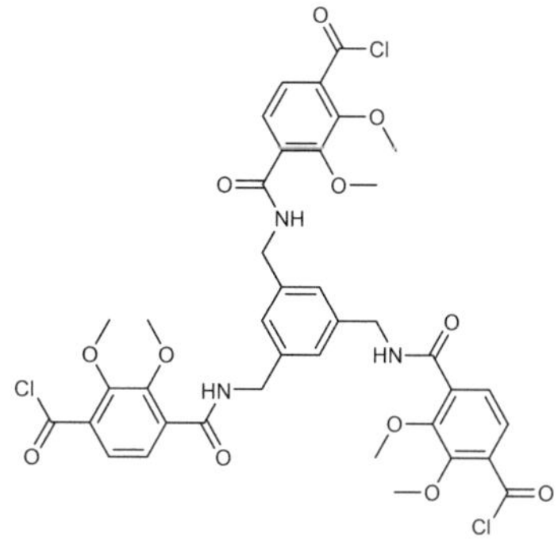

MIX
Papier aus verantwortungsvollen Quellen
Paper from responsible sources
FSC® C105338

If you have any concerns about our products,
you can contact us on
ProductSafety@springernature.com

In case Publisher is established outside the EU,
the EU authorized representative is:
Springer Nature Customer Service Center GmbH
Europaplatz 3, 69115 Heidelberg, Germany

Printed by Libri Plureos GmbH
in Hamburg, Germany